Universality in Chaos

Second Edition

CLASSICS
Illustrated
FEATURING IDENTITIES BY THE WORLD'S FASTEST THEORISTS
GRAF VON CHAOS
IN:
DAS EWIG-WEIBLICHE ZIEHT UNS HINAN
APROVED BY THE CHAOTIC CODE g(x) AUTHORITY
HOW STRANGELY TURBULEN
COMING SOON: THE ASTOUNDING HOLLYWOOD SPECTACULAR
!!!
P.C.

Universality in Chaos

Second Edition

a reprint selection
compiled and introduced by

Predrag Cvitanović
Niels Bohr Institute
Copenhagen

The classification of the constituents of a chaos, nothing less is here essayed.

Herman Melville
Moby Dick chapter 32

Published in 1989 by
Taylor & Francis Group
270 Madison Avenue
New York, NY 10016

Published in Great Britain by
Taylor & Francis Group
2 Park Square
Milton Park, Abingdon
Oxon OX14 4RN

Printed in the United States of America on acid-free paper
10 9 8 7 6 5 4 3 2

International Standard Book Number-10: 0-85274-260-6 (Softcover)
International Standard Book Number-13: 973-0-85274-260-0 (Softcover)
First published 1984, Second edition, 1989

Library of Congress Cataloging-in-Publication Data

Catalog record is available from the Library of Congress

Taylor & Francis Group
is the Academic Division of Informa plc.

Visit the Taylor & Francis Web site at
http://www.taylorandfrancis.com

Contents

Part 3: Theory

Part 4: Noise

Part 7: Beyond the one-dimensional theory

Part 8: Recent developments

Part 9: References

Preface to the Second Edition

The first edition of this reprint selection has origin in those dim times in the past when there was no chaos in physics, or at least not in the minds of most physicists. The people who did this kind of thing were meteorologists, geometers, astronomers, biologists, plasma physicists, field theorists, ..., and there was a clear need for a reprint selection, as the seminal references were scattered far and wide. In those days it was still possible to get a fair overview of the infant field by reading leisurely a few fundamental articles. Already then I found it necessary to narrow the scope to the exciting new development of that time, the universal aspects of chaotic motions. Since then the chaos literature has turned into a veritable torrent, and no fair overview of the field can be attempted within the confines of a single volume. The articles added to the second edition (Part 8 in this selection) reflect to some extent the status of the universality theory today. A particularly fruitful recent development has been the reinjection of techniques of statistical mechanics into the description of dynamically generated strange sets (here represented by the article by Halsey *et al*). Though the methods are old and the original mathematics literature deeper than the recent physics reworking, what once seemed arcane mathematics has turned out to be a very useful tool for the description of observed strange sets, and winged expressions like 'f of α' have by now become a part of our conceptual vocabulary.

P. Cvitanović
Niels Bohr Institute
Copenhagen
May 1989

Preface to the First Edition

This reprint selection presents some of the recent developments in the study of the chaotic behavior of deterministic systems. The problem, posed in its most general form, is old and appears under many guises: Why are clouds the way they are? Is the solar system stable? What determines the structure of turbulence in liquids, the noise in electronic circuits, the stability of plasma in a tokomak? The subject, defined so broadly, could not possibly be covered in a single reprint selection. This selection concentrates on the universal aspects of chaotic motions: those qualitative and quantitative predictions which apply to large classes of (often very different) physical systems. The selection can be divided into roughly four parts. The first part offers a general introduction to deterministic chaos and universality. The second part presents some of the experimental evidence for universality in transitions to turbulence. The third part concentrates on the theoretical investigations of the universality ideas, and the last part gives a glimpse of the further developments stimulated by the success of the one-dimensional universality theory.

This selection originates from a NORDITA reprint selection prepared together with Mogens Høgh Jensen in the fall of 1981. I am grateful to Mogens and to the NORDITA staff, in particular Nils Robert Nilsson, for their help with this project. I thank Harry L Swinney, J Doyne Farmer, David Ruelle, Albert Libchaber, Yves Pomeau, Robert H G Helleman, David Rand, Robert MacKay and Stellan Ostlund for their suggestions and criticisms. And last, but not least, I thank Mitchell J Feigenbaum for teaching me almost all that I know about universality in chaos, and all that I know about Schubert.

P. Cvitanović
NORDITA
August 1983

Note: The reference list for articles referred to in the introduction and the comments to reprint selections is placed at the end of this volume.

Introduction

Universality in Chaos

Predrag Cvitanović[1]
Nordita, Blegdamsvej 17,
DK-2100 Copenhagen Ø

The often repeated statement, that given the initial conditions we know what a deterministic system will do far into the future, is false. Poincaré (1892) knew it was false, and we know it is false, in the following sense: given infinitesimally different starting points, we often end up with wildly different outcomes. Even with the simplest conceivable equations of motion, almost any non-linear system will exhibit chaotic behaviour. A familiar example is turbulence.

Turbulence is _the_ unsolved problem of classical physics. However, recent developments have greatly increased our understanding of turbulence, and given us new concepts and modes of thought that we hope will have far reaching repercussions in many different fields (solid state physics, hydrodynamics, plasma physics, chemistry, quantum optics, biology, meteorology, acoustics, mechanical engineering, elementary particle physics, mathematics, fishery[2], astrophysics, cosmology, electrical engineering and so on).

The developments that we shall describe here are one of those rare demonstrations of the unity of physics. The key discovery was made by a physicist not trained to work on problems of turbulence. In the fall of 1975 Mitchell Feigenbaum, an elementary particle theorist, discovered a universality in one-dimensional iterations. At the time the physical implications of the discovery were rather unclear. During the next few years, however, numerical and theoretical studies established this universality in a number of models in various dimensions. Finally, in 1980, the universality theory passed its first test in an actual turbulence experiment.

The discovery was that large classes of non-linear systems exhibit transitions to chaos which are _universal_ and _quantitatively measurable_. This advance can be compared to past advances in the theory of solid state phase transitions; for the first time we can predict and measure "critical exponents" for turbulence. But the breakthrough consists not so much in discovering a new set of scaling numbers, as in developing a new way to do physics. Traditionally we use regular motions (harmonic oscillators, plane waves, free particles, etc.) as zeroth-order approximations to physical systems, and account for weak non-linearities perturbatively. We think of a dynamical system as a smooth system whose evolution we can follow by integrating

----

1. Lectures given at the XXII-nd Cracow School of Theoretical Physics, Zakopane, June 1982. Sections 1 to 4 were written in collaboration with Mogens Høgh Jensen (Cvitanović and Høgh Jensen 1982). Published in _Acta Physica Polonica_, vol _A65_ (April 1984).

2. Of special interest to our Icelandic colleagues.

a set of differential equations. The universality theory seems to tell us that the zeroth-order approximations to strongly non-linear systems should be quite different. They show an amazingly rich structure which is not at all apparent in their formulation in terms of differential equations. However, these systems do show self-similar structures which can be encoded by universality equations of a type which we will describe here. To put it more succinctly, junk your old equations and look for guidance in clouds' repeating patterns.

In these lectures we shall reverse the chronology, describing first an actual turbulence experiment, then a numerical experiment, and finally explain the observations using the universality theory. We will try to be intuitive and concentrate on a few key ideas, referring you to the literature for more detailed expositions[3]. Even though we illustrate it by turbulence, the universality theory is by no means restricted to the problems of fluid dynamics. The key concepts of phase-space trajectories, Poincaré maps, bifurcations, and local universality are common to all non-linear dynamical systems. The essence of this subject is incommunicable in print; intuition is developed by computing. We urge the reader to carry through a few simple numerical experiments on a desktop computer, because that is probably the only way to start perceiving order in chaos.

1. Onset of turbulence

We start by describing schematically the experiment of Libchaber and Maurer (1980) (a nice description has been given by Libchaber and Maurer(1981)[4]). In this type of experiment a liquid contained in a small box is heated from the bottom. The salient points are:

1. There is a controllable parameter, the Rayleigh number, which is proportional to the temperature difference between the bottom and the top of the cell. (Rayleigh number describes the stability of a convective flow (see Velarde and Normand 1980).)

2. The system is dissipative. Whenever the Rayleigh number is increased, one waits for the transients to die out.

For small temperature gradients there is a heat flow across the cell, but the liquid is static. At a critical temperature a convective flow sets in. The hot liquid rises in the middle, the cool liquid flows down at the sides, and two convective rolls appear:

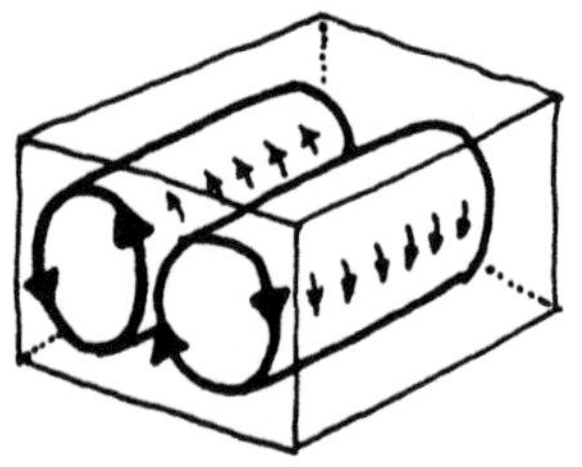

Fig.1.1

----

3. The most thorough exposition available is the Collet and Eckmann (1980a) monograph. We also recommend Hu (1982), Crutchfield, Farmer and Huberman (1982), Eckmann (1981) and Ott (1981).

4. See p.109 this selection.

As the temperature difference is increased further, the rolls become unstable in a very specific way - a wave starts running along the roll:

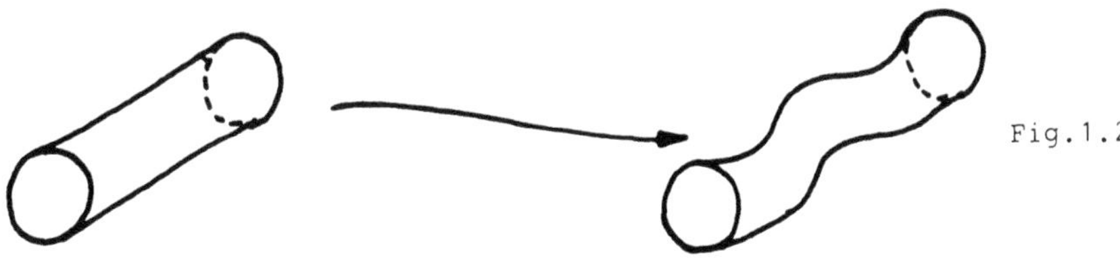
Fig.1.2

As the warm liquid is rising on one side of the roll, while cool liquid is descending down the other side, the position and the sideways velocity of the ridge can be measured with a thermometer:

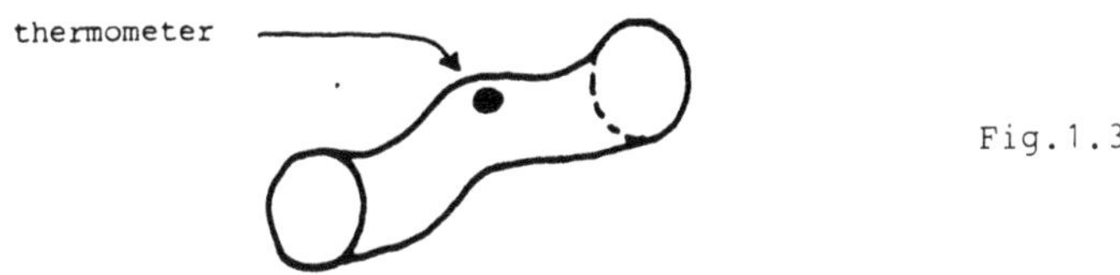

Fig.1.3

One observes a sinusoid:

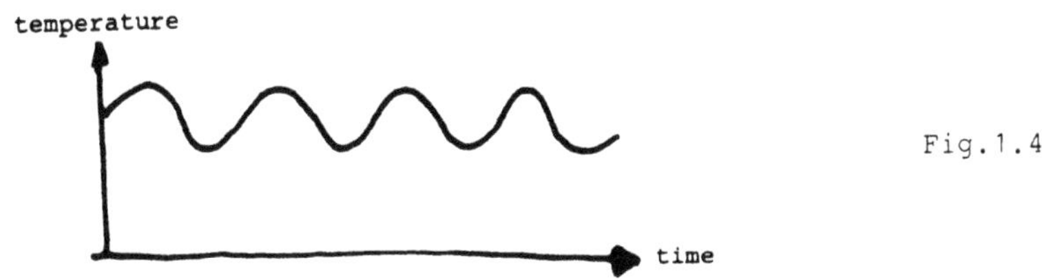

Fig.1.4

The periodicity of this instability suggests two other ways of displaying the measurement:

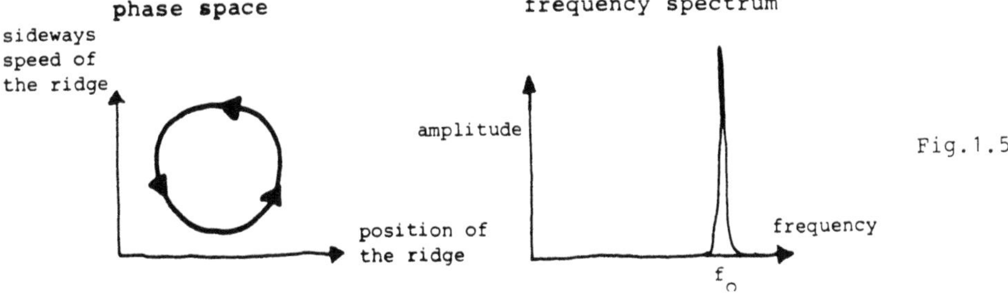

Fig.1.5

Now the temperature difference is increased further. After the stabilisation of the phase-space trajectory, a new wave is observed superimposed on the original sinusoidal instability. The three ways of looking at it (real time, phase space, frequency spectrum) are:

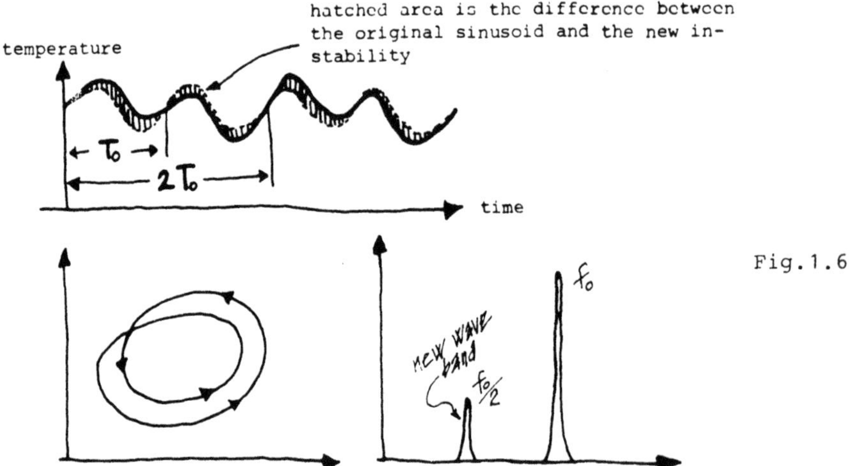

Fig.1.6

A coarse measurement would make us believe that T_o is the periodicity; however, a closer look reveals that the phase-space trajectory misses the starting point at T_o, and closes on itself only after $2T_o$. If we look at the frequency spectrum, a new wave band has appeared at half the original frequency. Its amplitude is small, because the phase-space trajectory is still approximately a circle with periodicity T_o.

Now, as one increases the temperature very slightly, a fascinating thing happens - the phase-space trajectory undergoes a very fine splitting:

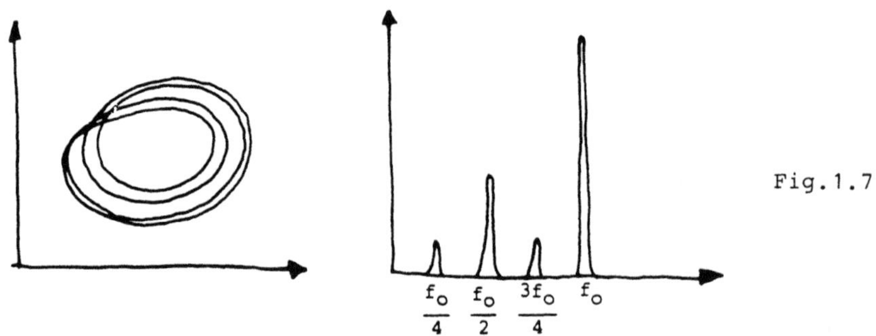

Fig.1.7

We see that there are three scales involved here. Looking casually, we see a circle with period T_o; looking a little closer, we see a pretzel ⓐ with period $2T_o$; and looking very closely, we see that the trajectory closes on itself only after $4T_o$. The same information can be read off the frequency spectrum; the dominant frequency is f_o (the circle), then $f_o/2$ (the pretzel), and finally, much weaker $f_o/4$ and $3f_o/4$.

The experiment now becomes very difficult. A minute increase in the temperature gradient causes the phase-space trajectory to split on an even finer scale, with the periodicity 2^3T_o. If the noise were not killing us, we would expect these splittings to continue, yielding a trajectory with finer and finer detail, and a frequency spectrum

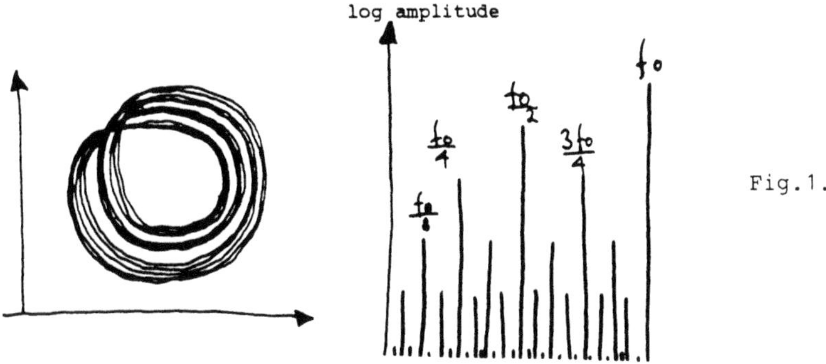

Fig.1.8

with families of ever weaker frequency components. For a critical value of the Rayleigh number, the periodicity of the system is $2^{\infty}T_O$, and the convective rolls have become turbulent (this is weak turbulence - the rolls persist, wiggling irregularly). The ripples which are running along them show no periodicity, and the spectrum of idealized, noise-free experiment contains infinitely many subharmonics:

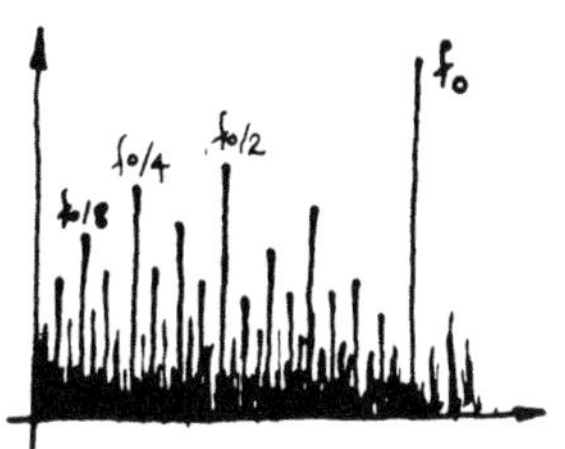

Fig.1.9

If one increases the temperature gradient beyond this critical value, there are further surprises: we refer you to Libchaber and Maurer (1981). We now turn to a numerical simulation of a simple non-linear oscillator in order to start understanding why the phase-space trajectory splits in this peculiar fashion.

2. Onset of chaos in a numerical experiment

In the experiment that we have just described, limited experimental resolution makes it impossible to observe more than a few bifurcations. Much longer sequences can be measured in numerical experiments; the non-linear oscillator studied by Arecchi and Lisi (1982) is a typical example:

$$\ddot{x} + k\dot{x} - x + 4x^3 = A\cos(\omega t) \tag{2.1}$$

The oscillator is driven by an external force of frequency ω, with amplitude A and the natural time unit $T_O = 2\pi/\omega$. The dissipation is controlled by the friction coefficient k. Given the initial displacement and velocity one can easily follow numerically (by the Runge-Kutta method, for example) the phase-space trajectory of the system. Due to the dissipation it does not matter where one starts in the phase space; for a wide range of initial points the phase-space trajectory converges to a limit cycle (trajectory loops onto itself) which for some $k = k_o$ looks something like this (fig. 12a in Feigenbaum 1980a)[5]:

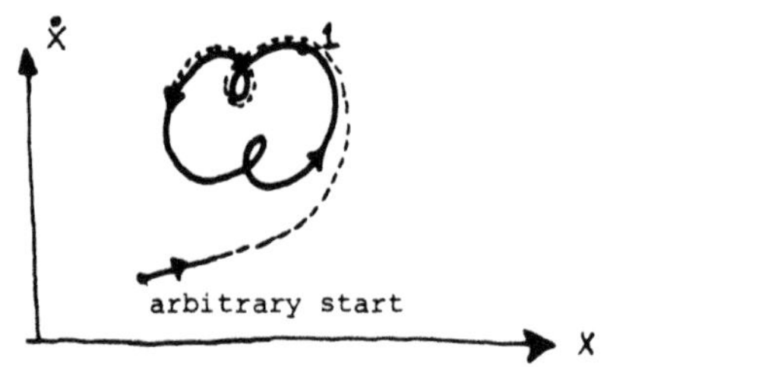

Fig.2.1

If it were not for the external driving force, the oscillator would have simply come to a stop; as it is, it is executing a motion forced on it externally, independent of the initial displacement and velocity. You can easily visualise this non-linear pendulum executing little backward jerks as it swings back and forth. Starting at the point marked 1, the pendulum returns to it after the unit period T_o.

However, as one decreases the friction, the same phenomenon is observed[6] as in the turbulence experiment; the limit cycle undergoes a series of period-doublings

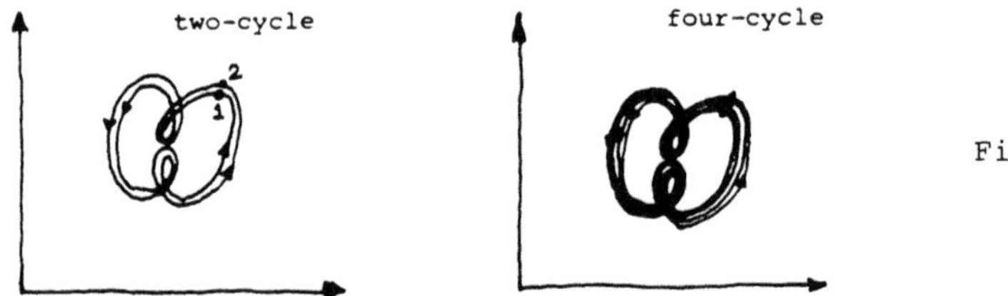

Fig.2.2

The trajectory keeps on nearly missing the starting point, until it hits it after exactly $2^n T_o$. The phase-space trajectory is getting increasingly hard to draw; however, the sequence of points 1, 2, ..., 2^n, which corresponds to the state of the oscillator at times T_o, $2T_o$, ..., $2^n T_o$, sits in a small region of the phase space, so we enlarge it for a closer look:

----

5. See p. 49 this selection.

6. If you have a desktop computer with graphics, you can easily do this experiment yourself. For example, if you take $k = 0.154$, $\omega = 1.2199778$ and $A = 0.1, 0.11, 0.114, 0.11437, \ldots$, you will observe bifurcations. There is nothing special about these parameter values; we give them just to help you with finding your first bifurcation sequence.

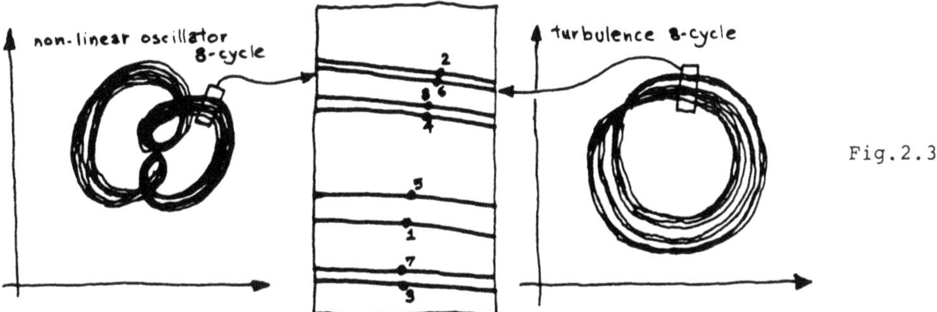

Fig.2.3

Globally the phase-space trajectories of the turbulence experiment and of the non-linear oscillator numerical experiment look very different. However, the above sequence of near misses is <u>local</u>, and looks roughly the same for both systems. Furthermore, this sequence of points lies approximately on a straight line

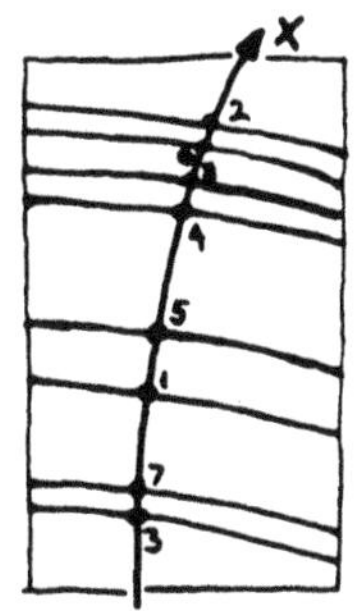

Fig.2.4

Let us concentrate on this line: this way of reducing the dimensionality of the phase space is often called a <u>Poincaré map</u>. Instead of staring at the entire phase-space trajectory, one looks at its points of intersection with a given surface. The Poincaré map contains all the information we need; from it we can read off when an instability occurs, and how large it is. One varies continuously the non-linearity parameter (friction, Rayleigh number, etc.) and plots the location of the intersection points; in the present case, the Poincaré surface is a line, and the result is a <u>bifurcation tree</u>:

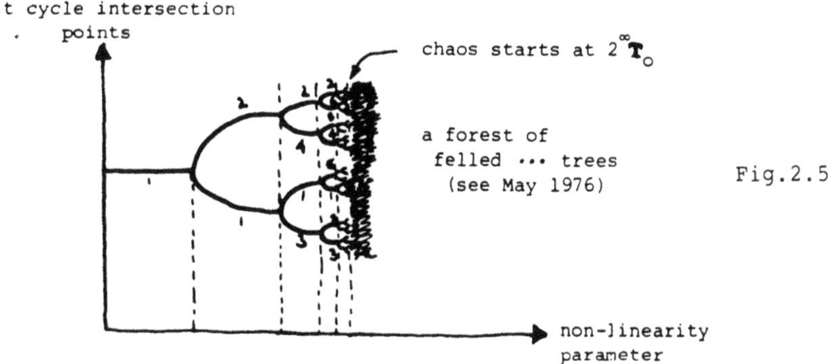

Fig.2.5

We already have some qualitative understanding of this plot. The phase-space trajectories we have drawn are localised (the energy of the oscillator is bounded) so the tree has a finite span. Bifurcations occur simultaneously because we are cutting a single trajectory; when it splits, it does so everywhere along its length. Finer and finer scales characterise both the branch separations and the branch lengths.

Feigenbaum's discovery[7] consists of the following quantitative observations:

1. The parameter convergence is universal (i.e., independent of the particular physical system):

$$\Delta_i/\Delta_{i+1} \to \delta = 4.6692... \quad \text{for } i \text{ large}$$

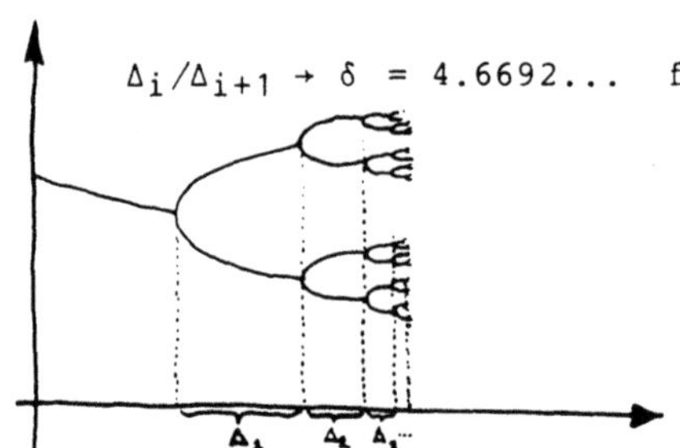

Fig.2.6

2. The relative scale of successive branch splittings is universal:

$$\varepsilon_i/\varepsilon_{i+1} \to \alpha = 2.5029... \quad \text{for } i \text{ large}$$

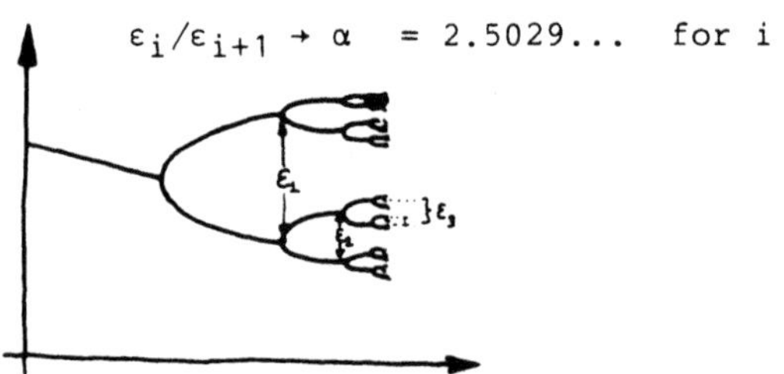

Fig.2.7

The beauty of this discovery is that if turbulence (chaos) is arrived at through an infinite sequence of bifurcations, we have two quantitative predictions:

1. The convergence of the critical Rayleigh numbers corresponding to the cycles of length 2, 4, 8, 16, ... is controlled by the universal convergence parameter $\delta = 4.6692016...$.

2. The splitting of the phase-space trajectory is controlled by the universal scaling parameter $\alpha = 2.50290787...$. As we have indicated in our discussion of the turbulence experiment, the relative heights of successive subharmonics measure this splitting and hence α.

----

7. The geometric parameter convergence was first noted by Myrberg (1958), and independently of Feigenbaum, by Grossmann and Thomae (1977). However, these authors have not emphasised the universality of δ.

These universal numbers are measured in a variety of experiments: we shall summarise the experimental situation in section 9.

You might think that this universality applies only to very simple, essentially one-dimensional systems (single pendulum, oscillations along a convective roll), but that is not true. For example, Franceschini and Tebaldi (1979)[8] have investigated numerically the following system:

$$\dot{x}_1 = -2x_1 + 4x_2 x_3 + 4x_4 x_5$$
$$\dot{x}_2 = -9x_2 + 3x_1 x_3$$
$$\dot{x}_3 = -5x_3 - 7x_1 x_2 + \underline{r}$$
$$\dot{x}_4 = -5x_4 - x_1 x_5$$
$$\dot{x}_5 = -x_5 - 3x_1 x_4$$

where $\underline{r}$ is the controllable external parameter. Within certain intervals of values of the parameter $\underline{r}$ ("Reynolds number" for the system) they found infinite sequences of period-doublings. Moreover, even though the phase space was ten-dimensional, any Poincaré map they tried yielded period-doublings along a one-dimensional line. The convergence parameter δ and the scaling number α they obtained agree with Feigenbaum's universal numbers. This particular system of equations is a truncation of the Navier-Stokes equations, and in the literature there are innumerable other examples of period-doublings in many-dimensional systems. A wonderful thing about this universality is that it does not matter much how close our equations are to the ones chosen by nature; as long as the model is in the same universality class (in practice this means that it can be modelled by a mapping of form (3.2)) as the real system, both will undergo a period-doubling sequence. That means that we can get the right physics out of very crude models, and this is precisely what we will do next. (Of course, we have no clue whether we are really in the same universality class.)

The reason why multidimensional <u>dissipative</u> systems become effectively one-dimensional is roughly this: for a dissipative system phase-space volumes shrink. They shrink at different rates in different directions; the direction of the slowest convergence defines a one-dimensional line which will contain the attractor (the region of the phase space to which the trajectory is confined at asymptotic times):

Fig.2.8

The real story is both more subtle and more interesting; the phase-space volume shrinks in some directions and grows and folds onto itself in others - so the attractor, while thin, can be very complicated. Nearby points on this attractor can have very different

----

8. See p.379 this selection.

histories. You should read Collet and Eckmann (1980a) and Collet, Eckmann and Koch (1980)[9] for a detailed description of how a dissipative system becomes one-dimensional.

What we have presented so far are a few experimental facts; we now have to convince you that they are universal. To do this, we shall have to talk about fish.

3. What does all this have to do with fishing?

Looking at the phase-space trajectories shown earlier, we observe that the trajectory bounces within a restricted region of the phase space. How does this happen? One way to describe this bouncing is to plot the (n+1)th intersection of the trajectory with the Poincaré surface as a function of the preceding intersection. Referring to fig.2.4 we find

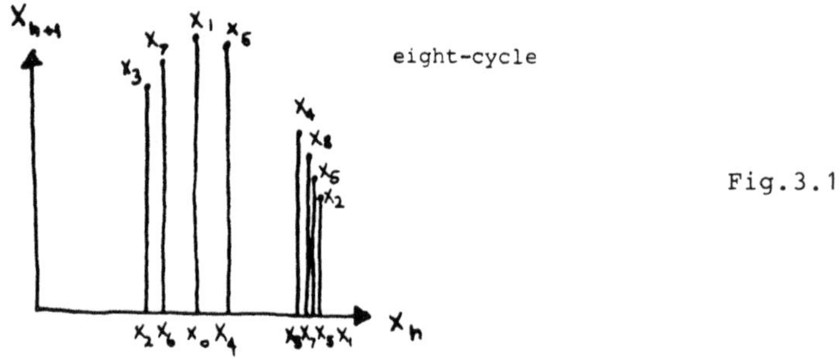

Fig.3.1

This is a Poincaré map (or return map) for the limit cycle. If we start at various points in the phase space (keeping the non-linearity parameter fixed) and mark all passes as the trajectory converges to the limit cycle, we trace an approximately continuous curve f(x)

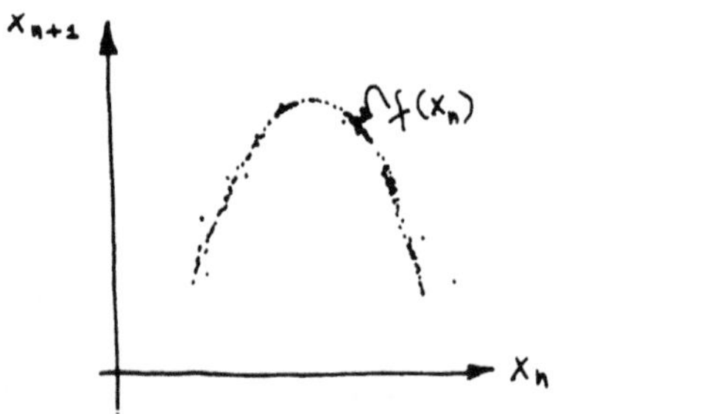

Fig.3.2

which gives the location of the trajectory at time $t+T_o$ as a function of its location at time t:

$$x_{n+1} = f(x_n) \tag{3.1}$$

The trajectory bounces within a trough in the phase space, and f(x) gives a local description of the way the trajectories converge to the

----

9. See p.353 this selection.

limit cycle. In principle we know $f(x)$, as we can measure it (see Simoyi, Wolf and Swinney (1982) for a construction of a return map in a chemical turbulence experiment[10]) or compute it from the equations of motion. The form of $f(x)$ depends on the choice of Poincaré map, and obtaining an analytic expression for $f(x)$ is difficult (see Gonzales and Piro (1983) for an example of an explicit return map), but we know what $f(x)$ should look like; it has to fall on both sides (to confine the trajectory), so it has a maximum. Around the maximum it looks like a parabola

$$f(x) = a_o+a_2(x-x_c)^2 +\ldots \tag{3.2}$$

like any sensible polynomial approximation to a function with a hump[11].

This brings us to the problem of a rational approach to fishery. By means of a Poincaré map we have reduced a continuous trajectory in phase space to one-dimensional iteration. This one-dimensional iteration has an analog in population biology, where $f(x)$ is interpreted as a population curve (the number of fish x_{n+1} in the given year as a function of the number of fish x_n the preceding year), and the bifurcation tree fig.2.5 has been studied in considerable detail. We recommend reviews by May (1976)[12] and Hoppensteadt (1978) for further reading.

The first thing we need to understand is the way in which a trajectory converges to a limit cycle. A numerical experiment will give us something like:

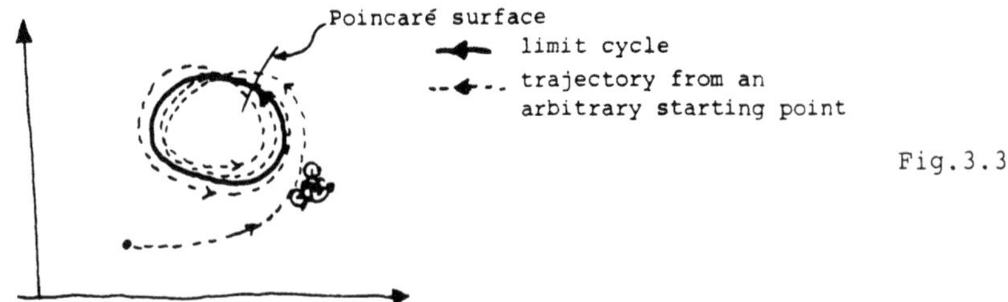

Fig.3.3

In the Poincaré map the limit trajectory maps onto itself

$$x^* = f(x^*)$$

Hence a limit trajectory corresponds to a <u>fixed point</u> of $f(x)$. Take a programmable pocket calculator and try to determine x^*. Type in a simple approximation[13] to $f(x)$, such as

$$f(x) = 1-Rx^2 \tag{3.3}$$

----

10. See p.164 this selection.

11. If $f(x)$ is not quadratic around the maximum, the universal numbers will be different - see Villela Mendés (1981) and Hu and Mao (1982b) for their values. According to Kuramoto and Koga (1982) such mappings can arise in chemical turbulence.

12. See p. 85 this selection.

13. This way of modelling dynamical systems was introduced by Lorenz (1964).

Here R is the non-linear parameter. Enter your guess x_1 and press the button. The number x_2 appears on the display. Is it a fixed point? Press the button again, and again, until $x_{n+1} = x_n$ to desired accuracy. Diagrammaticaly

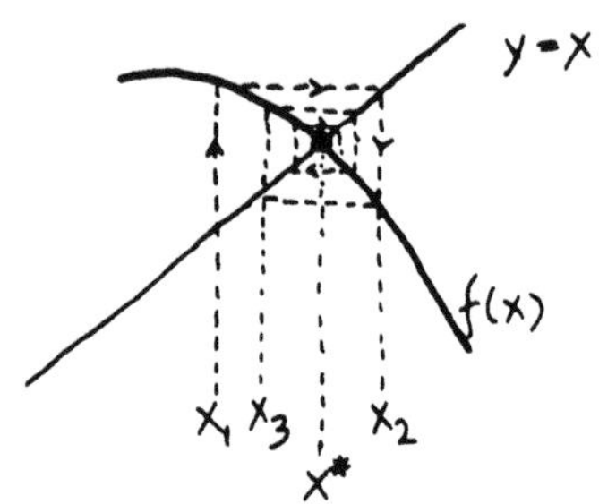

Fig.3.4

Note the tremendous simplification gained by the use of the Poincaré map. Instead of computing the entire phase-space trajectory by a numerical integration of the equations of motion, we are merely pressing a button on a pocket calculator.

This little calculation confirms one´s intuition about fishery. Given a fishpond, and sufficient time, one expects the number of fish to stabilise. However, no such luck - a rational fishery manager soon discovers that anything can happen from year to year. The reason is that the fixed point x^* need not be attractive, and our pocket calculator computation need not converge.

4. A universal equation

Why is the naive fishery manager wrong in concluding that the number of fish will eventually stabilise? He is right when he says that $x^* = f(x^*)$ corresponds to the same number of fish every year. However, this is not necessarily a stable situation. Reconsider how we got to the fixed point in fig.3.4. Starting with a sufficiently good guess, the iterates converge to the fixed point. Now start increasing gently the non-linearity parameter (Rayleigh number, the nutritional value of the pond, etc.). f(x) will slowly change shape, getting steeper and steeper at the fixed point,

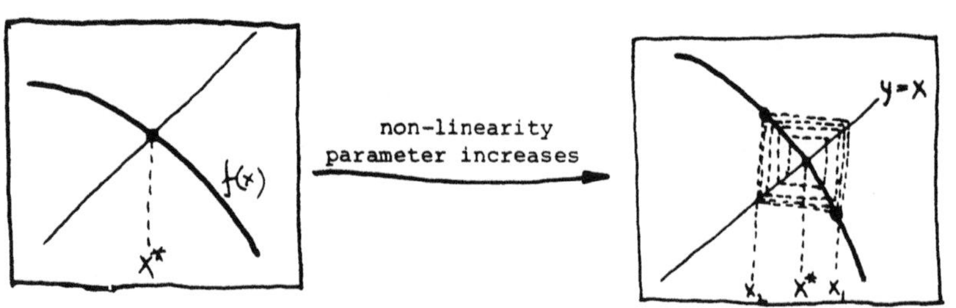

Fig.4.1

until the fixed point becomes unstable and gives birth to a cycle of two points[14]. This is precisely the first bifurcation observed in our experiments. This is the only gentle way in which our trajectory can become unstable (cycles of other lengths can be created, but that requires delicate fiddling with parameters; they are not generic). Now we return to the same point after every second iteration

$$x_i = f(f(x_i)) \quad i = 1, 2.$$

so the cycle points of f(x) are the fixed points of f(f(x)).

To study <u>their</u> stability, we plot f(f(x)) alongside f(x):

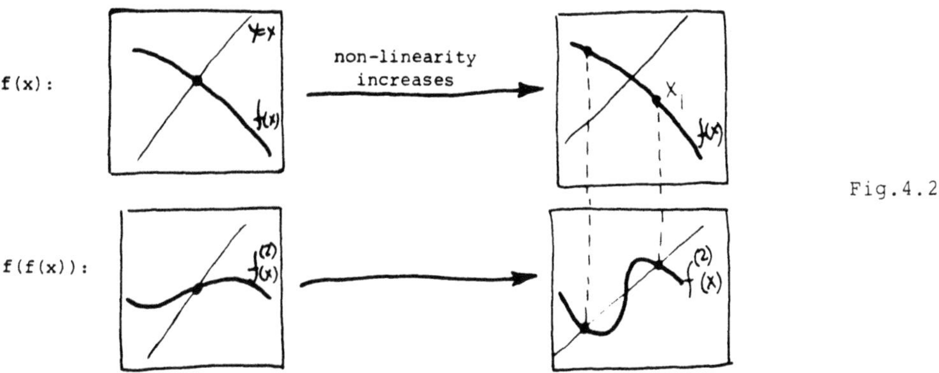

Fig.4.2

What happens as we continue to increase the "Rayleigh number"? f(x) becomes steeper at its fixed point, and so does f(f(x)). Eventually the magnitude of the slope at the fixed points of f(f(x)) exceeds one, and they bifurcate. Now the cycle is of length four, and we can study the stability of the fixed points of the fourth iterate. They too will bifurcate, and so forth. This is why the phase-space trajectories keep on splitting 2 → 4 → 8 → 16 → 32 ... in our experiments. The argument does not depend on the precise form of f(x), and therefore the phenomenon of successive period-doublings is <u>universal</u> (Metropolis,Stein and Stein, 1973)[15].

More amazingly, this universality is not only qualitative. In our analysis of the stability of fixed points we kept on magnifying the neighbourhood of the fixed point:

----

14. Program your pocket calculator to evaluate (3.3). Chose some value of R between 0 and 2, and x between -1 and 1. Type in the initial x, press the start button and read off the next x. Now press the start button again. The game consists in staring at the display, and looking for regularities in the sequences of iterates. Try also the following values of R: 1, 1.31070274134, 1.38154748443, 1.3979453597. Compute the next number in this series.

15. See p.187 this selection.

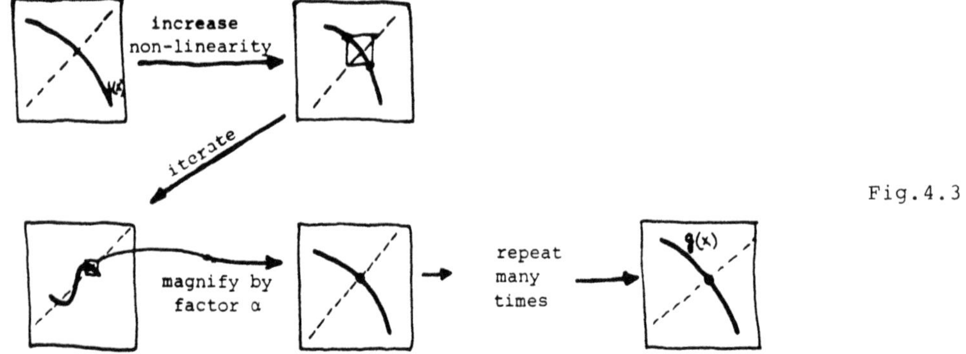

Fig.4.3

The neighbourhoods of successive fixed points look very much the same after iteration and rescaling. After we have magnified the neighbourhoods of fixed points many times, practically all information about the global shape of the starting function f(x) is lost, and we are left with a universal function g(x). Denote by T the operation indicated in fig.4.3: iterate twice and rescale by α (without changing the non-linearity parameter),

$$Tf(x) = -\alpha f(f(-x/\alpha)). \tag{4.1}$$

g(x) is self-reproducing under rescaling and iteration:

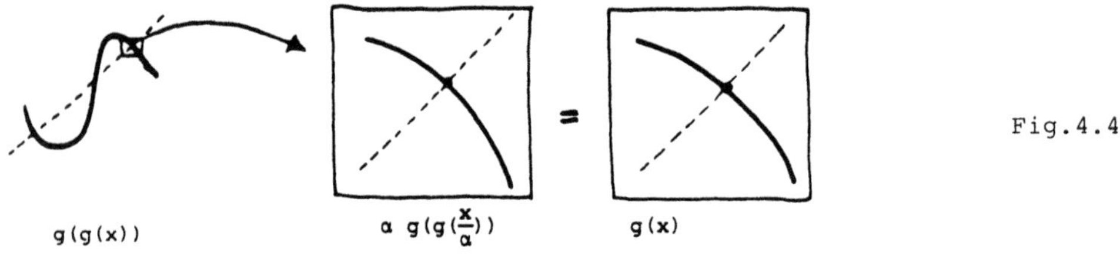

Fig.4.4

More precisely, this can be stated as the universal equation

$$g(x) = -\alpha g(g(-x/\alpha)), \tag{4.2}$$

which determines both the universal function g(x) and $\alpha = -1/g(1) = 2.50290787\ldots$ (with normalisation convention $g(0) = 1$).

If you arrive at g(x) the way we have, by successive bifurcations and rescalings, you can hardly doubt its existence. However, if you start with (4.2) as an equation to solve, it is not obvious what its solutions should look like. The simplest thing to do is to approximate g(x) by a finite polynomial and solve the universal equation numerically, by Newton's method (Feigenbaum 1979a). This way you can compute α and δ to much higher accuracy than you can ever hope to measure them to experimentally.

There is much pretty mathematics in universality theory. Despite its simplicity, nobody seems to have written down the universal equation[16] before 1976, so the subject is still young. We do not have a series expansion for α, or an analytic expression for g(x); the numbers that we have are obtained by boring numerical methods. So far, all we know is that g(x) exists (Lanford 1982)[17]. What is proved is that the Newton iteration converges, so we are no wiser for the result. In some situations the universal equation (4.2) has analytic solutions; we shall return to this in the discussion of intermittency (section 10). The universality theory has also been extended to iterations of complex polynomials (section 12).

To see why the universal function must be a rather crazy function, consider high iterates of f(x) for parameter values corresponding to 2-, 4- and 8-cycles:

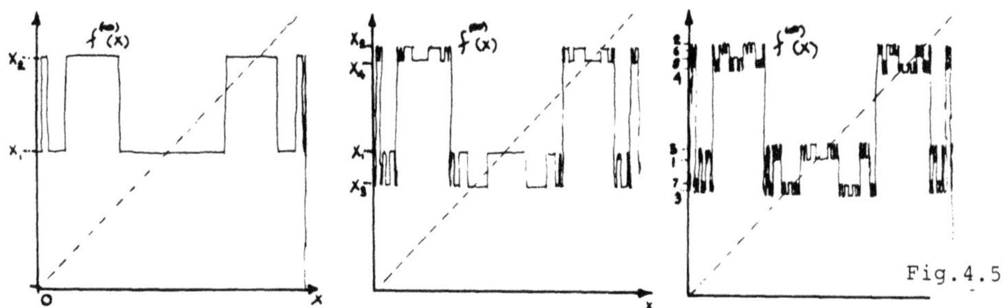

Fig.4.5

If you start anywhere in the unit interval and iterate a very large number of times, you end up in one of the cycle points. For the 2-cycle there are two possible limit values, so f(f(...f(x))) resembles a castle battlement. Note the infinitely many intervals accumulating at the unstable x = 0 fixed point. In a bifurcation of the 2-cycle into the 4-cycle each of these intervals gets replaced by a smaller battlement. After infinitely many bifurcations this becomes a fractal (i.e., looks the same under any enlargement), with battlements within battlements on every scale. Our universal function g(x) does not look like that close to the origin, because we have enlarged that region by the factor α = 2.5029... after each period-doubling, but all the wiggles are still there; you can see them in Feigenbaum's (1978)[18] plot of g(x). For example, (4.2) implies that if x^* is a fixed point of g(x), so is αx^*. Hence g(x) must cross the lines y = x and y = -x infinitely many times. It is clear that while around the origin g(x) is roughly a parabola and well approximated by a finite polynomial, something more clever is needed to describe the infinity of g(x)'s wiggles further along the real axis and in the complex plane.

All this is fun, but not essential for understanding the physics of the onset of chaos. The main thing is that we now understand where the universality comes from. We start with a complicated

----

16. The universal equation was introduced by the author, in collaboration with M. J. Feigenbaum (1978). Coullet and Tresser (1978a,b) have proposed similar equations.

17. See p.245 this selection.

18. See p. 67 this selection.

many-dimensional dynamical system. A Poincaré map reduces the problem from a study of differential equations to a study of discrete iterations, and dissipation reduces this further to a study of one-dimensional iterations (now we finally understand why the phase-space trajectory in the turbulence experiment undergoes a series of bifurcations as we turn the heat up!). The successive bifurcations take place in smaller and smaller regions of the phase space. After n bifurcations the trajectory splittings are of order $\alpha^{-n} = (0.399...)^n$ and practically all memory of the global structure of the original dynamical system is lost:

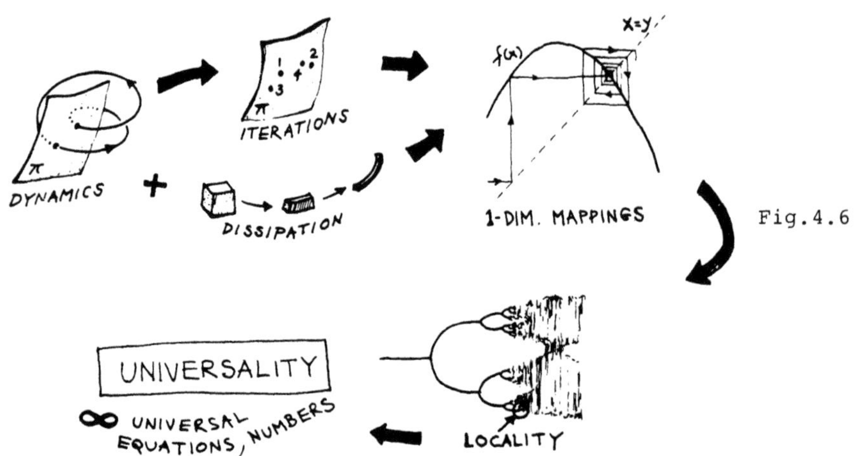

Fig.4.6

The asymptotic self-similarities can be encoded by universal equations. The physically interesting scaling numbers can be quickly estimated by simple truncations of the universal equations (May and Oster 1980, Derrida and Pomeau 1980, Helleman 1980a[19], Hu 1981). The full universal equations are designed for accurate determinations of universal numbers; as they have built-in rescaling, the round-off errors do not accumulate, and the only limit on the precision of the calculation is the machine precision of the computer.

Anything that can be extracted from the asymptotic period-doubling regime is universal[20]; the trick is to identify those universal features that have a chance of being experimentally measurable. We will discuss several such extensions of the universality theory in the remainder of this introduction.

----

19. See p.458 this selection.

20. Derrida, Gervois and Pomeau (1979) have extracted a great many metric universalities from the asymptotic regime. Grassberger (1981) has computed the Hausdorff dimension of the asymptotic attractor. Lorenz (1980) and Daido (1981b) have found a universal ratio relating bifurcations and reverse bifurcations. A number of other universal quantities are discussed in this reprint selection.

5. The unstable manifold

Feigenbaum's delta:

$$\delta = \lim_{n\to\infty} \frac{r_{n-1} - r_n}{r_n - r_{n+1}} = 4.6692016... \qquad (5.1)$$

is the universal number of the most immediate experimental import: it tells us that in order to reach the next bifurcation we should increase the Rayleigh number (or friction, or whatever the controllable parameter is in the given experiment) by about one fifth of the preceding increment. Which particular parameter is being varied is largely a question of experimental expedience; if r is replaced by another parameter R = R(r), then the Taylor expansion

$$R(r) = R(r_\infty)+(r-r_\infty)R'(r_\infty)+(r-r_\infty)^2R''(r_\infty)/2 + ...$$

yields the same asymptotic delta

$$\delta \simeq \frac{R(r_{n-1}) - R(r_n)}{R(r_n) - R(r_{n+1})} = \frac{r_{n-1} - r_n}{r_n - r_{n+1}} + O(\delta^{-n}) \qquad (5.2)$$

providing, of course, that $R'(r_\infty)$ is non-vanishing (the chance that a physical system just happens to be parametrised in such a way that $R'(r_\infty) = 0$ is nil).

In deriving the universal equation (4.2) we were intentionally sloppy, because we wanted to introduce the notion of encoding self-similarity by universal equations without getting entangled in too much detail. We obtained a universal equation which describes the self-similarity in the x-space, under iteration and rescaling by α. However, the bifurcation tree fig.2.5 is self-similar both in the x-space and the parameter space: each branch looks like the entire bifurcation tree. We will exploit this fact to construct a universal equation which determines both α and δ.

Let T* denote the operation of iterating twice, rescaling x by α, shifting the non-linearity parameter to the corresponding value at the next bifurcation[21], and rescaling it by δ:

$$T^*f_{R_n + \Delta_n p}(x) = -\alpha_n f^{(2)}_{R_n + \Delta_n(1 + p/\delta_n)}(-x/\alpha_n) \qquad (5.3)$$

Here R_n is a value of the non-linearity parameter for which the limit cycle is of length 2^n, Δ_n is the distance to R_{n+1}, $\delta_n = \Delta_n/\Delta_{n+1}$, p provides a continuous parametrisation, and we appologise that there are so many subscripts. T* operation encodes the self-similarity of the bifurcation tree[22] (fig.2.5):

----

21. More precisely, the value of the nonlinearity parameter with the same stability, i.e. the same slope at the cycle points.

22. Collet and Eckmann (1980a) give a very nice illustration of this self-similarity in Fig.I.28 of their monograph.

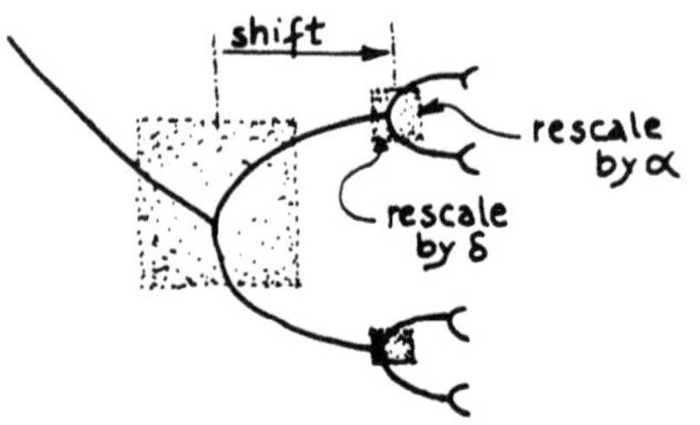

Fig.5.1

For example, if we take the fish population curve f(x) (3.3) with R value corresponding to a cycle of length 2^n, and act with T*, the result will be a similar cycle of length 2^n, but on a scale α times smaller. If we apply T* infinitely many times, the result will be a universal function with a cycle of length 2^n:

$$g_p(x) = (T^*)^\infty f_{R+p\Delta}(x) \tag{5.4}$$

If you can visualise a space of all functions with quadratic maximum, you will find the following picture helpful:

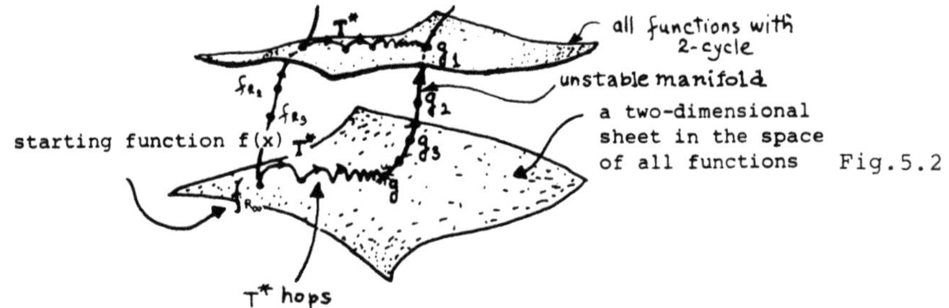

Fig.5.2

Each transverse sheet is a manifold consisting of functions with 2^n-cycle of given stability. T* moves us across this transverse manifold toward g_p.

$g_p(x)$ is invariant under the self-similarity operation T*, so it satisfies a universal equation:

$$g_p(x) = -\alpha g_{1+p/\delta}(g_{1+p/\delta}(-x/\alpha)) \tag{5.5}$$

p parametrises a one-dimensional continuum family of universal functions[23]. Our first universal equation (4.2) is the fixed point of the above equation:

$$p^* = 1+p^*/\delta \tag{5.6}$$

and corresponds to the asymptotic 2^∞-cycle[24].

----

23. This elegant formulation of universality is due to Vul and Khanin (1982) and Goldberg, Sinai and Khanin (1983). I have learned it from M.J. Feigenbaum. The family of universal functions parametrised by p is called the unstable manifold because T-operation (4.1) drives p away from the fixed point value $g(x) = g_{p^*}(x)$.

24. This is not necessarily the only way to formulate universality; for example, Daido (1981a) has introduced a different set of universal equations.

You have probably forgotten by now, but we started this section promising a computation of δ. Feigenbaum (1979a) solved this problem by linearising the equations (5.5) around the fixed point p*. Close to the fixed point $g_p(x)$ does not differ much from g(x), so one can treat it as a small deviation from g(x):

$$g_p(x) = g(x) + (p-p^*)h(x)$$

Substitute this into (5.5), keep the leading term in p-p*, and use the universal equation (4.2). This yields a universal equation for δ:

$$g'(g(x))h(x) + h(g(x)) = -(\delta/\alpha)h(\alpha x) \qquad (5.7)$$

We already know g(x) and α, so this can be solved numerically by polynomial approximations, yielding $\delta = 4.6692016...$ (plus a whole spectrum of eigenvalues and eigenvectors h(x) (see Feigenbaum 1979a [25]).

Actually, one can do better with less work; T*-operation treats the coordinate x and the parameter p on the same footing, which suggests that we should approximate the entire unstable manifold by a double power series (Vul and Khanin 1982, Goldberg, Sinai and Khanin 1983)

$$g_p(x) = \sum_{j=0}^{N} \sum_{k=0}^{M} c_{jk} x^{2j} p^k \qquad (5.8)$$

The scale of x and p is arbitrary. We will fix it by the normalisation conditions

$$g_0(0) = 0 \qquad (5.9)$$

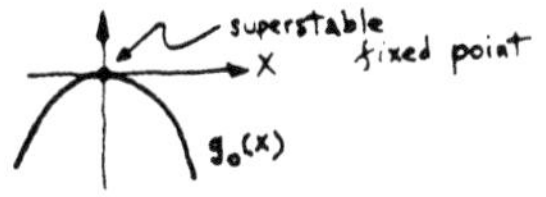

$$g_1(0) = 1$$
$$g_1(1) = 0 \qquad (5.10)$$

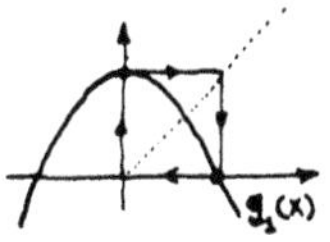

The first condition means that the origin of p corresponds to the superstable fixed point. The second condition sets the scale of x and p by the superstable 2-cycle. (Superstable cycles are the cycles which have the maximum of the map as one of the cycle points.) Start with any simple approximation to $g_p(x)$ which satisfies the above conditions (for example, $g_p(x) = p-x^2$). Apply the T*-operation (5.3) to it. This involves polynomial expansions in which terms of order higher than M and N in (5.8) are dropped. Now find by Newton's method the value of δ which satisfies normalisation (5.10). This is the only numerical calculation you have to do; the condition (5.10) automatically yields the value of α. The result is a new approximation to g_p. Keep applying T* until the coefficients in (5.8) repeat; this has moved the approximate g_p toward the unstable manifold along the transverse sheets indicated in fig.5.2. Computationally this is straightforward, the accuracy of the computation is limited only by computer precision, and at the end you will have α, δ and a polynomial approximation to the unstable manifold $g_p(x)$.

As δ controls the convergence of the high iterates of the initial

----

25. See p. 207 this selection.

mapping toward their universal limit g(x), it also controls the convergence of most other numbers toward their universal limits, such as the scaling number $\alpha_n = \alpha+O(\delta^{-n})$, or even δ itself, $\delta_n = \delta+O(\delta^{-n})$. As $1/\delta = 0.2141...$, the convergence is very rapid, and already after a few bifurcations the universality theory is good to a few per cent. This rapid convergence is both a blessing and a curse. It is a theorist's blessing because the asymptotic theory applies already after a few bifurcations; but it is an experimentalist's curse because a measurement of every successive bifurcation requires a fivefold increase in the experimental accuracy of the determination of the non-linearity parameter r.

6. Power spectra

We have stated that the physical significance of the scaling number α is that it sets the scale of trajectory splitting. That is true, but not good enough to make connection with experiments; our theory describes the splitting of the phase-space trajectories, while experimentalists (as we have discussed in section 1) usually measure the power spectrum. To construct the asymptotic spectrum, we need not only α (the splitting at the maximum of the return map), but also the splitting everywhere along the trajectory. This splitting is described by Feigenbaum's (1979b, 1980a[26]) scaling function $\sigma(t)$.

To estimate the shape of the power spectrum, consider the 1-dimensional iteration analogue of the phase-space and real time outputs of the turbulence experiment (fig.1.6):

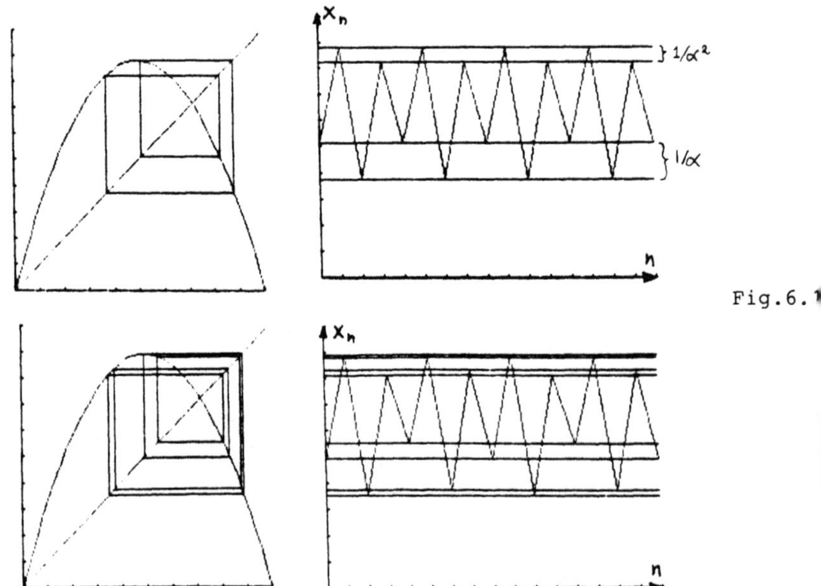

Fig.6.1

----

26. See p. 49 this selection.

(these were obtained by taking the non-linearity parameter in (3.3) corresponding to 4- and 8-cycles and iterating). With each period-doubling a new instability comes in, at twice the period, or half the frequency of the preceding instability. Its amplitude can be read off the above figures; very crudely, it is just $1/\alpha$ (the width of the trajectory splitting) times the previous amplitude. We can easily do better; it is obvious from the above figures that the trajectory splitting is dominated by two scales; half of the time the splitting is of order $1/\alpha$, and the other half is determined by the projection through the quadratic maximum, $1/\alpha^2$. The successive subharmonics in the power spectrum (amplitude squared) fall off like the mean-square average of the two dominant trajectory splitting scales:

$$\mu \simeq 2(1/\alpha^2 + 1/\alpha^4)^{-\frac{1}{2}} = 4.648... \qquad (6.1)$$

This rough estimate tells us that the scaling function $\sigma(t)$ is basically either α^{-1} or α^{-2}. (A plot of $\sigma(t)$ is given in Feigenbaum 1980b[27].) The splitting is taking place on a smaller and smaller scale, and we expect universality. A numerical calculation (Nauenberg and Rudnick 1981)[28] yields $\mu = 4.578...$. The actual power spectrum (a trivial calculation to do on a desktop computer) looks something like this:

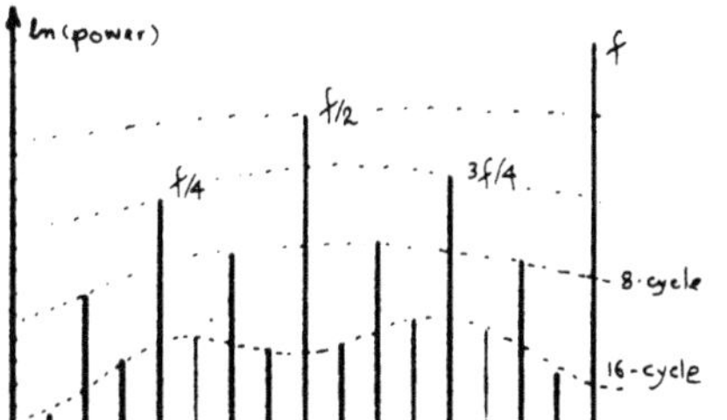

Fig.6.2

On average the subharmonics do drop by the predicted ratio, but they are strongly modulated.

Comparison with experiments (see section 9) requires some care. The problem is that one wants to use all subharmonics and higher harmonics seen in an experiment in computing the average (6.1), but the envelope of higher harmonics is not universal; it represents the deviation of the global phase-space trajectory (fig.1.6) from a perfect circle. A careful analysis of an experiment would utilise the spectrum observed in the first few bifurcations as an input for the calculation of the spectrum of the subsequent subharmonics.

7. Reverse bifurcations

So far we have concentrated on the sequence of bifurcations which arises as we smoothly destabilise a fixed point into a 2-cycle, 2-cycle into 4-cycle, and so on, until 2^∞-cycle. One can say that the system

----

27. See p. 74 this selection.

28. See p. 276 this selection.

has become turbulent in the sense that the time period has become infinite, but the trajectory is anything but chaotic; it follows a very strict itinerary. Chaotic motion would usually mean that nearby trajectories diverge exponentially (positive Lyapunov coefficients) or that the power spectrum is characterised by broad-band noise (we somehow forgot to mention what we mean by chaos in this introduction to chaos). What happens if we keep on increasing the non-linearity parameter?

The most surprising thing that happens is that the system does not necessarily get more chaotic (remember, we are turning up the heat in the turbulence experiment, so one would expect more and more turbulence). Instead, one observes an infinite number of parameter ranges (windows) for which the system becomes periodic again. The largest such window is the 3-window, whose emergence we can detect by searching for fixed points of f(f(f(x))). For the parameter value R = 1.750... in the mapping (3.3), f(f(f(x))) acquires three attractive fixed points, and f(x) acquires a 3-cycle:

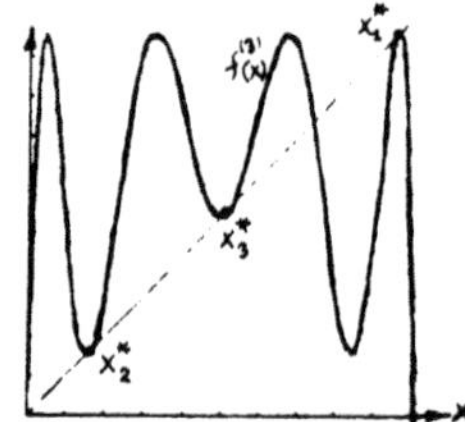

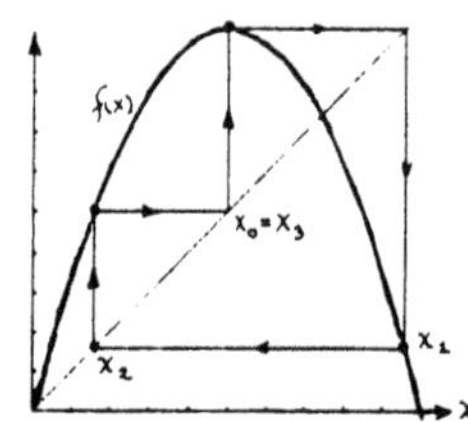

Fig.7.1

If we now smoothly increase R, the fixed points of f(f(f(x))) bifurcate as before: the 3-cycle goes into a 6-cycle which then bifurcates into a 12-cycle, etc., with the same asymptotic universal period-doubling numbers[29].

Metropolis, Stein and Stein (1973)[30] have discovered a qualitative universality in the relative ordering of the windows of different periods; the order does not depend on the map f(x), as long as f(x) has a differentiable maximum and falls off monotonically on both sides (see Collet and Eckmann (1980a) for a detailed discussion of universal ordering and itineraries of periodic orbits). Observation of the universal ordering in an experiment (such as Simoyi, Wolf and Swinney (1982)[31]) is strong evidence that the return map is one-dimensional; there is no ordering in higher dimensions where one can have co-existing cycles of different lengths and with different basins of attraction.

Periodic windows notwithstanding, the regime beyond 2^∞ parameter value is very chaotic. You can easily map out both the periodic windows and the chaotic bands if you increase R in (3.3) in small steps and plot a thousand or so iterates for each parameter increment:

----

29. There is a crisis ahead, though - see Grebogi, Ott and Yorke (1982).

30. See p.187 this selection.

31. See p.164 this selection.

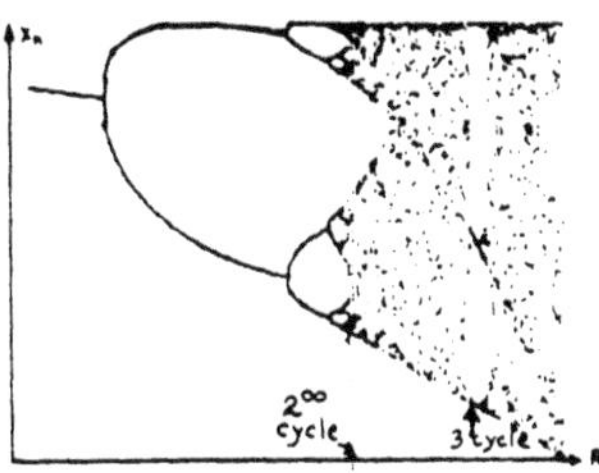

Fig.7.2

Ulam and von Neumann (1947) have actually used mapping (3.3) with R = 2 as a random number generator. At R = 2 the critical point (maximum of f(x)) is mapped into an unstable fixed point, which causes infinitesimally different initial conditions to result in wildly different sequences of iterates:

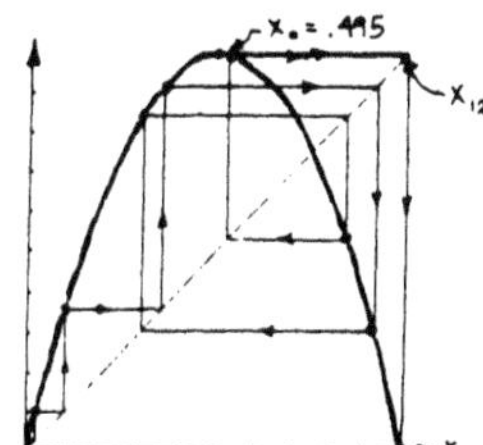

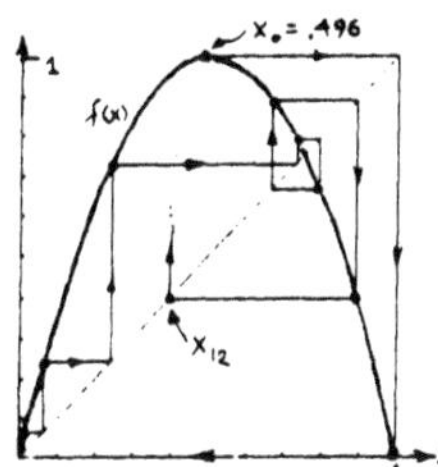

Fig.7.3

This is genuine deterministic chaos, as random as coin flipping (Grossmann and Thomae 1977[32], Ott 1981); iterates of any initial point (except for a set of measure zero) fill out the interval with a probability density independent of the initial point (see the plots of probability distributions in Collet and Eckmann 1980a). Another striking feature apparent in fig.7.2 is the existence of <u>reverse bifurcations</u>: the sequence of chaotic band-doublings which joins onto the bifurcation tree at R_∞. We can explain this sequence by the same tricks that we have used in our study of period-doublings. Plot f(f(x)) and look for the parameter value for which the critical point maps into the unstable fixed point:

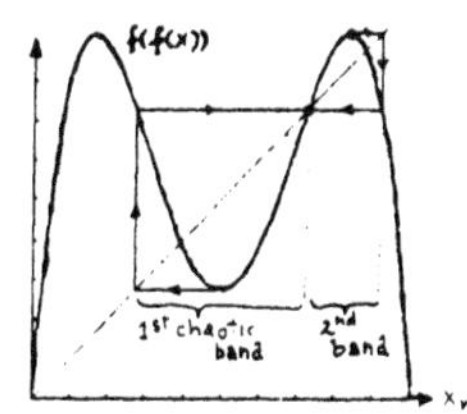

Fig.7.4

There are two chaotic bands, both similar to the original one. Now look at the second iterate of f(f(x)) and find the parameter value for which each of these bands has been split into a pair of chaotic bands. This yields a Misiurewicz (1981) sequence of chaotic band-doublings:

----

32. See p. 281 this selection.

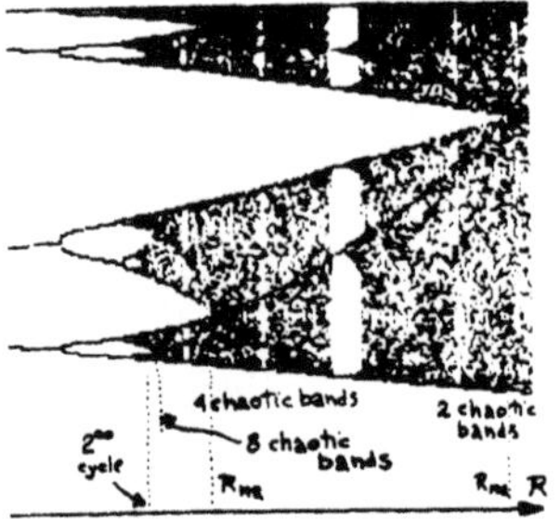

Fig.7.5

The Misiurewicz sequence is defined by the same self-similarity rule as the bifurcation sequence, and obeys the same universal equations (5.5), with the same universal convergence and scaling numbers α and δ. As the motion within the chaotic bands is aperiodic, the associated power spectrum will be characterised by broad-band noise. As was already observed by Lorenz (1964, 1979), the chaotic bands are a fat Cantor set in the parameter space, so this noise is appreciable. The universality of the asymptotic chaotic band-doublings suggests that we look for universal features in the noisy power spectrum. Using the self-similarity of the chaotic bands, Wolf and Swift (1981)[33], Huberman and Zisook (1981)[34], and Farmer (1981)[35] have shown that the total power of the broad-band spectrum scales as

$$N(R) = \text{const.}(R-R_\infty)^{1.5247..} \qquad (7.5)$$

The exponent can be estimated (Wolf and Swift 1981) in terms of α by the same kind of argument as the one that lead to (6.1).

Periodic windows and broad-band noise are seen in all experiments on period doublings. The above universal exponent $\beta = 1.5247...$ has been measured in several non-linear electronic circuit experiments (Testa, Pérez and Jeffries 1982[36], Yeh and Kao 1982a).

8. External noise

You must have asked yourself by now: "But what about the noise in a real physical experiment? Do bifurcations survive noise?" Indeed, one might worry that at the asymptotic times to which our theory applies, the noise might build up without bound and wipe out everything. However, as we are considering dissipative physical systems, the noise

----

33. See p. 305 this selection.

34. See p. 308 this selection.

35. See p. 311 this selection.

36. See p. 174 this selection.

gets damped as well. You can see that by iterating a quadratic return map - every time a noisy trajectory goes through the maximum, the noise width σ is reduced to σ^2:

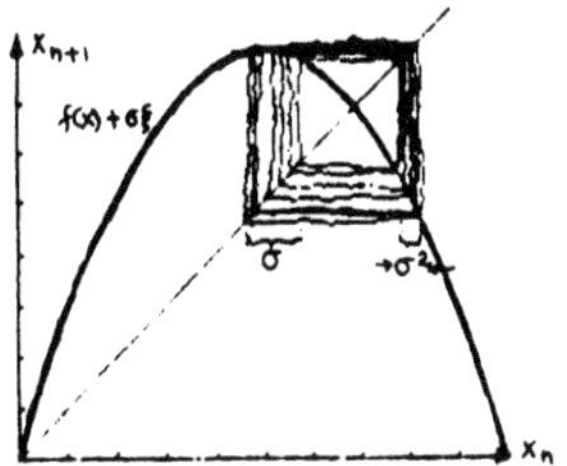

Fig.8.1

The noise does not get out of hand, but it does truncate the bifurcation sequences. As soon as the scale of trajectory splitting is smaller than the noise width, the bifurcations cannot be resolved any longer (Crutchfield and Huberman 1980)[37]. If α^{-n} (the scale of trajectory splittings) is of the order of the noise width, one can hope for at most n bifurcations. Actually, the bifurcations stop sooner than that, because the trajectory gets noisier with each bifurcation; after each period-doubling there is twice as much time for noise to accumulate.

As we know by now, anything that can be extracted from the aymptotic n regime is universal. Adding a weak noise term to mapping (3.1) is a good example of such universality

$$x_{n+1} = f(x_n)+\sigma\xi \tag{8.1}$$

$$\langle\xi\rangle = 0, \quad \langle\xi^2\rangle = 1$$

If σ is very small, the bifurcations proceed as before, until the scale of the nth bifurcation is comparable to the noise level. Hence we can iterate, rescale, etc. and the result will be close to the universal function (4.2):

$$g_n(x) = g(x) + \varepsilon_n \sigma\xi h(x)$$

Here n refers to a parameter p value corresponding to a 2^n-cycle, ε_n and ε_n is the noise at the nth level. As in (5.7) we can linearise around the fixed point, obtaining

$$\xi g'(g(x))h(x) + \xi' h(g(x)) = -(\varkappa/\alpha)\xi' h(x) \tag{8.2}$$

where ξ is the stochastic noise at the nth level, ξ' the noise at the next level, and $\varkappa = \varepsilon_{n+1}/\varepsilon_n$ is the factor by which the noise increases from one level to the next. Performing the average we obtain[38]:

$$\sqrt{[g'(g(x))h(x)]^2+h(g(x))^2} = \kappa/\alpha h(\alpha x) \tag{8.3}$$

(A more careful derivation can be found in Shraiman, Wayne and Martin

----

37. See p. 314 this selection.

38. Crutchfield, Nauenberg and Rudnick (1981), p. 321 this selection.

1981[39] and Feigenbaum and Hasslacher 1982). You can think of noise as the uncertainty in determining the non-linearity parameter; it measures the distance from the fixed point, just like δ. That is why the eigenvalue equation looks very much like (5.7), and can be solved by the same methods. As $\kappa^n\sigma = 1$ is the last resolvable bifurcation, the new universal number $\kappa = 6.61903...$ can be interpreted as the factor by which the noise should be increased to wipe out one bifurcation. This means that demands on an experimentalist eager to observe still another bifurcation are even greater than intimated hitherto; not only does she have to have higher phase-space and parameter resolution, but also lower noise.

Experimentalists are usually not in a position to vary noise over many decades, so κ is not as interesting physically as α and δ. Still, it has been measured in a few experiments on non-linear electronic circuits (Testa, Pérez and Jeffries 1982[40], Yeh and Kao 1982a).

9. Experiments

Sequences of bifurcations, reverse bifurcations, return maps, periodic windows and universal scaling numbers have been measured in many experiments in a wide variety of physical systems: hydrodynamic (water, helium, liquid mercury), optical (lasers), acoustical, electronic, biological (heart muscles), chemical (Belousov-Zhabotinsky) and so on. Numerical experiments are even more numerous, and they are not necessarily trivial; we include some of them in the tabulation of the experimental results to give you a feeling for what kind of precision one can hope for in measurements of universal numbers.

Table 9.1 summarises the experimental situation; the experimental evidence that sequences of bifurcations are common and characterised by the universal numbers is firm. However, accurate measurement of universal numbers is difficult, if not impossible, because each successive bifurcation requires a fivefold increase in the precision of determination of the controllable external parameter, a threefold increase in the experimental resolution of the phase-space trajectory, and twice the time needed for the transients to damp out. Inevitably the noise inherent in the system, long time drifts, and other experimental problems interrupt the bifurcation sequence - about the best that one can hope for is 4 to 5 bifurcations. There are also theoretical problems; in most of these experiments we do not have sufficient understanding of the underlying dynamics to be able to estimate how well they are approximated by the one-dimensional theory. Because of all this it is hard to know what to make of the errors quoted by experimentalists; in some cases the universal numbers measured from the first three bifurcations are so uncannily close to the predicted asymptotic values that one wishes that publication of theoretical predictions were outlawed.

To summarise, the one-dimensional universality is in good shape. It is theoretically sound and it has been tested in many experiments. We lack (and will lack for a long time) a global theory which would tell us what physical systems, and which parameter ranges exhibit transitions to chaos via infinite sequences of bifurcations.

----

39. See p. 317 this selection.

40. See p. 174 this selection.

TABLE 9.1. Period doublings, experimental. A summary of the experimental observations of period doublings. The numbers in brackets are estimates of experimental errors; 4.3(8) means 4.3±.8. A few numerical "experiments" are included to indicate the precision of measurement of universal numbers attainable in numerical simulations.

experiment	no. period doublings	δ	α	μ	σ	$\varkappa$
hydrodynamic:						
water[1]	2					
water[2]	4	4.3(8)		4(1)		
helium[3]	4	3.5(1.5)		4(?)		
mercury[4]	4	4.4(1)		5(1)		
Electronic:						
diode[5]	4	4.5(6)		6(?)		
diode[6]	5	4.3(1)	2.4(1)	O.K.		6.3(3)
transistor[7]	4	4.7(3)		O.K.		
Josephson simul.[8]	3	4.5(3)	2.7(2)		1.5(1)	5(2)
Laser:						
laser feedback[9]	3	4.3(3)	O.K.			
laser[10]	2					
laser[11]	3					
Acoustic:						
helium[12]	3					
helium[13]	3	4.8(6)		6(?)		
Chemical:						
B-Zh reaction[14]	3					
Computer:						
N-S truncation[15]	5	4.6(2)	2.5(1)			
Brusselator[16]	7	4.6(2)		4.77(3)	1.5(?)	
Theory:	∞	4.669..	2.503..	4.58..	1.52..	6.55..
equation no.		(5.1)	(4.2)	(6.1)	(7.5)	(8.3)

1. Gollub and Benson (1980).
2. Giglio, Musazzi and Perini (1981).
3. Libchaber and Maurer (1981).
4. Libchaber, Laroche and Fauve (1982).
5. Linsay (1982).
6. Testa, Pérez and Jefferies (1982).
7. Arecchi and Lisi (1982).
8. Yeh and Kao (1982).
9. Hopf, Kaplan, Gibbs and Shoemaker (1981).
10. Arecchi, Meucci, Puccioni and Tredicce (1982).
11. Weiss, Godone and Olafsson (1983).
12. Lauterborn and Cramer (1981).
13. Smith, Tejwani and Farris (1982).
14. Simoyi, Wolf and Swinney (1982).
15. Franceschini and Tebaldi (1979).
16. Kai (1981).

10. Intermittency

Period-doublings are rather common, but they are by no means the only way in which a deterministic system can reach chaos. Intermittency is another type of chaotic behaviour commonly observed in deterministic systems. It is characterised by long periods of regular motion interrupted by short chaotic bursts. For dissipative systems it too can be modelled by one-dimensional iterations (3.1). Consider the neighborhood of a fixed point of f(f(f(x))) as the non-linearity parameter sweeps through the critical value for the creation of the 3-cycle (fig.7.1):

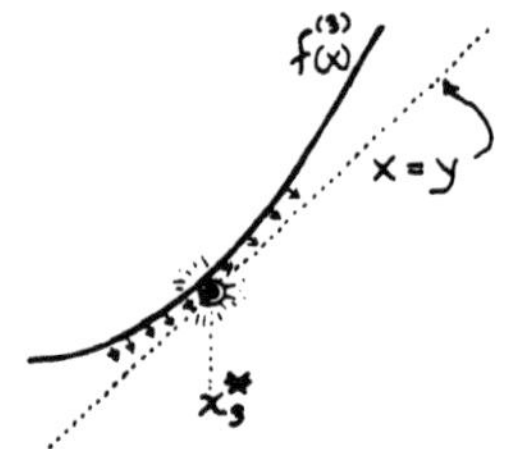

Fig.10.1

The new fixed point is created by tangent bifurcation. If you iterate f(f(f(x))) for a value of the non-linearity parameter just below the birth of the 3-cycle, you will note that the iterates accumulate in the neighbourhood of the 3-cycle points-to-be:

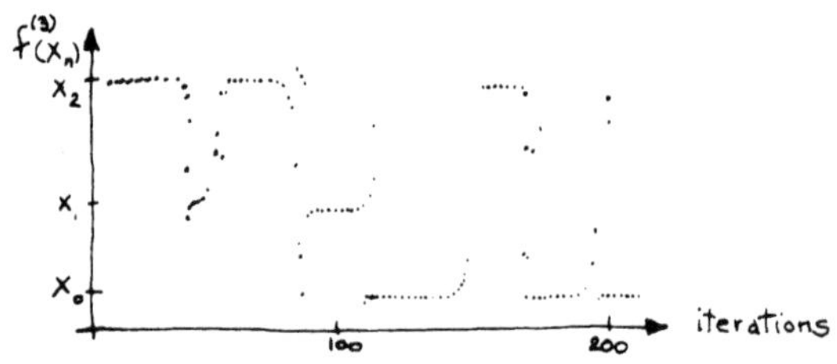

Fig.10.2

For many iterations the system is being fooled into believing that it is converging toward a fixed point, only to discover that the fixed point is not there after all; it then wanders away again, in the hope of finding a true fixed point:

Fig.10.3

This is typical intermittent behaviour. Arbitrarily small differences in initial conditions will result in totally different sequences of iterates, so the intermittent motion is chaotic. Can we make any quantitative predictions about it?

Pomeau and Manneville (1980) have discovered a scaling law which relates the average duration of regular motion to the parameter deviation from its tangent bifurcation value. Somewhat suprisingly, not only can this scaling can be described by universal equations of the same form as for the period-doubling sequences, but in this case the universal equations have simple analytic solutions (Cosnard 1981, Hirsch, Nauenberg and Scalapino 1982[41]).

The self-similarity in this case is almost trivial; close to the tangent fixed point, f(x) and f(f(x)) look very much the same; the only difference is that for f(f(x)) the steps in the iteration staircase are twice as long

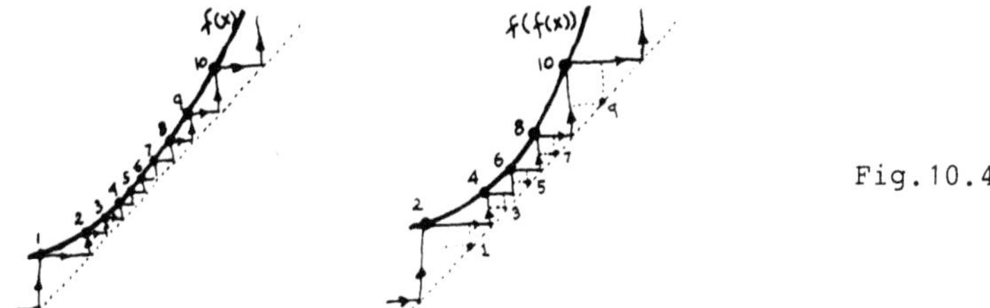

Fig.10.4

so one writes the self-similarity equation (4.2)

$$g(x) = \alpha g(g(x/\alpha)) \tag{10.1}$$

imposes tangency conditions

$$g(0) = 0 \tag{10.2}$$

$$g'(0) = 1 \tag{10.3}$$

and guesses $\alpha = 2$. As you can easily check, $g(x) = x/(1-ax)$, a arbitrary, solves the above equation. The equation is universal for the same reason as in the period-doubling case; it describes the neighbourhood of the tangent fixed point enlarged infinitely many times with the scaling factor $\alpha = 2$, so almost all memory of the original mapping is lost. The unstable manifold can be studied the same way as in the period-doubling case, by linearising as in (5.7); this time one obtains $\delta = 4$.

It is cute to see that the universal equations can be solved explicitly. More importantly, the above universal numbers α and δ have direct physical significance. The length of the periodic motion depends on how close f(x) is to the (incipient) tangent fixed point. The universal equation (10.1) says that if we move $1/\delta = 1/4$ times closer to the tangent fixed point, the iteration hangs $\alpha = 2$ times as long in the neighbourhood of the fixed point. Hence L, the average length of periodic motion, and R_c-R, the amount by which the parameter differs from the tangency value, are related by the Manneville and Pomeau (1979) scaling law

$$L = \text{const.}(R_c-R)^{-1/2} \tag{10.4}$$

Beyond this, one can play with everything we tried for period-doublings; add noise (Hirsch, Nauenberg and Scalapino 1982, Hirsch, Huberman and Scalapino 1982), change the power of the mapping

----

41. See p. 336 this selection.

(Hu and Rudnick 1982), and so on. Intermittency has been observed in chemical (Pomeau, Roux, Rossi, Bachelart and Vidal 1981[42]) and electronic (Jeffries and Pérez 1982, Yeh and Kao 1982a) experiments.

11. Universality in higher dimensions

The theory of the period-doubling route to chaos, as discussed above, is essentialy complete. A pretty mathematical theory is lacking, but we understand when and why period-doublings can occur, and we have solid experimental evidence that they are common in many physical systems. Where do we go from here?

Period-doubling is a one-dimensional theory; it describes dynamical systems for which the dissipation has effectively damped out all but one of the dynamical degrees of freedom. Sooner or later we have to face the real world: chaos in many-dimensional systems. The simplest way to postpone this unpleasant moment is to turn to the study of iterations of two-dimensional maps. Three types of two-dimensional mappings have been extensively studied: contractive, complex and conservative.

For some regions of parameter values the <u>contractive</u> mappings reduce to the one-dimensional theory, as argued at the end of section 2, and shown by numerical calculations by Derrida, Gervois and Pomeau (1979), Franceschini (1979), Franceschini and Tebaldi (1979)[43] and many others. Beyond this there is much more (such as Hénon´s 1976[44] strange attractor) that is chaotic, but so far not covered by the universality theory, and therefore beyond the scope of this introduction.

Iterations of functions of one <u>complex</u> variable can be viewed as mappings in two real dimensions. They have been extensively studied (Julia 1918, Fatou 1919, 1920, Myrberg 1958-1962, Brolin 1965, Mandelbrot 1980, 1982, Douady and Hubbard 1982, Douady 1982, Sullivan 1982, Manton and Nauenberg 1983). The universal function (4.2) has rich structure in the complex plane (Epstein and Lascoux 1981). In the complex plane the theory of period-doublings has been generalised to the theory of period n-tuplings (Golberg, Sinai and Khanin 1983, Cvitanović and Myrheim 1983). The universality theory for complex mappings is very beautiful, but we do not know whether it can be used to model any physical systems.

<u>Conservative</u> (Hamiltonian) mappings appear in many physical problems. In solid state physics conservative mappings arise in the study of commensurate-incommensurate transitions, this time not as Poincaré maps, but as iterations on physical lattices (Bak 1981). An effective theory of conservative chaos would have many important practical applications in a variety of problems, such as the design of plasma fusion devices and intersecting storage rings (Helleman 1981a).

----

42. See p. 167 this selection.

43. See p. 379 this selection.

44. See p. 341 this selection.

In a dissipative system transients die out and the trajectory settles into some low-dimensional attractor. In a conservative system transients never die out, and the trajectory (for example, an ion in a tokomak) keeps spiralling on forever. We have seen that an ordinary fish population curve leads to an amazingly rich structure. The phase-space structure of conservative systems is truly bewildering, and has fascinated physicists and mathematicians for many generations (see for example Berry 1978 and Helleman 1980a).

The discovery of the universality for one-dimensional iterations has prompted a search for period-doublings in two-dimensional conservative mappings. They have been discovered (Benettin, Cercignani, Galgani and Giorgilli 1980, Bountis 1981, Bak and Høgh Jensen 1982) and look something like this:

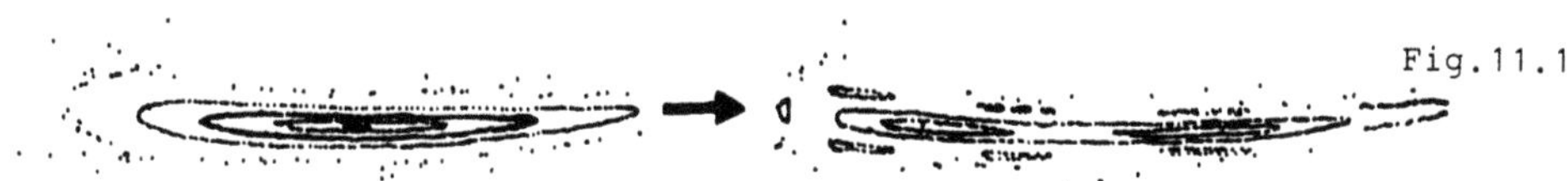

Fig.11.1

An elliptic fixed point turns hyperbolic and gives birth to a pair of new elliptic fixed points. Variation of the external parameter yields an infinite sequence of such period-doublings, ending in chaos. The universal scaling and convergence numbers can be computed as before; they are different for conservative and dissipative systems. Their size has been predicted by simple renormalisation group arguments (Derrida and Pomeau 1980), and they can be related smoothly to the one-dimensional universal numbers by varying the amount of dissipation in two-dimensional mappings (Helleman 1980a, 1980b, Hu 1981, Zisook 1981[45]). The universal equations for two-dimensional conservative period-doublings have been formulated and investigated by Greene, MacKay, Vivaldi and Feigenbaum (1981), Collet, Eckmann and Koch (1981), Eckmann, Koch and Wittwer (1982), MacKay (1983)[46] and for two-dimensional intermittency by Zisook (1982) and Zisook and Shenker (1982). As far as we know, there have been no experiments which probe this type of universality.

Another problem in which the universality ideas have had some success is the transitions to chaos for diffeomorphisms on the circle. Maps of this type arise in a variety of physical problems.

Hamiltonian mappings are one class of such problems. According to the KAM theorem (see Arnold 1978, Moser 1968), the phase-space trajectories in two dimensions are confined within KAM tori, and large-scale chaos sets in with the dissolution of these tori. Greene (1979) has shown that the winding number of the last surviving KAM is the golden mean. Shenker and Kadanoff (1982) have investigated the precise manner in which the last KAM dissolves and discovered that it does so by turning into a fractal. A way to study this transition is to neglect the radial variation of the torus, and model the angular variable by a map on the circle;

$$x_{n+1} = x_n + \Omega + k/(2\pi)\sin(2\pi x_n) \pmod 1 \qquad (11.1)$$

The circle mappings also arise in the study of dissipative

----

45. See p. 350 this selection.

46. See p. 412 this selection.

systems, such as the cylindrical Couette flow. After two Hopf bifurcations (a fixed point with inward spiralling stability has become unstable and outward spirals to a limit cycle) a system lives on a two-torus, executing quasi-periodic motion. The Poincaré map of this system can again be modelled by a mapping of type (11.1). The question is what happens next. In an influential paper Ruelle and Takens (1971) have proposed that chaos is reached via a three-torus (see Eckmann 1981[47] for a discussion). However, chaos can arise already on the two-torus, again just as the map is losing its invertibility.

Yet another situation in which the circle maps arise naturally is for periodically driven non-linear oscillators, such as the Duffing oscillator and models of the Josephson junction (Kautz 1981, Crutchfield and Huberman 1980, Levinsen 1982, D'Humiéres et al. 1982, Tomita 1982). Periodicity is imposed by the driving frequency, and the phase-lockings between the driving frequency and the intrinsic frequency of the oscillator can again be modelled by the circle map (11.1).

The transition to chaos occurs as the mapping (11.1) starts to lose invertibility, $k = 1$. The iterates for Ω corresponding to the golden mean winding number show self-similarity (Shenker 1982)[48]. It is not hard to see what this self-similarity is and to write down the corresponding universality equations (Rand et al. 1982, Ostlund et al. 1983, Feigenbaum, Kadanoff and Shenker 1982) - the method is the same as in the examples discussed above, but getting into this here would take too much space.

It is rather unlikely that the predictions of the universality theory for the golden mean winding number can be tested experimentally; the scaling numbers for the critical case differ by only few per cent from the scaling numbers for the trivial (invertible mapping) case[49].

Beyond circle maps and two dimensions, there is no end to the problems waiting for us. We have to understand how the attractors grow as more and more degrees of freedom go chaotic on us (Farmer 1982), how to do the quantum mechanics of classically chaotic systems (Berry 1978, Zaslavsky 1981, Pullen and Edmonds 1981), prove quark confinement as an effect of Yang-Mills turbulence (Matinyan, Savvidy and Ter-Arutyunyan-Savvidy 1981), write the universal equation for the brain (Feigenbaum, unpublished), and finally, understand why the clouds are the way they are.

ACKNOWLEDGEMENTS

This introduction has its origin in a Nordita lecture prepared together with Mogens Høgh Jensen (Cvitanović and Høgh Jensen 1982), to whom I am very much indebted. I would also like to thank Ulla Selmer, Jan Myrheim, Peter Scharbach, Jim Revill, Nils Robert Nilsson, Olivia Kaypro, and Mitchell Feigenbaum for their help and encouragement.

----

47. See p. 94 this selection.
48. See p.405 this selection.
49. (Author's note, April 1986) The golden mean universality has been tested in a beautiful experiment by J. Stavans, F. Heslot and A. Libchaber, Phys. Rev. Lett., 55, 596 (1985).

Part 1

Introductory Articles

The Mathematical Intelligencer **2** 126–37 (1980)

Strange Attractors*

David Ruelle

Introduction: Deterministic Systems with a Touch of Fantasy

Systems with an irregular, non periodic, "chaotic" time evolution are frequently encountered in physics, chemistry, and biology. Think for example of the smoke rising in still air from a cigarette. Oscillations appear at a certain height in the smoke column, and they are so complicated as to apparently defy understanding. Although the time evolution obeys strict deterministic laws, the system seems to behave according to its own free will. Physicists, chemists, biologists, and also mathematicians have tried to understand this situation. We shall see how they have been helped by the concept of *strange attractor*, and by the use of modern computers.

A strange attractor consists of a infinity of points, in the plane as shown on Figure 1A, or in m-dimensional space. These points correspond to the states of a chaotic system. Strange attractors are relatively abstract mathematical objects, but computers give them some life, and draw pictures of them. (See the illustrations, and note that the computer may mark only a finite number of points.) It may well be that the reader has access to a computer, and can reproduce some of the "experiments" described below.

The Description of Time Evolution: Dynamical Systems

We specify the state of a physical, chemical, or biological system by parameters $x_1, x_2, \ldots, x_m$. A chemical system for example would be described by the concentrations of various reactants. The parameters vary with time, and we denote by

$$x_1(t), x_2(t), \ldots, x_m(t)$$

their values at time t. For simplicity we shall consider first only integer values of t (time expressed in seconds, or in years). We shall come back later to the case of continuously varying time.

How do we determine the time evolution of the system, in other words its *dynamics*? We shall admit that the parameters specifying the system at time $t + 1$ are given functions of the parameters at time t. We may thus write

$$\left.\begin{aligned} x_1(t+1) &= F_1(x_1(t), x_2(t), \ldots, x_m(t)) \\ x_2(t+1) &= F_2(x_1(t), x_2(t), \ldots, x_m(t)) \\ &\ldots \\ x_m(t+1) &= F_m(x_1(t), x_2(t), \ldots, x_m(t)) \end{aligned}\right\} \quad (1)$$

We assume that the functions $F_1, F_2, \ldots, F_m$ are continuous and have continuous derivatives. This "technical" differentiability condition will be satisfied in our examples. We shall see later why it is important.

Given *initial values* $x_1(0), x_2(0), \ldots, x_m(0)$ for the parameters we can, using (1), compute $x_1(t), x_2(t), \ldots, x_m(t)$ successively for all positive integer times t. Thus, knowing the state of the system at time zero one may compute its state at time t. We say that the functions $F_1, F_2, \ldots, F_m$ determine a discrete time *dynamical system*. It is a *differentiable* dynamical system because we have assumed that the functions $F_1, F_2, \ldots, F_m$ have continuous derivatives.

An Example: The Hénon Attractor

Let us now examine a concrete case. Let $m = 2$, and write x, y instead of x_1, x_2. We are given

$$F_1(x, y) = y + 1 - ax^2$$
$$F_2(x, y) = bx$$

with $a = 1.4$ and $b = 0.3$. The relations (1) thus take the form

$$\left.\begin{aligned} x(t+1) &= y(t) + 1 - ax(t)^2 \\ y(t+1) &= bx(t) \end{aligned}\right\} \quad (2)$$

Given $x(0), y(0)$ we may compute $x(t)$ and $y(t)$ for $t = 1, 2, \ldots, 10{,}000$ for instance, keeping everywhere sixteen significant figures. Done by hand this calculation would take many months and, since its interest is not obvious, nobody undertook it. For a digital computer on

* Translated by the author from his French article published in *La Recherche* N° 108, Février 1980, with kind permission of La Recherche.

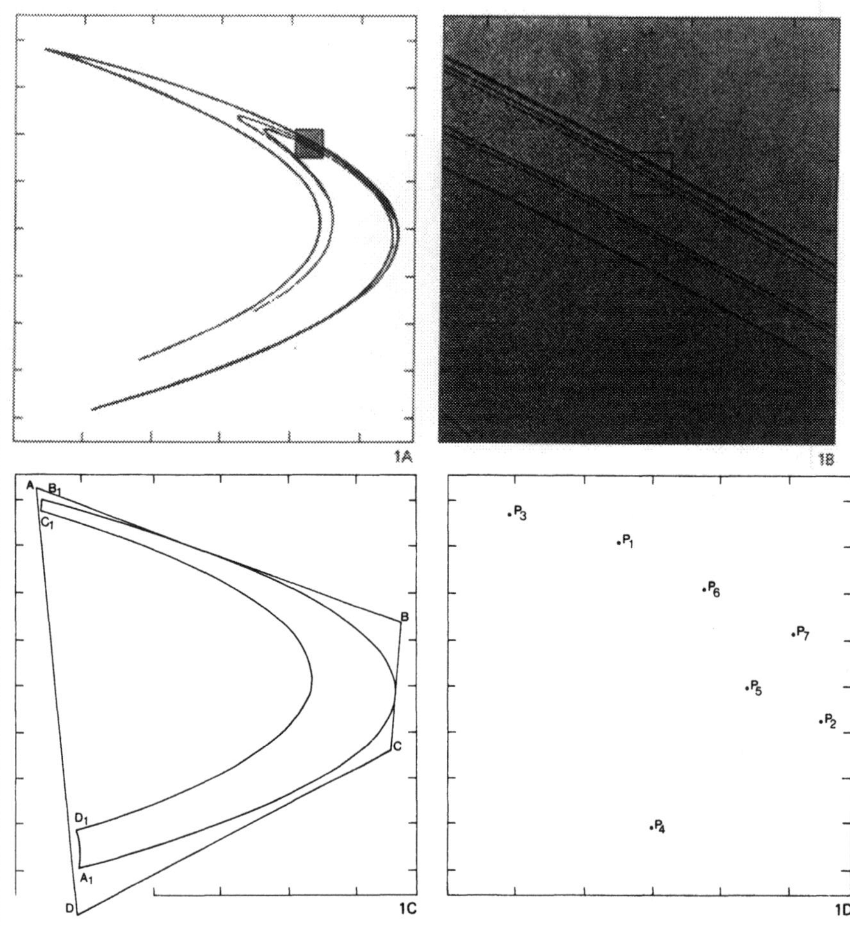

Figure 1. *The Hénon attractor.* A computer has been asked to mark points of coordinates $x(t)$, $y(t)$ for t going from 1 to 10,000. The point $(x(0), y(0))$ is given, and the following points are determined by

$$x(t+1) = y(t) + 1 - ax(t)^2, \quad y(t) = bx(t)$$

with $a = 1.4$ and $b = 0.3$. Figure 1A shows the result. The 10,000 points distribute themselves on a complex system of lines: the *Hénon attractor.* It is an example of a *strange attractor.* Magnification of the little square in Figure 1A yields 1B, and magnification of the little square in 1B would again yield a similar picture. Each new magnification resolves lines into more lines. The Hénon attractor is associated with a map of the plane which sends the point (x, y) to $(F_1(x, y), F_2(x, y))$, with $F_1(x, y) = y + 1 - ax^2$, $F_2(x, y) = bx$. In particular, the quadrilateral $ABCD$ of Figure 1C is mapped inside itself into $A_1B_1C_1D_1$. Notice that F_1, F_2 are polynomials, and therefore have continuous derivatives

$$\partial F_1/\partial x = -2ax \qquad \partial F_1/\partial y = 1$$

$$\partial F_2/\partial x = b \qquad \partial F_2/\partial y = 0$$

One can see that the surface of $A_1B_1C_1D_1$ is equal to three tenths of the surface of $ABCD$ (the factor $b = 0.3$ is given, up to sign, by the determinant of the above derivatives). In Figure 1D one has kept $b = 0.3$ but taken $a = 1.3$. The strange attractor disappears, and is replaced by the seven points of a periodic attractor.

the other hand, this boring and repetitive task is not a problem. Michel Hénon, of the observatory in Nice, did the first calculations with an HP-65 programmable pocket computer. He then went on to a more powerful machine (IBM 7040). That computer had a plotter, which marked on a sheet of paper the points with coordinates $x(t), y(t)$, for t ranging from 1 to 10,000. Figure 1A shows the picture obtained. Unexpectedly, the ten thousand points lie on a system of lines with complex structure. If the little square of Figure 1A is magnified, Figure 1B is obtained. If the square of Figure 1B were magnified, one would obtain again a similar picture, and so on, each magnification revealing lines which were not previously visible [1].

What happens if the initial point $(x(0), y(0))$ is changed? Well, for a "bad" choice $(x(t), y(t))$ will go to infinity (and in particular, leave the sheet of paper). For a "good" choice, $(x(1), y(1)), (x(2), y(2)), \ldots$, will rapidly get close to the "noodle" of Figure 1A, and the general aspect of this picture will be reproduced after a few thousand points have been marked.

Our "noodle" is the *Hénon attractor*. It is an example of a *strange attractor*. Let me mention, among other curiosities, that the attractor may suddenly disappear when the parameters a, b in (2) are changed. Taking for instance $a = 1.3$ and $b = 0.3$ one sees the points $(x(t), y(t))$ approaching, when t increases, a set of seven points $P_1, \ldots, P_7$ (Figure 1D). Instead of a strange attractor we now have a *periodic attractor* (of period 7).

In trying to understand the Hénon attractor, it is helpful to consider the map F of the plane to itself defined by (2). If X has coordinates x and y, $F(X)$ has coordinates

$$F_1(x, y) = y + 1 - ax^2, \quad F_2(x, y) = bx$$

Call X_t the point with coordinates $x(t), y(t)$. Then $X_1 = F(X_0)$, $X_2 = F(F(X_0))$, etc ..., X_t is obtained from X_0 by applying t times the map F. Figure 1C shows a quadrilateral $ABCD$, and its *image* $A_1B_1C_1D_1$ by F. This image is by definition the set of points $F(X)$ with X in the quadrilateral $ABCD$. Hénon has chosen the quadrilateral $ABCD$ in such a manner that it contains the image $A_1B_1C_1D_1$. Figure 1C shows that the quadrilateral is "folded in two" by the map F. If the initial point X_0 is in $ABCD$, then X_1 is in the image $A_1B_1C_1D_1$, and thus again in $ABCD$. All the points $X_1, X_2, \ldots, X_t, \ldots$ are therefore in the quadrilateral $ABCD$, and the Hénon attractor is also contained in that quadrilateral.

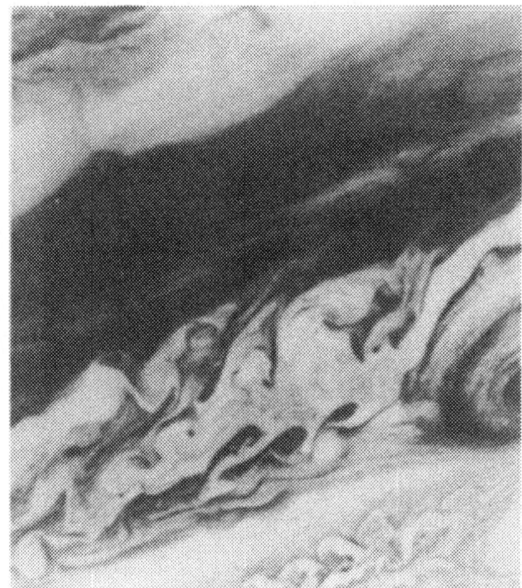

Smoke rising from a cigarette. – The atmosphere of Jupiter. Two of the many examples of systems whose evolution through time involves oscillations which can be described by strange attractors. (Clichés E. Rousseau & IPS)

Another Example: The Solenoid

We shall now examine an attractor in three dimensions, i.e., we shall take $n = 3$ in the formulae (1). Instead of writing explicit expressions for the functions F_1, F_2, F_3, we describe geometrically the map F of three-dimensional space to itself which they define. (This map F sends the point with coordinates x_1, x_2, x_3 to the point with coordinates $F_1(x_1, x_2, x_3), F_2(x_1, x_2, x_3), F_3(x_1, x_2, x_3)$). We suppose that F takes a ring A (the solid torus of Figure 2A), stretches it, makes it thinner, folds it, and places it in the manner drawn in Figure 2B. This figure shows both A, and its image $F(A)$ by the map F. The image $F(A)$ winds twice around the central hole of the ring A.

Starting from a point X_0 in the ring A, we write $X_1 = F(X_0), X_2 = F(X_1), \ldots$ Figure 2C shows the five thousand points $X_{51}, X_{52}, \ldots, X_{5050}$ (together with the set $F(A)$). A new strange attractor appears. Since the point X_0 is arbitrary in A, it is not in general on the attractor, but $X_1, X_2, X_3, \ldots$ get progressively closer to it. This is why we have marked the points starting at X_{51}. It is fascinating to observe the plotter (of the HP 9830A) draw the picture. About once per second a click is produced and a point is marked, in an apparently random manner. It takes a fairly long time before one can guess the final form of the attractor.

The attractor of Figure 2C has been called a *solenoid*. Indeed the picture is suggestive of electric wires around an axis. To understand this structure, note that the solenoid is contained not only in the ring A of Figure 2A, but also in its image $F(A)$ drawn in Figure 2C, and also in $F(F(A))$, $F(F(F(A)))$, ... The image $F(A)$ is the inside of a tube which winds twice around the central hole of A, $F(F(A))$ is in a thinner tube which winds four times around the hole, $F(F(F(A)))$ is still thinner and winds around eight times, etc. ... The solenoid is thus contained in very thin tubes winding around many times, and this explains how it looks.

Sensitive Dependence with Respect to Initial Conditions: How Errors Grow with Time

Remember that the parameters $x_1(t), x_2(t), \ldots, x_m(t)$ are supposed to describe a physical, chemical, or biological

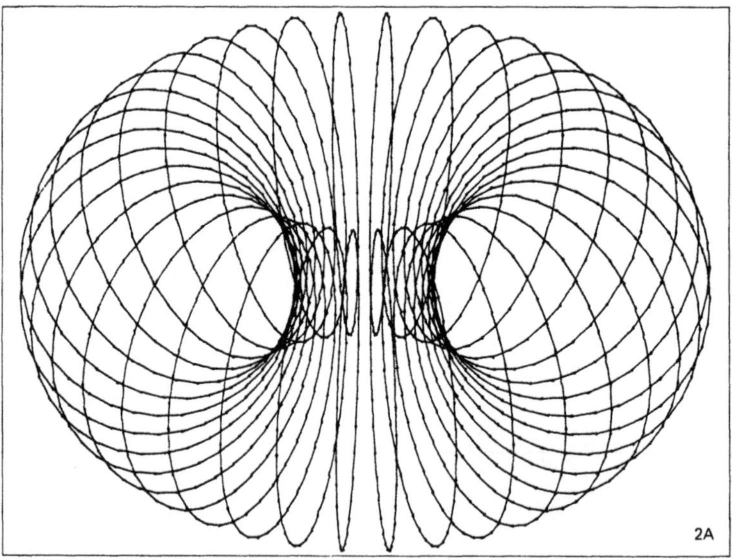

Figure 2. *The solenoid.* Figure 2A is a perspective view of a ring A in three dimensional space. A map F stretches A, makes it thinner, folds it, and places the image $F(A)$ inside A so that $F(A)$ turns twice around the central hole of A, as shown in Figure 2B. Figure 2C shows $F(A)$ again, and also 5,000 points successively defined by $X_{t+1} = F(X_t)$ starting from some initial point X_0. The 5,000 points produce a wiry structure. It is a new strange attractor, called *solenoid*.

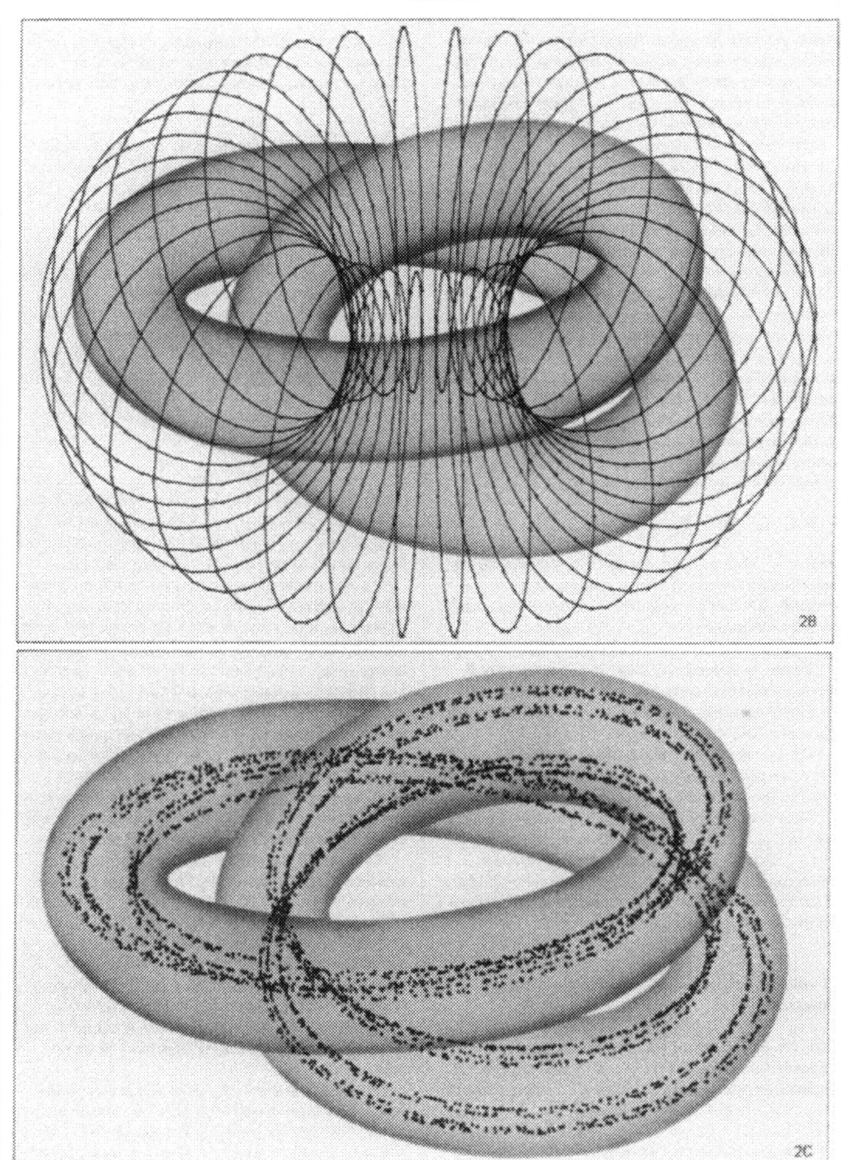
2B
2C

system at time t. We assume that the system has a deterministic time-evolution defined by the equations (1). With what precision can we predict the evolution if the choice of the initial values $x_1(0), x_2(0), \ldots, x_m(0)$ is slightly in error, as is always the case for experimental data? How will the error increase (or decrease) with increasing t? The answer will of course depend on the given functions $F_1, F_2, \ldots, F_m$, and on the initial values $x_1(0), x_2(0), \ldots, x_m(0)$. For the two strange attractors which we have examined (the Hénon attractor and the solenoid) a small error (or uncertainty) on the initial values gives an error (or uncertainty) at time t, which increases rapidly with t.

Let us verify this assertion for the Hénon attractor. We know that there is, around the attractor, a quadrilateral $ABCD$ such that the map F folds the quadrilateral in two. As Figure 1C shows, the folding in two is accompanied by stretching. Thus if X_t and X'_t correspond to initial data X_0 and X'_0 close to each other, the distance $d(X_t, X'_t)$ generally increases with t. At least this is the case as long as this distance remains small; when the distance from X_t to X'_t becomes of the order of the total size of the attractor it cannot increase any more. Numerically one finds

$$d(X_t, X'_t) \sim d(X_0, X'_0) . a^t \qquad (3)$$

with $a \approx 1.52$. Since $a > 1$, the factor a^t increases rapidly (exponentially) with t. Therefore *the error $d(X_t, X'_t)$ increases exponentially with time*. The rate of exponential increase is determined by a (or by its logarithm $\lambda = \ln a$ called characteristic exponent, here $\lambda \approx 0.42$).

We may argue similarly for the solenoid. The map F stretches a tube containing the solenoid and, because of this stretching, formula (3) remains valid, with a different choice of $a > 1$.

The exponential increase of errors described by formula (3) is expressed by saying that the dynamical system under consideration has *sensitive dependence on initial condition*.

Notice that to give a precise meaning to (3) we have to take $d(X_0, X'_0)$ "infinitesimal". The assumption that $F_1, F_2, \ldots, F_m$ have continuous derivatives is used here. Notice also that, for given X_0, there may be exceptional X'_0 for which the error does not grow as indicated by (3) (it may for instance decrease).

A Little Bit of Mathematics: A Definition of Strange Attractors

Let us come back to the general dynamical system described by the equations (1). We call F the map of m dimensional space to itself which sends X with coordinates $x_1, \ldots, x_m$ to $F(X)$ with coordinates $F_1(x_1, \ldots, x_m), \ldots, F_m(x_1, \ldots, x_m)$. We shall say that a bounded set A in m-dimensional space is a ***strange attractor*** for the map F if there is a set U with the following properties:

(a) U is an m-dimensional ***neighborhood*** of A, i.e., for each point X of A, there is a little ball centered at X and entirely contained in U. In particular A is contained in U.

(b) For every initial point X_0 in U, the point X_t with coordinates $x_1(t), \ldots, x_m(t)$ remains in U for positive t; it becomes and stays as close as one wants to A for t large enough. This means that A is ***attracting.***

(c) There is sensitive dependence on initial condition when X_0 is in U. This makes A a ***strange*** attractor.

In the case of the Hénon attractor one can take for U the quadrilateral $ABCD$ (Figure 1C), in the case of the solenoid one can take for U the solid torus A (Figure 2).

The above definition allows the practical determination of strange attractors in computer studies, but it is not quite complete mathematically. It is desirable also to impose the following condition.

(d) One can choose a point X_0 in A such that, arbitrarily close to each other point Y in A, there is a point X_t for some positive t. This *indecomposability condition* implies that A cannot be split into two different attractors.

It would also be necessary to make the notion of sensitive dependence on initial condition more precise. This however, leads to questions which are not too well understood. It must be said that the mathematical theory of strange attractors is difficult and, in part, still in its infancy. The solenoid is well understood, thanks to the work of Steve Smale [2] of Berkeley. By contrast, it has not been *proved* that Figures 1A and 1B do not just show a periodic orbit of very long period. The fact that the Hénon attractor exists as a strange attractor is for the time being a *belief* based on computer calculations! Perhaps our definition of strange attractors will have to be changed to adapt to more general situations. Do not take it too seriously.

It seems that the phrase "strange attractor" first appeared in print in a paper by Floris Takens (of Groningen) and myself [3]. I asked Floris Takens if he had created this remarkably successful expression. Here is his answer: "Did you ever ask God whether he created this damned universe? . . . I don't remember anything . . . I often create without remembering it . . ." The creation of strange attractors thus seems to be surrounded by clouds and thunder. Anyway, the name is beautiful, and well suited to these astonishing objects, of which we understand so little.

Besides strange attractors, we should remember that there are also non strange attractors. For instance ***attracting fixed points.*** The point A is an attracting fixed point if X_t gets arbitrarity close to A when t increases, provided X_0 is in a neighborhood U of A. In that case of course errors decrease when t increases, and there is no sensitive dependence on initial conditions. Attracting fixed points belong to the *periodic attractors,* which we have already met (Figure 1D). A periodic attractor has a finite number of points.

Attracting fixed points have been known for a long time. They describe an asymptotically stationary situation, i.e., for large t, X_t practically no longer depends on t. In the same manner the periodic attractors describe an asymptotically periodic situation. Scientists had got used to the notion that the asymptotic behavior of natural phenomena should be stationary, or perhaps periodic. Only recently did interest arise in the "chaotic" behavior, with sensitive dependence on initial condition, which occurs in many natural phenomena.

Strange Attractors in Nature

To describe the systems which they encounter, physicists, chemists, and biologists use equations of the type (1), or differential equations in the case of continuous time. One should not underestimate the amount of idealization implied by such a description. Certain parameters are selected as variables $x_1, \ldots, x_m$, others are ignored, and various simplifications are made. Idealization is a basic ingredient of all natural sciences, and a serious scientist must show that the natural system which he considers obeys deterministic laws of the type (1) with a good approximation. He may then look for strange attractors, either by the direct study of experimental results, or by computer simulation. In this manner, the "chaos" which occurs in certain phenomena becomes understandable, and it may be hoped that this understanding will lead to practical applications.

The study of "chaotic" or "turbulent" time evolutions in natural phenomena is now only at its beginnings. Progress is slow, due in part to experimental difficulties, in part to the insufficient development of the theory. In the absence of a satisfactory mathematical theory, computers play an important role in the interpretation of data.

We shall now discuss some examples of chaotic phenomena, and in particular the problem of fluid turbulence. In order to do this we shall have to use a continuous time t rather than a discrete time.

The Lorenz Attractor, and Meteorological Predictions

In order to define differentiable dynamical systems with continuous time we replace the equations (1) by differential equations

$$\left.\begin{aligned} \frac{d}{dt}x_1(t) &= G_1(x_1(t), \ldots, x_m(t)) \\ &\cdots \\ \frac{d}{dt}x_m(t) &= G_m(x_1(t), \ldots, x_m(t)) \end{aligned}\right\} \qquad (4)$$

If $G_1, \ldots, G_m$ satisfy certain conditions (existence of continuous derivatives, etc.) the equations (4) uniquely determine the functions $x_1(t), \ldots, x_m(t)$ of time t when the initial data $x_1(0), \ldots, x_m(0)$ are known. The equations (4) thus define a deterministic evolution with continuous time, just as the equations (1) defined a deterministic evolution with discrete time.

Let us take for example $m = 3$, and write $x_1(t) = x$, $x_2(t) = y$, $x_3(t) = z$. We consider the differential equations

$$\left.\begin{aligned} \frac{dx}{dt} &= -\sigma x + \sigma y \\ \frac{dy}{dt} &= -xy + rx - y \\ \frac{dz}{dt} &= xy - bz \end{aligned}\right\} \qquad (5)$$

with $\sigma = 10$, $b = 8/3$, and $r = 28$. Figure 3 shows the trajectory of the point (x, y, z) corresponding to the solution of these equations with initial condition (0, 0, 0). It appears that we have here again a strange attractor, and one can show that there is indeed sensitive dependence on initial condition.

The attractor of Figure 3 is the *Lorenz attractor*, named after Edward Lorenz, professor in the Meteorology department of the Massachusetts Institute of Technology. The equations (5) were indeed first written and studied by Lorenz [4]. These equations give an approximate description of a horizontal fluid layer heated from below. The warmer fluid formed at the bottom is lighter. It tends to rise, creating convection currents. If the heating is sufficiently intense, the convection takes place in an irregular, turbulent manner. This phenomenon takes place for instance in the earth atmosphere, and since it has sensitive dependence on initial condition, it is understandable that meteorologists cannot predict the state of the atmosphere with precision a long time in advance. The work of Ed Lorenz thus gives some theoretical excuse to the well-known unreliability of weather forecasts.

Fluid Turbulence: One of the Great Unsolved Problems of Theoretical Physics

Turbulence is a phenomenon easily produced by opening the tap over the bath tub or the kitchen sink. The nature

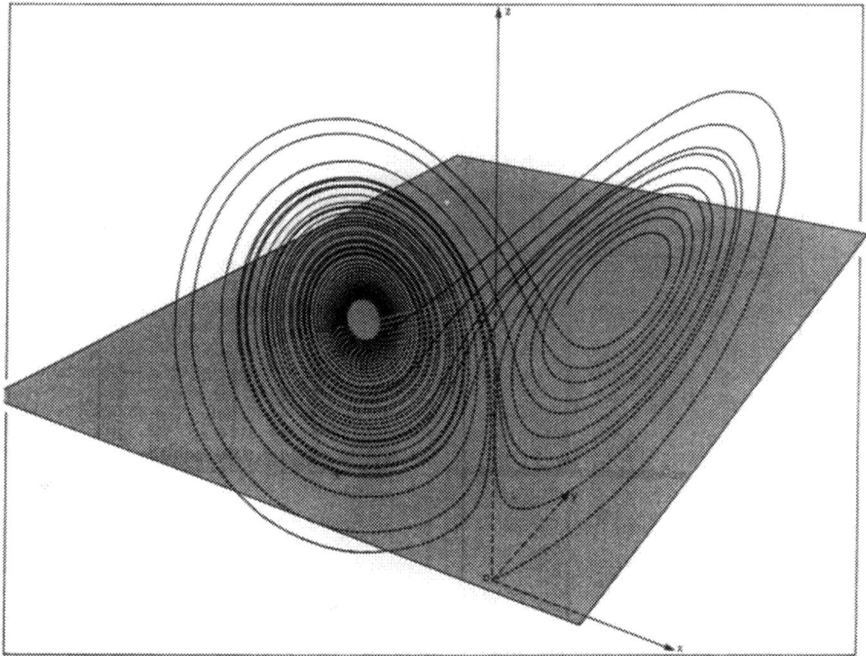

Figure 3. *The Lorenz attractor.* This beautiful figure has been obtained by Oscar Lanford, of Berkeley. It illustrates a new strange attractor, the ***Lorenz attractor,*** which is approached by the solutions of the Lorenz system of equations:

$$\frac{dx}{dt} = -10x + 10y, \quad \frac{dy}{dt} = -xz + 28x - y, \quad \frac{dz}{dt} = xy - \frac{8}{3}z.$$

Lanford has chosen the solution which starts from the origin (0, 0, 0) at time $t = 0$. It makes one loop to the right, then a few loops to the left, then to the right, and so on in irregular manner. One follows the solution here for fifty loops. The part below the plane $z = 27$ is drawn as a dotted line. If one would take, instead of (0, 0, 0), a nearby initial condition, the new solution would soon deviate from the old one, and the numbers of loops to the left and to the right would no longer be the same. There is ***sensitive dependence with respect to initial conditions.*** The Lorenz equations are suggested by a problem of atmospheric convection. Edward Lorenz has used the sensitive dependence on initial condition observed with the above equations to justify the imprecision of weather forecasting.

of turbulence remains however rather mysterious and controversial.

One may in principle describe the time evolution of a viscous fluid by equations of the form (4). The number m will have to be taken infinite, because the state of the fluid at a given instant of time requires an infinite number of variables for its description. We admit that there are no further problems, and write $X(t)$ and G instead of $x_1(t), x_2(t), \ldots$, and $G_1, G_2, \ldots$. The equations (4) can then be written in compact form as

$$\frac{d}{dt} X(t) = G_\mu(X(t)) \tag{6}$$

We have introduced a parameter μ in (6) to indicate the intensity of external action on the fluid. (If there is no external action, viscosity brings the fluid to rest, and there is no turbulence). In the example of the tap, μ might give the degree of opening of the tap. In the convection equations (5) of Lorenz, μ is replaced by r, which is proportional to the temperature difference between the top and the bottom of the fluid layer. In many hydrodynamical problems, the role of μ is taken by a parameter called *Reynolds number*.

If $\mu = 0$, i.e., if there is no external action, the fluid tends to a state of rest $X(t) = X_0$. This state corresponds to an attracting fixed point X_0 for our dynamical system.

For small μ one observes again a steady state $X(t) = X_\mu$. As μ is further increased, one often sees periodic oscillations in the fluid. This means that asymptotically

$$X(t) = f(\omega t)$$

where f is a function of period 2π and ω the frequency of the oscillations. This situation corresponds to a periodic attractor for continuous time, i.e., a circle or "attracting limit cycle". For sufficiently large μ, the fluid motion becomes irregular, chaotic: turbulence has set in.

When I became interested in turbulence, around 1970, Lorenz' paper of 1963 was not known to physicists and mathematicians. The most popular theory of turbulence was that of Lev D. Landau of Moscow [5]. According to this theory, the time evolution of a turbulent fluid is asymptotically given by

$$X(t) = f_k(\omega_1 t, \omega_2 t, \ldots, \omega_k t) \tag{7}$$

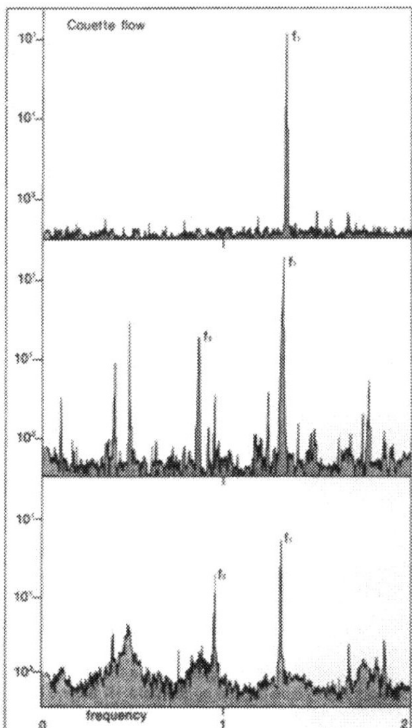

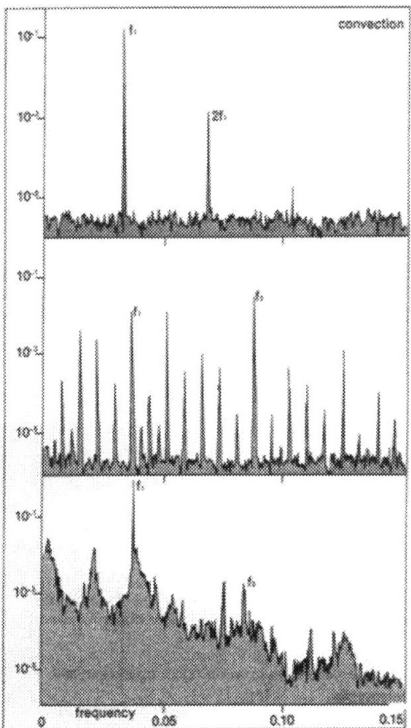

Figure 4. *Frequency spectra.* A frequency analysis of the time dependence of a phenomenon is possible, whether this dependence is periodic or not. One obtains thus a "frequency spectrum" giving the square of the amplitude associated with each frequency. The spectra on the left of the figure have been measured by R. Fenstermacher for the Couette flow (the interval between two coaxial circular cylinders is filled with fluid, and the inner cylinder is rotated at constant speed). The spectra on the right have been measured by S. Benson for a convective flow (a liquid layer is heated from below, the hot liquid formed below is lighter and rises, producing convection currents).

The different spectra shown correspond to different speeds of rotation (Couette) or different intensities of heating (convection). The spectra at the top contain isolated peaks corresponding to a certain frequency and its harmonics: the system is *periodic*. The spectra in the middle row exhibit several independent frequencies: the system is *quasi periodic*. The spectra at the bottom show some wide peaks on a background of *continuous spectrum*, this suggests that a strange attractor is present. Notice that the frequency spectra are shown with a logarithmic vertical scale.

where f_k is a periodic function of period 2π in each of its arguments, and $\omega_1, \omega_2, \ldots, \omega_k$ are independent frequencies. A function of t of the form (7) is called quasiperiodic. (One can see that the corresponding quasiperiodic attractor is a k-dimensional torus). A quasiperiodic function has a non periodic, irregulàr aspect, suggestive of turbulence. However a small change in initial conditions simply replaces $\omega_1 t, \ldots, \omega_k t$ by $\omega_1 t + \alpha_1, \ldots, \omega_k t + \alpha_k$ with small $\alpha_1, \ldots, \alpha_k$. There is thus no sensitive dependence on initial conditions.

It was tempting to appeal to strange attractors rather than quasiperiodic attractors to interpret turbulence. A mathematical argument against quasiperiodic attractors is their fragility. My attention had been drawn on this fragility, or absence of "structural stability" by the seminars of René Thom at the Institut des Hautes Etudes Scientifiques (Bures-sur-Yvette). By a small perturbation of (6) one can destroy a quasiperiodic attractor and, if $k \geqslant 3$, obtain a strange attractor. I had published this result with Floris Takens [3] in 1971, and we had on this occasion proposed the idea that turbulence is described by strange attractors. While structural stability may not be as important an aspect of things as we thought at the time, the connection between strange attractors and turbulence was a lucky idea.

It remained to be seen if strange attractors would give a better description of turbulence than quasiperiodic attractors. There is no direct experimental test of sensitive dependence on initial condition in hydrodynamics. One may however do a frequency analysis of the fluid velocity at a point, considered as a function of time. The function giving the square of the amplitude versus the frequency is called *frequency spectrum* (see Figure 4). For a quasiperiodic function the frequency spectrum is formed of discrete peaks at the frequencies $\omega_1, \ldots, \omega_k$ and their linear combinations with integer coefficients. By contrast if the time evolution is governed by a strange attractor one may obtain a continuous frequency spectrum.

It was known that the frequency spectrum of a turbulent fluid is continuous, but this fact was attributed to the accumulation of a large number of independent frequencies simulating, in the limit, a continuous spectrum. Recently (1974–75), delicate experiments performed by Guenter Ahlers at Bell Labs (Murray Hill, NJ), Jerry Gollub

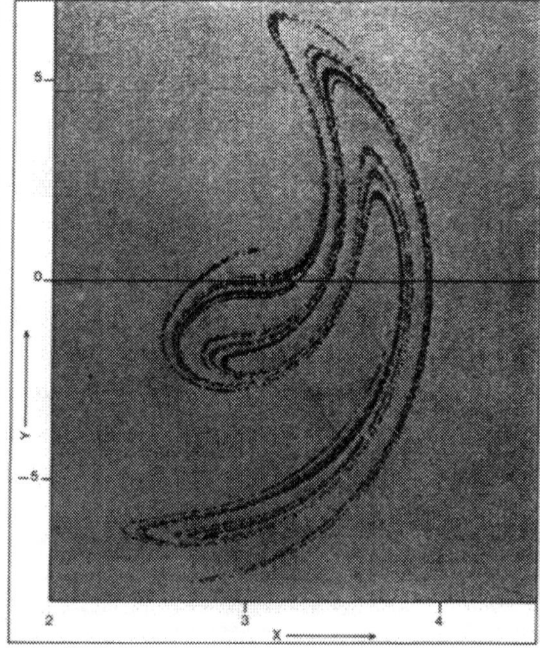

Figure 5. ***A Japanese attractor.*** **This picture shows a strange attractor invented by Yoshisuke Ueda, of Kyoto University. It is obtained by solving the differential equation**

$$\frac{d^2x}{dt^2} + k\frac{dx}{dt} + x^3 = B \cos t$$

for $k = 0.1$ and $B = 12$, and marking the points with coordinates $x(2n\pi)$, $\frac{d}{dt}x(2n\pi)$ for integer n (discrete time $t = 2n\pi$). Depending on the initial conditions one obtains either the above attractor, or a single point (attracting fixed point). Y. Ueda has studied strange attractors numerically for a number of years on analog and digital computers. Esthetically, his pictures are probably the finest obtained to this date.

and Harry Swinney at City College (New York) [6], and others, have shown that things happen differently. When one increases the parameter μ describing the system, the transition to the continuous spectrum characteristic of turbulence is rapid. There is no progressive accumulation of many independent discrete frequencies. So it seems that the onset of turbulence may well correspond to the appearance of strange attractors.

Other Chaotic Phenomena: Turbulence Everywhere

It should here be mentioned that frictionless mechanical systems (conservative systems) give rise neither to strange attractors, nor in fact to attractors at all. Actually, a theorem of mechanics, Liouville's theorem, asserts that time evolution preserves volumes in phase space. This prevents the volume contraction which occurs near an attractor. On the other hand, conservative systems often show sensitive dependence on initial condition.

The physico-chemical systems which give rise to strange attractors are the *dissipative systems*, i.e., those for which a "noble" form of energy (for instance mechanical, electrical, or chemical energy) is changed into heat [7]. These systems actually exhibit an interesting behavior only if they are constantly fed some noble energy, otherwise they go to rest.

One knows chemical reactions which are periodic in time (see inset). I asked in 1971 a chemist, specialist of these periodic reactions, if he thought that one would find chemical reactions with chaotic time dependence. He answered that if an experimentalist obtained a chaotic record in the study of a chemical reaction, he would throw away the record, saying that the experiment was unsuccessful. Things, fortunately, have changed, and we now have several examples of non periodic chemical reactions.

The magnetism of the earth perhaps gives an example of a strange attractor. It is known that the earth magnetic field reverses itself at irregular intervals. This phenomenon occurred at least sixteen times in the last four million years. Geophysicists have written "dynamo equations" with chaotic solutions which describe irregular changes of direction of a magnetic field. There is however as yet no quantitatively satisfactory theory.

Ecologists have studied non periodic models in population dynamics. If m species have, in the year $t + 1$, populations $x_1(t + 1), \ldots, x_m(t + 1)$ determined by the equations (1) in terms of the populations in the year t, one may expect strange attractors to occur. In fact, already for $m = 1$, the equation

$$x(t + 1) = Rx(t)\,(1 - x(t))$$

gives rise to nonperiodic behavior [8].

One imagines easily that strange attractors may play a role in economics, where periodic processes (economic cycles) are well-known. In fact, let us suppose that the macroeconomical evolution equations contain a parameter μ describing, say, the level of technological development. By analogy with hydrodynamics we would guess that for small μ the economy is in a steady state and that, as μ increases, periodic or quasiperiodic cycles may develop. For high μ chaotic behavior with sensitive dependence on initial condition would be present. This discussion is some

A Periodic Chemical Phenomenon: The Belousov-Zhabotinski Reaction

For about twenty years now, an oscillating reaction has been known to chemists. The oscillations have a period of the order of one minute, and continue for perhaps an hour, until the reagents are exhausted. If reagents are added continuously, while reaction products are removed, the oscillations proceed periodically forever. The reaction is, roughly speaking, the oxydization of malonate by bromate, catalyzed by Cerium. The experiment is fairly easy to realize: here is the recipe.

Malonic acid	0.3 M
Cerous nitrate	0.005–0.01 M
Sulfuric acid	3.0 M
Sodium bromate	0.05–0.01 M
Ferroin	a little

M means "molar", for instance sulfuric acid occurs at the concentration of 3 moles per liter. Ferroin is an oxidation reduction indicator (obtained by mixing in water a small amount of *o*-phenanthroline and ferrous sulfate). In practice one prepares one solution with part of the reagents (in water), and another solution with the rest of the reagents. The oscillating reaction starts when the two solutions are mixed. Perhaps the mathematical reader should be warned that diluting sulfuric acid produces heat and requires caution (see a chemistry text). The sulfuric solution should be allowed to cool before being mixed with the other solution, otherwise the oscillations will not be seen. During the reaction, the ferroin turns from blue to purple to red, making the oscillations visible. At the same time the Cerium ion changes from pale yellow to colorless, so that all kinds of hues are produced.

The Belousov-Zhabotinski reaction, which we just described, caused astonishment and some disbelief among chemists when it was discovered. Other periodic chemical reactions have now been discovered, in particular in systems of biological origin. One speculates on the physiological significance of these reactions, but little is really known with certainty.

what metaphorical, but its conclusions are suggestive, and a more detailed analysis may be useful.

To conclude this list of examples, let me mention a dynamical system of vital interest to everyone of us: the heart. The normal cardiac regime is periodic, but there are many nonperiodic pathologies (like ventricular fibrillation) which lead to the steady state of death. It seems that great medical benefit might be derived from computer studied of a realistic mathematical model which would reproduce the various cardiac dynamical regimes.

The application of the ideas which we have discussed often poses serious methodological problems. How does one maintain constant experimental conditions, and how does one make precise measurements? In any case, the recognition of the role of strange attractors in many problems is a great conceptual progress. The nonperiodic fluctuations of a dynamical system do not necessarity indicate an experiment spoilt by mysterious random forces; they often point to a strange attractor, which one may try to understand [9].

I have not spoken of the esthetic appeal of strange attractors. These systems of curves, these clouds of points suggest sometimes fireworks or galaxies, sometimes strange and disquieting vegetal proliferations. A realm lies there of forms to explore, and harmonies to discover.

References

1. M. Hénon. A two-dimensional mapping with a strange attractor. Commun. Math. Phys. *50*, 69–77 (1976). See also S. D. Feit. Characteristic exponents and strange attractors. Commun. Math. Phys. *61*, 249–260 (1978). J. H. Curry. On the Hénon transformation. Commun. Math. Phys. *68*, 129–140 (1979).
2. S. Smale. Differentiable dynamical systems. Bull. Amer. Math. Soc. *73*, 747–817 (1967).
3. D. Ruelle, F. Takens. On the nature of turbulence. Commun. Math. Phys. *20*, 167–192 (1971); *23*, 343–344 (1971). See also S. Newhouse, D. Ruelle, F. Takens. Occurrence of strange Axiom A attractors near quasiperiodic flows on T^m, $m > 3$. Commun. Math. Phys. *64*, 35–40 (1978).
4. E. N. Lorenz. Deterministic nonperiodic flow. J. Atmos. Sci. *20*, 130–141 (1963).
5. L. D. Landau, E. M. Lifshitz. *Fluid mechanics*. Pergamon, Oxford, 1959.
6. H. L. Swinney, J. P. Gollub. The transition to turbulence. Physics Today *31*, No. 8, 41–49 (1978).
7. I. Prigogine. *Introduction to thermodynamics of irreversible processes*. Wiley, New York, 1962. [Does not cover chaotic time evolutions. I prefer this little book on the physics of dissipative systems to later and more ambitions works of the same author].
8. R. May. Simple mathematical models with very complicated dynamics. Nature *261*, 459–467 (1976).
9. A wealth of information is contained in the proceedings of two conferences organized by the New York Academy of Sciences. *Bifurcation theory and applications in scientific disciplines*. Ann. N. Y. Acad. Sci. 316 (1979). *Nonlinear dynamics*. Ann. N. Y. Acad. Sci. (To appear).

David Ruelle
Institut des Hautes Etudes Scientifiques
F-91440 Bures-sur-Yvette, France

Editor's note. Two other enjoyable popular articles on deterministic chaos are Hofstadter (1981) and J Ford, *Physics Today* p 40 (April 1983).

Los Alamos Science **1** 4–27 (1980)

Universal Behavior in Nonlinear Systems

Mitchell J Feigenbaum

Universal numbers, $\delta = 4.6692016\ldots$ and $\alpha = 2.502907875\ldots$, determine quantitatively the transition from smooth to turbulent or erratic behavior for a large class of nonlinear systems.

There exist in nature processes that can be described as complex or chaotic and processes that are simple or orderly. Technology attempts to create devices of the simple variety: an idea is to be implemented, and various parts executing orderly motions are assembled. For example, cars, airplanes, radios, and clocks are all constructed from a variety of elementary parts each of which, ideally, implements one ordered aspect of the device. Technology also tries to control or minimize the impact of seemingly disordered processes, such as the complex weather patterns of the atmosphere, the myriad whorls of turmoil in a turbulent fluid, the erratic noise in an electronic signal, and other such phenomena. It is the complex phenomena that interest us here.

When a signal is noisy, its behavior from moment to moment is irregular and has no simple pattern of prediction. However, if we analyze a sufficiently long record of the signal, we may find that signal amplitudes occur within narrow ranges a definite fraction of the time. Analysis of another record of the signal may reveal the same fraction. In this case, the noise can be given a *statistical* description. This means that while it is impossible to say what amplitude will appear next in succession, it is possible to estimate the probability or likelihood that the signal will attain some specified range of values. Indeed, for the last hundred years disorderly processes have been taken to be statistical (one has given up asking for a precise causal prediction), so that the goal of a description is to determine what the probabilities are, and from this information to determine various behaviors of interest—for example, how air turbulence modifies the drag on an airplane.

We know that perfectly definite causal and *simple* rules can have statistical (or random) behaviors. Thus, modern computers possess "random number generators" that provide the statistical ingredient in a simulation of an erratic process. However, this generator does nothing more than shift the decimal point in a rational number whose repeating block is suitably long. Accordingly, it is possible to predict what the nth generated number will be. Yet, in a list of successive generated numbers there is such a seeming lack of order that all statistical tests will confer upon the numbers a pedigree of randomness. Technically, the term "pseudorandom" is used to indicate this nature. One now may ask whether the various complex processes of nature themselves might not be merely pseudorandom, with the full import of randomness, which is untestable, a historic but misleading concept. Indeed our purpose here is to explore this possibility. What will prove altogether remarkable is that some very simple schemes to produce erratic numbers behave *identically* to some of the erratic aspects of natural phenomena. More specifically, there is now cogent evidence that the problem of how a fluid changes over from smooth to turbulent flow can be solved through its relation to the simple scheme described in this article. Other natural problems that can be treated in the same way are the behavior of a population from generation to generation and the noisiness of a large variety of mechanical, electrical, and chemical oscillators. Also, there is now evidence that various Hamiltonian systems—those subscribing to classical mechanics, such as the solar system—can come under this discipline.

The feature common to these phenomena is that, as some external parameter (temperature, for example) is varied, the behavior of the system changes from simple to erratic. More precisely, for some range of parameter values, the system exhibits an orderly *periodic* behavior; that is, the system's behavior reproduces itself every *period* of time T. Beyond this range, the behavior fails to reproduce itself after T seconds; it almost does so, but in fact it requires *two* intervals of T to repeat itself. That is, the period has *doubled* to 2T. This new periodicity remains over some range of parameter values until another critical parameter value is reached after which the behavior *almost* reproduces itself after 2T, but in fact, it now requires 4T for reproduction. This process of successive period doubling recurs continually (with the range of parameter values for which the period is 2^nT becoming successively smaller as n increases) until, at a certain value of the parameter, it has doubled *ad infinitum*, so that the behavior is no longer periodic. Period doubling is then a characteristic route for a system to follow as it changes over from simple periodic to complex aperiodic motion. All the phenomena mentioned above exhibit period doubling. In the limit of aperiodic behavior, there is a unique and hence *universal* solution common to all systems undergoing period doubling. This fact implies remarkable consequences. For a given system, if we denote by Λ_n the value of the parameter at which its period doubles for the nth time, we find that the values Λ_n converge to Λ_∞ (at which the motion is aperiodic) *geometrically* for large n. This means that

$$\Lambda_\infty - \Lambda_n \propto \delta^{-n} \qquad (1)$$

for a fixed value of δ (the *rate* of onset of complex behavior) as n becomes large. Put differently, if we define

$$\delta_n \equiv \frac{\Lambda_{n+1} - \Lambda_n}{\Lambda_{n+2} - \Lambda_{n+1}}, \qquad (2)$$

δ_n (quickly) approaches the constant value δ. (Typically, δ_n will agree with δ to several significant figures after just a few period doublings.) What is quite remarkable (beyond the fact that there is always a geometric convergence) is that, for all systems undergoing this period doubling, the value of δ is *predetermined* at the universal value

$$\delta = 4.6692016 \ldots \quad . \tag{3}$$

Thus, this definite number must appear as a natural rate in oscillators, populations, fluids, and all systems exhibiting a period-doubling route to turbulence! In fact, most measurable properties of *any* such system in this aperiodic limit now can be determined, in a way that essentially bypasses the details of the equations governing each specific system because the theory of this behavior is universal over such details. That is, so long as a system possesses certain *qualitative* properties that enable it to undergo this route to complexity, its *quantitative* properties are determined. (This result is analogous to the results of the modern theory of critical phenomena, where a few qualitative properties of the system undergoing a phase transition, notably the dimensionality, determine *universal* critical exponents. Indeed at a *formal* level the two theories are identical in that they are fixed-point theories, and the number δ, for example, can be viewed as a critical exponent.) Accordingly, it is sufficient to study the simplest system exhibiting this phenomenon to comprehend the general case.

Functional Iteration

A random number generator is an example of a simple iteration scheme that has complex behavior. Such a scheme generates the next pseudorandom number by a definite transformation upon the present pseudorandom number. In other words, a certain function is reevaluated successively to produce a sequence of such numbers. Thus, if f is the function and x_0 is a starting number (or "seed"), then $x_0, x_1, \ldots, x_n, \ldots$, where

$$\begin{aligned} x_1 &= f(x_0) \\ x_2 &= f(x_1) \\ &\ \vdots \\ x_{n+1} &= f(x_n) \\ &\ \vdots \end{aligned} \tag{4}$$

is the sequence of generated pseudorandom numbers. That is, they are generated by *functional iteration*. The nth element in the sequence is

$$x_n = f(f(\ldots f(f(x_0)) \ldots)) \equiv f^n(x_0) , \tag{5}$$

where n is the total number of applications of f. [$f^n(x)$ is not the nth power of f(x); it is the nth *iterate* of f.] A property of iterates worthy of mention is

$$f^n(f^m(x)) = f^m(f^n(x)) = f^{m+n}(x) , \tag{6}$$

since each expression is simply m + n applications of f. It is understood that

$$f^0(x) = x . \tag{7}$$

It is also useful to have a symbol, $\circ$, for functional iteration (or composition), so that

$$f^n \circ f^m = f^m \circ f^n = f^{m+n} . \tag{8}$$

Now f^n in Eq. (5) is itself a definite and computable function, so that x_n as a function of x_0 is known in principle.

If the function f is *linear* as, for example,

$$f(x) = ax \tag{9}$$

for some constant a, it is easy to see that

$$f^n(x) = a^n x\,, \tag{10}$$

so that, for this f,

$$x_n = a^n x_0 \tag{11}$$

is the solution of the *recurrence relation* defined in Eq. (4),

$$x_{n+1} = ax_n\,. \tag{12}$$

Should $|a| < 1$, then x_n geometrically converges to zero at the rate 1/a. This example is special in that the linearity of f allows for the explicit computation of f^n.

We must choose a *nonlinear* f to generate a pseudorandom sequence of numbers. If we choose for our nonlinear f

$$f(x) = a - x^2\,, \tag{13}$$

then it turns out that f^n is a polynominal in x of order 2^n. This polynomial rapidly becomes unmanageably large; moreover, its coefficients are polynomials in a of order up to 2^{n-1} and become equally difficult to compute. Thus even if $x_0 = 0$, x_n is a polynomial in a of order 2^{n-1}. These polynomials are nontrivial as can be surmised from the fact that for certain values of a, the sequence of numbers generated for almost all starting points in the range $(a - a^2, a)$ possess *all* the mathematical properties of a random sequence. To illustrate this, the figure on the cover depicts the iterates of a similar system in two dimensions:

$$x_{n+1} = y_n - x_n^2$$

$$y_{n+1} = a - x_n\,. \tag{14}$$

Analogous to Eq. (4), a starting coordinate pair (x_0,y_0) is used in Eq. (14) to determine the next coordinate (x_1,y_1). Equation (14) is reapplied to determine (x_2,y_2) and so on. For some initial points, all iterates lie along a definite elliptic curve, whereas for others the iterates are distributed "randomly" over a certain region. It should be obvious that no explicit formula will account for the vastly rich behavior shown in the figure. That is, while the iteration scheme of Eq. (14) is trivial to specify, its nth iterate as a function of (x_0,y_0) is unavailable. Put differently, applying the simplest of *nonlinear* iteration schemes to itself sufficiently many times can create vastly complex behavior. Yet, precisely because the same operation is reapplied, it is conceivable that only a select few self-consistent patterns might emerge where the consistency is determined by the key notion of iteration and *not* by the particular function performing the iterates. These self-consistent patterns do occur in the limit of infinite period doubling and in a well-defined intricate organization that can be determined *a priori* amidst the immense complexity depicted in the cover figure.

The Fixed-Point Behavior of Functional Iterations

Let us now make a direct onslaught against Eq. (13) to see what it possesses. We want to know the behavior of the system after many iterations. As we already know, high iterates of f rapidly become very complicated. One way this growth can be prevented is to have the first iterate of x_0 be precisely x_0 itself.

Generally, this is impossible. Rather this condition *determines* possible x_0's. Such a self-reproducing point is called a *fixed point* of f. The sequence of iterates is then $x_0, x_0, x_0, \ldots$ so that the behavior is *static*, or if viewed as periodic, it has period 1.

It is elementary to determine the fixed points of Eq. (13). For future convenience we shall use a modified form of Eq. (13) obtained by a translation in x and some redefinitions:

$$f(x) = 4\lambda x(1 - x), \tag{15}$$

so that as λ is varied, $x = 0$ is always a fixed point. Indeed, the fixed-point condition for Eq. (15),

$$x^* = f(x^*) = 4\lambda x^*(1 - x^*), \tag{16}$$

gives as the two fixed points

$$x^* = 0, x_0^* = 1 - 1/4\lambda. \tag{17}$$

The maximum value of f(x) in Eq. (15) is attained at $x = \frac{1}{2}$ and is equal to λ. Also, for $\lambda > 0$ and x in the interval (0,1), f(x) is always positive. Thus, if λ is anywhere in the range [0,1], then any iterate of any x in (0,1) is also always in (0,1). Accordingly, in all that follows we shall consider only values of x and λ lying between 0 and 1. By Eq. (16) for $0 \leq \lambda < \frac{1}{4}$, only $x^* = 0$ is within range, whereas for $\frac{1}{4} \leq \lambda \leq 1$, both fixed points are within the range. For example, if we set $\lambda = \frac{1}{2}$ and we start at the fixed point $x_0^* = \frac{1}{2}$ (that is, we set $x_0 = \frac{1}{2}$), then $x_1 = x_2 = \ldots = \frac{1}{2}$; similarly if $x_0 = 0$, $x_1 = x_2 = \ldots = 0$, and the problem of computing the nth iterate is obviously trivial.

What if we choose an x_0 *not* at a fixed point? The easiest way to see what happens is to perform a graphical analysis. We graph $y = f(x)$ together with $y = x$. Where the lines intersect we have $x = y = f(x)$, so that the intersections are precisely the fixed points. Now, if we choose an x_0 and plot it on the x-axis, the ordinate of f(x) at x_0 is x_1. To obtain x_2, we must transfer x_1 to the x-axis before reapplying f. Reflection through the straight line $y = x$ accomplishes precisely this operation. Altogether, to iterate an initial x_0 successively,

1. move *vertically* to the graph of f(x),
2. move *horizontally* to the graph of $y = x$, and
3. repeat steps 1, 2, etc.

Figure 1 depicts this process for $\lambda = \frac{1}{2}$. The two fixed points are circled, and the first several iterates of an arbitrarily chosen point x_0 are shown. What should be obvious is that if we start from any x_0 in (0,1) ($x = 0$ and $x = 1$ excluded), upon continued iteration x_n will converge to the fixed point at $x = \frac{1}{2}$. No matter how close x_0 is to the fixed point at $x = 0$, the iterates diverge away from it. Such a fixed point is termed *unstable*. Alternatively, for almost all x_0 near enough to $x = \frac{1}{2}$ [in this case, all x_0 in (0,1)], the iterates converge towards $x = \frac{1}{2}$. Such a fixed point is termed *stable* or is referred to as an *attractor* of period 1.

Now, if we don't care about the *transient* behavior of the iterates of x_0, but only about some regular behavior that will emerge eventually, then knowledge of the stable fixed point at $x = \frac{1}{2}$ satisfies our concern for the *eventual* behavior of the iterates. In this restricted sense of eventual behavior, the existence of an attractor determines the solution *independently* of the initial condition x_0 provided that x_0 is within the *basin of attraction* of the attractor; that is, that it *is* attracted. The attractor satisfies Eq. (16), which is explicitly independent of x_0. This condi-

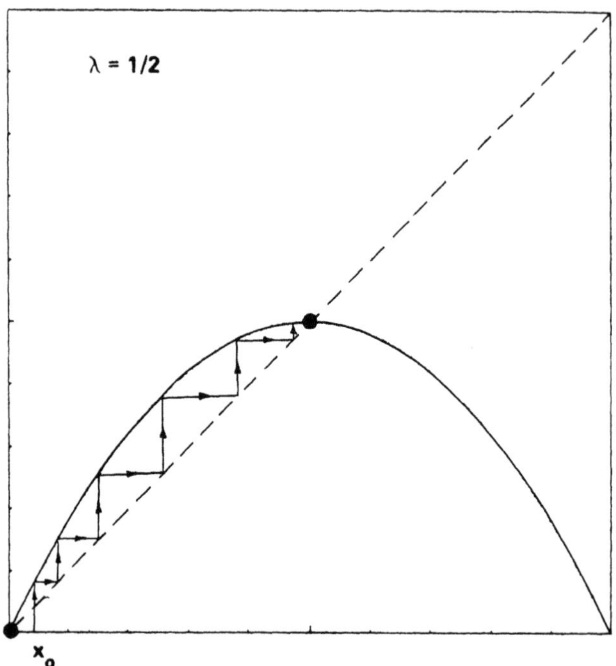

Fig. 1. Iterates of x_0 ***at*** $\lambda = 0.5$.

tion is the basic theme of universal behavior: if an attractor exists, the eventual behavior is independent of the starting point.

What makes $x = 0$ unstable, but $x = \frac{1}{2}$ stable? The reader should be able to convince himself that $x = 0$ is unstable because *the slope of $f(x)$ at $x = 0$ is greater than 1*. Indeed, if x^* is a fixed point of f and the derivative of f at x^*, $f'(x^*)$, is smaller than 1 in absolute value, then x^* is stable. If $|f'(x^*)|$ is greater than 1, then x^* is unstable. Also, only *stable* fixed points can account for the eventual behavior of the iterates of an arbitrary point.

We now must ask, "For what values of λ are the fixed points attracting?" By Eq. (15), $f'(x) = 4\lambda(1 - 2x)$ so that

$$f'(0) = 4\lambda \tag{18}$$

and

$$f'(x_0^*) = 2 - 4\lambda. \tag{19}$$

For $0 < \lambda < \frac{1}{4}$, only $x^* = 0$ is stable. At $\lambda = \frac{1}{4}$, $x_0^* = 0$ and $f'(x_0^*) = 1$. For $\frac{1}{4} < \lambda < \frac{3}{4}$, x^* is unstable and x_0^* is stable, while at $\lambda = \frac{3}{4}$, $f'(x^*_0) = -1$ and x_0^* also has become unstable. Thus, for $0 < \lambda < \frac{3}{4}$, the eventual behavior is known.

Period 2 from the Fixed Point

What happens to the system when λ is in the range $\frac{3}{4} < \lambda < 1$, where there are no attracting fixed points? We will see that as λ increases slightly beyond $\lambda = \frac{3}{4}$, f undergoes period doubling. That is, instead of having a stable cycle of period 1 corresponding to one fixed point, the system has a stable cycle of period 2; that is, the cycle contains two points. Since these two points are fixed points of the function f^2 (f applied twice) and since stability is determined by the slope of a function at its *fixed* points, we must now focus on f^2. First, we examine a graph of f^2 at λ just below $\frac{3}{4}$. Figures 2a and b show f and f^2, respectively, at $\lambda = 0.7$.

To understand Fig. 2b, observe first that, since f is symmetric about its maximum at $x = \frac{1}{2}$, f^2 is also symmetric about $x = \frac{1}{2}$. Also, f^2 must have a fixed point whenever f does because the second iterate of a fixed point is still that same point. The main ingredient that determines the period-doubling behavior of f as λ increases is the relationship of the slope of f^2 to the slope of f. This relationship is a consequence of the chain rule. By definition

$$x_2 = f^2(x_0) ,$$

where

$$x_1 = f(x_0),\ x_2 = f(x_1) .$$

We leave it to the reader to verify by the chain rule that

$$f^{2\prime}(x_0) = f'(x_0)f'(x_1) \tag{20}$$

and

$$f^{n\prime}(x_0) = f'(x_0)f'(x_1) \ldots f'(x_{n-1}) , \tag{21}$$

an elementary result that determines period doubling. If we start at a fixed point of f and apply Eq. (20) to $x_0 = x^*$, so that $x_2 = x_1 = x^*$, then

$$f^{2\prime}(x^*) = f'(x^*)f'(x^*) = [f'(x^*)]^2 . \tag{22}$$

Since at $\lambda = 0.7$, $|f'(x^*)| < 1$, it follows from Eq. (22) that

$$0 < f^{2\prime}(x^*) < 1 .$$

Also, if we start at the extremum of f, so that $x_0 = \frac{1}{2}$ and $f'(x_0) = 0$, it follows from Eq. (21) that

$$f^{n\prime}(\tfrac{1}{2}) = 0 \tag{23}$$

for all n. In particular, f^2 is extreme (and a minimum) at $\frac{1}{2}$. Also, by Eq. (20), f^2 will be extreme (and a maximum) at the x_0 that will iterate under f to $x = \frac{1}{2}$, since then $x_1 = \frac{1}{2}$ and $f'(x_1) = 0$. These points, the *inverses* of $x = \frac{1}{2}$, are found by going *vertically* down along $x = \frac{1}{2}$ to $y = x$ and then ***horizontally*** to $y = f(x)$. (Reverse the arrows in Fig. 1, and see Fig. 2a.) Since f has a maximum, there are *two* horizontal intersections and, hence, the two maxima of Fig. 2b. *The ability of f to have complex behaviors is precisely the consequence of its double-valued inverse*, which is in turn a reflection of its possession of an extremum. A monotone f, one that always increases, *always* has simple behaviors, whether or not the behaviors are easy to compute. A *linear* f is always monotone. The f's we care about always fold over and so are *strongly* nonlinear. This folding nonlinearity gives rise to universality. Just as linearity in any system implies a definite method of solution, folding nonlinearity in any system also implies a definite method of solution. In fact folding nonlinearity in the aperiodic limit of period doubling in any system is solvable, and many systems, such as various coupled

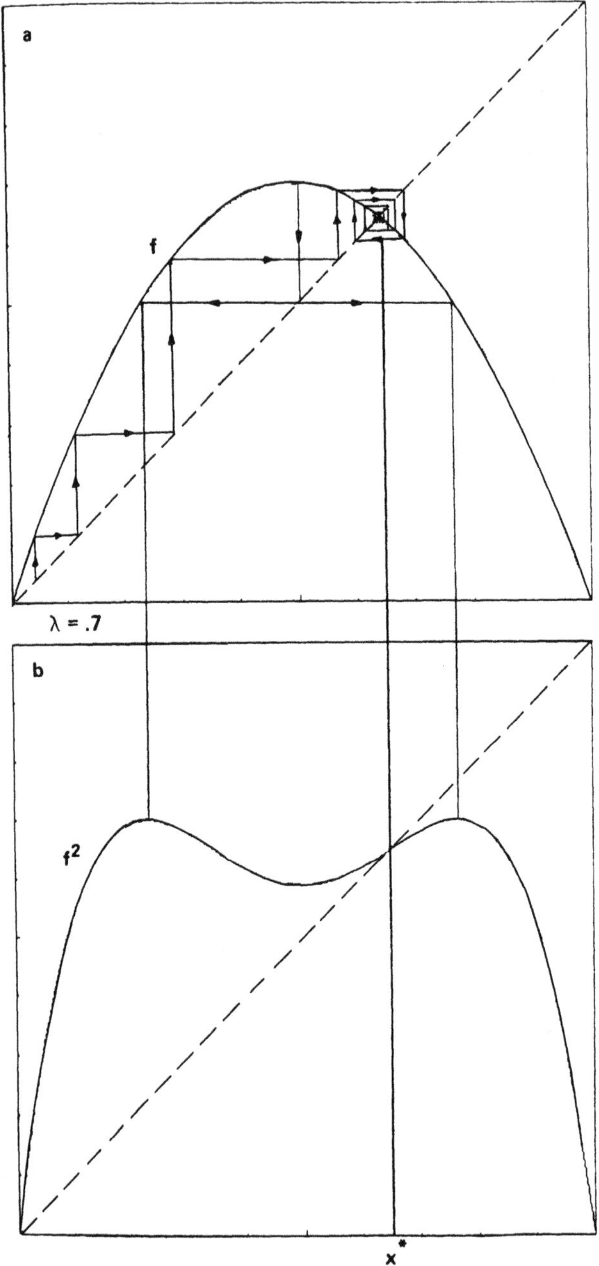

Fig. 2. $\lambda = 0.7$. x^* *is the stable fixed point. The extrema of* f^2 *are located in (a) by constructing the inverse iterates of* $x = 0.5$.

nonlinear differential equations, possess this nonlinearity.

To return to Fig. 2b, as $\lambda \rightarrow \frac{3}{4}$ and the maximum value of f increases to $\frac{3}{4}$, $f'(x^*) \rightarrow -1$ and $f^{2\prime}(x^*) \rightarrow +1$. As λ increases beyond $\frac{3}{4}$, $|f'(x^*)| > 1$ and $f^{2\prime}(x^*) > 1$, so that f^2 must develop two new fixed points beyond those of f; that is, f^2 will cross $y = x$ at two more points. This transition is depicted in Figs. 3a and b for f and f^2, respectively, at $\lambda = 0.75$, and similarly in Figs. 4a and b at $\lambda = 0.785$. (Observe the exceptionally slow convergence to x^* at $\lambda = 0.75$, where iterates approach the fixed point not geometrically, but rather with deviations from x^* inversely proportional to the square root of the number of iterations.) Since x_1^* and x_2^*, the new fixed points of f^2, are *not* fixed points of f, it must be that f sends one into the other:

$$x_1^* = f(x_2^*)$$

and

$$x_2^* = f(x_1^*) \ .$$

Such a *pair of points*, termed a *2-cycle*, is depicted by the limiting unwinding circulating square in Fig. 4a. Observe in Fig. 4b that the slope of f^2 is in excess of 1 at the fixed point of f and so is an unstable fixed point of f^2, while the two new fixed points have slopes smaller than 1, and so are *stable*; that is, every two iterates of f will have a point attracted toward x_1^* if it is sufficiently close to x_1^* or toward x_2^* if it is sufficiently close to x_2^*. This means that the sequence under f,

$$x_0, x_1, x_2, x_3, \ldots ,$$

eventually becomes arbitrarily close to the sequence

$$x_1^*, x_2^*, x_1^*, x_2^*, \ldots ,$$

so that this is a stable 2-cycle, or an *attractor of period 2*. Thus, we have observed for Eq. (15) the first period doubling as the parameter λ has increased.

There is a point of paramount importance to be observed; namely, f^2 has the same slope at x_1^* and at x_2^*. This point is a direct consequence of Eq. (20), since if $x_0 = x_1^*$, then $x_1 = x_2^*$, and vice versa, so that the product of the slopes is the same. More generally, if $x_1^*, x_2^*, \ldots, x_n^*$ is an n-cycle so that

$$x_{r+1}^* = f(x_r^*) \qquad r = 1, 2, \ldots, n-1$$

and

$$x_1^* = f(x_n^*) \, , \tag{24}$$

then *each* is a fixed point of f^n with identical slopes:

$$x_r^* = f^n(x_r^*) \qquad r = 1, 2, \ldots, n \tag{25}$$

and

$$f^{n\prime}(x_r^*) = f'(x_1^*) \ldots f'(x_n^*) \, . \tag{26}$$

From this observation will follow period doubling *ad infinitum*.

As λ is increased further, the minimum at $x = \frac{1}{2}$ will drop as the slope of f^2 through the fixed point of f increases. At some value of λ, denoted by λ_1, $x = \frac{1}{2}$ will become a fixed point of f^2. Simultaneously, the right-hand maximum will also become a fixed point of f^2. [By Eq. (26), both elements of the 2-cycle have slope 0.] Figures 5a and b depict the situation that occurs at $\lambda = \lambda_1$.

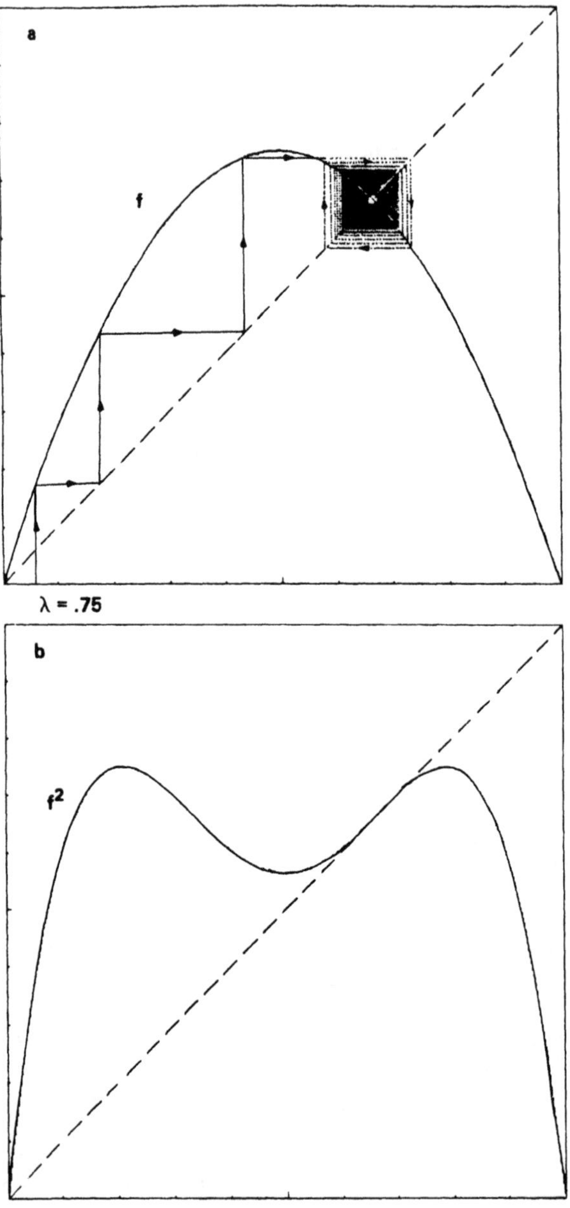

Fig. 3. λ = 0.75. (a) depicts the slow convergence to the fixed point. f^2 osculates about the fixed point.

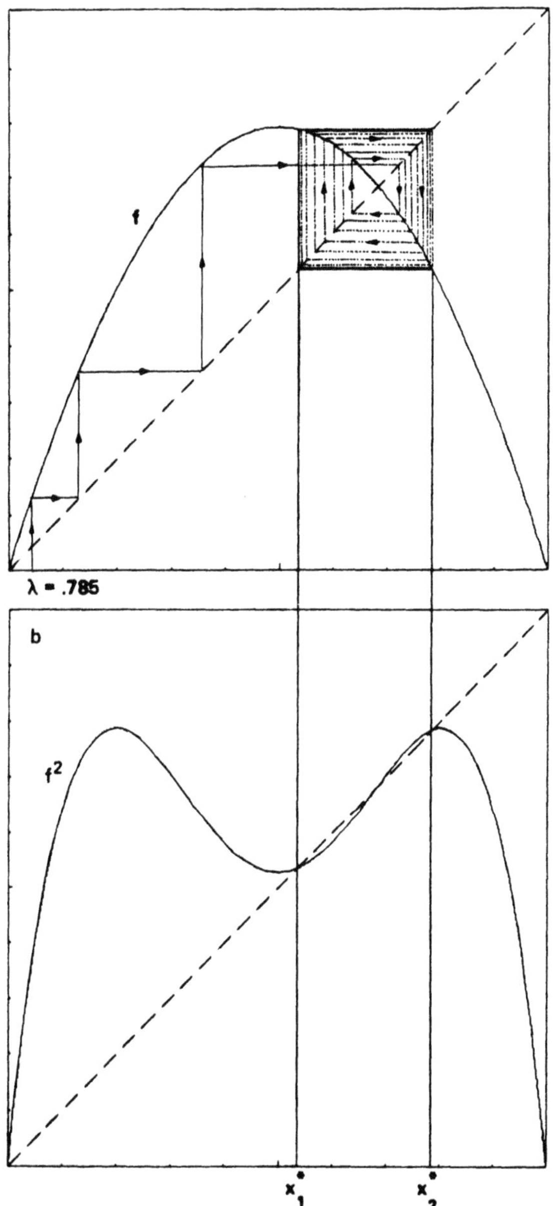

Fig. 4. λ = 0.785. (a) shows the outward spiralling to a stable 2-cycle. The elements of the 2-cycle, x_1^* and x_2^*, are located as fixed points in (b).

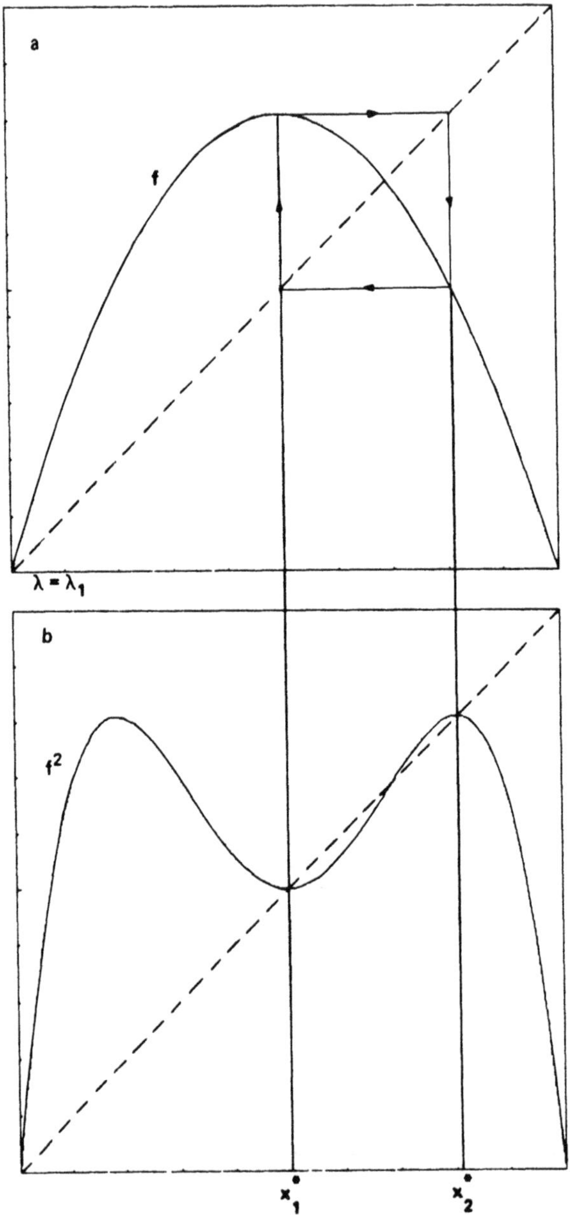

Fig. 5. $\lambda = \lambda_1$. *A superstable 2-cycle.* x_1^* *and* x_2^* *are at extrema of* f^2.

Period Doubling *Ad Infinitum*

We are now close to the end of this story. As we increase λ further, the minimum drops still lower, so that both x_1^* and x_2^* have negative slopes. At some parameter value, denoted by Λ_2, the slope at *both* x_1^* and x_2^* becomes equal to -1. Thus at Λ_2 the same situation has developed for f^2 as developed for f at $\Lambda_1 = \frac{3}{4}$. This transitional case is depicted in Figs. 6a and b. Accordingly, just as the fixed point of f at Λ_1 issued into being a 2-cycle, so too does *each* fixed point of f^2 at Λ_2 create a 2-cycle, which in turn is a 4-cycle of f. That is, we have now encountered the second period doubling.

The manner in which we were able to follow the creation of the 2-cycle at Λ_1 was to anticipate the presence of period 2, and so to consider f^2, which would resolve the cycle into a pair of fixed points. Similarly, to resolve period 4 into fixed points we now should consider f^4. Beyond being the fourth iterate of f, Eq. (8) tells us that f^4 can be computed from f^2:

$$f^4 = f^2 \circ f^2 .$$

From this point, we can abandon f itself, and take f^2 as the "fundamental" function. Then, just as f^2 was constructed by iterating f with itself we now iterate f^2 with itself. The manner in which f^2 reveals itself as being an iterate of f is the slope equality at the fixed points of f^2, which we saw imposed by the chain rule. Since the operation of the chain rule is "automatic," we actually needed to consider only the fixed point of f^2 nearest to $x = \frac{1}{2}$; the behavior of the other fixed point is slaved to it. Thus, at the level of f^4, we again need to focus on only the fixed point of f^4 nearest to $x = \frac{1}{2}$: the other *three* fixed points are similarly slaved to it. Thus, a recursive scheme has been unearthed. We now increase λ to λ_2, so that the fixed point of f^4 nearest to $x = \frac{1}{2}$ is again at $x = \frac{1}{2}$ with slope 0. Figures 7a and b depict this situation for f^2 and f^4, respectively. When λ increases further, the maximum of f^4 at $x = \frac{1}{2}$ now moves up, developing a fixed point with negative slope. Finally, at Λ_3 when the slope of this fixed point (as well as the other three) is again -1, each fixed point will split into a pair giving rise to an 8-cycle, which is now stable. Again, $f^8 = f^4 \circ f^4$, and f^4 can be viewed as fundamental. We define λ_3 so that $x = \frac{1}{2}$ again is a fixed point, this time of f^8. Then at Λ_4 the slopes are -1, and another period doubling occurs. Always,

$$f^{2^{n+1}} = f^{2^n} \circ f^{2^n} . \tag{27}$$

Provided that a constraint on the range of λ does not prevent it from decreasing the slope at the appropriate fixed point past -1, this doubling must recur *ad infinitum*.

Basically, the mechanism that f^{2^n} uses to period double at Λ_{n+1} is the same mechanism that $f^{2^{n+1}}$ will use to double at Λ_{n+2}. The function $f^{2^{n+1}}$ is constructed from f^{2^n} by Eq. (27), and similarly $f^{2^{n+2}}$ will be constructed from $f^{2^{n+1}}$. Thus, there is a definite operation that, by acting on functions, creates functions; in particular, the operation acting on f^{2^n} at Λ_{n+1}, (or better, f^{2^n} at λ_n) will determine $f^{2^{n+1}}$ at λ_{n+1}. Also, since we need to keep track of f^{2^n} only in the interval including the fixed point of f^{2^n} closest to $x = \frac{1}{2}$ and since this interval becomes increasingly small as λ increases, the part of f that generates this region is also the restriction of f to an increasingly small interval about $x = \frac{1}{2}$. (Actually, slopes of f at points farther away also matter, but these merely set a "scale," which will be

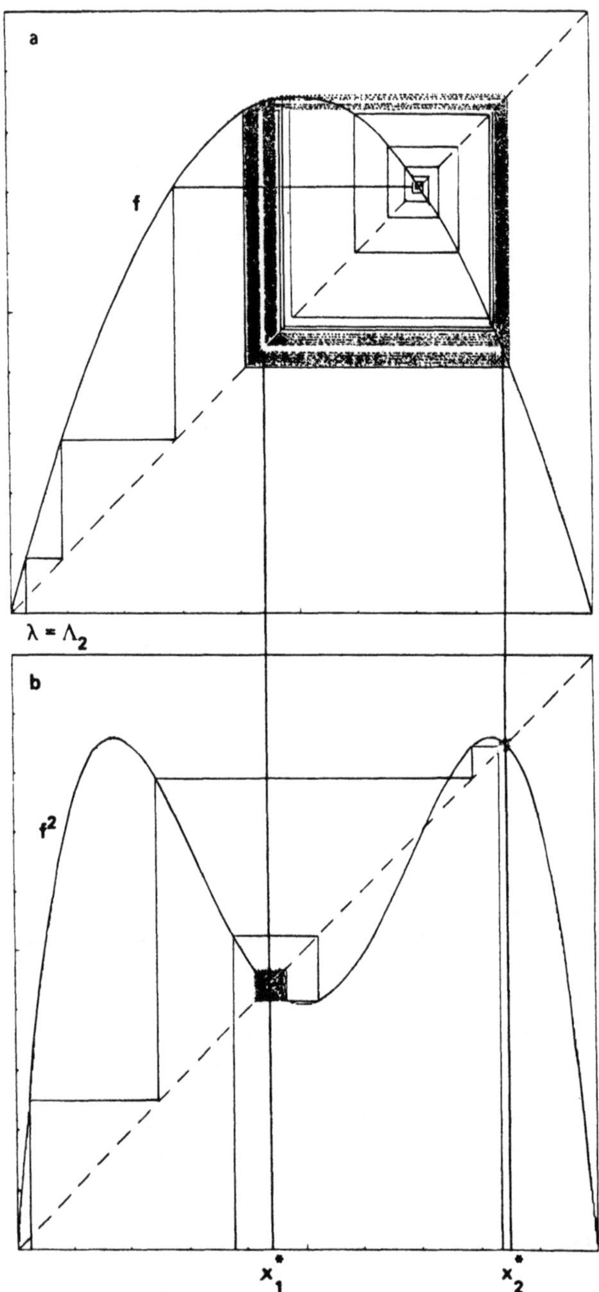

Fig. 6. $\lambda = \Lambda_2$. x_1^* ***and*** x_2^* ***in (b) have the same slow convergence as the fixed point in Fig. 3a.***

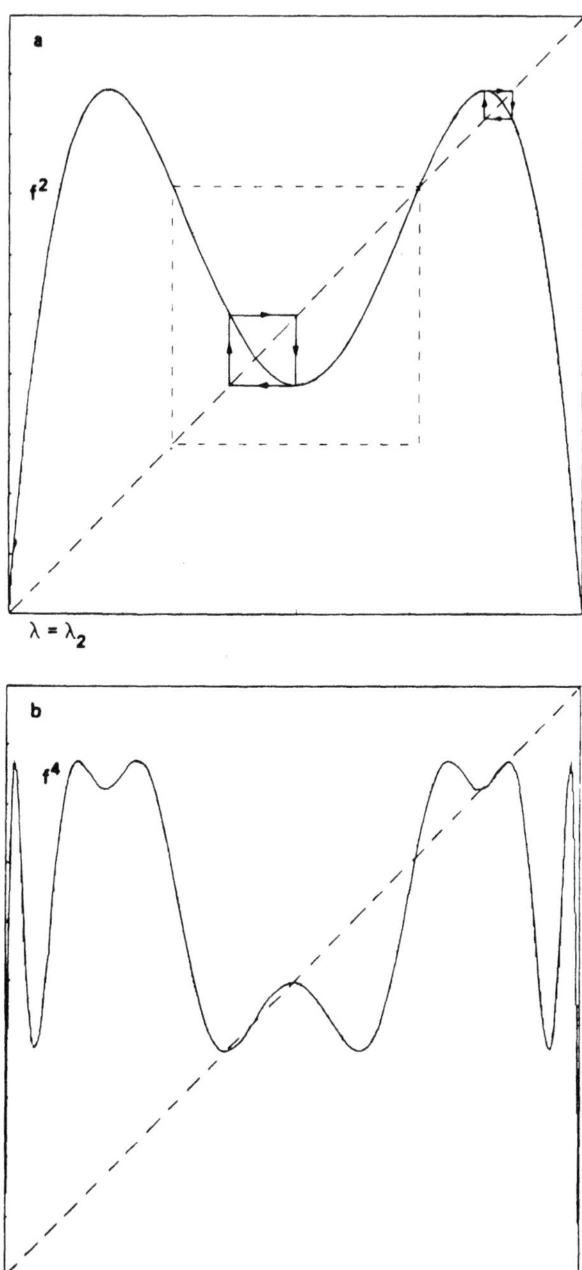

Fig. 7. ***$\lambda = \lambda_2$. A superstable 4-cycle. The region within the dashed square in (a) should be compared with all of Fig. 5a.***

eliminated by a rescaling.) The behavior of f away from $x = \frac{1}{2}$ is immaterial to the period-doubling behavior, and in the limit of large n only the *nature of f's maximum* can matter. This means that in the infinite period-doubling limit, all functions with a quadratic extremum will have identical behavior. [$f''(\frac{1}{2}) \neq 0$ is the generic circumstance.] Therefore, the operation on functions will have a *stable fixed point* in the space of functions, which will be the common universal limit of high iterates of any specific function. To determine this universal limit we must enlarge our scope vastly, so that the role of the starting point, x_0, will be played by an arbitrary *function*; the attracting fixed point will become a universal function obeying an equation implicating only itself. The role of the function in the equation $x_0 = f(x_0)$ now must be played by an *operation* that yields a new function when it is performed upon a function. In fact, the heart of this operation is the functional composition of Eq. (27). If we can determine the exact operator and actually can solve *its* fixed-point problem, we shall understand why a special number, such as δ of Eq. (3), has emerged independently of the specific system (the starting function) we have considered.

The Universal Limit of High Iterates

In this section we sketch the solution to the fixed-point problem. In Fig. 7a, a dashed square encloses the part of f^2 that we must focus on for all further period doublings. This square should be compared with the unit square that comprises all of Fig. 5a. If the Fig. 7a square is reflected through $x = \frac{1}{2}$, $y = \frac{1}{2}$ and then *magnified* so that the circulation squares of Figs. 4a and 5a are of equal size, we will have in each square a piece of a function that has the same kind of maximum at $x = \frac{1}{2}$ and falls to zero at the right-hand lower corner of the circulation square. Just as f produced this second curve of f^2 in the square as λ increased from λ_1 to λ_2, so too will f^2 produce another curve, which will be similar to the other two when it has been magnified suitably and reflected twice. Figure 8 shows this superposition for the first *five* such functions; at the resolution of the figure, observe that the last three curves are coincident. Moreover, the scale reduction that f^2 will determine for f^4 is based solely on the functional composition, so that if these curves for f^{2^n}, $f^{2^{n+1}}$, *converge* (as they obviously do in Fig. 8), the scale reduction from level to level will *converge to a definite constant.* But the width of each circulation square is just the distance between $x = \frac{1}{2}$ when it is a fixed point of f^{2^n} and the fixed point of f^{2^n} next nearest to $x = \frac{1}{2}$ (Figs. 7a and b). That is, asymptotically, *the separation of adjacent elements of period-doubled attractors is reduced by a constant value from one doubling to the next.* Also from one doubling to the next, this next nearest element *alternates* from one side of $x = \frac{1}{2}$ to the other. Let d_n denote the algebraic distance from $x = \frac{1}{2}$ to the nearest element of the attractor cycle of period 2^n, in the 2^n-cycle at λ_n. A positive number α scales this distance down in the 2^{n+1}-cycle at λ_{n+1}:

$$\frac{d_n}{d_{n+1}} \sim -\alpha\,. \tag{28}$$

But since rescaling is determined only by functional composition, there is some function that composed with itself will *reproduce* itself reduced in scale by −α. The function has a quadratic maximum at $x = \frac{1}{2}$, is symmetric about $x = \frac{1}{2}$, and can be scaled by hand to equal 1 at x =

½. Shifting coordinates so that $x = ½ \rightarrow x = 0$, we have

$$-\alpha g(g(x/\alpha)) = g(x) . \quad (29)$$

Substituting $g(0) = 1$, we have

$$g(1) = -\frac{1}{\alpha} . \quad (30)$$

Accordingly, Eq. (29) is a definite equation for a function g depending on x through x^2 and having a maximum of 1 at $x = 0$. There is a unique smooth solution to Eq. (29), which determines

$$\alpha = 2.502907875 \ldots . \quad (31)$$

Knowing α, we can predict through Eq. (28) a definite scaling law binding on the iterates of any scheme possessing period doubling. The law has, indeed, been amply verified experimentally. By Eq. (29), we see that the relevant operation upon functions that underlies period doubling is functional composition followed by magnification, where the magnification is determined by the fixed-point condition of Eq. (29) with the function g the fixed point in this space of functions. However, Eq. (29) does not describe a stable fixed point because we have not incorporated in it the parameter increase from λ_n to λ_{n+1}. Thus, g is not the limiting function of the curves in the circulation squares, although it is intimately related to that function. The full theory is described in the next section. Here we merely state that we can determine the limiting function and thereby can *determine the location of the actual elements of limiting 2^n-cycles.* We also have established that g is an unstable fixed point of functional composition, where the rate of divergence away from g is precisely δ of Eq. (3) and so is computable. Accordingly, there is a full theory that determines, in a precise quantitative way, the aperiodic limit of functional iterations with an *unspecified* function f.

Some Details of the Full Theory

Returning to Eq. (28), we are in a position to describe theoretically the universal scaling of high-order cycles and the convergence to a universal limit. Since d_n is the distance between $x = ½$ and the element of the 2^n-cycle at λ_n nearest to $x = ½$ and since this nearest element is the 2^{n-1} iterate of $x = ½$ (which is true because these two points were coincident before the nth period doubling began to split them apart), we have

$$d_n = f^{2^{n-1}}(\lambda_n, ½) - ½ . \quad (32)$$

For future work it is expedient to perform a coordinate translation that moves $x = ½$ to $x = 0$. Thus, Eq. (32) becomes

$$d_n = f^{2^{n-1}}(\lambda_n, 0) . \quad (33)$$

Equation (28) now determines that the rescaled distances,

$$r_n \equiv (-\alpha)^n d_{n+1} .$$

will converge to a definite finite value as $n \rightarrow \infty$. That is,

$$\lim_{n\rightarrow\infty} (-\alpha)^n f^{2^n}(\lambda_{n+1}, 0) \quad (34)$$

must exist if Eq. (28) holds.

However, from Fig. 8 we know something stronger than Eq. (34). When the nth iterated function is *magnified* by $(-\alpha)^n$, it converges to a definite function. Equation (34) is the value of this function at $x = 0$. After the magnification, the convergent functions are given by

$$(-\alpha)^n f^{2^n}(\lambda_{n+1}, x/(-\alpha)^n) .$$

Thus,

$$g_1(x) \equiv \lim_{n\rightarrow\infty} (-\alpha)^n f^{2^n}(\lambda_{n+1}, x/(-\alpha)^n) \quad (35)$$

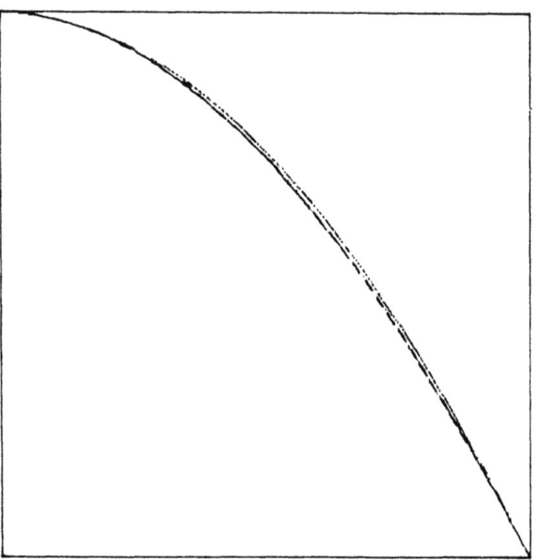

Fig. 8. The superposition of the suitably magnified dotted squares of $f^{2^{n-1}}$ at λ_n (as in Figs. 5a, 7a, ...).

is the limiting function inscribed in the square of Fig. 8. The function $g_1(x)$ is, by the argument of the restriction of f to increasingly small intervals about its maximum, the *universal* limit of all iterates of all f's with a quadratic extremum. Indeed, it is numerically easy to ascertain that g_1 of Eq. (35) is always the same function independent of the f in Eq. (32).

What is this universal function good for? Figure 5a shows a crude approximation of g_1 [n = 0 in the limit of Eq. (35)], while Fig. 7a shows a better approximation (n = 1). In fact, the extrema of g_1 near the fixed points of g_1 support circulation squares each of which contains two points of the cycle. (The two squares shown in Fig. 7a locate the four elements of the cycle.) That is, g_1 determines the location of elements of high-order 2^n-cycles near x = 0. Since g_1 is *universal*, we now have the amazing result that the location of the actual elements of highly doubled cycles is universal! The reader might guess this is a *very* powerful result. Figure 9 shows g_1 out to x sufficiently large to have 8 circulation squares, and hence locates the 15 elements of a 2^n-cycle nearest to x = 0. Also, the universal value of the scaling parameter α, obtained numerically, is

$$\alpha = 2.502907875 \ldots . \qquad (36)$$

Like δ, α is a number that can be *measured* [through an experiment that observes the d_n of Eq. (28)] in any phenomenon exhibiting period doubling.

If g_1 is universal, then of course its iterate g_1^2 also is universal. Figure 7b

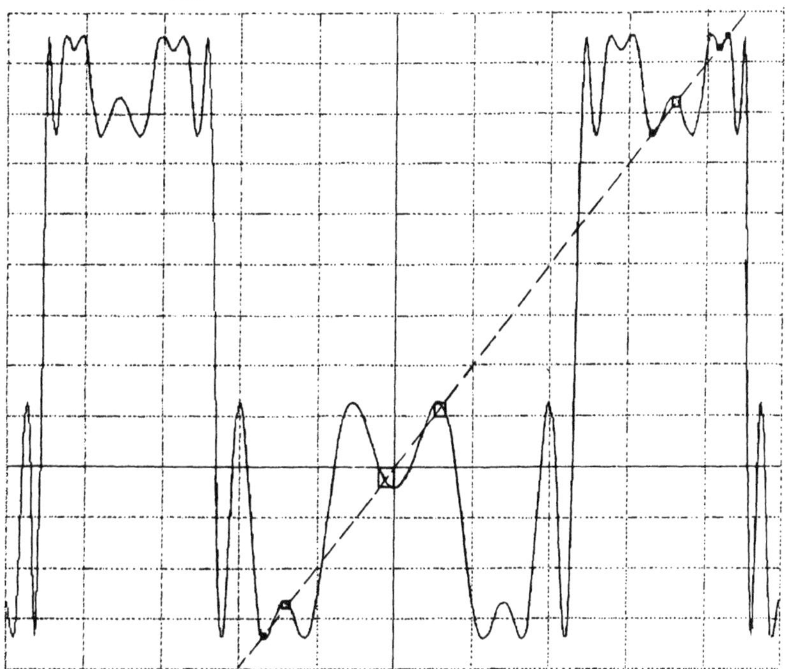

Fig. 9. The function g_1. The squares locate cycle elements.

depicts an early approximation to this iterate. In fact, let us define a new universal function g_0, obtained by scaling g_1^2:

$$g_0(x) \equiv -\alpha g_1^2(-x/\alpha)\,. \tag{37}$$

(Because g_1 is universal and the iterates of our quadratic function are all symmetric in x, both g_1 and g_0 are symmetric functions. Accordingly, the minus sign within g_1^2 can be dropped with impunity.) From Eq. (35), we now can write

$$g_0(x) = \lim_{n\to\infty}(-\alpha)^n f^{2^n}(\lambda_n, x/(-\alpha)^n)\,. \tag{38}$$

[We introduced the scaling of Eq. (37) to provide one power of α per period doubling, since each successive iterate of f^{2^n} reduces the scale by α].

In fact, we can generalize Eqs. (35) and (38) to a *family* of universal functions g_r:

$$g_r(x) = \lim_{n\to\infty}(-\alpha)^n f^{2^n}(\lambda_{n+r}, x/(-\alpha)^n)\,. \tag{39}$$

To understand this, observe that g_0 locates the cycle elements as the fixed points of g_0 at extrema; g_1 locates the same elements by determining two elements per extremum. Similarly, g_r determines 2^r elements about each extremum near a fixed point of g_r. Since each f^{2^n} is always magnified by $(-\alpha)^n$ for each r, the scales of all g_r are the same. Indeed, g_r for $r > 1$ looks like g_1 of Fig. 9, except that each extremum is slightly higher, to accommodate a 2^r-cycle. Since each extremum must grow by convergently small amounts to accommodate higher and higher 2^r-cycles, we are led to conclude that

A DISCOVERY

The inspiration for the universality theory came from two sources. First, in 1971 N. Metropolis, M. Stein, and P. Stein (all in the LASL Theoretical Division) discovered a curious property of iterations: as a parameter is varied, the behavior of iterates varies in a fashion independent of the particular function iterated. In particular for a large class of functions, if at some value of the parameter a certain cycle is stable, then as the parameter increases, the cycle is replaced successively by cycles of doubled periods. This period doubling continues until an infinite period, and hence erratic behavior, is attained.

Second, during the early 1970s, a scheme of mathematics called dynamical system theory was popularized, largely by D. Ruelle, with the notion of a "strange attractor." The underlying questions addressed were (1) how could a purely causal equation (for example, the Navier-Stokes equations that describe fluid flow) come to demonstrate highly erratic or statistical properties and (2) how could these statistical properties be computed. This line of thought merged with the iteration ideas, and the limiting infinite "cycles" of iteration systems came to be viewed as a possible means to comprehend turbulence. Indeed, I became inspired to study the iterates of functions by a talk on such matters by S. Smale, one of the creators of dynamical system theory, at Aspen in the summer of 1975.

My first effort at understanding this problem was through the complex analytic properties of the generating function of the iterates of the quadratic map

$$x_{n+1} = \lambda x_n(1 - x_n) .$$

This study clarified the mechanism of period doubling and led to a rather different kind of equation to determine the values of λ at which the period doubling occurs. The new equations were intractable, although approximate solutions seemed possible. Accordingly, when I returned from Aspen, I numerically determined some parameter values with an eye toward discerning some patterns. At this time I had never used a large computer—in fact my sole computing power resided in a programmable pocket calculator. Now, such machines are very slow. A particular parameter value is obtained iteratively (by Newton's method) with each step of iteration requiring 2^n iterates of the map. For a 64-cycle, this means 1 minute per step of Newton's method. At the same time as n increased, it became an increasingly more delicate matter to locate the desired solution. However, I immediately perceived the λ_n's were converging geometrically. This enabled me to predict the next value with increasing accuracy as n increased, and so required just one step of Newton's method to obtain the desired value. To the best of my knowledge, this observation of geometric convergence has never been made independently, for the simple reason that the solutions have always been performed automatically on large and fast computers!

That a geometric convergence occurred was already a surprise. I was interested in this for two reasons: first, to gain insight into my theoretical work, as already mentioned, and second, because a convergence rate is a number invariant under all smooth transformations, and so of mathematical interest. Accordingly, I spent a part of a day trying to fit the convergence rate value, 4.669, to the mathematical constants I knew. The task was fruitless, save for the fact that it made the number memorable.

At this point I was reminded by Paul Stein that period doubling isn't a unique property of the quadratic map, but also occurs, for example, in

$$x_{n+1} = \lambda \sin \pi x_n \ .$$

However, my generating function theory rested heavily on the fact that the nonlinearity was simply quadratic and not transcendental. Accordingly, my interest in the problem waned.

Perhaps a month later I decided to determine the λ's in the transcendental case numerically. This problem was even slower to compute than the quadratic one. Again, it became apparent that the λ's converged geometrically, and altogether amazingly, the convergence rate was the same 4.669 that I remembered by virtue of my efforts to fit it.

Recall that the work of Metropolis, Stein, and Stein showed that precise qualitative features are independent of the specific iterative scheme. Now I learned that precise quantitative features also are independent of the specific function. This discovery represents a complete inversion of accustomed ritual. Usually one relies on the fact that similar equations will have qualitatively similar behavior, but quantitative predictions depend on the details of the equations. The universality theory shows that qualitatively similar equations have the identical *quantitative* behavior. For example, a system of differential equations naturally determines certain maps. The computation of the actual analytic form of the map is generally well beyond present mathematical methods. However, should the map exhibit period doubling, then precise quantitative results are available from the universality theory because the theory applies independently of which map it happens to be. In particular, certain fluid flows have now been experimentally observed to become turbulent through period doubling (subharmonic bifurcations). From this one fact we know that the universality theory applies—and indeed correctly determines the precise way in which the flow becomes turbulent, without any reference to the underlying Navier-Stokes equations.

$$g(x) = \lim_{r\to\infty} g_r(x) \tag{40}$$

must exist. By Eq. (39),

$$g(x) = \lim_{n\to\infty}(-\alpha)^n f^{2^n}(\lambda_\infty, x/(-\alpha)^n)\,. \tag{41}$$

Unlike the functions g_r, $g(x)$ is obtained as a limit of f^{2^n}'s at a *fixed value* of λ. Indeed, this is the special significance of λ_∞; it is an isolated value of λ at which repeated iteration and magnification lead to a convergent function.

We now can write the equation that g satisfies. Analogously to Eq. (37), it is easy to verify that all g_r are related by

$$g_{r-1}(x) = -\alpha g_r(g_r(-x/\alpha))\,. \tag{42}$$

By Eq. (40), it follows that g satisfies

$$g(x) = -\alpha g(g(x/\alpha))\,. \tag{43}$$

The reader can verify that Eq. (43) is invariant under a magnification of g. Thus, the theory has nothing to say about absolute scales. Accordingly, we must fix this by hand by setting

$$g(0) = 1\,. \tag{44}$$

Also, we must specify the nature of the maximum of g at $x = 0$ (for example, quadratic). Finally, since g is to be built by iterating $a - x^2$, it must be both smooth and a function of x through x^2. With these specifications, Eq. (43) has a *unique* solution. By Eqs. (44) and (43),

$$g(0) = 1 = -\alpha g(g(0)) = -\alpha g(1)\,,$$

so that

$$\alpha = -1/g(1)\,. \tag{45}$$

Accordingly, Eq. (43) determines α together with g.

Let us comment on the nature of Eq. (43), a so-called functional equation. Because g is smooth, if we know its value at a finite number of points, we know its value to some approximation on the interval containing these points by any sufficiently smooth interpolation. Thus, to some degree of accuracy, Eq. (43) can be replaced by a finite coupled system of nonlinear equations. Exactly then, Eq. (43) is an infinite-dimensional, nonlinear vector equation. Accordingly, we have obtained the solution to one-dimensional period doubling through our infinite-dimensional, explicitly universal problem. Equation (43) must be infinite-dimensional because it must keep track of the infinite number of cycle elements demanded of any attempt to solve the period-doubling problem. Rigorous mathematics for equations like Eq. (43) is just beyond the boundary of present mathematical knowledge.

At this point, we must determine two items. First, where is δ? Second, how do we obtain g_1, the real function of interest for locating cycle elements? The two problems are part of one question. Equation (42) is itself an iteration scheme. However, unlike the elements in Eq. (4), the elements acted on in Eq. (42) are *functions*. The analogue of the function of f in Eq. (4) is the operation in function space of functional composition followed by a magnification. If we call this operation T, and an element of the function space ψ, Eq. (42) gives

$$T[\psi](x) = -\alpha\psi^2(-x/\alpha)\,. \tag{46}$$

In terms of T, Eq. (42) now reads

$$g_{r-1} = T[g_r]\,, \tag{47}$$

and Eq. (43) reads

$$g = T[g]\,. \tag{48}$$

Thus, g is precisely the fixed point of T. Since g is the limit of the sequence g_r, we can obtain g_r for large r by linearizing T about its fixed point g. Once we have g_r in the linear regime, the exact repeated application of T by Eq. (47) will provide g_1. Thus, we must investigate the stability of T at the fixed point g. However, it is obvious that T is *unstable* at g: for a large enough r, g_r is a point arbitrarily close to the fixed point g; by Eq. (47), successive iterates of g_r under T move away from g. How unstable is T? Consider a one-parameter family of functions f_λ, which means a "line" in the function space. For each f, there is an isolated parameter value λ_∞, for which repeated applications of T lead to convergence towards g [Eq. (41)]. Now, the function space can be "packed" with all the lines corresponding to the various f's. The set of all the points on these lines specified by the respective λ_∞'s determines a "surface" having the property that repeated applications of T to any point on it will converge to g. This is the surface of stability of T (the "stable manifold" of T through g). But through each point of this surface issues out the corresponding line, which is one-dimensional since it is parametrized by a single parameter, λ. Accordingly, T is *unstable* in only *one* direction in function space. Linearized about g, this line of instability can be written as the one-parameter family

$$f_\lambda(x) = g(x) - \lambda\, h(x)\,, \tag{49}$$

which passes through g (at $\lambda = 0$) and deviates from g along the unique direction h. But f_λ is just one of our transformations [Eq. (4)]! Thus, as we vary λ, f_λ will undergo period doubling, doubling to a 2^n-cycle at Λ_n. By Eq. (41), λ_∞ for the family of functions f_λ in Eq. (49) is

$$\lambda_\infty = 0\,. \tag{50}$$

Thus, by Eq. (1)

$$\lambda_n \sim \delta^{-n}\,. \tag{51}$$

Since applications of T by Eq. (47) iterate in the opposite direction (diverge away from g), it now follows that the rate of instability of T along h must be precisely δ.

Accordingly, we find δ and g_1 in the following way. First, we must linearize the operation T about its fixed point g. Next, we must determine the stability directions of the linearized operator. Moreover, we expect there to be precisely one direction of instability. Indeed, it turns out that infinitesimal deformations (conjugacies) of g determine *stable* directions, while a unique unstable direction, h, emerges with a stability rate (eigenvalue) precisely the δ of Eq. (3). Equation (49) at λ_r is precisely g_r for asymptotically large r. Thus g_r is known asymptotically, so that we have entered the sequence g_r and can now, by repeated use of Eq. (47), step down to g_1. All the ingredients of a full description of high-order 2^n-cycles now are at hand and evidently are universal.

Although we have said that the function g_1 universally locates cycle elements near $x = 0$, we must understand that it doesn't locate all cycle elements. This is possible because a finite distance of the scale of g_1 (for example, the location of the element nearest to $x = 0$) has been magnified by α^n for n diverging. Indeed, the distances from $x = 0$ of all elements of a 2^n-cycle, "accurately" located by g_1, are reduced by $-\alpha$ in the 2^{n+1}-cycle. However, it is obvious that some elements have no such scaling: because f(0) $= a_n$ in Eq. (13), and $a_n \rightarrow a_\infty$, which is a definite nonzero number, the distance from the origin of the element of the 2^n-cycle farthest to the right certainly has not been reduced by $-\alpha$ at each period doubling. This suggests that we must measure locations of elements on the far right with respect to the farthest right point. If we do this, we can see that these distances scale by α^2, since they are the images through the quadractic maximum of f at x = 0 of elements close to x = 0 scaling with $-\alpha$. In fact, if we image g_1 through the maximum of f (through a quadratic conjugacy), then we shall indeed obtain a new universal function that locates cycle elements near the right-most element. The correct description of a highly doubled cycle now emerges as one of universal local clusters.

We can state the scope of universality for the location of cycle elements precisely. Since $f(\lambda_1, x)$ exactly locates the two elements of the 2^1-cycle, and since $f(\lambda_1, x)$ is an approximation to g_1 [n = 0 in Eq. (35)], we evidently can locate both points exactly by appropriately sealing g_1. Next, near x = 0, $f^2(\lambda_2, x)$ is a better approximation to g_1 (suitably scaled). However, in general, the more accurately we scale g_1 to determine the smallest 2-cycle elements, the greater is the error in its determination of the right-most elements. Again, near x = 0, $f^4(\lambda_3, x)$ is a still better approximation to g_1. Indeed, the suitably scaled g_1 now can determine several points about x = 0 accurately, but determination of the right-most elements is still worse. In this fashion, it follows that g_1, suitably scaled, can determine 2^r points of the 2^n-cycle near x = 0 for $r \ll n$. If we focus on the neighborhood of one of these 2^r points at some definite distance from x = 0, then by Eq. (35) the larger the n, the larger the *scaled* distance of this region from x = 0, and so, the poorer the approximation of the location of fixed points in it by g_1. However, just as we can construct the version of g_1 that applies at the right-most cycle element, we also can construct the version of g_1 that applies at this chosen neighborhood. Accordingly, the universal description is set through an acceptable tolerance: if we "measure" f^{2^n} at some definite n, then we can use the actual location of the elements as foci for 2^n versions of g_1, each applicable at one such point. For all further period doubling, we determine the new cycle elements through the g_1's. In summary, the *more accurately we care to know the locations* of arbitrarily high-order cycle elements, the *more parameters we must measure* (namely, the cycle elements at some chosen order of period doubling). This is the sense in which the universality theory is asymptotic. Its ability to have serious predictive power is the fortunate consequence of the high convergence rate $\delta(\sim 4.67)$. Thus, typically after the first two or three period doublings, this asymptotic theory is already accurate to within several percent. If a period-doubling system is *measured* in its 4- or 8-cycle, its behavior throughout and symmetrically beyond the period-doubling regime also is determined to within a few percent.

To make precise dynamical predictions, we do not have to construct all the local versions of g_1; all we really need to know is the local *scaling* everywhere along the attractor. The scaling is $-\alpha$ at x = 0 and α^2 at the right-most element. But what is it at an arbitrary point? We can determine the scaling law if we order elements not by their location on the x-axis, but rather by their order as iterates of x = 0. Because the time sequence in which a process evolves is precisely this ordering, the result will be of immediate and powerful predictive value. It is precisely this scaling law that allows us to compute the spectrum of the onset of turbulence in period-doubling systems.

What must we compute? First, just as the element in the 2^n-cycle nearest to $x = 0$ is the element halfway around the cycle from $x = 0$, the element nearest to an arbitrarily chosen element is precisely the one halfway around the cycle from it. Let us denote by $d_n(m)$ the distance between the mth cycle element (x_m) and the element nearest to it in a 2^n-cycle. [The d_n of Eq. (28) is $d_n(0)$]. As just explained,

$$d_n(m) = x_m - f^{2n-1}(\lambda_n, x_m) . \quad (52)$$

However, x_m is the mth iterate of $x_0 = 0$. Recalling from Eq. (6) that powers commute, we find

$$d_n(m) = f^m(\lambda_n, 0) - f^m(\lambda_n, f^{2n-1}(\lambda_n, 0)) . \quad (53)$$

Let us, for the moment, specialize to m of the form 2^{n-r}, in which case

$$\begin{aligned} d_n(2^{n-r}) &= f^{2n-r}(\lambda_n, 0) - f^{2n-r}(\lambda_n, f^{2n-1}(\lambda_n, 0)) \\ &= f^{2n-r}(\lambda_{(n-r)+r}, 0) - f^{2n-r}(\lambda_{(n-r)+r}, f^{2n-1}(\lambda_n, 0)) . \end{aligned} \quad (54)$$

For $r \ll n$ (which can still allow $r \gg 1$ for n large), we have, by Eq. (39),

$$d_n(2^{n-r}) \sim (-\alpha)^{-(n-r)}[g_r(0) - g_r((-\alpha)^{n-r} f^{2n-1}(\lambda_n, 0))]$$

or

$$d_n(2^{n-r}) \sim (-\alpha)^{-(n-r)}[g_r(0) - g_r((-\alpha)^{-r+1} g_1(0))] . \quad (55)$$

The object we want to determine is the local scaling at the mth element, that is, the ratio of nearest separations at the mth iterate of $x = 0$, at successive values of n. That is, if the scaling is called σ,

$$\sigma_n(m) \equiv \frac{d_{n+1}(m)}{d_n(m)} . \quad (56)$$

[Observe by Eq. (28), the definition of α, that $\sigma_n(0) \sim (-\alpha)^{-1}$.] Specializing again to $m = 2^{n-r}$, where $r \ll n$, we have by Eq. (55)

$$\sigma(2^{n-r}) \sim \frac{g_{r+1}(0) - g_{r+1}((-\alpha)^{-r} g_1(0))}{g_r(0) - g_r((-\alpha)^{-r+1} g_1(0))} . \quad (57)$$

Finally, let us rescale the axis of iterates so that all 2^{n+1} iterates are within a unit interval. Labelling this axis by t, the value of t of the mth element in a 2^n-cycle is

$$t_n(m) = m/2^n . \quad (58)$$

In particular, we have

$$t_n(2^{n-r}) = 2^{-r} . \quad (59)$$

Defining σ along the t-axis naturally as

$$\sigma(t_n(m)) \sim \sigma_n(m) \qquad (\text{as } n \to \infty) ,$$

we have by Eqs. (57) and (59),

$$\sigma(2^{-r-1}) = \frac{g_{r+1}(0) - g_{r+1}((-\alpha)^{-r} g_1(0))}{g_r(0) - g_r((-\alpha)^{-r+1} g_1(0))} . \quad (60)$$

It is not much more difficult to obtain σ for all t. This is done first for rational t by writing t in its binary expansion:

$$t_{r_1 r_2 r_3 \ldots} = 2^{-r_1} + 2^{-r_2} + \ldots .$$

In the 2^n-cycle approximation we require σ_n at the $2^{n-r_1} + 2^{n-r_2} + \ldots$ iterate of the origin. But, by Eq. (8),

$$f^{2^{n-r_1} + 2^{n-r_2} + \ldots} = f^{2^{n-r_1}} \circ f^{2^{n-r_2}} \circ \ldots .$$

It follows by manipulations identical to those that led from Eq. (54) to Eq. (60) that σ at such values of t is obtained by replacing the individual g_r terms in Eq. (60) by appropriate iterates of various g_r's.

There is one last ingredient to the computation of σ. We know that $\sigma(0) = -\alpha^{-1}$. We also know that $\sigma_n(1) \sim \alpha^{-2}$. But, by Eq. (59),

$$t_n(1) = 2^{-n} \rightarrow 0 .$$

Thus σ is discontinuous at $t = 0$, with $\sigma(0 - \varepsilon) = -\alpha^{-1}$ and $\sigma(0 + \varepsilon) = \alpha^{-2}(\varepsilon \rightarrow 0^+)$. Indeed, since $x_{2^{n-r}}$ is always very close to the origin, each of these points is imaged quadratically. Thus Eq. (60) actually determines $\sigma(2^{-r-1} - \varepsilon)$, while $\sigma(2^{-r-1} + \varepsilon)$ is obtained by replacing each numerator and denominator g_r by its square. The same replacement also is correct for each multi-g_r term that figures into σ at the binary expanded rationals.

Altogether, we have the following results. $\sigma(t)$ can be computed for all t, and it is *universal* since its explicit computation depends only upon the universal functions g_r. σ is *discontinuous* at all the rationals. However, it can be established that the *larger* the number of terms in the binary expansion of a rational t, the smaller the discontinuity of σ. Lastly, as a finite number of iterates leaves t unchanged as $n \rightarrow \infty$, σ must be

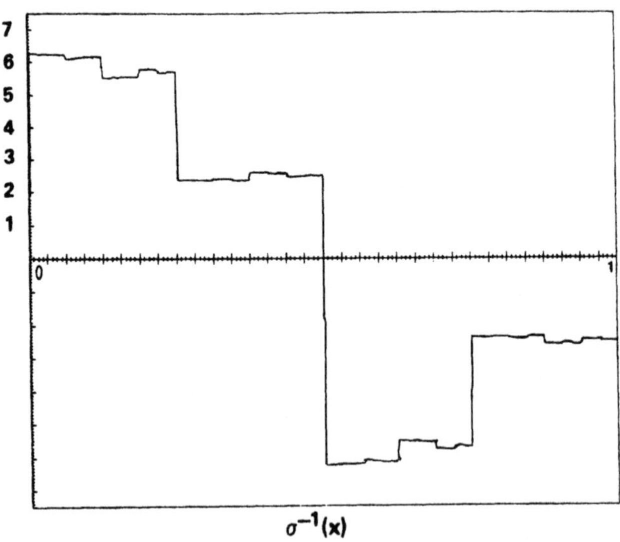

Fig. 10. The trajectory scaling function. Observe that $\sigma(x + 1/2) = -\sigma(x)$.

continuous except at the rationals. Figure 10 depicts $1/\sigma(t)$. Despite the pathological nature of σ, the reader will observe that basically it is constant half the time at α^{-1} and half the time at α^{-2} for $0 < t < \frac{1}{2}$. In a succeeding approximation, it can be decomposed in each half into two slightly different quarters, and so forth. [It is easy to verify from Eq. (52) that σ is periodic in t of period 1, and has the symmetry

$$\sigma(t + \tfrac{1}{2}) = -\sigma(t) .$$

Accordingly, we have paid attention to its first half $0 < t < \frac{1}{2}$.] With σ we are at last finished with one-dimensional iterates per se.

Universal Behavior in Higher Dimensional Systems

So far we have discussed iteration in *one* variable; Eq. (15) is the prototype. Equation (14), an example of iteration in two dimensions, has the special property of preserving areas. A generalization of Eq. (14),

$$x_{n+1} = y_n - x_n^2$$

and

$$y_{n+1} = a + bx_n \qquad (61)$$

with $|b| < 1$, contracts areas. Equation (61) is interesting because it possesses a

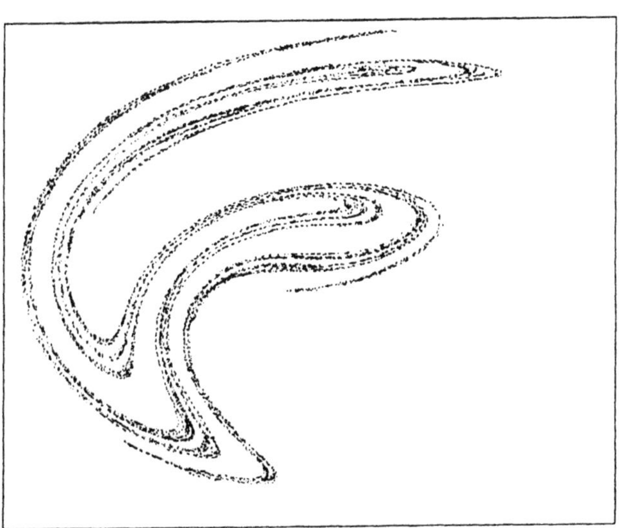

Fig. 11. The plotted points lie on the "strange attractor" of Duffing's equation.

so-called *strange attractor*. This means an attractor (as before) constructed by folding a curve repeatedly upon itself (Fig. 11) with the consequent property that two initial points very near to one another are, in fact, very far from each other when the distance is measured along the folded attractor, which is the path they follow upon iteration. This means that after some iteration, they will soon be far apart in actual distance as well as when measured along the attractor. This general mechanism gives a system highly sensitive dependence upon its initial conditions and a truly statistical character: since very small differences in initial conditions are magnified quickly, unless the initial conditions are known to *infinite precision*, all known knowledge is eroded rapidly to future ignorance. Now, Eq. (61) enters into the early stages of statistical behavior through period doubling. Moreover, δ of Eq. (3) is *again* the rate of onset of complexity, and α of Eq. (31) is again the rate at which the spacing of adjacent attractor points is vanishing. Indeed, the one-dimensional theory determines all behavior of Eq. (61) in the onset regime.

In fact, dimensionality is irrelevant. The same theory, the same numbers, etc. also work for iterations in N dimensions, provided that the system goes through period doubling. The basic process, wherever period doubling occurs *ad infinitum*, is functional composition from one level to the next. Accordingly, a modification of Eq. (29) is at the heart of the process, with composition on functions from N dimensions to N dimensions. Should the specific iteration function contract N-dimensional volumes (a dissipative process), then in general there is one direction of slowest contraction, so that after a number of iterations the process is effectively one-dimensional. Put differently, the one-dimensional solution to Eq. (29) is always a solution to its N-dimensional analogue. It is the relevant fixed point of the analogue if the iteration function is contractive.

Universal Behavior in Differential Systems

The next step of generalization is to include systems of differential equations. A prototypic equation is Duffing's oscillator, a driven damped anharmonic oscillator,

$$\ddot{x} + k\dot{x} + x^3 = b\sin 2\pi t\,. \tag{62}$$

The periodic drive of period 1 determines a natural time step. Figure 12a depicts a period 1 attractor, usually referred to as a *limit cycle*. It is an attractor because, for a range of initial conditions, the solution to Eq. (62) settles down to the cycle. It is period 1 because it repeats the same curve in every period of the drive. Figures 12b and c depict attractors of periods 2 and 4 as the friction or damping constant k in Eq. (62) is reduced systematically. The parameter values $k = \lambda_0, \lambda_1, \lambda_2, \ldots$, are the damping constants corresponding to the most stable 2^n-cycle in analogy to the λ_n of the one-dimensional functional iteration. Indeed, this oscillator's period doubles (at least numerically!) *ad infinitum*. In fact, by $k = \lambda_5$, the δ_3 of Eq. (2) has converged to 4.69. Why is this? Instead of considering the entire trajectories as shown in Fig. 12, let us consider only where the trajectory point is located every 1 period of the drive. The 1-cycle then produces only one point, while the 2-cycle produces a pair of points, and so forth. This *time-one map* [if the trajectory point is $(x,\dot{x})$ now, where is it one period later?] is by virtue of the differential equation a smooth and invertible func-

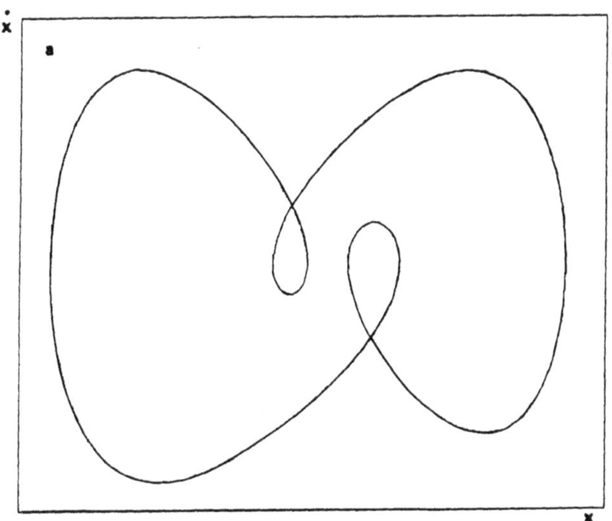

Fig. 12a. The most stable 1-cycle of Duffing's equation in phase space $(x,\dot{x})$.

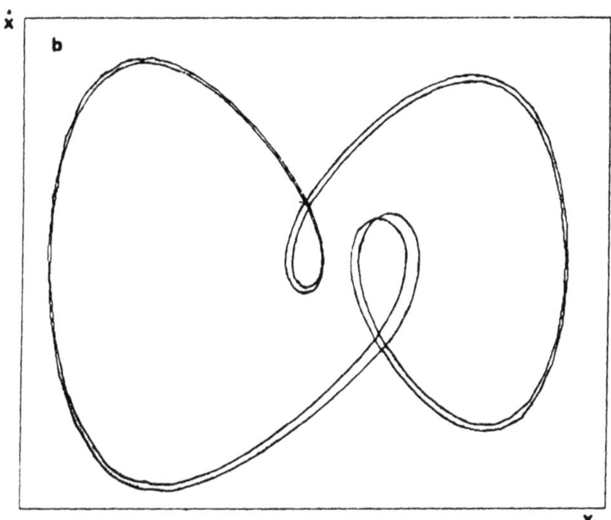

Fig. 12b. The most stable 2-cycle of Duffing's equation. Observe that it is two displaced copies of Fig. 12a.

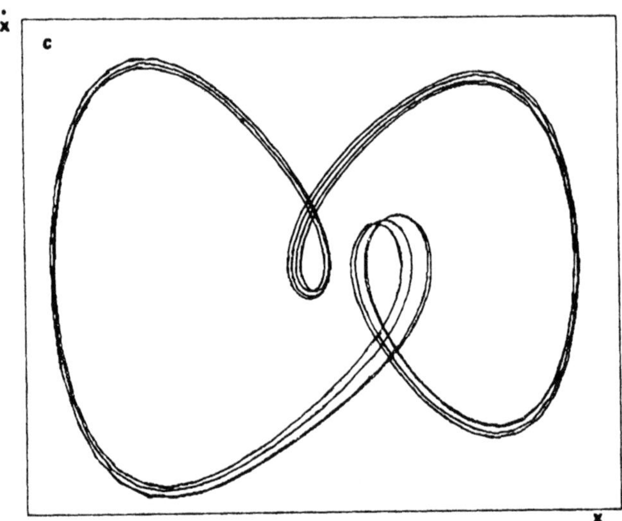

Fig. 12c. The most stable 4-cycle of Duffing's equation. Observe that the displaced copies of Fig. 12b have either a broad or a narrow separation.

tion in two dimensions. Qualitatively, it looks like the map of Eq. (61). In the present state of mathematics, little can be said about the analytic behavior of time-one maps; however, since our theory is universal, it makes no difference that we don't know the explicit form. We still can determine the complete quantitative behavior of Eq. (62) in the onset regime where the motion tends to aperiodicity. If we already know, by measurement, the precise form of the trajectory after a few period doublings, we can compute the form of the trajectory as the friction is reduced throughout the region of onset of complexity by carefully using the full power of the universality theory to determine the spacings of elements of a cycle.

Let us see how this works in some detail. Consider the time-one map of the Duffing's oscillator in the superstable 2^n-cycle. In particular, let us focus on an element at which the scaling function σ (Fig. 10) has the value σ_0, and for which the next iterate of this element also has the scaling σ_0. (The element is not at a big discontinuity of σ.) It is then intuitive that if we had taken our time-one examination of the trajectory at values of time displaced from our first choice, we would have seen the same scaling σ_0 for this part of the trajectory. That is, the differential equations will extend the map-scaling function continuously to a function along the entire trajectory so that, if two successive time-one elements have scaling σ_0, then the entire stretch of trajectory over this unit time interval has scaling σ_0. In the last section, we were motivated to construct σ as a function of t along an interval precisely towards this end.

To implement this idea, the first step is to define the analogue of d_n. We require the spacing between the trajectory at time t and at time $T_n/2$ where the period of the system in the 2^n-cycle is

$$T_n \cong 2^n T_0 . \quad (63)$$

That is, we define

$$d_n(t) \equiv x_n(t) - x_n(t + T_n/2) . \quad (64)$$

(There is a d for each of the N variables for a system of N differential equations.) Since σ was defined as periodic of period 1, we now have

$$d_{n+1}(t) \sim \sigma(t/T_{n+1}) d_n(t) . \quad (65)$$

The content of Eq. (65), based on the n-dependence arising solely through the T_n in σ, and not on the detailed form of σ, already implies a strong scaling prediction, in that the ratio

$$\frac{d_{n+1}(t)}{d_n(t)} ,$$

when plotted with t scaled so that $T_n = 1$, is a function *independent* of n. Thus if Eq. (65) is true for *some* σ, whatever it might be, then knowing $x_n(t)$, we can compute $d_n(t)$ and from Eq. (65) $d_{n+1}(t)$. As a consequence of periodicity, Eq. (64) for $n \rightarrow n + 1$ can be solved for $x_{n+1}(t)$ (through a Fourier transform). That is, if we have measured any chosen coordinate of the system in its 2^n-cycle, we can compute its time dependence in the 2^{n+1}-cycle. Because this procedure is recursive, we can compute the coordinate's evolution for all higher cycles through the infinite period-doubling limit. If Eq. (65) is true and σ not known, then by measurement at a 2^n-cycle and at a 2^{n+1}-cycle, σ could be *constructed* from Eq. (65), and hence all higher order doublings would again be determined. Accordingly, Eq. (65) is a very powerful result. However, we know much more. The universality theory tells us that period doubling is universal and that there is a ***unique*** function σ which, indeed, we have computed in the previous section. Accordingly, by ***measuring*** x(t) in some chosen 2^n-cycle (the higher the n, the more the number of effective parameters to be determined empirically, and the more precise are the predictions), we now can compute the entire evolution of the system on its route to turbulence.

How well does this work? The empirically determined σ [for Eq. (62)] of Eq. (65) is shown for $n = 3$ in Fig. 13a and $n = 4$ in Fig. 13b. The figures were constructed by plotting the ratios of d_{n+1} and d_n scaled respective to $T = 16$ in Fig. 13a and $T = 32$ in Fig. 13b. Evidently the scaling law Eq. (65) is being obeyed. Moreover, on the same graph Fig. 14 shows the empirical σ for $n = 4$ and the recursion theoretical σ of Fig. 10. The reader should observe the detail-by-detail agreement of the two. In fact, if we use Eq. (65) and the theoretical σ with $n = 2$ as empirical input, the $n = 5$ frequency spectrum agrees with the empirical $n = 5$ spectrum to within 10%. (The $n = 4$ determines $n = 5$ to within 1%.) Thus the asymptotic universality theory is correct *and* is already well obeyed, even by $n = 2$!

Equations (64) and (65) are solved, as mentioned above, through Fourier transforming. The result is a recursive scheme that determines the Fourier coefficients of $x_{n+1}(t)$ in terms of those of $x_n(t)$ and the Fourier transform of the (known) function $\sigma(t)$. To employ the formula accurately requires knowledge of the entire spectrum of x_n (amplitude *and* phase) to determine each coefficient

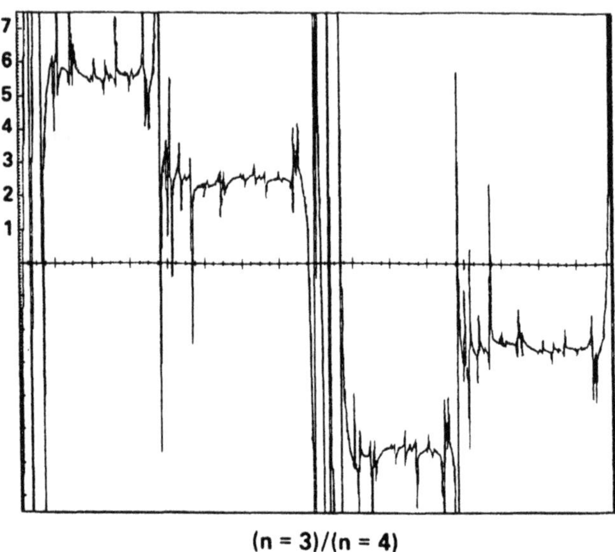

Fig. 13a. The ratio of nearest copy separations in the 8-cycle and 16-cycle for Duffing's equation.

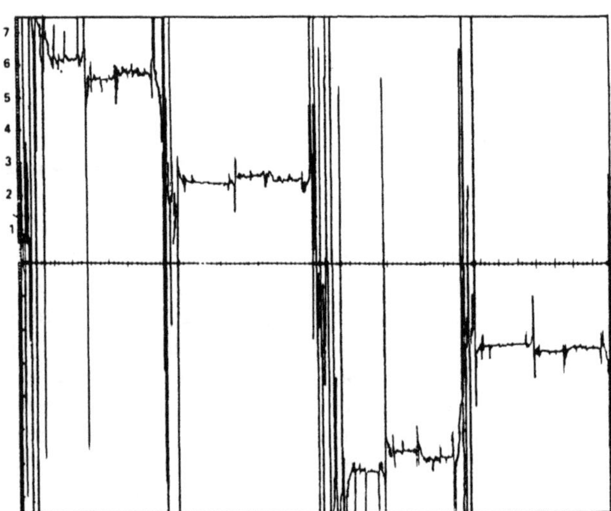

Fig. 13b. The same quantity as in Fig. 13a, but for the 16-cycle and 32-cycle. Here, the time axis is twice as compressed.

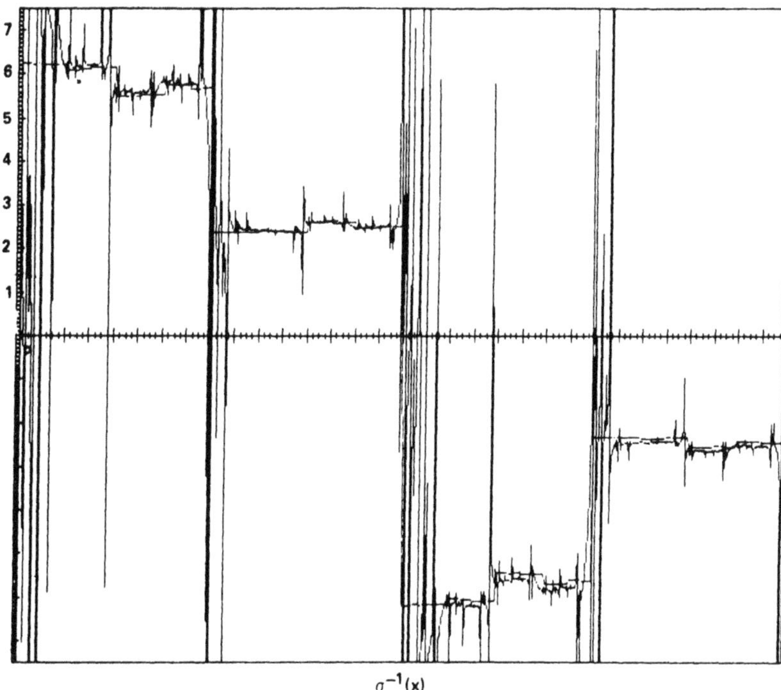

Fig. 14. Figure 13b overlayed with Fig. 10 compares the universal scaling function σ with the empirically determined scaling of nearest copy separations from the 16-cycle to the 32-cycle for Duffing's equation.

of x_{n+1}. However, the formula enjoys an approximate local prediction, which roughly determines the amplitude of a coefficient of x_{n+1} in terms of the amplitudes (alone) of x_n near the desired frequency of x_{n+1}.

What does the spectrum of a period-doubling system look like? Each time the period doubles, the fundamental frequency halves; period doubling in the continuum version is termed half-subharmonic bifurcation, a typical behavior of coupled nonlinear differential equations. Since the motion *almost* reproduces itself every period of the drive, the amplitude at this original frequency is high. At the first subharmonic halving, spectral components of the odd halves of the drive frequency come in. On the route to aperiodicity they saturate at a certain amplitude. Since the motion more nearly reproduces itself every two periods of drive, the next saturated subharmonics, at the odd fourths of the original frequency, are smaller still than the first ones, and so on, as each set of odd 2^nths comes into being. A crude approximate prediction of the theory is that whatever the system, the saturated amplitudes of each set of successively lower half-frequencies define a smooth interpolation located 8.2 dB *below* the smooth interpolation of the previous half-frequencies. [This is shown in Fig. 15 for Eq. (62).] After subharmonic bifurcations *ad infinitum*, the system is now no longer periodic; it has developed a continuous broad spectrum

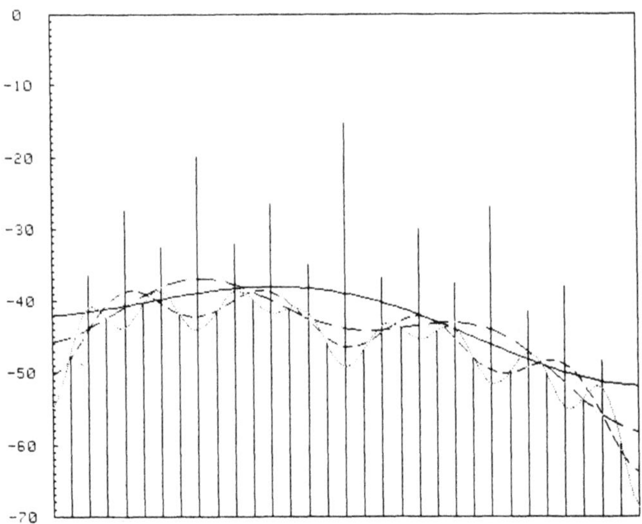

Fig. 15. The subharmonic spectrum of Duffing's equation in the 32-cycle. The dotted curve is an interpolation of the odd 32nd subharmonics. The shorter dashed curve is constructed similarly for the odd 16th subharmonics, but lowered by 8.2 dB. The longer dashed curve of the 8th subharmonics has been dropped by 16.4 dB, and the solid curve of the 4th subharmonics by 24.6 dB.

down to zero frequency with a definite internal distribution of the energy. That is, the system emerges from this process having developed the beginnings of broad-band noise of a determined nature. This process also occurs in the onset of turbulence in a fluid.

The Onset of Turbulence

The existing idea of the route to turbulence is Landau's 1941 theory. The idea is that a system becomes turbulent through a succession of instabilities, where each instability creates a new degree of freedom (through an indeterminate phase) of a time-periodic nature with the frequencies successively higher and incommensurate (*not* harmonics); because the resulting motion is the superposition of these modes, it is quasi-periodic.

In fact, it is experimentally clear that quasi-periodicity is incorrect. Rather, to produce the observed noise of rapidly decaying correlation the spectrum must become *continuous* (broad-band noise) down to zero frequency. The defect can be eliminated through the production of successive half-subharmonics, which then emerge as an allowable route to turbulence. If the general idea of a succession of instabilities is maintained, the new modes do *not* have indeterminate phases. However, only a small number of modes need be excited to produce the required spectrum. (The number of modes participating in the transition is, as of now, an open experimental question.) Indeed, knowledge of the phases of a small number of amplitudes at an early stage of period doubling suffices to determine the phases of the transition spectrum. What is important is that a purely causal system can and does possess essentially statistical properties. Invoking *ad hoc* statistics is unnecessary and generally incompatible with the true dynamics.

A full theoretical computation of the onset demands the calculation of successive instabilities. The method used traditionally is perturbative. We start at the static solution and add a small time-dependent piece. The fluid equations are linearized about the static solution, and the stability of the perturbation is studied. To date, only the first instability has been computed analytically. Once we know the parameter value (for example, the Rayleigh number) for the onset of this first time-varying instability, we must determine the correct form of the solution after the perturbation has grown large *beyond* the linear regime. To this solution we add a new time-dependent perturbative mode, again linearized (now about a time-varying, nonanalytically available solution) to discover the new instability. To date, the second step of the analysis has been performed only numerically. This process, in principle, can be repeated again and again until a suitably turbulent flow has been obtained. At each successive stage, the computation grows successively more intractable.

However, it is just at this point that the universality theory solves the problem; it works only after enough instabilities have entered to reach the asymptotic regime. Since just two such instabilities already serve as a good approximate starting point, we need only a few parameters for each flow to empower the theory to complete the hard part of the infinite cascade of more complex instabilities.

Why should the theory apply? The fluid equations make up a set of coupled field equations. They can be spatially Fourier-decomposed to an infinite set of coupled ordinary differential equations. Since a flow is viscous, there is some smallest spatial scale below which no significant excitation exists. Thus, the equations are effectively a finite coupled set of nonlinear differential equations. The number of equations in the set is completely irrelevant. The universality theory is generic for such a dissipative system of equations. Thus it is possible that the flow exhibits period doubling. If it does, then our theory applies. However, to prove that a given flow (or any flow) actually should exhibit doubling is well beyond present understanding. All we can do is experiment.

Figure 16 depicts the experimentally measured spectrum of a convecting liquid helium cell at the onset of turbulence. The system displays measurable period doubling through four or five levels; the spectral components at each set of odd half-subharmonics are labelled with the level. With n = 2 taken as asymptotic, the dotted lines show the crudest interpolations implied for the n = 3, n = 4 component. Given the small amount of ***amplitude*** data, the interpolations are perforce poor, while ignorance of higher odd multiples prevents construction of any significant interpolation at the right-hand side. Accordingly, to do the crudest test, the farthest right-hand amplitude was dropped, and the oscillations were smoothed away by averaging. The experimental results, −8.3 dB and −8.4 dB, are in surprisingly good agreement with the theoretical 8.2!

From this good experimental agree-

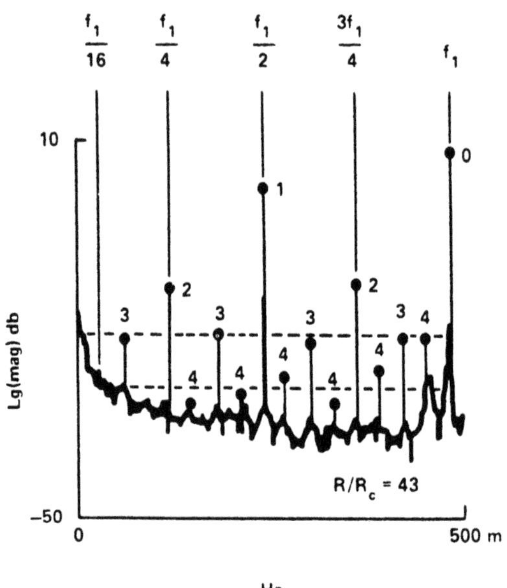

Fig. 16. The experimental spectrum (redrawn from Libchaber and Maurer) of a convecting fluid at its transition to turbulence. The dashed lines result from dropping a horizontal line down through the odd 4th subharmonics (labelled 2) by 8.2 and 16.4 dB.

ment and the many period doublings as the clincher, we can be confident that the measured flow has made its transition according to our theory. A measurement of δ from its fundamental definition would, of course, be altogether convincing. (Experimental resolution is insufficient at present.) However, if we work backwards, we find that the several percent agreement in 8.2 dB is an *experimental observation* of α in the system to the same accuracy. Thus, the present method has provided a theoretical calculation of the actual dynamics in a field where such a feat has been impossible since the construction of the Navier-Stokes equations. In fact, the scaling law Eq. (65) transcends these equations, and applies to the *true* equations, whatever they may be.

Nature **261** 459–67 (1976)

Simple mathematical models with very complicated dynamics

Robert M. May*

First-order difference equations arise in many contexts in the biological, economic and social sciences. Such equations, even though simple and deterministic, can exhibit a surprising array of dynamical behaviour, from stable points, to a bifurcating hierarchy of stable cycles, to apparently random fluctuations. There are consequently many fascinating problems, some concerned with delicate mathematical aspects of the fine structure of the trajectories, and some concerned with the practical implications and applications. This is an interpretive review of them.

THERE are many situations, in many disciplines, which can be described, at least to a crude first approximation, by a simple first-order difference equation. Studies of the dynamical properties of such models usually consist of finding constant equilibrium solutions, and then conducting a linearised analysis to determine their stability with respect to small disturbances: explicitly nonlinear dynamical features are usually not considered.

Recent studies have, however, shown that the very simplest nonlinear difference equations can possess an extraordinarily rich spectrum of dynamical behaviour, from stable points, through cascades of stable cycles, to a regime in which the behaviour (although fully deterministic) is in many respects "chaotic", or indistinguishable from the sample function of a random process.

This review article has several aims.

First, although the main features of these nonlinear phenomena have been discovered and independently rediscovered by several people, I know of no source where all the main results are collected together. I have therefore tried to give such a synoptic account. This is done in a brief and descriptive way, and includes some new material: the detailed mathematical proofs are to be found in the technical literature, to which signposts are given.

Second, I indicate some of the interesting mathematical questions which do not seem to be fully resolved. Some of these problems are of a practical kind, to do with providing a probabilistic description for trajectories which seem random, even though their underlying structure is deterministic. Other problems are of intrinsic mathematical interest, and treat such things as the pathology of the bifurcation structure, or the truly random behaviour, that can arise when the nonlinear function $F(X)$ of equation (1) is not analytical. One aim here is to stimulate research on these questions, particularly on the empirical questions which relate to processing data.

Third, consideration is given to some fields where these notions may find practical application. Such applications range from the abstractly metaphorical (where, for example, the transition from a stable point to "chaos" serves as a metaphor for the onset of turbulence in a fluid), to models for the dynamic behaviour of biological populations (where one can seek to use field or laboratory data to estimate the values of the parameters in the difference equation).

*King's College Research Centre, Cambridge CB2 1ST; on leave from Biology Department, Princeton University, Princeton 08540.

Fourth, there is a very brief review of the literature pertaining to the way this spectrum of behaviour—stable points, stable cycles, chaos—can arise in second or higher order difference equations (that is, two or more dimensions; two or more interacting species), where the onset of chaos usually requires less severe nonlinearities. Differential equations are also surveyed in this light; it seems that a three-dimensional system of first-order ordinary differential equations is required for the manifestation of chaotic behaviour.

The review ends with an evangelical plea for the introduction of these difference equations into elementary mathematics courses, so that students' intuition may be enriched by seeing the wild things that simple nonlinear equations can do.

First-order difference equations

One of the simplest systems an ecologist can study is a seasonally breeding population in which generations do not overlap[1–4]. Many natural populations, particularly among temperate zone insects (including many economically important crop and orchard pests), are of this kind. In this situation, the observational data will usually consist of information about the maximum, or the average, or the total population in each generation. The theoretician seeks to understand how the magnitude of the population in generation $t+1$, X_{t+1}, is related to the magnitude of the population in the preceding generation t, X_t: such a relationship may be expressed in the general form

$$X_{t+1} = F(X_t) \tag{1}$$

The function $F(X)$ will usually be what a biologist calls "density dependent", and a mathematician calls nonlinear; equation (1) is then a first-order, nonlinear difference equation.

Although I shall henceforth adopt the habit of referring to the variable X as "the population", there are countless situations outside population biology where the basic equation (1), applies. There are other examples in biology, as, for example in genetics[5,6] (where the equation describes the change in gene frequency in time) or in epidemiology[7] (with X the fraction of the population infected at time t). Examples in economics include models for the relationship between commodity quantity and price[8], for the theory of business cycles[9], and for the temporal sequences generated by various other economic quantities[10]. The general equation (1) also is germane to the social sciences[11], where it arises, for example, in theories of learning (where X may be the number of bits of information that can be remembered after an interval t), or in the propagation of rumours in variously structured societies (where X is the number of people to have heard the rumour after time t). The imaginative reader will be able to invent other contexts for equation (1).

In many of these contexts, and for biological populations in particular, there is a tendency for the variable X to increase from one generation to the next when it is small, and for it to decrease when it is large. That is, the nonlinear function $F(X)$ often has the following properties: $F(0)=0$; $F(X)$ increases monotonically as X increases through the range $0<X<A$ (with $F(X)$ attaining its maximum value at $X=A$); and $F(X)$ decreases monotonically as X increases beyond $X=A$. Moreover, $F(X)$ will usually contain one or more parameters which "tune" the severity of this nonlinear behaviour; parameters which tune the steepness of the hump in the $F(X)$ curve. These parameters will typically have some biological or economic or sociological significance.

A specific example is afforded by the equation[1,4,12-23]

$$N_{t+1} = N_t(a-bN_t) \tag{2}$$

This is sometimes called the "logistic" difference equation. In the limit $b=0$, it describes a population growing purely exponentially (for $a>1$); for $b\neq 0$, the quadratic nonlinearity produces a growth curve with a hump, the steepness of which is tuned by the parameter a. By writing $X=bN/a$, the equation may be brought into canonical form[1,4,12-23]

$$X_{t+1} = aX_t(1-X_t) \tag{3}$$

In this form, which is illustrated in Fig. 1, it is arguably the simplest nonlinear difference equation. I shall use equation (3) for most of the numerical examples and illustrations in this article. Although attractive to mathematicians by virtue of its extreme simplicity, in practical applications equation (3) has the disadvantage that it requires X to remain on the interval $0<X<1$; if X ever exceeds unity, subsequent iterations diverge towards $-\infty$ (which means the population becomes extinct). Furthermore, $F(X)$ in equation (3) attains a maximum value of $a/4$ (at $X=\frac{1}{2}$); the equation therefore possesses non-trivial dynamical behaviour only if $a<4$. On the other hand, all trajectories are attracted to $X=0$ if $a<1$. Thus for non-trivial

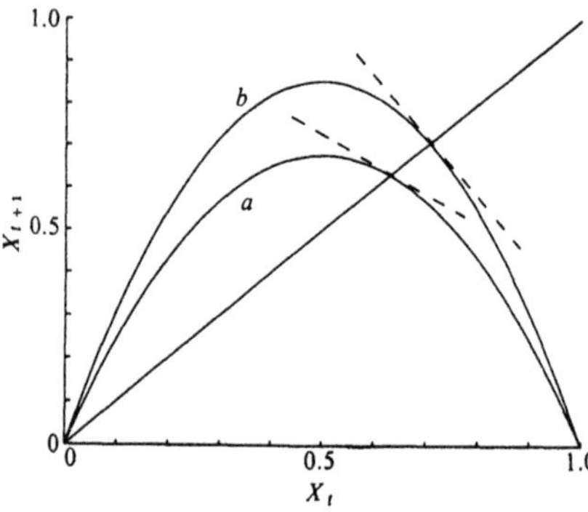

Fig. 1 A typical form for the relationship between X_{t+1} and X_t described by equation (1). The curves are for equation (3), with a = 2.707 (a); and a = 3.414 (b). The dashed lines indicate the slope at the "fixed points" where $F(X)$ intersects the 45° line: for the case a this slope is less steep than $-45°$ and the fixed point is stable; for b the slope is steeper than $-45°$, and the point is unstable.

dynamical behaviour we require $1<a<4$; failing this, the population becomes extinct.

Another example, with a more secure provenance in the biological literature[1,23-27], is the equation

$$X_{t+1} = X_t \exp[r(1-X_t)] \tag{4}$$

This again describes a population with a propensity to simple exponential growth at low densities, and a tendency to decrease at high densities. The steepness of this nonlinear behaviour is tuned by the parameter r. The model is plausible for a single species population which is regulated by an epidemic disease at high density[28]. The function $F(X)$ of equation (4) is slightly more complicated than that of equation (3), but has the compensating advantage that local stability implies global stability[1] for all $X>0$.

The forms (3) and (4) by no means exhaust the list of single-humped functions $F(X)$ for equation (1) which can be culled from the ecological literature. A fairly full such catalogue is given, complete with references, by May and Oster[1]. Other similar mathematical functions are given by Metropolis *et al.*[16]. Yet other forms for $F(X)$ are discussed under the heading of "mathematical curiosities" below.

Dynamic properties of equation (1)

Possible constant, equilibrium values (or "fixed points") of X in equation (1) may be found algebraically by putting $X_{t+1}=X_t=X^*$, and solving the resulting equation

$$X^* = F(X^*) \tag{5}$$

An equivalent graphical method is to find the points where the curve $F(X)$ that maps X_t into X_{t+1} intersects the 45° line, $X_{t+1}=X_t$, which corresponds to the ideal nirvana of zero population growth; see Fig. 1. For the single-hump curves discussed above, and exemplified by equations (3) and (4), there are two such points: the trivial solution $X=0$, and a non-trivial solution X^* (which for equation (3) is $X^*=1-[1/a]$).

The next question concerns the stability of the equilibrium point X^*. This can be seen[24,25,19-21,1,4] to depend on the slope of the $F(X)$ curve at X^*. This slope, which is illustrated by the dashed lines in Fig. 1, can be designated

$$\lambda^{(1)}(X^*) = [dF/dX]_{X=X^*} \tag{6}$$

So long as this slope lies between 45° and $-45°$ (that is, $\lambda^{(1)}$ between $+1$ and -1), making an acute angle with the 45° ZPG line, the equilibrium point X^* will be at least locally stable, attracting all trajectories in its neighbourhood. In equation (3), for example, this slope is $\lambda^{(1)}=2-a$: the equilibrium point is therefore stable, and attracts all trajectories originating in the interval $0<X<1$, if and only if $1<a<3$.

As the relevant parameters are tuned so that the curve $F(X)$ becomes more and more steeply humped, this stability-determining slope at X^* may eventually steepen beyond $-45°$ (that is, $\lambda^{(1)}<-1$), whereupon the equilibrium point X^* is no longer stable.

What happens next? What happens, for example, for $a>3$ in equation (3)?

To answer this question, it is helpful to look at the map which relates the populations at successive intervals 2 generations apart; that is, to look at the function which relates X_{t+2} to X_t. This second iterate of equation (1) can be written

$$X_{t+2} = F[F(X_t)] \tag{7}$$

or, introducing an obvious piece of notation,

$$X_{t+2} = F^{(2)}(X_t) \tag{8}$$

The map so derived from equation (3) is illustrated in Figs 2 and 3.

Population values which recur every second generation (that

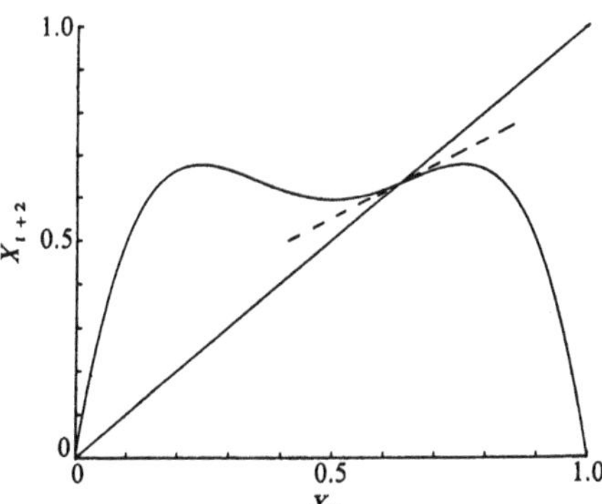

Fig. 2 The map relating X_{t+2} to X_t, obtained by two iterations of equation (3). This figure is for the case (*a*) of Fig. 1, $a = 2.707$: the basic fixed point is stable, and it is the only point at which $F^{(2)}(X)$ intersects the 45° line (where its slope, shown by the dashed line, is less steep than 45°).

is, fixed points with period 2) may now be written as X^*_2, and found either algebraically from

$$X^*_2 = F^{(2)}(X^*_2) \tag{9}$$

or graphically from the intersection between the map $F^{(2)}(X)$ and the 45° line, as shown in Figs 2 and 3. Clearly the equilibrium point X^* of equation (5) is a solution of equation (9); the basic fixed point of period 1 is a degenerate case of a period 2 solution. We now make a simple, but crucial, observation[1]: the slope of the curve $F^{(2)}(X)$ at the point X^*, defined as $\lambda^{(2)}(X^*)$ and illustrated by the dashed lines in Figs 2 and 3, is the square of the corresponding slope of $F(X)$

$$\lambda^{(2)}(X^*) = [\lambda^{(1)}(X^*)]^2 \tag{10}$$

This fact can now be used to make plain what happens when the fixed point X^* becomes unstable. If the slope of $F(X)$ is less than $-45°$ (that is, $|\lambda^{(1)}| < 1$), as illustrated by curve *a* in Fig. 1, then X^* is stable. Also, from equation (10), this implies $0 < \lambda^{(2)} < 1$ corresponding to the slope of $F^{(2)}$ at X^* lying between 0° and 45°, as shown in Fig. 2. As long as the fixed point X^* is stable, it provides the only non-trivial solution to equation (9). On the other hand, when $\lambda^{(1)}$ steepens beyond $-45°$ (that is, $|\lambda^{(1)}| > 1$), as illustrated by curve *b* in Fig 1, X^* becomes unstable. At the same time, from equation (10) this implies $\lambda^{(2)} > 1$, corresponding to the slope of $F^{(2)}$ at X^* steepening beyond 45°, as shown in Fig. 3. As this happens, the curve $F^{(2)}(X)$ must develop a "loop", and two new fixed points of period 2 appear, as illustrated in Fig. 3.

In short, as the nonlinear function $F(X)$ in equation (1) becomes more steeply humped, the basic fixed point X^* may become unstable. At exactly the stage when this occurs, there are born two new and initially stable fixed points of period 2, between which the system alternates in a stable cycle of period 2. The sort of graphical analysis indicated by Figs 1, 2 and 3, along with the equation (10), is all that is needed to establish this generic result[1,4].

As before, the stability of this period 2 cycle depends on the slope of the curve $F^{(2)}(X)$ at the 2 points. (This slope is easily shown to be the same at both points[1,20], and more generally to be the same at all k points on a period k cycle.) Furthermore, as is clear by imagining the intermediate stages between Figs 2 and 3, this stability-determining slope has the value $\lambda = +1$ at the birth of the 2-point cycle, and then decreases through zero towards $\lambda = -1$ as the hump in $F(X)$ continues to steepen. Beyond this point the period 2 points will in turn become unstable, and bifurcate to give an initially stable cycle of period 4. This in turn gives way to a cycle of period 8, and thence to a hierarchy of bifurcating stable cycles of periods 16, 32, 64, . . ., 2^n. In each case, the way in which a stable cycle of period k becomes unstable, simultaneously bifurcating to produce a new and initially stable cycle of period $2k$, is basically similar to the process just adumbrated for $k = 1$. A more full and rigorous account of the material covered so far is in ref. 1.

This "very beautiful bifurcation phenomenon"[22] is depicted in Fig. 4, for the example equation (3). It cannot be too strongly emphasised that the process is generic to most functions $F(X)$ with a hump of tunable steepness. Metropolis *et al.*[18] refer to this hierarchy of cycles of periods 2^n as the harmonics of the fixed point X^*.

Although this process produces an infinite sequence of cycles with periods 2^n ($n \to \infty$), the "window" of parameter values wherein any one cycle is stable progressively diminishes, so that the entire process is a convergent one, being bounded above by some critical parameter value. (This is true for most, but not all, functions $F(X)$: see equation (17) below.) This critical parameter value is a point of accumulation of period 2^n cycles. For equation (3) it is denoted a_c: $a_c = 3.5700\ldots$

Beyond this point of accumulation (for example, for $a > a_c$ in equation (3)) there are an infinite number of fixed points with different periodicities, and an infinite number of different periodic cycles. There are also an uncountable number of initial points X_0 which give totally aperiodic (although bounded) trajectories; no matter how long the time series generated by $F(X)$ is run out, the pattern never repeats. These facts may be established by a variety of methods[1,4,20,21,23]. Such a situation, where an infinite number of different orbits can occur, has been christened "chaotic" by Li and Yorke[20].

As the parameter increases beyond the critical value, at first all these cycles have even periods, with X_t alternating up and down between values above, and values below, the fixed point X^*. Although these cycles may in fact be very complicated (having a non-degenerate period of, say, 5,726 points before repeating), they will seem to the casual observer to be rather like a somewhat "noisy" cycle of period 2. As the parameter value continues to increase, there comes a stage (at $a = 3.6786$ . . for equation (3)) at which the first odd period cycle appears. At first these odd cycles have very long periods, but as the parameter value continues to increase cycles with smaller and smaller odd periods are picked up, until at last the three-point

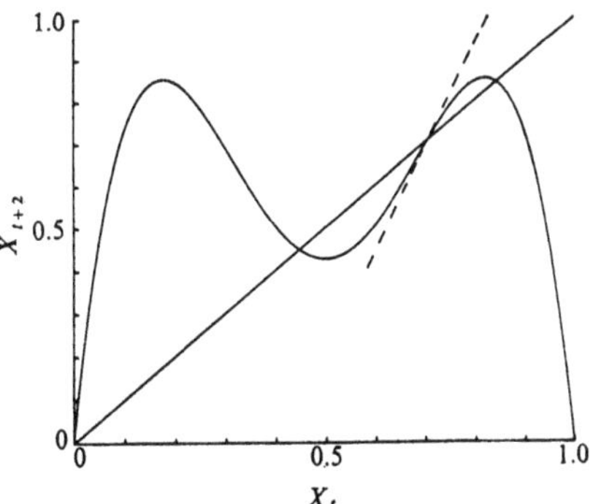

Fig. 3 As for Fig. 2, except that here $a = 3.414$, as in Fig. 1*b*. The basic fixed point is now unstable: the slope of $F^{(2)}(X)$ at this point steepens beyond 45°, leading to the appearance of two new solutions of period 2.

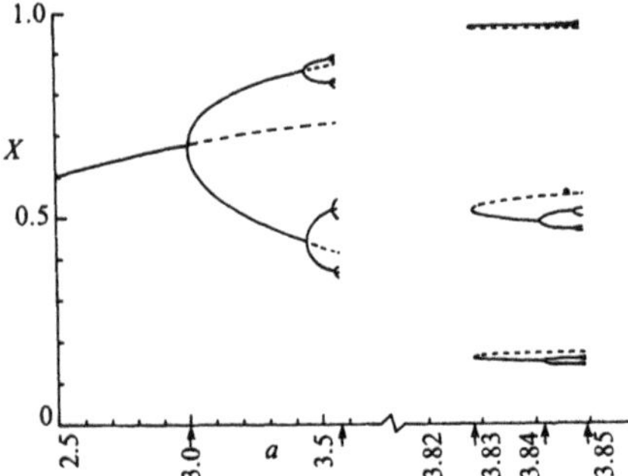

Fig. 4 This figure illustrates some of the stable (———) and unstable (— — — —) fixed points of various periods that can arise by bifurcation processes in equation (1) in general, and equation (3) in particular. To the left, the basic stable fixed point becomes unstable and gives rise by a succession of pitchfork bifurcations to stable harmonics of period 2^n; none of these cycles is stable beyond $a = 3.5700$. To the right, the two period 3 cycles appear by tangent bifurcation: one is initially unstable; the other is initially stable, but becomes unstable and gives way to stable harmonics of period 3×2^n, which have a point of accumulation at $a = 3.8495$. Note the change in scale on the a axis, needed to put both examples on the same figure. There are infinitely many other such windows, based on cycles of higher periods.

cycle appears (at a=3.8284 . . for equation (3)). Beyond this point, there are cycles with every integer period, as well as an uncountable number of asymptotically aperiodic trajectories: Li and Yorke[20] entitle their original proof of this result "Period Three Implies Chaos".

The term "chaos" evokes an image of dynamical trajectories which are indistinguishable from some stochastic process. Numerical simulations[12,15,21,23,25] of the dynamics of equation (3), (4) and other similar equations tend to confirm this impression. But, for smooth and "sensible" functions $F(X)$ such as in equations (3) and (4), the underlying mathematical fact is that for any specified parameter value there is one unique cycle that is stable, and that attracts essentially all initial points[22,29] (see ref. 4, appendix A, for a simple and lucid exposition). That is, there is one cycle that "owns" almost all initial points; the remaining infinite number of other cycles, along with the asymptotically aperiodic trajectories, own a set of points which, although uncountable, have measure zero.

As is made clear by Tables 3 and 4 below, any one particular stable cycle is likely to occupy an extraordinarily narrow window of parameter values. This fact, coupled with the long time it is likely to take for transients associated with the initial conditions to damp out, means that in practice the unique cycle is unlikely to be unmasked, and that a stochastic description of the dynamics is likely to be appropriate, in spite of the underlying deterministic structure. This point is pursued further under the heading "practical applications", below.

The main messages of this section are summarised in Table 1, which sets out the various domains of dynamical behaviour of the equations (3) and (4) as functions of the parameters, a and r respectively, that determine the severity of the nonlinear response. These properties can be understood qualitatively in a graphical way, and are generic to any well behaved $F(X)$ in equation (1).

We now proceed to a more detailed discussion of the mathematical structure of the chaotic regime for analytical functions, and then to the practical problems alluded to above and to a consideration of the behavioural peculiarities exhibited by non-analytical functions (such as those in the two right hand columns of Table 1).

Fine structure of the chaotic regime

We have seen how the original fixed point X^* bifurcates to give harmonics of period 2^n. But how do new cycles of period k arise?

The general process is illustrated in Fig. 5, which shows how period 3 cycles originate. By an obvious extension of the notation introduced in equation (8), populations three generations apart are related by

$$X_{t+3} = F^{(3)}(X_t) \tag{11}$$

If the hump in $F(X)$ is sufficiently steep, the threefold iteration will produce a function $F^{(3)}(X)$ with 4 humps, as shown in Fig. 5 for the $F(X)$ of equation (3). At first (for $a<3.8284$ . . in equation 3) the 45° line intersects this curve only at the single point X^* (and at X=0), as shown by the solid curve in Fig. 5. As the hump in $F(X)$ steepens, the hills and valleys in $F^{(3)}(X)$ become more pronounced, until simultaneously the first two valleys sink and the final hill rises to touch the 45° line, and then to intercept it at 6 new points, as shown by the dashed curve in Fig. 5. These 6 points divide into two distinct three-point cycles. As can be made plausible by imagining the intermediate stages in Fig. 5, it can be shown that the stability-determining slope of $F^{(3)}(X)$ at three of these points has a common value, which is $\lambda^{(3)}=+1$ at their birth, and thereafter steepens beyond +1: this period 3 cycle is never stable. The slope of $F^{(3)}(X)$ at the other three points begins at $\lambda^{(3)}=+1$, and then decreases towards zero, resulting in a stable cycle of period 3. As $F(X)$ continues to steepen, the slope $\lambda^{(3)}$ for this intially stable three-point cycle decreases beyond −1; the cycle becomes unstable, and gives rise by the bifurcation process discussed in the previous section to stable cycles of period 6, 12, 24, . . ., 3×2^n. This birth of a stable and unstable pair of period 3 cycles, and the subsequent harmonics which arise as the initially stable cycle becomes unstable, are illustrated to the right of Fig. 4.

Table 1 Summary of the way various "single-hump" functions $F(X)$, from equation (1), behave in the chaotic region, distinguishing the dynamical properties which are generic from those which are not

The function $F(X)$ of equation (1)	$aX(1-X)$	$X\exp[r(1-X)]$	aX; if $X<\frac{1}{2}$ $a(1-X)$; if $X>\frac{1}{2}$	λX; if $X<1$ λX^{1-b}; if $X>1$
Tunable parameter	a	r	a	b
Fixed point becomes unstable	3.0000	2.0000	1.0000*	2.0000
"Chaotic" region begins [point of accumulation of cycles of period 2^n]	3.5700	2.6924	1.0000	2.0000
First odd-period cycle appears	3.6786	2.8332	1.4142	2.6180
Cycle with period 3 appears [and therefore every integer period present]	3.8284	3.1024	1.6180	3.0000
"Chaotic" region ends	4.0000†	∞‡	2.000†	∞‡
Are there stable cycles in the chaotic region?	Yes	Yes	No	No

* Below this a value, $X = 0$ is stable.
† All solutions are attracted to $-\infty$ for a values beyond this.
‡ In practice, as r or b becomes large enough, X will eventually be carried so low as to be effectively zero, thus producing extinction in models of biological populations.

Table 2 Catalogue of the number of periodic points, and of the various cycles (with periods $k = 1$ up to 12), arising from equation (1) with a single-humped function $F(X)$

k	1	2	3	4	5	6	7	8	9	10	11	12
Possible total number of points with period k	2	4	8	16	32	64	128	256	512	1,024	2,048	4,096
Possible total number of points with non-degenerate period k	2	2	6	12	30	54	126	240	504	990	2,046	4,020
Total number of cycles of period k, including those which are degenerate and/or harmonics and/or never locally stable	2	3	4	6	8	14	20	36	60	108	188	352
Total number of non-degenerate cycles (including harmonics and unstable cycles)	2	1	2	3	6	9	18	30	56	99	186	335
Total number of non-degenerate, stable cycles (including harmonics)	1	1	1	2	3	5	9	16	28	51	93	170
Total number of non-degenerate, stable cycles whose basic period is k (that is, excluding harmonics)	1	–	1	1	3	4	9	14	28	48	93	165

There are, therefore, two basic kinds of bifurcation processes[1,4] for first order difference equations. Truly new cycles of period k arise in pairs (one stable, one unstable) as the hills and valleys of higher iterates of $F(X)$ move, respectively, up and down to intercept the 45° line, as typified by Fig. 5. Such cycles are born at the moment when the hills and valleys become tangent to the 45° line, and the initial slope of the curve $F^{(k)}$ at the points is thus $\lambda^{(k)} = +1$: this type of bifurcation may be called[1,4] a tangent bifurcation or a $\lambda = +1$ bifurcation. Conversely, an originally stable cycle of period k may become unstable as $F(X)$ steepens. This happens when the slope of $F^{(k)}$ at these period k points steepens beyond $\lambda^{(k)} = -1$, whereupon a new and initially stable cycle of period $2k$ is born in the way typified by Figs 2 and 3. This type of bifurcation may be called a pitchfork bifurcation (borrowing an image from the left hand side of Fig. 4) or a $\lambda = -1$ bifurcation[1,4].

Putting all this together, we conclude that as the parameters in $F(X)$ are varied the fundamental, stable dynamical units are cycles of basic period k, which arise by tangent bifurcation, along with their associated cascade of harmonics of periods $k2^n$, which arise by pitchfork bifurcation. On this basis, the constant equilibrium solution X^* and the subsequent hierarchy of stable cycles of periods 2^n is merely a special case, albeit a conspicuously important one (namely $k=1$), of a general phenomenon. In addition, remember[1,4,22,29] that for sensible, analytical functions (such as, for example, those in equations (3) and (4)) there is a unique stable cycle for each value of the parameter in $F(X)$. The entire range of parameter values ($1 < a < 4$ in equation (3), $0 < r$ in equation (4)) may thus be regarded as made up of infinitely many windows of parameter

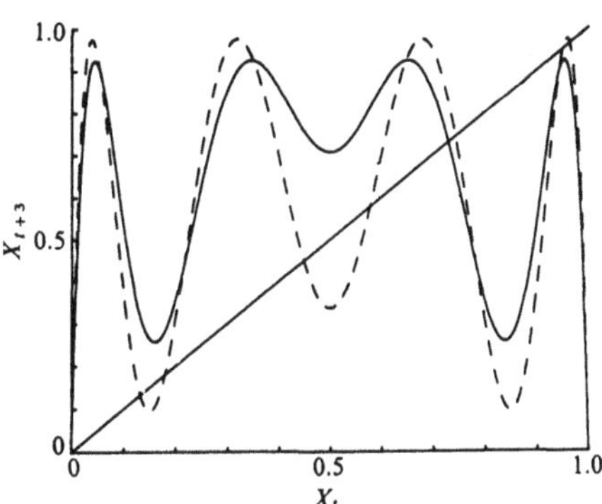

Fig. 5 The relationship between X_{t+3} and X_t, obtained by three iterations of equation (3). The solid curve is for $a = 3.7$, and only intersects the 45° line once. As a increases, the hills and valleys become more pronounced. The dashed curve is for $a = 3.9$, and six new period 3 points have appeared (arranged as two cycles, each of period 3).

values—some large, some unimaginably small—each corresponding to a single one of these basic dynamical units. Tables 3 and 4, below, illustrate this notion. These windows are divided from each other by points (the points of accumulation of the harmonics of period $k2^n$) at which the system is truly chaotic, with no attractive cycle: although there are infinitely many such special parameter values, they have measure zero on the interval of all values.

How are these various cycles arranged along the interval of relevant parameter values? This question has to my knowledge been answered independently by at least 6 groups of people, who have seen the problem in the context of combinatorial theory[16,30], numerical analysis[13,14], population biology[1], and dynamical systems theory[22,31] (broadly defined).

A simple-minded approach (which has the advantage of requiring little technical apparatus, and the disadvantage of being rather clumsy) consists of first answering the question, how many period k points can there be? That is, how many distinct solutions can there be to the equation

$$X^*_k = F^{(k)}(X^*_k)? \qquad (12)$$

If the function $F(X)$ is sufficiently steeply humped, as it will be once the parameter values are sufficiently large, each successive iteration doubles the number of humps, so that $F^{(k)}(X)$ has 2^{k-1} humps. For large enough parameter values, all these hills and valleys will intersect the 45° line, producing 2^k fixed points of period k. These are listed for $k \leqslant 12$ in the top row of Table 2. Such a list includes degenerate points of period k, whose period is a submultiple of k; in particular, the two period 1 points ($X=0$ and X^*) are degenerate solutions of equation (12) for all k. By working from left to right across Table 2, these degenerate points can be subtracted out, to leave the total number of non-degenerate points of basic period k, as listed in the second row of Table 2. More sophisticated ways of arriving at this result are given elsewhere[13,14,16,22,30,31].

For example, there eventually are $2^6 = 64$ points with period 6. These include the two points of period 1, the period 2 "harmonic" cycle, and the stable and unstable pair of triplets of points with period 3, for a total of 10 points whose basic period is a submultiple of 6; this leaves 54 points whose basic period is 6.

The 2^k period k points are arranged into various cycles of period k, or submultiples thereof, which appear in succession by either tangent or pitchfork bifurcation as the parameters in $F(X)$ are varied. The third row in Table 2 catalogues the total number of distinct cycles of period k which so appear. In the fourth row[14], the degenerate cycles are subtracted out, to give the total number of non-degenerate cycles of period k: these numbers must equal those of the second row divided by k. This fourth row includes the (stable) harmonics which arise by pitchfork bifurcation, and the pairs of stable–unstable cycles arising by tangent bifurcation. By subtracting out the cycles which are unstable from birth, the total number of possible stable cycles is given in row five; these figures can also be obtained by less pedestrian methods[13,16,30]. Finally we may subtract out the stable cycles which arise by pitchfork bifurcation, as harmonics of some simpler cycle, to arrive at the final

Table 3 A catalogue of the stable cycles (with basic periods up to 6) for the equation $X_{t+1} = aX_t(1-X_t)$

Period of basic cycle	*a* value at which: Basic cycle first appears	Basic cycle becomes unstable	Subsequent cascade of "harmonics" with period $k2^n$ all become unstable	Width of the range of *a* values over which the basic cycle, or one of its harmonics, is attractive
1	1.0000	3.0000	3.5700	2.5700
3	3.8284	3.8415	3.8495	0.0211
4	3.9601	3.9608	3.9612	0.0011
5(*a*)	3.7382	3.7411	3.7430	0.0048
5(*b*)	3.9056	3.9061	3.9065	0.0009
5(*c*)	3.99026	3.99030	3.99032	0.00006
6(*a*)	3.6265	3.6304	3.6327	0.0062
6(*b*)	3.937516	3.937596	3.937649	0.000133
6(*c*)	3.977760	3.977784	3.977800	0.000040
6(*d*)	3.997583	3.997585	3.997586	0.000003

row in Table 2, which lists the number of stable cycles whose basic period is k.

Returning to the example of period 6, we have already noted the five degenerate cycles whose periods are submultiples of 6. The remaining 54 points are parcelled out into one cycle of period 6 which arises as the harmonic of the only stable three-point cycle, and four distinct pairs of period 6 cycles (that is, four initially stable ones and four unstable ones) which arise by successive tangent bifurcations. Thus, reading from the foot of the column for period 6 in Table 2, we get the numbers 4, 5, 9, 14.

Using various labelling tricks, or techniques from combinatorial theory, it is also possible to give a generic list of the order in which the various cycles appear[1,13,16,22]. For example, the basic stable cycles of periods 3, 5, 6 (of which there are respectively 1, 3, 4) must appear in the order 6, 5, 3, 5, 6, 6, 5, 6: compare Tables 3 and 4. Metropolis *et al.*[16] give the explicit such generic list for all cycles of period $k \leqslant 11$.

As a corollary it follows that, given the most recent cycle to appear, it is possible (at least in principle) to catalogue all the cycles which have appeared up to this point. An especially elegant way of doing this is given by Smale and Williams[22], who show, for example, that when the stable cycle of period 3 first originates, the total number of other points with periods k, N_k, which have appeared by this stage satisfy the Fibonacci series, N_k=2, 4, 5, 8, 12, 19, 30, 48, 77, 124, 200, 323 for k=1, 2, . . ., 12: this is to be contrasted with the total number of points of period k which will eventually appear (the top row of Table 2) as $F(X)$ continues to steepen.

Such catalogues of the total number of fixed points, and of their order of appearance, are relatively easy to construct. For any particular function $F(X)$, the numerical task of finding the windows of parameter values wherein any one cycle or its harmonics is stable is, in contrast, relatively tedious and inelegant. Before giving such results, two critical parameter values of special significance should be mentioned.

Hoppensteadt and Hyman[31] have given a simple graphical method for locating the parameter value in the chaotic regime at which the first odd period cycle appears. Their analytic recipe is as follows. Let α be the parameter which tunes the steepness of $F(X)$ (for example, $\alpha=a$ for equation (3), $\alpha=r$ for equation (4)), $X^*(\alpha)$ be the fixed point of period 1 (the nontrivial solution of equation (5)), and $X_{max}(\alpha)$ the maximum value attainable from iterations of equation (1) (that is, the value of $F(X)$ at its hump or stationary point). The first odd period cycle appears for that value of α which satisfies[21,31]

$$X^*(\alpha) = F^{(3)}(X_{max}(\alpha)) \qquad (13)$$

As mentioned above, another critical value is that where the period 3 cycle first appears. This parameter value may be found numerically from the solutions of the third iterate of equation (1): for equation (3) it is[14] $a=1+\sqrt{8}$.

Myrberg[13] (for all $k \leqslant 10$) and Metropolis *et al.*[16]. (for all $k \leqslant 7$) have given numerical information about the stable cycles in equation (3). They do not give the windows of parameter values, but only the single value at which a given cycle is maximally stable; that is, the value of a for which the stability-determining slope of $F^{(k)}(X)$ is zero, $\lambda^{(k)}=0$. Since the slope of the k-times iterated map $F^{(k)}$ at any point on a period k cycle is simply equal to the product of the slopes of $F(X)$ at each of the points $X^*{}_k$ on this cycle[1,8,20], the requirement $\lambda^{(k)}=0$ implies that X=A (the stationary point of $F(X)$, where $\lambda^{(1)}=0$) is one of the periodic points in question, which considerably simplifies the numerical calculations.

For each basic cycle of period k (as catalogued in the last row of Table 2), it is more interesting to know the parameter values at which: (1) the cycle first appears (by tangent bifurcation); (2) the basic cycle becomes unstable (giving rise by successive pitchfork bifurcations to a cascade of harmonics of periods $k2^n$); (3) all the harmonics become unstable (the point of accumulation of the period $k2^n$ cycles). Tables 3 and 4 extend the work of May and Oster[1], to give this numerical information for equations (3) and (4), respectively. (The points of accumulation are not ground out mindlessly, but are calculated by a rapidly convergent iterative procedure, see ref. 1, appendix A.) Some of these results have also been obtained by Gumowski and Mira[32].

Practical problems

Referring to the paradigmatic example of equation (3), we can now see that the parameter interval $1<a<4$ is made up of a one-dimensional mosaic of infinitely many windows of a-values, in each of which a unique cycle of period k, or one of its harmonics, attracts essentially all initial points. Of these windows, that for $1<a<3.5700$ . . corresponding to k=1 and its harmonics is by far the widest and most conspicuous. Beyond the first point of accumulation, it can be seen from Table 3 that these windows are narrow, even for cycles of quite low periods, and the windows rapidly become very tiny as k increases.

As a result, there develops a dichotomy between the underlying mathematical behaviour (which is exactly determinable) and the "commonsense" conclusions that one would draw from numerical simulations. If the parameter a is held constant at one value in the chaotic region, and equation (3) iterated for an arbitrarily large number of generations, a density plot of the observed values of X_t on the interval 0 to 1 will settle into k equal spikes (more precisely, delta functions) corresponding to the k points on the stable cycle appropriate to this a-value. But for most a-values this cycle will have a fairly large period, and moreover it will typically take many thousands of generations before the transients associated with the initial conditions are damped out: thus the density plot produced by numerical simulations usually looks like a sample of points taken from some continuous distribution.

An especially interesting set of numerical computations are due to Hoppensteadt (personal communication) who has combined many iterations to produce a density plot of X_t for each one of a sequence of a-values, gradually increasing from 3.5700 . . to 4. These results are displayed as a movie. As can be expected from Table 3, some of the more conspicuous cycles do show up as sets of delta functions: the 3-cycle and its first few harmonics; the first 5-cycle; the first 6-cycle. But for most values of a the density plot looks like the sample function of a random process. This is particularly true in the neighbourhood of the a-value where the first odd cycle appears (a=3.6786 . .), and again in the neighbourhood of a=4: this is not surprising, because each of these locations is a point of accumulation of points of accumulation. Despite the underlying discontinuous changes in the periodicities of the stable cycles, the observed density pattern tends to vary smoothly. For example, as a increases toward the value at which the 3-cycle appears, the density plot tends to concentrate around three points, and it smoothly diffuses away from these three points after the 3-cycle and all its harmonics become unstable.

I think the most interesting mathematical problem lies in designing a way to construct some approximate and "effectively continuous" density spectrum, despite the fact that the exact density function is determinable and is always a set of delta functions. Perhaps such techniques have already been developed in ergodic theory[33] (which lies at the foundations of statistical mechanics), as for example in the use of "coarse-grained observers". I do not know.

Such an effectively stochastic description of the dynamical properties of equation (4) for large r has been provided[28], albeit by tactical tricks peculiar to that equation rather than by any general method. As r increases beyond about 3, the trajectories generated by this equation are, to an increasingly good approximation, almost periodic with period $(1/r)\exp(r-1)$.

The opinion I am airing in this section is that although the exquisite fine structure of the chaotic regime is mathematically fascinating, it is irrelevant for most practical purposes. What seems called for is some effectively stochastic description of the deterministic dynamics. Whereas the various statements about the different cycles and their order of appearance can be made in generic fashion, such stochastic description of the actual dynamics will be quite different for different $F(X)$: witness the difference between the behaviour of equation (4), which for large r is almost periodic "outbreaks" spaced many generations apart, versus the behaviour of equation (3), which for $a\to 4$ is not very different from a series of Bernoulli coin flips.

Mathematical curiosities

As discussed above, the essential reason for the existence of a succession of stable cycles throughout the "chaotic" regime is that as each new pair of cycles is born by tangent bifurcation (see Fig. 5), one of them is at first stable, by virtue of the way the smoothly rounded hills and valleys intercept the 45° line. For analytical functions $F(X)$, the only parameter values for which the density plot or "invariant measure" is continuous and truly ergodic are at the points of accumulation of harmonics, which divide one stable cycle from the next. Such exceptional parameter values have found applications, for example, in the use of equation (3) with a=4 as a random number generator[34,35]: it has a continuous density function proportional to $[X(1-X)]^{-\frac{1}{2}}$ in the interval $0<X<1$.

Non-analytical functions $F(X)$ in which the hump is in fact a spike provide an interesting special case. Here we may imagine spikey hills and valleys moving to intercept the 45° line in Fig. 5, and it may be that both the cycles born by tangent bifurcation are unstable from the outset (one having $\lambda^{(k)}>1$, the other $\lambda^{(k)}<-1$), for all $k>1$. There are then no stable cycles in the chaotic regime, which is therefore literally chaotic with a continuous and truly ergodic density distribution function.

One simple example is provided by

$$X_{t+1} = aX_t;\ \text{if } X_t<\tfrac{1}{2} \qquad X_{t+1} = a(1-X_t);\ \text{if } X_t>\tfrac{1}{2} \tag{14}$$

defined on the interval $0<X<1$. For $0<a<1$, all trajectories are attracted to X=0; for $1<a<2$, there are infinitely many periodic orbits, along with an uncountable number of aperiodic trajectories, none of which are locally stable. The first odd period cycle appears at $a=\sqrt{2}$, and all integer periods are represented beyond $a=(1+\sqrt{5})/2$. Kac[36] has given a careful discussion of the case a=2. Another example, this time with an extensive biological pedigree[1-5], is the equation

$$X_{t+1} = \lambda X_t;\ \text{if } X_t<1 \qquad X_{t+1} = \lambda X_t^{1-b};\ \text{if } X_t>1 \tag{15}$$

If $\lambda>1$ this possesses a globally stable equilibrium point for $b<2$. For $b>2$ there is again true chaos, with no stable cycles: the first odd cycle appears at $b=(3+\sqrt{5})/2$, and all integer periods are present beyond b=3. The dynamical properties of equations (14) and (15) are summarised to the right of Table 2.

The absence of analyticity is a necessary, but not a sufficient, condition for truly random behaviour[31]. Consider, for example,

$$X_{t+1} = (a/2)X_t;\ \text{if } X_t<\tfrac{1}{2} \qquad X_{t+1} = aX_t(1-X_t);\ \text{if } X_t>\tfrac{1}{2} \tag{16}$$

This is the parabola of equation (3) and Fig. 1, but with the left hand half of $F(X)$ flattened into a straight line. This equation does possess windows of a values, each with its own stable cycle, as described generically above. The stability-determining slopes $\lambda^{(k)}$ vary, however, discontinuously with the parameter a, and the widths of the simpler stable regions are narrower than for equation (3): the fixed point becomes unstable at a=3; the point of accumulation of the subsequent harmonics is at a=3.27 . .; the first odd cycle appears at a=3.44 . .; the 3-point cycle at a=3.67 . . (compare the first column in Table 1).

These eccentricities of behaviour manifested by non-analytical functions may be of interest for exploring formal questions in ergodic theory. I think, however, that they have no relevance to models in the biological and social sciences, where functions such as $F(X)$ should be analytical. This view is elaborated elsewhere[37].

As a final curiosity, consider the equation

$$X_{t+1} = \lambda X_t[1+X_t]^{-\beta} \tag{17}$$

Table 4 Catalogue of the stable cycles (with basic periods up to 6) for the equation $X_{t+1} = X_t\exp[r(1-X_t)]$

Period of basic cycle	r value at which: Basic cycle first appears	Basic cycle becomes unstable	Subsequent cascade of "harmonics" with period $k2^n$ *all* become unstable	Width of the range of r values over with the basic cycle, or one of its harmonics, is attractive
1	0.0000	2.0000	2.6924	2.6924
3	3.1024	3.1596	3.1957	0.0933
4	3.5855	3.6043	3.6153	0.0298
5(*a*)	2.9161	2.9222	2.9256	0.0095
5(*b*)	3.3632	3.3664	3.3682	0.0050
5(*c*)	3.9206	3.9295	3.9347	0.0141
6(*a*)	2.7714	2.7761	2.7789	0.0075
6(*b*)	3.4558	3.4563	3.4567	0.0009
6(*c*)	3.7736	3.7745	3.7750	0.0014
6(*d*)	4.1797	4.1848	4.1880	0.0083

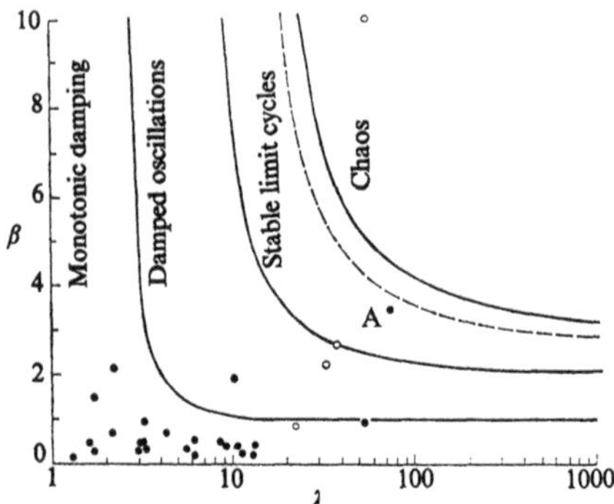

Fig. 6 The solid lines demarcate the stability domains for the density dependence parameter, β, and the population growth rate, λ, in equation (17); the dashed line shows where 2-point cycles give way to higher cycles of period 2^n. The solid circles come from analyses of life table data on field populations, and the open circles from laboratory populations (from ref. 3, after ref. 39).

This has been used to fit a considerable amount of data on insect populations[38,39]. Its stability behaviour, as a function of the two parameters λ and β, is illustrated in Fig. 6. Notice that for $\lambda < 7.39$ . . there is a globally stable equilibrium point for all β; for $7.39 \ldots < \lambda < 12.50 \ldots$ this fixed point becomes unstable for sufficiently large β, bifurcating to a stable 2-point cycle which is the solution for all larger β; as λ increases through the range $12.50 \ldots < \lambda < 14.77 \ldots$ various other harmonics of period 2^n appear in turn. The hierarchy of bifurcating cycles of period 2^n is thus truncated, and the point of accumulation and subsequent regime of chaos is not achieved (even for arbitrarily large β) until $\lambda > 14.77 \ldots$

Applications

The fact that the simple and deterministic equation (1) can possess dynamical trajectories which look like some sort of random noise has disturbing practical implications. It means, for example, that apparently erratic fluctuations in the census data for an animal population need not necessarily betoken either the vagaries of an unpredictable environment or sampling errors: they may simply derive from a rigidly deterministic population growth relationship such as equation (1). This point is discussed more fully and carefully elsewhere[1].

Alternatively, it may be observed that in the chaotic regime arbitrarily close initial conditions can lead to trajectories which, after a sufficiently long time, diverge widely. This means that, even if we have a simple model in which all the parameters are determined exactly, long term prediction is nevertheless impossible. In a meteorological context, Lorenz[15] has called this general phenomenon the "butterfly effect": even if the atmosphere could be described by a deterministic model in which all parameters were known, the fluttering of a butterfly's wings could alter the initial conditions, and thus (in the chaotic regime) alter the long term prediction.

Fluid turbulence provides a classic example where, as a parameter (the Reynolds number) is tuned in a set of deterministic equations (the Navier–Stokes equations), the motion can undergo an abrupt transition from some stable configuration (for example, laminar flow) into an apparently stochastic, chaotic regime. Various models, based on the Navier–Stokes differential equations, have been proposed as mathematical metaphors for this process[15,40,41]. In a recent review of the theory of turbulence, Martin[42] has observed that the one-dimensional difference equation (1) may be useful in this context. Compared with the earlier models[15,40,41], it has the disadvantage of being even more abstractly metaphorical, and the advantage of having a spectrum of dynamical behaviour which is more richly complicated yet more amenable to analytical investigation.

A more down-to-earth application is possible in the use of equation (1) to fit data[1,2,3,38,39,43] on biological populations with discrete, non-overlapping generations, as is the case for many temperate zone arthropods. Figure 6 shows the parameter values λ and β that are estimated[39] for 24 natural populations and 4 laboratory populations when equation (17) is fitted to the available data. The figure also shows the theoretical stability domains: a stable point; its stable harmonics (stable cycles of period 2^n); chaos. The natural populations tend to have stable equilibrium point behaviour. The laboratory populations tend to show oscillatory or chaotic behaviour; their behaviour may be exaggeratedly nonlinear because of the absence, in a laboratory setting, of many natural mortality factors. It is perhaps suggestive that the most oscillatory natural population (labelled *A* in Fig. 6) is the Colorado potato beetle, whose present relationship with its host plant lacks an evolutionary pedigree. These remarks are only tentative, and must be treated with caution for several reasons. Two of the main caveats are that there are technical difficulties in selecting and reducing the data, and that there are no single species populations in the natural world: to obtain a one-dimensional difference equation by replacing a population's interactions with its biological and physical environment by passive parameters (such as λ and β) may do great violence to the reality.

Some of the many other areas where these ideas have found applications were alluded to in the second section, above[5–11]. One aim of this review article is to provoke applications in yet other fields.

Related phenomena in higher dimensions

Pairs of coupled, first-order difference equations (equivalent to a single second-order equation) have been investigated in several contexts[4,44–46], particularly in the study of temperate zone arthropod prey–predator systems[2–4,23,47]. In these two-dimensional systems, the complications in the dynamical behaviour are further compounded by such facts as: (1) even for analytical functions, there can be truly chaotic behaviour (as for equations (14) and (15)), corresponding to so-called "strange attractors"; and (2) two or more different stable states (for example, a stable point and a stable cycle of period 3) can occur together for the same parameter values[4]. In addition, the manifestation of these phenomena usually requires less severe nonlinearities (less steeply humped $F(X)$) than for the one-dimensional case.

Similar systems of first-order ordinary differential equations, or two coupled first-order differential equations, have much simpler dynamical behaviour, made up of stable and unstable points and limit cycles[48]. This is basically because in continuous two-dimensional systems the inside and outside of closed curves can be distinguished; dynamic trajectories cannot cross each other. The situation becomes qualitatively more complicated, and in many ways analogous to first-order difference equations, when one moves to systems of three or more coupled, first-order ordinary differential equations (that is, three-dimensional systems of ordinary differential equations). Scanlon (personal communication) has argued that chaotic behaviour and "strange attractors", that is solutions which are neither points nor periodic orbits[48], are typical of such systems. Some well studied examples arise in models for reaction–diffusion systems in chemistry and biology[49], and in the models of Lorenz[15] (three dimensions) and Ruelle and Takens[40] (four dimensions) referred to above. The analysis of these systems is, by virtue of their higher dimensionality, much less transparent than for equation (1).

An explicit and rather surprising example of a system which

has recently been studied from this viewpoint is the ordinary differential equations used in ecology to describe competing species. For one or two species these systems are very tame: dynamic trajectories will converge on some stable equilibrium point (which may represent coexistence, or one or both species becoming extinct). As Smale[50] has recently shown, however, for 3 or more species these general equations can, in a certain reasonable and well-defined sense, be compatible with any dynamical behaviour. Smale's[50] discussion is generic and abstract: a specific study of the very peculiar dynamics which can be exhibited by the familiar Lotka-Volterra equations once there are 3 competitors is given by May and Leonard[51].

Conclusion

In spite of the practical problems which remain to be solved, the ideas developed in this review have obvious applications in many areas.

The most important applications, however, may be pedagogical.

The elegant body of mathematical theory pertaining to linear systems (Fourier analysis, orthogonal functions, and so on), and its successful application to many fundamentally linear problems in the physical sciences, tends to dominate even moderately advanced University courses in mathematics and theoretical physics. The mathematical intuition so developed ill equips the student to confront the bizarre behaviour exhibited by the simplest of discrete nonlinear systems, such as equation (3). Yet such nonlinear systems are surely the rule, not the exception, outside the physical sciences.

I would therefore urge that people be introduced to, say, equation (3) early in their mathematical education. This equation can be studied phenomenologically by iterating it on a calculator, or even by hand. Its study does not involve as much conceptual sophistication as does elementary calculus. Such study would greatly enrich the student's intuition about nonlinear systems.

Not only in research, but also in the everyday world of politics and economics, we would all be better off if more people realised that simple nonlinear systems do not necessarily possess simple dynamical properties.

I have received much help from F. C. Hoppensteadt, H. E. Huppert, A. I. Mees, C. J. Preston, S. Smale, J. A. Yorke, and particularly from G. F. Oster. This work was supported in part by the NSF.

1 May, R. M., and Oster, G. F., *Am. Nat.*, 110 (in the press).
2 Varley, G. C., Gradwell, G. R., and Hassell, M. P., *Insect Population Ecology* (Blackwell, Oxford, 1973).
3 May, R. M. (ed.), *Theoretical Ecology: Principles and Applications* (Blackwell, Oxford, 1976).
4 Guckenheimer, J., Oster, G. F., and Ipaktchi, A., *Theor. Pop. Biol.* (in the press).
5 Oster, G. F., Ipaktchi, A., and Rocklin, I., *Theor. Pop. Biol.* (in the press).
6 Asmussen, M. A., and Feldman, M. W., *J. theor. Biol.* (in the press).
7 Hoppensteadt, F. C., *Mathematical Theories of Populations: Demographics, Genetics and Epidemics* (SIAM, Philadelphia, 1975).
8 Samuelson, P. A., *Foundations of Economic Analysis* (Harvard University Press, Cambridge, Massachusetts, 1947).
9 Goodwin, R. E., *Econometrica*, **19**, 1–17 (1951).
10 Baumol, W. J., *Economic Dynamics*, 3rd ed. (Macmillan, New York, 1970).
11 See, for example, Kemeny, J., and Snell, J. L., *Mathematical Models in the Social Sciences* (MIT Press, Cambridge, Massachusetts, 1972).
12 Chaundy, T. W., and Phillips, E., *Q. Jl Math. Oxford*, **7**, 74–80 (1936).
13 Myrberg, P. J., *Ann. Akad. Sc. Fennicae*, *A*, I, No. 336/3 (1963).
14 Myrberg, P. J., *Ann. Akad. Sc. Fennicae*, *A*, I, No. 259 (1958).
15 Lorenz, E. N., *J. Atmos. Sci.*, **20**, 130–141 (1963); *Tellus*, **16**, 1–11 (1964).
16 Metropolis, N., Stein, M. L., and Stein, P. R., *J. Combinatorial Theory*, **15(A)**, 25–44 (1973).
17 Maynard Smith, J., *Mathematical Ideas in Biology* (Cambridge University Press, Cambridge, 1968).
18 Krebs, C. J., *Ecology* (Harper and Row, New York, 1972).
19 May, R. M., *Am. Nat.*, **107**, 46–57 (1972).
20 Li, T-Y., and Yorke, J. A., *Am. Math. Monthly*, **82**, 985–992 (1975).
21 Hoppensteadt, F. C., and Hyman, J. M. (Courant Institute, New York University: preprint, 1975).
22 Smale, S., and Williams, R. (Department of Mathematics, Berkeley: preprint, 1976).
23 May, R. M., *Science*, **186**, 645–647 (1974).
24 Moran, P. A. P., *Biometrics*, **6**, 250–258 (1950).
25 Ricker, W. E., *J. Fish. Res. Bd. Can.*, 11, 559–623 (1954).
26 Cook, L. M., *Nature*, **207**, 316 (1965).
27 Macfadyen, A., *Animal Ecology: Aims and Methods* (Pitman, London, 1963).
28 May, R. M., *J. theor. Biol.*, **51**, 511–524 (1975).
29 Guckenheimer, J., *Proc. AMS Symposia in Pure Math.*, *XIV*, 95–124 (1970).
30 Gilbert, E. N., and Riordan, J., *Illinois J. Math.*, **5**, 657–667 (1961).
31 Preston, C. J. (King's College, Cambridge: preprint, 1976).
32 Gumowski, I., and Mira, C., *C. r. hebd. Séanc. Acad. Sci., Paris*, **281a**, 45–48 (1975); **282a**, 219–222 (1976).
33 Layzer, D., *Sci. Am.*, **233**(6), 56–69 (1975).
34 Ulam, S. M., *Proc. Int. Congr. Math.1950, Cambridge, Mass.; Vol. II*, pp. 264–273 (AMS, Providence R.I., 1950).
35 Ulam, S. M., and von Neumann, J., *Bull. Am. math. Soc.* (abstr.), **53**, 1120 (1947).
36 Kac, M., *Ann. Math.*, **47**, 33–49 (1946).
37 May, R. M., *Science*, **181**, 1074 (1973).
38 Hassell, M. P., *J. Anim. Ecol.*, **44**, 283–296 (1974).
39 Hassell, M. P., Lawton, J. H., and May, R. M., *J. Anim. Ecol.* (in the press).
40 Ruelle, D., and Takens, F., *Comm. math. Phys.*, **20**, 167–192 (1971).
41 Landau, L. D., and Lifshitz, E. M., *Fluid Mechanics* (Pergamon, London, 1959).
42 Martin, P. C., *Proc. Int. Conf. on Statistical Physics, 1975, Budapest* (Hungarian Acad. Sci., Budapest, in the press).
43 Southwood, T. R. E., in *Insects, Science and Society* (edit. by Pimentel, D.), 151–199 (Academic, New York, 1975).
44 Metropolis, N., Stein, M. L., and Stein, P. R., *Numer. Math.*, **10**, 1–19 (1967).
45 Gumowski, I., and Mira, C., *Automatica*, **5**, 303–317 (1969).
46 Stein, P. R., and Ulam, S. M., *Rosprawy Mat.*, **39**, 1–66 (1964).
47 Beddington, J. R., Free, C. A., and Lawton, J. H., *Nature*, **255**, 58–60 (1975).
48 Hirsch, M. W., and Smale, S., *Differential Equations, Dynamical Systems and Linear Algebra* (Academic, New York, 1974).
49 Kolata, G. B., *Science*, **189**, 984–985 (1975).
50 Smale, S. (Department of Mathematics, Berkeley: preprint, 1976).
51 May, R. M., and Leonard, W. J., *SIAM J. Appl. Math.*, **29**, 243–253 (1975).

Reviews of Modern Physics **53** 643–54 (1981)

Roads to turbulence in dissipative dynamical systems

J.-P. Eckmann

Département de Physique Théorique, Université de Genève, 1211 Genève 4, Switzerland

Three scenarios leading to turbulence in theory and experiment are outlined. The respective mathematical theories are explained and compared.

CONTENTS

I. INTRODUCTION

Every physicist is exposed early in his career to solvable dynamical problems, for example, the harmonic oscillator and the Kepler problem. One also learns that a damped pendulum reaches its equilibrium position, and one learns how to find the exponential functions describing the approach to this equilibrium. Quite soon, one becomes aware that not all dynamical problems are explicitly solvable, even allowing for solutions in terms of the more complicated transcendental functions. This situation may occur for systems with few degrees of freedom, (i.e., few dynamical variables), and without external noise. In addition, it is not restricted to Hamiltonian problems, but appears as well for dynamical systems with internal friction, called dissipative dynamical systems. The reason for this difficulty is the fact that *dynamical problems with regular equations may have solutions which behave irregularly in time*.

We would like to understand, in the absence of explicit solutions, more about the qualitative aspects of these irregular solutions. There is no general classification of dynamical systems which is sufficiently fine to account for all possible types of erratic behavior of their solutions, and even such simple systems as a forced pendulum with friction are exceedingly hard to analyze. One would nevertheless like to find similarities among, and predictions for, various dynamical systems.

The aim here is to present an approach to the understanding of irregular (or nearly irregular) phenomena, which has been relatively successful recently.[1] To avoid any misunderstanding, I must insist that this approach does not reach any conclusions about such matters as the beautiful turbulences on Jupiter or the dynamics of the Niagara falls.[2] Rather, by setting more modest aims, I describe here examples of relatively simple, but nevertheless aperiodic behavior, and put them in perspective. In this view, systems exhibiting this behavior are still sufficiently irregular to be called turbulent, and in fact some of their aspects are found in (irregular) convection of fluids. All forms of aperiodicity (even very weak ones) are of interest, but the words *aperiodic*, *erratic*, *chaotic*, and (*weakly*) *turbulent* will be used interchangeably for any of these forms.

The approach I describe has its roots in the general study of deterministic differential equations which are supposed to model the physical (chemical, . . .) system under investigation (Smale, 1967). Throughout, we shall suppose that the system depends on an external controllable parameter and that for some value of the parameter its dynamical behavior is well understood (e.g., the system could have only a stable equilibrium state, or a stationary solution). As the parameter is changed from this value, the qualitative behavior of the system may change, too. After a finite or infinite succession of such changes the system may present erratic behavior in the sense that its time evolution may be quite unpredictable on large time scales, or it may show broad-band spectrum or may not be periodic any more. Some systems may show features of a stochastic process,[3] although no external noise source is present in the dynamical equations.

II. DISSIPATIVE SYSTEMS AND THEIR ATTRACTORS

In order to describe our main topic, we need an adequate language for describing deterministic evolution equations. Typical behavior will be described in terms of the attractors of a system. The evolution equations, for fixed value of the parameter, will be assumed

[1] In a way, this approach can be viewed as a concretization of some aspects of Thom's (1972) catastrophe theory.

[2] For a discussion of "fully developed turbulence," see, for example, Monin and Yaglom (1975).

[3] Good expository references about these aspects are Bowen (1975) and Lanford (1978).

throughout to be of one of two types, namely,

$$\frac{d}{dt}\mathbf{x}(t)=\mathbf{F}(\mathbf{x}(t)), \qquad (1)$$

or

$$\mathbf{x}_{n+1}=\mathbf{F}(\mathbf{x}_n). \qquad (2)$$

Here $\mathbf{x}$ is a vector in $\mathbb{R}^m$, $m \geq 1$ and each of its components describes a "mode" or a coordinate. When $\mathbf{F}$ will depend on a parameter, we shall denote it by μ and write $\mathbf{F}_\mu$. Typical examples of dynamical systems of the form of Eq. (1) are listed in Table I.

We shall describe later how Eq. (2) appears naturally in applications; in any case, the simple dynamical system (discrete iteration) which is defined by

$$x_{n+1}=f(x_n),$$

where $x_n \in \mathbb{R}$, $n=0,1,2,\ldots$ and f: $\mathbb{R}\to\mathbb{R}$ is continuous, often serves as a guiding tool (Collet and Eckmann, 1980). Here, one should think of n as the (discrete) time.

It is well known that in Hamiltonian dynamics Liouville's theorem asserts that the flow $t\to\mathbf{x}(t)$ preserves volumes in phase space. If we denote by $\mathbf{x}(\mathbf{y},t)$ the solution of Eq. (1) with initial condition $\mathbf{x}(\mathbf{y},t=0)=\mathbf{y}$, and if

$$\sum_{i=1}^{m}\frac{\partial F_i}{\partial x_i}(\mathbf{x})=0,$$

then the flow preserves volumes locally. On the other hand, for systems with internal friction, called *dissipative systems*, such as the last three examples in Table I, the flow contracts volumes, i.e.,

$$\sum_{i=1}^{m}\frac{\partial F_i}{\partial x_i}(\mathbf{x})<0,$$

or (equivalently)

$$\sum_{i=1}^{m}\frac{\partial \dot{x}_i(\mathbf{y},t)}{\partial y_i}<0,$$

where $\dot{\mathbf{x}}=d\mathbf{x}/dt$.

We shall deal exclusively with dissipative systems, and we start now with the description of their attractors. Assume there is a finite volume V in state space ($\mathbb{R}^m$) such that if $\mathbf{y}\in V$ then $T^t\mathbf{y}=\mathbf{x}(\mathbf{y},t)$ is in V for all $t>0$. Since the flow T^t decreases volumes, the sets T^tV decrease as $t\to\infty$ to a set

TABLE I. Dynamical systems and their phase-space coordinates.

System[a]	Interpretation of coordinates
Hamiltonian mechanics	Coordinates p,q in phase space
Particle accelerators	Deviations from ideal trajectory
Hydrodynamics	Fourier modes of velocity field (not position of molecules)
Chemical reactions	Concentrations
Electrical circuits	Currents, voltages

[a]Some introductory references are Siegel and Moser (1976), Hagedorn (1957), Foias and Temam (1979), Nicolis and Prigogine (1977), and Brayton and Moser (1964).

$$W=\bigcap_{t>0}T^tV$$

(of zero volume). Thus every solution curve starting at some $\mathbf{y}\in V$ approaches W as $t\to\infty$. We can alternately say that if $\mathbf{y}\in V\backslash W$ then $\mathbf{y}$ is *transient* and the curve $T^t\mathbf{y}$ will for some sufficiently large t definitively depart from $\mathbf{y}$ and converge to W. This is in sharp contrast with the situation encountered in nondissipative closed systems, where almost all curves $T^t\mathbf{y}$ return infinitely often arbitrarily close to their initial state $\mathbf{y}$. We shall not discuss the question of transience, although this is an interesting subject. Therefore we consider only systems which have attained some sort of "internal equilibrium." In other words, we analyze the motion on W or on parts of W, assuming the orbits which tend to W but are not in it behave similarly to those in W, at least after a sufficient lapse of time. These parts of W will be called *attractors*, and studying attractors only amounts to neglecting transient behavior. Before reading the definition of attractors, it should be kept in mind that there is no universal agreement about what the best definition should be [see, for example, Newhouse (1980b), Shub (1980), Lanford (1981)].

Definition. An *attractor* for the flow T^t is a compact set X satisfying

(1) X is invariant under T^t: $T^tX=X$.

(2) X has a shrinking neighborhood, i.e., there is an open neighborhood U of X, $U\supset X$ such that $T^tU\subset U$ for $t>0$ and $X=\bigcap_{t>0}T^tU$.

This definition excludes *repellors*—for example, an isolated fixed point $\mathbf{x}$, $T^t\mathbf{x}=\mathbf{x}$, in whose neighborhood there is for every $\varepsilon>0$ a $\mathbf{y}$ with $|\mathbf{y}-\mathbf{x}|<\varepsilon$, which escapes away from $\mathbf{x}$, i.e., $|T^t\mathbf{y}-\mathbf{x}|$ grows (relatively) large. A repellor $\mathbf{x}$ would be in W, but not in X. We are not interested in repellors, since from an experimental point of view only attractors can play a role. Many points behave like the points on attractors, but only few behave like a repellor; a repellor is a generalization of an unstable equilibrium point or of a saddle point.

A good definition of an attractor needs another ingredient which generalizes the description of k separate stable equilibria to k separate attractors. This is achieved by the following requirement.

(3) The flow T^t on X is *recurrent* and *indecomposable*. Recurrent means T^t is nowhere transient on X: If U is an open set in V and if $U\cap V\neq\varnothing$, then there are arbitrarily large values for t such that $T^t\mathbf{x}\in X\cap U$ when $\mathbf{x}\in X\cap U$. Indecomposable means that X cannot be split into two nontrivial closed invariant pieces.

In the simplest dynamical systems the situation might be as shown in Fig. 1. There are two attractors, x_1 and x_2, which are stable fixed points. There basins of attraction are respectively the left and right sides of the line L. The line L is attracted by x_3, which is not an attractor, since it also has an unstable direction. It is a saddle point. With our previous definitions, $W=\{x_1,x_2,x_3\}$.

If X is an attractor, its *basin of attraction* is defined to be the set of initial points $\mathbf{x}$ such that $T^t\mathbf{x}$ approaches X as $t\to\infty$.

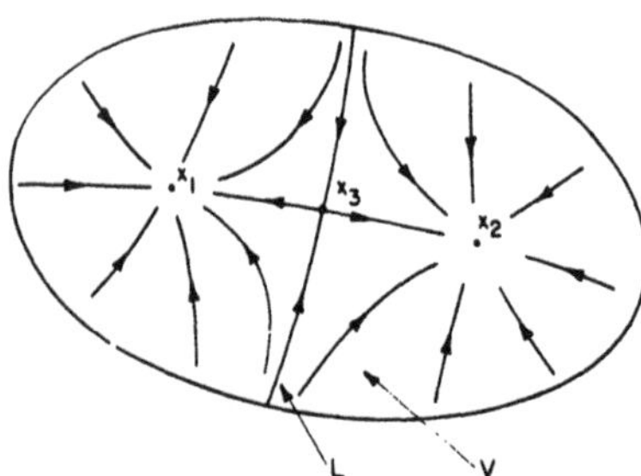

FIG. 1. Phase portrait illustrating two stable (x_1, x_2) and one unstable (x_3) fixed point.

It is now time to point out some misconceptions which could arise from the simple picture of Fig. 1.

(1) Although T^t contracts volumes, it *need not contract lengths*. If we take snapshots of T^t at $t=0,1,2$, say, we may have the picture shown in Fig. 2(a) but could also get that of Fig. 2(b) or even that of Fig. 2(c). In particular, even if all points in V converge to a single attractor X, one still may find that points which are arbitrarily close initially may get macroscopically separated on the attractor after sufficiently large time intervals. This property is called *sensitive dependence on initial conditions*. It is *not* excluded for area-contracting flows, i.e., it can, and will, occur in dissipative dynamical systems. An attractor exhibiting this property will be called a *strange attractor*. Strange attractors are neither periodic points nor periodic orbits. Observe, however, that there exists a large variety of attractors which are neither trivial (i.e., they are neither periodic orbits nor fixed points) and which are not strange attractors. All of them seem to present more or less pronounced chaotic features. We shall call the motion on any nontrivial attractor weakly

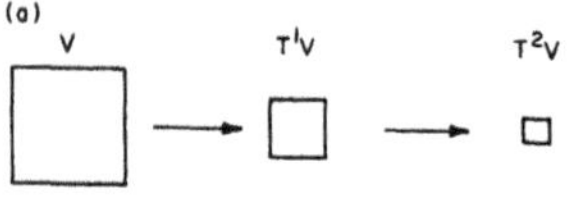

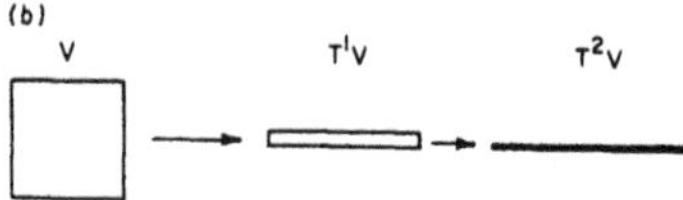

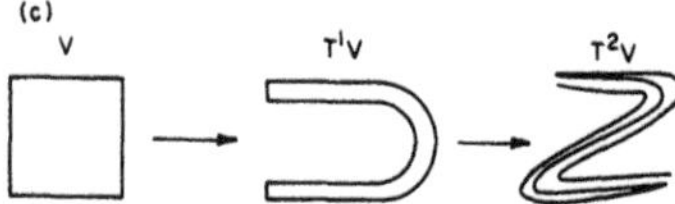

FIG. 2. (a) Contraction of volume in phase space. (b) Contraction of volume in phase space, with stretching of length. (c) Contraction of volume, stretching of length, and folding.

turbulent, erratic, etc., independently of whether or not the attractor is strange.

(2) Even simple dynamical systems may have an *infinity* of distinct attractors. As an example, it has been shown [Newhouse, 1980a; see also Levi, to appear] that the iterative scheme of Hénon

$$\begin{pmatrix} x_n \\ y_n \end{pmatrix} \to \begin{pmatrix} x_{n+1} \\ y_{n+1} \end{pmatrix} = \begin{pmatrix} 1 + y_n - a x_n^2 \\ b x_n \end{pmatrix}$$

has an infinity of attractors at some values of a near 1.15357 and $b=0.3$. The attractors correspond to periodic points of higher and higher period, which may be numerically indistinguishable from a strange attractor. Incidentally, it is believed that for some values of a and b the above system does have a strange attractor, but this has not been proved so far.[4]

(3) Basins of attraction may be complicated, even if the attractors are simple. A very old example[5] is the following: Consider the map

$$z_{n+1} = z_n - \frac{z_n^3 - 1}{3z_n^2},$$

defined on $\mathbb{C} \setminus \{0\}$. This is the Newton algorithm for finding the roots of $z^3=1$. It has three stable fixed points $z=1$, $\exp(i2\pi/3)$, $\exp(-i2\pi/3)$, with domains of attraction $\mathfrak{D}_1, \mathfrak{D}_2, \mathfrak{D}_3$. One can show that the boundary points of $\mathfrak{D}_1, \mathfrak{D}_2, \mathfrak{D}_3$ coincide. So these three regions must be highly interlaced.

III. THE PROBLEM OF CLASSIFYING ATTRACTORS. SCENARIOS

In the spirit of the preceding discussion, one should arrive at a description of the nontransient behavior of dynamical systems by classifying their attractors and the motion on them. This aim is clearly felt throughout the literature on dynamical systems. One is, however, far from any complete classification of attractors, or even from a canonical choice of adequate classification criteria. What I present here is a more modest approach which will lead to a *description of some nontrivial attractors, which have the additional feature that they arise as modifications of trivial attractors as an external parameter is changed.*[6] Thus, instead of considering a single problem, we deal with a one-parameter family of problems:

$$\frac{d}{dt}\mathbf{x}(t) = \mathbf{F}_\mu(\mathbf{x}(t)), \quad \mathbf{x}(0) = \mathbf{y}$$

or

$$\mathbf{x}_{n+1} = \mathbf{F}_\mu(\mathbf{x}_n), \quad \mathbf{x}_0 = \mathbf{y}.$$

The parameter μ, in the list of Table I, can be thought of as the strength of a driving force, the amount of friction, the amount of chemicals added per time unit, etc. It is assumed that μ stays fixed during the whole duration of an experiment. We are interested in the *changes of the attractors* as the parameter is varied.

---

[4]A partial answer is in Misiurewicz (1980).

[5]I have heard this from F. Sergeraert.

[6]This procedure has been advocated in Ruelle and Takens (1971).

In general, the attractor changes smoothly for small variations of the parameter. For example, a fixed point may move a little bit as the parameter is varied, or a stable limit cycle may change its shape and/or the time needed to complete a cycle (see Fig. 3).

Sometimes, however, the topological nature of the attractor may change as the parameter crosses a point μ_B. One calls this a *bifurcation* point. For example, in Fig. 4 the stable fixed point at μ_1 changes to a stable limit cycle at μ_2 (plus an unstable fixed point). Quite often a bifurcation is prompted by the crossing of eigenvalues of the linearized flow at the fixed point (or periodic orbit) through the unit circle when the parameter passes through μ_B.

A first bifurcation may be followed by further bifurcations, and we may ask what happens when a certain sequence of bifurcations has been encountered. In principle there is an infinity of further possibilities, but, in some sense to be specified, not all of them are equally probable. The more likely ones will be called *scenarios*, and below we shall examine three prominent scenarios which have had theoretical and experimental success. One should hope that further relevant scenarios will be found in the future.

We are now going to look at the nature of the prediction which can be made with the help of scenarios, since this may be a somewhat unfamiliar way of reasoning. But it appears that this kind of argument has the most promising chances of illuminating the nature of chaotic behavior. The statement of a scenario always takes the form "if... then...," i.e., if certain things happen to the attractor as the parameter is varied, then certain other things are likely to happen as the parameter is varied further. The mathematical meaning of "likely" may depend on the scenario and will be described below for each of the scenarios. But what does likely mean in a physical context? I do not intend to go to any philosophical depth but, rather, take a pragmatic stand. (1) One never knows exactly which equation (i.e., which **F**) is relevant for the description of a given physical system. (2) When an experiment is repeated, the equations may have slightly changed (e.g., the gravitational effects change on the earth by the motion of the moon). (3) The equation under investigation is one among several, all of which are very close to each other. (4) If among these there are many which satisfy the conclusions of the scenario, then we will say that if we perform an actual experiment, it will be probable that the conclusions of the scenario apply.

In general, a scenario deals with the description of a few attractors. On the other hand, a given dynamical system may have many attractors. Therefore, *several scenarios may evolve concurrently in different regions of phase space*. There is thus no contradiction if several scenarios occur in a given physical system, depending on how the initial state of the system is prepared. In addition, the relevant parameter ranges may overlap, and while the basins of attraction for different scenarios must be disjoint, they may be interlaced.

FIG. 3. Phase portraits illustrating stable limit cycles.

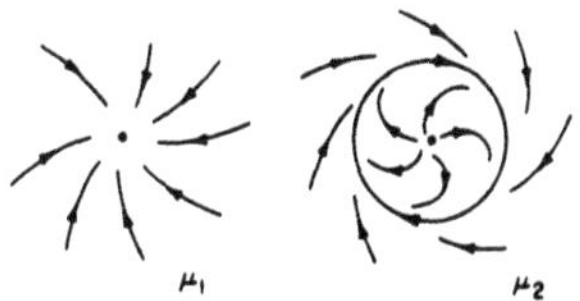

FIG. 4. Phase portraits illustrating Hopf bifurcation.

It is implicit in the preceding discussion that *a scenario does not describe its domain of applicability*. We have already stated that a scenario consists of an "if" part and a "then" part, which should be a statement that something is likely to happen. But there is no attempt being made to say how probable the "if" part is; such statements must be found by other, maybe more specific, theories. Therefore, if the hypotheses of a scenario do not apply, nothing is falsified and there is no contradiction, but no prediction is being made. Finally it should be stressed that while scenarios intend to describe roads to turbulence, no claim is made that this is the only way to find turbulence. Turbulence also occurs elsewhere, e.g., in the Niagara falls.

Let us recapitulate the main advantages and handicaps of the procedure.

(1) The turbulence described in the scenarios which have been found so far is a simple form of temporal aperiodicity, whose appearance is well under control. It has not been possible, so far, to find scenarios which lead to the rich spatiotemporal structure of fully developed turbulence, but nothing excludes in principle finding such scenarios.

(2) The theory is completely general, but it cannot describe its domain of applicability.

(3) The main field of study for scenarios is deterministic evolution equations, leading to stochastic behavior, whose occurrence *does not need any external noise source. Any external noise should be thought of as an additional complication.*[7]

The description of scenarios will be uniform, so that differences and similarities may appear more clearly. After a *mathematical description*, the scenario will be described in more simple-minded terms, followed by *interpretation, experimental evidence*, and a short description of the *influence of external noise*. Since there seems to be a general interest in such external noise, a final section will be devoted to a summary of the known results for the various scenarios. Table II at the end will summarize the results.

IV. THE RUELLE-TAKENS-NEWHOUSE SCENARIO

A. Description

This scenario is the oldest one, if we disregard the Landau scenario (see below for a discussion of why this

[7]For other formulations of this point of view, see Lanford (1981), Ruelle (1980), or Lorenz (1963).

TABLE II. Summary of the three scenarios discussed in this paper.

SCENARIO	Ruelle-Takens-Newhouse	Feigenbaum	Pomeau-Manneville
Typical bifurcations	Hopf	Pitchfork	(inverse) Saddle-node
Bifurcation diagram (s = stable, u = unstable).			
Eigenvalues of linearization in complex plane as μ is varied			
Main phenomenon	After 3 bifurcations strange attractor "probable"	Infinite cascade of period doublings with universal scaling of parameter values $\mu_i - \mu_\infty \sim (4.6692)^{-i}$	Intermittent transition to chaos. Laminar phase lasts $\sim(\mu-\mu_c)^{-1/2}$
Measurement	Power spectrum, correlation	Power spectrum subharmonics ~13.5 db below preceding level	Real-time measurements
Small noise	no influence	high periods disappear (noise level must go down by 6.62 to see one more period doubling)	time of laminarity scales as $(\mu-\mu_c)^{-1/2} T(\sigma/(\mu-\mu_c)^{3/4})$ for noise of standard deviation σ

is an inadequate scenario) (Ruelle and Takens, 1971).

In abstract mathematical terms, the situation is as follows.

Theorem (Newhouse, Ruelle, Takens, 1978).[8] *Let v be a constant vector field on the torus $T^n = R^n/Z^n$. If $n \geq 3$, every C^2 neighborhood of v contains a vector field v' with a strange Axiom A attractor. If $n \geq 4$, we may take C^∞ instead of C^2.*

For the definition of Axiom A vector fields, see Smale (1967).

B. Assumptions

It is now easy to describe an "if" for a scenario which implies the conditions of the theorem and hence its conclusion.

Assume a system $\dot{\mathbf{x}} = F_\mu(\mathbf{x})$ has a steady-state solution $\mathbf{x}_\mu$ for $\mu < \mu_c$. *Assume* further that this steady-state solution loses its stability through a *Hopf bifurcation* (Ruelle and Takens, 1971) (i.e., a pair of complex eigenvalues of

$$A_{ij} = \left.\frac{\partial F_\mu^{(i)}}{\partial x_j}\right|_{x = x_\mu}$$

crosses the imaginary axis, or $\exp A_{ij}$ has eigenvalues crossing the unit circle). This means that the steady state (a constant flow or an equilibrium) becomes oscillatory; we may say that some mode has been destabilized. *Assume* that this happens three times in succession, and that the three newly created modes are essentially independent [see Ruelle and Takens (1971) for details]. Thus the "if" part of the scenario is as shown in Fig. 5. Under all these assumptions, the scenario of Ruelle-Takens asserts: *A strange attractor may occur.* Its occurrence is "likely" in the following sense.

C. Interpretation

In the space of all differential equations, some equations have strange attractors; others have none. Those which do form a set which contains a subset which is *open* in the C^2 topology. The closure of this open set contains the constant vector fields on the torus T^3.

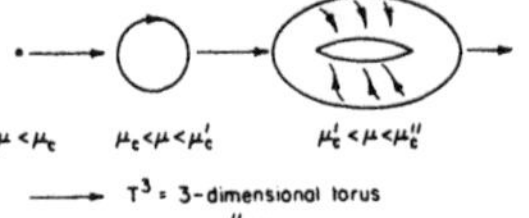

FIG. 5. Three critical values of the parameter μ_c, μ_c', μ_c'', and the associated motion in phase space.

[8]Ruelle and Takens's original work (1971) needed four dimensions. This was reduced to three by using an idea of Plykin

If a property of differential equations holds in an open set, then if we vary the coefficients of the differential equations sufficiently little, the property continues to hold. Thus the strangeness of the attractor is stable under small perturbations of the dynamical system; in other words, it is not exceptional. We can compare this with the *Landau scenario* (Landau and Lifshitz, 1959, III, Sec. 103), which assumes that the flow on the three-torus (and in fact on all n-tori which appear after further bifurcations) is the constant velocity flow. This is a much more stringent requirement than the one of the Ruelle-Takens scenario. While the latter is fulfilled on an open set of vector fields, the former does not hold on any open set of vector fields and is not even generic, i.e., it does not hold on any countable intersection of dense open sets (called a residual set). But genericity is perhaps a minimal way of saying that something is likely, and thus the Landau scenario is not likely. (In particular, if two properties are generic, they hold simultaneously on a residual set, and residual sets are more or less the weakest possibility for this simultaneity property to hold.)

Returning to the Ruelle-Takens scenario, we add a word of caution. While it is true that the set of vector fields with strange attractor is open near the constant vector fields, this does not mean that this set is large in the measure theoretic sense. We can visualize the situation in the space of vector fields near the constant vector fields as in Fig. 6.

D. Experimental evidence and its measurement

In order to describe how the appearance of the scenario manifests itself in measurements and to show the measurable consequences of the presence of strange attractor, let us reformulate the scenario: *If a system undergoes three Hopf bifurcations, starting from a stationary solution, as a parameter is varied, then it is likely that the system possesses a strange attractor with sensitivity to initial conditions after the third bifurcation.*

The *power spectrum* of such a system will exhibit one, then two, and possibly three independent basic frequencies. When the third frequency is about to appear, simultaneously some broad-band noise will appear if there is a strange attractor. This we interpret as chaotic, turbulent evolution of the system. Experiments have been performed on the formation of Taylor vortices between rotating cylinders and the Rayleigh-Bénard convection (see Figs. 7 and 8; for a review, from which these figures are taken, see Swinney and Gollub, 1978). They can be interpreted in the sense of the Ruelle-Takens-Newhouse scenario. It should also be stressed that measurements of time correla-

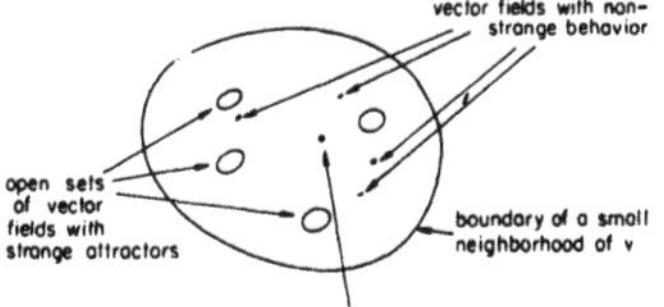

FIG. 6. Measure theoretic situation for the Ruelle-Takens-Newhouse scenario.

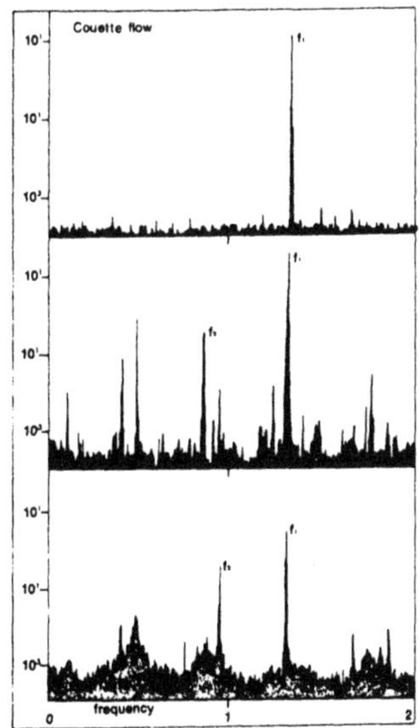

FIG. 7. Power spectrum of velocity in rotating cylinders driven at three different speeds.

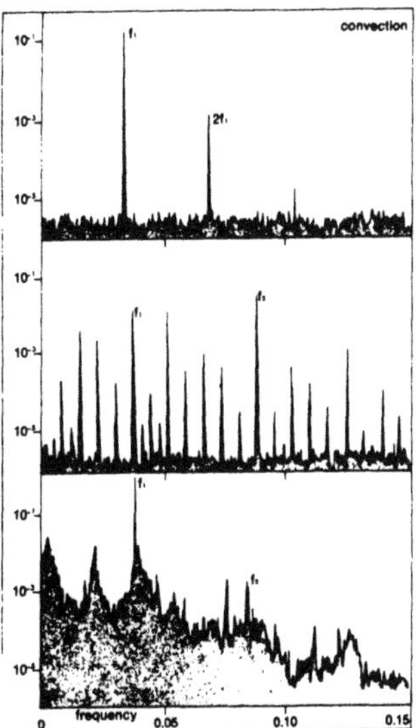

FIG. 8. Power spectrum of heat transport at different heating in Rayleigh-Bénard convection.

tions (measures of k-tuples $x_l, x_{l+1}, \ldots, x_{l+k-1}$ as a function of l) are very useful indicators about flows in general (Takens, 1980; Roux *et al.*, 1980), and allow one in some sense to reconstruct the dynamical system.

E. The influence of noise

The Ruelle-Takens scenario is *not* destroyed by the addition of small external noise to the evolution equations. This result, which is somewhat counterintuitive, will be explained in more detail in the final section. In effect, the chaos of the scenario is so strong that order cannot be accidentally established by small noise terms, much like a very attracting fixed point is locally not much altered by noise, and globally there is at most a small probability to change stochastically from one basin of attraction to another (Kifer, 1974; Ventsel and Freidlin, 1970).

V. THE FEIGENBAUM SCENARIO

A. Description

We start with the description of a general framework. Assume we are in the presence of a one-parameter family of vector fields v_μ in $\mathbb{R}^m$ (we conjecture that the results extend to the case $m=\infty$), where μ is the parameter. Assume each v_μ has a periodic orbit, and assume there is a piece of hyperplane of dimension $m-1$, transversal to this periodic orbit, for which the Poincaré map P_μ can be defined (Fig. 9). The scenario will make predictions about these Poincaré maps and hence for the corresponding flow.[9]

Now fix m. Two objects, Φ_m and W_m, whose existence is asserted by a mathematical theory, will be of fundamental importance in describing the scenario, namely, there is a neighborhood D_m of $[0,1]\times\{0\}^{m-2}$ in $\mathbb{C}^{m-1}$ and on this neighborhood an analytic function Φ_m: $\mathbb{C}^{m-1}\to\mathbb{C}^{m-1}$ whose restriction to $\mathbb{R}^{m-1}$ is real. In the space of analytic functions on D_m (with, for example, the sup norm) there is an open disk W_m of codimension one, containing Φ_m. The existence of the two objects Φ_m and W_m is assured through an extension of Feigenbaum's original theory (Feigenbaum, 1978, 1979a) ($m=2$, one-dimensional maps) by Collet, Eckmann, and Lanford (1980) and Collet, Eckmann, and Koch (1981).

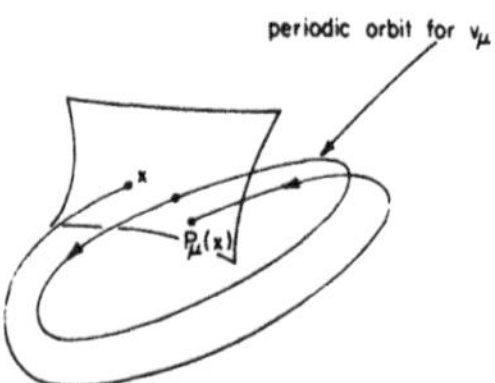

FIG. 9. Phase portrait illustrating Poincaré section of v_μ.

[9]These ideas were first explained in Eckmann (1980). See also Collet and Eckmann (1980).

B. Assumptions

The scenario assumes that P_μ extends to an analytic function on D_m and that the curve $\mu\to P_\mu$ transversally crosses W_m near Φ_m.

Under these hypotheses one can assert

(1) The family P_μ has an infinite sequence of period doubling bifurcations of stable periodic orbits at parameter values μ_1 (period 1-2), μ_2 (period 2-4), ..., μ_{j+1} (period 2^j-2^{j+1}) (the sequence might only start at some high j).

(2) $\lim_{j\to\infty}\mu_j=\mu_\infty$ exists.

(3) At $\mu=\mu_\infty$, P_μ has an aperiodic attractor (a stable periodic orbit of "period 2^∞"). The action on the attractor is ergodic, but not mixing (in particular, there is no sensitive dependence on initial conditions).

(4) There is a universal number $\delta=4.66920\ldots$ such that

$$\lim_{j\to\infty}\frac{1}{j}\log|\mu_j-\mu_\infty|=-\log\delta\,.$$

One even has

$$|\mu_j-\mu_\infty|\sim\text{const}\,\delta^{-j}\quad\text{as } j\to\infty\,.$$

C. Remarks

(1) The bifurcations of the orbit structure of P_μ are *pitchfork bifurcations*, i.e., a stable fixed point loses its stability and gives rise to a stable periodic orbit as the parameter is changed. This corresponds to a crossing of one eigenvalue of the tangent map DP_μ through -1 (Fig. 10).

(2) One can show that any suitable property (such as bifurcation) which can be described by a coordinate independent codimension 1 surface in the space of functions on D_m will double its spatial structure in phase space in the same way as the periodic orbits, i.e., it will split in 2, 4, 8, ... pieces. Typically, such surfaces are given by a single functional relation, e.g., fixing the value of a derivative at a fixed point.

(3) A similar scenario exists for area-preserving (=Hamiltonian) maps of the plane to itself, but with $8.721\ldots$ as the universal constant instead of $\delta=4.66920\ldots$ (Collet, Eckmann, and Koch, 1980; Greene *et al.*, 1981).

(4) The scenario can be somewhat extended under the assumption of very strong friction. This has the effect of making the situation very similar to the case of maps of the interval to itself. Then one can show that if the system has transitions from periods 1 to 2 and 2 to 4 at values μ_1 and μ_2, respectively, a stable period 3 with a large basin of attraction near

$$\mu=\frac{(\delta\mu_2-\mu_1)}{(\delta-1)}-\frac{\delta(\mu_1-\mu_2)0.803}{(\delta-1)}$$

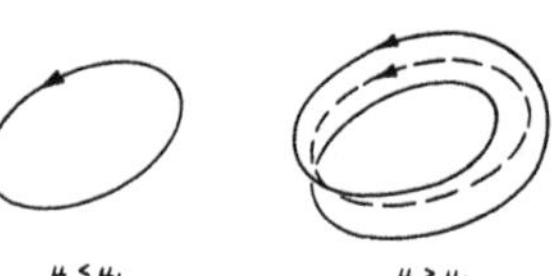

FIG. 10. Example of a pitchfork bifurcation for a flow.

can be expected.

(5) After the cascade of period doublings, one expects beyond the accumulation point μ_∞ an *inverse* cascade of *noisy* periods.

The physical interpretation of the Feigenbaum scenario can be brought to a more appealing form than for the Ruelle-Takens scenario, because the statement deals with *all* curves which cross W_∞ transversally. On the other hand, it is only a statement about a very small parameter range, and point (B.4) describes nothing more than a critical index.

D. Interpretation

In an experiment, if one observes subharmonic bifurcations at μ_1, μ_2, then, according to the scenario, it is very probable for a further bifurcation to occur near $\mu_3 = \mu_2 - (\mu_1 - \mu_2)/\delta$, where $\delta = 4.66920\ldots$. In addition, if one has seen three bifurcations, a fourth bifurcation becomes more probable than a third after only two, etc. At the accumulation point, one will observe aperiodic behavior, but no broad-band spectrum.

E. Experimental evidence

This scenario is extremely well tested on numerical and physical grounds. The period doublings have by now been observed in most current low dimensional dynamical systems (Hénon map, Lorenz equations, forced oscillator with friction, etc). Experiments with liquid helium have confirmed the predictions.

F. Measurement

In all numerical examples, the bifurcations are found by a direct analysis of the orbits and of their stability. The experiments on liquid helium produce power spectra. Feigenbaum has given a nice prediction of how the power spectrum evolves as a function of the parameter (see Fig. 11). At each successive bifurcation a new frequency is born. The mean of the squares of the new amplitudes is then expected to rise until it stops about 13.5 db below the level of its predecessors (Feigenbaum, 1979b, 1980; Nauenberg and Rudnick, 1981; Collet, Eckmann, and Thomas, 1981).

The measured power spectrum of Libchaber and Maurer (1980)[10] for the heat transport by convection of liquid helium, heated from below, shows a sequence of period doubling bifurcations. The power goes down by about 10 db per doubling, but the apparent discrepancy with the prediction of the scenario may be ascribed to not yet having reached the asymptotic regime (Fig. 12). The prediction (5) above has recently been seen by Libchaber (1981) [Fig. 12(c)].

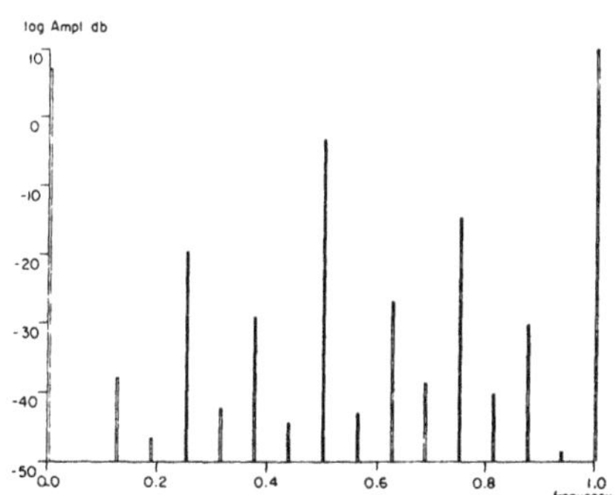

FIG. 11. Numerical prediction of the shape of the power spectrum.

G. The influence of noise (Crutchfield *et al.*, 1980)

Again we postpone a detailed description of the influence of noise. Since the structure of the periodic orbit must acquire finer and finer length scales as the parameter approaches μ_∞, it is clear that even very small noise will eventually play a role. There exist estimates on the relation between the noise level and the maximal period which can be observed. This is of course related to the power spectrum described above.

VI. THE POMEAU-MANNEVILLE SCENARIO

A. Description

This scenario (Pomeau and Manneville, 1980; Manneville and Pomeau, 1980) has been—correctly—termed *transition to turbulence through intermittency*. Its mathematical status is somewhat less satisfactory than that of the two other scenarios presented here. This is because the parameter region the scenario intends to describe contains an infinity of (very long) stable periods, and because there is no mention as to when the "turbulent" regime is reached or what the exact nature of this turbulence is. We nevertheless examine it here because of its esthetic and conceptual beauty.

While the two other scenarios have been associated with Hopf bifurcations (Ruelle-Takens) and pitchfork bifurcations (Feigenbaum), this one is associated with a "saddle node bifurcation," i.e., the collision of a stable and an unstable fixed point which then both disappear (into complex fixed points).

The general idea is best explained for the simple example of a one-parameter family of iterated maps on the unit interval, $x_{n+1} = f_\mu(x_n)$. We take $f_\mu(x) = 1 - \mu x^2$, which for $\mu \in [0,2]$ maps $[-1,1]$ into itself. The function $f_\mu^3 = f_\mu \circ f_\mu \circ f_\mu$ can be shown to have a saddle node for $\mu = \frac{7}{4}$. For $\mu > 1.75$, f_μ^3 has a stable periodic orbit of period three, and an unstable one nearby. The two collide at $\mu = 1.75$, and both have then eigenvalue 1. See Fig. 13.

For μ slightly below 1.75, the local picture near $x=0$ is shown in Fig. 14. It can be shown that if $\mu - 1.75 = \mathcal{O}(\varepsilon)$ then a typical orbit will need $\mathcal{O}(\varepsilon^{-1/2})$ iterations to cross a fixed small x interval around $x \sim 0$. As long as the orbit is in this small interval, an observer will have the impression of seeing a periodic orbit of period three. Once one has left the small interval, the iterations of the map will look rather like those of a chaotic map [a consequence of a

[10]See Collet and Eckmann (1980), pp. 39 and 42 for a list of tests. In particular, beautiful experiments on liquid helium were performed by Libchaber and Maurer (1980).

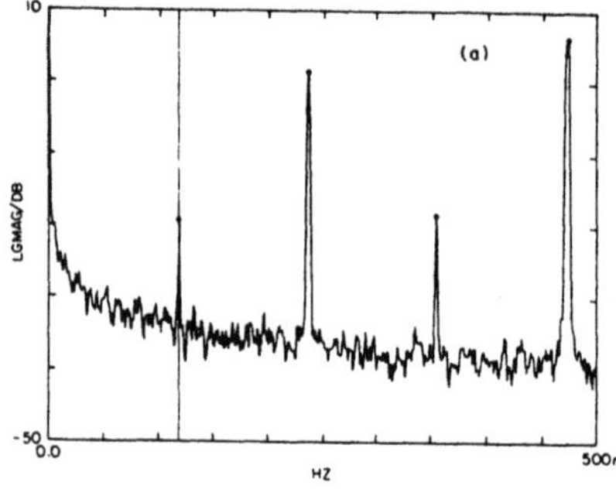

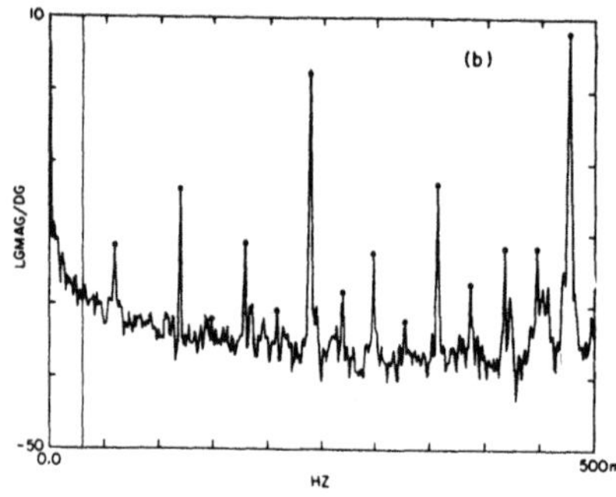

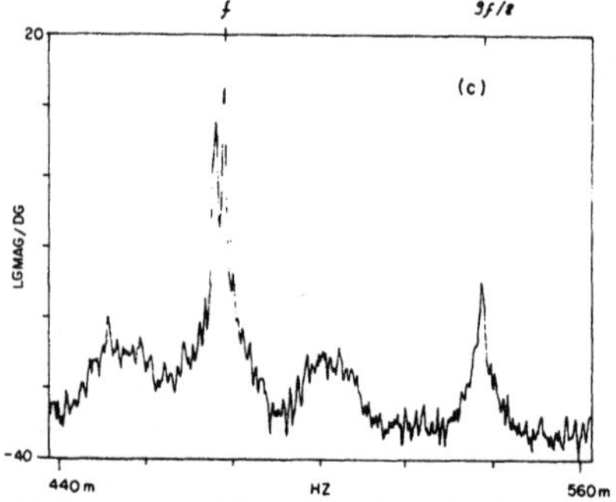

FIG. 12. Power spectra for two values of heating. (c) Observation of the noisy period 8.

result of Misiurewicz; see Collet and Eckmann (1980), Theorem 5.2.2]. Thus this map can be called intermittently turbulent (see Fig. 15).

The problem with this argument comes in the splitting into two regions. It is true that the iterated map may have sensitivity to initial conditions for $x \in$ small intervals around contact points. But this destabilizing effect may be lost whenever one passes near the contact point. In fact, we conjecture that this will happen for an infinity of parameter values near to, and just below $\mu = 1.75$. For these parameter values, one will have (very long) stable periods, but no chaos. On the other hand, we also conjecture that a modification of the proof of Jakobson (1980) would show that truly aperiodic behavior with sensitivity to initial conditions occurs for a set of parameter values of positive Lebesque measure near 1.75.

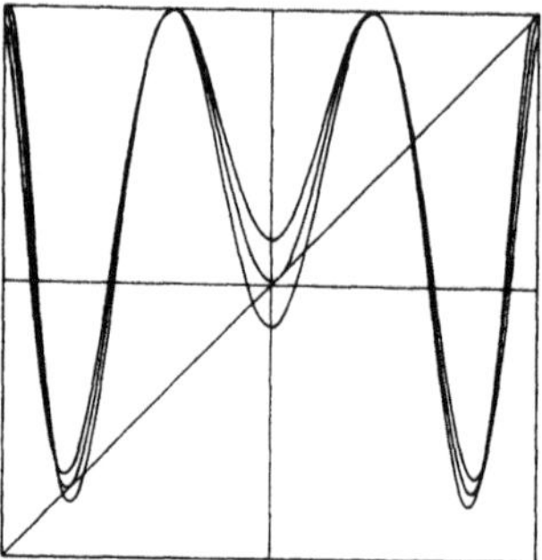

FIG. 13. Graph of f_μ^3 for three values of μ.

B. Assumptions

We can now formulate a reasonable version of this scenario for general dynamical systems.

Assume a one-parameter family of dynamical systems has Poincare maps close to a one-parameter family of maps of the interval, and that these maps have a stable and unstable fixed point which collide as the parameter is varied. Then, as the parameter is varied further to μ from the critical parameter value μ_c, one will see intermittently turbulent behavior of random duration, with laminar phases of mean duration $\sim(|\mu-\mu_c|^{-1/2})$ in between.

C. Interpretation

The difficulty with this scenario is that it does not have any clear-cut precursors, because the unstable fixed point which is going to collide with the stable fixed point (respectively periodic orbit) may not be visible. One can think of two ways out of this problem. The first would be that increasingly long transients can be observed before the two fixed points (periodic orbits) collide. The second kind of precursor is a cascade of *inverse* pitchfork bifurcations, and, at the "end" of this, the intermittent transition to turbulence (Collet and Eckmann, 1980).

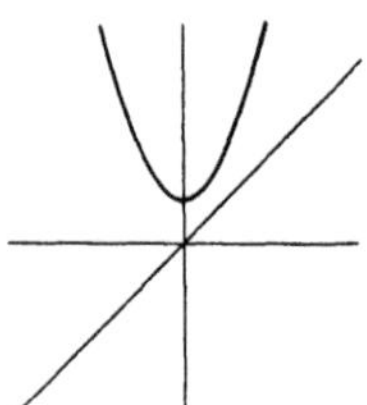

FIG. 14. Graph of f_μ^3 in the vicinity of the origin.

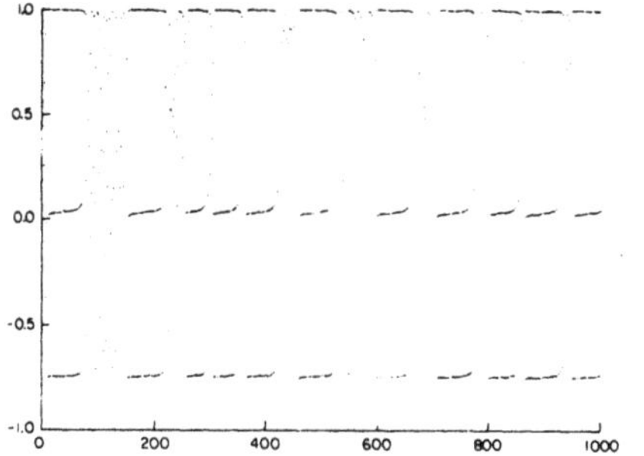

FIG. 15. Graph describing $f_\mu^n(0)$ as a function of n in the neighborhood of $\mu \simeq 1.75$, and indicating the existence of an intermittent turbulence.

D. Experimental evidence

Pomeau and Manneville based their work on observations for the Lorenz system. Intermittent transitions to turbulence can be seen in many physical experiments. The only ones which seem to agree with the scenario described above are those of Maurer and Libchaber (1980), Bergé *et al.* (1980), and Pomeau *et al.* (1981). They exhibit intermittent transition to aperiodic behavior, but more work needs to be done to show that these are really instances of the scenario described above.

E. Measurement

We have already discussed the difficulties of detecting the scenario. We add here only that one should not look at power spectra in this case, but rather at real-time measurements.

F. The influence of noise

As the parameter value at which the two fixed points collide is a critical point, the influence of noise is relevant. This has been first exhibited by Mayer-Kress and Haken (1981). A more detailed analysis of the tunneling through the region of contact shows that certain scaling relations hold between the noise level and the distance from the critical parameter value (Eckmann *et al.*, 1981).

VII. THE INFLUENCE OF EXTERNAL NOISE ON SCENARIOS

It seems to be a widespread opinion that external noise is relevant

(a) for the appearance of (even weak) turbulence and chaotic behavior and

(b) for the form, amplitude, and spectrum of the turbulence, once it has appeared.

The foregoing discussion of attractors and of the scenarios should have shown that this opinion is wrong for case (a)—*ergodic behavior is possible, and quite common, for dynamical equations without external noise*. In this section, we shall examine case (b) and see that the nature of chaotic systems may be totally insensitive to small external noise. The systems most sensitive to noise seem to be deterministic systems near transition (bifurcation) points.

This insensitivity to noise is surprising and at first sight counterintuitive. It has been discovered by Kifer (1974), whose work is an extension of a paper by Ventsel and Freidlin (1970). Kifer's theorem states that for a dynamical system with an Axiom A attractor, which has an invariant measure ν, the following is true: Given any reasonable small noise, going to zero with σ, consider the corresponding invariant measure ν_σ. [Under suitable assumptions, the measures ν and ν_σ are given, for discrete mappings f as follows:

$$\int d\nu(x)h(x) = \lim_{n\to\infty} \frac{1}{n} \sum_{k=0}^{n-1} h(f^k(y))$$

for Lebesque-almost every y, and every continuous h.] The density of the measure ν_σ, given a noise with transition probability $\rho_\sigma(x, y)$ [and an iteration scheme $x_{n+1} = f(x_n) + \xi_\sigma(x_n)$, where ξ_σ is a random variable with density $\rho_\sigma(x_n, .)$] satisfies

$$\nu_\sigma(x) = \int \rho_\sigma(f(y), x - f(y))\nu_\sigma(y)dy .$$

Theorem (Kifer, 1974). *ν_σ converges weakly to ν as $\sigma \to 0$ (i.e., all expectation values of bounded observables converge).*

This tells us, then, that if the noise is sufficiently small, the corresponding probability distributions (ν and ν_σ) are as close to each other in the weak-* topology as we wish. This result is astonishing, because any nontrivial (strange) Axiom A attractor is full of hyperbolic points, and one could think that a small random deviation might get amplified away from any deterministic path. But the celebrated "shadowing lemma" leads to a different conclusion. With high probability, the sample paths of the problem with external noise follow *some* orbit of the deterministic problem arbitrarily closely. This bounds ν_σ by ν (up to small errors). On the other hand, the central limit theorem shows that ν is bounded by ν_σ: For every deterministic orbit, there are many sample paths which follow it rather closely.

We next discuss the influence of noise on the Feigenbaum scenario. It is known (Collet, Eckmann, and Lanford, 1980; Collet *et al.*, 1981; Feigenbaum, 1978, 1979a) that the smallest scales of the period 2^n are of approximate size $O(\lambda^{2n})$, with $\lambda = .3995..$ (another universal constant). Thus it is obvious that even small noise can wipe out the finest structures of the orbit, and hence the orbit itself, provided n is sufficiently large. The question then is how large the noise may be if we want to see a period 2^n. Crutchfield *et al.*, (1980) give a heuristic argument with the following conclusion. Denote, for each k, by ξ_k the independent random variables with mean zero and density ρ. Let f_μ be a one-parameter family of maps of the interval, with μ so chosen that the accumulation of period doublings is at $\mu = \mu_\infty = 0$. Consider the stochastic iteration equation

$$x_{k+1} = f_\mu(x_k) + \xi_k ,$$

and define $\nu_{\mu,\rho}$, the corresponding invariant density. Then one has the approximate identity

$$\lambda \nu_{\mu\delta,\kappa\cdot\rho\cdot\kappa}(\lambda x) \sim \nu_{\mu,\rho}(x),$$

with $\lambda = 0.39953\ldots$, $\delta = 4.66920\ldots$, $\kappa = 6.619\ldots$. In words, in order to see twice the period, the noise must have a variance about κ times smaller. [Note that this is very close to the ratio of the amplitudes between a frequency and its subharmonic, which has been estimated by Feigenbaum (1979b) to be about 6.60... .]

In the Pomeau-Manneville scenario, the influence of noise can be modeled as follows (Eckmann *et al.*, 1981). In the "laminar" region, i.e., when the iteration steps are small, one can model the iteration scheme

$$x_{n+1} = x_n + x_n^2 + \varepsilon + \sigma \xi_n,$$

with ξ_n independent stochastic variables, by the stochastic differential equation

$$dx = (x^2 + \varepsilon')dt + \sigma' d\omega,$$

where ω is white noise, and $\varepsilon' = \varepsilon$, $\sigma' = \sigma \operatorname{Exp}(\xi^2)^{1/2}$. The estimated time to cross the laminar region is then easily seen to be a stopping time for the differential equation, and an analysis of its solution shows that the fraction of time spent in the laminar region scales approximately as $\varepsilon^{-1/2} T(\sigma'/\varepsilon^{3/4})$, where T is a universal function.

See Table II for a summary of these three scenarios.

ACKNOWLEDGMENT

I thank L. Thomas and A. Libchaber for many helpful comments.

REFERENCES

Arnold, V., 1976, ***Méthodes mathêmaques de la mécanique classique*** (Editions Mir, Moscow).

Bergé, P., M. Dubois, P. Manneville, and Y. Pomeau, 1980, J. Phys. (Paris) Lett. **41**, L341.

Bowen, R., 1975, ***Equilibrium states and the ergodic theory of Anosov diffeomorphisms***, Lecture Notes in Mathematics (Springer, Berlin), Vol. 470.

Brayton, R. K., and J. K. Moser, 1964, Q. Appl. Math. **22**, 1, 81.

Collet, P., and J.-P. Eckmann, 1980, ***Iterated maps on the interval as dynamical systems***, Progress in Physics (Birkhäuser, Basel), Vol. 1.

Collet, P., J.-P. Eckmann, and H. Koch, 1980, Physica D, in press.

Collet, P., J.-P. Eckmann, and H. Koch, 1981, J. Stat. Phys. **25**, 1.

Collet, P., J.-P. Eckmann, and O. Lanford, 1980, Commun. Math. Phys. **76**, 211.

Collet, P., J.-P. Eckmann, and L. Thomas, 1981, Commun. Math. Phys., in press.

Crutchfield, J., M. Nauenberg, and J. Rudnick, 1980, Phys. Rev. Lett. **46**, 933.

Eckmann, J.-P., 1980, in ***Bifurcation phenomena in Mathematical Physics and Related Topics***, edited by C. Bardos and D. Bessis (Reidel, Dordrecht), p. 115.

Eckmann, J.-P., L. Thomas, and P. Wittwer, 1981, J. Phys. A, in press.

Feigenbaum, M. J., 1978, J. Stat. Phys. **19**, 25.

Feigenbaum, M. J., 1979a, J. Stat. Phys. **21**, 669.

Feigenbaum, M. J., 1979b, Phys. Lett. A **74**, 375.

Feigenbaum, M. J., 1980, Commun. Math. Phys. **77**, 65.

Foias, C., and R. Temam, 1979, J. Math. Pures Appl. **58**, 339.

Greene, J. M., R. S. MacKay, F. Vivaldi, and M. Feigenbaum, 1981, Physica D, in press.

Hagedorn, R., 1957, CERN Report No. 57-1.

Jakobson, M., in ***Proceedings of the International Conference on Dynamical Systems***, Northeastern University, 1980 (Commun. Math. Phys., in press).

Kifer, J. I., 1974, Math. USSR Izvestija **8**, 1083.

Landau, L., and E. Lifshitz, 1959, ***Fluid Mechanics*** (Pergamon, Oxford).

Lanford, O. E., 1978, ***Qualitative and Statistical Theory of Dissipative Systems***, CIME Course, 1976 (Liguori Editore, Napoli).

Lanford, O. E., 1981, ***Strange Attractors and Turbulence***, Topics in Applied Physics (Springer, Berlin), Vol. 45, p. 7.

Levi, M., Mem. Am. Math. Soc., in press.

Libchaber, A., 1981, in Nato Study, Nonlinear Phenomena and Phase Transitions, edited by T. Riste, in press.

Libchaber, A., and J. Maurer, 1980, J. Phys. (Paris), Colloq. **41**, C3, 51.

Lorenz, E. N., 1963, J. Atmos. Sci. **20**, 130.

Manneville, P., and Y. Pomeau, 1980, Physica D **1**, 219.

Maurer, J., and A. Libchaber, 1980, J. Phys. (Paris) Lett. **41**, L515.

Mayer-Kress, G., and H. Haken, 1981, J. Phys. A **82**, 151.

Misiurewicz, M., 1980, Ann. N.Y. Acad. Sci., Vol. 357.

Monin, A. S., and A. M. Yaglom, 1975, ***Statistical Fluid Mechanics*** (MIT, Cambridge, Mass.), Vol. 2.

Nauenberg, M., and J. Rudnick, 1981, Preprint, University of California, Santa Cruz, UCSC 80/137.

Newhouse, S., 1980a, Ann. N.Y. Acad. Sci., Vol. 357.

Newhouse, S., 1980b, ***Lectures on Dynamical Systems***, Progress in Mathematics (Birkhäuser, Boston), Vol. 8.

Newhouse, S., D. Ruelle, and F. Takens, 1978, Commun. Math. Phys. **64**, 35.

Nicolis, G., and Y. Prigogine, 1977, ***Self-Organization in Nonequilibrium Systems*** (Wiley, New York).

Plykin, R., 1974, Math. USSR Sbornik **23**, 233.

Pomeau, Y., and P. Manneville, 1980, Commun. Math. Phys. **77**, 189.

Pomeau, Y., J. C. Roux, A. Rossi, S. Bachelart, and C. Vidal, 1981, J. Phys. Lett. **42**, 271.

Roux, J. C., A. Rossi, S. Bachelart, and C. Vidal, 1980, Phys. Lett. A **77**, 391.

Ruelle, D., 1980, La Recherche **108**, 132.

Ruelle, D., and F. Takens, 1971, Commun. Math. Phys. **20**, 167.

Shub, M., 1980, ***Stabilité globale des systèmes dynamiques*** Astérisque, Vol. 56 (Société Math. Française, Paris).

Siegel, C. L., and J. K. Moser, 1971, ***Lectures on Celestial Mechanics*** (Springer, Berlin).

Smale, S., 1967, Bull. Am. Math. Soc. **73**, 747.

Swinney, H. L., and H. P. Gollub, 1978, Phys. Today **31** (8), 41.

Takens, F., 1980, Preprint No. 7907, Dept. of Mathematics University, Groningen.

Thom, R., 1972, ***Stabilité Structurelle et Morphogenèse*** (Benjamin, Reading).

Ventsel, A. D., and M. I. Freidlin, 1970, Russ. Math. Surveys **25**, 1.

FURTHER BIBLIOGRAPHY

Here is a list of references which can be useful for further study of the subject matter of this paper. Although they are not directly relevant to our principal subject, they contain a wealth of further ideas. The references marked with an asterisk (*) contain important bibliographical sections.

Catastrophe theory

*Poston, T., and I. Stewart, 1978, *Catastrophe theory and its applications* (Pitman, London).
Thom, R., 1977, *Stabilité structurelle et morphogenèse* (Inter Editions, Paris).
Zeeman, E. C., 1977, *Catastrophe theory: selected papers, 1972–1977* (Addison-Wesley, London).

Differentiable dynamical systems

Abraham, R. H., 1978, *Foundations of mechanics* (Benjamin, Reading).
Abraham, R. H., and C. D. Shaw, 1980, University of California, Santa Cruz.
*Helleman, R. H. G., 1980, in *Fundamental problems in statistical mechanics*, edited by E. G. D. Cohen (North-Holland, Amsterdam), p. 165.
Shaw, R., 1978, Z. Naturforschung, in press.
Smale, S., 1967, Bull. Am. Math. Soc. **73**, 747.

Maps of the interval

Bowen, R., 1979, Commun. Math. Phys. **69**, 1.
Guckenheimer, J., 1979, Commun. Math. Phys. **70**, 133.
Hofbauer, F., 1979, Preprint, Inst. f. Mathematik, University, Wien.
Jakobson, M. V., Commun. Math. Phys., in press.
Jonker, L., 1979, Proc. London Math. Soc. **39**, 428.
Lasota, A., and J. A. Yorke, 1973, Am. Math. Soc. Transl. **183**, 481.
Li, T., and J. A. Yorke, 1972, Amer. Math. Monthly **82**, 985.
Lorenz, E. N., 1979, *On the Prevalence of Aperiodicity in Simple Systems*, Lecture Notes in Mathematics (Springer, Berlin), Vol. 755.
Matsumoto, S., 1980.
*May, R. M., 1976, Nature **261**, 459.
Metropolis, M., M. L. Stein, and P. R. Stein, 1973, J. Comb. Theory **15**, 25.
Milnor, J., and P. Thurston, 1977, Preprint, Dept. of Mathematics, Princeton University, Princeton.
Misiurewicz, M., Publ. Math. IHES, Vol. 53, in press.
*Nitecki, Z., to appear, *Topological Dynamics on the Interval. Ergodic Theory and Dynamical Systems, II*, Progress in Mathematics (Birkhäuser, Boston).
Ruelle, D., 1977, Commun. Math. Phys. **55**, 47.
Šarkovskii, A., 1964, Ukr. Mat. Zh. **16**, 61.
Singer, D., 1978, SIAM J. Appl. Math. **35**, 260.
Stefan, P., 1977, Commun. Math. Phys. **54**, 237.
Ulam, S. M., and J. v. Neumann, 1947, Bull. Am. Math. Soc. **53**, 1120.

Period-doubling bifurcations

Benettin, G., C. Cercignani, L. Galgani, and A. Giorgilli, 1980, Lett. Nuovo Cimento **28**, 1.
Benettin, G., L. Galgani, and A. Giorgilli, 1980, Lett. Nuovo Cimento **29**, 163.
Bountis, T., Physica D, in press.
Campanino, M., H. Epstein, and D. Ruelle, 1980, Topology, in press; Commun. Math. Phys. **79**, 261.
Coullet, P., and J. Tresser, 1978, C. R. Acad. Sci. **287**, 577; and J. Phys. (Paris) C **5**, 25.
Derrida, B., A. Gervois, and Y. Pomeau, 1979, J. Phys. (Paris) A **12**, 269.
Derrida, B., and Y. Pomeau, Phys. Lett. **80A**, 217.
Lanford, O. E., III, 1980, in *Mathematical Problems in Theoretical Physics*, Lecture Notes in Physics (Springer, Berlin), Vol. 116, p. 103.
McLaughlin, J. B., 1981, J. Stat. Phys. **24**, 375.

Intermittency

Mayer-Kress, G., and H. Haken, 1981, Phys. Lett. A **82**, 151.

Influence of noise

Crutchfield, J. P., J. D. Farmer, and B. A. Huberman, 1980, Preprint, University of California, Santa Cruz.
Shraiman, B., C. E. Wayne, and P. C. Martin, 1981, Phys. Rev. Lett. **46**, 935.

Miscellaneous

Feit, S., 1978, Commun. Math. Phys. **61**, 249.
Francheschini, V., 1980, J. Stat. Phys. **22**, 397.
Francheschini, V., and C. Tebaldi, 1979, J. Stat. Phys. **21**, 707.
*Gollub, J. P., and S. V. Benson, 1980, J. Fluid Mech. **100**, 449.
Maurer, J., and A. Libchaber, 1979, J. Phys. (Paris) Lett. **40**, L419.
**Nonlinear Dynamics*, edited by R. Helleman Sci., Vol. 357.
Ruelle, D., 1980, Ann. N.Y. Acad. Sci., Vol. 357, p. 1.
Vidal, C., and J. C. Roux, 1980, La Recherche.
Vidal, C., J. C. Roux, S. Bachelart, and A. Rossi, 1980, Ann. N.Y. Acad. Sci., Vol. 357, p. 377.

Editor's note. We also recommend review articles by Hu (1982), Ott (1981) and Crutchfield, Farmer and Huberman (1982).

Part 2

Experiments

Nonlinear Phenomena at Phase Transitions and Instabilities (ed T Riste)
259–86 (1982)

A RAYLEIGH BÉNARD EXPERIMENT: HELIUM IN A SMALL BOX

A. Libchaber and J. Maurer

Ecole Normale Supérieure, Groupe de Physique des Solides

24 rue Lhomond, 75231 Paris Cedex 05, France

This is a limited excursion in the field of hydrodynamical instabilities, in itself an infinite domain of research. It is first restricted to a Rayleigh Benard experiment, and we will study the case of a small Prandtl number fluid ($0.4 < P < 1$). To simplify the problem some more we shall restrict ourselves to the geometry of a small rectangular box with two or three convective rolls present. This somewhat artificial case allows us to truncate the degrees of freedom of the system and thus to define some simple bifurcations to turbulence.

Thermal convection provides a simple experimental example of a non linear physical system. We are dealing with a fluid confined in a box as compared to experiments with unconfined flows where the ordered structure moves in space and time. The control parameter is the temperature difference between top and bottom plates which can be highly stabilized. Local thermal bolometers allow an easy measurement of the temperature profile and temperature oscillations.

A low Prandtl number fluid has the following simple property. Above convection and up to the onset of the first time dependent instability, the so-called oscillatory instability, the convective structure has a two dimensional shape. For Prandtl numbers larger than one the convective structure acquires a three dimensional character before any bifurcation to a time dependent state (cross rolls, zig zag, skewed varicose).

Let us finally try to explain why our experiment is performed in a small confined geometry. In a large cell, G. Ahlers[1] showed that immediately above convection, a low frequency noise is present. We believe that in large cells with many convective rolls the wavenumber selection[2] leads to a perpetual motion of the convective structure which never achieves a state with a stable pattern. In our first experiment[3] with a cylindrical cell of aspect ratio $\Gamma = R/d = 6$ (R : radius, d : height) two distinct time dependent phenomena were observed and are shown on figure 1. As one increases the Rayleigh number above the critical Rayleigh number for convection (R_c), a low frequency noise starts at about twice the convection threshold ($R/R_c \simeq 2$). For $R/R_c \sim 3$ a distinct higher frequency mode appears, the oscillatory instability.

We are thus faced with two distinct phenomena. Using a two rolls convective cell we can freeze out the effect of the wavenumber selection and have a stable structure up to the onset of the oscillatory instability.

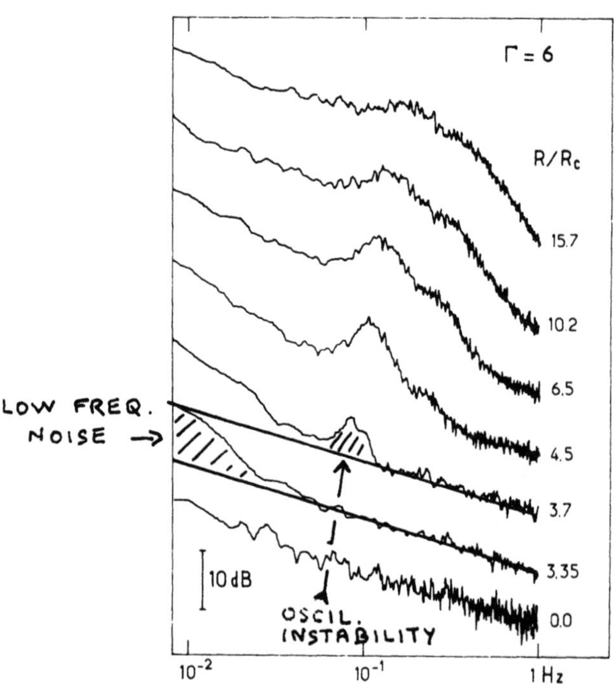

Fig. 1. Power spectra as a function of frequency. Both coordinates are in logarithmic scale. The curves have been translated along the vertical axis for clarity. The lower curve shows the 1/f character of the bolometer.

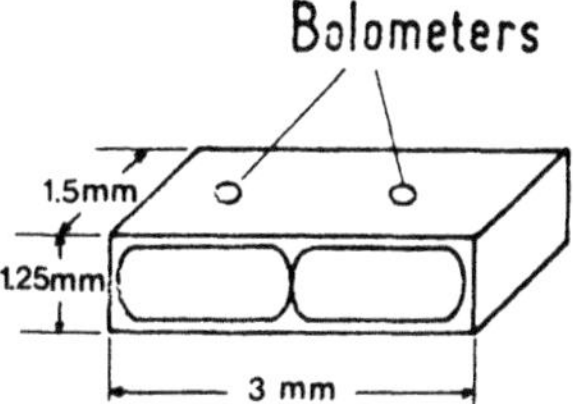

Fig. 2. Schematic of the experimental cell with the position of the local bolometers.

I. THE EXPERIMENT

This is just an outline of the experimental set-up and we refer to previous publications for a more detailed presentation. For our confined cell we use a rectangular geometry instead of a circular one to lift the degeneracy of the convective roll pattern. We thus get a non ambiguous pattern with rolls perpendicular to the largest lateral side, as shown by Stork and Müller[4]. A typical sample shape and dimension is presented on Fig. 2.

The fluid used is liquid Helium (^{4}He). We change the Prandtl number by varying the temperature in a range from 2.5 to 4.5°K and the pressure from 1 to 5 atmospheres. We refer to the NBS chart[5] for the values of the Prandtl number and all the relevant physical constants of the fluid. The Prandtl number can be easily changed from 0.48 to 1.

Given our typical cell height, $d \cong 1$ mm, the onset of convection corresponds to a ΔT in the millikelvin range. The thermal time constant for heat diffusion from bottom to top plate is around twenty seconds.

$$\tau \cong \frac{d^2}{\kappa} \cong 20 \text{ s} \qquad\qquad \kappa \sim 5\ 10^{-4} \text{ cm}^2/\text{sec}$$

An important experimental character of our set-up is that the thermal conductivity of the fluid is very close to the thermal conductivity of the lateral boundaries. In contrast the top and bottom plates have a thermal conductivity extremely high compared to the fluid. This is one of the advantages of low temperature.

Table of thermal conductivities

Bottom plate : high purity copper	∿ 100 Watt/cm °K
Top plate : sapphire crystal	∿ 10 Watt/cm °K
Helium	∿ 1 to 5 10^{-4} Watt/cm °K
Lateral side walls : araldite, nylon, teflon	∿ 1 to 5 10^{-4} Watt/cm °K

The overall set-up is shown on Fig. 3.

We use two bolometers to measure the local temperature as shown on Fig. 2. The bolometers are inserted in the top sapphire plate. They are cut and machined from an Allen-Bradley resistance (the

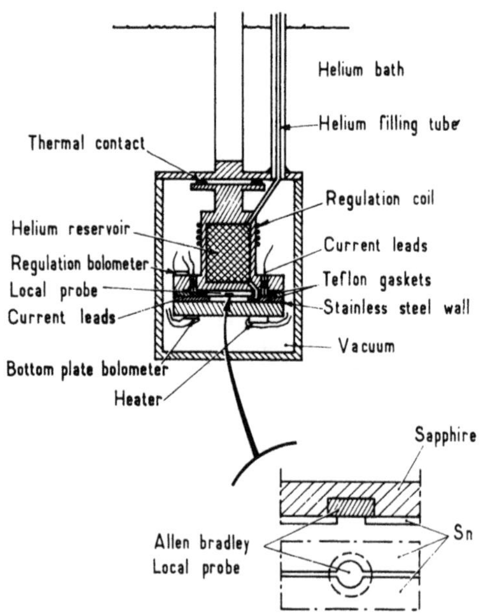

Fig. 3 The upper part is the detail of the experimental system. The lower part is an enlarged view of the local probe seen both in a transverse cut and from the bottom. Dimensions are not conserved.

formula for the variation of the resistance with temperature is

$$\text{Log } R + \frac{C}{\text{Log } R} = A + \frac{B}{T} .)$$

The effective dimension of a bolometer is less than 0.1 mm. The noise factor of the bolometers has a 1/f overall variation in the frequency range where the experiment is performed (refer to Fig. 1 for $R/R_c = 0.0$).

We use a balanced bridge at helium temperature and a lock-in detection to record the temperature signal from each bolometer.

The last important point concerning the experiment is the thermal regulation and the way the temperature constraint ΔT is applied. In our set-up the top plate only is highly regulated, within 10^{-6} K. The whole system is under very high cryogenic vacuum. We heat the bottom plate with a regulated power supply. The sample geometry is such that the thermal conductivity through the helium liquid is very small compared to the thermal conductivity through the lateral boundary. If we use an electronic image let us say the sample impedance is larger, by two orders of magnitude, than the parallel impedance of the lateral walls. We are thus applying a "constant voltage source" which means that we impose a stable temperature difference ΔT. This point if often overlooked in many experiments and it is quite a crucial one.

II. BASICS OF CONVECTION : BUSSE[6] THEORY

The Oberbeck-Boussinesq approximation is used, this means that all fluid properties are assumed constant, with the exception of the temperature dependence of the density, taken into account in the gravity term. To get dimensionless numbers one uses d as a length scale, d^2/κ as a time scale, $(T_2 - T_1)/R$ as a temperature scale.

The governing equations become then :

$$P^{-1}\left(\frac{\partial}{\partial t}\vec{v} + (v.\nabla)\vec{v}\right) = -\nabla p + \theta\vec{k} + \nabla^2\vec{v} \qquad (1)$$

$$\nabla.\vec{v} = 0 \qquad (2)$$

$$\frac{\partial}{\partial t}\theta + (v.\nabla)\theta = R\,\vec{k}.\vec{v} + \nabla^2\theta \qquad (3)$$

where $\vec{k}$ is a vertical unit vector opposed to gravity, θ the deviation from the static temperature distribution, ∇p all terms that can be written in gradient form, and where R and P are the Rayleigh number and the Prandtl number.

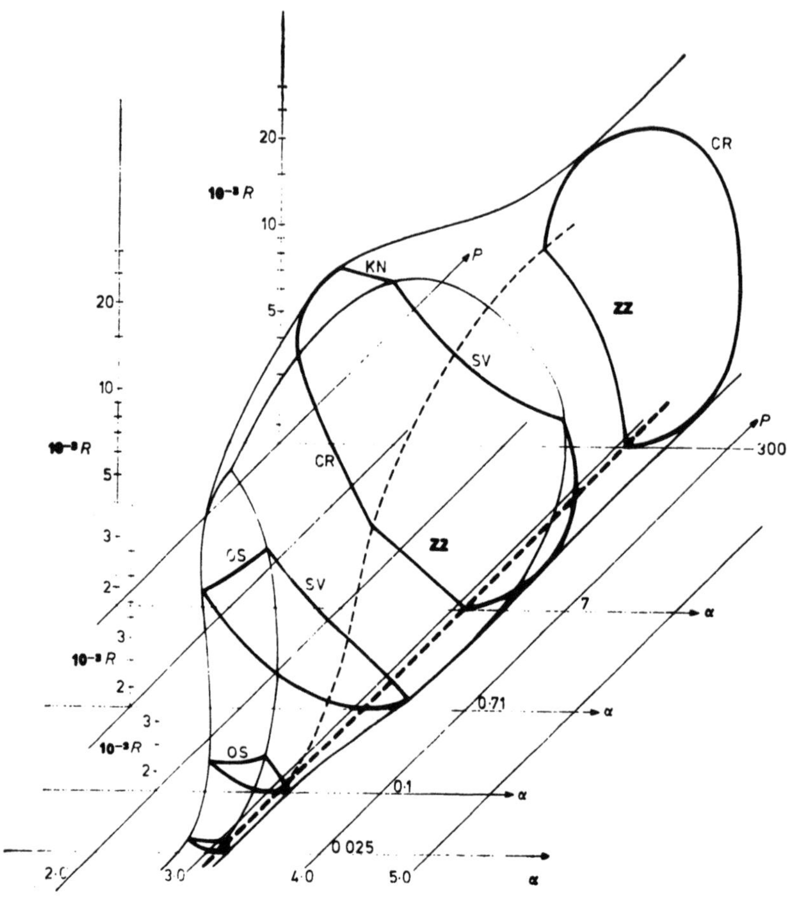

Fig. 4 Region of stable convection rolls in the R.P.α space. OS is the oscillatory instability, SV the skewed varicose, CR the cross roll, KN the knot and ZZ the zigzag one (from ref. 6).

For a good physical interpretation of the onset of convection and the meaning of the Rayleigh number see Velarde and Normand[7] article in Scientific American.

The onset of convection defines a critical Rayleigh number and a critical wavenumber for a fluid layer of infinite horizontal extent

$R_c = 1707$ $\alpha_c = 3.117$

The wavenumber is defined as $\alpha = \frac{2\pi d}{\lambda}$ where λ is the horizontal

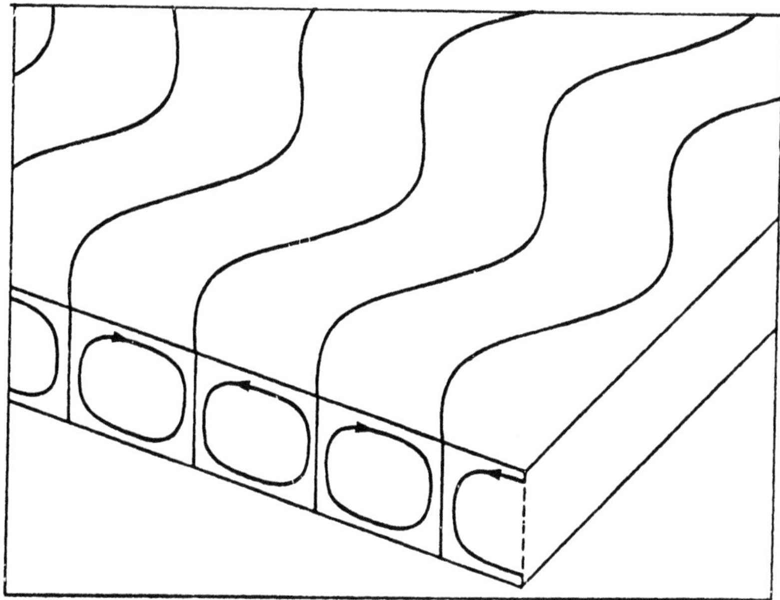

Fig. 5 Sketch of the oscillatory instability (from reference 6).

dimension of the cross section of two adjacent rolls (they define a unit cell as they rotate in opposite directions).

The fact that $\alpha_c \sim \Pi$ as $\lambda_c \sim 2d$ indicates that convective cells with a nearly square cross section are preferred at threshold. The physical reason for this choice is as follows[6] : for $\alpha < \alpha_c$ the potential energy released by the vertical motion is too small compared to viscous dissipation in the horizontal motion. For $\alpha > \alpha_c$ heat conduction between up and down fluid diminishes the buoyancy force.

Whereas for the onset of convection the Rayleigh number is the only relevant parameter, the Prandtl number enters insofar as the nonlinear properties of convection are concerned. Thus, depending on the Prandtl number, various instabilities are observed above the convection threshold. A beautiful analysis of the various possible convective structures has been performed by Busse[6], and is called the "Busse balloon". It is shown on Fig. 4.

In the experimental work reported here ($P < 0.7$) two instabilities will be of concern : the oscillatory instability and the skewed varicose.

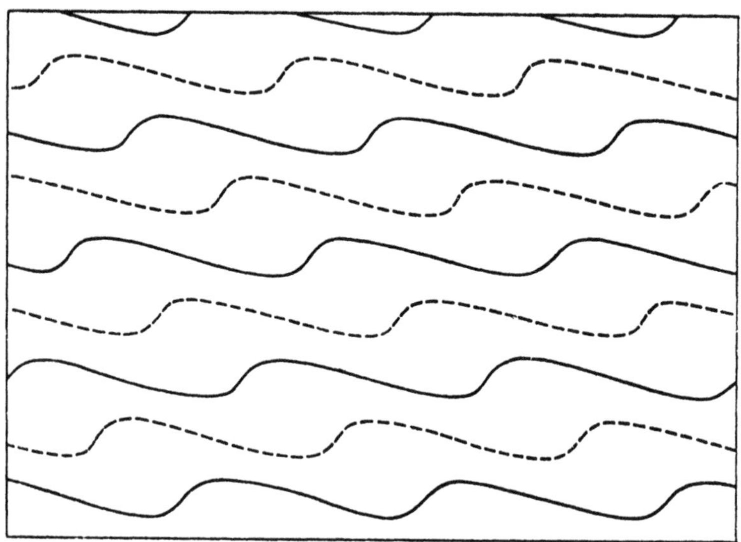

Fig. 6 Distortion of a horizontal pattern of convection rolls by the skewed varicose instability. Full and broken line represent the up and down going motion of the fluid (ref. 6).

For a very small P, the relevant nonlinear term is the momentum advection one, the $(v.\nabla)v$ term of equation (1). Thus starting from bidimensional rolls at R_c, the first instability which appears is a time dependent one, called the oscillatory instability. It consists of a transverse, time dependent, oscillation of the convective rolls, which propagates along the roll axis (Fig. 5).

As one increases P the nonlinear advection term in the heat equation (3), the $(v.\nabla)\theta$ term, can no longer be neglected. Thermal boundary layers at the top and bottom plates, perturb the perfect bidimensional up and down motion of the fluid. This leads to the skewed varicose instability which consists of a static periodic thickening and thinning of the convective rolls (Fig. 6).

III. THE WAVENUMBER SELECTION : EFFECT OF SIDE WALLS AND THE DYNAMICS OF A 3 ROLLS TO 2 ROLLS TRANSITION

Above the onset of convection a band of possible wavenumbers is allowed for the convective structure, in contrast with usual situations in physics where a well defined mode is selected. This effect may be relevant to Ahlers[1] observation that in large aspect

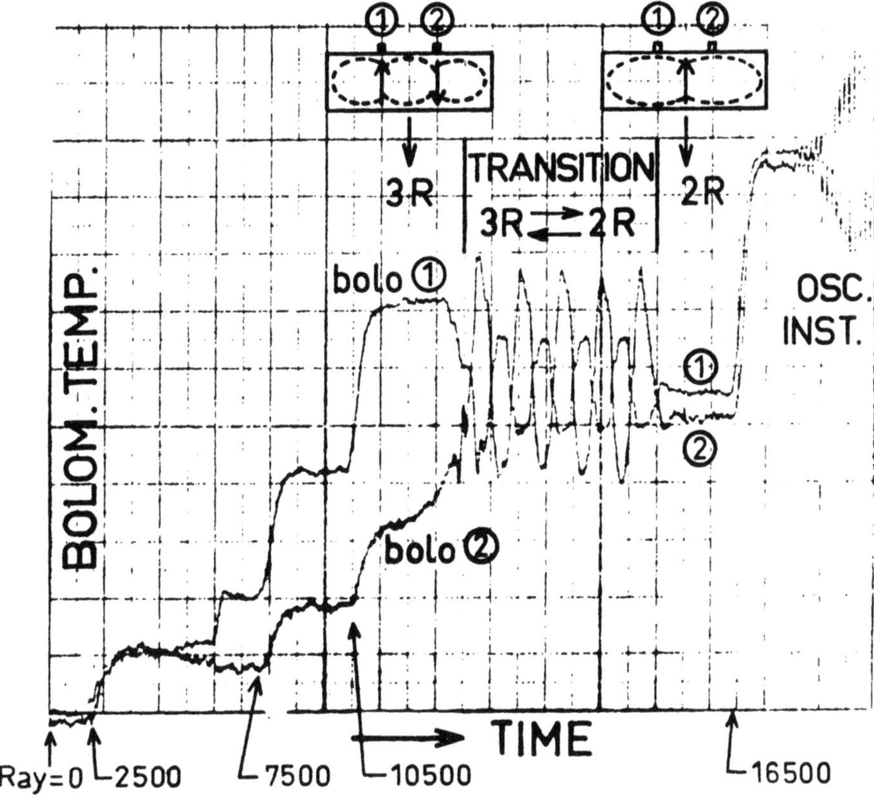

Fig. 7 Recordings of the local temperature seen by the bolometers as one varies the Rayleigh number stepwise. The regions labelled 3R and 2R correspond respectively to a three rolls and two rolls pattern. The fast oscillation is the oscillatory instability. The slow oscillation corresponds to the sloshing of the rolls pattern.

ratio cells, a low frequency noise appears above the onset of convection. In this context the experimental observation by Koschmieder[8], that beyond the onset of convection, there is a general tendency for a wavelength increase of the roll pattern, is important. From a theoretical point of view this problem has been addressed recently by Cross et al[9], Pomeau and Zaleski[10].

In a long and careful paper Cross et al have tried to answer the important question of the effect of sidewalls on the wavelength selection of the convective structure. They consider 2D motion in a large rectangular cell. The main result is that the presence of sidewalls, no matter how distant, severely restricts the possible wavevectors that can occur. Specifically the band of available q about q_o is reduced from a size $|q| \sim ((R - R_o)/R_o)^{1/2}$ to $|q| \sim (R - R_o)/R_o$. The result is strongly dependent on the Prandtl number P_o, and also on the sidewall conductivity as compared to the liquid one. For a small Prandtl number they do find a wavelength increase of the roll pattern.

Going back to the experiment with a small cell and a small P, as the Rayleigh number increases, one reaches a transition[11] from 3 rolls to 2 rolls.

Fig. 7 warrants some explanation, and is inserted here for pedagogical reasons. It is a direct laboratory recording of the local temperature seen by the bolometers as a function of time for five different values of the Rayleigh number, applied stepwise. In a three rolls pattern one bolometer is at a higher temperature than the other one as it experiences the arrival of hot fluid, the second one being at a location where cold fluid is going down. In a two rolls pattern they experience the same D.C. temperature. Thus a direct observation of the recording shows that one starts with a three rolls state at low Rayleigh number and that for R_a = 10500 the bolometers see a sloshing motion of the rolls before the system transits to a two rolls state. Then for a higher Rayleigh number the oscillatory instability sets in.

The conclusion is thus twofold. First as one increases the Rayleigh number we do observe a wavelength increase of the roll pattern, which in our quantized size cell leads to a transition from three or two rolls. Second this transition is a dynamic one with an instability region where the rolls move back and forth without finding a stable position.

As far as this instability is concerned we have found various possible trajectories. We show in Fig. 8 one of the typical recordings by one bolometer ; the second one sees the same temperature oscillations but in phase opposition as in Fig. 7. The main oscillation period is of about 100 sec (10 mHz) which scales with the diffusion time along the sample $\tau = (d^2/\kappa)(L/d)^2$. It starts for R/R_c close to three. The limit cycle can be qualitatively analyzed as follows.

3 rolls → one damped oscillation to 2 rolls → 3 reversed rolls

It looks as if when the three rolls pattern is destabilized a

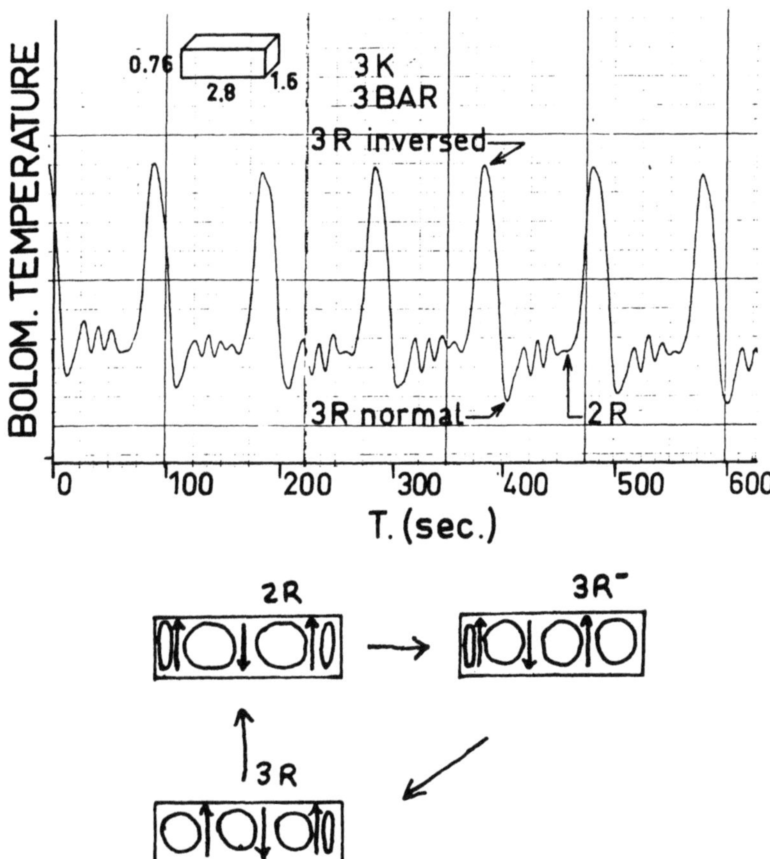

Fig. 8 Recordings of the sloshing motion of the rolls and schematic of the rolls motion in the cell with the limit cycle (3R and 2R mean three and two main rolls).

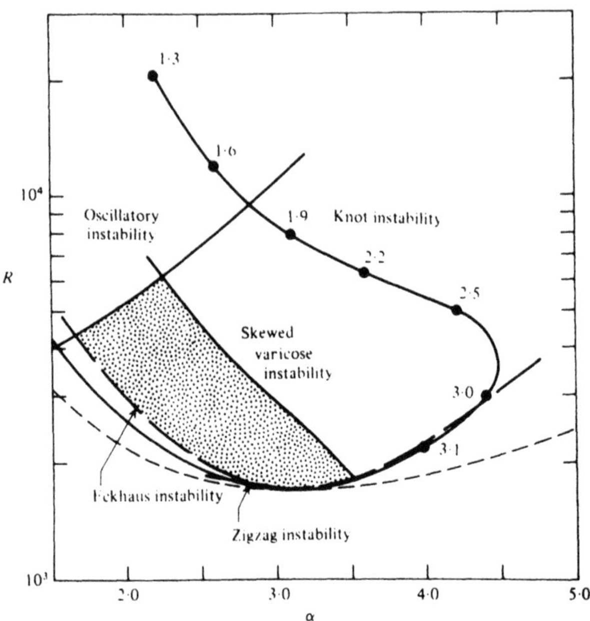

Fig. 9 Stability boundaries of convection rolls (ref. 12).

fourth roll germinates at one lateral boundary, and this leads to the following consecutive pattern.

In the next chapter we will start to study the routes to chaos for restricted geometries with two convective rolls present. We will see that the bifurcations to turbulence can be simple analyzed in a model with two or three oscillators present. But one should keep in mind that for larger aspect ratio cells, and even with three rolls present, the routes are more complex and should always include the wavelength selection problem which leads to very low frequencies oscillations or noise.

IV. THE OSCILLATORY INSTABILITY

From now on we shall reduce our study to the case of a confined geometry with two convective rolls present. A typical experimental recording of the oscillatory instability was shown on Fig. 7 for R_a = 16500. It represents the first bifurcation to a time dependent state in a two rolls geometry.

The theory for the stability boundary of the oscillatory instability as a function of the Rayleigh number and Prandtl number is due to Busse and Clever[12]. Fig. 9 is taken from their paper.

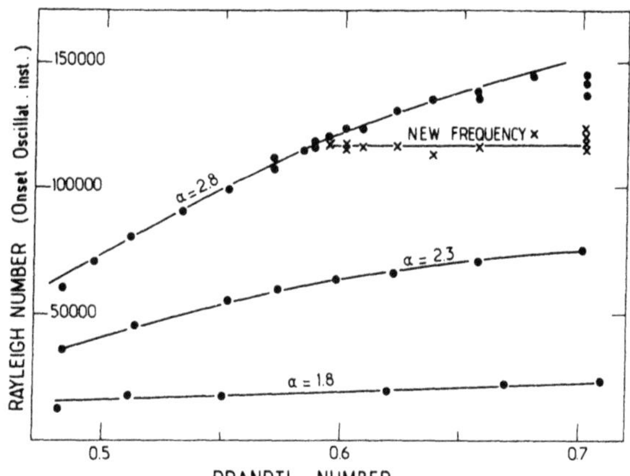

Fig. 10 Onset of the oscillatory instability (black dots). The crosses correspond to the onset of a new time dependent oscillation which we associate to the presence of the skewed varicose instability.

Whereas the onset of convection is related to a critical Rayleigh number. the onset of the oscillatory instability is related to a critical Reynolds number of the convective flow

$$\frac{vL}{\nu} = R_{e\ cr}$$

Taking for the velocity the vertical convective velocity, for L the rolls wavelength λ one gets

$$\frac{\kappa}{d}\,(R - R_c)^{1/2}\,\frac{\lambda}{\nu} = R_{e\ cr}$$

$$\boxed{\frac{(R - R_c)^{1/2}}{\alpha\, P_r} = R_{e\ cr}}$$

Thus the critical Rayleigh number for the onset of the oscillatory instability increases with the wavenumber and the Prandtl number as shown on Fig. 9.

The experimental results[13] for a cell with two rolls are shown on Fig. 10. In the experiment we define somewhat arbitrarily the wavenumber by $\alpha = 2\Pi d/L$ where L is the largest lateral side. We used three different samples. Busse analysis is relevant to a sample of infinite lateral dimensions whereas our experiment is in a confined geometry. Nevertheless the agreement is qualitatively good. The

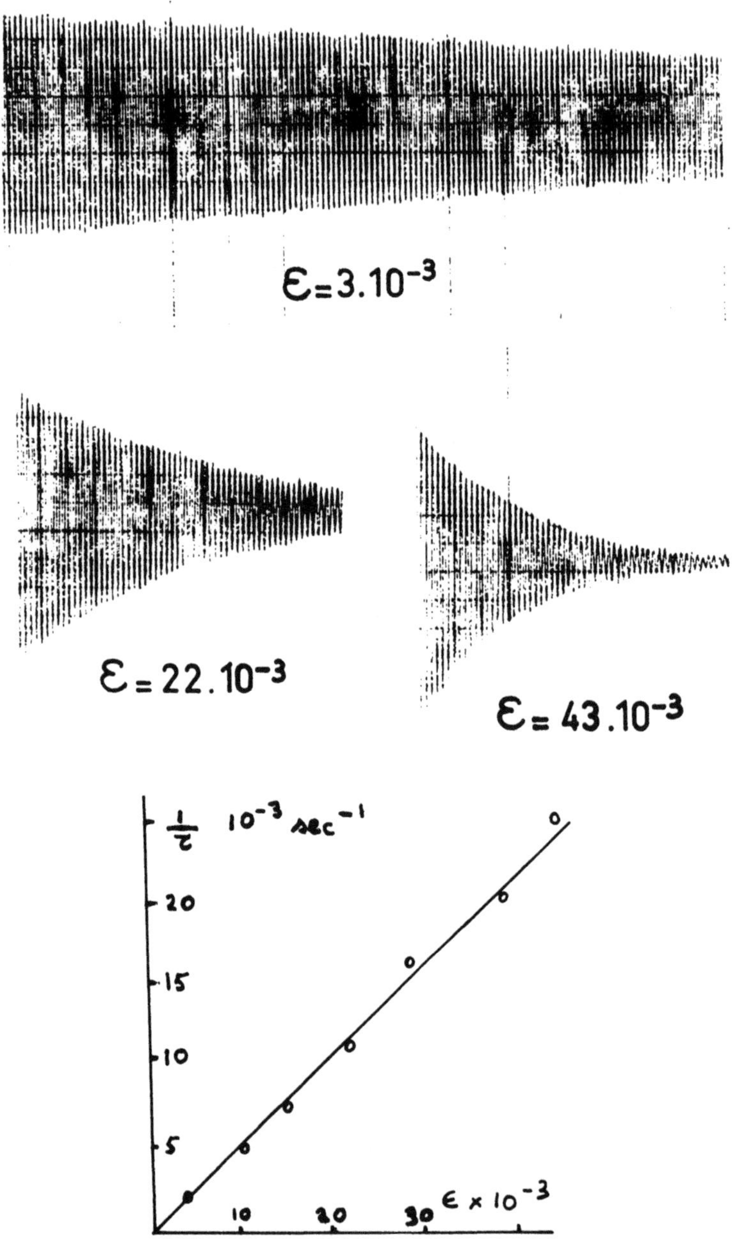

Fig. 11 The oscillatory instability : critical slowing down

main difference introduced by the lateral walls seems to be an increase by about an order of magnitude of the value of the Rayleigh number at onset of the oscillatory instability.

Before the onset of the oscillatory instability the vorticity field is horizontal. A vertical component of the vorticity is introduced by this instability. Some general considerations developed by

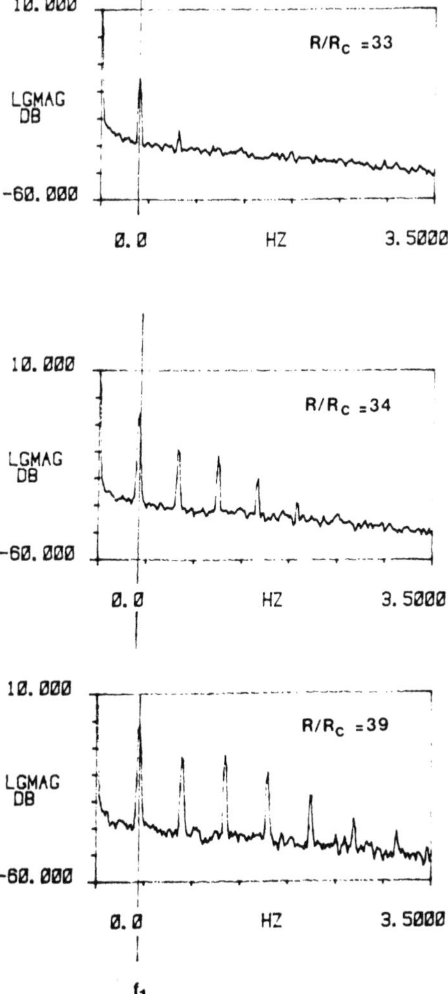

Fig. 12 Evolution of the frequency spectrum of the oscillatory instability as a function of R/R_c, for the sample corresponding to Fig. 2.

Siggia and Zippelius[14] are relevant now. The presence of a vertical vorticity term, whether it comes from the presence of a defect in the structure, a curvature of the rolls caused by the lateral boundaries, or the onset of an oscillating mode, may lead to a time dependent motion of the convective structure. In fact large scale changes in pattern require a vertical vorticity.

The onset of the oscillatory instability shows the characteristic slowing down of the Landau Hopf bifurcation. We test it as follows.

Starting from a Rayleigh number just above the onset of the oscillatory instability, R_0, we decrease the Rayleigh number by various amounts. We then get exponential decreases of the amplitude of the oscillation. Plotting the inverse of the relaxation time, inferred from this exponential decrease, as a function of ε ($\varepsilon = 1 - R/R_0$) we get a linear dependence.

This is shown on Fig. 11 for a sample of $\alpha = 2.75$ and $P_r = 0.48$.

V. TWO OSCILLATORS : ENTRAINMENT AND LOCKING

Starting from the onset of the oscillatory instability, the harmonic content of the oscillator (called from now on f_1) increases with Rayleigh number[15] as shown on Fig. 12.

For a further increase of the control parameter a new bifurcation appears associated to a second oscillator of frequency f_2. There is no theory as for the physical origin of this new mode. Its frequency is experimentally related to the largest lateral size, decreasing as L/d increases. This oscillator may be related to an instability of the boundary layers at the top and bottom plate.

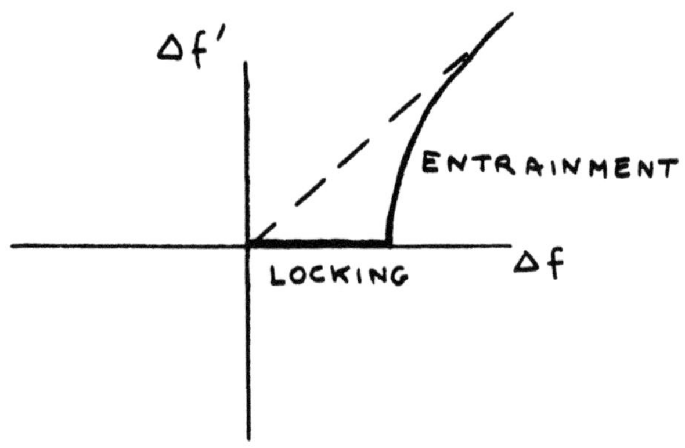

Fig. 13 Theoretical curve of entrainment and lock-in[16]

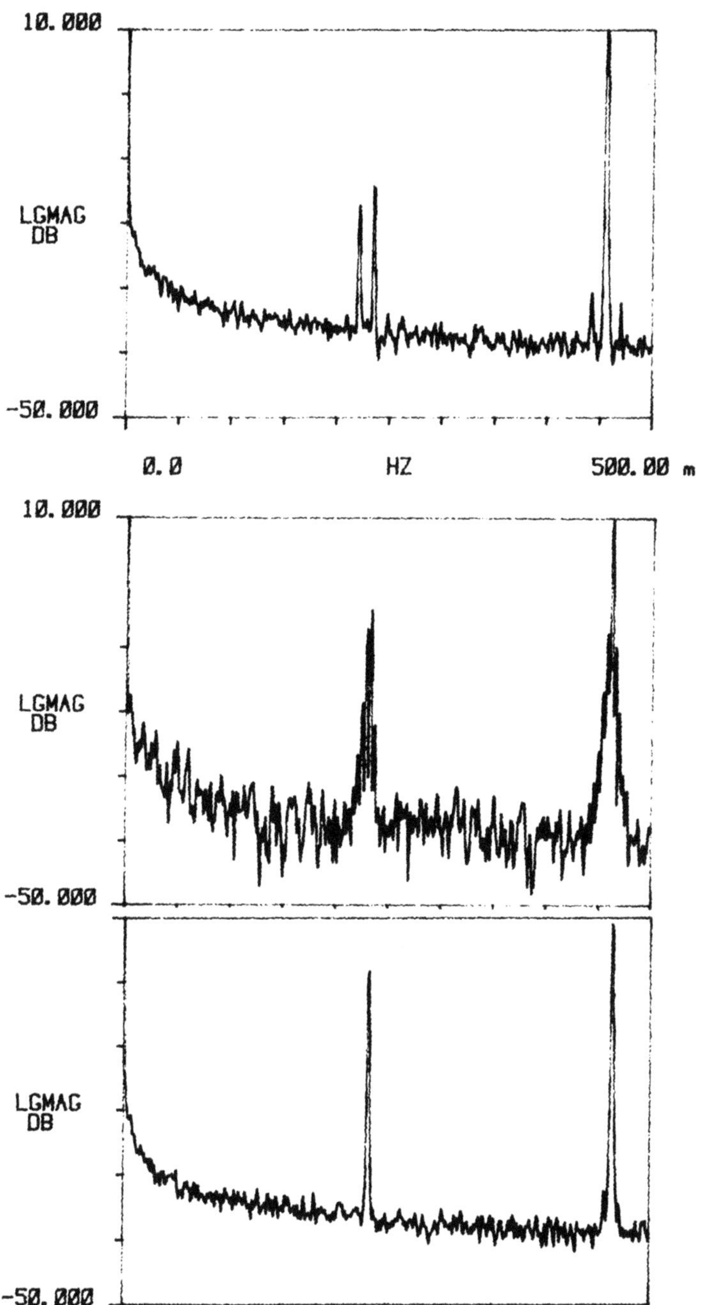

Fig. 14 Typical scenario : two noncommensurate frequencies, entrainment, locking.

First f_2 is noncommensurate with f_1, defining an invariant torus.

There is then a critical Rayleigh number where f_2 interacts with f_1, leading to an entrainment followed by a locking state[16].

A typical scenario is shown on Fig. 14 (for the sample of Fig. 2). The locking state observed shows no hysteresis and may be compared to the relevant theories in dynamical systems[17]. For a general approach to the theory of locking see Stratonovich et al[18].

In this scenario only one locking state was observed, but in various experiments with cells with largest aspect ratio (lower wavenumber for the rolls) a cascade of locking states exists. Flaherty and Hoppenstead have discussed extensively such regimes for the equation[19]

$$\ddot{y} + \frac{1}{\varepsilon}(y^2 - 1)\dot{y} + y = \frac{B}{\varepsilon}\cos \omega t$$

See also for a beautiful study of coupled relaxation oscillators the work of Gollub et al[20], and for a Rayleigh Benard experiment in water, Gollub and Benson study[21]. The results of one of our experiments[11] is shown on Fig. 15 with two clear lock-in for a frequency ratio 6.5 and 7 showing hysteresis and many other lock-in with no hysteresis

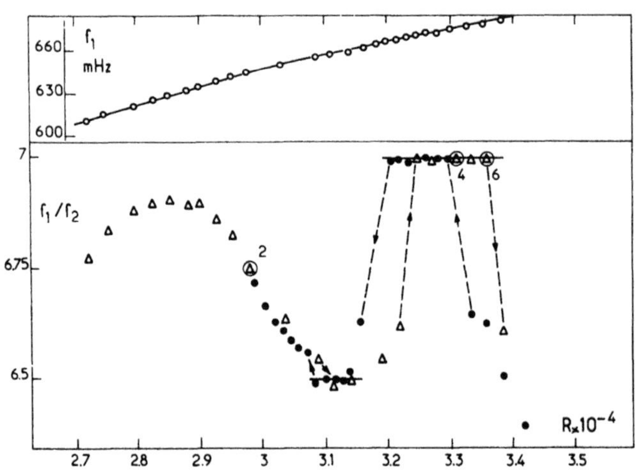

Fig. 15 Top curve : evolution of the frequency f_1.Bottom curve : the various lock-in states.

VI. THE ROUTES TO TURBULENT CONVECTION

In small boxes many routes to turbulent convection have been observed : Gollub and Benson[21] in water ($2.5 < P_r < 5$), Bergé and Dubois[22] in silicone oil ($P_r \sim 130$), Ahlers and Behringer[23] in small cylindrical cells of Helium. In all the studies the general consensus is that the presence of two oscillators is enough to reach a turbulent state. This is in some contradiction with the old picture of turbulence by Landau. It is closer to Lorenz, Ruelle and Takens[24] picture where chaos may occur when only few modes are present.

In our experiments the routes to turbulence are strongly dependent on the Prandtl number. In fact the important factor is the presence or not of static instabilities before we reach the two oscillators regime. More specifically it depends on whether the convective structure, before any time dependent phenomena, has a 2D or 3D character. Summing up our results :

Low P_r (2D character) $f_1 \to f_1, f_2 \to$ locking $\to$ period multiplication

High P_r (3D character) $\begin{cases} f_1 \to f_1, f_2 \to \text{intermittency to chaos} \\ f_1 \to f_1, f_2 \to f_1, f_2, f_3 + \text{chaos} \end{cases}$

We will now detail the two main routes to chaos, i.e. the period doubling cascade and intermittency.

VI.A. The Period Doubling Bifurcation to Chaos, Feigenbaum[25] Scheme

A good general approach to this scenario is given by Collet and Eckmann book[26]. A more physical approach is to present it as a cascade of parametric amplification. In this respect one should note the pionnering work of Rayleigh[27] on parametric amplification.

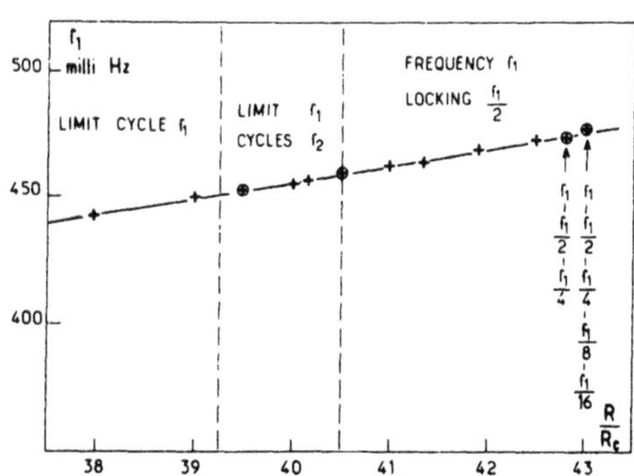

Fig. 16 Evolution of the period doubling cascade

Finally in this topic two textbooks are quite relevant : Landau mechanics and Bender and Orzag[28] on advanced mathematical methods.

Let us come back to our experimental results and pursue the scenario shown on Fig. 14. If we increase the Rayleigh number the subharmonic f/4 appears and then a cascade of subharmonics up to f/16 with a low frequency noise superposed. The evolution is shown on Fig. 16.

The Fourier spectrum of the last two steps of the cascade are shown on Fig. 17. The experimental observations are as follows :

- There is an accumulation point (called R_∞) in Rayleigh number

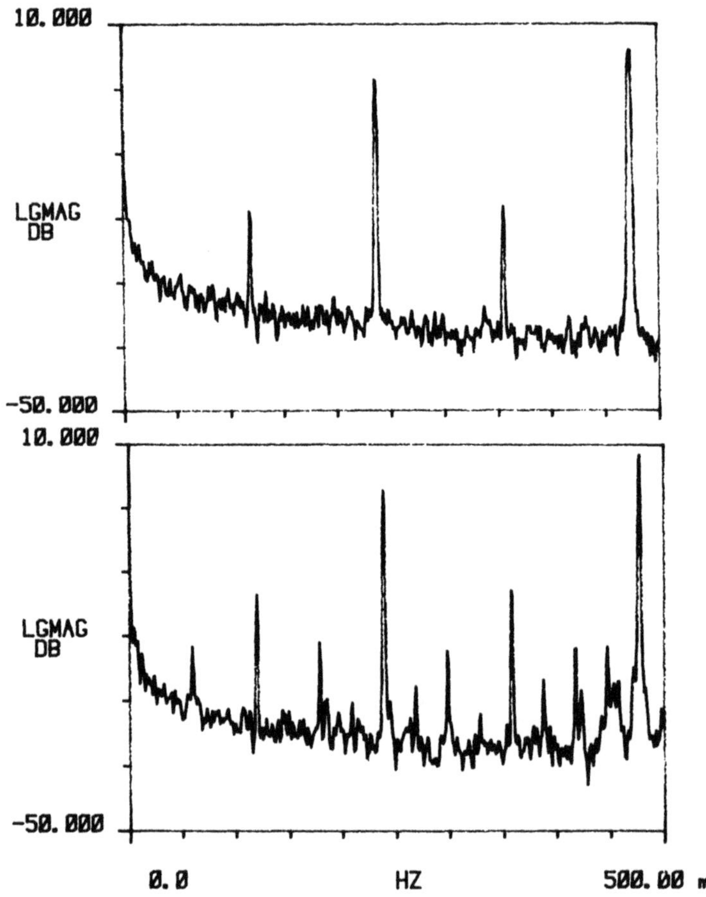

Fig. 17 The last two steps of the cascade

for the cascade but we were unable to observe subharmonics below f/16.

- There is a decrease of the power spectrum of every subharmonic by about 10 dB.

- The low frequency noise has a finite bandwidth and starts at the end of the cascade only.

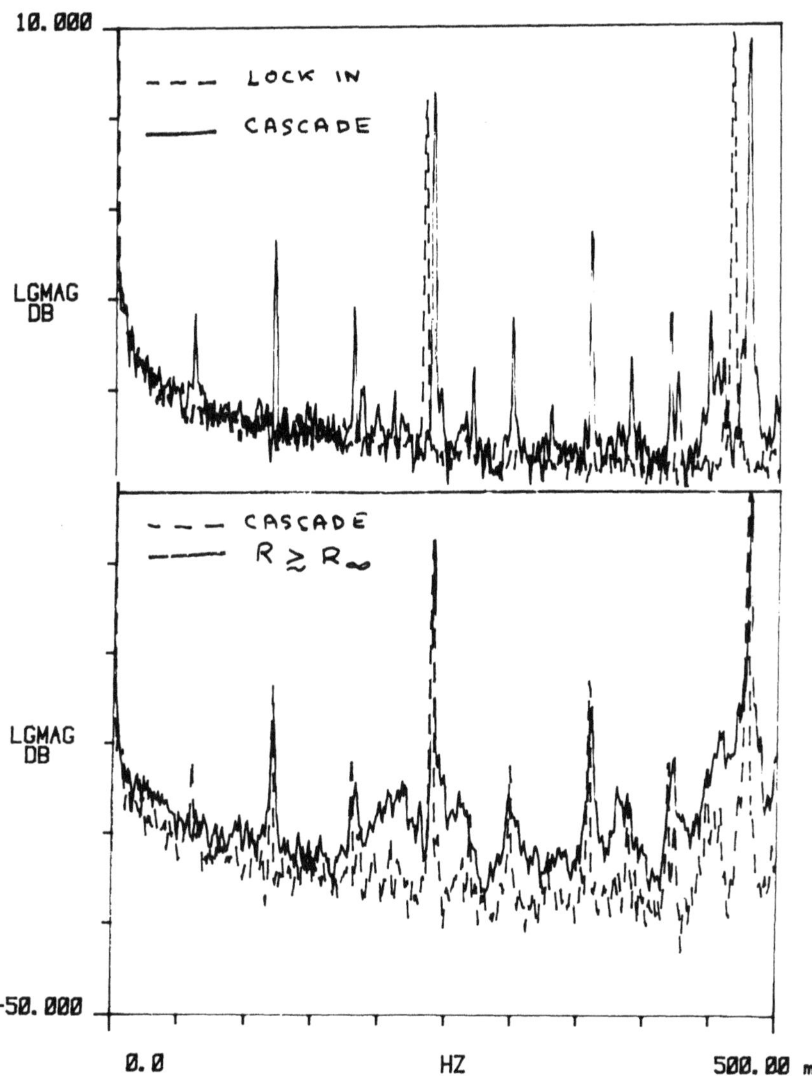

Fig. 18 The abrupt increase of the noise power above R_∞

To show clearly this effect we present two recordings in Fig. 18. In the first one we have superposed two spectra, one for the locking state f_1 and $f_{1/2}$, the other one being the final cascade. One can see that the base line is unaffected and that the noise is just barely modulating the various peaks. In the second one we have superposed the cascade and a recording for a very small increase of the Rayleigh number beyond it. There is a very large increase in the noise power.

There are two essential predictions of the theory.

- The Rayleigh number for the onset of each subharmonic should follow a geometric progression

$$\frac{R_n - R_{n+1}}{R_{n+1} - R_{n+2}} \xrightarrow[n\to\infty]{} \delta = 4.6692$$

- The peaks of the successive subharmonics should decrease by a constant amount in dB as $n \to \infty$.

The experimental precision on δ is quite bad. We get

$\delta = 3.5 \pm 1.5$

Two calculations[29] have proposed a number for the ratio of the successive subharmonics. If we define ϕ_n/ϕ_{n+1} as the ratio of the subharmonic power one gets

	Feigenbaum	Nauenberg Rudnick	This experiment
$\dfrac{\phi_n}{\phi_{n+1}}$	43	21	7 to 12

There is quite a discrepancy but let us recall that the numbers are asymptotic values for large n, whereas in the experiment we measure only the three first bifurcations.

Let us close this part by pointing out that recently Giglio et al[30] in a similar experiment in water have also measured the period doubling cascade with a better precision.

Another prediction of the theory is that beyond the accumulation point, a mirror image of the cascade should exist[31,26]. In other words the cascade is observed starting from the quiescent state of the fluid and increasing the heating. But the same one should be observed starting from the turbulent state and decreasing the heating. We show on the four next power spectra this phenomena. (see pages 24 and 25).

If we plot, within our precision, the results as a function of the frequency f_1, which is proportional to the Rayleigh number one gets

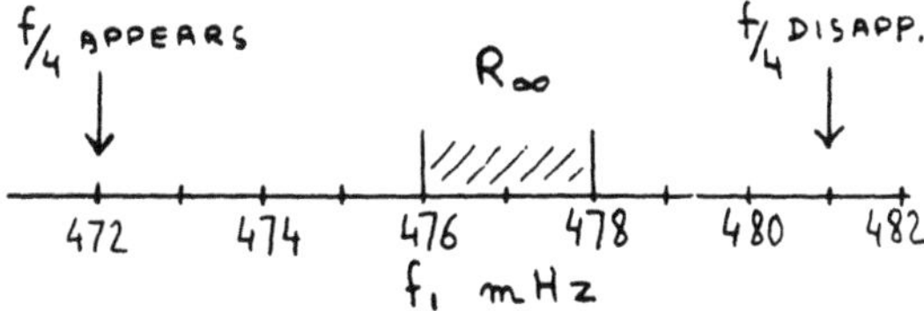

To conclude let us present an experimental recording of the cascade by plotting on an XY recorder the signals from one bolometer as a function of the signal from the other bolometer (let us recall that in our experiment we have a bolometer on top of each convecting roll). In those recordings we have just filtered the high harmonic contents of the frequency f_1.

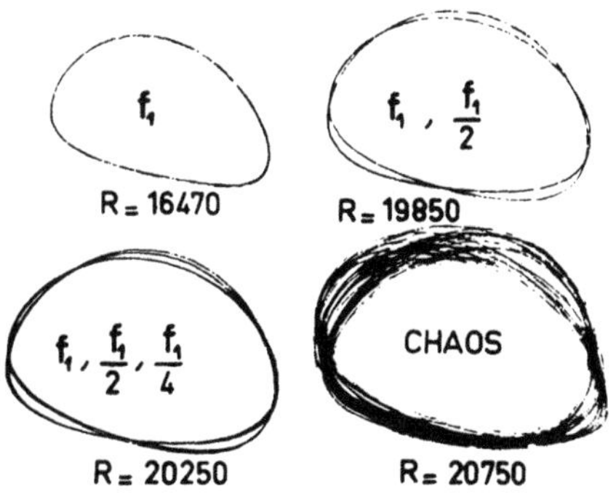

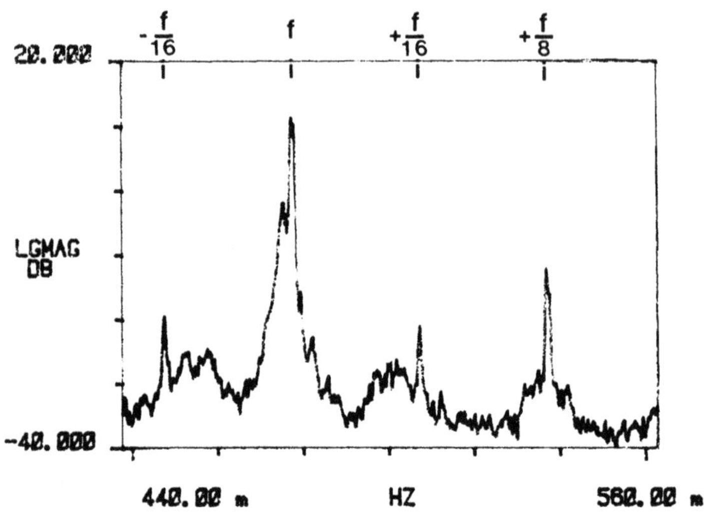

A : above the accumulation point. All the subharmonics are present plus noise.

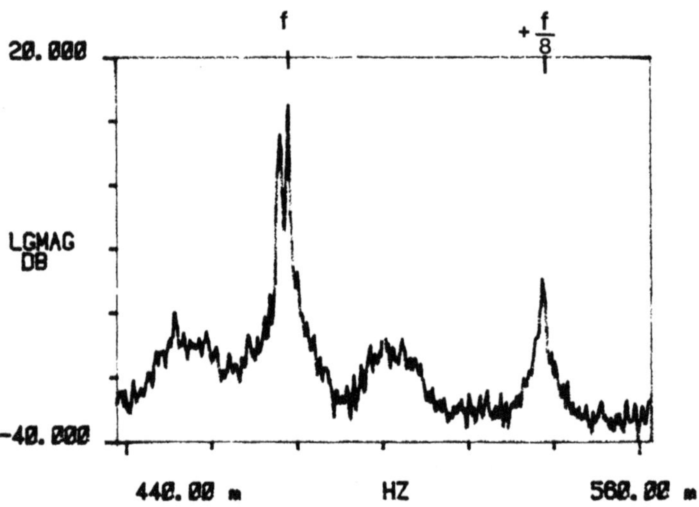

B : a slight increase in Rayleigh number $\frac{f}{16}$ disappears.

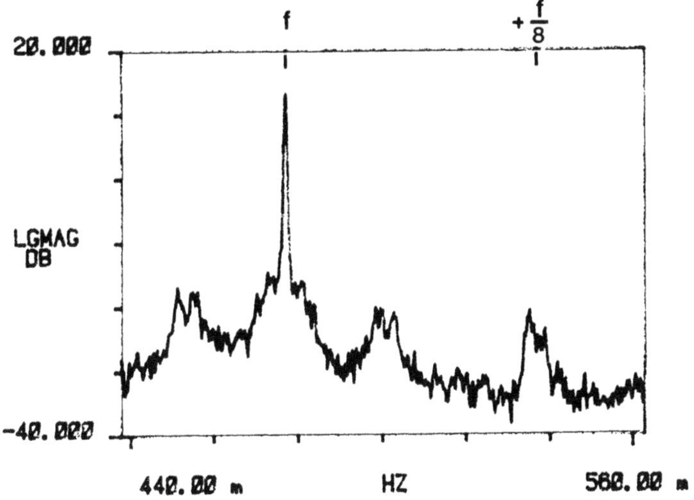

C : new increase in R_a. $\frac{f}{8}$ disappears.

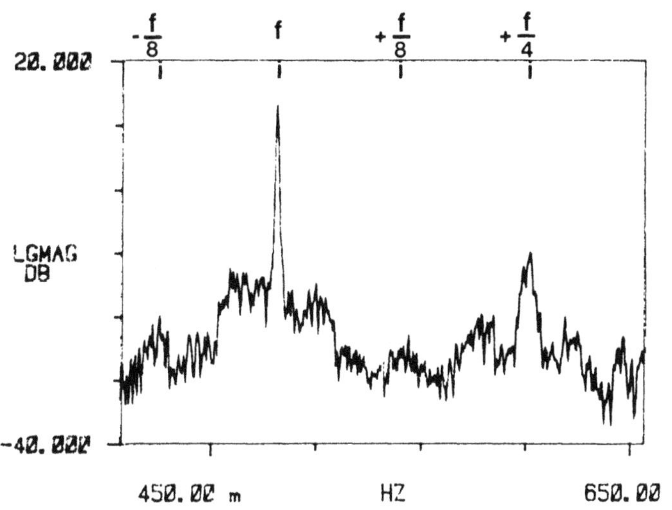

D : larger increase in R_a. $\frac{f}{4}$ disappears. Note that the bandwidth has been changed.

Conclusion

- The qualitative picture proposed by Feigenbaum seems to be correct. Quantitatively there are discrepancies, may be associated to the fact that we observe only the very first bifurcations.

- One of the main difficulties of this experiment is that one often bifurcates to other frequency divisions like the division by 3 or 5. We have observed clearly the following sequences

$$f \to \frac{f}{2} \to \frac{f}{6}$$

$$f \to \frac{f}{2} \to \frac{f}{4} \to \frac{f}{20}$$

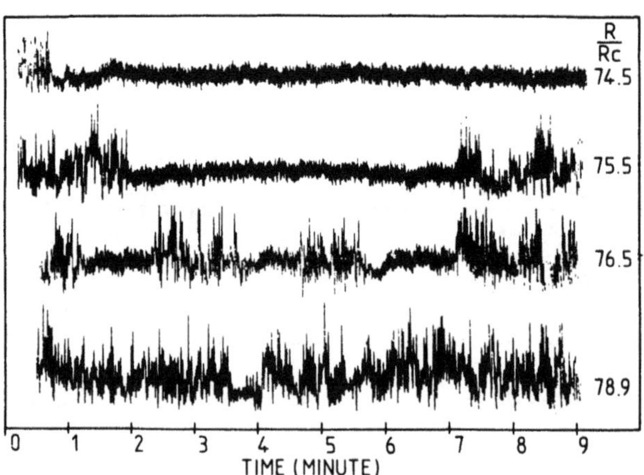

Fig. 19 Intermittent transition to turbulence for a cell $\alpha = 2.7$ $P_r = 0.62$.

VI.B. Intermittency as a Route to Turbulence

Manneville and Pomeau[32] have shown that an intermittent transition to chaos exists for the Lorenz model. Experiments by Gollub[21] in water and Bergé[33] in silicon oil have shown intermittent transitions to chaos.

In our experiments in helium an intermittent transition to chaos is measured[13] at high P and high wavenumber. We first reach a state with two oscillators with uncommensurate frequencies. The transition to turbulence is intermittent for a range of Rayleigh numbers $74.5 < R/R_c < 79$.

Defining the onset of noise bursts at $R/R_c = 74.4$ we find that the laminar time intervals without noise diverge like $\tau \simeq e^{-\beta}$, $1 < \beta < 1.5$. The data are shown on Fig. 19.

REFERENCES

1. G. Ahlers, R.W. Walden, Phys. Rev. Lett., 44:445 (1980).
2. J. Wesfreid, V. Croquette, Phys. Rev. Lett., 45:634 (1980).
3. A. Libchaber, J. Maurer, Journal Phys. Lettres, 39:369 (1978).
4. K. Stork, U. Müller, J. Fluid Mech., 71:231 (1975), 54:599 (1972).
5. R.D. Mc Carthy, Thermophysical Properties of ^{4}He, Bur. Stand. Tech. Note n° 631 (1972).
6. F.H. Busse, Report Progr. Physics 41:1929 (1978).
7. M. Velarde, C. Normand, Scientific American, 243:78 (July 1980).
8. E.L. Koschmieder, Adv. Chemical Physics, 26:177 (1974).
9. M.E. Cross, P.G. Daniels, P.C. Hohenberg, E. Siggia, Phys. Rev. Lett., 45:898 (1980).
10. Y. Pomeau, S. Zaleski, C. R. Acad. Sc. Paris, 290 (série B):505 (1980).
11. J. Maurer, A. Libchaber, J. Physique Lettres, 40:419 (1979).
12. B.H. Busse, R.M. Clever, J. Fluid Mech., 91:319 (1979).
13. J. Maurer, A. Libchaber, J. Phys. Lettres, 41:515 (1980).
14. E. Siggia, A. Zippelius, "Dynamics of Defects in Rayleigh-Benard Convection", preprint.
15. A. Libchaber, J. Maurer, J. de Physique, Coll. C3, 41:51 (1980).
16. A. Adler, Proc. I.R.E., 34:351 (1946).
17. G. Iooss, Math. Studies, 36, New York (1979).
    G. Iooss, W.F. Langford, Annals of the New York Academy of Sciences, Vol. 327 (1980).
18. R.L. Stratonovich, Topics in the Theory of Random Noise, Gordon and Breach (1967).
    W.E. Lamb Jr., Phys. Rev., 134:429 (1964).
    B. Van der Pol, Phil. Mag., 3:65 (1927).
19. J.E. Flaherty, F.C. Hoppensteadt, Study Appl. Math., 58:5 (1978).
20. J.P. Gollub, E.J. Romer, J.E. Socolar, Journ. Stat. Phys., 23:321 (1980).

21. J.P. Gollub, S.V. Benson, J. Fluid Mech., 100:449 (1980).
22. M. Dubois, Colloque Pierre Curie, Paris (1980).
M. Dubois, P. Bergé, J. Physique, 42:167 (1981).
23. G. Ahlers, R.L. Behringer, Phys. Rev. Lett., 40:712 (1978).
24. L. Landau, E. Lifshitz, Fluid Mechanics, Chapt. 3, Pergamon, Oxford (1959).
D. Ruelle, F. Takens, Comm. Math. Phys., 20:167 (1971).
E.N. Lorenz, J. Atmos. Sci., 20:130 (1978).
25. M.J. Feigenbaum, Phys. Lett., 74A:375 (1979) ; Comm. Math. Phys., 77:65 (1980).
P. Coullet, C. Tresser, A. Arneodo, Phys. Lett., 72A:268 (1979).
26. J. Collet, J.P. Eckmann, "Iterated Maps of the Interval as Dynamical Systems", Birkhaüser (1980).
27. Lord Rayleigh, "The Theory of Sound", Vol. I, Chapt. 3, Dover (1945).
28. C. Bender, S. Orzag, "Advanced Math. Methods for Scientists", McGraw Hill (1978).
29. M. Nauenberg, J. Rüdnick, "University and the Power Spectrum at the Onset of Chaos", preprint.
30. M. Giglio, S. Musazzi, U. Perini, "Transition to Chaos via a Well Defined Ordered Sequence of Period Doubling", preprint.
31. S. Grossmann, S. Thomae, Z. Naturforsch., 32a:1353 (1977).
S. Thomae, S. Grossman, "Correlations and Spectra of Periodic Chaos Generated by the Logistic Parabola", preprint.
A. Wolf, J. Swift, "Universal Power Spectra for the Reverse Bifurcation Sequence", preprint.
B. Huberman, A. Zisook, Phys. Rev. Lett., 46:626 (1981).
32. P. Manneville, Y. Pomeau, Phys. Lett., 75A:1 (1979).
Y. Pomeau, P. Manneville, Physica, D1:219 (1980).
33. P. Bergé, M. Dubois, P. Manneville, Y. Pomeau, J. Phys. Lettres, 41:341 (1980).

Editor's note. We also recommend the excellent review article by Fenstermacher, Swinney and Gollub (1979).

Le Journal de Physique–Lettres **43** L-211–L-216 (1982)

Period doubling cascade in mercury, a quantitative measurement

A. Libchaber, C. Laroche and S. Fauve

Groupe de Physique des Solides de l'Ecole Normale Supérieure,
24, rue Lhomond, 75231 Paris Cedex 05, France

(*Reçu le 21 décembre 1981, accepté le 12 février 1982*)

Résumé. — Observation de la cascade de doublement de période dans une expérience de Rayleigh-Bénard sur le mercure. La cellule expérimentale a un rapport d'aspect $\Gamma = 4$ et comporte quatre rouleaux convectifs. Un champ magnétique de 270 G est appliqué le long de l'axe des rouleaux. Le nombre de Feigenbaum mesuré est $\delta = 4{,}4$. Le rapport des sous harmoniques successifs est de l'ordre de 14 dB pour les sous harmoniques les plus bas mesurés.

Abstract. — Observation of the period doubling cascade in a Rayleigh-Bénard experiment in mercury. The experimental cell has an aspect ratio $\Gamma = 4$ and contains four convective rolls. A DC magnetic field of 270 G is applied along the convective roll axis. The measured Feigenbaum number is $\delta = 4.4$. The ratio of the successive subharmonics is of the order of 14 dB for the lowest measured subharmonics.

The period doubling cascade as a route to chaos is now well documented theoretically [1] as well as experimentally [2]. We present here new results on a Rayleigh-Bénard experiment in a cell of liquid mercury with an aspect ratio larger than in previously reported works on helium or water (four convective rolls). The very good signal to noise ratio of the experiment allows a precise determination of the Feigenbaum number and of the power ratio of the successive subharmonics. In this experiment, a DC magnetic field, applied along the convective roll axis, introduces an extra damping of the oscillators which favours the period doubling cascade.

1. **An experimental dynamical system with controlled dissipation.** — At low Prandtl number, the relevant instability of two-dimensional convective rolls, which leads to time dependent convection and chaos, is the oscillatory instability [3]. It was shown [2] in experiments of convection in liquid helium that for a range of Prandtl numbers and wavenumbers of the convective pattern, one route to chaos is a period doubling cascade, which follows the Feigenbaum scenario [1]. It is customary to compare this type of experiments to a one-dimensional mapping. But this mapping relates to strongly dissipative modes. A more realistic mapping is therefore a two-dimensional one with two parameters, a constraint and a viscosity (area contraction in phase space). The

simplest quadratic one is the Henon mapping [4] which reduces for infinite contraction to a one-dimensional mapping. In a way, we want to find a physical system depending on those two parameters. In a Rayleigh-Bénard experiment, the constraint is the Rayleigh number R. At low Prandtl number, the damping of the modes associated with the oscillatory instability depends on the Prandtl number and the wavenumber of the convective pattern [3]. But it is not easy to control such parameters experimentally, and furthermore, they do not affect selectively the oscillatory instability. On the contrary, a horizontal magnetic field, parallel to the convective roll axis, will add a quantitative damping to the oscillatory instability, as we have recently shown [5]. This increased damping of the oscillators will favour the period doubling cascade as a route to chaos [6].

2. **Experimental apparatus.** — Our experimental system has been described in detail elsewhere [5, 7]. We use a parallelepipedic cell of aspect ratio $\Gamma = 4$ (dimensions $7 \times 7 \times 28$ mm) with, in the convective state, four rolls parallel to the shorter side wall (see reference five for visualization). The top and bottom boundaries consist of two thick copper plates. The lateral boundaries are made of plexiglass. In the centre of the bottom plate, a small NTC (negative temperature coefficient) thermistor is located on a 2 mm diameter hole and is adjacent to the convective fluid. The temperature signal, given by the bolometer, is analysed by a 5 420 A H. P. digital analyser. An

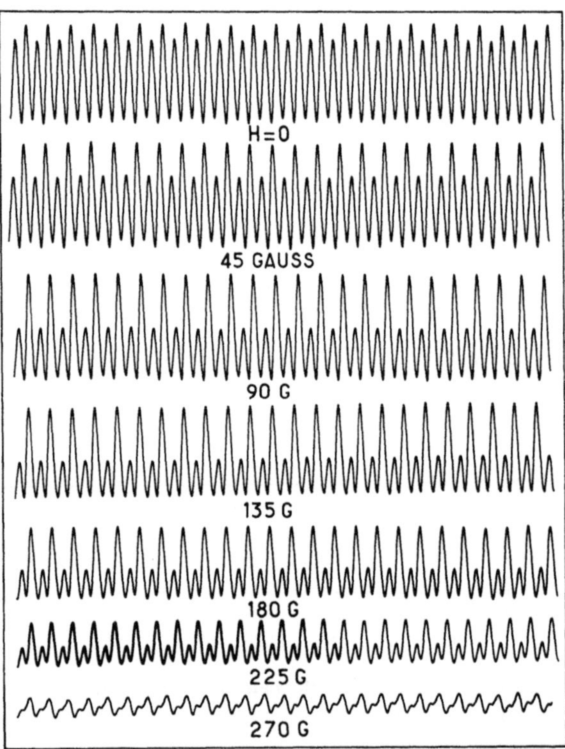

Fig. 1. — Effect of a magnetic field parallel to the roll axis. The Rayleigh number is $R/R_c = 2.9$ at zero field. The two oscillators are locked ($f_1/f_2 = 2$), the amplitude of oscillator f_1 about 20 dB larger than the amplitude of oscillator f_2 for $H = 0$. As H increases, the two oscillators tend to the same amplitude and keep their locking ratio 2.

electromagnet provides a uniform horizontal magnetic field. The experimental convective cell is placed in a vacuum chamber.

3. **Effect of a DC magnetic field, parallel to the roll axis, on the oscillating state.** — Defining the Rayleigh number for the onset of convection as R_c, the first time dependent instability observed is the oscillatory instability, its onset being at $R \simeq 2\ R_c$. As previously reported [7], the frequency of the oscillatory mode increases linearly with the Rayleigh number. It allows, by measuring the frequency, a precise determination of each bifurcation point. Increasing slightly the Rayleigh number, a second frequency f_2 appears in the temperature spectrum which, within a very small range of R, locks with f_1, the ratio being $f_1 = 2\ f_2$.

We then apply a DC magnetic field parallel to the roll axis up to a field of 270 G. We present, in figure 1, the effect of the magnetic field on the direct local temperature recording. Two distinct regimes appear.

For $H > 100$ G, the evolution is the one described in our previous study [5]. The frequency of the oscillatory instability increases and so does the damping of the mode. This will tend at a higher field to a complete inhibition of the oscillatory instability.

But for $H < 100$ G, and the effect is already noticeable around 10 G, the relative oscillating strength of the two locked oscillators changes. This sensitivity to a very small field is surprising.

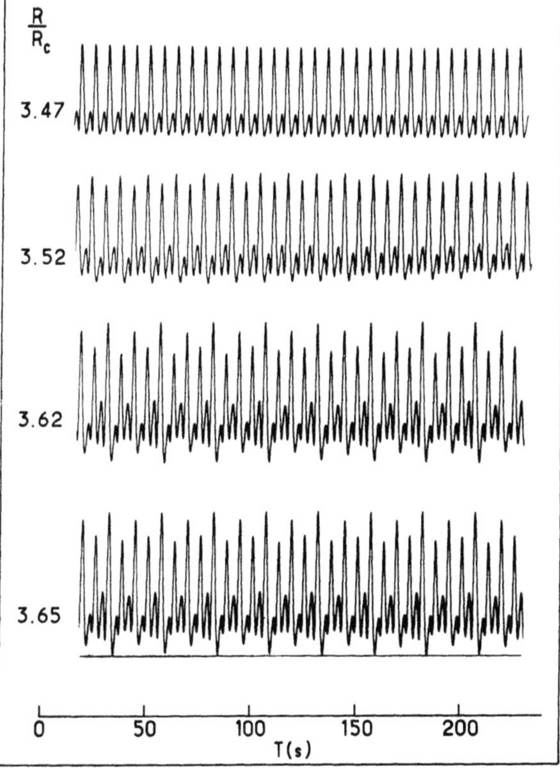

Fig. 2. — Direct time recordings of temperature for various stages of the period doubling cascade showing the onset of $f/4$ ($R/R_c = 3.52$), $f/8$ ($R/R_c = 3.62$), $f/16$ ($R/R_c = 3.65$).

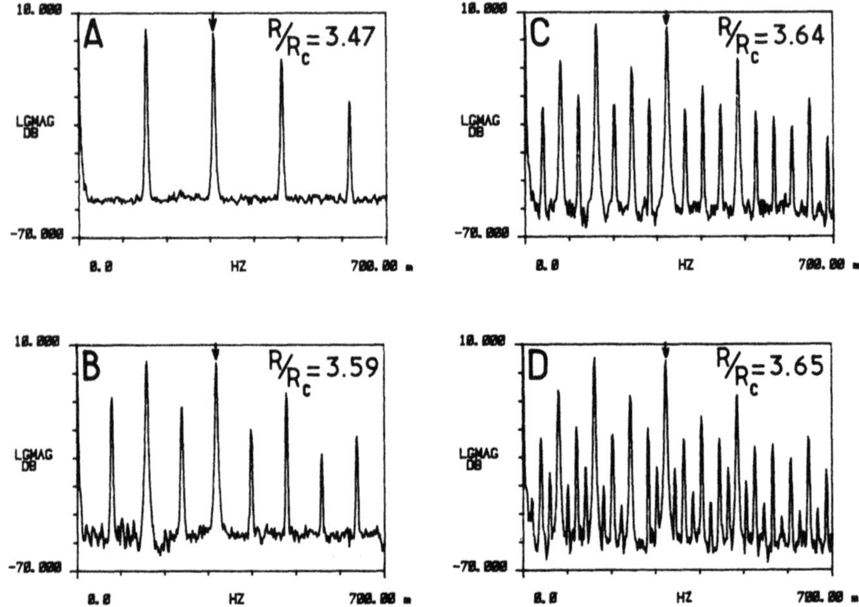

Fig. 3. — The Fourier spectrum. Arrows indicate the peak at the frequency f_1.

The net result of this phenomenon is that, at 270 G, the two modes are highly damped and have about the same oscillating strength, whereas at zero field the mode f_1 has an amplitude about 20 dB larger than the mode f_2.

4. **The period doubling cascade.** — From now on, we keep a constant magnetic field $H = 270$ G. A typical recording is shown in figure 2 for $R/R_c = 3.47$. Let us note that it shows a striking similarity with a recent Cray machine simulation done by Upson *et al.* [8] on a Rayleigh-Bénard experiment in a small box. The corresponding Fourier spectrum is shown in figure 3A.

As we increase the Rayleigh number, a period doubling cascade develops up to $f/32$, well resolved as far as the onset values up to $f/16$.

In table I, we present the various onset values of the cascade of pitchfork bifurcations. In figure 2, the temperature recordings are presented, showing the development of the subharmonics $f/4$, $f/8$ and $f/16$, for $R/R_c = 3.52$, 3.62 and 3.65. Their respective Fourier spectra are presented in figures 3B, C, D for values of R/R_c close to the preceding ones.

Table I

Onset of bifurcations	R/R_c		
—	—		
$f/4$	3.485		$\mu_{4/8} = 3.5$ (~ 11 dB)
$f/8$	3.618 3	$\delta = 4.4 \pm 0.1$	
$f/16$	3.648 6		$\mu_{8/16} = 5$ (14 dB)

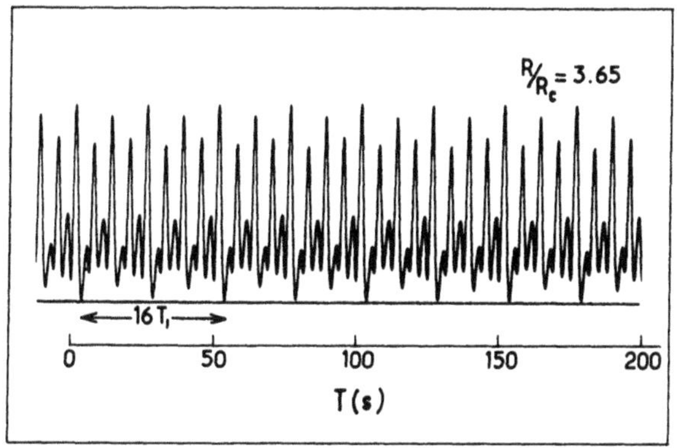

Fig. 4. — Enlargement of the recording showing the development of the cascade.

If we compute the Feigenbaum number δ for the last three bifurcations, we get

$$\delta = \frac{R_8 - R_4}{R_{16} - R_8} = 4.4 \pm 0.1 \,.$$

This is to be compared with the theoretical asymptotic value [1] $\delta = 4.669...$

We can also compute the ratio of the successive subharmonics amplitude called μ.

This ratio is measured directly on the temperature recordings. We show in figure 4 an enlargement of the temperature signal after the $f/16$ bifurcation and in figure 5 an enlargement of its Fourier spectrum. The last value of μ measured is $\mu \sim 5$ to be compared with theoretical values between 4.58 and 6.5 (the first one is given by Nauenberg and Rudnick [9], the second one by Feigenbaum [1]). This measurement of μ led to some confusion in the past, which can be understood if we look at the Fourier spectrum in figures 3D and 5. It is clear there that the odd harmonics of $f/16$ have an amplitude which is modulated and depends on the order of the harmonics.

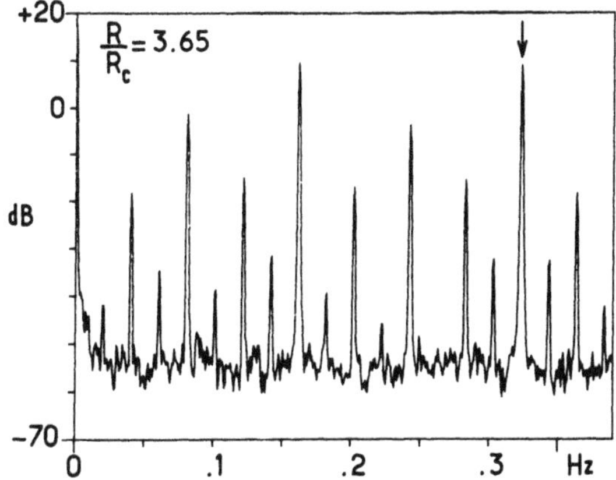

Fig. 5. — Enlargement of the Fourier spectrum corresponding to figure 4.

It is thus not clear how to calculate the ratio from the Fourier spectrum. The numbers we give in table I are derived from the direct recording.

We have also observed beyond the accumulation point the inverse cascade. Its study and the subtle phenomena of locking windows in the chaotic state will be presented elsewhere.

Let us add some final remarks. Recent computer experiments have shown the role of the breakdown of some spatial symmetries on the onset of chaos [10]. In this respect, the presence of a DC magnetic field plays an important role, such as, for example, introducing an anisotropic effective viscosity. Clearly, for other values of the field and especially at very high fields, other routes to chaos will be found. We must also stress the fact that a very small magnetic field affects the oscillating modes which is somewhat surprising.

References

[1] Feigenbaum, M. J., *Phys. Lett.* **75A** (1979) 375.
Collet, P. and Eckmann, J. P., *Iterated maps on the interval as dynamical systems* (Birkhaüser, Boston) 1980.

[2] Maurer, J. and Libchaber, A., *J. Physique Lett.* **40** (1979) L-419.
Libchaber, A. and Maurer, J., *J. Physique Colloq.* **41** (1980) C3-51.
Giglio, M., Muzzati, S. and Perini, U., *Phys. Rev. Lett.* **47** (1981) 243.
Linsay, P. S., *Phys. Rev. Lett.* **47** (1981) 1349.

[3] Busse, F. H., *Rep. Progr. Phys.* **41** (1978) 1929.

[4] Henon, M., *Commun. Math. Phys.* **77** (1976) 50.

[5] Fauve, S., Laroche, C. and Libchaber, A., *J. Physique Lett.* **42** (1981) L-455.

[6] Arneodo, A., Coullet, P., Tresser, C., Libchaber, A., Maurer, J. and D'Humières, D., *About the observation of the uncompleted cascade*, preprint.

[7] Fauve, S. and Libchaber, A., *Chaos and order in nature*, H. Haken Ed., (Springer) 1981.

[8] Upson, C. D., Gresho, P. M., Sani, R. L., Chan, S. T. and Lee, R. L., Lawrence Livermore Lab., Preprint UCRL 85555.

[9] Nauenberg, M. and Rudnick, J., UCSC preprint.

[10] Lipps, F. B., *J. Fluid Mech.* **75** (1976) 113.
Mc Laughlin, F. B. and Orszag, S. A., preprint.

Physical Review Letters **35** 927–30 (1975)

Onset of Turbulence in a Rotating Fluid*

J. P. Gollub†‡ and Harry L. Swinney
Physics Department, City College of the City University of New York, New York, New York 10031
(Received 17 July 1975)

Light-scattering measurements of the time-dependent local radial velocity in a rotating fluid reveal three distinct transitions as the Reynolds number is increased, each of which adds a new frequency to the velocity spectrum. At a higher, sharply defined Reynolds number all discrete spectral peaks suddenly disappear. Our observations disagree with the Landau picture of the onset of turbulence, but are perhaps consistent with proposals of Ruelle and Takens.

Thirty years ago, Landau proposed[1] that the turbulent state of a fluid results from a large number of discrete transitions or bifurcations, each of which causes the velocity field to oscillate with a different frequency f_i, until for sufficiently large i the motion appears chaotic, although the time correlation functions $C(\tau)$ of the velocity field do not strictly go to zero as $\tau \to \infty$. The Landau picture has been presumed applicable to a large class of systems, including the rotating fluid that we have studied. Systems in a second class (which we will not mention further) exhibit inverted bifurcations, where the transition to turbulence is hysteretic, and usually no periodic regime precedes the onset of chaotic behavior.

The Landau picture has been challenged by Ruelle and Takens,[2] who propose on the basis of abstract mathematical arguments that the motion should be aperiodic with exponentially damped correlation functions after three or four bifurcations to time-dependent states. Recently McLaughlin and Martin[3] have performed numerical calculations on a truncated set of equations applicable to Rayleigh-Bénard convection, and they found a sharp transition to aperiodic behavior following a periodic regime, in qualitative agreement with the arguments of Ruelle and Takens.

A great variety of periodic and chaotic states have been observed in past experiments on rotating[4] and convecting[5,6] systems. These experiments have not examined the onset of aperiodicity in sufficient detail to distinguish between what we term the Landau and Ruelle-Takens pictures. In contrast, Ahlers[7] has recently observed and characterized a sharp transition to aperiodic behavior in sensitive heat-flux measurements on convecting liquid helium; however, the periodic states which presumably precede the transition were not observed.

We present here the first detailed measurements of a *local* property that shows a sequence of periodic regimes followed by a sharp and reversible transition to an aperiodic state, as defined by the vanishing of all discrete spectral peaks (or equivalently, the decay of the time correlation functions). Specifically, we have studied the radial velocity in a fluid rotating between concentric cylinders. The observed behavior clearly contradicts the Landau model of the onset of turbulence.

In our experiments the fluid (water) was confined between an inner rotating stainless-steel cylinder of radius $r_1 = 2.224$ cm and a stationary precision-bore glass tube of inner radius $r_2 = 2.540$ cm. The gap d was uniform to within 1% over its entire length. The fluid height in the cell was 6.25 cm and the cell temperature was 27.5 ± 0.1°C. The ten-cycle average of the rotation period T was constant to within 0.3%.

The local radial velocity V_r was observed by an optical heterodyne technique using an optical arrangement described elsewhere.[8] The scattering volume was located at the center of the gap between the cylinders, and its largest dimension was 150 μm, about 0.05 of the gap. Thus the observations are essentially local measurements, and no significant spatial averaging is involved. The time-dependent frequency of the photocurrent oscillation, which is proportional to $V_r(t)$, was measured for 1024 adjacent sampling intervals of 5×10^{-4} to 5×10^{-1} sec.

In the discussion to follow the rotation rate is expressed in terms of a reduced Reynolds number $R^* = R/R_T$, where $R = 2\pi r_1 d/\nu T$ (ν is the kinematic viscosity) and $R_T = 2501$ is the value of R at the onset of aperiodic motion.

We now describe the sequence of transitions which are observed reversibly as R is varied. The first instability (the Taylor instability) occurs at $R^* \approx 0.051$, and results in a time-independent toroidal roll pattern that has been extensively studied.[4,8] The radial velocity is periodic in the axial coordinate z, with wavelength 0.79 cm.

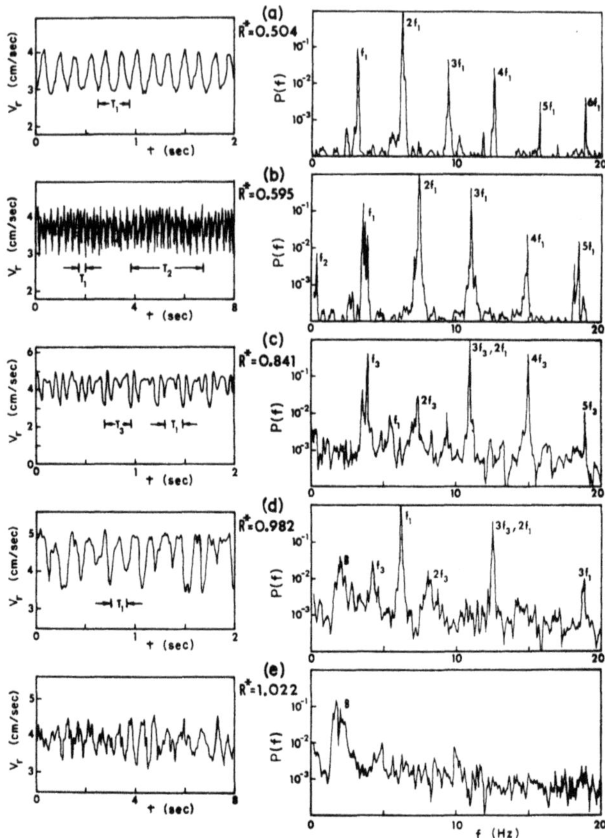

FIG. 1. Time dependence of the radial velocity and corresponding power spectra $P(f)$ [with units cm^2 sec^{-2} Hz^{-1}, normalized so that $\int_0^{25\,\mathrm{Hz}} P(f)\,df = \langle(\Delta V_r)^2\rangle$] for different reduced Reynolds numbers $R^* = R/R_T$.

Our scattering volume was always positioned at or near one of the maxima in $V_r(z)$, and these locations persisted well into the aperiodic regime.

The first transition to a periodic state occurs at $R^* = 0.064$, where transverse waves (with four wavelengths around the annulus) are superimposed on the toroidal vortices.[9] These waves, which have been previously observed visually,[4] manifest themselves as an oscillation at a frequency f_1 in our measurements, as shown in Fig. 1(a) for $R^* = 0.504$. The frequency f_1 scales with $R^* \propto T^{-1}$, as Figs. 1(a)–1(d) illustrate; hence the dimensionless frequency $f_1^* \equiv f_1 T$ is constant. The range in R^* of this and the subsequent time-dependent states is summarized in Fig. 2. The power spectrum $P(f)$ of the radial velocity, shown on a logarithmic scale in Fig. 1(a), contains strong peaks at the frequency f_1 and its harmonics nf_1, and the 0.05-Hz linewidth of these sharp peaks is determined only by the length of the data segment. The background noise in $P(f)$, which comprises only 2.3% of the total spectral power, is frequency independent to at least 100 Hz. This is mostly instrumental noise, with a magnitude of 10^{-4} cm^2 sec^{-2} Hz^{-1}, which corresponds to $\langle(\Delta V_r)^2\rangle^{1/2} = 0.05$ cm/sec.

When R^* is increased to 0.54 ± 0.01, a second time-dependent instability occurs, and a new frequency f_2 is visible as a low-frequency modulation of the radial velocity [Fig. 1(b)]. The corresponding power spectrum shows that f_2, though weaker than f_1, is still nearly 2 orders of magnitude above the noise level. The frequency f_2 is a transverse (i.e., axial) disturbance, as is f_1, but its precise nature is unknown. It does interact with the mode at f_1 to produce a splitting of some of the nf_1 lines. Further increase in rotation rate causes f_2 to decrease in frequency until it is no longer visible at $R^* = 0.78 \pm 0.04$ (see Fig. 2). Associated with this decrease is a gradual increase in the background noise level to about 10^{-3} cm^2 sec^{-2} Hz^{-1} in the range $0 < f < 60$ Hz, as in Fig. 1(c). This background now represents real noise in the fluid, but the peaks, which still contain

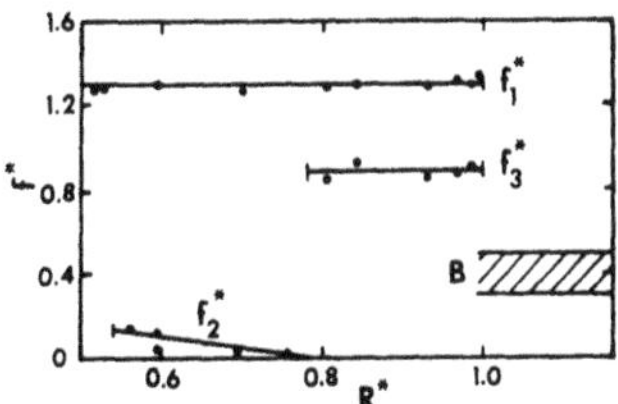

FIG. 2. The dimensionless frequencies $f_i^* = f_i T$ as a function of R^*. The solid lines are to guide the eye, and the vertical bars demarcate the regions in which the f_i are present (except that the lower bound for f_1 is $R^* = 0.064$). The fact that $f_1^* = 1.30$ and $f_3^* = 0.87$ are constant indicates that f_1 and f_3 scale with rotation rate, whereas f_2 does not.

90% of the power, remain sharp. A new frequency f_3 (and its harmonics) appears at $R^* = 0.78 \pm 0.03$, and this is also visible in Fig. 1(c). Note that f_3 appears only after f_2 has disappeared. Since f_3 is two-thirds of f_1, the behavior of Fig. 1(c) is periodic except for the additive noise.

Figure 1(d) shows the behavior just below the transition at $R^* = 1$. The qualitative features are unchanged, and the reduction in the amplitude of the f_3 peak is caused by a small change in the vertical position of the scattering volume. The peaks in $P(f)$ (which still contain 90% of the spectral power) remain sharp, and the corresponding correlation function $C(\tau) = \langle V_r(t) V_r(t+\tau)\rangle$ oscillates periodically without any detectable decay for time lags τ up to 10 sec.

At $R^* = 1$ $(R = R_T = 2501)$ a dramatic change occurs, as shown in Fig. 1(e). The sharp peaks at nf_1 and nf_3 disappear completely, leaving a broad doublet B that contains 60% of the power. While B was incipient at $R^* = 0.982$, it contained only 5% of the power there. Higher-resolution spectra confirm that B is in fact a broad doublet with a total linewidth of about 1 Hz. The disappearance of the sharp peaks is equally apparent in the correlation function, which now decays to zero in a few seconds. The velocity data of Fig. 1(e) show erratic and noisy behavior, but since there is also some noise in 1(d), one must examine $P(f)$ or $C(\tau)$ to see the qualitative distinction. A further increase in rotation rate to $R^* = 1.16$ produced no further qualitative changes, although $P(f)$ broadens substantially.

The transition at $R^* = 1$ is sharp, reversible, and nonhysteretic to within a resolution $\delta R^* = 0.01$. Although the behavior for $R^* > 1$ is independent of the total sample height L, there is some variation with L for $R^* < 1$. However, we always detect three basic frequencies (f_1^*, f_2^*, and f_3^*) followed by a sharp onset of aperiodicity.

Our observation of a sharply defined Reynolds number at which the correlation function $C(\tau)$ decays to zero and the discrete peaks in the power spectrum $P(f)$ disappear represents the first clear demonstration that the Landau picture of the onset of turbulence is wrong. The observed behavior seems to be of the general type described by Ruelle and Takens, in which a few nonlinearly coupled modes are sufficient to produce an aperiodic motion. However, there exists no specific theoretical model applicable to this experiment.

Many questions remain unanswered. The arguments of Ruelle and Takens are quite general, and seem to apply to all systems which exhibit normal bifurcations. For these systems how universal is the behavior we have observed? What physical assumptions are inherent in the arguments of Ruelle and Takens? Finally, what is the sequence of events describing the loss of *spatial* correlation of the velocity fluctuations?

It is a pleasure to acknowledge helpful discussions with H. Z. Cummins, W. Davidon, J. Gersten, J. B. McLaughlin, and W. A. Smith.

---

*Work supported by the National Science Foundation.
†Work performed while on leave from the Physics Department, Haverford College, Haverford, Pa. 19041.
‡J.P.G. gratefully acknowledges the support of the National Oceanic and Atmospheric Administration.

[1]L. Landau, C. R. (Dokl.) Acad. Sci. URSS 44, 311 (1944); L. D. Landau and E. M. Lifshitz, *Fluid Mechanics* (Pergamon, London, England, 1959).

[2]D. Ruelle and F. Takens, Commun. Math. Phys. 20, 167 (1971). Also see R. Bowen and D. Ruelle, to be published; D. Ruelle, to be published.

[3]J. B. McLaughlin and P. C. Martin, Phys. Rev. Lett. 33, 1189 (1974), and Phys. Rev. A 12, 186 (1975).

[4]See, e.g., D. Coles, J. Fluid Mech. 21, 385 (1965); H. A. Snyder, Int. J. Non-Linear Mech. 5, 659 (1970); R. J. Donnelly and R. W. Schwarz, Proc. Roy. Soc. London, Ser. A 283, 531 (1965).

[5]See, e.g., R. Krishnamurti, J. Fluid Mech. 60, 285 (1973); G. E. Willis and J. W. Deardorf, J. Fluid Mech. 44, 661 (1970).

[6]Recent volume-averaged neutron-scattering-intensity measurements on a convecting liquid crystal are provocative but difficult to interpret. See H. B. Møller and T. Riste, Phys. Rev. Lett. 34, 996 (1975).

[7]G. Ahlers, Phys. Rev. Lett. 33, 1185 (1974).

[8]J. P. Gollub and M. H. Freilich, Phys. Rev. Lett. 33, 1465 (1974).

[9]The variety of axial and azimuthal wavelengths observed by Coles were avoided by using a shorter cell and always exceeding $R^* = 1$ before taking data.

Physical Review Letters **47** 243–6 (1981)

Transition to Chaotic Behavior via a Reproducible Sequence of Period-Doubling Bifurcations

Marzio Giglio, Sergio Musazzi, and Umberto Perini
Centro Informazioni Studi Esperienze, Societá per Azioni, I-20100 Milano, Italy

(Received 1 May 1981)

We present the results of measurements of the vertical and horizontal temperature gradients in a Rayleigh-Benard cell. By an appropriate preparation of the initial state, the system can be brought into a single-frequency oscillatory regime. Further stepwise increase in the imposed temperature gradient makes the system go through a reproducible sequence of period-doubling bifurcations up to $f_1/16$. The Feigenbaum δ and μ universal numbers are determined.

PACS numbers: 47.25.-c

A recent theory by Feigenbaum[1,2] suggests that nonlinear systems which can be led into chaotic behavior via a sequence of period-doubling bifurcations will exhibit universal behavior. As the stress parameter λ is increased, the bifurcation points λ_n occur in such a way that the ratio $\delta_n = (\lambda_{n+1} - \lambda_n)/(\lambda_{n+2} - \lambda_{n+1})$ will approach the universal number $\delta = 4.669...$. Furthermore, an appropriately defined ratio of the amplitudes of the Fourier components of neighboring subharmonics will also approach a universal number $\mu = 6.574...$.

Experimental evidence supporting the applicability of the Feigenbaum picture to real physical systems is not very rich. Libchaber and Maurer[3-5] have reported spectra obtained in a low-Prandtl-number Rayleigh-Benard system showing remarkable agreement with the regular subharmonic amplitude decrease predicted by the theory. Because of the lack of resolution in R/R_c, the number δ could only be estimated. Gollub and Benson[6] have reported on the observation of period-doubling bifurcations in an intermediate-Prandtl-number Rayleigh-Benard system. Again the spectra shown exhibit the regular decrease in subharmonic amplitudes.

In this Letter we report on results obtained in a low-aspect-ratio Rayleigh-Benard cell filled with water. We have found a technique for the preparation of the initial state of the system which leads to a very reproducible sequence of period-doubling bifurcations up to $f_1/16$. We determine the first three terms in the δ_n sequence. Also we determine the value of the number μ for a few spectra close to bifurcation points. The overall experimental evidence is in favor of the Feigenbaum picture.

The inner volume of the convection cell is 25 mm wide, 15 mm long, and 7.9 mm high. The lateral boundaries are four 5-mm-thick glass plates. They are glued between two aluminum blocks, whose temperature difference is controlled to within 2 mK (ΔT_{chaos} is close to 7 K). Measurements of the vertical and horizontal temperature gradients averaged along a horizontal line parallel to the short side, roughly at midheight, and close (3 mm) to the sidewall, are obtained by a laser-beam–deflection technique. Deflections along both axes are determined simultaneously to an accuracy of a few microradians with the aid of a two-axis solid-state quadrant sensor (typical angular oscillations are close to 1 mrad). The two signals can be plotted one against the other on an XY recorder (pen recorder or digital oscilloscope). Alternatively, each signal can be fed to a fast-Fourier-transform analyzer for spectral analysis.

An important point we must make is related to the manipulations we perform in order to prepare the initial state of the system. If the temperature difference is increased in small steps starting from zero, and observations are made under stationary conditions at each fixed temperature difference, we notice that the system has a quite complicated behavior and the route to chaos does not seem to be uniquely defined. We have not attempted to collect data under these conditions and therefore we cannot make any meaningful comparison with the results obtained by other authors[6] under similar conditions. We have found, however, that if we suddenly apply a large temperature difference (larger than ΔT_{chaos}), and then we rapidly come back to smaller ΔT values, the system starts oscillating very regularly. If the temperature difference is then changed in small steps, the system can be brought into dynamical states which are very reproducible. Each run (which lasts typically two weeks) consists therefore of one initial quenching operation followed by a sequence of small variations of the

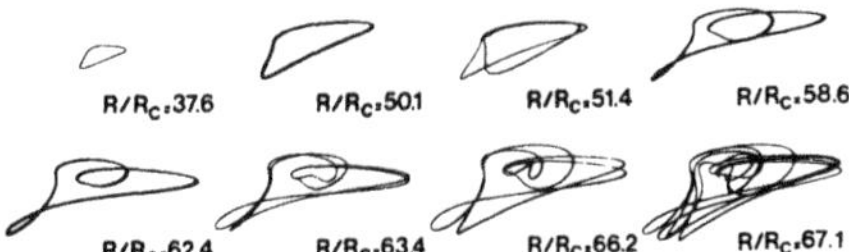

FIG. 1. Plots of vertical temperature gradients vs horizontal temperature gradients for different values of R/R_c.

temperature difference. Each measurement is performed at fixed ΔT (ΔT is our stress parameter λ). Data collected over three runs show very good consistency, and the details of the initial quenching operations seem to have little influence. It appears that the quenching technique brings the system into a phase-space basin from which a Feigenbaum-type route to chaos is accessible.

The actual planform of the instability is rather unclear. Crude whole-field shadowgraph observations indicate that the temperature-gradient oscillations are generated by propagating features which regularly pass under the exploring beam. It seems unlikely that we are dealing with oscillations of a simple spatial structure like in the case of the oscillatory instability studied by Libchaber and Maurer.[3,4] In Fig. 1, we report a sequence of temperature-gradient orbits observed at different values of ΔT (R/R_c values are reported, where R is the Rayleigh number and R_c is the value at threshold for convection). Splittings leading to the appearance of $f_1/8$ can be easily identified in the sequence. We point out that each orbit has been retraced at least fifteen times. Also, orbits obtained at the same R/R_c but on different days are virtually superimposable, and this justifies our previous comments on the general reproducibility of the data.

We must point out, however, that orbits can appear in a rather different form. Immediately after a bifurcation, orbits split into two closely lying replicas that eventually become more separated and distorted as λ is further increased. By recording the orbit traces over extended periods of time, we noticed that the separation of the newly split orbits did not remain constant. Indeed, the orbits execute a very slow oscillatory motion passing back and forth through each other. The orbits never quite retrace themselves, thus indicating that the oscillatory motion occurs at an incommensurate frequency. This effect is quite noticeable after the λ_3 and λ_4 bifurcations. Somewhere in between λ_3 and λ_4 the oscillations disappear, and the orbits resume their stable form as indicated at the end of the sequence in Fig. 1. Beyond λ_4, however, we have never been able to observe stable sixteenfold orbits.

The above observation is essential in order to understand some features of the spectra shown in Fig. 2. The spectra refer to the signals of the horizontal temperature gradient recorded after the λ_2, λ_3, and λ_4 bifurcations. The position of the fundamental frequency f_1 is indicated by the arrow. At $R/R_c = 62.6$ the $f_1/4$ is already present, and all the frequency components are in the form of sharp peaks. At $R/R_c = 66.2$ the emerging $f_1/8$ appears in the form of a finely divided doublet, with separation close to $f_1/38$. Notice that all the other features are sharp, as expected as a consequence of the orbit oscillatory behavior (orbit oscillations introduce an almost 100% modulation on the emerging subharmonic and its odd multiples). At $R/R_c = 67.4$ the emerging subharmonic $f_1/16$ is in reality an even more widely spaced doublet with separation close to $f_1/19$ (twice that of the previous bifurcations). Lack of graphical resolution prevents close examination of the detail, but expanded scale plots show the splitting

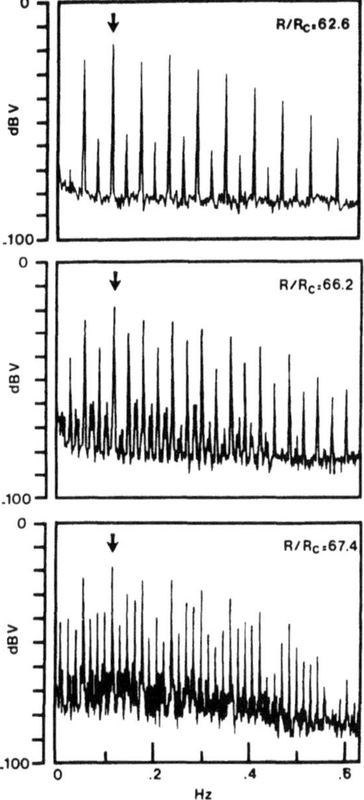

FIG. 2. Horizontal temperature-gradient spectra obtained close to λ_2, λ_3, and λ_4. The arrow indicates the position of f_1.

in a very unambiguous way. Notice that at this stage the splitting is barely larger than the frequency of the new subharmonic.

At this stage, we can attempt to estimate the Feigenbaum universal number μ. Unfortunately, the interpolation and averaging procedure necessary to construct $S(i)$ according to the original Feigenbaum scheme is fairly complex and difficult to use in analyzing our data. (It is best applicable when a large number of bifurcations has occurred.) We have, therefore, taken for $S(i)$ the geometric average of the odd multiples of the 2^ith subharmonic, and we define $\mu_{n,i}$ as the ratio $S(i)/S(i+1)$ evaluated immediately after the λ_n bifurcations.[7] If the average is taken up to $2f_1$, we obtain $\mu_{3,1}=4.1$, $\mu_{4,1}=3.8$, and $\mu_{4,2}=3.8$. An average up to $4f_1$ yields $\mu_{3,1}=4.0$, $\mu_{4,1}=4.2$, and $\mu_{4,2}=3.6$. These numbers are appreciably lower than the value reported in the introduction but in slightly better agreement with the value $\mu \sim 5.0$ estimated by Feigenbaum[8] when one takes for $S(i)$ the geometric average. We can also compare our results with the prediction $\mu=4.58$ put forward by Nauenberg and Rudnick[9] who take for $S(i)$ the rms integrated spectrum over all odd subharmonic multiples. When analyzed in this way, our data give $\mu_{3,1}=3.3$, $\mu_{4,1}=3.0$, and $\mu_{4,2}=4.0$ (averaged up to $4f_1$). Experimental results seem invariably smaller than the theoretical predictions.

Since the locations of the first four bifurcations are known, we can calculate the first three values of the δ_n sequence. They are $\delta_0=1.35$, $\delta_1=3.16$, and $\delta_2=3.53$. Estimates for δ_n can also be obtained by presenting the data in a different way. From the location of the bifurcations we can estimate R_{chaos}. We report in Fig. 3 the behavior of f_1 as a function of $\epsilon=(R_{chaos}-R)/R_{chaos}$, and we indicate the location of the bifurcations on this scale. Since the plot is logarithmic, the spacing of the bifurcations should approach a constant length (this is a consequence of the fact that the λ_n are described by a geometric sequence). In Fig. 3 we also report the length of the Feigenbaum ratio (F.R.) for comparison. The actual values thus determined are $\delta_0=2$, $\delta_1=3.3$, $\delta_2=3.6$, and $\delta_3=4.3$ and typical error bars can be estimated from the figure.

Beyond the last bifurcation we have indications that the system is deviating from the Feigenbaum picture. It is tempting to say that the crossover between the last subharmonic frequency and orbit oscillation frequency is heralding the premature termination of the sequence.

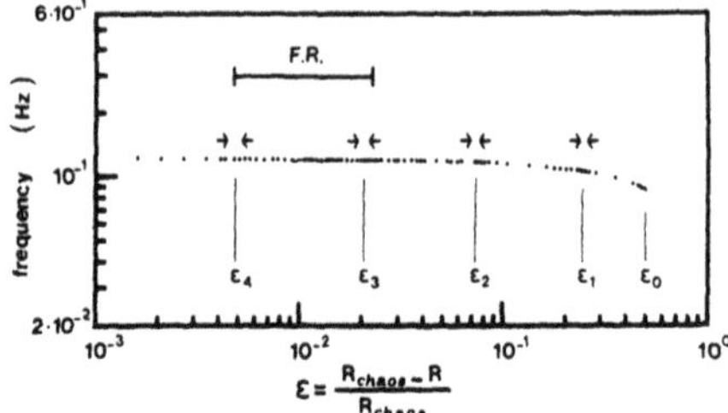

FIG. 3. Log-log plot of f_1 as a function of $\epsilon=(R_{chaos}-R)/R_{chaos}$. The data have been obtained over three runs. The location of bifurcations is reported together with error bars indicating the ϵ ranges outside which the presence (or absence) of a new subharmonic could be unambiguously assigned.

We are greatly indepted to A. Libchaber and M. Feigenbaum for illuminating discussions. Thanks are also due to F. Busse, V. Degiorgio, D. Cannell, H. Cummins, and to M. Corti for the loan of the analyzer. This work was supported by the Consiglio Nazionale delle Ricerche–Centro Informazioni Studi Esperienze, Contract No. 80.00016.02.

[1]M. J. Feigenbaum, Phys. Lett. 74A, 375 (1979).

[2]M. J. Feigenbaum, Commun. Math. Phys. 77, 65 (1980).

[3]J. Maurer and A. Libchaber, J. Phys. (Paris), Lett. 41, L515 (1980).

[4]A. Libchaber and J. Maurer, J. Phys. (Paris), Colloq. 41, C3-51 (1980).

[5]During the preparation of this manuscript, further data have been presented by A. Libchaber, in Proceedings of the NATO Advanced Study Institute on Nonlinear Phenomena at Phase Transitions and Instabilities (to be published). The data refer to the behavior above ΔT_{chaos} and show the ordered and symmetric disapperance of the various subharmonics as one moves away from ΔT_{chaos}.

[6]J. P. Gollub and S. V. Benson, J. Fluid Mech. 100, 449 (1980).

[7]The values for $\mu_{4,1}$ and $\mu_{4,2}$ have not been calculated from the last spectrum shown in Fig. 2. Indeed, the reported spectrum is already somewhat beyond λ_4, and it has been chosen because the doublets around $f_1/16$ are more clearly observable. If μ values are calculated on this spectrum, the values deviate substantially from the theoretical ones and this is a sign that at this stage the sequence is no longer followed (see remark at the end of the paper).

[8]M. J. Feigenbaum, private communication.

[9]M. Nauenberg and J. Rudnick, to be published.

Le Journal de Physique–Lettres **41** L-341–L-345 (1980)

Intermittency in Rayleigh-Bénard convection

P. Bergé, M. Dubois, P. Manneville and Y. Pomeau (*)

DPh-G/PSRM, CEN-Saclay, Boîte Postale 2, 91190 Gif-s/Yvette, France

(*Reçu le 13 mars 1980, accepté le 13 juin 1980*)

Résumé. — Nous présentons la première évidence expérimentale d'une transition vers la turbulence par intermittence dans une expérience de convection à nombre de Prandtl élevé et en géométrie confinée. Il est possible d'interpréter qualitativement cette transition à partir d'un modèle qui se déduit de résultats récents obtenus sur le système de Lorenz.

Abstract. — Experimental evidence of a transition to turbulence *via* intermittency is reported for a convection experiment of a high Prandtl number fluid in confined geometry. This transition is qualitatively understood in terms of a model derived from recent results on the Lorenz system.

1. **Introduction.** — Turbulence can appear in a system after a small number of bifurcations. This is now a well established fact both from experimental evidences [1] and from theoretical considerations [2]. A typical example is given by the Rayleigh-Bénard convective instability of a fluid heated from below, particularly in the case of confined geometries i.e. when the ratio of the horizontal extension to the height of the fluid layer, the so-called *aspect ratio*, is small enough. Increasing the Rayleigh number R_a which is the relevant control parameter, one observes a cascade of instabilities, each step of which adds a new complication to the convective behaviour. After the onset of the stationary convection at R_a^c, the first manifestation of time dependence generally occurs in the form of an oscillatory regime characterized by a remarkable time periodic behaviour of all the convective parameters [3, 4], velocity components [5] and temperature fluctuations [6]. Fourier analysis reveals a unique frequency f_1 (and its harmonics). From this point on, the different routes to turbulence begin to diverge according to the convective pattern and to the Prandtl number essentially. One possible way involves the appearance of a second frequency f_2 in the spectrum [1*c*, 7, 13]. In this paper we shall report on another route which involves only one fundamental frequency f_1 and which has been observed in a high Prandtl number fluid. Temporal chaos sets in continuously and manifests itself under the form of (apparently) random *bursts* interrupting quite regular *laminar phases*. A similar *intermittent* phenomenon has also been observed in low Prandtl number fluids [6*b*, 7].

Since the work of Lorenz [8] and Ruelle [2], theories of the transition to stochasticity in dynamical systems used to relate chaotic temporal behaviour with global unpredictability of otherwise deterministic systems (i.e. systems governed by ordinary differential equations or iterative transformations which are locally deterministic), this unpredictability being the result of the intrinsic instability of trajectories which belong to *strange attractors* [9]. The formal framework of these theories based on genericity arguments is in fact sufficiently versatile to allow for different possible transitions. One of these is the transition *via intermittency* just sketched above. In the following we shall present an interpretation of experimental findings which relates them directly to our present knowledge of the subject gained mainly by studying the Lorenz [10] system and other even simpler models [11].

2. **Experimental results.** — The convective fluid is silicone oil with a viscosity $\nu = 0.1$ St at 25 °C and Prandtl number 130. The cell is rectangular with height $d = 2$ cm, horizontal aspect ratios $\Gamma_x = L_x/d = 2$ and $\Gamma_y = L_y/d = 1.2$; for details on the set-up see reference [13]. Two types of measurements have been performed simultaneously : i) laser Doppler anemometry gives us the local vertical velocity component and ii) a differential interferometric method provides us vertical temperature gradients ($\nabla_z T$) projected on the *xz* and *yz* planes ; in this latter case the information being no longer local but integrated along a

(*) DPh-T, Orme des Merisiers.

direction gives an overall picture of the structure at any time.

Even in small cells such as the one used here, several different convective structures are possible at high Rayleigh number. In particular, we have observed two types of structure, which can be schematically described as the superposition of two rolls along the X-direction with either one or with two rolls along the Y-direction. In the first type of structure (one Y roll), the time-dependent phenomena are related to periodic instabilities in the cold boundary layer leading to the presence of one, then two frequencies in the corresponding spectra of the convective behaviour [12]. Results we present in the following are all related to the second structure which, though not exactly symmetrical, can be schematically described by two rolls along the X-direction and two rolls along the Y-direction. For that kind of structure the mechanism responsible for the time dependent regime (Fig. 1A) which sets in at $R_a/R_a^c = 250$, is the periodic formation, advection and thermal relaxation of a hot (or cold) droplet in a corner of the box [5]. The corresponding spectrum consists of the fundamental frequency f_1 and its harmonics only. When the Rayleigh number increases, the subharmonic $f_1/2$ appears in the spectrum, which corresponds to the fact that one droplet over two contains a greater thermal perturbation. We have to note that, all our experimental data show that, whatever the observed route to turbulence, there is a one to one correspondence between the particular route followed and the corresponding structure.

Increasing further the Rayleigh number, one observes a qualitatively new behaviour : from time to time, the amplitude of the velocity oscillations — see figure 1B where the velocity is measured near the centre of the cell — is strongly perturbed and then relaxes toward the *normal* oscillatory regime (that we call *laminar phase* for convenience). The time lag between two perturbations seems to be random but, when the Rayleigh number increases beyond the threshold R_a^I for this intermittent behaviour at $R_a^I/R_a^c \sim 290$, i) the mean duration of laminar phases shortens and ii) the statistical dispersion of these durations increases. Looking at the interferometric images one does not see any fundamental deviation from the basic picture sketched above, except perhaps small fluctuations of the overall circuit followed by a *normal* droplet during its growth and advection. Strong perturbations of the amplitude of the oscillations are associated with the occurrence of *giant* droplets which have roughly the same fate as *normal* droplets. The structure being more strongly distorted then relaxes to the regular oscillatory regime. That the giant droplet may result from a cumulative effect of fluctuations on normal droplets will be analysed further. At even larger Rayleigh number ($R_a/R_a^c > 310$) the local intermittent perturbations show a more chaotic nature, resembling bursts of turbulence (Fig. 1C) and the corresponding spectrum of the velocity fluctuations shows a *noise* at low frequencies superimposed to the somewhat enlarged peaks given by the oscillations.

3. **Theoretical interpretation.** — A detailed quantitative interpretation is clearly out of reach, even from the simplified point of view of dynamical systems with few degrees of freedom which are believed to contain most of the mechanics of the transition to turbulence in confined geometry. Indeed the drastic simplifications concerning the geometry, the boundary conditions, the order of truncation, etc... impede the construction of a realistic differential system relevant to the experiment reported above. Anyway, this unknown realistic dynamical systems should share some generic features with already well studied models and it has

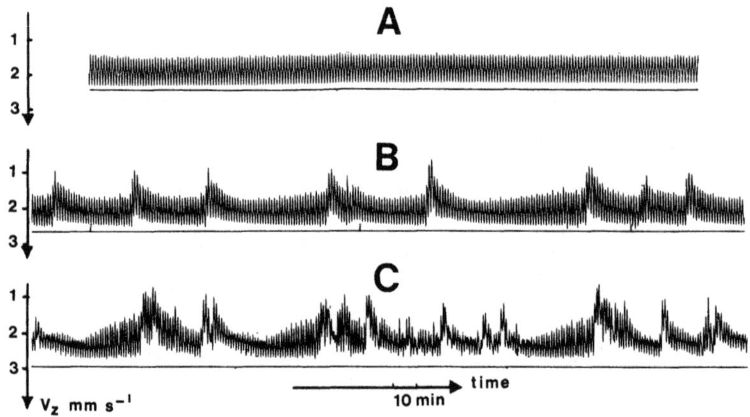

Fig. 1. — Time dependence of the vertical velocity component v_z measured at the centre of the cell ; A : $R_a/R_a^c = 270$; B : $R_a/R_a^c = 300$; C : $R_a/R_a^c = 335$ where $R_a^c = 1\,700$ is the critical Rayleigh number at the onset of convection in infinite geometry.

seemed more promising, i) to gain as much information as possible from these models and more especially from the Lorenz system [8, 10] and ii) to build abstract models displaying these generic features most susceptible of *explaining* the experimental observations [11].

In the Lorenz system [10], as well as in the present experimental case, one follows the destabilization of a limit cycle and the Poincaré map [14] is the best tool to understand the phenomenon. It consists of a series of successive intersections of the trajectories in phase space with a well chosen (hyper)-surface, which gives a kind of stroboscopic vision of the motion. From a general point of view, the structure of a strange attractor corresponding to an aperiodic regime is quite complicated. Known examples are the product of a manifold by a Cantor-like set [16]. If one can forget the transverse cantorian fine structure, then one can restrict oneself to consider a *reduced* Poincaré map which takes the form of a simple iteration on the coordinates which parametrize the manifold. Such a circumstance precisely occurs in the Lorenz model and allows for the quite simple picture of the transition to turbulence [10] summarized in figure 2.

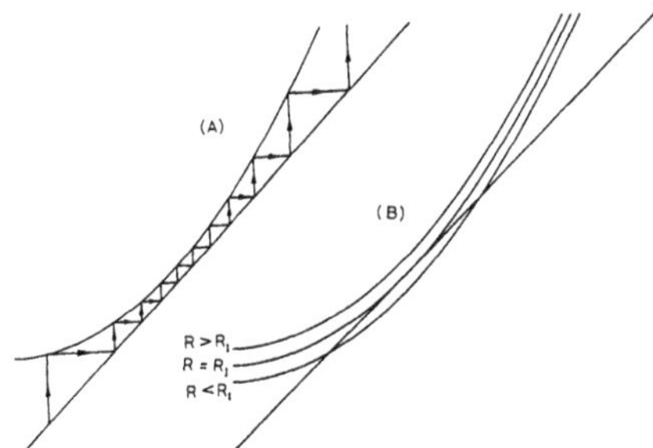

Fig. 2. — A stable limit cycle is associated with a fixed point of the Poincaré map. For the Lorenz model [11] it turns out that in the vicinity of this fixed point the graph of the reduced map is roughly parabolic and that increasing the control parameter r comes to translating the curve upwards (Fig. 2A). Then below the intermittency threshold $R_I = 166.07$, one has two fixed points one stable the other unstable. They collapse at $R = R_I$ and disappear completely above R_I. In the intermittent regime $R > R_I$, the system is allowed to visit from time to time the ghost of the limit cycle; to this *laminar phase* of the motion corresponds a slow drift of the successive iterates in a kind of channel which opens between the first bissectrix and the parabolic curve (Fig. 2B). Then the iterates are expelled from the vicinity of the ghost of the fixed point of the *channel* (the *turbulent burst*) and further recycled toward the intrance.

The Poincaré map is a theoretician's tool, the disposal of which we do not have in a concrete experiment. However we can get a similar stroboscopic vision of the motion (capable of pointing out a trend in the evolution of fluctuations affecting the periodic motion of the droplets) in examining a clearly identifiable feature of the oscillatory behaviour. Such a feature can be for example the minimum M of the velocity variations at a point x_0 recorded in figure 1, which would correspond to a projection of the Poincaré map with the surface of section given by the condition $\frac{d}{dt} v_z(x_0, t) = 0$. The procedure then consists in plotting a given minimum M_n against the precedent one M_{n-1} or, more judiciously since a subharmonic bifurcation has already taken place, against the minimum M_{n-2}. Figure 3 displays two typical *laminar*

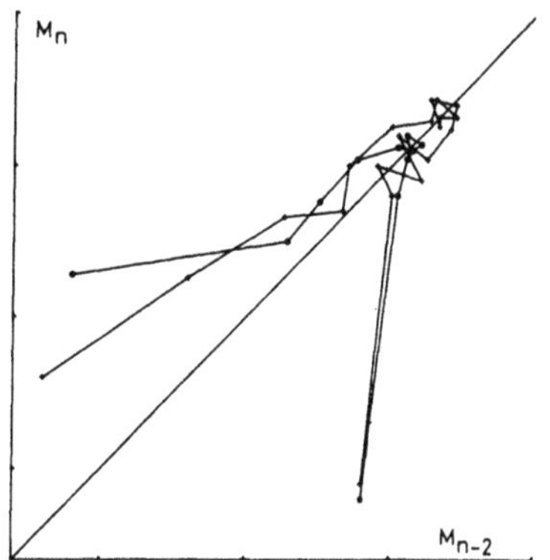

Fig. 3. — *Reduced Poincaré map* extracted from experimental results of figure 1. A minimum M_n of v_z measured at the centre of the cell is plotted against M_{n-2}. Two typical sequences beginning at the end of a *turbulent burst* and ending at the beginning of the next one are plotted, one long (+) and the other short (.); a stable limit cycle would correspond to a clustering of points at a given place on the first bissectrix. Here the upper branches of the cusp shaped graphs correspond to the relaxation of the structure toward what resembles a limit cycle while the lower branches describe the escape phase.

phases. It confirms the view of a behaviour which is deterministic in the short term, since the points organize themselves along a definite pattern hardly disturbed by experimental noise. Now let us assume that intermittency here is of the simplest conceivable type, namely that which occurs in the Lorenz model. The point to understand is then the cusp shape of the *Poincaré map* when compared to figure 2. Noticing that in the Lorenz model the convergence towards the ghost of the limit cycle is rather rapid (see Fig. 2 of Ref. [10]) while here it is much slower (see Fig. 1), gives a hint to construct a model which allows for a qualitative understanding of the behaviour depicted in figure 3 and of its generic meaning. The idea is then to add explicitly a weakly stable direction to the unstable one. The starting point will be the well known *Baker's Transform* of the square [16] $[0,1] \times [0,1]$ defined by $x \to 2x$, $y \to y/2$ for $x < 1/2$ and $x \to 2x - 1$, $y \to (y+1)/2$ for $x > 1/2$. The divergence of nearby trajectories and the sensitivity to initial conditions, typical of a *mixing* system and required for unpredictability, are contained in the expanding transformation on x (which then parametrizes the unstable manifold) while the transformation on y, which insured the aera preserving property, has now to be modified to account for the contraction in the stable manifold direction [15]. The transition from the stable limit cycle (fixed point of the Poincaré map) to the aperiodic trajectories is then obtained by a continuous deformation of the expanding map $x \to 2x$. Finally we retain :

$$x \to 2x, y \to ky \quad \text{for } x < 1/2$$

$$x \to 2x - 1 - \alpha \sin 2\pi x + \beta \sin 4\pi x, \quad y = ky + 1 - k$$
for
$$x > 1/2$$

with $k < 1/2$ to insure contraction, $\beta = 0.1$ and α variable around 0.25 ; the critical value is $\alpha_c \simeq 0.247$, the fixed points in the map disappear for $\alpha > \alpha_c$. A slightly more complicated (though equally generic) transformation on x would even allow to reproduce the sequence, subharmonic bifurcation $\to$ intermittency, which then gets a general meaning. Two typical trajectories in phase space are presented in figure 4. If the experiment could give a quantity directly linked to the x-coordinate (unstable direction) one would obtain the picture of figure 2, but this is not the case and one rather follows a projection of the motion along another direction. One easily sees from figure 4 that the projection in a direction of the form $u = y + \varepsilon x$ ($\varepsilon > 0$ and small) will lead to an iteration, the graph of which presents a striking similarity with experimental results, more specifically the existence of the cusp and of abortive visits to the ghost of the fixed point (see Fig. 5).

4. **Conclusion.** — Of course a complete quantitative account is out of reach but on the basis of quite general arguments, theory touches on several other points such as the mean duration of laminar phases or the decrease of correlation functions. Experiments are underway to further check these predictions. Finally recent abstract theories could appear at first sight only as conjectures on how things happen but contributions such as the work presented in this paper tend to confirm their general validity up to a quasi-quantitative level [17].

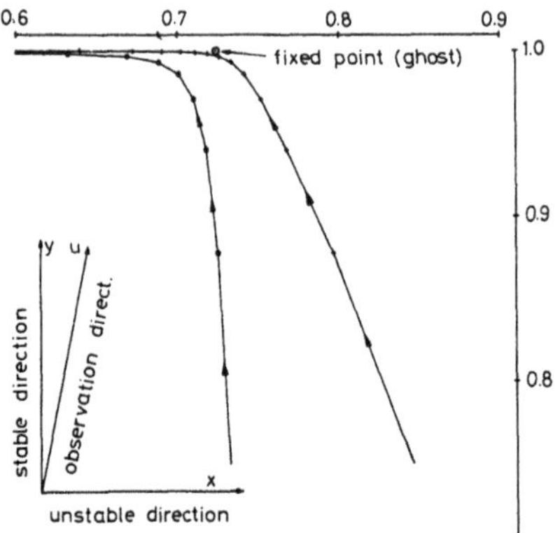

Fig. 4. — Two typical trajectories visiting the ghost of the fixed point of the modified *Baker's Transform* at $x \simeq 0.726$, $y = 1$. The first one (+) passes in its immediate vicinity; we call the second one (.) an abortive visit since it is much shorter.

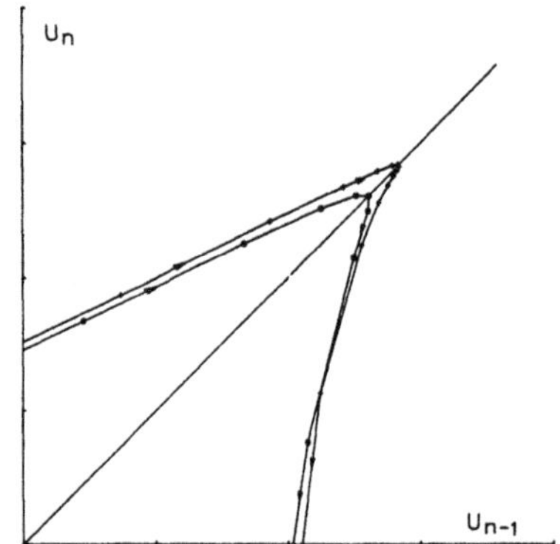

Fig. 5. — The projection of the mapping in a direction $u = y + \alpha x$ (here $\alpha = 0.2$) for the two trajectories of figure 4 should be compared with the experimental ones (Fig. 3). The general aspect is similar and forgetting the experimental dispersion, one can even distinguish between the long laminar phase (+) and the abortive one (.).

References

[1] *a*) AHLERS, G. and BEHRINGER, R. P., *Prog. Theor. Phys. Suppl.* **64** (1978) 186 ;
b) FENSTERMACHER, P. R., SWINNEY, H. L., BENSON, S. A., GOLLUB, J. P., *Ann. N.Y. Acad. Sci.* **316** (1979) 652.
c) MAURER, J. and LIBCHABER, A., *J. Physique Lett.* **40** (1979) L-419.
d) BERGÉ, P., in *Dynamical Critical Phenomena and Related Topics*, C. P. Enz ed. (Springer Verlag, Berlin) 1979.

[2] RUELLE, D., TAKENS, F., *Commun. Math. Phys.* **20** (1971) 167.

[3] KRISHNAMURTI, R., *J. Fluid Mech.* **60** (1973) 285.

[4] BUSSE, F. H., WHITEHEAD, J. A., *J. Fluid Mech.* **66** (1974) 67.

[5] BERGÉ, P., DUBOIS, M., *J. Physique Lett.* **40** (1979) L-505.

[6] *a*) LIBCHABER, A., MAURER, J., *J. Physique Lett.* **39** (1978) L-369 ;
b) LIBCHABER, A., MAURER, J. (private communication).

[7] Gollub, J. P., Benson, S. A., *Phys. Rev. Lett.* **41** (1978) 948 and submitted to *J. Fluid Mech.*
[8] Lorenz, E. N., *J. Atmos. Sci.* **20** (1963) 130.
[9] Pomeau, Y., in *Intrinsic Stochasticity in Plasmas*, G. Laval and D. Gresillon ed. (Editions de Physique, Orsay) 1979.
[10] Manneville, P. and Pomeau, Y., *Phys. Lett.* **75A** (1979) 1 and to be published in *Physica D* (Non Linear Physics).
[11] Pomeau, Y. and Manneville, P., cf. ref. [10] and to be published in *Commun. Math. Phys.*
[12] Dubois, M. and Bergé, P., *Phys. Lett. A* **76** (1980) 53.
[13] Dubois, M. and Bergé, P., *J. Fluid Mech.* **85** (1978) 641.
[14] Iooss, G., *Bifurcation of Maps and Applications*, Math. Studies n° 36 (North Holland, Amsterdam and N.Y.) 1979.
[15] Mandelbrot, B., *Fractals Form Chance and Dimension* (Freeman and Co, San Francisco) 1977.
[16] Arnold, V. I., Avez, A., *Problèmes Ergodiques de la Mécanique Classique* (Gauthier-Villars, Paris) 1967.
[17] Cf. for example Feigenbaum, M. J., *Phys. Lett.* **74A** (1979) 375.

Physics Letters **77A** 391–3 (1980)

REPRESENTATION OF A STRANGE ATTRACTOR FROM AN EXPERIMENTAL STUDY OF CHEMICAL TURBULENCE

J.C. ROUX, A. ROSSI, S. BACHELART and C. VIDAL
Centre de Recherche Paul Pascal, Domaine Universitaire,
33405 Talence Cédex, France

Received 9 April 1980

We show that in Belousof-Zhabotinsky reaction the experimentally observed turbulence can be depicted by a three dimensional "Strange Attractor".

In a previous study of the transition to turbulence in the Belousof-Zhabotinsky reaction we showed how a regular flow becomes more and more chaotic [1]. The method used at that time was a frequency analysis (Fast Fourier Transform) of the time variations of the concentration of one selected species (Ce^{4+} measured by its optical density at 340 nm) involved in this reaction. The appearance of a broad band in the frequency spectra was considered as the indication of the transition to turbulence as it was done in hydrodynamic studies [2]. However, to measure how turbulent a regime is, neither this broad band, because of its irregular shape, nor the autocorrelation time, because of its complicated behavior [1], can be used. The lack of a measure for this phenomenon leaves us with only qualitative descriptions whose precision and variety are thus of prime importance.

The aim of the present work is to provide such a new representation.

From a theoretical view point, turbulence is described by surveying the trajectories in the phase space. A system is turbulent when these trajectories describe a so-called strange attractor [3]. In the case of the Henon attractor [4] the calculated power spectra [5] show the appearance of a broad band at the same time as the bifurcation toward the strange attractor, thus giving an indication of the equivalence of the descriptions (see also ref. [11]).

In the study of chemical turbulence a three dimensional representation of the attractor would be obtained by the simultaneous recording of three selected signals. This procedure raises many difficulties in an experiment and, up to now, we have only succeeded in simultaneously measuring two independent intermediate species (see later). However, to prevent these difficulties two procedures have been recently proposed. Starting with the recording of the time variations of one signal ($x(t)$) one can reconstruct a finite dimensional phase space picture. The heuristic idea underlying these conjectures is that any set of independent quantities calculated at a given time from the time variation record will define an attractor diffeomorphically equivalent to the true one [6]. Thus we can obtain this representation of the attractor by choosing for the coordinates of a point in such

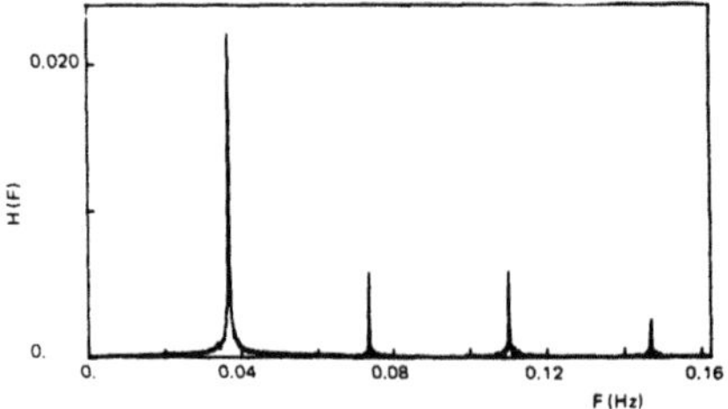

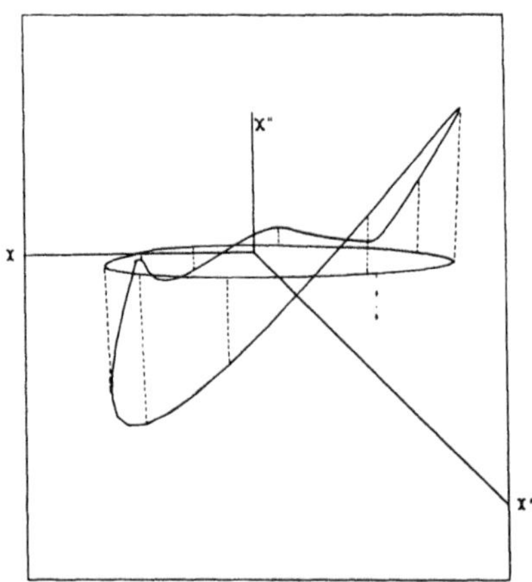

Fig. 1. Periodic regime; mean residence time 4.57 mn; $T = 40°C$. Concentrations of reagent in the reactor before any reaction. $[CH_2(COOH)_2]_0 = 8 \times 10^{-2}$ mole.l^{-1}; $[NaBrO_3]_0 = 3.6 \times 10^{-2}$ mole.l^{-1}; $[Ce_2(SO_4)_3]_0 = 2.5 \times 10^{-4}$ mole.l^{-1}; $[H_2SO_4]_0 = 1.44$ mole.l^{-1}. Arbitrary units on the axis.

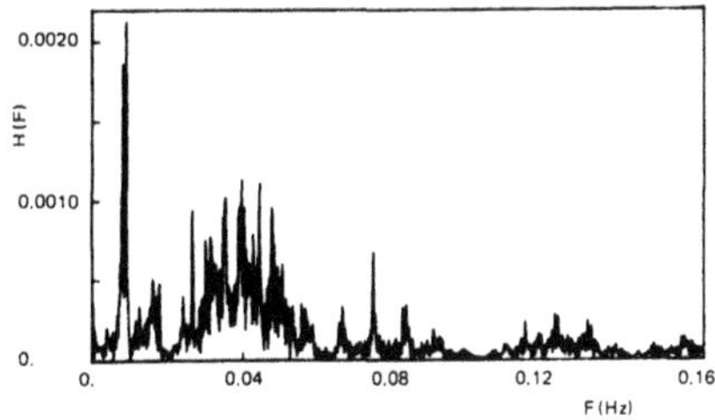

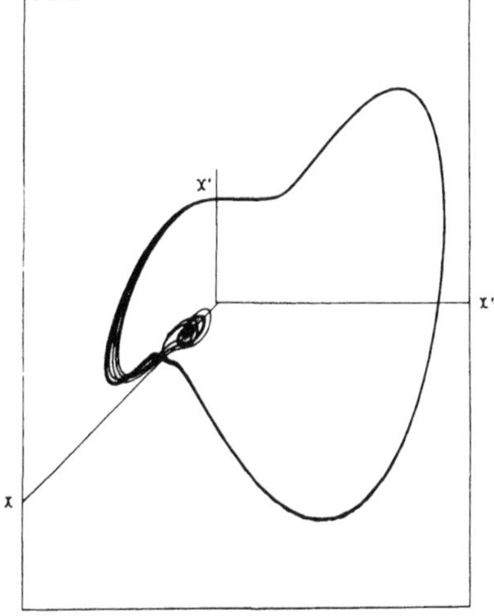

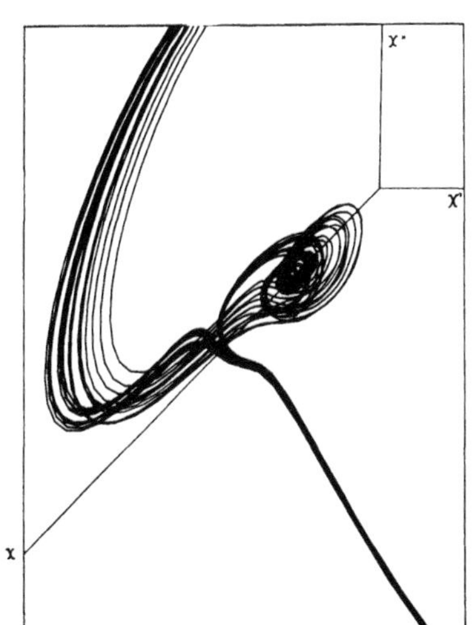

Fig. 2. Turbulent regime; mean residence time 5.60 mn. Same experimental conditions as in fig. 1.

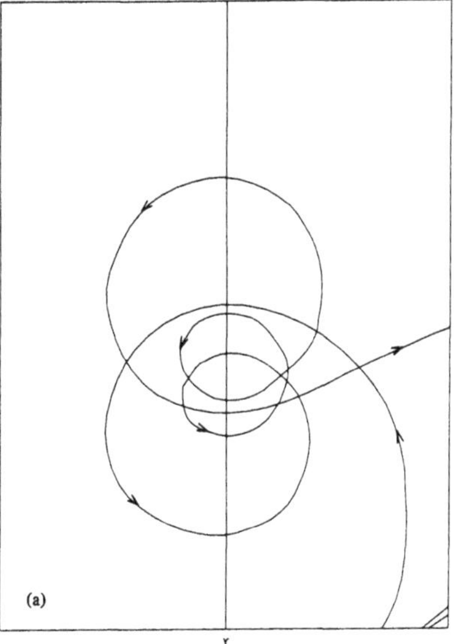

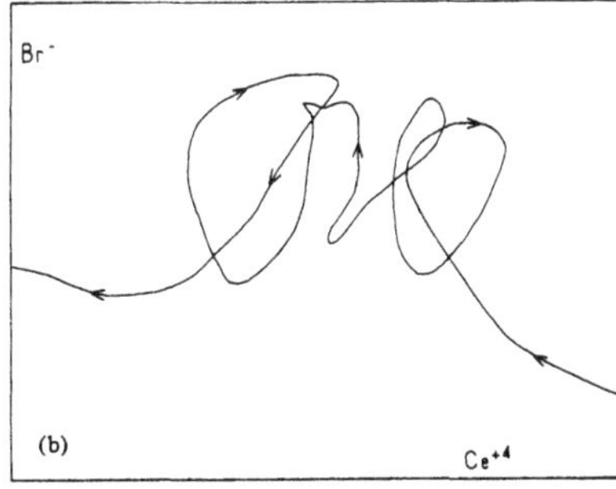

Fig. 3. (a) Projection of part of the trajectories shown in fig. 2 on the $x, \dot{x}$ plane. (b) Simultaneously recorded variations of the Bromide and Ce^{4+} concentration for the experimental conditions of fig. 2.

a space either $x(t)$, $x(t + \tau_1)$, $x(t + \tau_2)$ (τ_1, τ_2 independent and arbitrary numbers) [7] or $x(t)$, $\dot{x}(t)$, $\ddot{x}(t)$ [6]. We present here the results obtained by this last method when applied to the time variation of $[Ce^{4+}]$ in the Belousof-Zhabotinsky reaction.

Figure 1 and 2 describe for the experimental conditions quoted (differing only by the removal time of the reactor) the frequency spectrum and the three dimensional representation for two typical cases: periodic regime (fig. 1), and turbulent regime (fig. 2). One can see, at least on the enlarged view in fig. 2 that the trajectories look very much like those shown by Rössler [8] or others and considered as characteristic of a strange attractor. At first sight, however, it seems that the trajectories shown in fig. 2 exhibit some kind of regularity: when a trajectory emerges from the eddy centered on the x axis it will not visit the eddy on the next turn but only on the second one, and this situation is repeated again and again. Nevertheless the "sensitivity to initial condition " [9] is preserved: from the knowledge of the representative point at a given time we cannot infer its position some time later (long enough to be sure it enters the eddy at least one time) because the residence time in the eddy depends on the trajectory choosen in it and because this choice appears to be unforseeable.

A projection on the plane $x\,\dot{x}$ of part of the trajectories is shown in fig. 3 together with a two dimensional record of the simultaneous variations of $[Ce^{4+}]$ and $[Br^-]$ (measured by a specific bromide electrode). The main parts of the trajectories in both cases are not shown in order to make the comparison easier but they do not present any salient difference. Such a comparison indicates that the geometrical characteristics of the attractor are preserved in the $\dot{x}, \ddot{x}$ plot.

A "good" description of the attractor is thus obtained in the $x, \dot{x}, \ddot{x}$ representation; a qualitatively comparable picture would be reached in a $x(t)$, $x(t+\tau_1), x(t+\tau_2)$ coordinates system. We hope that this qualitative description will provide a better understanding of what we call chemical turbulence. A "measure" of turbulence by computing LYAPUNOF exponent on these trajectories is under progress in our laboratory. These results will be published elsewhere [10].

Special thanks to P. Hanusse for computational assistance in the three dimensional graphical representations.

[1] C. Vidal, J.C. Roux, A. Rossi, S. Bachelart, C.R. Acad. Sci. Paris 289C (1979) 73;
C. Vidal, J.C. Roux, S. Bachelart, A. Rossi, Annals of N.Y. Acad. Sci., to be published.
[2] P.R. Fenstermacher, H.L. Swinney, J.P. Gollub, J. Fluid Mech. 94 (1979) 103.
[3] D. Ruelle, F. Takens, Commun. Math. Phys. 20 (1971) 167.
[4] M. Henon, Commun. Math. Phys. 50 (1976) 69.
[5] J.H. Curry, preprint (1979).
[6] N.H. Packard, J.P. Crutchfield, J.D. Farmer, R.S. Shaw, Submitted to Phys. Rev. Letters (1979).
[7] D. Ruelle, private communication.
[8] O.E. Rössler, in: Structural stability in physics (Springer Verlag, Berlin, 1979) p. 290.
[9] D. Ruelle, Annals of N.Y. Acad. Sci. 316 (1978) 408.
[10] S. Bachelart, A. Rossi, J.C. Roux, C. Vidal, to be published.
[11] J. Crutchfield, D. Farmer, N. Packard, R. Shaw, C. Jones, R.J. Donnelly, Phys. Letters 76A (1980) 1.

Journal of Chemical Physics **74** 6171–7 (1981)

Chaos in the Belousov–Zhabotinskii reaction

J. L. Hudson and J. C. Mankin

Department of Chemical Engineering, University of Virginia, Thornton Hall, Charlottessville, Virginia 22901

(Received 22 December 1980; accepted 10 February 1981)

Experimental results obtained with the Belousov–Zhabotinskii reaction in a continuous stirred reactor are analyzed in order to distinguish between periodic and chaotic behavior. Power spectra and stereoscopic three-dimensional state space trajectories are presented for both periodic oscillations and for chaos. A next amplitude map is constructed for the chaotic case and from it a Liapunov characteristic exponent of $+0.62$ is calculated. The rate of divergence of trajectories is illustrated by following the behavior as predicted from the next amplitude map of 10 points initially very close together.

INTRODUCTION

There has been considerable interest recently in sustained oscillations in chemically reacting systems. These oscillations can be periodic in which case the concentrations of some species undergo regular variations with time or they can be nonperiodic in which case the reactor never approaches a globally attracting limit cycle. This latter condition has been termed chemical chaos.

Many of these studies have been carried out with the Belousov–Zhabotinskii reaction since it has been shown through experiments[1,2] and numerical calculations[3] that the reaction can exhibit a variety of oscillation types in an open reactor. In recent experimental studies apparent chaotic behavior has been observed with this reaction.[4–8] Chaotic behavior has also been observed in catalytic reactions,[9–11] in an enzyme reaction,[12] and in the decomposition of $S_2O_4^{2-}$.[13]

Rössler has made theoretical studies on chaos using abstract chemical kinetics.[14] This work has led to theoretical studies of chaos in more realistic chemical systems including both homogeneous[15–18] and surface catalytic[19,20] reactions.

In the experimental studies on chaos in chemical systmes the results have been largely restricted to the measurement of some dependent variable as a function of time. However, power spectra and autocorrelation functions have recently been presented for the Belousov–Zhabotinskii reaction by Vidal *et al.*[8] and next amplitude maps have been constructed by Olsen and Degn for an enzyme reaction[12] and by Tomita and Tsuda[21] for some experimental results of Hudson *et al.*[5] Furthermore, Tomita and Tsuda have developed a model for chaotic behavior based on their return map. Roux *et al.* have presented three dimensional state space plots for the Belousov–Zhabotinskii reaction.[22]

In this paper we present analyses of experimental results on the Belousov–Zhabotinskii reaction in a continuous reactor which illustrates the chaotic nature of the oscillations. We compare power spectra and three dimensional state trajectories for chaotic and periodic flows. We construct a next amplitude map which can reproduce chaotic behavior. From the next amplitude map we calculate a positive Liapunov characteristic exponent and demonstrate how neighboring trajectories diverge and how information is thus lost. In this manner we give further evidence that the experimentally obtained behavior is true chemical chaos.

EXPERIMENTS

A continuous stirred reactor with requisite feed system, temperature control, and measurement was used for the experiments. The apparatus is an improved version of that used in a previous publication.[5] The reactor volume is 26.4 ml and the temperature is 25 °C. The malonic acid, sodium bromate, sulfuric acid, and cerous ion (as cerous sulfate octahydrate) are fed with a four channel peristaltic pump to four separate inlet ports in the bottom of the reactor. The mixed feed concentrations are 0.3, 0.14, 0.2, and 0.001 M, respectively. (The concentrations in the stored solutions were four times these values.) Stock sodium bromate contains bromide ion impurity such that the mixed feed bromide ion concentration is approximately 3×10^{-6} M as measured by a specific ion electrode. The effect of Br^- impurity in the feed stream was discussed in an earlier presentation.[6] The reactor is stirred at 2000 rpm with a motor driven glass stirrer and has four small Plexiglas baffles on the wall. The liquid exits through a small gap surrounding the stirring rod at the top of the reactor. A platinum wire electrode and a bromide ion electrode are used to follow the course of the reaction. Data are taken both with a strip chart recorder and in digital form through direct hookup to a microcomputer.

RESULTS

Typical results are presented in Figs. 1(a), 1(b), and 1(c), where the output from the bromide ion electrode is shown as a function of time for a two peak periodic oscillation, nonperiodic behavior, and a three peak periodic oscillation, respectively. These three types of behavior are a subset of the 11 types which can occur at various residence times for the same inlet concentrations and temperature.[5] The three oscillations shown in Figs. 1(a), 1(b), and 1(c) occur at adjacent ranges of residence time, i.e., starting with a two peak oscillation, an increase in flow rate will result in a change or bifurcation to nonperiodic behavior and then

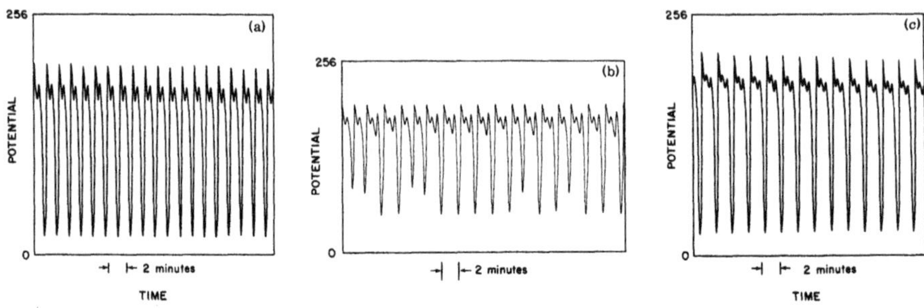

FIG. 1. Recording from bromide ion electrode; $T = 25\,°C$; (a) residence time = 6.35 min; (b) residence time = 6.01 min; (c) residence time = 5.68 min.

to a three peak periodic oscillation. The flow rates at which these bifurcations occur are reproducible and there is no hysteresis.

The electrode potentials in the figures and throughout the paper are expressed in units of 0 to 255, corresponding to the output of the 8 bit analog to digitial converter used to interface the electrodes to the microcomputer. This range covers a range of electrode potentials of 136 mV for Pt and 64 mV for Br^-, with the zeros of the output ranges at approximately 920 and 140 mV, respectively.

The question to be answered is whether or not the behavior shown in Fig. 1(b) is truly chaotic. Some information can be gained by an examination of the Fourier transform of the data. Power spectra of the data from Figs. 1(a), 1(b), and 1(c) are shown in Figs. 2(a), 2(b), and 2(c), respectively. It is seen in Figs. 2(a) and 2(c) that the transform of the periodic oscillations results in the expected sharp peaks and associated harmonics. Sharp peaks do not appear, however, in the power spectra of the nonperiodic oscillation as shown in Fig. 2(b).

Three dimensional state space trajectories are useful both for increasing the understanding of the type of behavior being observed and also for comparison to theories on chaotic flows. A minimum of three dimensions is necessary to generate chaos. Since three dimensions is the minimum number to yield chaos but the maximum number which can be displayed, we have made stereoplots of the data from Figs. 1(a), 1(b), and 1(c). These are shown in Figs. 3(a), 3(b), and 3(c), respectively. Since we have measured only two variables, viz., the bromide ion electrode potential (Br^-) and the platinum electrode potential (Pt), we have generated a third variable by differentiating the platinum potential with respect to time ($\dot{Pt}$). The derivative is calculated as the change in the above defined units per second. The view for the stereoplots of Fig. 3 is from an angle of 116.6° from the positive Pt axis toward the positive Br^- axis and from an angle of 23.75° below the Pt–Br^- plane. The observation point is at a distance from the origin of three times the length of the axes and the two views have a separation of 6°.

For a third order system the use of any three independent quantities will determine the behavior. Packard *et al.*[23] have recently shown that replacing the original variables $x(t)$, $y(t)$, and $z(t)$ in a set of equations by $x(t)$, $\dot{x}(t)$, $\ddot{x}(t)$ where the dot denotes differentiation does not alter the topological characteristics of the flow. In fact, it seems reasonable to use any combination of variables, time derivatives of these variables, and, as is done below, time delayed variables. There is, of course, the limitation that these variables must be independent.

Figure 4 is another stereoplot using the same data as Fig. 3(b). The view is the same as that of Fig. 3. In this case, however, the third variable employed is Pt(t-10 sec), i.e., a time delay is used in one variable to generate a third. The resulting trajectory seems to have some of the characteristics of the "screw" chaos described by Rössler.[24] Perhaps this type of flow should be called "dunce cap" or "cone" chaos since, unlike a screw which is tapered only at the end, the trajectories wind continuously outwards toward the base before being reinjected to the region near the tip. In our case the trajectory would make one or two loops around the dunce cap before being reinjected. However, we cannot say for certain that the chaos being observed is the same as that described theoretically by Rössler. A plane at a constant value of Pt(t-10 sec) passed thourgh the upper left-hand corner of Fig. 4 results in intersections of the plane and the trajectories which are almost on a single line. (This is not the case for the reinjection curves of the right-hand portion of the figure.) This would not occur in general for screw or dunce cap chaos. It may be that our dunce cap is simply rather short, but the answer awaits further experimentation.

The trajectories presented by Roux *et al.*[22] were based on [Ce^{4+}] and its first two time derivatives. Their results resemble Fig. 3 in that in both cases the trajectories have two parts: a large loop having a portion where the individual trajectories are close together, and a smaller region where more complicated behavior

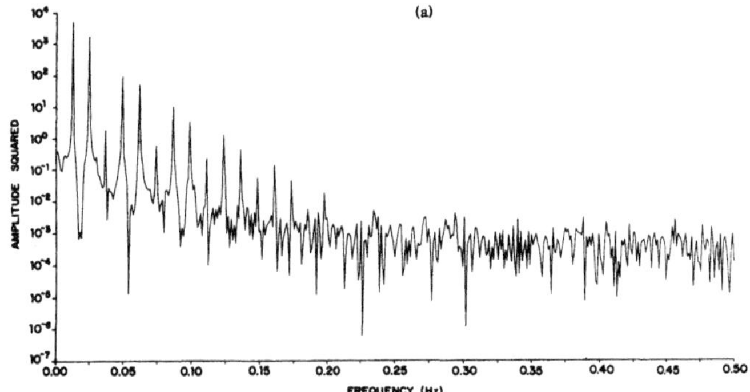

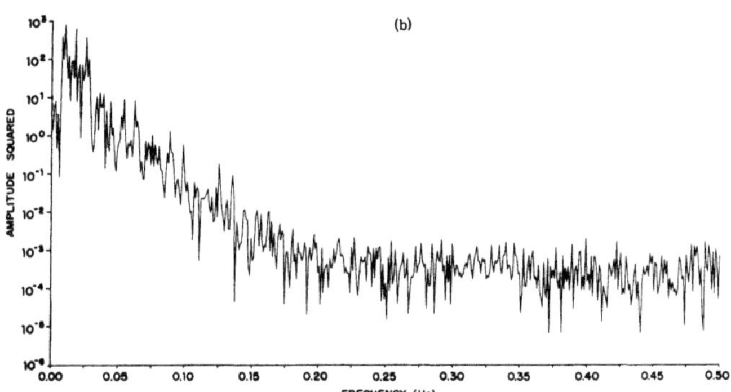

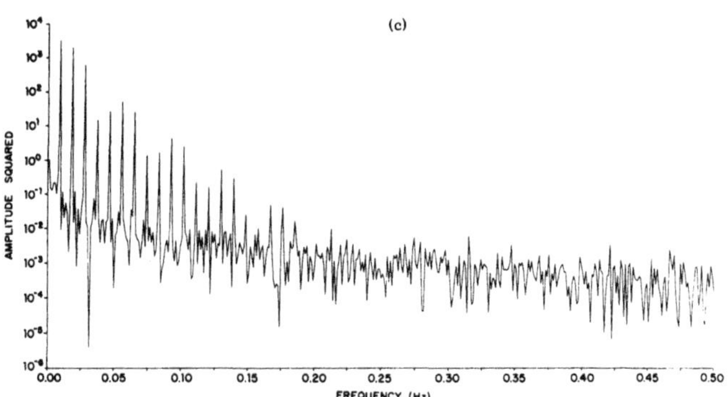

FIG. 2. Power spectra; (a), (b), (c) correspond to Figs. 1(a), 1(b), and 1(c), respectively.

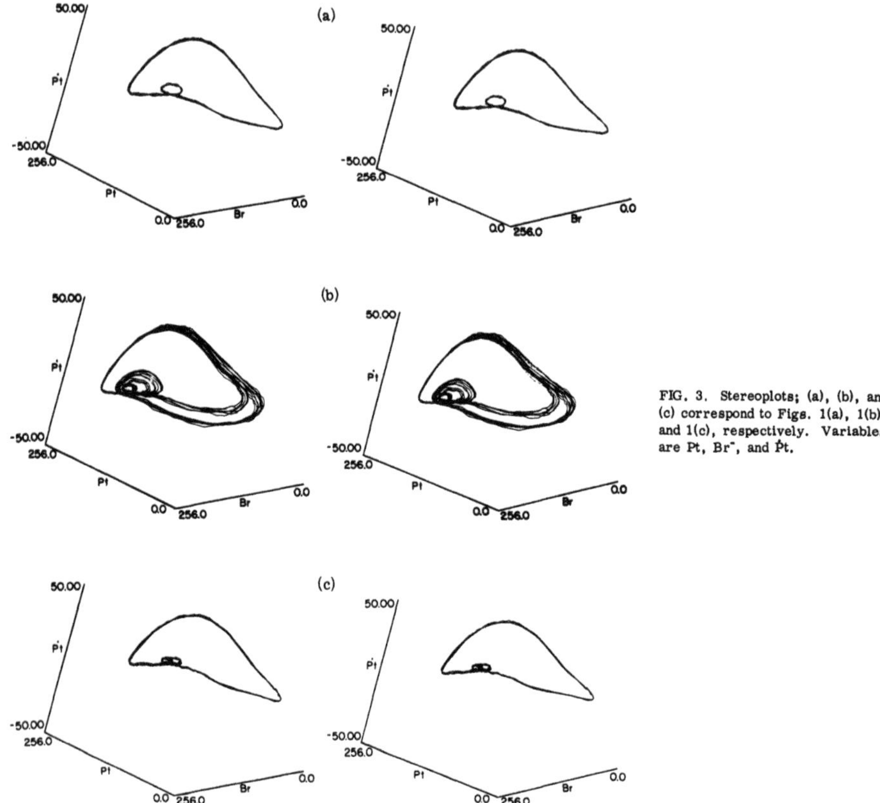

FIG. 3. Stereoplots; (a), (b), and (c) correspond to Figs. 1(a), 1(b), and 1(c), respectively. Variables are Pt, Br^-, and $\dot{P}t$.

occurs. However, in their case the smaller region is not visited after every cycle through the large loop.

We now turn to the construction of maps and the calculation of the Liapunov characteristic exponent, both of which help to characterize the nature of the chaotic trajectories. A map can be constructed from Fig. 3(b) by passing a surface through the flow and noting the successive traverses of the surface. Since the trajectories of Fig. 3(b) remain approximately on a two dimensional manifold, the resulting map is almost one dimensional. It is convenient to construct the return map by employing the minima in the bromide ion potential and this is done in Fig. 5, where the minima at the $(n+1)$st iterate are shown as a function of the value at the nth iterate. This figure is similar to that of Tomita and Tsuda,[21] who constructed a return map using some of our earlier data[5] which were then available only in analog form as a strip chart recording. It is noted that the figure is remarkably one dimensional which in itself is an indication that the chaotic nature of this system is due to chemistry and fluid flow and not to some random fluctuation in the experiment.

A mathematical representation of the curve in Fig. 5 was developed by fitting the data as follows:

$$y = \begin{cases} 175 - 4.42[1.44 - (x - 172)]^{1/3} e^{0.0126x}, & x \geq 172, \\ 175 - 4.42[1.44 + (172 - x)]^{1/3} e^{0.0126x}, & 172 > x \geq 150, \\ 175 - 182(175 - x)^{2.04} e^{0.0918x}, & 150 > x \geq 60, \\ 191 - 0.375x, & x < 60, \end{cases} \qquad (1)$$

where for convenience we write $y = \mathrm{Br}(\min)_{n+1}$ and $x = \mathrm{Br}(\min)_n$. For Br (min) greater than 60, the function was derived by using a form similar to that used by Tomita and Tsuda using constants such that

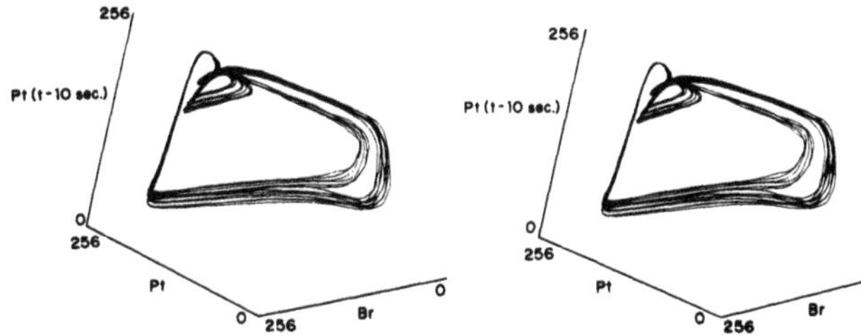

FIG. 4. Stereoplot of chaotic behavior. Variables are Pt, Br^-, and Pt(t − 10 sec).

(a) The function is continuous [except at Br^-(min) = 60].

(b) The position of the inflection point is consistent with the data.

(c) The minimum and its location are consistent with the data.

(d) The function value and derivative at Br^-(min) = 80 equal those of a line through the data at that point.

For Br^-(min) below 60, the map is approximated by a line through the data in this region. The line described by Eq. (1) is shown in Fig. 5.

One of the characteristics of chaotic behavior is that trajectories in state space diverge in at least one direction. For example, for the type of flow shown in Fig. 3(b) two adjacent trajectories will diverge in a direction within the two dimensional manifold given sufficient time, but will converge toward the manifold. This is is contrast to a state with an attracting limit cycle where all trajectories in the neighborhood of the limit cycle converge. The Liapunov characteristic exponent is a measure of the rate of divergence or convergence of trajectories[25] and thus can be used to differentiate between chaotic and periodic behavior. It is defined as

$$\bar{\lambda} = \lim_{n\to\infty} \frac{1}{n} \sum_{i}^{n} \log_2 \left| \frac{dy}{dx} \right| , \qquad (2)$$

where (dy/dx) is the slope of the return map in Fig. 5 evaluated at the bromide ion potential minimum of the nth iterate; the log base 2 is used for reasons given below. As discussed by Shaw,[25] the Liapunov characteristic exponent is invariant, i.e., a number of return maps such as Fig. 5 can be made from Fig. 3 by cutting the surface in various ways. Each of these return maps would generate the same value of the Liapunov characteristic exponent. We use bromide ion potential minima since the spread of the trajectories is greatest there and experimental error is thus reduced.

The value of the Liapunov characteristic exponent obtained from Eq. (2) and the return map of Eq. (1) is

$$\bar{\lambda} = 0.62. \qquad (3)$$

The fact that $\bar{\lambda}$ has a value greater than zero implies that the trajectories diverge and that the behavior of the system is chaotic. However, it should be noted that the value of the Liapunov characteristic exponent was found to be very sensitive to the parameters of Eq. (1). In order to investigate this sensitivity somewhat, we varied the slope and intercept of the linear portion of the fit. We varied the slope by ± 30% and the midpoint of the line by about one unit and calculated a 5 × 5 matrix of $\bar{\lambda}$'s. Nineteen of these $\bar{\lambda}$ were in the range 0.47 to 0.65 although six of the values were negative. There was no regular pattern to the results although raising the midpoint to its largest value resulted in the most negative values of $\bar{\lambda}$. It is thus seen that slight changes in the return map of Eq. (1) can have a significant effect on the nature of the series of Br^-(min) predicted from it; the result can even be a periodic flow $\bar{\lambda}(< 0)$.

There are three Liapunov characteristic exponents for trajectories in three dimensional space, and Eq. (3) gives the value of the largest of these exponents. The fact that the value of this largest exponent is greater than zero means that the trajectories diverge in one direction. The direction in this case is that along the (Br^-) axis. It would be of interest to know the rate of divergence in a direction along the two dimensional manifold within whose neighborhood the trajectories are confined as noted above. If such a two

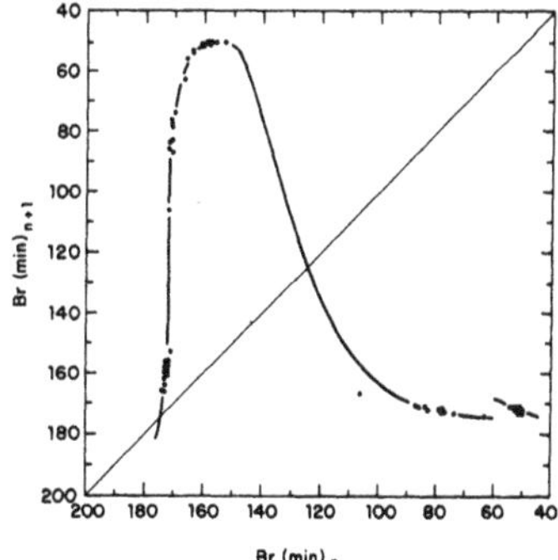

FIG. 5. Return map. Minimum in bromide ion electrode potential vs the value at the previous minimum.

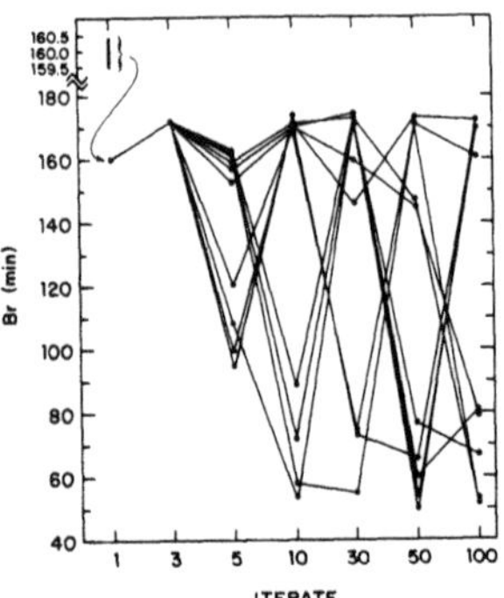

FIG. 6. Divergence of neighboring minima.

dimensional manifold were planar, the Liapunov characteristic exponent found for divergence within it would be that given by Eq. (3) since carrying out a linear transformation of the axes of Fig. 4 does not affect the calculation of $\bar{\lambda}$. The curvature of the manifold will affect the calculation somewhat. Nevertheless, the Liapunov characteristic exponent for a direction along the manifold is expected to be approximately 0.62. On the other hand, the value of the direction normal to the manifold is expected to be less than zero since trajectories probably converge in that direction.

We now know that the trajectories diverge in one direction. An interesting consequence of this fact is that information is lost with time, or, equivalently, with each iteration. Say, for example, that the value of some variable such as Br^- is known at some time to within some experimental error or uncertainty. Consider two different values of Br^- close enough together such that within experimental uncertainty they can be considered to be the same point. Trajectories beginning at these two points will diverge, or, equivalently, the uncertainty will grow and evantually all information is lost, which means that there is no more information on where the trajectory is in state space. Shaw[25] has noted that infornation is lost in N iterates, where N is given by

$$N = \text{bits information}/\bar{\lambda}\,. \tag{4}$$

The bits of information refers to the information in a measurement. For example, in our laboratory we are using an 8 bit analog to digital converter for data acquisition so that the absolute upper limit, even with a hypothetical perfect experiment, would be 8 bits or one part in 256. Thus, the upper limit on the number of iterates which can be made before all information is lost is

$$N = 8/0.62 \cong 13 \text{ iterates.} \tag{5}$$

This loss of information can perhaps be better seen in the following manner: Consider the values of Br^- (min) all within one part in 256, i.e., all within the uncertainty range of our experiments. Choose ten values from 159.5 to 160.4 on the 0 to 255 scale. All ten numbers round off to 160. We consider subsequent values of Br^- (min) for trajectories starting at each of these 10 points. These values of Br^-(min) are calculated using Eq. (1) and the results are shown in Fig. 6. At the third iterate the Br^-(min) are still so close together that they still appear to be a single point to the accuracy of the graph. At the fifth iterate the divergence is evident and by the 10th iterate the values of Br^- (min) fill the entire space. This is in reasonable agreement with the calculation shown above in which it was predicted that information would be lost in 13 iterates.

DISCUSSION

We have demonstrated that the apparent nonperiodic behavior that we have observed in a continuous stirred tank reactor is most likely true chemical chaos. The data were recorded by means of direct hookup to a microcomputer and were analyzed in order to compare the characteristics of the experimental system to those of theoretical systems. One such possible comparison was between the stereoplots for the experimental data and those of Rössler derived from mathematical models. These stereoplots showed that the type of return map constructed by Tomita and Tsuda is the most powerful for this experimental system. Of perhaps greatest significance, however, is the fact that apparently for the first time for a real chemical system the Liapunov characteristic exponent and the rate of divergence of trajectories have been calculated. Divergence of trajectories is one of the most distinctive characteristics of chaotic behavior; the fact that a positive exponent was found and that successive minima were seen to diverge are strong evidence that true chaotic behavior has been observed.

[1]K. R. Graziani, J. L. Hudson, and R. A. Schmitz, Chem. Eng. J. **12**, 9 (1976).
[2]M. Marek and E. Svobodova, Biophys. Chem. **3**, 263 (1975)
[3]K. Showalter, R. M. Noyes, and K. Bar-Eli, J. Chem. Phys. **69**, 2514 (1978).
[4]R. A. Schmitz, K. R. Craziani, and J. L. Hudson, J. Chem. Phys. **67**, 3040 (1977).
[5]J. L. Hudson, M. Hart, and D. Marinko, J. Chem. Phys. **71**, 1601 (1979).
[6]J. L. Hudson, D. Marinko, and C. Dove, Discussion meeting, "Kinetics of Physicochemical Oscillations," Aachen, September 1979.
[7]O. E. Rössler and K. Wegmann, Nature (London) **271**, 89 (1978).
[8]C. Vidal, J.-C. Rous, S. Bachelart, and A. Rossi, "Experimental Study of the Transition of Turbulence in the Belousov–Zhabotinsky Reaction," preprint (1980).
[9]J. E. Zuniga and D. Luss, J. Catal. **53**, 312 (1978).
[10]R. A. Schmitz, G. T. Renola, and P. C. Garrigan, Ann. N. Y. Acad. Sci. **316**, 638 (1979).
[11]Z. Kurtanjek, M. Sheintuch, and D. Luss, Discussion meeting, "Kinetics of Physicochemical Oscillations," Aachen, September 1979.
[12]L. F. Olsen and H. Degn, Nature (London) **267**, 177 (1977).
[13]Aaron B. Corbet and David M. Mason, Discussion meeting, "Kinetics of Physiochemical Oscillations," Aachen, September 1979.

[14]O. E. Rössler, Z. Naturforsch. Teil A **31**, 259 (1979).
[15]Kazuhisa Tomita and Ichiro Tsude, Physics Lett. A **71**, 489 (1979).
[16]Jack S. Turner, Discussion meeting, "Kinetics of Physiochemical Oscillations," Aachen, September 1979.
[17]K.-D. Willamowski and O. E. Rössler, Discussion meeting, "Kinetics of Physicochemical Oscillations," Aachen, September 1979.
[18]J. J. Tyson, J. Math. Biol. **5**, 351 (1978).
[19]R. A. Schmitz, G. T. Renola, and A. P. Ziodas, in *Dynamics and Modelling of Reactive Systems* (Academic, New York, 1980).
[20]K. R. Jensen and W. H. Ray, Discussion meeting, "Kinetics of Physicochemical Oscillations," Aachen, September 1979.
[21]Kazuhisa Tomita and Ichiro Tsuda, Prog. Theor. Phys. **64**, 1138 (1980).
[22]J. C. Roux, A. Rossi, S. Bachelart, and C. Vidal, Phys. Lett. A **77**, 391 (1980).
[23]N. H. Parkard, J. P. Crutchfield, J. D. Farmer, and R. S. Shaw, Phys. Rev. Lett. **45**, 712 (1980).
[24]Otto E. Rössler, Z. Naturforsch Teil A **31**, 1664 (1976).
[25]Robert Shaw, "Strange Attractors, Chaotic Behavior, and Information Flow," preprint (1979).

Physical Review Letters **49** 245–8 (1982)

One-Dimensional Dynamics in a Multicomponent Chemical Reaction

Reuben H. Simoyi, Alan Wolf, and Harry L. Swinney
Department of Physics, The University of Texas, Austin, Texas 78712
(Received 10 May 1982)

Experiments on the Belousov-Zhabotinskii reaction in a stirred flow reactor reveal behavior that is strikingly similar to that generated by one-dimensional maps with a single extremum. In particular, a period-doubling sequence is observed that leads to a regime containing both chaotic and periodic states. Within the experimental resolution the ordering of the periodic states is in accord with the theory of one-dimensional maps.

PACS numbers: 05.70.Ln, 47.70.Fw, 64.60.-i, 82.20.-w

We have conducted experiments on a complex chemically reacting system (with about 25 chemical species) which exhibits, as a function of the flow rate of the chemicals through the reactor, a sequence of periodic and chaotic states that is in good agreement with that exhibited by unimodal (single-extremum) one-dimensional (1D) maps. From the data we have constructed 1D maps that correspond to the different periodic and chaotic states.

A decade ago Metropolis, Stein, and Stein[1] showed that unimodal maps, $x_{n+1} = \lambda f(x_n)$, exhibit universal (map-independent) dynamics as a function of the bifurcation parameter λ. Analysis of higher-dimensional systems has led to the conjecture that, if such a system were to exhibit a period-doubling sequence, then the dynamics of the system would be similar to that of a 1D map.[2,3] Indeed, period-doubling sequences have been discovered in recent experiments[4] on a variety of physical systems, and the observed behavior for at least the first few doublings has been in accord with the theory for 1D maps. However, 1D maps were not obtained in any of those experiments, and the rich dynamical structure that 1D maps exhibit beyond the period-doubling sequence has been observed only in the experiments of Testa, Pérez, and Jeffries[4] on the simplest nonlinear physical system that has been studied, an electrical oscillator with three degrees of freedom. We will now review the properties of 1D maps and then present the results of our experiments.

1D maps.[5]—A method called symbolic dynamics can be used to show that the dynamics of unimodal 1D maps of the interval $[0, 1]$ is exhausted by the periodic states of the "U (universal) sequence" of Metropolis, Stein, and Stein[1] and the chaotic states of the "reverse bifurcation sequence" of Lorenz.[6] The theory uses only the unimodal property of the map to deduce the nature of the states and the order in which they appear as a function of the bifurcation parameter λ. Feigenbaum and others, making the additional assumption that the map has a quadratic extremum, have obtained detailed predictions for the scaling of various dynamical quantities.[2,5] We will confine our discussion to the results of symbolic dynamics theory since it is the ordering and nature of the states and not their scaling properties that have been determined in our experiments.

We begin with the mechanics of map iteration.[1,5] For a given value of λ one picks any initial condition (except for a set of measure zero) and iterates the map until transient behavior disappears. Further behavior of the sequence $\{x_n\}$ can be either periodic or chaotic. For the purpose of categorizing periodic states we may restrict our attention to the iterates of the point $\bar{x}$, where $f(\bar{x})$ is the extremum of the map.[1] If the nth iterate of $\bar{x}$ falls to the right of $\bar{x}$, then the nth character of a descriptive character string is set to "R"; otherwise it is set to "L." Thus, for example, the 4-cycle in Fig. 1 is described by the string "RLR" where a character for the initial condition $\bar{x}$ (neither R nor L) is omitted. Periodic states may be uniquely classified also by the order in which points on the x_n axis are visited. For the example in Fig. 1 the iteration pattern can be seen to be 2-0-3-1.

Consider the dynamics of a map as a function of λ. For small λ the map has a fixed point (1-cycle). If λ is increased the 1-cycle eventually loses its stability to a 2-cycle in a pitchfork bifurcation. There exists an infinite sequence of such period-doubling transitions that converges to a 2^∞-cycle at finite $\lambda = \lambda_c$.[2]

The dynamics past λ_c is very complex.[1,5] Fundamentals of all integer periods (the first fundamental was the 1-cycle) appear and undergo their own complete period-doubling sequences. Thus, for example, there is a 3-cycle and its "harmonics" (3×2^n-cycles for all positive n) for some interval in λ. The larger the integer, the larger

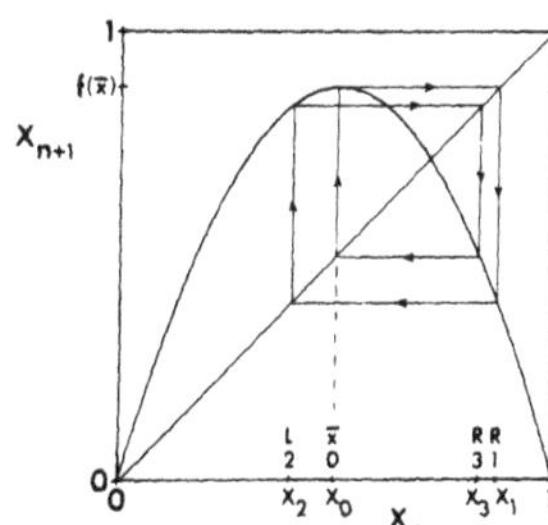

FIG. 1. The map $x_{n+1} = \lambda x_n(1-x_n)$ with $\lambda = 3.498\,561\,7$ exhibits a 4-cycle of the type RLR.

the number of allowed states; for example, there are three distinct 5-cycles (RLRR, RLLR, and RLLL) and 27 distinct 9-cycles. In Table I we list in order of increasing λ some of the periodic states of period less than 11, along with their "RL..." strings and iteration patterns. The full U sequence consists of the extension of this table to all allowed periodic states. Each allowed pattern occurs only once, and at any given λ not more than one periodic state is stable.

Additional structure in the region past λ_c was described by Lorenz.[6] In any system of finite resolution there exist gaps between various period-doubling sequences. These contain chaotic (intrinsically noisy) "reverse bifurcation sequences" which appear at the end of each period-doubling sequence; these sequences show period *halving* with increasing λ back down to the appropriate fundamental. While the chaotic states do not exist for intervals in λ, there is a finite probability of encountering a chaotic state.[5]

Experimental methods.[7]—We have conducted experiments on the Belousov-Zhabotinskii reaction in a well-stirred reactor as a function of the flow rate of the chemicals through the reactor (with input chemical concentrations held fixed). The time dependence of the concentration of one of the chemicals, the bromide ion, was measured with a specific ion probe, as described previously.[7]

Results.—The states listed in Table I are in fact the *observed* states. Time series records for several of the states are shown in Fig. 2. Presumably, unobserved U-sequence states exist over flow rate ranges too small to resolve in our experiments; indeed, our data files contain many short segments corresponding to U-sequence states not included in Table I—the table lists only states that were observed in several runs.

The ordering of the states in Table I was difficult to determine definitively because the pump had to be recalibrated for each run and not every state was observed in a given run, and because even with our high signal-to-noise ratio, a state identified as periodic in records of finite length could in some cases be its reverse-sequence counterpart (our data are *consistent* with the identification of all states as being periodic). Also, a slow drift in flow rate, characteristic of peristaltic pumps, resulted in data files that sometimes contained two (occasionally more) different periodic states; however, the periodic states observed in the same file were always found to be close together in the U sequence—this in itself is *strong* evidence for the existence of the U sequence in the chemical system.

The data taken as a whole support the ordering given in Table I. Further confirmation of the U sequence is provided by the 1D maps described in the following section.

Phase portraits and 1D maps.—The well-stirred Belousov-Zhabotinskii reaction may be described by the instantaneous concentrations of about 25 chemicals. It is not feasible to monitor all these quantities and thus determine the phase-space behavior of the system. For many purposes, however, embedding theorems[8] justify the use of a *single* chemical concentration, $B(t_i)$ $(i = 1, \ldots,$

TABLE I. Some elements of the U sequence.

Period	Sequence	Pattern
1	...	0
2	R	0-1
2×2	RLR	2-0-3-1
$2^2\times 2$	RLR^3LR	2-6-0-4-3-7-5-1
10	RLR^3LRLR	2-8-6-0-4-3-9-5-7-1
6	RLR^3	2-0-4-3-5-1
5	RLR^2	2-0-4-3-1
3	RL	2-0-1
2×3	RL^2RL	2-5-3-0-4-1
9	RL^2RLR^2L	2-8-5-3-0-6-4-7-1
5	RL^2R	2-3-0-4-1
4	RL^2	2-3-0-1
2×4	RL^3RL^2	2-6-3-7-4-0-5-1

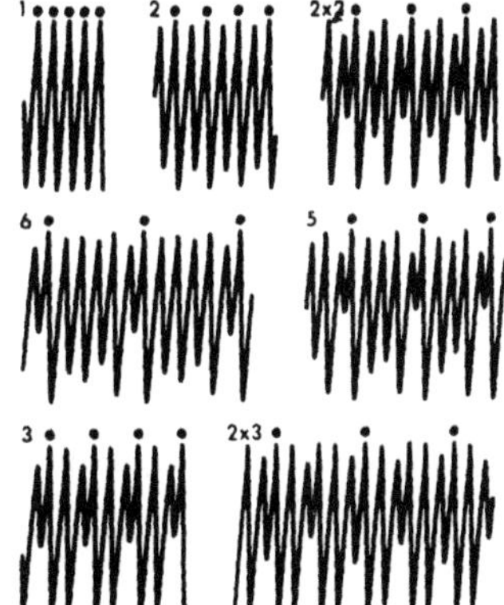

FIG. 2. Observed bromide-ion potential time series with periods τ (115 s), 2τ, $2\times 2\tau$, 6τ, 5τ, 3τ, and $2\times 3\tau$; the dots above the time series are separated by one period.

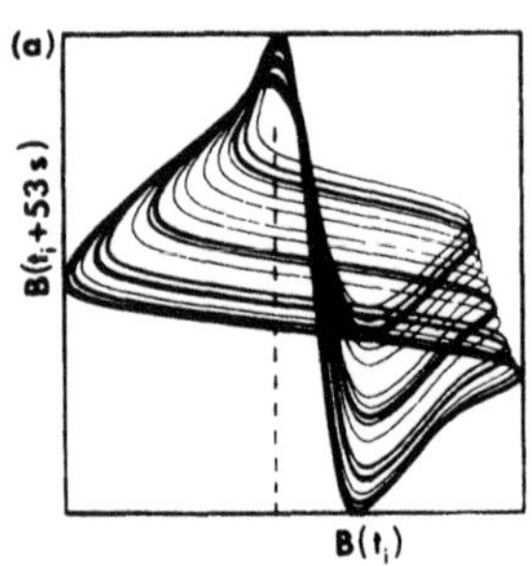

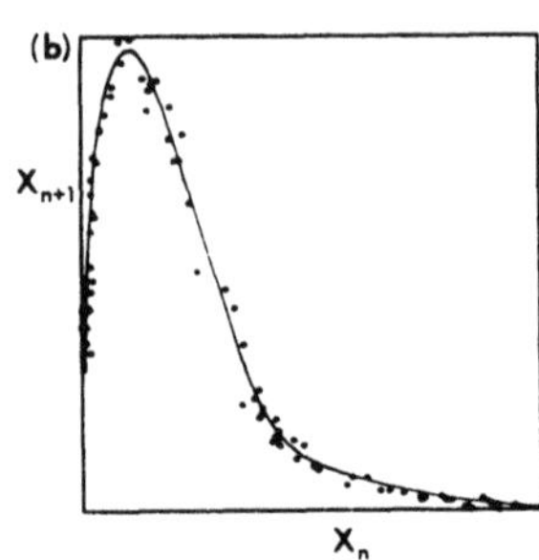

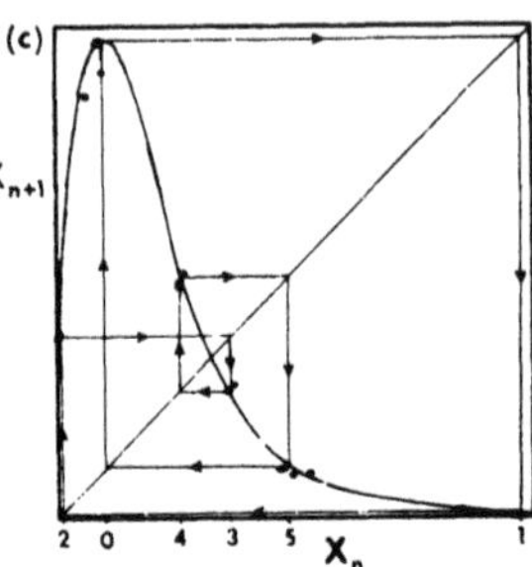

FIG. 3. (a) A 2D projection of a 3D phase portrait for a chaotic state. (b) A 1D map constructed from the data in (a) (see text). (c) A 1D map for the nearby 6-cycle RLR^3. In (b) and (c) the curves are drawn to guide the eye.

∞), to construct an m-dimensional phase portrait with the vectors $\{B(t_i), B(t_i+T), \ldots, B(t_i+(m-1)T)\}$, for sufficiently large m (and for almost any time delay T).

In Fig. 3(a) we show a 2D projection of a 3D phase portrait constructed with the third axis normal to the page. Our studies of the resulting strange attractor (an attracting set in phase space with the property that infinitesimally separated trajectories exponentially diverge on the average) suggest that it is essentially two-dimensional and that a 3D construction of the phase portrait is adequate for our system.[7,9] The connection between continuous motions on the attractor and a unimodal map is provided by the Poincaré section, the intersection of an $(m-1)$-dimensional hypersurface with "positively" directed orbits in m space. The intersections of our sheetlike attractor with a plane normal to the page [through the dashed line in Fig. 3(a)] lie approximately along a parametrizable curve, not on a higher-dimensional set.[10] Thus within this resolution the parameter values at successive intersections provide a sequence $\{x_n\}$ which defines a 1D map, as shown in Fig. 3(b). The shape of the map evolves slowly with flow rate, and so its shape in the periodic regions is known from the chaotic maps for nearby flow rates. For example, the map for a 6-cycle is shown in Fig. 3(c). The iteration pattern, 2-0-4-3-5-1, can be read from the map.

Conclusions.—In the parameter range studied here the Belousov-Zhabotinskii reaction exhibits the U sequence of 1D maps. The iteration patterns and, within the experimental resolution, the order of occurrence of the periodic states are in accord with the theory for 1D maps. To our knowledge these observations provide the first example of a physical system with many degrees of freedom that can be modeled in detail by a 1D map.

We acknowledge collaboration with J. C. Roux (who discovered the period-doubling sequence described here), J. S. Turner, W. D. McCormick, M. Kilgore, and J. Swift, and helpful discussions with M. J. Feigenbaum. This research was supported by National Science Foundation Grant No. CHE79-23627 and Robert A. Welch Foundation Grants No. F-805 and No. F-767.

---

[1]N. Metropolis, M. L. Stein, and P. R. Stein, J. Combinatorial Theory Ser. A 15, 25 (1973); J. Guckenheimer, Invent. Math. 39, 165 (1977).

[2]M. J. Feigenbaum, Commun. Math. Phys. 77, 65 (1980), and Los Alamos Sci., Summer, 1980, p. 4.

[3]P. Collet, J. P. Eckmann, and H. Koch, J. Stat. Phys. 25, 1 (1981).

[4]J. Maurer and A. Libchaber, J. Phys. (Paris), Lett. 40, L419 (1979); J. P. Gollub and S. V. Benson, J. Fluid Mech. 100, 449 (1980); H. M. Gibbs, F. A. Hopf, D. L. Kaplan, and R. L. Shoemaker, Phys. Rev. Lett. 46, 474 (1981); M. Giglio, S. Musazzi, and U. Perini, Phys. Rev. Lett. 47, 243 (1981); W. Lauterborn and E. Cramer, Phys. Rev. Lett. 47, 1445 (1981); P. S. Linsay, Phys. Rev. Lett. 47, 1349 (1981); R. Keolian, L. A. Turkevich, S. J. Putterman, I. Rudnick, and J. Rudnick, Phys. Rev. Lett. 47, 1133 (1981); C. W. Smith, M. J. Tejwani, and D. A. Farris, Phys. Rev. Lett. 48, 492 (1982); J. Testa, J. Pérez, and C. Jeffries, Phys. Rev. Lett. 48, 714 (1982).

[5]P. Collet and J.-P. Eckmann, *Iterated Maps of the Interval as Dynamical Systems* (Birkhäuser, Boston, 1980).

[6]E. N. Lorenz, Ann. N. Y. Acad. Sci. 357, 282 (1980); see also A. Wolf and J. Swift, Phys. Lett. 83A, 184 (1981).

[7]J. S. Turner, J. C. Roux, W. D. McCormick, and H. L. Swinney, Phys. Lett. 85A, 9 (1981); J. C. Roux, J. S. Turner, W. D. McCormick, and H. L. Swinney, in *Nonlinear Problems: Present and Future*, edited by A. R. Bishop, D. K. Campbell, and B. Nicolaenko (North-Holland, Amsterdam, 1982), p. 409. In the present experiment the residence time ranges from 0.9 to 1.0 h.

[8]H. Whitney, Ann. Math. 37, 645 (1936); F. Takens, in *Lecture Notes in Mathematics, Volume 898*, edited by D. A. Rand and L.-S. Young (Springer, Berlin, 1981), p. 366.

[9]J. C. Roux and H. L. Swinney, in *Nonlinear Phenomena in Chemical Dynamics*, edited by C. Vidal and A. Pacault (Springer, Berlin, 1981), p. 38.

[10]Strictly speaking, the Poincaré section for a chaotic strange attractor in a 3D construction must have dimension greater than unity because of the fractal nature of the attractor [see B. Mandelbrot, *Fractals, Form, Chance, and Dimension* (Freeman, San Francisco, 1977)].

Le Journal de Physique–Lettres **42** L-271–L-273 (1981)

Intermittent behaviour in the Belousov-Zhabotinsky reaction

Y. Pomeau

CEN Saclay, BP n° 2, 91190 Gif sur Yvette and GPS, Laboratoire de Physique de l'ENS, 24, rue Lhomond, 75231 Paris, France

J. C. Roux, A. Rossi, S. Bachelart and C. Vidal

Centre de Recherche Paul-Pascal, 33405 Talence, France

(*Reçu le 6 mars 1981, accepté le 11 mai 1981*)

Résumé. — L'une des routes possibles vers la turbulence est la transition intermittente. Avant la transition vers la turbulence, des oscillations régulières existent et au-dessus des oscillations apparemment régulières sont interrompues au hasard par des bruits de grande amplitude. Ce comportement se manifeste dans la réaction de Bélousov-Zhabotinsky. D'autre part une analyse des mesures montre que l'on a affaire dans ce cas à une intermittence de type 1.

Abstract. — One of the possible routes to turbulence is the intermittent transition. Below the onset of turbulence regular oscillations exist, and above the onset seemingly regular oscillations are interrupted randomly by large amplitude bursts. This behaviour shows up in the Belousov-Zhabotinsky reaction. Moreover an analysis of the measurements indicates that a so-called « type 1 » transition takes place in this case.

The possibility of turbulence in chemical kinetics, as suggested by Ruelle [1], has been verified recently [2, 3] for the Belousov-Zhabotinsky (B-Z) reaction. We report here the discovery of an intermittent behaviour in this reaction. Such a behaviour has been seen during the transition to turbulence in thermo-convection experiments [4] and in studies of ordinary equations [5].

In our experiment, the B-Z reaction takes place in an open well stirred tank reactor. Due to stirring, no spatial pattern can develop and the concentrations of the chemical species evolve according to the non linear equations of chemical kinetics, as derived from the mass action law; this deterministic behaviour is sufficient to produce chaos. The experimental conditions are : reacting volume 28 ml, temperature 39.6 °C, concentration of reagents before reaction in $\text{mol} \cdot \text{l}^{-1}$ (standard analytical reagent grade without further purification) :

$$NaBrO_3 = 1.8 \times 10^{-3}, \quad CH_2(COOH)_2 = 5.6 \times 10^{-3},$$
$$Ce_2(SO_4)_3 = 5.8 \times 10^{-4}, \quad H_2SO_4 = 1.5 .$$

A peristaltic pump feeds the reactor at constant adjustable rate. The chemical reaction in the reactor is monitored by the optical density at 340 nm. As this wavelength is absorbed [6] by the Ce^{4+} ion only, the signal is a sort of « pure quantity » and has been preferred to the redox potential that depends on the concentration of several chemical species.

As shown in figure 1*a*, when the mean residence time of chemicals in the reactor is 100 min., the optical density oscillates regularly. At higher fluxes (residence time 76 min. in figure 1*b*) the time record changes in a specific manner : seemingly stable oscillations exist which are interrupted from time to time and at random by large peaks. Such a transition from stable periodic behaviour to oscillations interrupted by random bursts denotes an intermittent transition to turbulence [5]. The present transition is well described by « type 1 intermittency » [7], shortly described hereafter. Although detailed modelizations of the B-Z kinetics have been proposed [8], we have not tried to relate them to our observations, since most features of the intermittent transition are model independent.

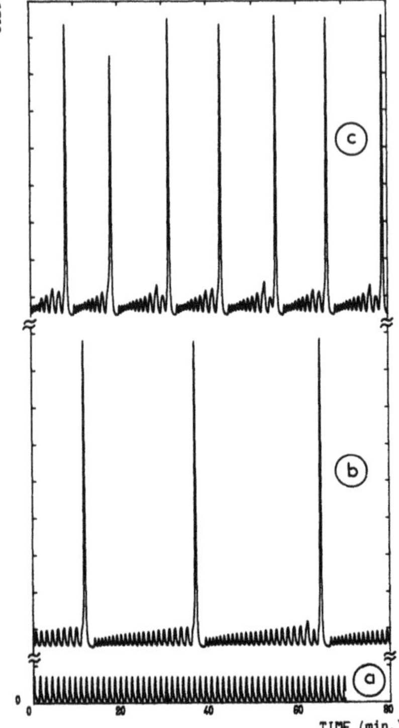

Fig. 1. — $[Ce^{4+}]$ oscillations recorded as a function of time : *a*) residence time 100 min. ; *b*) residence time 76 min. ; *c*) residence time 35 min.

Following an idea already used [4*b*] in studying the Rayleigh-Bénard thermoconvection, we consider the peak to peak amplitude of the « regular » oscillations between two bursts; this gives ordered sequence of numbers $X_1 \ldots X_n, X_{n+1} \ldots$ In type 1 intermittency, these numbers are connected to each other by a finite difference equation. This reads [7] after suitable choice of normalizations and origins in the generic form :

$$X_{n+1} = X_n + \varepsilon + X_n^2 . \tag{1}$$

ε is the control parameter, a smooth function of the residence time. For $\varepsilon < 0$ (in particular for a residence time of 100 min.), $X = \pm (-\varepsilon)^{1/2}$ are two fixed points of the iteration : $-(-\varepsilon)^{1/2}$ (resp. $+(-\varepsilon)^{1/2}$) being stable (resp. unstable). The stable fixed point corresponds to the stable oscillations before the intermittent transition (Fig. 1*a*). If ε is positive, the fixed points of (1) vanish and a small channel is created [7] between the first bissectrix [in the Cartesian plane (X_n, X_{n+1})] and the representative curve $X_{n+1}(X_n)$. This curve is locally approximated by a parabola, as implied by equation (1). Far away from the region $X \sim 0$, this local form is no longer valid ; however this large distance behaviour affects the structure of the large bursts only, that we do not consider here.

For $\varepsilon > 0$, starting from a negative X, the successive X_n's, as given by equation (1), drift slowly through the channel toward positive values. Plotting $X_{n+1}(X_n)$, as given by the time records of the B-Z reaction in the intermittent conditions, one can recognize this drift process (Fig. 2). Furthermore, near $\varepsilon = 0+$, one may replace equation (1) by a differential equation [7], n being taken as continuous. This yields :

$$X(n) \simeq \varepsilon^{1/2} \operatorname{tg}(n\varepsilon^{1/2}) .$$

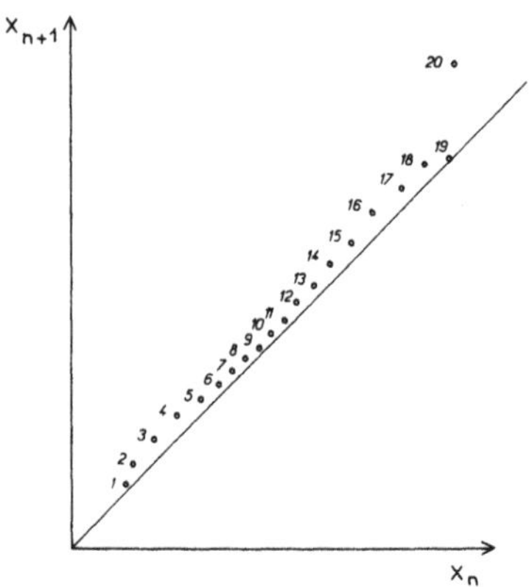

Fig. 2. — Peak to peak amplitude of the « regular » oscillations between two bursts (residence time : 76 min.).

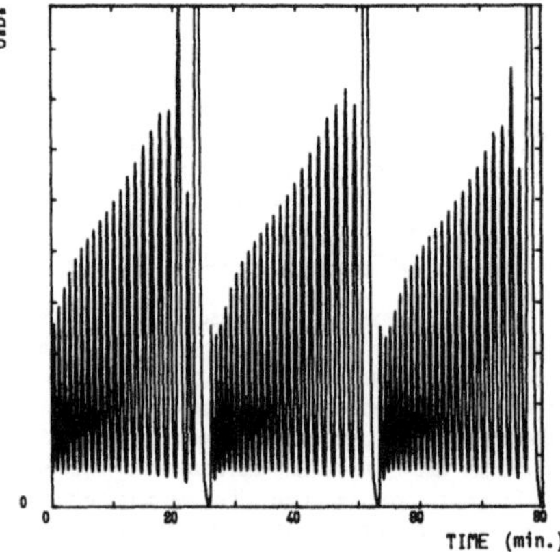

Fig. 3. — Enlarged view of the oscillations (residence time : 76 min.).

This approximation breaks down when $n\varepsilon^{1/2}$ tends to $\pm \pi/2$. However, the tangent like shape of the local maxima can be recognized in the records (Fig. 3).

In real life experiments, it is in principle impossible to conclude surely about the stochastic or non stochastic nature of a process recorded during a finite time. Nevertheless we present hereafter more plausible arguments for our interpretation of this transition :

(i) at values of the residence time still lower than these reported here, the behaviour becomes more and more chaotic. The duration and structure of the oscillations between two bursts fluctuate more and more (Fig. 1*c*). This agrees with the idea that chaos is already present just beyond the onset of intermittency, even though the records look quite regular then.

(ii) the frequency of occurrence of large bursts does not seem to be locked with the frequency of fast oscillations. The number of oscillations between two bursts varies randomly, with a probability distribution shown in figure 4. This distribution is typical of « type 1 intermittency » : the time needed to drift through the channel is bounded from above. This time can fluctuate to lower values only. A two dimensional iteration scheme, as the one proposed in references [4, 6] yields a probability distribution similar to the one of figure 4.

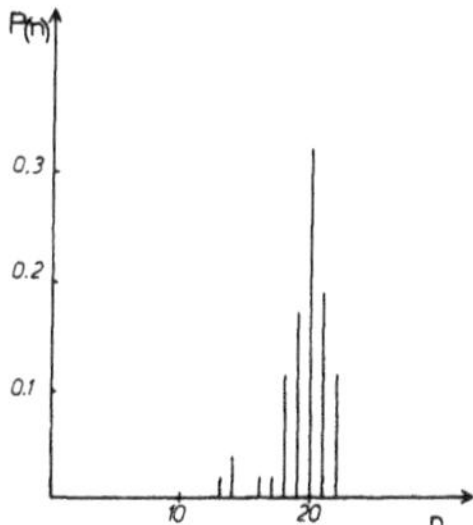

Fig. 4. — Probability distribution of the number of oscillations between two bursts (residence time : 76 min.).

To conclude, all our experimental data are consistent with a transition to turbulence *via* type 1 intermittency.

References

[1] RUELLE, D., *Trans. N. Y. Acad. Sci.* **35** (1973) 66.

[2] *a*) VIDAL, C., ROUX, J. C., ROSSI, A. and BACHELART, S., *C.R. Hebd. Séan. Acad. Sci.* Paris **C289** (1979) 73-76 and *Ann. N. Y. Acad. Sci.* (1980), in press.

b) ROUX, J. C., ROSSI, A., BACHELART, S. and VIDAL, C., *Experimental observation of complex dynamical behaviour*, accepted for publication in *Physica D*.

[3] ROUX, J. C., ROSSI, A., BACHELART, S. and VIDAL, C., *Phys. Lett. A* **77** (1980) 391-393.

[4] *a*) MAURER, J., LIBCHABER, A., *J. Physique Lett.* **41** (1980) L-515-L-518.

b) BERGÉ, P., DUBOIS, M., MANNEVILLE, P., POMEAU, Y., *J. Physique Lett.* **41** (1980) L-341-L-346.

[5] POMEAU, Y., MANNEVILLE, P., *Physica D* **1** (1980) 219-226.

[6] VIDAL, C., ROUX, J. C., ROSSI, A., *J. Am. Chem. Soc.* **102** (1980) 1241-1245.

[7] POMEAU, Y., MANNEVILLE, P., *Comm. Math. Phys.* **74** (2) (1980) 189-197.

[8] TYSON, J. J., « The Belousov-Zhabotinsky reaction », *Lecture Notes in Biomathematics*, vol. **10** (Springer Verlag, Berlin) 1976.

Physical Review Letters **49** 1217–20 (1982)

Experimental Evidence of Subharmonic Bifurcations, Multistability, and Turbulence in a Q-Switched Gas Laser

F. T. Arecchi,[a] R. Meucci, G. Puccioni, and J. Tredicce
Istituto Nazionale di Ottica, Firenze, Italy

(Received 9 August 1982)

Subharmonic bifurcations, generalized multistability, and chaotic behavior were found experimentally in a Q-switched CO_2 laser operating at 10.6 μm. Jumps between two strange attractors lead to a low-frequency ($1/f$ type) divergence in the power spectrum. This is the first experimental evidence of these phenomena in a quantum-optical molecular system. A theoretical model is also presented whose results are in good agreement with the experimental data.

PACS numbers: 05.40.+j, 05.70.Ln, 42.50.+q

Over the past years, the laser has been a test bench for many conjectures in nonequilibrium statistical mechanics. Measurements of photon statistics around laser threshold suggested a set of analogies between stationary quantum optical devices and thermodynamical phase transitions.[1] Furthermore, evidence of transient anomalous fluctuations[2] introduced the new concept of nonstationary statistics, later extended to other physical systems.[1,3,4] Recently, interest has arisen in higher-order bifurcations of quantum optical dynamical systems, which may eventually lead to chaos. Theoretical models have been formulated,[5] and an experiment has been performed on a hybrid electro-optical device.[6]

Here we report the first experimental evidence of multiple subharmonic bifurcations, eventually leading to chaos, in a quantum optical molecular system.[7] Specifically, we show these effects in a Q-switched CO_2 laser operating at the 10.6 μm $P(20)$ line. Furthermore we give evidence of the coexistence of independent basins of attraction in the phase space (generalized multistability) which may lead in suitable situations to the appearance of low-frequency divergences in the power spectrum ($1/f$ noise).[8] We have chosen a CO_2 laser system because the relaxation time $1/\gamma_\parallel$ of the population inversion is much larger ($1/\gamma_\parallel = 0.4$ ms) than the memory time $1/\gamma_\perp$ of the induced dipoles[9] ($1/\gamma_\perp = 10^{-8}$ s), thus reducing the single-mode dynamics to the coupling between two degrees of freedom, namely photon population n and molecular population inversion Δ. Introducing within the cavity a time-dependent perturbation by an electro-optical modulator driven by a sinusoidal frequency, we have a nonautonomous differential system in n and Δ amounting to the crucial three degrees of freedom that have been shown to be the necessary condition for the onset of chaos.[10] As shown later, the relevant range of modulation frequencies is correlated to the $\gamma_\parallel$ value; hence the choice of CO_2 laser has put the working frequency range in the easily accessible 50–130-kHz region.

The coupled field-molecules equations for a single-mode laser lead, after adiabatic elimination of the polarization, to the following rate equations:

$$\dot{\Delta} = R - Gn\Delta - \gamma_\parallel\Delta, \quad \dot{n} = 2Gn\Delta - K(t)n, \tag{1}$$

where

$$G = \omega\mu^2/\hbar\epsilon_0\gamma_\perp V = 0.25\times10^{-4}\ \mathrm{s}^{-1} \tag{2}$$

is the coupling constant (SI units) including the frequency ω and the dipole matrix element μ of the $P(20)$ transition and the collisional broadening rate $\gamma_\perp$, R is the pump rate, V the cavity volume, and

$$K(t) = K_1(1 + m\cos\Omega t) \tag{3}$$

is the cavity damping rate, modulated by the inserted electro-optical device. A linear perturbation analysis around the steady-state values

$$\bar{\Delta} = K_1/2G, \quad \bar{n} = 2R/K_1 - \gamma_\parallel/G, \tag{4}$$

yields a linearized frequency value $\Omega_0/2\pi \simeq 43$ kHz for the following typical parameters: spontaneous emission rate $\gamma_{sp} = 1.3\times10^3\ \mathrm{s}^{-1}$, and $K_1 = 3\times10^7\ \mathrm{s}^{-1}$ (corresponding to a cavity length of 2 m with losses of 20% per pass), and for a pumping rate R such that $\bar{n} \simeq \gamma_\parallel/G \simeq 10^8$ (corresponding to a dc power output of 50 μW). We select for the two modulation parameters m and Ω ranges which have been shown to be relevant in previous studies of driven nonlinear oscillators.[8,11] Specifically, we set K_1 consistently below the maximum damping rate $K_0 = GR/\gamma_\parallel$ compatible with the fixed pump rate R. It is easily

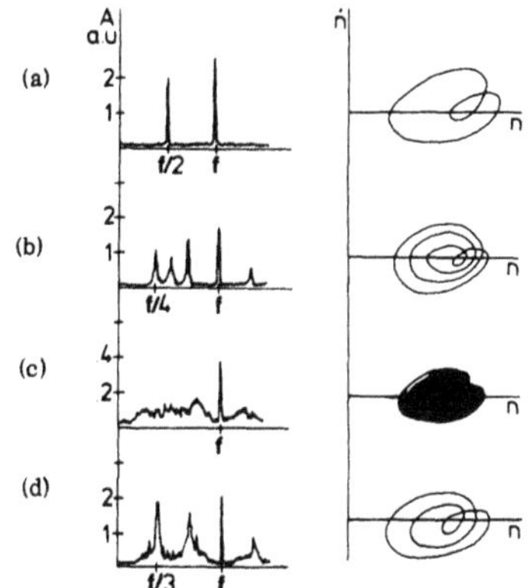

FIG. 1. Experimental phase-space portraits ($\dot{n}-n$) (right side) and the corresponding frequency spectra (left side) for different modulation frequencies f. (a) $f = 62.75$ kHz. Period-two limit cycle and corresponding $f/2$ subharmonic. (b) $f = 63.80$ kHz. Period-four limit cycle and $f/4$ subharmonic. (c) $f = 64.00$ kHz. The phase-space portrait shows a strange attractor (the oscilloscope spot could not resolve single windings). The power spectrum is a quasicontinuous one with a small peak at the modulation frequency (see the scale change with respect to previous figures). (d) $f = 64.1$[illegible] kHz. Period-three limit cycle and $f/3$ subharmonic.

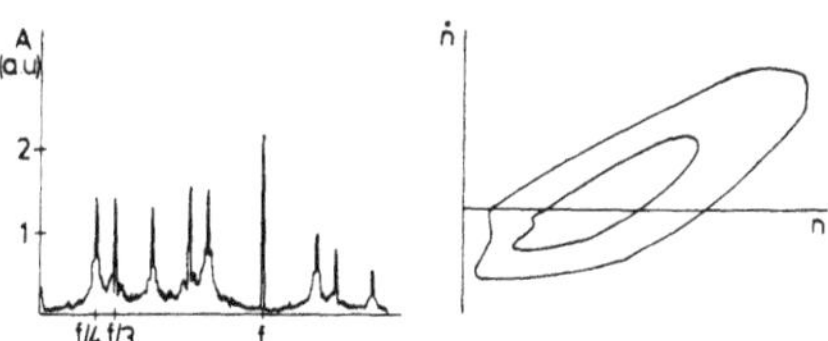

FIG. 2. $f = 63.85$ kHz. Experimental evidence of generalized multistability (coexistence of two independent attractors). The power spectrum shows that those attractors correspond to $f/3$ and $f/4$ subharmonic bifurcations, respectively; in phase space, the multipole windings merged within the thickness of the phase portrait contour.

shown that the linearized eigenfrequency is

$$\Omega_0 \simeq (\gamma_{\parallel} K_1)^{1/2}, \tag{5}$$

provided we choose $K_1 \approx K_0/2$ and a modulation depth m sufficiently small to have

$$(1/K_1)(dK/dt) = \Omega m < \gamma_{\parallel}. \tag{6}$$

The driving frequency $f = \Omega/2\pi$ is chosen to vary in the region from $\Omega_0/2\pi$ on, that is, from 40 to 150 kHz. As a consequence, $m < \gamma_{\parallel}/\Omega \simeq 10^{-2}$. We have explored modulation values between 1% and 5%. A complete state diagram would yield the dynamical features for all possible values of the modulation parameters m and Ω. However, the strip $m = 1\%-5\%$ does not display m dependence; therefore we limit ourselves to giving experimental results at $m = 1\%$ for various Ω values.

The experimental setup consists of a CO_2 laser carefully stabilized against thermal and acoustic disturbances, with the discharge current stabilized better than $1/10^3$. No long-term stabilization was necessary. The electro-optical modulator was a CdTe, antireflex-coated, 6-cm-long crystal, with an absorption less than 0.2%. The laser cavity includes also a $\lambda/4$ plate and a beam expander, both coated to limit the total losses per pass to 20%. The laser output is detected on a fast (2.5-ns rise time) pyroelectric detector whose current, proportional to the photon number $n(t)$, is sent together with its time derivative $\dot{n}(t)$ to an x-y scope, in order to have the phase-space portrait $(n, \dot{n})$. The detector is also sent to

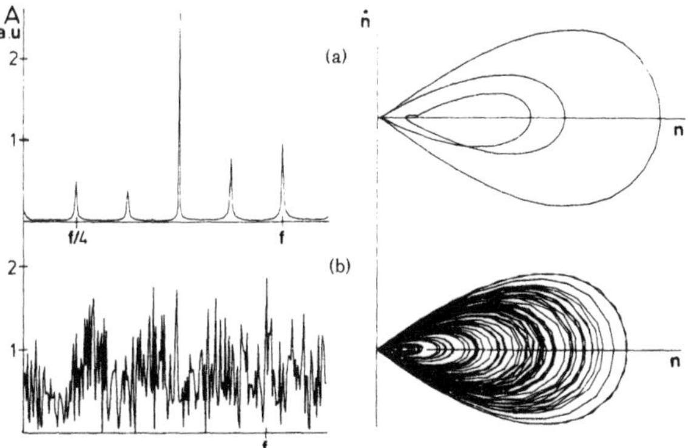

FIG. 3. Computer plots for the parameter values $\gamma_{\parallel} = 10^3\ s^{-1}$, $K_1 = 7\times10^7\ s^{-1}$, $m = 2.0\times10^{-2}$, $GR = 2.0\times10^{11}$. (a) $f = 64.33$ kHz. Subharmonic bifurcation $f/4$, as in the experiment of Fig. 1(b). (b) $f = 78.8$ kHz, $m = 3\times10^{-2}$. Strange attractor and broad spectrum corresponding to a chaotic solution.

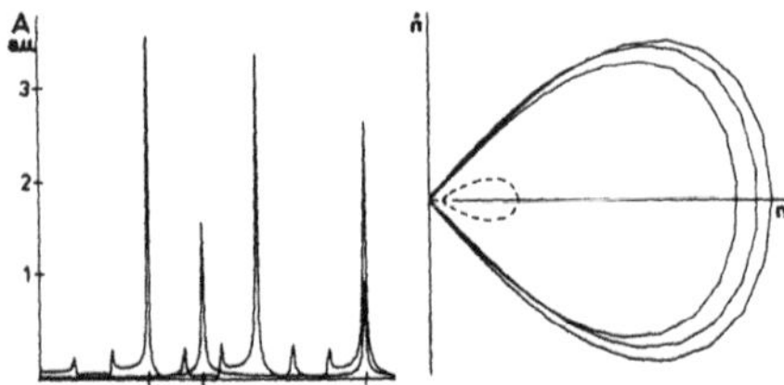

FIG. 4. Theoretical generalized bistability; $f = 119.0$ kHz, $m = 2.0\times 10^{-2}$. The phase-space portrait shows the existence of two independent attractors, corresponding to the subharmonic frequencies $f/2$ (dashed line) and $f/3$ (continuous line); relative spectra are superimposed. It must be noted that one attractor remains inside the other as in the experiment of Fig. 2. If initial conditions are properly changed, a third attractor is found with a subharmonic frequency $f/10$ (not plotted for the sake of simplicity). Initial conditions: $n_0 = 4\times 10^8$, $\dot{n}_0 = 0$ (dashed), $n_0 = 2\times 10^5$, $\dot{n}_0 = -2\times 10^6$ (continuous).

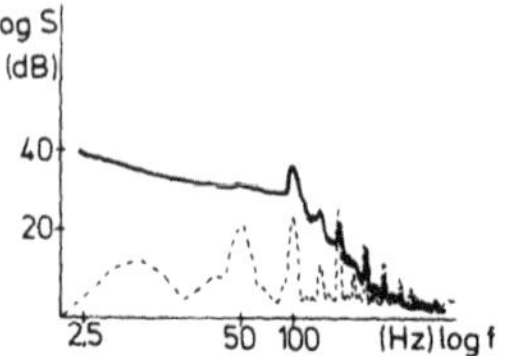

FIG. 5. Experimental power spectra in the case of two attractors, stable (dashed line), and strange (solid line).

a Rockland spectrum analyzer to measure the power spectra. The limited range (up to 100 kHz) of the spectrum analyzer has limited the frequency range explored thus far. We show later that interesting bifurcations are also expected in the 130-kHz domain for which we do not have experimental data.

In Fig. 1 we show experimental data in a narrow region between 62.7 and 64.25 kHz where various bifurcations occur. This region is limited above and below by wide intervals with stable single-period limit cycles. Fig. 1(a) shows the $f/2$ bifurcation at $f = 62.7$ kHz, Fig. 1(b) the $f/4$ case for $f = 63.8$ kHz; Fig. 1(c) shows the strange attractor and a broad-band spectrum for $f = 64.0$ kHz; and Fig. 1(d) shows the $f/3$ case for $f = 64.2$ kHz. As one can see, in this interval there is no Feigenbaum sequence.[12] Furthermore at $f = 63.85$ kHz a new feature appears, namely the coexistence of two independent stable attractors, one of period 4 ($f/4$) and the other of period 3 ($f/3$) (Fig. 2). Notice that the driving frequency is intermediate between those of Figs. 1(b) and 1(d). Therefore the chaotic behavior of Fig. 1(c) can be considered as a merging of those two attractors. This bistable situation has nothing to do with the common optical bistability[13] where two dc output amplitude values appear for a single dc driving amplitude. We call this coexistence of two attractors "generalized bistability."

In Figs. 3 and 4 we report the theoretical equivalents of Figs. 1 and 2, respectively, obtained by computer solution of Eqs. (1) with parameter values in the ranges of the experiment. More details appear in the figure captions. The theoretical data also cover another interesting region at a driving frequency about twice that chosen the experiments. Preliminary observations in that frequency range show also a qualitative agreement between the computer solution and the experiments.

We conclude with a new experimental feature whose connection with turbulence has recently been shown.[8] As stated in Ref. 8, $1/f$ type low-frequency divergences, with power spectra as $f^{-\alpha}$ ($\alpha = 0.6-1.2$), appear when the following conditions are fulfilled: (i) There are at least two basins of attraction; (ii) the attractors have become strange and any random noise (always present in a macroscopic system) acts as a bridge, triggering jumps between them. These jumps have the $f^{-\alpha}$ feature. In the region of bistability (see Fig. 2) we have increased the modulator amplitude m up to the point where the two attractors have become strange. Figure 5 shows the sudden increase in the low-frequency spectrum. The divergent part has a power-law behavior $f^{-\alpha}$ with $\alpha \approx 0.6$. The theory of these new phenomena is under investigation.

We thank P. Poggi for technical assistance. This work was supported in part by the Consiglio Nazionale delle Ricerche through contract with the Istituto Nazionale di Ottica.

---

[a]Also with the University of Florence, I-50125 Florence, Italy.

[1]F. T. Arecchi, in *Order and Fluctuations in Equilibrium and Non-Equilibrium Statistical Mechanics*, edited by G. Nicolis, G. Dewel, and J. W. Turner (Wiley, New York, 1981).

[2]F. T. Arecchi, V. Degiorgio, and B. Querzola, Phys. Rev. Lett. 19, 1168 (1967).

[3]M. Suzuki, J. Stat. Phys. 16, 447 (1977).

[4]F. T. Arecchi and A. Politi, Phys. Rev. Lett. 45, 1219 (1980).

[5]K. Ikeda, H. Daido, and O. Akimoto, Phys. Rev. Lett. 45, 709 (1980); T. Yamada and R. Graham, Phys. Rev. Lett. 45, 1322 (1980).

[6]A. H. Gibbs, F. A. Hopf, D. L. Kaplan, and R. L.

Shoemaker, Phys. Rev. Lett. 46, 474 (1981).

[7]Observation of chaos in pulsed Xe laser has been reported by N. B. Abraham, M. D. Coleman, M. Maeda, and J. C. Wesson, Appl. Phys. B 28, 169 (1982).

[8]F. T. Arecchi and F. Lisi, Phys. Rev. Lett. 49, 94 (1982).

[9]W. W. Duley, *CO_2 Lasers* (Academic, New York, 1976), pp. 57–72.

[10]E. N. Lorenz, J. Atmos. Sci. 20, 130 (1963). Our modulation implements the third degree of freedom. It would be unnecessary for a medium without adiabatic elimination; see H. Haken, Phys. Lett. 53A, 77 (1975).

[11]J. P. Crutchfield and B. A. Huberman, Phys. Rev. Lett. 43, 1743 (1979).

[12]M. J. Feigenbaum, Phys. Lett. 74A, 375 (1979), and Commun. Math. Phys. 77, 65 (1980).

[13]H. M. Gibbs, S. L. McCall, and T. N. C. Venkatesan, Phys. Rev. Lett. 19, 1135 (1976).

Editor's note. See also Hopf, Kaplan, Gibbs and Shoemaker (1982) and Weiss, Godone and Olafsson (1983). There are also a number of experiments on bifurcations in acoustical systems, such as Lauterborn and Cremer (1981), Smith, Tejwani and Farris (1982) and Keolian, Turkevich, Putterman, Rudnick and Rudnick (1981).

According to the last authors, the first $f/2$ subharmonic was observed by Faraday (1831), in shallow-water waves. Rayleigh (*Phil. Mag.* **15** 229 (1883), **16** 50 (1883), **24** 145 (1887)) confirmed experimentally Faraday's results, and noted subharmonic response in driven nonlinear oscillators. Pedersen (1935) observed $f/2$ and $f/4$ subharmonics in strongly driven loudspeakers. Neppiras (1968) observed subharmonics in acoustically cavitating fluids.

Physical Review Letters **48** 714–7 (1982)

Evidence for Universal Chaotic Behavior of a Driven Nonlinear Oscillator

James Testa, José Pérez, and Carson Jeffries
Materials and Molecular Research Division, Lawrence Berkeley Laboratory, and Department of Physics, University of California, Berkeley, California 94720
(Received 8 January 1982)

A bifurcation diagram for a driven nonlinear semiconductor oscillator is measured directly, showing successive subharmonic bifurcations to $f/32$, onset of chaos, noise band merging, and extensive noise-free windows. The overall diagram closely resembles that computed for the logistic model. Measured values of universal numbers are reported, including effects of added noise.

PACS numbers: 05.40.+j, 05.20.Dd, 47.25.-c

Our purpose is to report detailed measurements on a driven nonlinear semiconducting oscillator and to make quantitative comparisons with the predictions of a simple model of period-doubling bifurcation as a route to chaos,[1-3] which stems from earlier work in topology.[4] There is surprising agreement, lending support to the belief and the hope that some nonlinear systems can be approximately understood by a universal model, as has been suggested by some experiments.[5,6] This upsurge of interest in nonlinear behavior has been triggered by the remarkable result that deterministic computer iterations of such a simple nonlinear recursion relation as the logistic equation

$$x_{n+1} = \lambda x_n(1 - x_n) \qquad (1)$$

yield exceedingly complex pseudorandom or chaotic behavior.[2,3] The results are best summarized by a bifurcation diagram[7-9]: a scatter plot of the iterated value $\{x_n\}$ versus the control parameter λ, which shows that as λ is increased $\{x_n\}$ displays a series of pitchfork bifurcations at λ_n, with period doubling by 2^n, $n = 1, 2, \ldots$. These converge geometrically, as $\lambda_c - \lambda_n \propto \delta^{-n}$, to the onset of chaos at λ_c, where $\{x_n\}$ becomes aperiodic; in the chaotic regime, $\lambda > \lambda_c$, noise bands merge and there exist narrow periodic windows in a specific order and pattern.[4] This model is quantified by universal numbers as $n \to \infty$: $\delta = 4.669\ldots$, and the pitchfork scaling parameter $\alpha = 2.502\ldots$, first computed by Feigenbaum. Other universal numbers characterize the spectral power density[10,11] and effects of noise.[8,12]

Our experimental system is a series *LRC* circuit driven by a controlled oscillator, described by $L\ddot{q} + R\dot{q} + V_c = V_d(t) = V_0 \sin(2\pi f t)$, where V_c is the voltage across a Si varactor diode (type 1N953 supplied by TRW Company), which is the nonlinear element. Under reverse voltage, $V_c = q/C$, where $C \simeq C_0/[1 + V_c/0.6]^{0.5}$, $C_0 \simeq 300$ pF; under forward voltage the varactor behaves like a normal conducting diode. The coil inductance $L = 10$ mH, the resistance $R = 28\ \Omega$. At low values of V_0, the system behaves like a high-Q resonant circuit at $f_{res} = 93$ kHz; as V_0 is increased, the resonant frequency shifts upward and the Q is lowered. It is not our intention to solve the intractable nonlinear differential equations for this system[13] but rather to do extensive and novel measurements designed to compare its behavior as fully as possible with the simple logistic model. We fix f near f_{res}, vary the driving voltage V_0, and measure the varactor voltage $V_c(t)$. We assume a correspondence between V_0 and λ and between V_c and x of Eq. (1).

A real-time display, e.g., Fig. 1, of $V_c(t)$ and $V_0(t)$ on a dual-beam oscilloscope, with V_0 as a

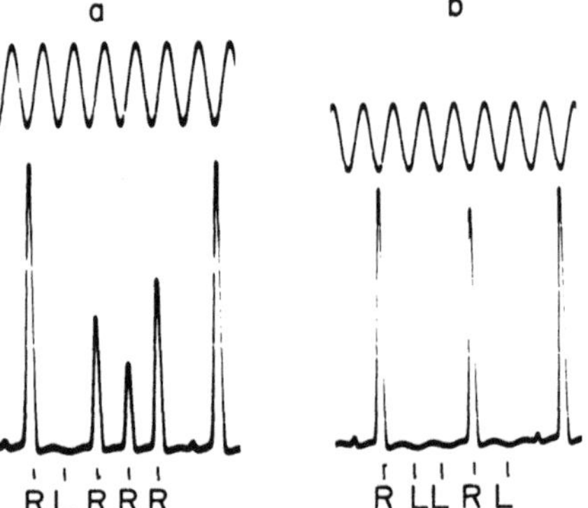

FIG. 1. (a) The varactor voltage $V_c(t)$ and the driving oscillator voltage $V_d(t)$ (upper) for period 6 window at 2.073 V; the pattern is RLRRR, and describes the sequence of visitation of the oscillator to its states according to whether it is to the right or left of zero, following the notation of Ref. 4. (b) Period 6 window at 3.338 V, with different pattern RLLRL.

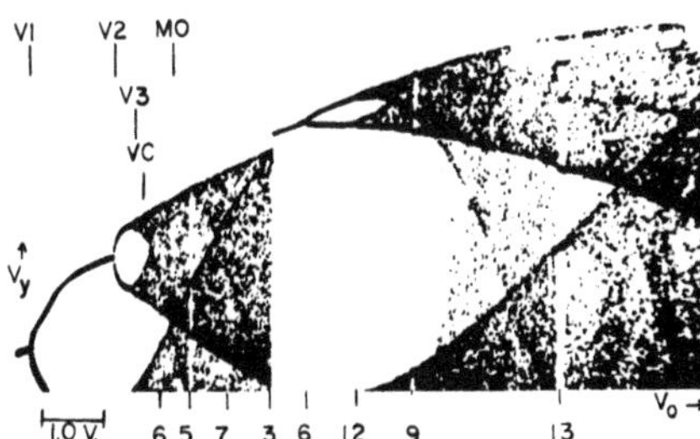

FIG. 2. Bifurcation diagram V_y vs V_0 at $f = 96.85$ kHz, showing thresholds V_1, V_2, and V_3 for periods 2, 4, and 8; threshold for chaos V_c; band merging M_0; and windows of periods 6, 5, 7, 3, 6, 12, 9, and 13. The veiled lines are peaks in the spectral density in the chaotic regime.

parameter, clearly revealed threshold values V_{0n} for bifurcation; the bifurcation subharmonics $f/2^n$ up to $f/16$; and the pattern of visitation of the oscillator to its stable points. The data shown at two different windows in the chaotic regime, both for period-6 orbits, show different patterns, as expected.[4] During the diode conducting half cycle, V_c is compressed toward the zero line; in the reverse half cycle, V_c has a set of discrete values, which correspond to the upper half of the bifurcation diagram.

To analyze V_c, a window comparator was constructed which selected components between V_y and $V_y + \Delta V$, $\Delta V \simeq 10$ mV. A vertical scan of V_y simultaneously with a slower horizontal scan of V_0 on an oscilloscope yielded Figs. 2 and 3, the first measured bifurcation diagram for a physical system showing subharmonic sequences. It has a striking resemblance to the computed diagram,[7,8] including bifurcation thresholds, onset of chaos, band merging, noise-free windows, and the subtle veiled structure, corresponding to regions of high probability.[8] The diagram allows a direct

TABLE I. Measured thresholds at 99 kHz.

Period	Threshold V_0 rms volts	Comments
2	0.639	Threshold for periodic bifurcation
4	1.567	
8	1.785	
16	1.836	
32	1.853	
Chaos	1.856	Onset of noise
12	1.901	Window
24	1.902	
6	2.073	Window
12	2.074	
5	2.353	Window
10	2.363	
7	2.693	Window
14	2.696	
3	3.081	Wide Window
6	3.338	
12	3.711	
24	3.821	
9	4.145	Window
18	4.154	

measurement of the number α; from the expanded region, Fig. 4, the ratio of the pitchfork splittings is directly measured in a series of ten similar measurements:

$$\alpha = 2.41 \pm 0.1 . \tag{2}$$

The diagram shows at least five noise-free windows, which bifurcate within the window: From Fig. 2 and Table I, at $V_0 = 3.081$ V, a noise-free window of period 3 appears, which bifurcates to periods 6, 12, and 24 before onset of chaos again.

The power spectral density of $V_c(t)$ was measured with a spectrum analyzer with 40 dB dynamic range, which showed the expected subharmonics $\frac{1}{2}, \frac{1}{4}, \frac{3}{4}; \frac{1}{8}, \frac{3}{8}, \frac{5}{8}, \frac{7}{8}$, etc., rather symmetrically displayed about $f/2$. The data shown in Fig. 5 were obtained with a more sensitive spectrum analyzer with 85 dB of dynamic range, sensitiv-

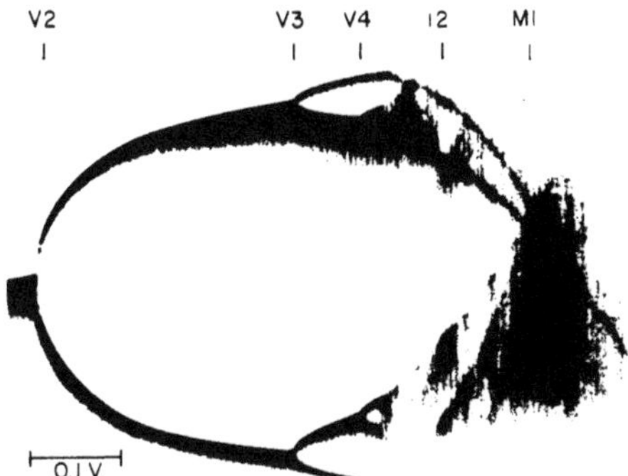

FIG. 3. Expansion of a region of Fig. 2, showing bifurcation thresholds V_2, V_3, and V_4; window of period 12; and band merging M_1.

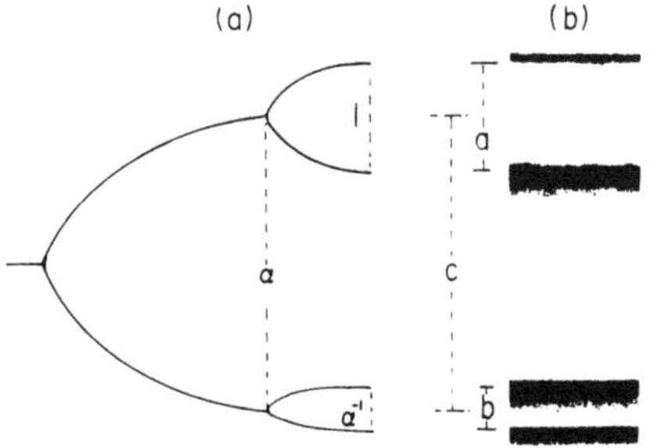

FIG. 4. (a) Schematic of universal metric scaling of pitchfork bifurcation, determined by α (Ref. 2). (b) Data for period 16 between V_4 and V_5, which yield the values $\alpha = a/b = 2.35$ and $\alpha = c/a = 2.61$.

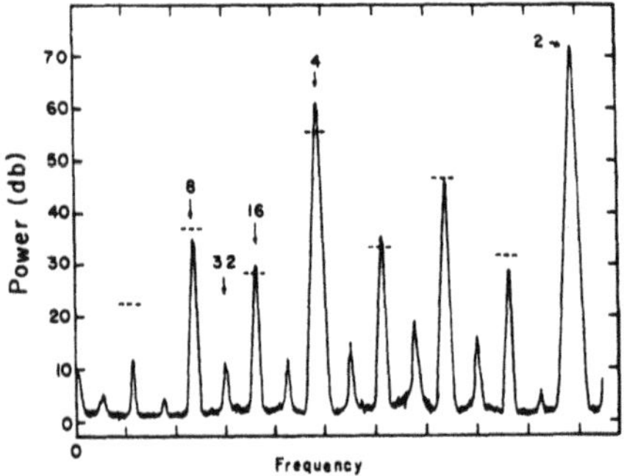

FIG. 5. Power spectral density (dB) vs frequency for $f = 98$ kHz, dynamic range 70 dB, showing subharmonics to $f/32$. The components agree with prediction (dashed bars, Ref. 14) within 2 dB rms deviation, except for the peak at $f/16$.

ity of 300 nV, and range $f=0$ to 50 kHz $\geq f/2$, thus allowing observation of spectral components 95 dB below V_0 at f. Figure 5 shows periodic subharmonics to $f/32$ at V_0 just below the threshold for chaos V_{0c}; the predicted values of the individual spectral components are shown.[14] It is predicted[10] that the average heights of the peaks for a period is $10 \log 20.963 = 13.21$ dB below the previous period; the data are consistent with this, although the region between $f/2$ and f is not available for exact averaging. Spectral analysis showed other noise-free windows (60 dB above noise) at periods 12, 6, 5, 7, and 9, at thresholds listed in Table I; all show bifurcations within the window. The entire V_0 sequence of Table I, identified by period and pattern, is consistent with the universal U sequence of Metropolis, Stein, and Stein[4] (who limit computation to period ≤ 11). From the first four threshold voltages V_{0n} we calculate the convergence rate

$$\delta_1 = \frac{V_{02} - V_{01}}{V_{03} - V_{02}} = 4.257 \pm 0.1\,;$$
$$\delta_2 = \frac{V_{03} - V_{02}}{V_{04} - V_{03}} = 4.275 \pm 0.1\,. \tag{3}$$

We observed the effect on the system of adding a random noise voltage $V_n(t)$ to $V_a(t)$. The bifurcation diagram and the power spectra were observed as $|V_n|$ was increased: periods 16, 8, 4, and 2 were successively obliterated at $V_n = 10$, 62, 400, and 2500 $\mathrm{mV_{rms}}$, respectively, yielding an average value

$$\kappa = 6.3 \tag{4}$$

for the noise voltage factor required to reduce by one the number of observable bifurcations.

To summarize, Table II compares our measured values with predicted values for some universal numbers. There is overall reasonable

TABLE II. Measured and predicted values for universal numbers.

Number		Measured	Predicted
δ_1	Eq. (3)	4.26 ± 0.1	4.751[a]
δ_2	Eq. (3)	4.28 ± 0.1	4.656[a]
δ_1	Period 3 window	0.69 ± 0.1	0.979[a]
δ_2	Period 3 window	3.38 ± 0.1	4.429[a]
α		2.41 ± 0.1	2.502[b]
ϵ		6.3 ± 0.3	6.55[c]
Average spectral power ratio		11 to 15 dB	13.61 dB[d]

[a] Computed from Eq. (1); cf. asymptotic limit 4.669, Ref. 2.
[b] Ref. 2.
[c] Ref. 12.
[d] Ref. 10.

quantitative agreement between the data and the logistic model. The likely cause for some discrepancy in δ is that the data cannot be taken in the asymptotic limit $n \to \infty$. These are first direct measurements for α and κ. The strong similarity between the predicted and the observed bifurcation diagram gives further support to the utility of simple models as a key to chaotic behavior of nonlinear systems. The measurement of a bifurcation diagram is a powerful method for assessing the degree to which a particular physical system will follow this route, or other routes[14]; it is not yet known how to predict this in advance.

We thank J. Rudnick, M. Nauenberg, J. P. Crutchfield, M. P. Klein, and H. A. Shugart for helpful conversations. This work was supported by the Director, Office of Energy Research, Office of Basic Energy Sciences, Materials Sciences Division of the U. S. Department of Energy under Contract No. W-7405-ENG-48.

[1] R. M. May, Nature (London) 261, 459 (1976).
[2] M. J. Feigenbaum, J. Stat. Phys. 19, 25 (1978).
[3] P. Collet and J.-P. Eckmann, *Iterated Maps on the Interval as Dynamical Systems* (Birkhauser, Boston, 1980).
[4] N. Metropolis, M. L. Stein, and P. R. Stein, J. Comb. Theory, Ser. A 15, 25 (1973).
[5] A. Libchaber and J. Maurer, J. Phys. (Paris), Colloq. 41, C3-51 (1980); M. Giglio, S. Musazzi, and U. Perini, Phys. Lett. 47, 243 (1981).
[6] P. S. Linsay, Phys. Rev. Lett. 47, 1349 (1981), first reported period doubling in a varactor oscillator, similar to the system studied here; however, our experimental methods differ.

[7]Collet and Eckmann, Ref. 3, pp. 26, 38, and 44.
[8]J. P. Crutchfield, J. D. Farmer, and B. A. Huberman, to be published.
[9]S. Grossman and S. Thomas, Z. Naturforsch. 32A, 1353 (1977).
[10]M. Nauenberg and J. Rudnick, Phys. Rev. B 24, 493 (1981).
[11]B. A. Huberman and A. B. Zisook, Phys. Rev. Lett. 46, 626 (1981).
[12]J. Crutchfield, M. Nauenberg, and J. Rudnick, Phys. Rev. Lett. 46, 933 (1981).
[13]However, B. A. Huberman and J. P. Crutchfield, Phys. Rev. Lett. 43, 1743 (1979), have computed solutions for an anharmonic oscillator with a restoring force $\propto x-4x^3$.
[14]J.-P. Eckmann, Rev. Mod. Phys. 53, 643 (1981).

Editor's note. The first electronic experiment was performed by Linsay (1981). See also Jeffries and Pérez (1982), Yeh and Kao (1982a, b) and Cascais, Dilao and Noronha da Costa (1983).

Science **214** 1350–3 (1981)

Phase Locking, Period-Doubling Bifurcations, and Irregular Dynamics in Periodically Stimulated Cardiac Cells

Abstract. *The spontaneous rhythmic activity of aggregates of embryonic chick heart cells was perturbed by the injection of single current pulses and periodic trains of current pulses. The regular and irregular dynamics produced by periodic stimulation were predicted theoretically from a mathematical analysis of the response to single pulses. Period-doubling bifurcations, in which the period of a regular oscillation doubles, were predicted theoretically and observed experimentally.*

The phase of neural and cardiac oscillators can be reset by a single brief depolarizing or hyperpolarizing stimulus (*1–3*). Experimental determination of the dependence of the phase shift on the phase of the autonomous cycle at which the stimulus was delivered allows computation of a mathematical function called the Poincaré map (*4*). Analysis of the Poincaré map is carried out to predict the response to periodic stimulation (*1, 2, 4*). This work provides experimental confirmation of a recent theoretical prediction (*4*) that period-doubling bifurcations and irregular dynamics (*5*) should be observable in periodically stimulated oscillators.

The preparation has been described in detail (*6*). Briefly, apical portions of heart ventricles of 7-day-old embryonic chicks were dissociated into their component cells in 0.05 percent trypsin. The cells were transferred to a flask containing tissue culture medium (818A with a potassium concentration of 1.3 m*M*), which was placed on a gyratory shaker. Spheroidal aggregates (100 to 200 μm in diameter) of electrically coupled cells that beat spontaneously with a period between 0.4 and 1.3 seconds form after 48 to 72 hours of gyration. Experiments were performed on aggregates in the same culture medium at 35°C under a gas mixture of 5 percent CO_2, 10 percent O_2, and 85 percent N_2. Intracellular electrical recordings were made with glass microelectrodes filled with 3*M* KCl (resistance, 20 to 60 megohms). Current pulses were delivered through the same electrode and measured with a virtual ground circuit. Impalements were maintained for 2 to 5 hours. This report presents results for two aggregates out of ten studied.

Consider the response of an aggregate to a single current pulse delivered δ msec after the upstroke of the action potential (Fig. 1A). The length of the cycle immediately preceding the perturbation is called τ, and the phase ϕ of the cycle at which the stimulus was delivered is $\phi = \delta/\tau$, $0 \leq \phi < 1$. Control cycles with the phase labeled are shown in Fig. 1B. The cycle time of the perturbed cycle (the time from the upstroke immediately preceding the stimulus to the next upstroke) is called T. A stimulus was delivered after every ten beats, with δ increased by 10 msec each time. In Fig. 1C the normalized perturbed cycle length T/τ is plotted for two different preparations. In a single preparation, an increase

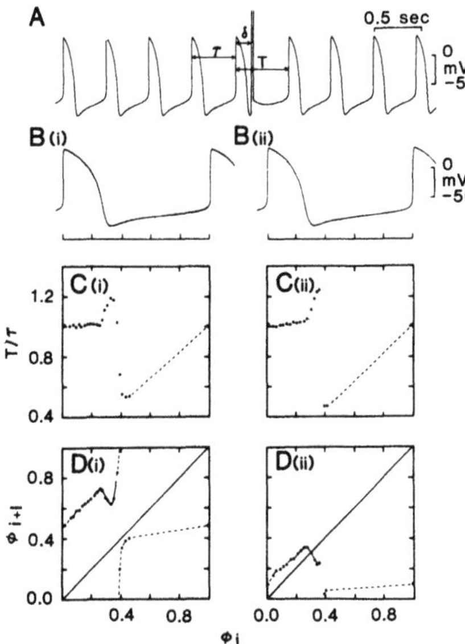

Fig. 1. (A) Transmembrane potential from an aggregate as a function of time, showing spontaneous electrical activity and effect of a 20-msec, 9-nA depolarizing pulse delivered at an interval of 160 msec following the action potential upstroke. The stimulus artifact is an off-scale vertical deflection following the fifth action potential. This early depolarizing stimulus prolongs the time at which the next action potential occurs. In (B to D), parts (ii) show results from this aggregate (aggregate 2), while parts (i) are from aggregate 1, taken from a different culture. (B) Membrane voltage as a function of phase ϕ, $0 \leq \phi < 1$. (C) Phase-resetting data, showing the normalized length T/τ of the perturbed cycle as a function of ϕ. (i) Pulse duration 40 msec, pulse amplitude 5 nA; (ii) pulse duration 20 msec, pulse amplitude 9 nA. For approximately $0.4 < \phi < 1.0$ the action potential upstroke occurs during the stimulus artifact and hence the perturbed cycle length cannot be exactly determined. The dashed line represents a linear interpolation that approximates the data. During collection of these data, the average control interbeat intervals ($\pm$ 1 standard deviation) were (i) $\bar{\tau} = 515 \pm 5.7$ msec and (ii) $\bar{\tau} = 434 \pm 5.5$ msec. (D) Poincaré maps computed from Eq. 1 and the data in Fig. 1C; (i) $t_s = 250$ msec, (ii) $t_s = 480$ msec. The dashed line represents a linear interpolation used in iterating the Poincaré map; the solid line through the data points is a quartic fit for $0.22 < \phi_i < 0.37$.

in stimulus intensity produces a transition from a continuous function such as that shown in Fig. 1C(i) to an apparently discontinuous function such as that shown in Fig. 1C(ii) (*7*).

Current pulses were periodically injected into the aggregate with period t_s ($100 \leq t_s \leq 700$ msec). For most t_s values in this range, phase-locked patterns result. A pattern is called an N:M phase-locked pattern if it is periodic in time and if for every N stimuli there are M action potentials, with action potentials occurring at M different times in the stimulus cycle. At first t_s was varied in 50-msec steps to sketch the boundaries of the major (2:1, 1:1, and 2:3) phase-locked regions. Then the intermediate regions were sampled by varying t_s in 10-msec steps. We first describe our experimental observations and then offer an interpretation based on an analysis of the Poincaré map.

Characteristic zones of regular and irregular dynamics were seen in both aggregates. Transitions occur at approximately the same values of $t_s/\bar{\tau}$, where $\bar{\tau}$ is the average control interbeat interval. For $0.55 < t_s/\bar{\tau} < 1.05$, 1:1 phase locking [Fig. 2A(ii)] is found. When t_s is decreased to $0.40 < t_s/\bar{\tau} < 0.55$ (zone α), dynamics analogous to the clinically observed Wenckebach phenomenon (*8*) are present. This phenomenon is characterized by a gradual prolongation of the time between a stimulus and the subsequent action potential until an action potential is skipped. This can occur in an irregular fashion (Fig. 2B) as well as in $m + 1$:m phase-locked patterns. As t_s is decreased below $0.4\,\bar{\tau}$, 2:1 phase locking is observed [Fig. 2A(i)].

For $1.05 < t_s/\bar{\tau} < 1.15$ (zone β), the ratio of stimulus frequency to action potential frequency is 1, but the stimulus no longer occurs at one fixed phase of the aggregate cycle as it does in 1:1 locking. Instead, the stimulus falls at two or more phases of the cycle. The dynamics in this narrow zone are highly variable and phase-locked patterns, when

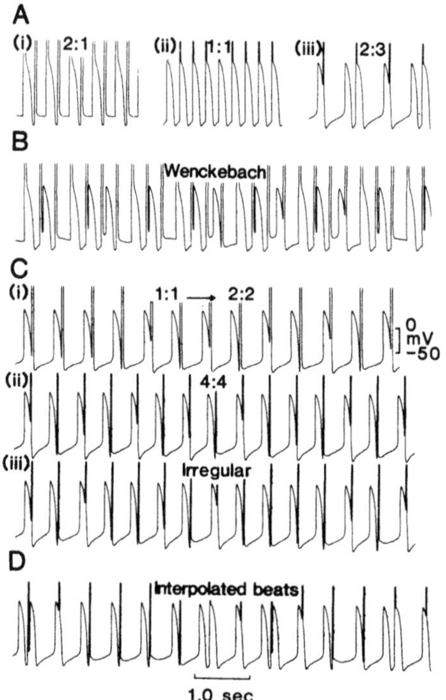

Fig. 2. Representative transmembrane recordings from both aggregates showing the effects of periodic stimulation with the same pulse durations and amplitudes as in Fig. 1C. (A) Stable phase-locked patterns: (i) 2:1 (aggregate 1, t_s = 210 msec); (ii) 1:1 (aggregate 2, t_s = 240 msec); (iii) 2:3 (aggregate 2, t_s = 600 msec). (B) Dynamics in zone α: irregular dynamics displaying the Wenckebach phenomenon (aggregate 1, t_s = 280 msec). (C) Dynamics in zone β: (i) 1:1 phase locking spontaneously changing to 2:2 phase locking (aggregate 1, t_s = 550 msec). During 2:2 phase locking there are two distinct phases of the cycle at which the stimuli fall. (ii) 4:4 phase locking (aggregate 2, t_s = 490 msec). There are four distinct phases of the cycle at which the stimuli fall. (iii) Irregular dynamics with one action potential in each stimulus cycle (aggregate 2, t_s = 490 msec). There is a narrow range of phases in which the stimuli fall. (D) Dynamics in zone γ: irregular dynamics displaying extra interpolated beats (aggregate 2, t_s = 560 msec).

they exist, are typically not maintained for long stretches of time. For example, Fig. 2C(i) shows a transition from a 1:1 to a 2:2 phase-locked pattern which spontaneously occurred during stimulation with a fixed frequency. Similarly, brief stretches of 4:4 phase locking [Fig. 2C(ii)] and irregular dynamics [Fig. 2C(iii)] can both be observed at t_s = 490 msec. Such transitions may be due to slow drifts in the intrinsic frequency of the aggregate during stimulation.

For $1.15 < t_s/\bar{\tau} < 1.35$ (zone γ) there are irregular patterns with extra or interpolated beats (Fig. 2D). Further increase in t_s leads to a 2:3 phase-locked pattern [Fig. 2A(iii)].

As the stimulus strength increases, the widths of zones α and γ decrease. However, zone β is widest at intermediate stimulus strength. The two examples in this report were selected because there is a relatively broad β zone.

The experimentally derived curves shown in Fig. 1C can be used to predict the effects of periodic stimulation (*1, 2, 4*). If ϕ_i is the phase of the oscillator immediately before the *i*th stimulus, then

$$\phi_{i+1} = 1 - f(\phi_i) + \phi_i + \frac{t_s}{\bar{\tau}} \qquad (\text{mod } 1) \tag{1}$$

where $f(\phi)$ gives the normalized perturbed interbeat interval (T/τ) as a function of ϕ. The relation

$$\phi_{i+1} = g(\phi_i, t_s) \tag{2}$$

defined by Eq. 1 is called the Poincaré map (Fig. 1D). For t_s = 0, the Poincaré map corresponds to the "new phase–old phase" phase-resetting curve, also called the phase transition curve (*1, 3, 4*). Starting from some initial phase ϕ_0, Eq. 1 can be iterated to compute the sequence of phases $\phi_0, \phi_1, \phi_2, \ldots$ If $\phi_N = \phi_0$ and $\phi_j \neq \phi_0$ for $0 < j < N$, then the Poincaré map has a cycle $\phi_0, \phi_1, \ldots, \phi_{N-1}, \phi_N = \phi_0$ of period N. If

$$\prod_{i=0}^{N-1} \left| \frac{\partial g}{\partial \phi} \bigg|_{\phi_i} \right| < 1 \qquad (3)$$

then the period N cycle is stable, and the corresponding $N{:}M$ phase-locked pattern is stable with a period of Nt_s (*4*). A stable pattern is maintained despite small variations in either the oscillator itself or the stimulus parameters. The value of M is given by

$$M = \sum_{i=1}^{N} \left[1 - f(\phi_i) + \frac{t_s}{\bar{\tau}} \right] \qquad (4)$$

To iterate the Poincaré map, the function $f(\phi_i)$ in Eq. 1 was approximated by linear interpolation between the data points of Fig. 1C. There is close agreement (*9*) between the experimentally observed ranges of t_s that give simple phase-locked patterns (Fig. 3A) and those theoretically predicted from iteration of the Poincaré map (Fig. 3B). The Poincaré map also predicts the existence of irregular dynamics (dynamics that are not phase locked) in zones α, β, and γ. There are complex changes in the phase-locking patterns as t_s is changed within these zones. We explicitly compute these changes in zone β and briefly discuss the dynamics in the other two zones.

Numerical analysis of the Poincaré maps for t_s in zone β shows that for any ϕ_i in the interval (0.22, 0.37) all iterates of ϕ_i remain in this invariant interval. Moreover, this interval attracts iterates of ϕ_i for all ϕ_0 outside this interval. To determine the theoretically predicted dynamics in zone β, the Poincaré map in the invariant interval was fit to a quartic polynomial by a least-squares method. Numerical iteration of the Poincaré map in zone β shows a sequence of period-doubling bifurcations and irregular dynamics (Fig. 3C) as t_s is increased (*5*). In the zone of irregular dynamics, the phases of the stimuli are restricted to the shaded regions of Fig. 3C (*10*). The irregular dynamics result from a deterministic iterative process, with no added stochastic terms.

We do not experimentally observe all the bifurcations theoretically computed in Fig. 3C (*11*). However, we propose

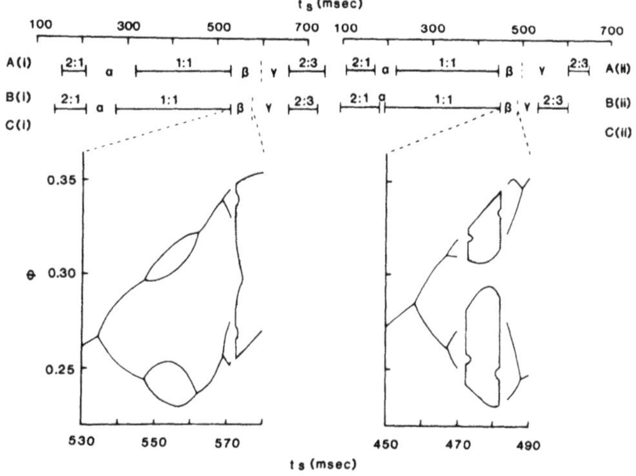

Fig. 3. Experimentally determined and theoretically computed dynamics. Parts (i) refer to aggregate 1, parts (ii) to aggregate 2. (A) Experimentally determined dynamics: there are three major phase-locking regions (2:1, 1:1, 2:3) and three zones of complicated dynamics labeled α, β, and γ (see text). (B) Theoretically predicted dynamics; note agreement with (A). (C) Theoretically predicted dynamics in zone β: curves give phase or phases in the cycle at which the stimuli fall during 1:1, 2:2, and 4:4 locking; stippled regions show the range of phases in which the stimulus falls during irregular dynamics.

that the transition shown in Fig. 2C(i) from 1:1 phase locking (period of repeating pattern = 550 msec) to 2:2 phase locking (period of repeating pattern = 1100 msec) corresponds to the period-doubling bifurcation theoretically computed at $t_s \simeq 535$ msec [Fig. 3C(i)]. The 4:4 phase-locked pattern of Fig. 2C(ii) and the irregular dynamics of Fig. 2C(iii) provide further evidence for the correspondence between the experimental observations and the theoretical computations [Fig. 3C(ii)].

In zones α and γ complex behavior is also observed experimentally and predicted theoretically. For example, in zone α, the dynamics experimentally seen at $t_s = 280$ msec (Fig. 2B) are very similar to those predicted at $t_s = 250$ msec from the Poincaré map in Fig. 1D(i). The extra beats characteristic of zone γ (Fig. 2D) are also predicted.

This work has implications for the understanding of normal and pathologic behavior in cardiac tissue. The experimental work supports previous studies showing that periodically forced oscillators display phase-locked dynamics that are similar to clinically observed cardiac dysrhythmias (*2, 4, 12*). Moreover, the work suggests novel explanations for the genesis of 2:2 rhythms (*13*) and irregular dysrhythmias (*4*).

We have observed behavior that we propose arises as a consequence of period-doubling bifurcations in these experiments. Thus, exotic dynamic behavior that was previously seen in mathematical studies and in experiments in the physical sciences (*5*) may in general be present when biological oscillators are periodically perturbed.

MICHAEL R. GUEVARA
LEON GLASS, ALVIN SHRIER
Department of Physiology, McGill University, Montreal, Canada H3G 1Y6

References and Notes

1. D. H. Perkel, J. H. Schulman, T. H. Bullock, G. P. Moore, J. P. Segundo, *Science* **145**, 61 (1964); T. Pavlidis, *Biological Oscillators: Their Mathematical Analysis* (Academic Press, New York, 1973); H. M. Pinsker, *J. Neurophysiol.* **40**, 527 (1977); *ibid.*, p. 544.
2. G. K. Moe, J. Jalife, W. J. Mueller, B. Moe, *Circulation* **56**, 968 (1977); J. Jalife and G. K. Moe, *Am. J. Cardiol.* **43**, 761 (1979); S. Scott, thesis, State University of New York, Buffalo (1979).
3. A. T. Winfree, *Science* **197**, 761 (1977); *The Geometry of Biological Time* (Springer-Verlag, New York, 1980); J. Jalife and C. Antzelevitch, *Science* **206**, 695 (1979); T. Sano, T. Sawanobori, H. Adaniya, *Am. J. Physiol.* **235**, H379 (1978).
4. M. R. Guevara and L. Glass, *J. Math. Biol.*, in press.
5. A period-doubling bifurcation is the doubling of the period of an oscillation due to parametric changes. Period-doubling bifurcations and "chaotic" dynamics are observed in simple mathematical models and in experiments in the physical sciences: T. Y. Li and J. A. Yorke, *Am. Math. Mon.* **82**, 985 (1975); R. M. May, *Nature (London)* **261**, 459 (1976); M. C. Mackey and L. Glass, *Science* **197**, 287 (1977); J. P. Gollub, T. O. Brunner, B. G. Danly, *ibid.* **200**, 48 (1978); K. Tomita and T. Kai, *Prog. Theor. Phys. Suppl.* *64* (1978), p. 280; R. H. G. Helleman, Ed., *Nonlinear Dynamics* (New York Academy of Sciences, New York, 1980); Editorial, *Phys. Today* **34**, 17 (March 1981).
6. R. L. DeHaan and L. J. DeFelice, *Theor. Chem.* **4**, 181 (1978); J. R. Clay and R. L. DeHaan, *Biophys. J.* **28**, 377 (1979); J. R. Clay and A. Shrier, *J. Physiol. (London)* **312**, 471 (1981).
7. There are many questions concerning the continuity properties of the graphs in Fig. 1, C and D. It is difficult to demonstrate experimentally that the points in Fig. 1C(ii) actually display a discontinuity. However, in this and many other preparations, repeated stimulation at the apparent discontinuity did not show evidence for intermediate values of T/τ. On theoretical grounds, discontinuities in the plots in Fig. 1C are expected [M. Kawato, *J. Math. Biol.* **12**, 13 (1981)]. However, the plots in Fig. 1D would not be expected to show discontinuities if one constructed the Poincaré map using the eventual phase [M. Kawato, cited above; E. N. Best, *Biophys. J.* **27**, 87 (1979)].
8. W. J. Mandel, Ed., *Cardiac Arrhythmias: Their Mechanisms, Diagnosis and Management* (Lippincott, Philadelphia, 1980).
9. There are many potential reasons for discrepancy between the predictions and experimental results. (i) The use of the Poincaré map is an approximation since it maps the state of a many-dimensional dynamical system to a single variable, the phase. (ii) The stimulation can lead to secondary electrophysiological changes in the aggregate such as changes in intrinsic frequency. (iii) There is beat-to-beat fluctuation in the autonomous cycle length in the absence of stimulation (*6*), and this "noise" tends to decrease the widths of the major stable phase-locked

zones (compare Fig. 3A with Fig. 3B).

10. In the irregular region there are strong limitations on the phase at which the stimulus occurs. Compare Fig. 3C in this report with figure 3 in E. N. Lorenz, *Ann. N.Y. Acad. Sci.* **357**, 282 (1980).
11. "Noise" tends to destroy stable phase-locked patterns that either are complex or exist over very small ranges of t_s [L. Glass, C. Graves, G. A. Petrillo, M. C. Mackey, *J. Theor. Biol.* **86**, 455 (1980); J. P. Crutchfield and B. A. Huberman, *Phys. Lett.* **A77**, 407 (1980); R. Guttman, L. Feldman, E. Jakobsson, *J. Membr. Biol.* **56**, 9 (1980)]. Thus, there are fundamental limitations to the experimental observability of such patterns.
12. B. van der Pol and J. van der Mark, *Philos. Mag.* **6**, 763 (1928); S. D. Moulopoulos, N. Kardaras, D. A. Sideris, *Am. J. Physiol.* **208**, 154 (1965); J. V. O. Reid, *Am. Heart J.* **78**, 58 (1969); F. A. Roberge and R. A. Nadeau, *Can. J. Physiol. Pharmacol.* **42**, 695 (1969); D. A. Sideris and S. D. Moulopous, *J. Electrocardiol.* **10**, 51 (1977); C. R. Katholi, F. Urthaler, J. Macy, T. N. James, *Comp. Biomed. Res.* **10**, 529 (1977).
13. R. Langendorff, *Am. Heart J.* **55**, 181 (1958).
14. Supported by grants from the Canadian Heart Foundation and the Natural Sciences and Engineering Research Council of Canada. M.R.G. is a recipient of a predoctoral traineeship from the Canadian Heart Foundation. We thank D. Colizza for technical assistance and M. C. Mackey and A. T. Winfree for helpful conversations.

3 August 1981; revised 13 October 1981

Part 3

Theory

Journal of Combinatorial Theory **15** 25–44 (1973)

On Finite Limit Sets for Transformations on the Unit Interval

N. METROPOLIS, M. L. STEIN, AND P. R. STEIN*

University of California, Los Alamos Scientific Laboratory, Los Alamos, New Mexico 87544

Communicated by G.-C. Rota

Received June 8, 1971

An infinite sequence of finite or denumerable limit sets is found for a class of many-to-one transformations of the unit interval into itself. Examples of four different types are studied in some detail; tables of numerical results are included. The limit sets are characterized by certain patterns; an algorithm for their generation is described and established. The structure and order of occurrence of these patterns is universal for the class.

1. Introduction. The iterative properties of 1-1 transformations of the unit interval into itself have received considerable study, and the general features are reasonably well understood. For many-to-one transformations, however. the situation is less satisfactory, only special and fragmentary results having been obtained to date [1, 2]. In the present paper we attempt to bring some coherence to the problem by exhibiting an infinite sequence of finite limit sets whose structure is common to a broad class of non 1-1 transformations of [0, 1] into itself. Generally speaking, the limit sets we shall construct are not the only possible ones belonging to an arbitrary transformation in the underlying class. Nevertheless, our sequence—which we shall call the "U-sequence"—constitutes perhaps the most interesting family of finite limit sets in virtue of the universality of their structure and of their order of occurrence. With regard to infinite limit sets we shall have little to say. There is reason to believe, however, that for a non-vacuous (in fact, infinite) subset of the class of transformations considered here, our construction—suitably extended to the limit of "periods of infinite length"—is exhaustive in the

* Work performed under the auspices of the U.S. Atomic Energy Commission under contract W-7405-ENG-36.

probabilistic sense, namely, that with "probability 1" every limit set belongs to the U-sequence.

2. We begin by describing the class of transformations to which our construction applies, but make no claim that the conditions imposed are strictly necessary. The description is complicated by our attempt to exclude, insofar as is possible, certain finite limit sets, not belonging to the U-sequence, whose existence and structure depend on detailed properties of the particular transformation in question.

All our transformations will be of the form

$$T_\lambda(x) : x' = \lambda f(x),$$

where x' denotes the first iterate of x (*not* the derivative!) and λ varies in a certain open interval to be specified below. The fundamental properties of $f(x)$ will be:

A.1. $f(x)$ is continuous, single-valued, and piece-wise $C^{(1)}$ on $[0, 1]$, and strictly positive on the open interval, with $f(0) = f(1) = 0$.

A.2. $f(x)$ has a unique maximum, $f_{max} \leqslant 1$, assumed either at a point or in an interval. To the left or right of this point (or interval) $f(x)$ is strictly increasing or strictly decreasing, respectively.

A.3. At any x such that $f(x) = f_{max}$, the derivative exists and is equal to zero.

We allow the possibility that $f(x)$ assumes its maximum in an interval so as to include certain broken-linear functions with a "flat top" (cf. example (3.4) below).

In addition to the properties (A) we need some further conditions which will serve to define the range of the parameter λ:

B. Let $\lambda_{max} = 1/f_{max}$. Then there exists a λ_0 such that, for $\lambda_0 < \lambda < \lambda_{max}$, $\lambda f(x)$ has only two fixed points, the origin and $x_F(\lambda)$, say, both of which are repellent. For functions of class $C^{(1)}$ this means simply that

$$\left.\lambda \frac{df}{dx}\right|_{x=0} > 1 \qquad\qquad (\lambda_0 < \lambda \leqslant \lambda_{max}).$$
$$\left.\lambda \frac{df}{dx}\right|_{x=x_F(\lambda)} < -1$$

For piece-wise $C^{(1)}$ functions the generalization of these conditions is obvious.

The above conditions are sufficient to guarantee the existence of the

U-sequence; that they are not necessary can be shown by various examples, but we shall not pursue this matter here.

$f(x)$ as defined above clearly has the property that its piecewise derivative is less than 1 in absolute value in some interval N which includes that for which $f(x) = f_{\max}$. In order to exclude certain unwanted finite limit sets, we append the following condition:

C. $f(x)$ is convex in the interval N; at every point $x \notin N$, the piece-wise derivative of $f(x)$ is greater than 1 in absolute value.

Unfortunately, property (C) is not sufficient to exclude all finite limit sets not given by our construction; to achieve this end it might be necessary to restrict the underlying class of transformations rather drastically. We shall return to this point in Section 4 below.

It will simplify the subsequent discussion to make the non-essential assumption that $f(x)$ assumes its maximum at the point $x = \frac{1}{2}$ (or, if the function assumes its maximum in an interval, that the interval includes $x = \frac{1}{2}$). In the sequel we shall make this assumption without further comment. A particular iterate x' will then be said to be of "type L" or of "type R" according as $x' < \frac{1}{2}$ or $x' > \frac{1}{2}$, respectively. Given an initial point x_0, the minimum distinguishing information about the sequence of iterates $T_\lambda^{(k)}(x_0)$, $k = 1, 2, \ldots$, will consist in a "pattern" of R's and L's, the k-th letter giving the relative position of the k-th iterate of x_0 with respect to the point $x = \frac{1}{2}$. The patterns turn out to play a fundamental role in our construction; they will be discussed in detail in the following sections.

3. Let us give some simple examples of the class of transformations we are considering:

$$Q_\lambda(x):\quad x' = \lambda x(1-x)$$
$$3 < \lambda < 4 \tag{3.1}$$

$$S_\lambda(x):\quad x' = \lambda \sin \pi x$$
$$\lambda_0 < \lambda < 1 \qquad (\text{with } .71 < \lambda_0 < .72) \tag{3.2}$$

$$C_\lambda(x):\quad x' = \lambda W(3 - 3W + W^2), \qquad W \equiv 3x(1-x)$$
$$\lambda_0 < \lambda < \frac{64}{63} \qquad (\text{with } .872 < \lambda_0 < .873) \tag{3.3}$$

In the last two examples, more precise limits for λ_0 are available, but they are not important for our discussion. All these examples are convex functions of class $C^{(\infty)}$ which are, moreover, symmetric about $x = \frac{1}{2}$.

With regard to the existence of the U-sequence, these restrictions are in no way essential. As will be remarked in Section 4, however, these examples happen to belong to the subclass for which our construction does exhaust all finite limit sets.

As a further example, consider the broken-linear mapping:

$$L_\lambda(x; e): \quad x' = \frac{\lambda}{e} x, \qquad 0 \leqslant x \leqslant e,$$

$$x' = \lambda, \qquad e \leqslant x \leqslant 1 - e, \tag{3.4}$$

$$x' = \frac{\lambda}{e}(1 - x), \qquad 1 - e \leqslant x \leqslant 1,$$

$$\text{with} \quad 1 - e < \lambda < 1.$$

Here e is a parameter characterizing the width 1-2e of the maximum, and may be chosen to have any value in the range $0 < e < \frac{1}{2}$. It remains fixed as λ is varied, and different choices of e yield distinct transformations.

4. The finite limit sets of our class of transformations—and, in particular, of the four special transformations given above—are attractive periods of order $k = 2, 3,\ldots$. (We exclude the case $k = 1$ by invoking property (B).) The reader will recall that an "attractive period of order k" is a set of k periodic points x_i, $i = 1, 2,\ldots, k$, with $T_\lambda(x_i) = x_{i+1}$ (in some order). Each of these is a fixed point of the k-th power of $T_\lambda : T_\lambda^{(k)}(x_i) = x_i$, for which, moreover, the (piece-wise) derivative satisfies

$$\left| \frac{dT_\lambda^{(k)}}{dx} \right|_{x=x_i} < 1.$$

(By the chain rule, the slope is the same at all points in the period.) As a consequence of this slope condition, there exists for each x_i an attractive neighborhood $n(x_i)$ such that for any $x^* \in n(x_i)$ the sequence of iterates $T_\lambda^{(jk)}(x^*), j = 1, 2,\ldots,$ will converge to x_i. Periodic points which do not satisfy this slope condition (more precisely, for which the absolute value of the derivative is greater than 1) have no attractive neighborhood; they are consequently termed repellent (or unstable). These points belong to what is sometimes called "the set of exceptional points," a set of measure zero in the interval which plays no role in a discussion of limit sets.

The sequence of finite periods which we shall exhibit will be characterized *inter alia* by the following property:

J. For every period belonging to the U-sequence there is a period point whose attractive neighborhood includes the point $x = \frac{1}{2}$.

Now it follows from a theorem of G. Julia [3] that, if $T_\lambda(x)$ is the restriction to [0, 1] of some function analytic in the complex plane whose derivative vanishes at a single point in the interval, then the only possible finite limit sets ($k > 1$) are those with the property (J). The transformations (3.1) through (3.3) are clearly of this type, so that, with respect to finite limit sets, the U-sequence will exhaust all possibilities for them. That Julia's criterion is not necessary is shown by example (3.4); in this case there cannot be any attractive periods which do not have a period point lying in the region $e \leqslant x \leqslant 1 - e$. Such a period, however, clearly is of the type described by property (J), and hence belongs to the U-sequence.

Taking our clue from property (J), we now investigate the solutions λ of the equation:

$$T_\lambda^{(k)}(\tfrac{1}{2}) = \tfrac{1}{2}. \tag{4.1}$$

The corresponding periodic limit sets will be attractive, since the slope of $\lambda f(x)$ at $x = \frac{1}{2}$ is zero by hypothesis (property (A.3)). By way of example, we choose $k = 5$. Then for each of the four transformations of Section 3 there are precisely three distinct solutions of equation (4.1). The three patterns—common to all four transformations—are:

$$\tfrac{1}{2} \to R \to L \to R \to R \to \tfrac{1}{2},$$
$$\tfrac{1}{2} \to R \to L \to L \to R \to \tfrac{1}{2},$$
$$\tfrac{1}{2} \to R \to L \to L \to L \to \tfrac{1}{2}.$$

Omitting the initial and final points as understood, we write these patterns in the simplified form:

$$\begin{aligned}&\mathrm{RLR}^2,\\ &\mathrm{RL}^2\mathrm{R},\\ &\mathrm{RL}^3.\end{aligned} \tag{4.2}$$

In accordance with this convention, a pattern with $k - 1$ letters R or L will be said to be of "length k."

These solutions are clearly ordered on the parameter λ. In Table I we give the full set of solutions of (4.1), through $k = 7$, for all four special transformations; in the broken-linear case we choose $e = .45$. The numerical values of λ were found by a simple iterative technique (the "binary chopping process"); although they are given to only seven decimal digits, they are actually known to approximately twice that precision. Of course, once a particular λ has been found, the corresponding pattern can be generated by direct iteration.

TABLE I

			Values of λ_i			
i	k_i	P_i	$Q_\lambda(x)$	$S_\lambda(x)$	$C_\lambda(x)$	$L_\lambda(x; .45)$
1	2	R	3.2360680	.7777338	.9325336	.6581139
2	4	RLR	3.4985617	.8463822	.9764613	.7457329
3	6	RLR^3	3.6275575	.8811406	.9895107	.7806832
4	7	RLR^4	3.7017692	.9004906	.9955132	.8031673
5	5	RLR^2	3.7389149	.9109230	.9990381	.8180892
6	7	RLR^2LR	3.7742142	.9213346	1.0024311	.8318799
7	3	RL	3.8318741	.9390431	1.0073533	.8645337
8	6	RL^2RL	3.8445688	.9435875	1.0083134	.8858150
9	7	RL^2RLR	3.8860459	.9568445	1.0111617	.8977794
10	5	RL^2R	3.9057065	.9633656	1.0123766	.9085993
11	7	RL^2R^3	3.9221934	.9687826	1.0132699	.9187692
12	6	RL^2R^2	3.9375364	.9735656	1.0140237	.9278274
13	7	RL^2R^2L	3.9510322	.9782512	1.0146450	.9361518
14	4	RL^2	3.9602701	.9820353	1.0149542	.9462185
15	7	RL^3RL	3.9689769	.9857811	1.0152122	.9564172
16	6	RL^3R	3.9777664	.9892022	1.0154974	.9635343
17	7	RL^3R^2	3.9847476	.9919145	1.0156711	.9702076
18	5	RL^3	3.9902670	.9944717	1.0157727	.9775473
19	7	RL^4R	3.9945378	.9966609	1.0158320	.9846165
20	6	RL^4	3.9975831	.9982647	1.0158621	.9903134
21	7	RL^5	3.9993971	.9994507	1.0158718	.9957404

We note that the set of 21 patterns and its λ-ordering is common to all four transformations. This remains true when we extend our calculations through $k = 15$. As k increases, the total number of solutions of (4.1) becomes large, as indicated in Table II. Thus for $k \leqslant 15$ there is a total of 2370 distinct solutions of equation (4.1). In the appendix we give a complete list of ordered patterns for $k \leqslant 11$.

The fact that these patterns and their λ-ordering are a common property of four apparently unrelated transformations (note that they are not connected by ordinary conjugacy, a relation which will be discussed in Section (6) suggests that the pattern sequence is a general property of a wide class of mappings. For this reason we have called this sequence of patterns the U-sequence where "U" stands (with some exaggeration) for

TABLE II

k ...	2	3	4	5	6	7	8	9	10	11	12	13	14	15
Number of solutions ...	1	1	2	3	5	9	16	28	51	93	170	315	585	1091

"universal." In the next section we shall state and prove a logical algorithm which generates the U-sequence for any transformation having the properties (A) and (B) of Section 2. In the present section we confine ourselves to describing what might be called the "λ-structure" of the limit sets associated with the patterns of the U-sequence. No proofs are included, since the results given here will not be used in the proof of our main theorem.

As constructed, the patterns of the U-sequence correspond to distinct solutions of equation (4.1); they are attractive k-periods containing the point $x = \frac{1}{2}$ and possessing the property (J). It is clear by continuity that, given any solution λ (with finite k) and its associated pattern $P_k(\lambda)$, then for sufficiently small $\epsilon > 0$ there will exist periodic limit sets with the same pattern for all $\bar{\lambda}$ in the interval $\lambda - \epsilon \leqslant \bar{\lambda} \leqslant \lambda + \epsilon$. In other words, each period has a finite "λ-width." It is also clear that there exist critical values $m_1(\lambda)$ and $m_2(\lambda)$ such that, for $\bar{\lambda} < \lambda - m_1$ and $\bar{\lambda} > \lambda + m_2$, the pattern $P_k(\lambda)$ does *not* correspond to an attractive period of $T_{\bar{\lambda}}(x)$.

Consider now for simplicity the case in which the transformation is $C^{(1)}$, and take m_1 and m_2 to be boundary values such that for $\lambda - m_1 < \bar{\lambda} < \lambda + m_2$ the periodic limit set with pattern $P_k(\lambda)$ is attractive. As $\bar{\lambda}$ varies in this interval, the slope $dT_{\bar{\lambda}}^{(k)}/dx$ at a period point varies continuously from $+1$ to -1, the values ± 1 being assumed at the boundary points. It is natural to ask: what happens if $\bar{\lambda}$ lies just to the left or just to the right of the above interval? The question as to what the limit sets look like if $\bar{\lambda} = \lambda - m_1 - \delta$ (δ small) is a difficult one; the conjectured behavior will be described in Section 6, but rigorous proof is lacking. For $\bar{\lambda} = \lambda + m_2 + \delta$ we are in better case. As shown in Section 5, corresponding to any solution λ of (4.1) and its associated pattern there exists an infinite sequence of solutions $\lambda < \lambda^{(1)} < \lambda^{(2)} < \cdots < \lambda^{(\infty)}$ with associated patterns $H^{(1)}(\lambda^{(1)}), H^{(2)}(\lambda^{(2)}),\ldots,$ called "harmonics," with the property that they exhaust all possible solutions λ^* in the interval $\lambda \leqslant \lambda^* \leqslant \lambda^{(\infty)}$. The sequence of harmonics of a given solution is a set of periods of order $2^m k$, $m = 1, 2,\ldots,$ with contiguous λ-widths and well-defined pattern structure; no other periods of the U-sequence can exist for any λ^* in the given interval (harmonics have been encountered before in a more restrictive context: cf. reference 4). From the construction of Section 5 it will be obvious that $\lambda^{(\infty)}$ exists as a right-hand limit; the question as to the nature of the limit sets for $\lambda^* = \lambda^{(\infty)} + \epsilon$ remains open (but cf. Conjecture 2 of Section 6).

In order to prepare the ground for the discussion in the next section, we give here the following formal

DEFINITION 1. Let $P = \mathrm{RL}^{\alpha_1}\mathrm{R}^{\alpha_2}\mathrm{L}^{\alpha_3} \cdots$ be a pattern corresponding to some solution of (4.1). Then the (first) harmonic of P is the pattern $H = P\mu P$, where $\mu = L$ if P contains an odd number of R's, and $\mu = R$ otherwise.

For example, the pattern RLR^2 has the harmonic $H = RLR^2LRLR^2$, while for RL^2R we have $H = RL^2R^3L^2R$.

Naturally, the construction of the harmonic can be iterated, so that one may speak of the second, third,..., m-th harmonic, etc. When necessary, we shall write $H^{(j)}$ to denote the j-th harmonic.

In addition to the harmonic H of a pattern P, there is another formal construct which will be used in the sequel:

DEFINITION 2. The "antiharmonic" A of a pattern P is constructed analogously to the harmonic H except that $\mu = \mathrm{L}$ when P contains an even number of R's, while $\mu = \mathrm{R}$ otherwise.

Thus in passing from a pattern to its harmonic the "R-parity" changes, while for the antiharmonic the parity remains the same. The antiharmonic is a purely formal construct and never corresponds to any periodic limit set; the reason for this will become clear in the next section. Note that, like that of the harmonic, the antiharmonic construction can be iterated to any desired order.

5. We begin by defining the "extension" of a pattern:

DEFINITION 3. The H-extension of a pattern P is the pattern generated by iterating the harmonic construction applied j times to P, where j increases indefinitely.

DEFINITION 4. The A-extension of P is the pattern $A^{(j)}(P)$, where j increases indefinitely. Here $A^{(j)}(P)$ denotes the j-th iterate of the antiharmonic.

In these definitions we avoid writing $j \to \infty$, in order to avoid raising questions concerning the structure of the limiting pattern. In practice, all that will be required is that j is "sufficiently large."

We are now in a position to state

THEOREM 1. *Let K be an integer. Consider the complete ordered sequence of solutions of* (4.1) *and their associated patterns for* $2 \leqslant k \leqslant K$. *Let λ_1 be any such solution with pattern P_1 and length k_1, and let $\lambda_2 > \lambda_1$ be the "adjacent" solution (i.e., the next in order) with pattern P_2 and length k_2.*

Form the H-extension of P_1 and the A-extension of P_2. Reading from left to right, the two extensions $H(P_1)$ and $A(P_2)$ will have a maximal common leading subpattern P^* of length k^*, so that we may write

$$H(P_1) = P^*\mu_1 \ldots, \quad A(P_2) = P^*\mu_2 \ldots, \quad \mu_1 \neq \mu_2,$$

where μ_i stands for one of the letters L *or* R.

Case 1. $k^* \geqslant 2k_1$. *Then the solution λ^* of lowest order such that $\lambda_1 < \lambda^* < \lambda_2$ is the harmonic of P_1.*

Case 2. $k^* < 2k_1$. *Then the solution λ^* of lowest order such that $\lambda_1 < \lambda^* < \lambda_2$ corresponds to the pattern P^* of length k^* ($> K$ necessarily).*

A simple consequence of this theorem is the following:

COROLLARY. *Let $|k_1 - k_2| = 1$ in Theorem* 1. *Then the lowest order solution λ^* with $\lambda_1 < \lambda^* < \lambda_2$ has length $k^* = 1 + \max(k_1, k_2)$.*

This follows from the theorem on noting that all patterns have the common leading subpatterns (not maximal!) RL; therefore, in forming the extensions, the first disagreement will indeed come at the indicated value of k^*.

We give some examples of the application of Theorem 1.

Example 1. Take $K = 9$. Reference to the table in the appendix shows that patterns #12 and #14 are adjacent. We have

$$P_1 = \mathrm{RLR}^4,\ k_1 = 7,\ H(P_1) = \mathrm{RLR}^4\mathrm{LRLR}^4\ldots,$$
$$P_2 = \mathrm{RLR}^4\mathrm{LR},\ k_2 = 9,\ A(P_2) = \mathrm{RLR}^4\mathrm{LRLRL}\ldots,$$

so $P^* = \mathrm{RLR}^4\mathrm{LRLR}$, with $k^* = 11$, as verified by the table.

Example 2. Again take $K = 9$. Patterns #16 and #19 are adjacent and $P_1 = \mathrm{RLR}^2$ with $k_1 = 5$; here $k^* \geqslant 2k_1$. Therefore, by Case 1, the lowest order solution between the two patterns is the harmonic of P_1, namely, $\mathrm{RLR}^2\mathrm{LRLR}^2$, as given in the table (pattern #17).

To prove Theorem 1 we must first introduce some new concepts. Consider the transformation:

$$T_m(x): x' = \lambda_{\max} f(x) \tag{5.1}$$

This transformation maps [0, 1] *onto* itself; hence, for any point in the interval, the inverses of all orders exist. Let us restrict ourselves for the moment to the point $x = \frac{1}{2}$ and its set (2^k in number) of k-th order inverses. At each step in constructing a k-th order inverse we have the free choice of taking a point on the right or on the left. For example, designating the point $x = \frac{1}{2}$ by the letter O, a possible inverse of order 5 would be represented by the sequence of letters

$$\mathrm{RLR^2O}, \tag{5.2}$$

which is to be read from *right to left*. Let us call a sequence like (5.2), when read from right to left, a "5-th order inverse path of the point $x = \frac{1}{2}$." Note that (5.2) is precisely the pattern associated with the first solution of equation (4.1) for $k = 5$. Another possible inverse path of the same order would be $\mathrm{L^2R^2O}$, but this clearly does not correspond to any solution of (4.1).

Choosing a particular k-th order inverse path of $x = \frac{1}{2}$, let us call the numerical value of the corresponding k-th inverse the "coordinate" of the path. Obviously, no path whose coordinate is less than $\frac{1}{2}$ can correspond to a pattern associated with a solution of (4.1). In order to achieve a 1-1 correspondence between a subclass of inverse paths and our periodic patterns we introduce the concept of a "legal inverse path," which we abbreviate as "l.i.p."

DEFINITION 5. For the transformation $T_m(x)$ (cf. (5.1)), an l.i.p. of order k is a k-th order inverse path of $x = \frac{1}{2}$ whose coordinate x_k has the greatest numerical value of any point on the path.

In other words, of all the inverses constituting the path, the coordinate (i.e., the k-th inverse) lies farthest to the right. Note that any inverse path of $x = \frac{1}{2}$ can be inversely extended to an l.i.p. by appending on the left some suitable sequence, e.g., the sequence RL^α with α sufficiently large. Now consider the transformation $T_\lambda(x)$ corresponding to $T_m(x)$ with $\lambda < \lambda_{\max}$. As λ decreases, the original l.i.p. is deformed into an inverse path with varying coordinate $x_k(\lambda)$, but with the same pattern. By continuity, there clearly exists a λ^* for which

$$T_{\lambda^*}(\tfrac{1}{2}) = x_k(\lambda^*);$$

this in turn implies that for $\lambda = \lambda^*$ there exists a solution of equation (4.1) with the same pattern as that of the original l.i.p. On the other hand, for an inverse path (with, say, an R-type coordinate) which is *not* an l.i.p. the cycle will close on some intermediate point of the path (farther to the right than $x_k(\lambda)$), so that the path cannot be further inverted; this means that the original pattern cannot correspond to a solution of (4.1). Thus we have proved

LEMMA 1. *There is a* 1-1 *correspondence between the set of l.i.p.'s and the patterns associated with the solutions of equation* (4.1).

We note that the l.i.p.'s are naturally ordered on the values of their coordinates. By Lemma 1, any true statement about the pattern structure and coordinate ordering of the set of l.i.p.'s corresponds to a true statement about the pattern structure and λ-ordering of the set of solutions of (4.1).

Given some l.i.p. of order k, we construct an *inverse extension* $I(P)$ of the path according to the prescription $I(P) = P\mu PO$, where μ is R or L. Obviously, one choice corresponds to the harmonic, the other to the antiharmonic (Definitions 1 and 2). We can therefore speak of the harmonic or antiharmonic of an l.i.p. as well as of a pattern. Now, because of the monotonicity property (A.2) it follows that, given any two points, taking the left-hand inverse of both points preserves their relative order, while taking the right-hand inverse reverses it. A simple argument shows that $x_A < x < x_H$, where x is the coordinate of some l.i.p. and x_A, x_H are the coordinates of its antiharmonic and harmonic, respectively. This explains why the harmonic of an l.i.p. is again an l.i.p.. while the antiharmonic is not (and hence can never correspond to an attractive period of the U-sequence).

One final concept, the "projection" of an interval, will be of value in the subsequent discussion.

DEFINITION 6. Choose any two points x_1, x_2 in (0, 1); they define some interval I. Let $\bar{P}$ be an arbitrary sequence of R's and L's with $k - 1$ letters in all. Now, for some $T_m(x)$, construct the inverse paths $\bar{P}x_1$ and $\bar{P}x_2$. The coordinate x_1^* and x_2^* of these two paths define a new interval I^*, called the "projection under $\bar{P}$ of I." Because the defining inverse paths are of length k, we refer to it as a k-th order projection. (If we wish to exhibit explicitly the end-points x_1, x_2 of the interval I, we write I(x_1, x_2); in contrast to the usual notation for an interval, no ordering is implied.)

Proof of Theorem 1. It is clear that, if two intervals I, I^* are related by a k-th order projection, then for any point $x \in I^*$ we have $T_m^{(k)}(x) \in I$.

Consider now any l.i.p. with pattern PO and coordinate x_1. Its harmonic is again an l.i.p., with pattern PμPO and coordinate x_H, μ being either R or L depending on the R-parity of P. If x_μ is the point corresponding to the choice μ, then this construction shows that the interval $I^*(x_1, x_H)$ is the (k-th order) projection under P of the interval $I(\frac{1}{2}, x_\mu)$. Now any point x in the interior of I^* must map into the interior of I, and the end-points must map into end-points. Thus no inverse path of $x = \frac{1}{2}$—which is one of the end-points of I—can have a coordinate x^* satisfying $x_1 < x^* < x_H$. Precisely the same argument can be made for the antiharmonic. This proves

LEMMA 2. *Let* PO *be some l.i.p. with coordinate* x_1. *Form the* $H^{(j)}$-*extension of* P, *with coordinate* $x_H^{(j)}$, *and the* $A^{(j)}$-*extension of* P *with coordinate* $x_A^{(j)}$. *We then have* $x_A^{(j)} < x_1 < x_H^{(j)}$. *The intervals* $I^*(x_1, x_H^{(j)})$ *and* $I(x_1, x_A^{(j)})$ *do not contain the coordinate of any inverse path of* $x = \frac{1}{2}$.

The left-hand interval $I^*(x_1, x_A^{(j)})$ is of no significance for the limit sets of $T_\lambda(x)$; in fact, for λ a solution of (4.1) this interval shrinks to zero (and for $\lambda^* < \lambda$, neither the harmonic nor the antiharmonic exists). The right-hand, interval, however, is important. Using Lemma 2 and Lemma 1 we immediately derive

LEMMA 3. *If* λ_1 *is a solution of equation* (4.1) *and* λ_H *is the solution corresponding to its harmonic, then there does not exist any solution* λ^* *of* (4.1) *with the property* $\lambda_1 < \lambda^* < \lambda_H$.

Iterating this argument, we verify the statement of Section 4 that the sequence of harmonics is contiguous, i.e., that harmonics are always adjacent.

The adjacency property of harmonics serves to prove Case 1 of Theorem 1. We now proceed to Case 2.

Given some K, let (P_1, x_1, k_1) and (P_2, x_2, k_2) be two adjacent l.i.p.'s with $x_1 < x_2$ and $K + 1 < 2k_1$. Form the H-extension of P_1 and the A-extension of P_2; these can be written in the form

$$\begin{aligned} H(P_1) &= P^*\mu_1 \cdots \\ A(P_2) &= P^*\mu_2 \cdots \end{aligned} \qquad (\mu_1 \neq \mu_2).$$

The coordinates x_H and x_A define an interval I^* which is a projection of $I(x_{\mu_1}, x_{\mu_2})$; clearly, I^* is contained in the original interval $I(x_1, x_2)$. Since I contains the point $x = \frac{1}{2}$, there must exist an inverse path of $\frac{1}{2}$, P*O, with coordinate x^* satisfying $x_1 < x^* < x_2$. But P*O must be an l.i.p. since it is a leading subpattern of the interated harmonic of P_1. Moreover, by the adjacency assumption, its length k^* must necessarily be greater than k_1 or k_2. On the other hand, P*O is the shortest pattern for which an interval with non-zero content exists. Invoking Lemma 1, we see that the proof of the theorem is complete.

The formulation of a practical algorithm, using the results of Theorem 1, needs little comment. Given the complete U-sequence for $k \leqslant K$, one generates the sequence for $K + 1$ by inserting the appropriate pattern of length $K + 1$ between every two (non-harmonic) adjacent patterns whenever the theorem permits it. The pattern $R(k = 2)$ remains the lowest pattern; as is easily shown, for any k the last pattern is always of the form RL^{k-2}, and this is simply appended to the list. As previously mentioned, the algorithm has been checked (to $k \leqslant 15$) for the four special transformations of Section 3 by actually finding the corresponding solutions of equation (4.1)—a simple process in which there are no serious accuracy limitations.

We remark here that the combinatorial problem of enumerating all l.i.p.'s of a given length k has been solved [5]; the number of patterns turns out to be just the number of symmetry types of primitive periodic sequences (with two "values" or letters allowed) under the cyclic group C_k (so that the full symmetry group is $C_k \times S_2$, where S_2 is the symmetric group on two letters). For k a prime, this number is simple, and turns out to be given by the expression

$$\frac{1}{k}(2^{k-1} - 1).$$

We encountered this enumeration problem previously (cf. reference 4, Table 1); at that time we were not aware of the work of Gilbert and Riordan [5].

6. In this final section we collect some observations and conjectures concerning the nature of limit sets not belonging to the U-sequence, ending with a few remarks on the relation of conjugacy.

(1) *Other finite limit sets.* As remarked in the introduction, it does not seem possible to exclude "anomalous" limit sets without seriously restricting the underlying class of transformations. To convince the reader that such anomalous periods can in fact exist we give a simple example:

Let us alter the special transformation (3.4) in the following way (we take $e = .45$):

$$\left.\begin{aligned} x' &= 4.5\lambda x, && 0 \leqslant x \leqslant .2 \\ x' &= \lambda(.4x + .82), && .2 \leqslant x \leqslant .45 \\ x' &= \lambda, && .45 \leqslant x \leqslant .55 \\ x' &= \frac{\lambda}{.45}(1 - x), && .55 \leqslant x \leqslant 1 \end{aligned}\right\} \quad (.55 < \lambda < 1). \tag{6.1}$$

Then, in addition to the U-sequence (with λ values different from those of the original transformation), there exists an attractive 2-period in the range $\lambda_1 < \lambda \leqslant 1$ with

$$\lambda_1 = \tfrac{1}{2} + \tfrac{1}{2}\sqrt{.19}.$$

Note that the 2-period remains attractive even for $\lambda = 1$. While the anomalous periods do not affect the existence of the U-sequence, they do cause additional partitioning of the unit interval because their existence implies that there is a set of points (with non-zero measure) whose sequence of iterated images will converge to the periods in question.

These anomalous periods, however, differ radically from those belonging to the U-sequence in that they do not possess the property (J). This in turn means that the slope at a period point is strongly bounded away from zero. Thus, at least for transformations with the property (C), it seems reasonable to conjecture that such periods cannot have arbitrary length and still remain attractive. Hence we make

Conjecture 1. *For transformations with properties* (A), (B) *and* (C), *the anomalous attractive periods constitute at most a finite sequence.*

(2) *Infinite limit sets.* For simplicity we consider the case in which there are no anomalous periods, e.g., functions covered by Julia's theorem (or some valid extension thereof). We assign to each period of the U-sequence a λ-measure equal to its λ-width. The question is then: is the λ-measure of the full U-sequence equal to $\lambda_{\max} - \lambda_0$? Or, put otherwise, is there a set of non-zero measure in the interval $(\lambda_0, \lambda_{\max})$ such that the sequence of iterates of $x = \frac{1}{2}$ does *not* converge to a member of the U-sequence? Numerical experiments with the four special transformations of Section 3 together with some heuristic arguments based on the iteration of the algorithm of Theorem 1 leads us to make the modest

CONJECTURE 2. *For an infinite subclass of transformations with properties* (A) *and* (B), *the* λ*-measure of the U-sequence is the whole* λ *interval.*

(3) *A limiting case.* Take the transformation $L_\lambda(x; e)$ of Section 3 and set $e = \frac{1}{2}$. We then have

$$L_\lambda(x; \tfrac{1}{2}): \begin{array}{ll} x' = 2\lambda x, & 0 \leqslant x \leqslant \frac{1}{2} \\ x' = 2\lambda(1-x), & \frac{1}{2} \leqslant x \leqslant 1 \end{array} \qquad (\tfrac{1}{2} < \lambda < 1). \tag{6.2}$$

Although we cannot speak of attractive periods in this case (since the slope of the function is nowhere less than 1 in absolute value), it is still of interest to investigate the corresponding solutions of equation (4.1). These turn out to be a subset of the U-sequence in which the 2-period, all harmonics, and all patterns algorithmically generated from the harmonics and adjacent nonharmonics, are absent. The count through $k \leqslant 15$ is given in Table III, which may be compared with Table II.

TABLE III

k ...	3	4	5	6	7	8	9	10	11	12	13	14	15
Number of solutions	1	1	3	4	9	14	27	48	93	163	315	576	1085

One can explain this behavior by saying that, as the width 1-2e of the flat-top shrinks to zero, the harmonics and harmonic-generated periods "coalesce" in structure with their fundamentals. This provides another illustration of the nature of the harmonics outlined in Section 4.

(4) *Conjugacy.* Two transformations $f(x)$, $g(x)$ on [0, 1] are said to be conjugate to each other if there exists a continuous, 1-1 mapping $h(x)$ of [0, 1] onto itself such that

$$g(x) = hf[h^{-1}(x)], \qquad x \in [0, 1]. \tag{6.3}$$

If $f(x)$ and $g(x)$ are themselves 1-1, the question of the existence of an $h(x)$ satisfying (6.3) is settled by a theorem of Schreier and Ulam [6]. When $f(x)$, $g(x)$ are not homeomorphisms, very little is known about the existence or nonexistence of a conjugating function $h(x)$.

The importance of (6.3) for our purpose is that the attractive nature of limit sets is preserved under conjugacy; in particular, if $T_\lambda(x)$ possess the U-sequence, then so does every conjugate of it. Clearly, our class of transformations must be invariant under conjugation by the set of all continuous, 1-1 functions $h(x)$ on $[0, 1]$. (Incidentally, we now see why our special choice of the point $x = \frac{1}{2}$ is no restriction, since it can be shifted by conjugation with an appropriate $h(x)$.)

It has long been known [7] that the parabolic transformation (3.1) with $\lambda = \lambda_{max} = 4$ is conjugate to the broken-linear transformation (6.2) with $\lambda = 1$, the conjugating function being

$$h(x) = \frac{2}{\pi} \sin^{-1} (\sqrt{x}).$$

In general, no such pairwise conjugacy exists for the four special transformations of Section 3. For particular choices of the parameters this can be shown by making the following simple test. If $f(x_0) = x_0$ and $g(x_1) = x_1 (x_0, x_1 \neq 0)$, then a short calculation shows that

$$\left.\frac{df(x)}{dx}\right|_{x=x_0} = \left.\frac{dg(x)}{dx}\right|_{x=x_1} \tag{6.4}$$

It is easily established that (6.4) does not hold in general for any pair of our special transformations.

In view of the existence of the U-sequence, it is of interest to speculate whether there is not some well-defined but less restrictive equivalence relation that will serve to replace conjugacy (for one such suggestion-due to S. Ulam—see the remarks in reference 1, p. 49). Of course, Theorem 1 itself provides such an equivalence relation, albeit not a very useful one:

Let $T_{1\lambda}(x)$, $T_{2\mu}(x)$ be two transformations with properties (A) and (B). Then there exists a mapping function M_{12} such that $M_{12}(\lambda) = \mu$, the domain of M being the union of the λ-widths of the U-sequence for T_1 and the range being the union of the μ-widths of the U-sequence for T_2.

Since at present nothing whatsoever is known about these mappings M_{ij}, the above correspondence amounts to nothing more than a restatement of the existence of the U-sequence itself.

Appendix

The following table gives the complete ordered set of patterns associated with the U-sequence for $K \leqslant 11$; i is a running index, K gives the pattern length, and $I(K)$ indicates the relative order of periods of given length K. The ordering corresponds to the λ-ordering of solutions of equation (4.1).

i	K	I(K)	Pattern	i	K	I(K)	Pattern
1	2	1	R	41	10	9	RL^2RLR^3L
2	4	1	RLR	42	7	3	RL^2RLR
3	8	1	RLR^3LR	43	10	10	$RL^2RLRLRL$
4	10	1	RLR^3LRLR	44	11	15	$RL^2RLRLRLR$
5	6	1	RLR^3	45	9	7	RL^2RLRLR
6	10	2	RLR^5LR	46	11	16	RL^2RLRLR^3
7	8	2	RLR^5	47	10	11	RL^2RLRLR^2
8	10	3	RLR^7	48	11	17	RL^2RLRLR^2L
9	11	1	RLR^8	49	8	5	RL^2RLRL
10	9	1	RLR^6	50	11	18	RL^2RLRL^2RL
11	11	2	RLR^6LR	51	5	2	RL^2R
12	7	1	RLR^4	52	10	12	$RL^2R^3L^2R$
13	11	3	RLR^4LRLR	53	11	19	$RL^2R^3L^2RL$
14	9	2	RLR^4LR	54	8	6	RL^2R^3L
15	11	4	RLR^4LR^3	55	11	20	$RL^2R^3LR^2L$
16	5	1	RLR^2	56	10	13	$RL^2R^3LR^2$
17	10	4	RLR^2LRLR^2	57	11	21	$RL^2R^3LR^3$
18	11	5	RLR^2LRLR^3	58	9	8	RL^2R^3LR
19	9	3	RLR^2LRLR	59	11	22	RL^2R^3LRLR
20	11	6	$RLR^2LRLRLR$	60	10	14	RL^2R^3LRL
21	7	2	RLR^2LR	61	7	4	RL^2R^3
22	11	7	RLR^2LR^3LR	62	10	15	RL^2R^5L
23	9	4	RLR^2LR^3	63	11	23	RL^2R^5LR
24	11	8	RLR^2LR^5	64	9	9	RL^2R^5
25	10	5	RLR^2LR^4	65	11	24	RL^2R^7
26	8	3	RLR^2LR^2	66	10	16	RL^2R^6
27	10	6	RLR^2LR^2LR	67	11	25	RL^2R^6L
28	11	9	$RLR^2LR^2LR^2$	68	8	7	RL^2R^4
29	3	1	RL	69	11	26	RL^2R^4LRL
30	6	2	RL^2RL	70	10	17	RL^2R^4LR
31	9	5	RL^2RLR^2L	71	11	27	$RL^2R^4LR^2$
32	11	10	$RL^2RLR^2LR^2$	72	9	10	RL^2R^4L
33	10	7	RL^2RLR^2LR	73	11	28	$RL^2R^4L^2R$
34	11	11	RL^2RLR^2LRL	74	6	3	RL^2R^2
35	8	4	RL^2RLR^2	75	11	29	$RL^2R^2LRL^2R$
36	11	12	RL^2RLR^4L	76	9	11	RL^2R^2LRL
37	10	8	RL^2RLR^4	77	11	30	$RL^2R^2LRLR^2$
38	11	13	RL^2RLR^5	78	10	18	RL^2R^2LRLR
39	9	6	RL^2RLR^3	79	11	31	RL^2R^2LRLRL
40	11	14	RL^2RLR^3LR	80	8	8	RL^2R^2LR

i	K	I(K)	Pattern
81	11	32	$RL^2R^2LR^3L$
82	10	19	$RL^2R^2LR^3$
83	11	33	$RL^2R^2LR^4$
84	9	12	$RL^2R^2LR^2$
85	11	34	$RL^2R^2LR^2LR$
86	10	20	$RL^2R^2LR^2L$
87	7	5	RL^2R^2L
88	10	21	$RL^2R^2L^2RL$
89	11	35	$RL^2R^2L^2RLR$
90	9	13	$RL^2R^2L^2R$
91	11	36	$RL^2R^2L^2R^3$
92	10	22	$RL^2R^2L^2R^2$
93	11	37	$RL^2R^2L^2R^2L$
94	4	2	RL^2
95	8	9	RL^3RL^2
96	11	38	$RL^3RL^2R^2L$
97	10	23	$RL^3RL^2R^2$
98	11	39	$RL^3RL^2R^3$
99	9	14	RL^3RL^2R
100	11	40	RL^3RL^2RLR
101	10	24	RL^3RL^2RL
102	11	41	$RL^3RL^2RL^2$
103	7	6	RL^3RL
104	11	42	$RL^3RLR^2L^2$
105	10	25	RL^3RLR^2L
106	11	43	RL^3RLR^2LR
107	9	15	RL^3RLR^2
108	11	44	RL^3RLR^4
109	10	26	RL^3RLR^3
110	11	45	RL^3RLR^3L
111	8	10	RL^3RLR
112	11	46	$RL^3RLRLRL$
113	10	27	RL^3RLRLR
114	11	47	RL^3RLRLR^2
115	9	16	RL^3RLRL
116	11	48	RL^3RLRL^2R
117	10	28	RL^3RLRL^2
118	6	4	RL^3R
119	10	29	$RL^3R^3L^2$
120	11	49	$RL^3R^3L^2R$
121	9	17	RL^3R^3L
122	11	50	$RL^3R^3LR^2$
123	10	30	RL^3R^3LR
124	11	51	RL^3R^3LRL
125	8	11	RL^3R^3
126	11	52	RL^3R^5L
127	10	31	RL^3R^5
128	11	53	RL^3R^6
129	9	18	RL^3R^4
130	11	54	RL^3R^4LR
131	10	32	RL^3R^4L
132	11	55	$RL^3R^4L^2$
133	7	7	RL^3R^2
134	11	56	$RL^3R^2LRL^2$
135	10	33	RL^3R^2LRL
136	11	57	RL^3R^2LRLR
137	9	19	RL^3R^2LR
138	11	58	$RL^3R^2LR^3$
139	10	34	$RL^3R^2LR^2$
140	11	59	$RL^3R^2LR^2L$
141	8	12	RL^3R^2L
142	11	60	$RL^3R^2L^2RL$
143	10	35	$RL^3R^2L^2R$
144	11	61	$RL^3R^2L^2R^2$
145	9	20	$RL^3R^2L^2$
146	11	62	$RL^3R^2L^3R$
147	5	3	RL^3
148	10	36	RL^4RL^3
149	11	63	RL^4RL^3R
150	9	21	RL^4RL^2
151	11	64	$RL^4RL^2R^2$
152	10	37	RL^4RL^2R
153	11	65	RL^4RL^2RL
154	8	13	RL^4RL
155	11	66	RL^4RLR^2L
156	10	38	RL^4RLR^2
157	11	67	RL^4RLR^3
158	9	22	RL^4RLR
159	11	68	RL^4RLRLR
160	10	39	RL^4RLRL

i	K	I(K)	Pattern
161	11	69	RL^4RLRL^2
162	7	8	RL^4R
163	11	70	$RL^4R^3L^2$
164	10	40	RL^4R^3L
165	11	71	RL^4R^3LR
166	9	23	RL^4R^3
167	11	72	RL^4R^5
168	10	41	RL^4R^4
169	11	73	RL^4R^4L
170	8	14	RL^4R^2
171	11	74	RL^4R^2LRL
172	10	42	RL^4R^2LR
173	11	75	$RL^4R^2LR^2$
174	9	24	RL^4R^2L
175	11	76	$RL^4R^2L^2R$
176	10	43	$RL^4R^2L^2$
177	11	77	$RL^4R^2L^3$
178	6	5	RL^4
179	11	78	RL^5RL^3
180	10	44	RL^5RL^2
181	11	79	RL^5RL^2R
182	9	25	RL^5RL
183	11	80	RL^5RLR^2
184	10	45	RL^5RLR
185	11	81	RL^5RLRL
186	8	15	RL^5R
187	11	82	RL^5R^3L
188	10	46	RL^5R^3
189	11	83	RL^5R^4
190	9	26	RL^5R^2
191	11	84	RL^5R^2LR
192	10	47	RL^5R^2L
193	11	85	$RL^5R^2L^2$
194	7	9	RL^5
195	11	86	RL^6RL^2
196	10	48	RL^6RL
197	11	87	RL^6RLR
198	9	27	RL^6R
199	11	88	RL^6R^3
200	10	49	RL^6R^2
201	11	89	RL^6R^2L
202	8	16	RL^6
203	11	90	RL^7RL
204	10	50	RL^7R
205	11	91	RL^7R^2
206	9	28	RL^7
207	11	92	RL^8R
208	10	51	RL^8
209	11	93	RL^9

References

1. P. R. Stein and S. M. Ulam, Non-linear transformation studies on electronic computers, *Rozprawy Mat.* **39** (1964), 1–66.
2. O. W. Rechard, Invariant measures for many-one transformations, *Duke Math. J.* **23** (1956), 477.
3. G. Julia, Mémoire sur l'itération des functions rationelles, *J. de Math. Ser.* 7, **4** (1918), 47–245. The relevant theorem appears on p. 129 ff.
4. N. Metropolis, M. L. Stein, and P. R. Stein, Stable states of a non-linear transformation, *Numer. Math.* **10** (1967), 1–19.
5. E. N. Gilbert and J. Riordan, Symmetry types of periodic sequences, *Illinois J. Math.* **5** (1961), 657.
6. J. Schreier and S. Ulam, Eine Bemerkung über die Gruppe der topologischen Abbildung der Kreislinie auf sich selbst, *Studia Math.* **5** (1935), 155–159.
7. See reference 1, p. 52. The result is due to S. Ulam and J. von Neumann.

Editor's note. One-dimensional maps as models of dynamical systems were introduced by Lorenz (1964). The results of Metropolis, Stein and Stein were put on a rigorous footing in a beautiful article by Guckenheimer (1977). The mathematics of itineraries is discussed in detail in Collet and Eckmann (1980a).

Journal of Statistical Physics **21** 669–706 (1979)

The Universal Metric Properties of Nonlinear Transformations

Mitchell J. Feigenbaum[1]

Received May 29, 1979

The role of functional equations to describe the exact local structure of highly bifurcated attractors of $x_{n+1} = \lambda f(x_n)$ independent of a specific f is formally developed. A hierarchy of universal functions $g_r(x)$ exists, each descriptive of the same local structure but at levels of a cluster of 2^r points. The hierarchy obeys $g_{r-1}(x) = -\alpha g_r(g_r(x/\alpha))$, with $g = \lim_{r\to\infty} g_r$ existing and obeying $g(x) = -\alpha g(g(x/\alpha))$, an equation whose solution determines both g and α. For r asymptotic

$$g_r \sim g - \delta^{-r} h \qquad (*)$$

where $\delta > 1$ and h are determined as the associated eigenvalue and eigenvector of the operator $\mathscr{L}$:

$$\mathscr{L}[\psi] = -\alpha[\psi(g(x/\alpha)) + g'(g(x/\alpha))\psi(-x/\alpha)]$$

We conjecture that $\mathscr{L}$ possesses a unique eigenvalue in excess of 1, and show that this δ is the λ-convergence rate. The form (*) is then continued to all λ rather than just discrete λ_r and bifurcation values Λ_r and dynamics at such λ is determined. These results hold for the high bifurcations of any fundamental cycle. We proceed to analyze the approach to the asymptotic regime and show, granted $\mathscr{L}$'s spectral conjecture, the stability of the g_r limit of highly iterated λf's, thus establishing our theory in a local sense. We show in the course of this that highly iterated λf's are conjugate to g_r's, thereby providing some elementary approximation schemes for obtaining λ_r for a chosen f.

KEY WORDS: Recurrence; bifurcation; attractor; universal; functional equations; scaling; conjugacy; spectrum of linearized operator.

Work performed under the auspices of the U.S. Energy Research and Development Administration.

[1] Theoretical Division, Los Alamos Scientific Laboratory, University of California, Los Alamos, New Mexico.

1. INTRODUCTION

In a previous paper[1] (hereafter referred to as I), a viewpoint was advanced that detailed information about large stability sets of a recursion relation

$$x_{n+1} = \lambda f(x_n) \tag{1}$$

is available independent of the exact form of f for a wide class of functions. A heuristic argument (corroborated by computer computation) was offered to the effect that appropriate functional equations, free of reference to the recursion equation, furnish all this detailed quantitative information. Specifically, the exact distribution of points of large limit cycles of the recursion equation within local clusters is determined by a certain universal function $g^*(x)$ obeying a functional equation we conjectured to exist, but only approximately could specify. A parameter α implicated in that equation, and presumably determined by it collaterally with $g^*(x)$, plays the role of a fundamental scale factor: upon bifurcation of a high-order cycle, the points of the bifurcated cycle are identically distributed, save for a reduction in scale by the factor α. Another fundamental parameter δ, the convergence rate of a variety of universal details, was crudely determined from $g^*(x)$.

In the present paper we shall vindicate these conjectures in exhibiting an exact equation determining α and a universal function g closely related to g^* of I. Indeed, two functions $g^*(x)$ and $-\alpha g^*(g^*(x/\alpha))$ were discussed in I; these are the first two (g_1 and g_0, respectively) of an infinite sequence of functions $g_r(x)$ linked by the shift operation

$$g_{r-1}(x) = -\alpha g_r(g_r(x/\alpha)) \tag{2}$$

The equation

$$g(x) = -\alpha g(g(x/\alpha)) \tag{3}$$

is obeyed by $g(x) = \lim_{r\to\infty} g_r(x)$, a function determining the local distribution of infinite clusters of elements of *all* the infinite attractors of (1). We then proceed to determine $g_r(x)$ for large r in terms of an auxiliary function $h(x)$ obeying a functional equation implicating δ and determining both $h(x)$ and δ. Utilizing (2), one can then step down to determine from a g_r for $r \gg 1$ the g^* of I. Thus, as conjectured in I, the entire local structure of high-order stability sets of (1) is determined in a framework liberated from (1).

With the structure of the infinite limiting attractors laid out in Sections 2 and 3, we investigate in Section 4 the asymptotic approach to this structure. The equation obeyed by h is a linear functional eigenvalue equation, whose eigenvalues in excess of 1 lead to convergence of g_r to g; the eigenvalues bounded by 1 in absolute value represent potential instabilities.

However, in analyzing the large-n approach to a g_r, we discover that exactly these eigenvalues lead to convergence, so that in the infinite-n limit, g_r possesses no unstable components. In this fashion, though, the large eigenvalues destroy convergence to g_r. We discover, however, that the choice of the λ_r dictated by the recursion equation exactly suppresses this instability *providing* there is a *unique* eigenvalue in excess of 1. This eigenvalue is δ and we conjecture its uniqueness. Proof of this conjecture would constitute a local proof of universality. (At present we have only computer corroboration.)

Finally, in Section 5 we discuss techniques for the solution of the fundamental functional equations and various approximation schemes.

2. THE SEQUENCE $\{g_r\}$ OF UNIVERSAL FUNCTIONS AND THE BASIC FUNCTIONAL EQUATION

As heuristically argued in I, defining

$$g(\lambda, x) \equiv \lambda f(x)$$

where f possesses a differentiable zth-order maximum at $x = 0$,

$$f(0) - f(x) \propto |x|^z \qquad z > 1 \text{ for } |x| \text{ small}$$

we have

$$(-\alpha)^n g^{(2^n)}(\lambda_{n+1}, x/\alpha^n) \sim \mu g^*(x/\mu) \tag{4}$$

for large n, where $g^{(n)}(x)$ is the nth iterate of g:

$$g^{(2)}(x) \equiv g(g(x)); \qquad g^{(n+1)}(x) \equiv g(g^{(n)}(x))$$

and μ depends upon the specific form of f. Rescaling g^* on both height and width by μ removes all vestige of the specific form of f, and is accomplished through the definition of an absolute scale:

$$g^*(0) = 1$$

We understand this absolute rescaling implicitly, and simply write (4) as

$$(-\alpha)^n g^{(2^n)}(\lambda_{n+1}, x/\alpha^n) \sim g^*(x) \tag{5}$$

The value of λ, λ_{n+1} is determined by the condition

$$g^{(2^n)}(\lambda_n, 0) = 0, \qquad g^{(2^{n'})}(\lambda_n, 0) \neq 0 \qquad \text{for } n' < n$$

g^* accordingly describes a two-point cycle near $x = 0$, since $g^*(0) = 1$ and $g^*(1) = 0$ (see Fig. 1). With g^* the limit of (5) as $n \to \infty$, and universal for

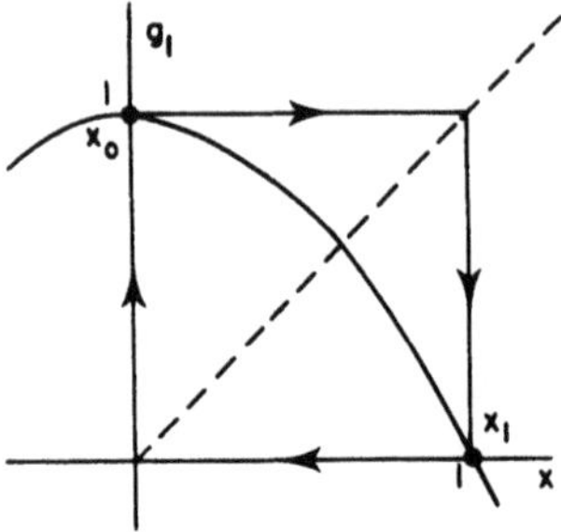

Fig. 1. The function g_1 normalized to $g_1(0) = 1$; the 2-cycle $x_0 \to x_1 \to x_0 \to x$, etc., is indicated.

all f of fixed z, $g^*(g^*(x))$ is itself universal, possessing 0 and 1 as fixed points. $g^*(g^*)$ is also a limit of highly iterated g's:

$$\begin{aligned} g^*(g^*(x)) &\sim (-\alpha)^n g^{(2^n)}(\lambda_{n+1}, g^{(2^n)}(\lambda_{n+1}, x/\alpha_n)) \\ &= (-\alpha)^n g^{(2n+1)}(\lambda_{n+1}, x/\alpha^n) \end{aligned}$$

or,

$$\begin{aligned} g_0 \equiv -\alpha g^*(g^*(x/\alpha)) &\sim (-\alpha)^{n+1} g^{(2^{n+1})}(\lambda_{n+1}, x/\alpha^{n+1}) \\ &\sim (-\alpha)^n g^{(2^n)}(\lambda_n, x/\alpha^n) \end{aligned} \tag{6}$$

Thus, $g^*(x)$ is obtained from $g_0(x)$ by increasing λ into the next bifurcation; conversely, g_0 describes a one-cycle near $x = 0$ (Fig. 2). Both g_0 and $g^* \equiv g_1$ describe the identical local structure of the elements of a large 2^n-cycle near $x = 0$, but at different "magnifications." Generalizing,

$$g_r(x) \equiv \lim_{n\to\infty} (-\alpha)^n g^{(2^n)}(\lambda_{n+r}, x/\alpha^n) \tag{7}$$

again describes the identical local structure, but now at a level of magnification such that 2^r elements of the cycle are clustered about the central bump (Fig. 3). From the definition (7),

$$g_{r-1}(x) = -\alpha g_r(g_r(x/\alpha)) \tag{8}$$

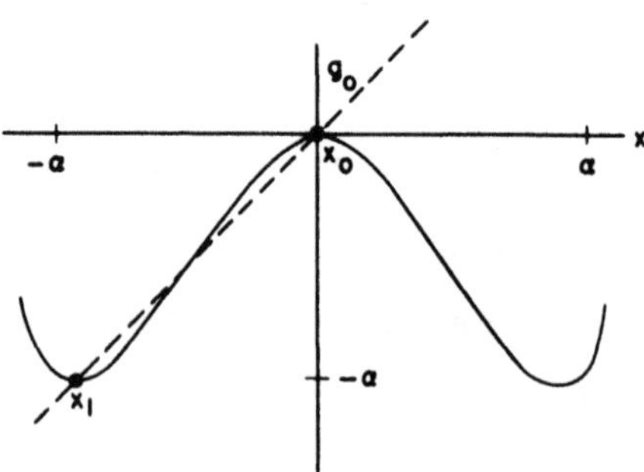

Fig. 2. The function g_0 corresponding to g_1 normalized as in Fig. 1; the locations of x_0 and x_1 are magnified by $-\alpha$.

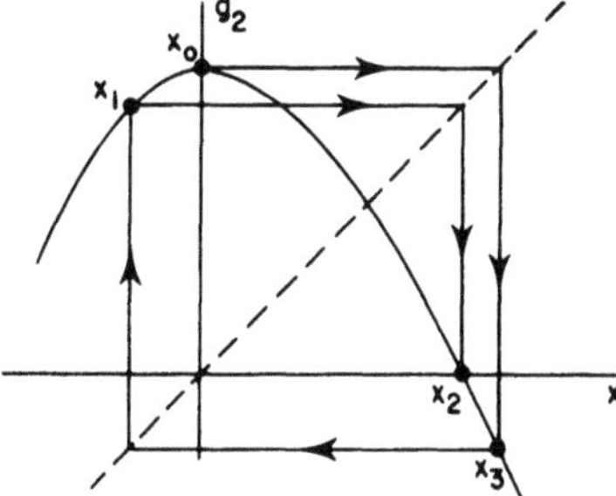

Fig. 3. The function g_2; the 4-cycle $x_0 \to x_3 \to x_1 \to x_2 \to x_0$, etc., is indicated. x_0 and x_1 are as in Figs. 1 and 2, but now *reduced* from Fig. 1 by $-\alpha$.

(Universality implies that all g_r are symmetric functions.) Given any g_r, simply iterating produces all other g_m for $m < n$, and each contains the identical information.

When r is so large that n can become large and yet much smaller than r, then

$$g^{(2^n)}(\lambda_{n+r+1}) \approx g^{(2^n)}(\lambda_{n+r})$$

since $\lambda_m \to \lambda_\infty$ and $\lambda_{n+r+1} \approx \lambda_{n+r}$: n must be of the order of r before any error in λ_{n+r} can become significant. Alternatively, the central bump suffers very slight distortion to accommodate the infinite attractor when it already accommodates a very large attractor. That is, we intuitively conjecture that the limit

$$\lim_{r\to\infty} g_r(x) \equiv g(x)$$

exists. This granted, (8) implies that g satisfies

$$g(x) = -\alpha g(g(x/\alpha)) \tag{9}$$

Qualitatively, $g(x)$ looks like the curve of Fig. 3. Yet, $g(x)$ contains different information from $g_r(x)$ for any finite r: quite simply, any two stability points located by g_r possess a minimum separation, which is not true for g. Rather, g represents a different level of universal distribution of stability points: the entirety of $g_1(x)$ is collapsed to a point at the level of g. (This is again a reflection of the Cantor set-like nature of highly bifurcated stability sets—indeed, infinitely bifurcated.)

An alternate definition of g in the n-limit sense is

$$g(x) = \lim_{n\to\infty} (-\alpha)^n g^{(2^n)}(\lambda_\infty, x/\alpha^n) \tag{10}$$

since λ_∞ is a finite, perfectly definite value of λ. It is clear from (10) why we succeeded in obtaining an exact equation for g: the λ-shifting which frustrated our attempt at an exact equation for g^* in I is here absent. There is, however, a strong price to be exacted for this grace: unlike the hoped-for equation in I, (9) must be recursively unstable. To understand this, let us rederive (9) from (10). Define

$$\tilde{g}_n(x) \equiv (-1)^n \beta_n g^{(2^n)}(\lambda_\infty, x/\beta_n) \tag{11}$$

or

$$(1/\beta_n)\tilde{g}_n(\beta_n x) = (-1)^n g^{(2^n)}(\lambda_\infty, x)$$

Then

$$\begin{aligned}(-1)^n g^{(2^{n+1})}(\lambda_\infty, x) &= (-1)^n g^{(2^n)}(\lambda_\infty, (-1)^n g^{(2^n)}(\lambda_\infty, x)) \\ &= \frac{1}{\beta_n}\tilde{g}_n((-1)^n \beta_n g^{(2^n)}(\lambda_\infty, x)) \\ &= \frac{1}{\beta_n}\tilde{g}_n(\tilde{g}_n(\beta_n x))\end{aligned}$$

or,

$$-\frac{1}{\beta_{n+1}}\tilde{g}_{n+1}(\beta_{n+1}x) = \frac{1}{\beta_n}\tilde{g}_n(\tilde{g}_n(\beta_n x))$$

or, with $\beta_{n+1}/\beta_n \equiv \alpha_n$,

$$\tilde{g}_{n+1}(x) = -\alpha_n \tilde{g}_n(\tilde{g}_n(x/\alpha_n)) \tag{12}$$

Setting an absolute scale

$$\tilde{g}_n(0) = 1 \qquad \text{for all } n$$

(12) implies that $g_n(x)$ determines α_n:

$$\tilde{g}_{n+1}(0) = -\alpha_n \tilde{g}_n(\tilde{g}_n(0))$$

or

$$1 = -\alpha_n \tilde{g}_n(1) \tag{13}$$

Accordingly, choosing a $\tilde{g}_0(x)$ satisfying $\tilde{g}_0(0) = 1$ and possessing a zth order maximum at $x = 0$, we can use (12) and (13) to recursively generate $\tilde{g}_n(x)$. Should $\tilde{g}_n(x) \to g(x)$, then

$$g(x) = -\alpha g(g(x/\alpha))$$

with $\alpha = \lim_{n\to\infty} \alpha_n$.

Apart from manipulations, the regimen of (12) and (13) is simply a machine to perform the (attempted) limit of (10) starting with a $\tilde{g}_0$ that is essentially $\lambda_\infty f(x)$, or more exactly

$$\tilde{g}_0(x) = f(\lambda_\infty x) \tag{14}$$

[So that $\tilde{g}_0(0) = 1$, we have rescaled on height and width by λ_∞: (9) is invariant to such rescaling.] Since λ_∞ depends upon f,

$$\tilde{g}_0(x) = f(\lambda_\infty(f)x) \tag{15}$$

Indeed, for any f of our class, $\tilde{g}_0$ as given by (15) must result in $\tilde{g}_n \to g$. *However*, for

$$\tilde{g}_0(x) = f(ax), \qquad a \neq \lambda_\infty$$

it must be *impossible* for $\tilde{g}_n$ to converge: unless $a = \lambda_\infty$ for some harmonic sequence, there is no infinite attractor and no sequence of g_r's converging to g. For example, if we choose

$$\tilde{g}_0(x) = 1 - ax^2$$

then unless a is chosen at special isolated values, the $\tilde{g}_n$ will not converge. Rather, a could in general be a value λ_m; after a number of iterations, $\tilde{g}_n$ would, by definition, be a g_r [Eq. (7)] approximately. Since (12) is (8) (with some rescaling), successive iterations would move toward g_0 rather than g and then divergently away. Figure 4 represents a suggestive picture of the situation. That is, the fixed point g is repellent, and unless $\tilde{g}_0$ is correctly chosen so that the $\tilde{g}_n$ will "aim" into g, they will at first approach $g(\lambda_r \simeq \lambda_\infty$ so far as $g^{(2^n)}$ is concerned until $n \sim r$), but then diverge away from it along the "path" of decreasing g_r's. That is, (9) in general defines a recursively unstable problem.

This instability can, however, be turned to excellent advantage. Since an arbitrary $\tilde{g}_0$ will lead to divergence, a "good" $\tilde{g}_0$ must already be a good approximation to g. With $g(0) = 1$, (9) implies that

$$g(1) = -1/\alpha$$

By (14) one should then estimate

$$f(\lambda_\infty(f)) = \tilde{g}_0(1) \simeq -1/\alpha \tag{16}$$

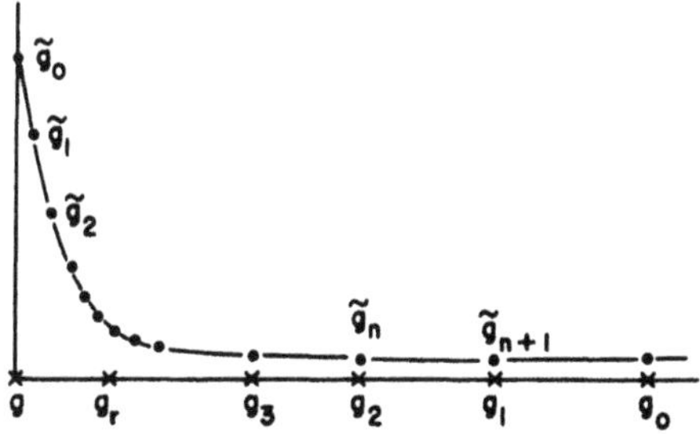

Fig. 4. The sequence of g_r's is indicated as points along the x axis. The iterates of $\tilde{g}_0$ are shown at first approaching g, the trajectory spending many iterations in the vicinity of g (or g_r for large r), but ultimately diverging away near low-lying g_r's.

that is, the instability of (9) provides an estimation of $\lambda_\infty(f)$ for any suitable f. The closer an f is to g (in some sense), the better the estimate. For example, consider

$$x_{n+1} = \lambda(1 - 2x_n^2)$$

or

$$f = 1 - 2x^2$$

and

$$f(\lambda_\infty x) = 1 - 2\lambda_\infty^2 x^2$$

With $\alpha = 2.5029\cdots$ for $z = 2$, we obtain from $1 - 2\lambda_\infty^2 \simeq -1/\alpha$ a value of $\lambda_\infty \simeq 0.8365$, to be compared to $\lambda_\infty = 0.8370$ for the limit of 2^n-cycles. In Section 5 we shall pursue this idea to obtain a technique for solving (9).

3. THE INFINITESIMAL λ SHIFT AND CONVERGENCE

Increasing λ from λ_n to λ_{n+1} maps g_r into g_{r+1} for all r. Calling this operation R, the λ shift, we write

$$R(g_r) = g_{r+1} \tag{17}$$

In I, R was applied to g_0 to produce g_1. Equation (8), written as

$$g_{r-1} = -\alpha g_r(g_r(x/\alpha)) \equiv L(g_r)$$

accomplishes the operation inverse to R. The combined operation

$$B \equiv L \cdot R$$

is the bifurcation transformation of I, which serves as an identity on the sequence $\{g_r\}$:

$$B(g_r) = g_r, \qquad r = 0, 1, \ldots \tag{18}$$

or, each g_r is a fixed point of the transformation B. We select g_1 by imposing the conditions

$$g_1(0) = 1, \qquad g_1(1) = 0 \tag{19}$$

and our universality conjecture is phrased in this language by saying the fixed point g_1 of B is *stable*, so that if any $\tilde{g}_0$ satisfying (19) with a zth-order maximum is chosen,

$$\tilde{g}_{n+1} = B[\tilde{g}_n]$$

will result in

$$\tilde{g}_n \rightarrow g_1$$

for the zth-order maximum universal g_1. The empirical computer evidence for universality, together with the instability of (12), means that R stabilizes B, as R reduces trivially to the identity only for the fixed-point g. Indeed, our approximate modelings of R in I resulted in recursively stable functional equations. We now determine R restricted to operation on g_r's in the limit of infinite r. The interchangeability of r and n in (7)

$$(-\alpha)^n g^{(2^n)}(\lambda_{n+r}, x/\alpha_n) \sim (-\alpha)^{n-s} g^{(2^{n-s})}(\lambda_{n+(r-s)}, x/\alpha^{n-s}), \qquad n - s \gg 1$$

together with the shifting operators implies that our study shall provide information about large-n convergence properties, and so determine δ as well.

We want to compute

$$\delta g^{(2^n)}(\lambda_{n+r}, x) \equiv g^{(2^n)}(\lambda_{n+r+1}, x) - g^{(2^n)}(\lambda_{n+r}, x) \tag{20}$$

Defining

$$\delta g_r(x) \equiv (-\alpha)^n \, \delta g^{(2^n)}(\lambda_{n+r}, x/\alpha^n) \tag{21}$$

(20) becomes

$$\delta g_r(x) = g_{r+1}(x) - g_r(x) = (R - 1)(g_r) \tag{22}$$

Substituting (22) in (8),

$$\begin{aligned} g_r(x) &= -\alpha(g_r + \delta g_r)[g_r(x/\alpha) + \delta g_r(x/\alpha)] \\ &= -\alpha g_r(g_r(x/\alpha)) - \alpha g_r'(g_r(x/\alpha))\, \delta g_r(x/\alpha) \\ &\quad - \alpha\, \delta g_r(g_r(x/\alpha)) + O((\delta g_r)^2) \\ &= g_{r-1}(x) - \alpha[\delta g_r(g_r(x/\alpha)) + g_r'(g_r(x/\alpha))\, \delta g_r(x/\alpha)] + O((\delta g_r)^2) \end{aligned} \tag{23}$$

or,

$$\delta g_{r-1}(x) = -\alpha[\delta g_r(g_r(x/\alpha)) + g_r'(g_r(x/\alpha))\, \delta g_r(x/\alpha)] + O((\delta g_r)^2) \tag{24}$$

Since $g_r \to g$, $\delta g_r \to 0$, and in the limit of infinite r

$$\begin{aligned} \delta g_{r-1}(x) &\sim -\alpha[\delta g_r(g_r(x/\alpha)) + g_r'(g_r(x/\alpha))\, \delta g_r(x/\alpha)] \\ &\sim -\alpha[\delta g_r(g(x/\alpha)) + g'(g(x/\alpha))\, \delta g_r(x/\alpha)] \end{aligned} \tag{25}$$

Separating $\delta g_r(x)$ as

$$\delta g_r(x) = \eta_r h(x) \tag{26}$$

with $\eta_r \to 0$ as $r \to \infty$, we obtained a closed equation for $h(x)$—the generator of infinitesimal λ shifts—and an equation for η_r:

$$\eta_{r-1} = \delta \eta_r \tag{27}$$

and

$$h(x) = -(\alpha/\delta)[h(g(x/\alpha)) + g'(g(x/\alpha))h(x/\alpha)] \tag{28}$$

In fact, (28) represents a rederivation of (28) of I, where (28) of I was an approximate realization of R applied to g_0, the approximation consisting of "mild" λ shifting, which becomes rigorous in the present context, and in this context, involving g and not g_1. Given $g(x)$ and α obeying (9), (28) determines both $h(x)$ and δ and defines a recursively stable equation. We return to this in the last section.

Equation (27) is trivially solved:

$$\eta_r = \delta^{-r} \tag{29}$$

so that

$$\delta g_r(x) = \delta^{-r} h(x) \tag{30}$$

or,

$$g_{r+1}(x) - g_r(x) = \delta^{-r} h(x), \qquad r \gg 1 \tag{31}$$

Summing (31) from $r = r_0$ to ∞, we obtain

$$g_{r_0}(x) = g(x) - \frac{\delta^{-r_0}}{1 - \delta^{-1}} h(x), \qquad r_0 \gg 1 \tag{32}$$

so that $g_r \to g$ (asymptotically) geometrically at the rate δ.

We now show that the δ determined by (28) and (9) is the δ of I: the argument is that of I resulting in (13) made exact. By (30),

$$\delta g_r(x) = \delta^{-1}\, \delta g_{r-1}(x)$$

which, by (21) reads for $x = 0$

$$(-\alpha)^n\, \delta g^{(2^n)}(\lambda_{n+r}, 0) = \delta^{-1}(-\alpha)^{n+1}\, \delta g^{(2^{n+1})}(\lambda_{n+r}, 0)$$

or

$$\delta g^{(2^{n-r})}(\lambda_n, 0) = -(\alpha/\delta)\, \delta g^{(2^{n-(r-1)})}(\lambda_n, 0)$$

or,

$$\delta g^{(2^s)}(\lambda_n, 0) = -(\alpha/\delta)\, \delta g^{(2^{s+1})}(\lambda_n, 0) \qquad \text{for } 1 \ll s \ll n$$

Thus, for n very large, the change in $g^{(2^r)}$ is the constant multiple $-\delta/\alpha$ of the change in $g^{(2^r-1)}$ induced by increasing λ_n to λ_{n+1} for all r except for the very small (initial transient) and very large (the bottom of the g_r sequence). Accordingly,

$$(-1)^n\, \delta g^{(2^n)}(\lambda_n, 0) \sim (\delta/\alpha)^n\, \delta g^{(1)}(\lambda_n, 0) = (\lambda_{n+1} - \lambda_n)(\delta/\alpha)^n$$

in the sense of logarithms. Since

$$\begin{aligned}(-\alpha)^n\, \delta g^{(2^n)}(\lambda_n, 0) &= (-\alpha)^n g^{(2^n)}(\lambda_{n+1}, 0) - (-\alpha)^n g^{(2^n)}(\lambda_n, 0)\\ &= g_1(0) - g_0(0) \sim 1\end{aligned}$$

we have

$$\lambda_{n+1} - \lambda_n \sim \delta^{-n}$$

logarithmically, or $(\lambda_{n+1} - \lambda_n)/(\lambda_{n+2} - \lambda_{n+1}) \to \delta$ as $n \to \infty$, which is the original definition of δ in I.

Combining (9) and (28), we can obtain $g(x)$, $h(x)$, α, and δ. By (32) we next obtain g_r for large r, and then by repeated application of (8) obtain low-lying g_r's. We have thus succeeded in determining all local quantitative properties of all highly bifurcated (and infinite) attractors of (1) in a framework independent of (1), and its unspecified $f(x)$. *The theory of high-order attractors is fully posed in a functional equation framework*, and represents the common residue of all specifically posed recursion equations $x_{n+1} = \lambda f(x_n)$.

It is important to make two observations pertaining to Eq. (28) at this point: one concerning the uniqueness of δ and the other concerning the linearity of (28) to any scaling of h.

Equation (32) can be derived by setting

$$g_r(x) = g(x) + \eta_r(x)$$

substituting in (8), and expanding to first order in η: we are simply analyzing the manner of approach of g_r to g. The separation of (26), namely,

$$\eta_r(x) = \eta_r h(x) = \delta^{-r} h(x)$$

demonstrates that

$$\lim_{r \to \infty} g_r = g$$

provided the eigenvalue δ of (28) is strictly greater than 1. In fact, it is easy to see that $\delta = 1$ with $h(x) = g(x) - xg'(x)$ exactly satisfies (28). To see the significance of this solution, observe that

$$g(x) - xg'(x) = (1 - x\, d/dx)g(x)$$

is exactly the generator of infinitesimal magnifications:

$$\begin{aligned}(1 + \mu)g(x/1 + \mu) &= (1 + \mu)g(x - x\mu) + O(\mu^2)\\ &= g(x) + \mu(g(x) - xg'(x)) + O(\mu^2)\end{aligned}$$

However, the magnifications comprise a degeneracy group of (9): if $g(x)$ obeys (9), then so too does $\mu g(x/\mu) \equiv g_\mu(x)$:

$$g_\mu\left(g_\mu\left(\frac{x}{\alpha}\right)\right) = \mu g\left(\frac{1}{\mu}\mu g\left(\frac{x}{\alpha\mu}\right)\right) = \mu g\left(g\left(\frac{x/\mu}{\alpha}\right)\right) = -\frac{1}{\alpha}\mu g\left(\frac{x}{\mu}\right) = -\frac{1}{\alpha}g_\mu(x)$$

Thus, the r-independent piece of λ_r corresponding to the eigenvalue $+1$ simply represents a convergence of g_r to a suitably magnified g; by *choosing*

$g(0) = 1$, this freedom is eliminated, and the r-independent piece of η_r set to 0.

We shall see in the next section that a spectrum of eigenvalues bounded by 1 in absolute value also exists. Anticipating some of that discussion, it turns out that g_r is orthogonal to the span of this part of the spectrum, so that only the large (convergence-producing) eigenvalues matter here.

Observe that (28) is linear in h, so that if $h(x)$ is a solution, so too is $\mu h(x)$ for any μ. That is, $h(0)$, say, is free. By (32), with $g(0) = 1$ by convention, this leaves $g_r(0)$ free in the asymptotic-r regime. However, a definite choice of $h(0)$ is necessary to ensure that $g_0(0) = 0$. A different choice of $h(0)$ would, for no number of iterations of (8), result in a g_r satisfying $g_r(0) = 0$.

It is easy to comprehend the meaning of other choices of $h(0)$. Since

$$g_r(x) = g(x) - \delta^{-r}h(x)$$

if $\bar{h}(x)$ [for a definite $\bar{h}(0)$] guarantees that $g_0(0) = 0$, then

$$\bar{h}_1(x) \equiv \delta\bar{h}(x)$$

produces a $g_1(x)$ such that $g_1(0) = 0$: that is, by increasing $h(0)$ we need perform fewer iterations to obtain a g_r satisfying $g_r(0) = 0$. Differently put, the absolute size of $h(x)$ is logarithmically periodic with period $\log \delta$: if

$$\bar{h}(x) \Rightarrow g_0(0) = 0$$

then

$$\delta^n\bar{h}(x) \Rightarrow g_n(0) = 0$$

All this means is that as $\log \bar{h}(0)$ is increased by $\log \delta$, one has moved through an entire bifurcation. Choices of $h(0) \neq \delta^n\bar{h}(0)$ determine a sequence of g_r's whose λ's are chosen *not* at λ_n's, but rather at intermediate values of λ between λ_n and λ_{n+1}. In particular, there is choice of $h(0) \equiv H(0)$ such that the λ's are the *bifurcation* values Λ_n. That is, our results determine the behavior of stability points not just at those values of λ_n such that $\bar{x}$ is an element, but indeed the entire behavior as λ is *continually* increased to λ_∞. The reason is simple: since $\lambda_n \sim \lambda_\infty - \delta^{-n}$, $\delta^{-n} \sim \lambda_\infty - \lambda_n$, and $g_r(x)$ is

$$g_r(x) = g(x) - (\lambda_\infty - \lambda_n)h(x)$$

and so,

$$g_\lambda(x) \equiv g(x) - (\lambda_\infty - \lambda)h(x)$$

is the continuation from discrete λ to continuous λ. Deviations of λ from

λ_∞ are most naturally measured logarithmically to the base δ: the bifurcation values Λ_n obey

$$\lambda_\infty - \Lambda_n \sim \delta^{-n}$$

as do the λ widths of a given harmonic

$$\Lambda_{n+1} - \Lambda_n \sim \delta^{-n}$$

A given "kind" of cycle recurs in the next harmonic periodic logarithmically. Moreover, had we started with, say, a stable three-cycle, bifurcated to the six-cycle, and considered $\lambda_n \equiv \lambda$ such that the 3×2^n cycle is stable and includes $\bar{x}$, then $g^{(3)}(\lambda, x)$ about $\bar{x}$ describes a two-cycle and

$$g_r(x) = (-\alpha)^n g^{(3\times 2^n)}(\lambda_{n+r}, x/(-\alpha)^n)$$

more generally are of the same character as the g_r's obtained from the harmonics of the two-cycle. Clearly (8) is again obeyed, leading to (9). With α and g unique solutions to (9) for a fixed z, we now realize that the entirety of the above treatment carries over *un*changed in every way to the structure of *every* highly bifurcated cycle of (1) no matter from which fundamental the bifurcations are obtained. That is, the local description of stability points at both the isolated-point and infinite-cluster level as well as α and δ are unique for every highly bifurcated cycle of (1) independent of f for any fixed z. Thus, in the so-called "chaotic" regime of (1) where most values of λ correspond to high bifurcation of a high-order fundamental, the local description of the attractor is essentially that of the g's.

We mention in passing that once g_r has been continued to a continuous index, Eq. (8) in the form

$$g_{r-s}(x) = (-\alpha)^s g_r^{(2^s)}(x/\alpha^s)$$

defines the notion of a continuous interaction, since every ingredient of the equation has received a natural continuation, save for the 2^s iterations.

4. THE APPROACH TO THE $\{g_r\}$ FIXED POINT

At this point we return to the fundamental question of the large-n limit of (7). In general f is not symmetric about $\bar{x}$ (although universality implies that even for an asymmetric f the g_r's must be symmetric) and the correct form of (7) is

$$g_r(x) = \lim(-\alpha)^n g^{(2^n)}(\lambda_{n+r}, x/(-\alpha)^n) \qquad (33)$$

Defining

$$g_{r,n}(x) \equiv (-\alpha)^n g^{(2^n)}(\lambda_{n+r}, x/(-\alpha)^n) \qquad (34)$$

we can immediately verify that

$$g_{r-1,n+1}(x) = -\alpha g_{r,n}(g_{r,n}(-x/\alpha)) \qquad (35)$$

Viewing (8) as a fixed point (in n) of (35), we could establish our theory in at least a local sense by ascertaining the stability of (8). Thus, we are led to consider

$$g_{r,n}(x) = g_r(x) + \eta_{r,n}(x) \tag{36}$$

and attempt to show that

$$\lim_{n\to\infty} \eta_{r,n}(x) = 0$$

Substituting (36) in (35), with (8) valid at the fixed point, we have in linear approximation (in η)

$$\eta_{r-1,n+1}(x) = -\alpha[\eta_{r,n}(g_r(x/\alpha)) + g_r'(g_r(x/\alpha))\eta_{r,n}(-x/\alpha)] \tag{37}$$

[Neglecting the n index, and replacing $g_r \to g$, we find that (37) reduces to (25): we shall have more to say about the eigenvalues of (28).] Since the $g_r(x)$ have been normalized to $g(0) = 1$, in general

$$\lim_{n\to\infty} g_{r,n} \neq g_r$$

but rather $g_{r,n}$ will approach a suitably magnified g_r. [We could alternatively have defined $\eta_{r,n} = g_{r,n} - \mu_{r+n}g_r(x/\mu_{r+n})$ with $\lim_{t\to\infty}\mu_t = \mu \neq 1$.] We shall account for this with $\eta^*_{r,n} \sim \mu(g_r - xg_r')$ a piece of $\eta_{r,n}$ to be determined from (37). Also, by defining

$$\eta_{r,n} \equiv \psi_{r,n+r}; \qquad n + r \equiv t$$

we find that (37) becomes

$$\psi_{r-1,t}(x) = -\alpha[\psi_{r,t}(g_r(x/\alpha)) + g_r'(g_r(x/\alpha))\psi_{r,t}(-x/\alpha)]$$

so that (37) itself is insufficient to determine any $n + r$ dependence; rather,

$$\eta_{r,0} = g_{r,0} - g_r = \lambda_r f - g_r \tag{38}$$

shall serve as initial data to fix $\eta_{r,n}$ uniquely.

We proceed to solve (37) by an artificial quadrature. Setting

$$\eta_{r,n} \equiv \eta^1_{r,n} + \eta^2_{r,n} \tag{39}$$

we find that (37) becomes

$$\begin{aligned}\eta^1_{r-1,n+1}(x) + \alpha[\eta^2_{r,n}(g_r(x/\alpha)) + g_r'(g_r(x/\alpha))\eta^1_{r,n}(-x/\alpha) + \eta^2_{r,n}(g_r(x/\alpha))] \\ = -\{\eta^2_{r-1,n+1}(x) + \alpha g_r'(g_r(x/\alpha))\eta^2_{r,n}(-x/\alpha)\} \equiv 0\end{aligned}$$

thereby defining η^2. That is (38) is replaced with the pair of equations

$$\eta^1_{r-1,n+1}(x) = -\alpha[\eta^1_{r,n}(g_r(x/\alpha)) + g_r'(g_r(x/\alpha))\eta^1_{r,n}(-x/\alpha) + \eta^2_{r,n}(g_r(x/\alpha))] \tag{40}$$

and

$$\eta^2_{r-1,n+1}(x) = -\alpha g_r'(g_r(x/\alpha))\eta^2_{r,n}(-x/\alpha) \tag{41}$$

However, by (8)

$$g'_{r-1}(x) = -g_r'(g_r(x/\alpha))g_r'(x/\alpha)$$

so that (41) becomes

$$\frac{\eta^2_{r-1,n+1}(x)}{g'_{r-1}(x)} = \alpha \frac{\eta^2_{r,n}(-x/\alpha)}{g_r'(x/\alpha)}$$

Since $g_r(x)$ is symmetric,

$$g_r'(-x/\alpha) = -g_r'(x/\alpha)$$

so that

$$\frac{\eta^2_{r-1,n+1}(x)}{g'_{r-1}(x)} = -\alpha \frac{\eta^2_{r,n}(-x/\alpha)}{g_r'(-x/\alpha)} \tag{42}$$

Defining

$$f_{r,n}(x) \equiv \frac{\eta^2_{r,n}(x)}{g_r'(x)} \tag{43}$$

we find that (39) reads

$$f_{r-1,n+1}(x) = -\alpha f_{r,n}(-x/\alpha)$$

with solution

$$f_{r,n}(x) = (-\alpha)^n f_{r+n,0}(x/(-\alpha)^n) \equiv (-\alpha)^n F_{r+n}(x/(-\alpha)^n)$$

so that we have for η^2

$$\eta^2_{r,n}(x) = (-\alpha)^n F_{r+n}(x/(-\alpha)^n)g_r'(x) \tag{44}$$

With (44), (40) now reads

$$\begin{aligned}\eta^1_{r-1,n+1} = &-\alpha\{\eta^1_{r,n}(g_r'(x/\alpha)) + g_r(g_r'(x/\alpha)) \\ &\times [\eta^1_{r,n}(-x/\alpha) + (-\alpha)^n F_{r+n}(g_r(x/\alpha)/(-\alpha)^n)]\}\end{aligned} \tag{45}$$

Setting

$$\eta^1_{r,n}(x) \equiv \eta^0_{r,n}(x) - (-\alpha)^n F_{r+n}(g_r(x)/(-\alpha)^n)$$

we can immediately verify from (45) that

$$\eta^0_{r-1,n+1}(x) = -\alpha[\eta^0_{r,n}(g_r(x/\alpha)) + g_r(g_r'(x/\alpha))\eta^0_{r,n}(-x/\alpha)]$$

that is, if $\eta^0_{r,n}$ obeys (37), then for any F_{r+n}, so too does

$$\eta_{r,n}(x) = \eta^0_{r,n}(x) + [(-\alpha)^n F_{r+n}(x/(-\alpha)^n)g_r'(x) - (-\alpha)^n F_{r+n}(g_r(x)/(-\alpha)^n)] \tag{46}$$

so that we have obtained a particular solution of (37). We now regard $\eta^0_{r,n}(x)$ to be a homogeneous transient to the F solution which we utilize to meet initial data. Specifically, we absorb all antisymmetric parts of $f(x)$ into F, leaving $\eta^0_{r,0}(x)$ and hence all $\eta^0_{r,n}(x)$ symmetric, and so obeying the "intrinsic" equation

$$\eta^0_{r-1,n+1}(x) = -\alpha[\eta^0_{r,n}(g_r(x/\alpha)) + g_r'(g_r(x/\alpha))\eta^0_{r,n}(x/\alpha)] \tag{47}$$

η^0 is viewed as built exclusively of g_r's with minimal dependence on the initial $f(x)$. [It will be seen that the decomposition of (39)–(41) is determined and not at all artificial if η^1 and η^2 are respectively taken to be the symmetric and antisymmetric parts of η.]

We now utilize the initial data (38):

$$\lambda_r f(x) - g_r(x) = \eta_{r,0}(x) = \eta^0_{r,0}(x) + F_r(x)g_r'(x) - F_r(g_r(x)) \tag{48}$$

With an overbar denoting symmetry and a circumflex denoting antisymmetry, (48) reads

$$\lambda_r \hat{f}(x) = \bar{F}_r(x)g_r'(x) \tag{49}$$

and

$$\lambda_r \bar{f}(x) - g_r(x) = \eta^0_{r,0}(x) + \hat{F}_r(x)g_r'(x) - \hat{F}_r(g_r(x)) - \bar{F}_r(g_r(x)) \tag{50}$$

By (49), $\bar{F}_r(x)$ is nonvanishing only when f is asymmetric, in which case it can absorb all the antisymmetry.

Let us specialize to the case $\hat{f} \neq 0$. We then have from (49)

$$\bar{F}_r(x) = \frac{\lambda_r}{g_r'(x)}\hat{f}(x)$$

and so, by (46), a piece of $\eta_{r,n}$ of the form

$$\eta^*_{r,n}(x) = \lambda_{r+n}\left\{(-\alpha)^n \frac{\hat{f}(x/(-\alpha)^n)}{g'(x/(-\alpha)^n)} g_r'(x) - (-\alpha)^n \frac{\hat{f}(g_r(x)/(-\alpha)^n)}{g'_{r+n}(g_r(x)/(-\alpha)^n)}\right\}$$

$$\underset{n\to\infty}{\sim} \lambda_\infty\left\{(-\alpha)^n \frac{\hat{f}(x/(-\alpha)^n)}{g'(x/(-\alpha)^n)} g_r'(x) - (-\alpha)^n \frac{\hat{f}(g_r(x)/(-\alpha)_n)}{g'(g_r(x)/(-\alpha)^n)}\right\}$$

With $1 - \bar{f}(x) \propto |x|^z$, $\hat{f}(x) \propto |x|^{z+\epsilon} \operatorname{sgn} x$ ($\epsilon > 0$: otherwise f not extreme at $x = 0$), and $g'(x) \propto |x|^{z-1} \operatorname{sgn} x$,

$$(\hat{f}/g')(x) \propto |x|^{1+\epsilon}$$

and so

$$\eta^*_{r,n}(x) \underset{n\to\infty}{\sim} (-1)^n\alpha^{-n\epsilon}(|x|^{1+\epsilon}g_r'(x) - |g_r(x)|^{1+\epsilon}) \to 0$$

Thus, the g_r fixed point is stable against antisymmetric perturbations. As an example, if

$$f(x) = 1 - ax^2 - bx^3$$

then $\epsilon = 1$ and $\eta_{r,n}$ converge to zero geometrically at the rate $-\alpha^{-1}$, in perfect agreement with the computer data for this f, since the $\eta_{r,n}$ convergence rate is exactly the α_n convergence rate:

$$\alpha_n^{(r)} \equiv -\frac{g^{(2^n)}(\lambda_{n+r,0})}{g(2^{n+1})(\lambda_{n+1,r+0})} = \alpha\frac{g_{r,n}(0)}{g_{r,n+1}(0)}$$

$$\approx \alpha\left\{1 + \frac{1}{g_r(0)}[\eta_{r,n}(0) - \eta_{r,n+1}(0)]\right\}$$

Accordingly, we consider now only symmetric f's so that

$$F_r(x) = \hat{F}_r(x)$$

and

$$\lambda_r f(x) - g_r(x) = \eta^0_{r,0}(x) + \hat{F}_r(x)g_r'(x) - \hat{F}_r(g_r(x)) \tag{51}$$

As a first observation, following the parenthetic remark below Eq. (37),

$$\eta_r(x) \equiv \hat{F}_r(x)g'(x) - \hat{F}_r(g(x))$$

with

$$\hat{F}_r(x) = \alpha^{-r}\hat{F}_0(\alpha^r x)$$

must obey

$$\eta_{r-1}(x) = -\alpha[\eta_r(g(x/\alpha)) + g'(g(x/\alpha))\eta_r(x/\alpha)]$$

For monomials,

$$\hat{F}_0(x) = |x|^z \operatorname{sgn} x, \qquad \hat{F}_{r-1}(x) = \alpha^{-z+1}\hat{F}_r(x)$$

and so

$$-\alpha[h^{(z)}(g(x/\alpha)) + g'(g(x/\alpha))h^{(z)}(x/\alpha)] = \alpha^{-z+1}h^{(z)}(x)$$

with

$$h^{(z)}(x) = |g(x)|^z \operatorname{sgn}(g(x)) - g'(x)|x|^z \operatorname{sgn} x$$

That is, the eigenvalue of (28) can assume any positive value less than or equal to 1 in addition to the value $\delta > 1$. These eigenvalues represent potential instabilities of the convergence of g_r to g. However, they are unexcited in every g_r exactly because they provide stable convergence of $g_{r,n}$ to g_r: The g_r meet no initial conditions save for their convergence to g. The potentially hazardous parts $h^{(z)}$ of a g_r are all shed in the approach of $g_{r,n}$ (for any suitable f) to g_r. That is,

$$g = \lim_{r\to\infty} g_r$$

must exist since all unstable eigenvalues are exactly those that vanish in the formation of g_r from $g_{r,n}$.

We return to (51) and now ask whether the $\hat{F}$'s can span the initial data, in which case $\eta^0 \equiv 0$ and our theory is complete. In fact, an $\eta_{r,n}$ built wholly from $\hat{F}$ must produce convergence at a rate $(-\alpha)^{1-z}$ for

$$\hat{F}_r(x) = \mu_r x + a_r x^z + \cdots \text{ higher order}$$

For example, if $f = 1 - zx^2 + \cdots$, the smallest value of z would be 3, providing a convergence rate α^{-2}. However, $\alpha_n^{(1)} \to \alpha$ at the rate δ^{-1} in this case. Since $\delta < \alpha^2$, only the δ rate survives asymptotically, and leaves the question as to how it enters $\eta_{r,n}$. This suggests that η^0 might not vanish in general, and so we examine the situation more carefully.

Returning to (35), define

$$r + n \equiv t, \qquad g_{r,n} \equiv \psi_{r+n,n} = \psi_{t,n}$$

so that

$$\psi_{t,n+1}(x) = -\alpha\psi_{t,n}(\psi_{t,n}(-x/\alpha)) \tag{52}$$

We next expand ψ about g:

$$\psi_{t,n} \equiv g + \omega_{t,n} \tag{53}$$

so that

$$\omega_{t,n+1}(x) = -\alpha[\omega_{t,n}(g(x/\alpha)) + g'(g(x/\alpha))\omega_{t,n}(-x/\alpha)] \equiv \mathscr{L}[\omega_{t,n}] \tag{54}$$

in linear approximation. Equation (54) is our familiar shift equation written in the form of (12). We already know a variety of eigenvalues of $\mathscr{L}$:

$$\psi_\rho(x) \equiv g^\rho(x) - x^\rho g'(x) \Rightarrow \mathscr{L}[\psi_\rho] = (-\alpha)^{1-\rho}\psi_\rho$$

and

$$h(x)\colon \quad \mathscr{L}[h] = \delta h$$

Expanding $\omega_{t,n}$ along these eigenvectors, we have

$$\omega_{t,n} = \sum_\rho c_t{}^\rho(-\alpha)^{n(1-\rho)}(g^\rho - x^\rho g') + c_t\, \delta^n h + \omega_{t,n}^0 \tag{55}$$

Observe at this point a strong similarity to (46), and yet with the difference that (55) possesses an isolated h piece plus

$$\sum_\rho c_t{}^\rho(-\alpha)^{n(1-\rho)}(g^\rho - x^\rho g') = (-\alpha)^n F_t(g/(-\alpha)^n) - (-\alpha)^n F_t(x/(-\alpha)^n)g'$$

This is the same form of particular solution as in (46) but constructed from g rather than g_r. Now, if $c_t\delta^n = c_{r+n}\delta^n \sim \delta^{-r}$, then the piece $c_t\delta^n h$ is a fixed (in n) perturbation about g. Now, $\mathscr{L}$ accounts for only first-order perturbative effects. One would obtain a second-order correction by expanding about $g + c_t\delta^n h$. However, this is the form of expansion about g_r that would modify the $\psi_\rho(x)$ to be

$$\psi_\rho \to g_r{}^\rho - x^\rho g_r{}' = (g^\rho - x^\rho g') - \delta^{-r}(\rho g^{\rho-1}h - x^\rho h')$$

Thus, assuming $c_t\delta^n \sim \delta^{-r}$, the correct form of (55) including all first-order $c_t{}^\rho$ dependence is

$$\omega_{t,n} = \sum_\rho c_t{}^\rho(-\alpha)^{n(1-\rho)}(g^\rho - x^\rho g') + c_t\delta^n\left(h + \sum_\rho c_t{}^\rho(-\alpha)^{n(1-\rho)}(\rho g^{\rho-1}h - x^\rho h')\right) + \omega^0_{t,n} \qquad (56)$$

which possesses extra h dependence beyond the $F_t(g_r)$ terms than does (46): even with $\omega^0_{t,n} = 0$, (56) has already included a transient piece $\eta^{0'}_{r,n}$, which is still constructed from the span of $h \oplus \{\psi\}$. We now pose the (strong) conjecture that $\omega^0_{t,n} = 0$:

Conjecture. The spectrum of the operator $\mathscr{L}$ is δ and $(-\alpha)^{1-\rho}$, $\rho \leqslant 1$, and, moreover, the spectrum is complete.

(We possess computational evidence for this conjecture at least for $z = 2$ and 4, which we discuss in Section 5.) Accordingly, we have

$$\omega_{t,n} = \sum_\rho c_t{}^\rho(-\alpha)^{n(1-\rho)}(g^\rho - x^\rho g') + c_t\delta^n\left(h + \sum_\rho c_t{}^\rho(-\alpha)^{n(1-\rho)} \times (\rho g^{\rho-1}h - x^\rho h')\right) \qquad (57)$$

or

$$g_{r,n} = g + \sum_\rho c^\rho_{r+n}(-\alpha)^{n(1-\rho)}(g^\rho - x^\rho g') + c_{r+n}\delta^n\left(h + \sum_\rho c^\rho_{r+n}(-\alpha)^{n(1-\rho)}(\rho g^{\rho-1}h - x^\rho h')\right) \qquad (58)$$

in linear approximation.

It is easy to extend (58) to an exact solution of (35), since the first-order terms are exactly the generators of conjugacy transformations connected to the identity. Defining

$$F_t(x) = \sum_\rho c_t{}^\rho x^\rho$$

we find that (58) becomes

$$g_{r,n} = (g + c_{r+n}\delta^n h) + (-\alpha)^n F_{r+n}((-\alpha)^{-n}(g + c_{r+n}\delta^n h)) - (-\alpha)^n(g + c_{r+n}\delta^n h)' F_{r+n}(x/(-\alpha)^n) \qquad (59)$$

Defining

$$S_{t,n}(x) = x + (-\alpha)^n F_t(x/(-\alpha)^n)$$

we have that (59) constitutes the leading approximation to

$$g_{r,n} = S_{r+n,n} \circ (g + c_{r+n}\delta^n h) \circ S^{-1}_{r+n,n} \tag{60}$$

However, defining $c_t = \delta^{-t} d_t$, we have

$$g + c_{r+n}\delta^n h = g + d_{r+n}\delta^{-r} h \equiv \tilde{g}_{r,r+n} + O(\delta^{-2r})$$

which obeys

$$\tilde{g}_{r-1,t}(x) = -\alpha \tilde{g}_{r,t}(\tilde{g}_{r,t}(x/\alpha))$$

to first order for any choice of d_t, converging to g as $r \to \infty$ for t fixed. That is, (58) is the leading approximation to

$$g_{r,n} = S_{r+n,n} \circ \tilde{g}_{r,r+n} \circ S^{-1}_{r+n,n} \tag{61}$$

which is easily seen to exactly satisfy (35). However, while (58) is *compatible* with the solution (61), (61) is not the *general* solution to (35) containing (58) as its linear approximation. That is, if (35) is stable about the $\tilde{g}_{r,t}$ fixed point, the linear approximation becomes a conjugacy transformation upon $\tilde{g}_{r,t}$, deviations from conjugacy vanishing in the higher order transient.

According to the discussion on p. 680, the functions $\tilde{g}_{r,t}$ all converge to g as $r \to \infty$, but differ in the small-r regime: only $d_t = -1$ (for the properly normalized h) will lead to $\tilde{g}_{0,t}(0) = 0$. Equation (61) is correct for large n; setting $r = 0$, it reads

$$g_{0,n}(x) = S_{n,n} \circ g_{0,n} \circ S^{-1}_{n,n}$$

Since $S_{t,n}$ is connected to the identity,

$$g_{0,n}(0) = 0 \Rightarrow \tilde{g}_{0,n}(0) = 0$$

However, $g_{0,n}(0)$ *must* vanish for all n:

$$g_{0,n}(0) = (-\alpha)^n g^{(2^n)}(\lambda_n, 0) = 0$$

by the recursion-equation definition of λ_n. But, if $\tilde{g}_{0,t}(0) = 0$ for all t, then $d_t = -1$ for all t. Thus, the recursion-defined values of λ_r determine $c_t = \delta^{-t} d_t = -\delta^{-t}$, which, when entered in (58), yields

$$\begin{aligned} g_{r,n} = g &+ \sum_\rho c^\rho_{r+n}(-\alpha)^{n(1-\rho)}(g^\rho - x^\rho g') \\ &- \delta^{-r}\left(h + \sum_\rho c^\rho_{r+n}(-\alpha)^{n(1-\rho)}(\rho g^{\rho-1} h - x^\rho h')\right) \end{aligned} \tag{62}$$

so that the potentially divergent δ^n terms have been stabilized. Moreover, for n large, all terms decay with powers of $-\alpha$ save for $\rho = 1$:

$$g_{r,n} \underset{n\to\infty}{\sim} g + c_{r+n}^1(g - xg') - \delta^{-r}(h + c_{r+n}^1(h - xh'))$$

or, with $\mu_t \equiv 1 + c_t^{\,1}$,

$$g_{r,n} \sim \mu_t(g - \delta^{-r}h)(x/\mu_t)$$

or

$$g_{r,n} \sim \mu_{r+n}g_r(x/\mu_{r+n}) \tag{63}$$

a magnification of g_r. Thus, our conjecture implies the local stability of the g_r fixed point of (35). [Conversely, the one parameter λ could be adjusted to cancel the potentially growing δ mode; had $\mathscr{L}$ possessed several growing eigenvalues, it is difficult to see how this cancellation could be arranged. Also, although the conjugacy generators produce convergence at rates $(-\alpha)^{1-\rho}$, we can see from (63) how $\mu_t \to \mu_\infty$ can produce a different convergence scheme for α_n.] We have not investigated any higher order stability questions, and apart from some approximation schemes and computational methods which we shall discuss in the next section, have nothing further to say about the ingredients of a nonlocal proof.

5. APPROXIMATIONS AND METHODS OF SOLUTION

All infinite attracters are locally determined by the hierarchy (8),

$$g_{r-1}(x) = -\alpha g_r(g_r(x/\alpha))$$

As previously described, (8) is solved by first computing g and α through (9),

$$g(x) = -\alpha g(g(x/\alpha))$$

with g_r for asymptotic r given by (32),

$$g_r \sim g - \delta^{-r}h$$

where h and δ are obtained through (28),

$$-\alpha[h(g(x/\alpha)) + g'(g(x/\alpha))h(x/\alpha)] = \delta h(x), \qquad \delta > 1$$

To any desired accuracy, an r_0 is chosen such that (32) provides g_r for all $r \geqslant r_0$, and g_r for $r < r_0$ determined from g_{r_0} through (8). In particular $h(0)$ is fixed through the requirement that $g_0(0) = 0$.

We now seek an approximate equation for g_r for a fixed r that bypasses the above asymptotic ansatz. The virtue of such an equation is that it must define a recursively stable scheme for obtaining g_r. As we shall see, the approximate formula of I shall appear to linear approximation.

By (8),

$$g_1(x) = -\alpha g_2(g_2(x/\alpha)) \tag{64}$$

Relation (32) provides g_r up to first order in δ^{-r}. Since $\delta \to \infty$ as $z \to \infty$, for large enough z

$$g_1 \simeq g - \delta^{-1}h \qquad \text{and} \qquad g_2 \simeq g - \delta^{-2}h \tag{65}$$

will be arbitrarily accurate. Accordingly,

$$g_2 - g_1 \simeq \delta^{-1}(1 - \delta^{-1})h$$

or

$$g_2 \simeq g_1 + \delta^{-1}(1 - \delta^{-1})h \tag{66}$$

Substituting (66) in (64), we find

$$g_1(x) = -\alpha\{g_1(g_1(x/\alpha)) + (1 - \delta^{-1})\delta^{-1}[h(g_1(x/\alpha)) + g_1'(g_1(x/\alpha))h(x/\alpha)] + O(\delta^{-2})\}$$

or, by (65),

$$g_1(x) = -\alpha g_1(g_1(x/\alpha)) + (1 - \delta^{-1})(-\alpha\delta^{-1})[h(g(x/\alpha)) + g'(g(x/\alpha))h(x/\alpha)] + O(\delta^{-2})$$

which, by (28), is

$$g_1(x) = -\alpha g_1(g_1(x/\alpha)) + (1 - \delta^{-1})h(x) + O(\delta^{-2}) \tag{67}$$

Thus, to leading order in δ^{-1},

$$g_0(x) = -\alpha g_1(g_1(x/\alpha)) \simeq g_1(x) - (1 - \delta^{-1})h(x) = g(x) - h(x) \tag{68}$$

That is, as $z \to \infty$, the asymptotic form (32) becomes arbitrarily accurate for all $r \geqslant 0$. Since $g_0(0) = 0$, (68) produces

$$h(0) \simeq g(0) = 1$$

in this limit, a

$$g_1(0) \simeq 1 - \delta^{-1} \tag{69}$$

[To appreciate this estimate, for $z = 2$, $g_1(0) \doteq 0.733$, in comparison with $1 - \delta^{-1} \doteq 0.786$.] Defining

$$g^*(x) \equiv (1 - \delta^{-1})^{-1} g_1((1 - \delta^{-1})x)$$

and

$$h^*(x) \equiv h((1 - \delta^{-1})x)$$

we can write (67) as

$$g^*(x) = -\alpha g^*(g^*(x/\alpha)) + h^*(x) + O(\delta^{-2}) \tag{70}$$

Since

$$-\alpha[h(g_1(x/\alpha)) + g_1'(g_1(x/\alpha))h(x/\alpha)] = \delta h(x) + O(\delta^{-1})$$

we also have

$$h^*(x) = -(\alpha/\delta)[h^*(g^*(x/\alpha)) + g^{*\prime}(g^*(x/\alpha))h^*(x/\alpha)] + O(\delta^{-2}) \qquad (71)$$

Equations (70) and (71) are exactly the approximate equations of I, since

$$h^*(0) = 1 \Rightarrow g^*(0) = 1 \qquad \text{by (70)}$$

while the definition of g_1 implies that

$$g^*(1) = 0$$

Since (70) and (71) constitute equations for g_1, their natural recursion forms ($g^*, h^* \to g_n^*, h_n^*$ on the right-hand sides and g_{n+1}^*, h_{n+1}^* on the left-hand sides) accomplish the recursion

$$g_{1,n+1} = f(g_{1,n})$$

which is stable.

We now exhibit a computational technique for solving (9) based on the observation about Eq. (16). The recursion form of (9) given by (12),

$$\tilde{g}_{n+1}(x) = -\alpha_n \tilde{g}_n(\tilde{g}_n(x/\alpha_n))$$
$$\tilde{g}_n(0) \equiv 1 \qquad \text{for all } n \Rightarrow \alpha_n = -[\tilde{g}_n(1)]^{-1}$$

must be convergent to g if $\tilde{g}_0$ is appropriately chosen. Thus, if f is any function of our class, there is a value λ_∞ such that

$$x_{n+1} = \lambda_\infty f(x_n)$$

will determine the infinite attractor bifurcated from the 2-cycle, in which case Eq. (15),

$$\tilde{g}_0(x) = f(\lambda_\infty x)$$

will lead to convergence. However, the strong instability of (12) requires that λ_∞ be known to very high precision in order that its high iterates will be accurate approximates of g. As a rough estimate, $\tilde{g}_0(x)$ should approximate g, and so (16),

$$f(\lambda_\infty) = \tilde{g}_0(1) \simeq g(1) = -\alpha^{-1}$$

provides an estimate for λ_∞. It is elementary to obtain better estimates. Clearly,

$$\alpha_0 = -[\tilde{g}_0(1)]^{-1} = -[f(\lambda_\infty)]^{-1}$$

and so

$$\tilde{g}_1(x) = [f(\lambda_\infty)]^{-1} f\{\lambda_\infty f(\lambda_\infty x f(\lambda_\infty))\} \qquad (72)$$

But now, $\tilde{g}_1(x)$ is a better estimate of g and so

$$\tilde{g}_1(1) \simeq -\alpha^{-1}$$

will provide a better estimate of λ_∞. Accordingly, with α known accurately, we could, for any f of our class, determine high-accuracy estimates of $\lambda_\infty(f)$. With α unknown, we could seek to collaterally determine it with λ_∞ by setting

$$\tilde{g}_1(1) \simeq \tilde{g}_0(1) \tag{73}$$

which by (16) and (72) provides an equation purely for λ_∞. Evidently, by successively setting $\tilde{g}_{n+1}(1) \simeq \tilde{g}_n(1)$ more accurate estimations are obtained. We now show that this can be turned into a highly convergent scheme for λ_∞. It is immediate to see that (12) can be "solved" as

$$\tilde{g}_n(x) = (-1)^n \alpha_{n-1}\alpha_{n-2} \cdots \alpha_0 \tilde{g}_0^{(2n)}(x/\alpha_{n-1} \cdots \alpha_0)$$

so that

$$\tilde{g}_n(0) = (-1)^n \alpha_{n-1} \cdots \alpha_0 \tilde{g}_0^{(2n)}(0) \tag{74}$$

Since $\alpha_n \to \alpha$ if $\tilde{g}_0(x) = f(\lambda_\infty x)$ for the exact λ_∞,

$$\tilde{g}_0^{(2n)}(0) \sim (-\alpha)^{-n} \tag{75}$$

Also, by (74)

$$\alpha_n = -\tilde{g}_0^{(2n)}(0)/\tilde{g}_0^{(2n+1)}(0) \tag{76}$$

so that, with the definition

$$\xi_n \equiv \tilde{g}_0^{(2n-1)}(0)\tilde{g}_0^{(2n+1)}(0) - [\tilde{g}_0^{(2n)}(0)]^2 \tag{77}$$

one has

$$\frac{\xi_{n+1}}{\xi_n} = \frac{1}{\alpha_{n+1}\alpha_n}\frac{\alpha_{n+1} - \alpha_n}{\alpha_n - \alpha_{n-1}} \equiv \frac{1}{\alpha_{n+1}\alpha_n}\rho_n \tag{78}$$

with ρ_n the α convergence rate. With

$$\tilde{g}_0(x) = f(ax) \tag{79}$$

and a chosen exactly at $\lambda_\infty(f)$, $\alpha_n \to \alpha$ and so $\rho_n < 1$. Since $\alpha > 1$ for all $z > 1$ ($\alpha \to \infty$ as $z \to 1$ and $\alpha \to 1$ as $z \to \infty$), ξ_n converges to zero α^2 times faster than $\alpha_n \to \alpha$. Accordingly, if one sets $\xi_n = 0$ for each n, an equation for a results (whose solution is a λ_∞ of f) that is exact to linear order in the error of the estimation. (For $f = 1 - 2x^2$, $\xi_n = 0$ yields λ_∞ for the 2-cycle fundamental to $2n$ significant figures and α_n to n significant figures.) Had we wanted λ_∞ for a 3-cycle, $\tilde{g}_0(x) = f^{(3)}(\lambda_\infty x)$ provides the starting $\tilde{g}$, and similarly *all* λ_∞ of a chosen f are rapidly determined (with of course the same α resulting). Once λ_∞ is determined, the iterates of $f(\lambda_\infty x)$ converge toward g until the error in λ_∞ is sufficiently magnified to cause divergence. (A 25-significant-figure estimate of λ_∞ provides a g obeying (9) to one part in 10^{-14} on [0, 1]. As shall follow, one can do significantly better far more quickly.)

We now consider a Newton's-method scheme of solution of (9) which shall lead into deeper considerations of the spectral problem of $\mathscr{L}$, and simple hand computations of λ_r's to several significant figures.

Regarding $g(x)$ on a compact interval—say [0, 1]—as a matrix of its values at N points x_i *together* with an interpolation scheme (of at least zth order to protect a zth-order g), (9) evaluated at the N points x_i becomes a set of N coupled nonlinear equations for the N quantities $g(x_i)$. Accordingly, one can perform an N-dimensional Newton's-method recursion to obtain the $g(x_i)$ from an initial estimate. However, high-precision estimates of g require a high-order interpolation scheme upon a large-N matrix, leading to an inaccurate inversion. Schematically, one writes

$$\bar{g}(x) = g(x) + \delta g(x), \qquad \bar{\alpha} = \alpha + \delta\alpha \tag{80}$$

where $\bar{g}$ and $\bar{\alpha}$ satisfy (9) and g and α serve as an approximate solution. We insert (80) in (9) and expand about the approximate values to first order in δg and $\delta\alpha$. By setting

$$g(0) \equiv \bar{g}(0) = 1$$

in the approximate solution, we have $\delta g(0) = 0$, which determines $\delta\alpha$ in terms of the $\delta g(x)$'s, so that we have a linear equation for $\delta g(x)$ alone. Expressions like $\delta g(g(x/\alpha))$ appear: the equation is evaluated at each x_i and $\delta g(g(x_i/\alpha))$ is expressed through the interpolation procedure in terms of linear combinations of $\delta g(x_i)$. The equations are then inverted to obtain $\delta g(x_i)$ and the procedure iterated. Convergence is slow and precision-limited.

However, the matrix of $g(x_i)$ and interpolation scheme simply constitute a certain parametrization of $g(x)$. Accordingly, one can perform the method with far simpler parametrizations. In particular, setting

$$g_N(x) = 1 + \sum_{i=1}^{N} G_i x^{zi}, \qquad \delta g_N(x) = \sum_{i=1}^{N} \delta G_i x^{zi}$$

and evaluating the linear approximate equation at N points (say $x_m = m/N$) produces an N-dimensional linear system again to be inverted. {In fact, for $z = 2$, precision limitations occurred for $N = 14$, determining α and g consistent with (9) to within 10^{-20} on [0, 1] in 10 sec of CDC 6600 time.} The solution obtained (the G_i) of course provides a very rapidly computable g for any further usage.

With α and g determined, we now face the determination of δ and h from the solution of (28),

$$\mathscr{L}[\psi] = -\alpha[\psi(g(x/\alpha)) + g'(g(x/\alpha)\psi(-x/\alpha)] = \lambda\psi(x)$$

In light of the previous discussion, this is an infinite-dimensional linear eigenvalue problem, which we shall study in a finite-dimensional approximation. That is, set

$$\psi(x) = \sum_{n=0}^{N-1} \psi_n x^{zn} \qquad (\psi_0 \equiv 1) \tag{81}$$

so that (28) evaluated at N points x_i becomes

$$\sum_{n=0}^{N-1} \{\lambda x_i^{zn} + \alpha[g(x_i/\alpha)]^{zn} + \alpha g'(g(x_i/\alpha))(-x_i/\alpha)^{zn}\}\psi_n = 0 \tag{82}$$

or

$$(\lambda X_{in} - \tilde{L}_{in})\psi_n = 0 \qquad \text{(summation convention)} \tag{83}$$

with (83) defining from (82) the $N \times N$ matrices X and $\tilde{L}$. The matrix X is invertible, and so

$$(X^{-1}\tilde{L})_{in}\psi_n \equiv L_{in}\psi_n = \lambda\psi_n \tag{84}$$

Accordingly, the eigenvalues λ are determined from

$$\det(\lambda X - \tilde{L}) = 0 \tag{85}$$

producing N eigenvalues in the N-dimensional approximating space.

The computational results are highly interesting. Starting with $N = 1$ and $x_1 = 0$ there simply results

$$\lambda = -\alpha - \alpha g'(1) = \alpha^z - \alpha$$

which is an approximate formula for δ with $h(1) = h(0) = 1$, asymptotically accurate as $z \to \infty$. Setting $N = 2$ with $x_1 = 0$ and $x_2 = 1$ results in a larger eigenvalue more nearly δ and a smaller one quite close to $\lambda = 1$, with corresponding eigenvectors approximating h and $\psi_1 = g - xg'$. Increasing N and evaluating at equally spaced points in [0, 1] produces more accurate determinations of δ and various $(-\alpha)^{1-\rho}$, $\rho = 1, 2,\ldots$: *δ is the solitary eigenvalue of L greater than* 1, at least in the two cases we studied, $z = 2$ and $z = 4$. {At $N = 14$ we obtain δ and h to 20 places consistent with (28) on [0, 1] and agreeing to the 14 places of our best recursion data.} That is, the spectrum of $\mathscr{L}$ restricted to these discrete linear systems is comprised of the conjugacy eigenvalues smaller than or equal to 1 in absolute value, plus a solitary larger one equal to δ. The eigenvalues are always *nondegenerate*, so that *L is complete* despite its nonsymmetric form. That is, defining L^*, the adjoint of L (the transpose in the present context), and adjoint eigenvectors

$$L^*\psi_\lambda^* \equiv \psi_\lambda^* L = \lambda\psi_\lambda^* \tag{86}$$

normalized to

$$(\psi_{\lambda'}, \psi_\lambda) \equiv \sum_n \psi^*_{\lambda',n}\psi_{\lambda,n} = \delta_{\lambda\lambda'} \tag{87}$$

L can be spectrally decomposed:

$$L_{mn} = \sum_\lambda \lambda\psi_{\lambda,m}\psi^*_{\lambda,n} \tag{88}$$

or symbolically,

$$L = \sum_\lambda \lambda\psi_\lambda\psi_\lambda^*$$

so that

$$L^\rho = \sum_\lambda \lambda^\rho\psi_\lambda\psi_\lambda^*$$

The condition to be met for $L^\rho\phi$ to not contain any eigenvector $\psi_{\bar\lambda}$ is then

$$(\psi_{\bar\lambda}^*, \phi) = 0 \tag{89}$$

We now consider universality in the light of this framework. With $\tilde g_0(x) = f(\lambda_\infty x)$,

$$\tilde g_0^{(n)}(x) = \lambda_\infty^{-1}(\lambda_\infty f)^{(n)}(\lambda_\infty x)$$

or

$$(\lambda_\infty f)^{(n)}(x) = \lambda_\infty \tilde g_0^{(n)}(x/\lambda_\infty) \tag{90}$$

which is simply a magnification by λ_∞ of $\tilde g_0^{(n)}(x)$. Defining the recursion

$$\tilde g_{n+1}(x) = -\alpha\tilde g_n(\tilde g_n(x/\alpha)), \qquad \tilde g_0(x) = \lambda_\infty f(x) \tag{91}$$

will cause $\tilde g_n \to g$, where

$$\tilde g_n(x) = (-\alpha)^n \tilde g_0^{(2^n)}(x/\alpha^n)$$

Defining $\eta_n \equiv \tilde g_n - g$, then

$$\eta_{n+1} = \mathscr{L}[\eta_n]$$

in linear approximation. Expanding η_n along eigenvectors of $\mathscr{L}$, we have by the orthogonality (77)

$$\eta_n(x) = \sum_\lambda \lambda^n(\psi_\lambda^*, \eta_0)\psi_\lambda(x) \tag{92}$$

Since $\eta_n \to 0$, the solitary growing mode corresponding to $\lambda = \delta$ must be unexcited. Calling h^* the adjoint eigenvector of eigenvalue δ, this means

$$(h^*, \eta_0) = 0 \tag{93}$$

However,

$$\eta_0 = \tilde{g}_0 - g = \lambda_\infty f - g$$

so that (93) becomes

$$\lambda_\infty(f) = (h^*, g)/(h^*, f) \tag{94}$$

That is, the λ_∞ of the 2-cycle receives the interpretation as the unique value of λ to extinguish the diverging mode of $\mathscr{L}$. [For a fundamental cycle of order s, $\tilde{g}_0$ must be $(\lambda_\infty f)^{(s)}$, so that λ_∞ is no longer multiplicative, although (93) will still provide an equation to determine λ_∞.] For example, for $z = 2$ and $N = 2$

$$g \simeq 1 + g_1 x^2, \qquad h^* \simeq 1 + h_1{}^* x^2, \qquad f \simeq 1 + f_1 x^2$$

and (94) becomes

$$\lambda_\infty(f) \simeq (1 + g_1 h_1{}^*)/(1 + f_1 h_1{}^*) \tag{95}$$

To be a good estimate, η_0 must be in the linear domain, so that f should be "nice." Thus $f = 2x^2$ determines through (95) a 0.1% estimate of λ_∞. Provided f is nice, once g_n and $h_n{}^*$ are determined, (95) allows for 5-sec estimates of $\lambda_\infty(f)$.

In view of the computer spectral evidence, h^* is the unique eigenvector to all conjugacy-generator eigenvectors. This is important in the application of (94): iterates of $\lambda_\infty f$ converge not to g, but to $\mu g(x/\mu)$. In writing $\eta_0 = \lambda_\infty f - g$, we never specified the normalization of g. Indeed, it is irrelevant: h^* is computed from the g normalized to $g(0) = 1$. Since h^* is orthogonal to all conjugacy generators of g,

$$(h^*, \mu g(x/\mu)) = (h^*, g)$$

for all μ (in linear approximation) and so (94) is correct for g with fixed normalization. Moreover, the conjugacy problem of g is solved: if

$$f = \psi \circ g \circ \psi^{-1}$$

for some ψ connected to the identity, then it must follow that

$$(h^*, f) = (h^*, g)$$

and conversely. (Clearly our spectral conjecture is quite strong.) This leads to another method of estimating λ_∞:

$$(\lambda_\infty f - g, h^*) = 0 \Rightarrow \lambda_\infty f \sim g \qquad \text{(conjugacy)}$$

Thus, should λ_∞ satisfy a *necessary* condition for conjugacy, $\lambda_\infty f$ must be conjugate to g. The condition is elementary: if

$$g(x^*) = x^* \qquad \text{and} \qquad \lambda f(\xi_\lambda{}^*) = \xi_\lambda{}^*$$

then

$$g \sim \lambda_\infty f \Rightarrow g'(x^*) = \lambda_\infty f(\xi^*_{\lambda_\infty})$$

But $g'(x^*)$ is a fixed value for fixed z, and so upon calculation of the fixed point of λf an estimate of λ_∞ is had:

$$\lambda_\infty(f) = g'(x^*)/f'(\xi^*_{\lambda_\infty})$$

Again for $f = 1 - 2x^2$, $f = \sin \pi x$, $f = x - x^3$, and other "nice" f's, a 0.1% estimate is obtained for λ_∞. "Nice" here means that f is "close" to conjugate to g.

We now extend these ideas to the $g_{r,n}$ recursion to provide another proof that

$$\lambda_r \sim \lambda_\infty - \mu\delta^{-r}$$

with μ now determined by the same simple kind of estimates (and to equal precision) as was λ_∞. Moreover, we shall demonstrate how the convergence rate of α_n is computable and equal to δ. Repeating Eq. (62),

$$g_{r,n} = g + \sum_\rho c^\rho_{r+n}(-\alpha)^{n(1-\rho)}(g^\rho - x^\rho g')$$
$$- \delta^{-r}\left(h + \sum_\rho c^\rho_{r+n}(-\alpha)^{n(1-\rho)}(\rho g^{\rho-1}h - x^\rho h')\right)$$

it is clear that the $g_{r,n}$ for large n are fixed by determining the $c_t{}^\rho$ from initial data. Thus, for $n = 0$

$$g_{r,0} = \lambda_r f(x) = g + \sum_\rho c_r{}^\rho(g^\rho - x^\rho g') - \delta^{-r}\left(h + \sum_\rho c_r{}^\rho(\rho g^{\rho-1}h - x^\rho h')\right) \tag{96}$$

Recall that

$$\psi_\rho \equiv g^\rho - x^\rho g'$$

is the eigenvector of $\mathscr{L}$ corresponding to $\lambda = (-\alpha)^{1-\rho}$ and so orthogonal to h^*. Defining

$$\rho g^{\rho-1}h - x^\rho h' \equiv h_\rho \tag{97}$$

and projecting (96) on h^*, we have

$$\lambda_r(h^*, f) = (h^*, g) - \delta^{-r}\left(1 + \sum_\rho c_r{}^\rho(h^*, h_\rho)\right) \tag{98}$$

or

$$\lambda_r = \frac{(h^*, g)}{(h^*, f)} - \frac{1 + \sum_\rho c_r{}^\rho(h^*, h_\rho)}{(h^*, f)}\delta^{-r} \equiv \lambda_\infty - \mu_r\delta^{-r}$$

Thus,

$$\lambda_\infty = (h^*, g)/(h^*, f)$$

as before and

$$\mu_r/\lambda_\infty = \left[1 + \sum_\rho c_r{}^\rho(h^*, h_\rho)\right]\Big/(h^*, g) \tag{99}$$

Projecting next upon $\psi_{\bar{\rho}}^*$, we have

$$\lambda_r(\psi_{\bar{\rho}}^*, f) = (\psi_{\bar{\rho}}^*, g) + c_r^{\bar{\rho}} - \delta^{-r} \sum_{\rho} c_r^{\rho}(\psi_{\bar{\rho}}^*, h_\rho) \tag{100}$$

In particular, as $r \to \infty$,

$$\lambda_\infty(\psi_{\bar{\rho}}^*, f) = (\psi_{\bar{\rho}}^*, g) + c_\infty^{\bar{\rho}}$$

so that

$$\lim_{r \to \infty} c_r^{\rho} = c_\infty^{\bar{\rho}} = \lambda_\infty(\psi_{\bar{\rho}}^*, f) - (\psi_{\bar{\rho}}^*, g) \tag{101}$$

exists and is finite to meet initial data. Accordingly, for large r we have

$$\lambda_r \sim \lambda_\infty - \mu\delta^{-r}$$

with

$$\frac{\mu}{\lambda_\infty} = \frac{1 + \sum_\rho c_\infty^{\rho}(h^*, h_\rho)}{(h^*, g)}$$

$$= \frac{1 - \sum_\rho (\psi_\rho^*, g)(h^*, h_\rho) + \lambda_\infty \sum_\rho (\psi_\rho^*, f)(h^*, h_\rho)}{(h^*, g)} \tag{102}$$

Accordingly, μ/λ_∞ is also available and easily computed for small N quite accurately.

We next obtain a sum rule for c_r^{ρ}. Setting $x = 0$ in (96), we have

$$\lambda_r = 1 + \sum_\rho c_r^{\rho} - \delta^{-r} h(0)\left(1 + \sum_\rho c_r^{\rho}\right) \tag{103}$$

which as $r \to \infty$ becomes

$$\lambda_\infty = 1 + \sum_\rho c_\infty^{\rho} \tag{104}$$

Since

$$\sum_\rho \rho c_r^{\rho} < \infty$$

by (103),

$$c_r^{\rho} < 1/\rho^{2+\epsilon}$$

for large ρ. Accordingly, truncation of the ρ sum allows high-accuracy estimates. Setting $N = 2$ so that only $\lambda = \delta$ and $\lambda = 1$ contribute, one has the rough result

$$\lambda_\infty \simeq 1 + c_\infty^{1} \tag{105}$$

By (62),

$$g_{r,n} \underset{n\to\infty}{\sim} g + c^1_{r+n}(g - xg') - \delta^{-r}(h + c^1_{r+n}(h - xh'))$$

$$\simeq (1 + c^1_{r+n})(g - \delta^{-r}h)(x/1 + c^1_{r+n}) \tag{106}$$

or,

$$g_{r,n} \xrightarrow[n\to\infty]{} (1 + c_\infty{}^1)g_r(x/1 + c_\infty{}^1) \tag{107}$$

With the rough estimate (105), this reads

$$g_{r,n} \xrightarrow[n\to\infty]{} \lambda_\infty g_r(x/\lambda_\infty)$$

Also, assuming a rapid n approach, and setting $r \to \infty$, we have

$$g_{\infty,n} \simeq \lambda_\infty g(x/\lambda_\infty)$$

so that $g_{\infty,n} \simeq g_{\infty,0}$ is the estimate

$$\lambda_\infty f(x) \simeq \lambda_\infty g(x/\lambda_\infty)$$

or

$$g(x) \simeq f(\lambda_\infty x)$$

Thus we realize that all our approximation schemes produce estimates of the same accuracy. Next, Eqs. (98) and (100) for $N = 2$ are

$$\lambda_r(h^*, f) \simeq (h^*, g) - \delta^{-r}(1 + c_r{}^1(h^*, h_1))$$

and

$$\lambda_r(\psi_1{}^*, f) \simeq (\psi_1{}^*, g) + c_r{}^1(1 - \delta^{-r}(\psi_1{}^*, h_1))$$

The ratio of these equations produces

$$c_r{}^1 - c_\infty{}^1 \propto \delta^{-r} \tag{108}$$

Together with (106), we then have

$$g_{r,n} - g_{r,\infty} \sim \delta^{-n}$$

in fixed r, providing the mechanism for $\alpha_n \to \alpha$ at the rate δ. We are unsure as to why $\alpha_n \to \alpha$ at a rate $\delta' \neq \delta$ for $z > 2$, especially since the spectrum of L for $z = 4$ possesses δ as the unique growing eigenvalue. Presumably, higher order transients can here decay at a rate below that of the "aymptotic" features discussed here. But for this one defect, the above techniques explain to good accuracy every detail of all our recursion data.

6. AFTERWORD

The preceding parts of this paper were contained in a preprint first circulated in November 1976. This paper is incomplete insofar as the uniqueness of an appropriate solution to (9) as well as the basic spectral conjecture remain unproven. Failing to publish it immediately (because it was not self-contained), I allowed it to hover in a limbo while I anticipated some measure of success at a proof, foreshadowing its content in the final section of its predecessor.[(1)]

Early in 1979, I was informed that an effort by Collet *et al.*[(2),2] has succeeded in this task. These authors have proven existence and uniqueness of the appropriate solution to the fundamental equation (9) and verified the spectral conjecture of $\mathscr{L}$. (This demonstration is, so far, restricted to $z = 1 + \epsilon$ with ϵ small.) Accordingly, the theory presented here is now well-founded, although no extension beyond the local stability of the fixed point is expected in the immediate future.

At this time, I should like to mention another effort. In the special case $z = 1 + \epsilon$, ϵ small, it is easy to approximately solve (9) and the spectral problem of $\mathscr{L}$ since α^{-1} is perturbatively small. This result first appeared in a work by Derrida *et al.*[(4)] (DGP). In this and another interesting paper by these authors,[(5)] the work of Metropolis *et al.*[(6)] (MSS) has been significantly elaborated upon through the discovery of an "internal symmetry" of the MSS sequences which allows organizations of these sequences in manners approaching λ_∞ from *above* rather than from below along the harmonics. I will here briefly explore the connection of one aspect of their work with the present work.

There exists a unique fundamental 4-cycle above the λ_∞ of the 2-cycle. Related to the pattern of this 4-cycle by the operation of DGP is a fundamental 8-cycle, below the 4-cycle and closest to λ_∞. Similarly, for each n there is a fundamental 2^n-cycle below the 2^{n-1}-cycle and closest to and above λ_∞. Denoting the parameter value of these cycles that are superstable by $\hat{\lambda}_n$, we have

$$\hat{\lambda}_2 > \hat{\lambda}_3 > \cdots > \hat{\lambda}_n > \cdots > \lambda_\infty \tag{109}$$

DGP observe that

$$\hat{\lambda}_n - \lambda_\infty \propto \delta^{-n}$$

with the same δ as for the harmonics of the 2-cycle. It is easy to see why this can be so. For the harmonics, the functions g_r were constructed, with

$$g_r \sim g - \delta^{-r} h$$

In this form, the coefficient of h is *negative*. Indeed, this is required for the harmonics to guarantee that $g_0(0) = 0$. However, the term in h is perturbative about the fixed point g, so that nothing in the local analysis requires this negative coefficient. Indeed, for an appropriate *positive* coefficient the

[2] See Ref. 3 for a preview.

phenomena described above are explained, with the g's constructed determining the elements of these cycles.

To see how these phenomena are described, write

$$\lambda f = \lambda_\infty f + (\lambda - \lambda_\infty) f \equiv G_0 + (\lambda - \lambda_\infty) H_0 \tag{110}$$

Iterating 2^n times, and keeping terms to order $\lambda - \lambda_\infty$, we obtain

$$(\lambda f)^{2^n} = G_n + (\lambda - \lambda_\infty) H_n + O((\lambda - \lambda_\infty)^2) \tag{111}$$

where

$$G_n = (\lambda_\infty f)^{2^n}$$

Defining

$$\begin{aligned} (-\alpha)^n (\lambda f)^{2^n} (x/(-\alpha)^n) &\equiv f_n \\ (-\alpha)^n G_n (x/(-\alpha)^n) &\equiv g_n \\ (-\alpha)^n H_n (x/(-\alpha)^n) &\equiv h_n \end{aligned} \tag{112}$$

we can write (111) as

$$f_n(x) = g_n(x) + (x - \lambda_\infty) h_n(x) \tag{113}$$

where

$$\begin{aligned} h_{n+1}(x) &= -\alpha[h_n(g_n(-x/\alpha)) + g_n'(g_n(-x/\alpha)) h_n(-x/\alpha)] \\ &\equiv \mathscr{L}_n[h_n(x)] \end{aligned}$$

and

$$h_n = \mathscr{L}_{n-1} \mathscr{L}_{n-2} \cdots \mathscr{L}_0 f \tag{114}$$

By the definition of λ_∞ and g,

$$g_n \to g, \qquad \mathscr{L}_n \to \mathscr{L} \qquad \text{as } n \to \infty$$

Accordingly, (114) becomes

$$h_n \sim c(f) \delta^n h$$

and (113) reads

$$f_n \sim g + c(f)(\lambda - \lambda_\infty) \delta^n h \tag{115}$$

Equation (115) is approximately correct so long as $(\lambda - \lambda_\infty)^n$ is small, which is the case when

$$|\lambda - \lambda_\infty| \delta^n \leqslant \text{small constant}$$

With λ_n chosen as usual to determine the superstable 2^n-cycle harmonic of the 2-cycle,

$$f_n \to g_0$$

Since $g_0(0) = 0$ and $g(0), h(0) \neq 0$ evidently

$$(\lambda_n - \lambda_\infty) \delta^n \sim 1$$

again establishing δ as the λ convergence rate. If an n-independent *finite* condition on the f_n can more generally be maintained, δ will again be the convergence rate and the corresponding limit of the f_n will be given by (115).

Accordingly, consider determining λ_n by the condition that

$$(\lambda_{nf})^{2^n}(\xi_n) = \xi_n$$
$$D(\lambda_{nf})^{2^n}(\xi_n) = \mu \qquad \text{(independent of } n\text{)}$$

where ξ_n is the fixed point closest to $x = 0$. By (112) these conditions transcribe to

$$f_n(x_n) = x_n, \qquad f_n'(x_n) = \mu, \qquad x_n = (-\alpha)^n\xi_n$$

As $n \to \infty, f_n \to f_\mu, x_n \to x_\mu$, by (115)

$$x_\mu \sim g(x_\mu) + c(f)(\lambda_n - \lambda_\infty)\delta^n h(x_\mu) \tag{116}$$
$$\mu \sim g'(x_\mu) + c(f)(\lambda_n - \lambda_\infty)\delta^n h'(x_\mu) \tag{117}$$

Denoting the fixed point of g by $\hat{x}$, and the slope of g at $\hat{x}$ by $\hat{\mu}$, then from (116) and (117) we immediately obtain the approximation

$$\lambda_n \sim \lambda_\infty + \delta^{-n}(\hat{\mu} - \mu)/c|h'(\hat{x})| \tag{118}$$

for $\mu \simeq \hat{\mu}$.

Thus, so long as $\mu \neq \hat{\mu}$, the corresponding λ_n converge to λ_∞ at the rate δ. (At $\mu = \hat{\mu}$, $\lambda_n \to \lambda_\infty$ faster than geometric at the rate δ.) Also by (118), the coefficient of h in f_μ, by (115), changes sign at $\mu = \hat{\mu}$: for $|\mu| < |\hat{\mu}|$, f_μ is a g_r or its continuous analog (for example, at bifurcation values) as described at the end of Section 3; for $|\mu| > |\hat{\mu}|$, $\lambda_n \to \lambda_\infty$ from *above*, and evidently the harmonics are not under consideration. For example, with $\hat{\lambda}_n$ of (109)

$$f_n = (-\alpha)^n(\hat{\lambda}_{n+1}f)^{2^n}(x/(-\alpha)^n)$$

corresponds to a limiting value of $|\mu| > |\hat{\mu}|$ and $\hat{\lambda}_n \to \lambda_\infty$ from above at rate δ, as was to be demonstrated. Accordingly, the fixed point g is the "organizing center" for all attractors with $\lambda \to \lambda_\infty$ whether from above or below λ_∞.

As a final comment, it is perhaps worthy to point out the resemblance of the theory presented to the renormalization-group notions of Wilson.[7] Essentially, the function g_r determine elements of infinitely bifurcated attractors at various levels of magnification, belying a self-similarity of their distribution; this structure is precisely reproduced through the operations of composition and rescaling $\mathscr{T}$, resulting in the next lower g_r. The function g itself is the fixed point of $\mathscr{T}$, while the g_r lie on the one-dimensional unstable manifold through g along h; δ and h indeed were determined by linearizing $\mathscr{T}$ about g. More generally, applied to any f, $\mathscr{T}$ can be viewed as a renormalization-group transformation with self-similarity (critical behavior) determined by the fixed point g. Viewing the parameter λ as temperature, λ_∞ is the critical point and δ emerges as a critical exponent. More intuitively, an analog of Kadanoff's block-spin notion is also available. Thus, consider the superstable 2^n cycles starting at $n = 0$, for which there is a single point at $x = 0$. For $n = 1$, this point is split into one at $x = 0$ again, and another point x_1 to the right. For $n = 2$, $x = 0$ again splits, with x_2 nearest to $x = 0$ and to the left, while x_1 splits into a more closely spaced pair with centroid roughly at x_1. By the definition of α, $x_1 \simeq -\alpha x_2$. As n increases, each point splits into a pair with the element nearest to $x = 0$ located $-\alpha$ times nearer to $x = 0$ than its predecessor. It is thus clear that if each closely spaced pair is replaced by a point at its centroid (viewing at lower resolution), then the same set of points about $x = 0$ is reproduced, but with all distances $-\alpha$ times larger. Accordingly, spin-blocking has here the analog of functional composition, while the following volume rescaling is here, rather than a geometrical factor of 2, now a dynamically determined factor of α. In this way, the theory presented in this work may be viewed as an instance arising mathematically of the renormalization-group notions of statistical mechanics.

APPENDIX

We include here some numerical results, useful for normal ($z = 2$) recursive calculations.

A1. $g(x) = 1 + \sum_{i=1}^{7} g_i x^{2i}$, determining g to ten significant figures as [0, 1]:

$$
\begin{aligned}
g_1 &= 1.527632997 \\
g_2 &= 1.048151943 \times 10^{-1} \\
g_3 &= 2.670567349 \times 10^{-2} \\
g_4 &= -3.527413864 \times 10^{-3} \\
g_5 &= 8.158191343 \times 10^{-5} \\
g_6 &= 2.536842339 \times 10^{-5} \\
g_7 &= -2.687772769 \times 10^{-6} \\
&\Rightarrow -g'(1) = \alpha = 2.502907876
\end{aligned}
$$

A2. From the above, $g(x^*) = x^*$ for $x^* = 0.5493052461$ and $g'(x^*) = -1.601191328$, which is required for the estimate

$$\lambda_\infty(f) \simeq g'(x^*)/f'(\xi^*)$$

where ξ^* satisfies

$$\lambda_\infty f(\xi^*) = \xi^*$$

For example, with $f = x(1 - x)$,

$$\xi^* = 1 - \lambda_\infty^{-1} \quad \text{and} \quad \lambda_\infty f'(\xi^*) = -\lambda_\infty + 2$$

so that

$$-\lambda_\infty + 2 \simeq g'(x^*)$$

or $\lambda_\infty \simeq 3.60119$, to be compared with the correct result $\lambda_\infty = 3.56995$.

A3. In order for g_r to be computed, one needs $h(x)$ normalized to $h(0) = 1$ together with the correct $h(0)$ to ensure that $g_0(0) = 0$. Regarding $r = 6$ as asymptotic, we have

$$h(0) = 1.318707$$

and a parametrization of similar accuracy to Section A1 is

$$h(x) = h(0)\left(1 + \sum_{i=1}^{6} h_i x^{2i}\right)$$

with

$$\begin{aligned} h_1 &= -3.256513712 \times 10^{-1} \\ h_2 &= -5.055393508 \times 10^{-2} \\ h_3 &= 1.455982806 \times 10^{-2} \\ h_4 &= -8.810422078 \times 10^{-4} \\ h_5 &= -1.062170276 \times 10^{-4} \\ h_6 &= 1.983988805 \times 10^{-5} \end{aligned}$$

Iterating (8), g_1 or g_0 is obtained for estimates of the locations of elements of a highly bifurcated cycle near $x = 0$. Observe that since

$$g_{r-s}(x) = (-\alpha)^s g_r^{(2^s)}(x/(-\alpha)^s)$$

with $s = r = 6$,

$$g_0(x) = (-\alpha)^6 g_6^{(2^6)}(x/\alpha^6) \simeq (-\alpha)^6 (g - \delta^{-6} h)^{(2^6)}(x/\alpha^6)$$

so that g and h restricted to $[0, 1]$ provide g_0 or $[0, \alpha^6]$, thereby determining many elements near $x = 0$.

A4. Solving the eigenvalue problem of L for $N = 2$, we have

$$\delta \simeq 4.6736, \qquad \lambda_1 \simeq 0.9880$$

to be compared with

$$\delta = 4.6692, \qquad \lambda_1 = 1.0000$$

The corresponding eigenvectors and adjoint eigenvectors (unnormalized) are $\left[\psi = \begin{pmatrix} a \\ b \end{pmatrix} \Rightarrow \psi(x) = a + bx^2\right]$

$$h = \begin{pmatrix} 1 \\ -0.3644 \end{pmatrix}, \qquad \psi_1 = \begin{pmatrix} 1 \\ 1.1082 \end{pmatrix}$$

$$h^* = \begin{pmatrix} 1 \\ -0.9024 \end{pmatrix}, \qquad \psi_1{}^* = \begin{pmatrix} 1 \\ 2.7444 \end{pmatrix}$$

Writing g as

$$g \simeq \begin{pmatrix} 1 \\ g_1 + \frac{1}{3}g_2 \end{pmatrix} = \begin{pmatrix} 1 \\ -1.4927 \end{pmatrix}$$

it is trivial to estimate $\lambda_\infty \simeq (h^*, g)/(h^*, f)$. For example, with $f(x) = 1 - 2x^2$, $f = \begin{pmatrix} 1 \\ -2 \end{pmatrix}$ and $\lambda_\infty(f) \simeq 0.8368$. Also, $h_1 = \begin{pmatrix} 1 \\ 0.3644 \end{pmatrix}$ and $(h^*, h_1) \simeq 0.5051$ (properly normalized), so that by (102)

$$\frac{\mu}{\lambda_\infty} \simeq \frac{1 - (\psi_1{}^*, g)(h^*, h_1) + \lambda_\infty(\psi_1{}^*, f)(h^*, h_1)}{(h^*, g)} \simeq 0.6851$$

or $\mu \simeq 0.5733$, to be compared with $\mu(f) = 0.5981$. While it is true that $N = 3$ significantly improves this result, this is already quite accurate and trivial to obtain.

ACKNOWLEDGMENTS

The first exact equation, Eq. (9) of the text, together with the scheme of solution incorporating Eq. (91), was obtained by Predrag Cvitanović in discussion and collaboration with the author. This result proved to be seminal in the construction of the theory presented. The author has profited from discussion with R. Menikoff, and thanks his colleagues, especially D. Campbell and F. Cooper, for critical interest.

REFERENCES

1. Mitchell J. Feigenbaum, *J. Stat. Phys.* **19**:25 (1978).
2. P. Collet, J.-P. Eckmann, and O. E. Lanford III, Universal Properties of Maps on an Interval, in draft.
3. P. Collet and J.-P. Eckmann, Bifurcations et Groupe de Renormalisation, IHES/P/78/250.

4. B. Derrida, A. Gervois, and Y. Pomeau, Universal Metric Properties of Bifurcations of Endomorphisms, Saclay preprint (1977).
5. B. Derrida, A. Gervois, and Y. Pomeau, Iterations of Endomorphisms on the Real Axis and Representation of Numbers, Saclay preprint (1977).
6. N. Metropolis, M. L. Stein, and P. R. Stein, *J. Combinatorial Theory* **15**:25 (1973).
7. K. Wilson and J. Kogut, *Phys. Rep.* **12C**:75 (1974).

Editor's note. This paper was submitted to *Advances in Mathematics* in November 1976 and rejected. Feigenbaum (1978) was submitted to *SIAM Journal of Applied Mathematics* in April 1977 and rejected in October 1977. The geometrical parameter convergence had already been noted by Myrberg (1958–64) and the value of δ was independently determined by Grossman and Thomae (1977). Coulet and Tresser (1978) have also formulated universal equations. There are many examples of universality which have been published since 1978, some of them included in this reprint selection. A nice article which should have been included, but was cut because of space limitations, is Grassberger (1981).

Bulletin of the American Mathematical Society **6** 427–34 (1982)

A COMPUTER-ASSISTED PROOF OF THE FEIGENBAUM CONJECTURES

BY OSCAR E. LANFORD III[1]

1. Introduction. Let $\mathcal{M}$ denote the space of continuously differentiable even mappings ψ of the interval $[-1, 1]$ into itself such that

M1. $\psi(0) = 1$,

M2. $x\psi'(x) < 0$ for $x \neq 0$.

M2 says that ψ is strictly increasing on $[-1, 0)$ and strictly decreasing on $(0, 1]$, so $\mathcal{M}$ is a space of mappings which are unimodal in a strict sense.

Condition M1 says that the unique critical point 0 is mapped to 1. We want to consider ψ's which map 1 slightly – but not too far – to the left of 0. It may then be possible to find nonoverlapping intervals I_0 about 0 and I_1 near 1 which are exchanged by ψ. Technically, we proceed as follows: Write a for $-\psi(1) = -\psi^2(0)$ and b for $\psi(a)$; we suppress from the notation the dependence of a and b on ψ. Define $\mathcal{D}(T)$ to be the set of all ψ's in $\mathcal{M}$ such that:

D1. $a > 0$,

D2. $b > a$,

D3. $\psi(b) \leqslant a$.

The two intervals $I_0 = [-a, a]$ and $I_1 = [b, 1]$ are then nonoverlapping and ψ maps I_0 into I_1 and vice versa. If $\psi \in \mathcal{D}(T)$, then $\psi \circ \psi|_{I_0}$ has a single critical point, which is a minimum. By making the change of variables $x \longrightarrow -ax$, we replace I_0 by $[-1, 1]$ and the minimum by a maximum, i.e., if we define

$$T\psi(x) = -\frac{1}{a}\,\psi \circ \psi(-ax) \quad \text{for } x \in [-1, 1]$$

then $T\psi$ is again in $\mathcal{M}$. Thus, T defines a mapping of $\mathcal{D}(T)$ into $\mathcal{M}$. (In general, $T\psi$ need not lie in $\mathcal{D}(T)$. If a is small, then $T\psi(1)$ will be approximately 1 so $T\psi$ will not satisfy D1. On the other hand, if $\psi(b)$ is near a, then $T\psi(1)$ will be near -1 from which it follows that $T\psi$ does not satisfy D2.)

M. Feigenbaum [6] has proposed an explanation for some universal features displayed by infinite sequences of period doubling bifurcations based on some conjectures about T. We will not review has argument here; a version with due regard for mathematical technicalities may be found in Collet and Eckmann [3],

---

Received by the editors October 27, 1981.

1980 *Mathematics Subject Classification.* Primary 58F14.

[1]The author gratefully acknowledges the financial support of the Stiftung Volkswagenwerk for a visit to the IHES during which this paper was written, and the continuing financial support of the National Science Foundation (Grant MCS 78-06718).

Collet, Eckmann and Lanford [4], or in Lanford [8]. The purpose of this note is to announce a proof of essentially all of these conjectures and to indicate the kind of analysis used.

2. Statement of results.

THEOREM 1. *There exists a function g, analytic and even on* $\{z \in \mathbf{C}: |z| < \sqrt{8}\}$ *whose restriction to* $[-1, 1]$ *is a fixed point for* T. *The Schwarzian derivative of g is negative on* $[-1, 1]$.

Let Ω denote $\{z \in \mathbf{C}: |z^2 - 1| < 2.5\}$ and write

$\mathfrak{H}$ for the Banach space of even functions bounded and analytic on Ω, real on real points, equipped with the supremum norm.

$\mathfrak{H}_0$ for the subspace of $\mathfrak{H}$ consisting of those functions vanishing (to second order) at 0.

$\mathfrak{H}_1$ for $\mathfrak{H}_0 + 1$.

PROPOSITION 2. *There is an open neighborhood* V *of g in* $\mathfrak{H}_1$ *such that*
Every $\psi \in V$ *is in* $\mathcal{D}(T)$ *(i.e., its restriction to* $[-1, 1]$ *is).*
If $\psi \in V$, $T\psi \in \mathfrak{H}_1$.
T *is infinitely differentiable as a mapping from* V *into* $\mathfrak{H}_1$,
The derivative $DT(\psi)$ *is compact operator on* $\mathfrak{H}_0$ *for each* $\psi \in V$.

THEOREM 3. $DT(g)$ *is hyperbolic on* $\mathfrak{H}_0$ *with one-dimensional expanding subspace; the expanding eigenvalue* δ *is positive.*

In other words: The spectrum of $DT(g)$ does not intersect the unit circle, and the part of the spectrum outside the unit circle consists of a single simple positive eigenvalue δ.

It then follows from invariant manifold theory that T admits locally invariant local stable and local unstable manifolds, of codimension one and dimension one respectively. Because of the noninvertibility of T, we do not construct global stable and unstable manifolds; we will let W_s and W_u denote respectively some particular local stable and local unstable manifolds.

Let Σ_0 denote the bifurcation surface for the simple period-doubling bifurcation. By this we mean the following: Any ψ in M has exactly one fixed point x_0 in $[0, 1]$; Σ_0 then denotes $\{\psi \in M: \psi'(x_0) = -1; (\psi \circ \psi)'''(x_0) < 0\}$. As a one-parameter family of ψ's crosses Σ_0 (in the appropriate direction) the fixed point x_0 loses stability in favor of an attracting orbit of period 2.

THEOREM 4. *There is a positive integer j and an element* g_j^* *of* W_u *such that* $T^j g_j^* \in \Sigma_0$.

Except for the difficulties in defining a global unstable manifold, we could formulate this theorem by saying that the unstable manifold crosses Σ_0. We would like to know more, viz., that the crossing is transversal. This – properly formulated – is almost certainly true, but we have not proved it.

Let $\psi_\mu^{(0)}(x)$ denote the quadratic mapping $1 - \mu x^2$.

THEOREM 5. *There is a positive integer j and a parameter value μ_∞ (between 1.4011550 and 1.4011554) such that $\psi_\mu^{(0)}$ is in $\mathcal{D}(T^j)$ for μ sufficiently near to μ_∞ and such that the curve $T^j\psi_\mu^{(0)}$ crosses W_s transversally at $\mu = \mu_\infty$.*

Except for technicalities, this says that $\psi_\mu^{(0)}$ crosses the stable manifold transversally at μ_∞.

3. Remarks on the method of proof. The heart of the proof is a set of complicated numerical estimates proved rigorously with the aid of a computer. To formulate these estimates, we have first to establish some notation. We will work, initially, not in $\mathfrak{H}_1$ but in a subspace equipped with a stronger norm. The idea is that we want to write ψ as

$$\psi(x) = 1 - x^2 h(x^2)$$

and to use the l^1 norm for the Taylor coefficients of h at 1. Formally, given an element (u, v) of $\mathbf{R} \oplus l^1$, we associate with it an element ψ of $\mathfrak{H}_1$ by

$$\psi(x) = 1 - x^2\left\{u/10 + \sum_{n=1}^{\infty} v_n\left(\frac{(x^2-1)}{2.5}\right)^n\right\}. \tag{3.1}$$

We denote the set of ψ's obtained in this way by A , and we equip A with the norm $|u| + \Sigma|v_n|$. Note that A contains any element of $\mathfrak{H}_1$ which is analytic on the closure of Ω. (Of course, $\mathbf{R} \oplus l^1$ could have been identified with l^1, but we have singled out the u component – and introduced the factor of 10 in the formula (3.1) for $\psi(x)$ – for convenience later on.) For the remainder of this section, the norm of an element of A will always mean the norm of l^1 type just introduced.

The first step is to choose an explicit polynomial ψ_0 which will turn out to be a good approximate fixed point. We will take ψ_0 to be the polynomial of degree 20 defined by the first ten terms of the series given in Table 1 below. It can be checked without difficulty that

For any $\psi \in A$ with $\|\psi - \psi_0\| < .01$, $T\psi \in A$

T is infinitely differentiable from $\{\|\psi - \psi_0\| < .01\}$ to A.

For any ψ in this ball, $DT(\psi)$ is a compact operator on A.

Identifying A with $\mathbf{R} \oplus l^1$, we can represent $DT(\psi)$ as a matrix

$$\begin{pmatrix} \alpha(\psi) & \beta(\psi) \\ \gamma(\psi) & \delta(\psi) \end{pmatrix}$$

with $\alpha \in \mathbf{R}; \beta \in (l^1)^*; \gamma \in l^1; \delta \in L(l^1, l^1)$. In this notation, we can formulate

ESTIMATE 1. If $\|\psi - \psi_0\| < .01$, then

$$|\alpha - 4.669| < .148; \quad \|\beta\| < .560; \quad \|\gamma\| < .756; \quad \|\delta\| < .719.$$

These bounds imply that the inequality

$$[\alpha(\psi) - 1]\,[1 - \|\delta(\psi)\|] > \|\beta(\psi)\| \cdot \|\gamma(\psi)\| \tag{3.2}$$

holds uniformly on the ball of radius .01 about ψ_0. If T has a fixed point g in this ball, then hyperbolicity of $DT(g)$ acting on A follows readily from (3.2).

To prove the existence of a fixed point, we use a variant of Newton's method. Instead of studying

$$\psi \mapsto \psi - (DT(\psi) - \mathbb{1})^{-1}[T\psi - \psi],$$

we replace $(DT(\psi) - \mathbb{1})^{-1}$ by the approximation

$$J = \begin{pmatrix} \frac{1}{3.669} & 0 \\ 0 & -\mathbb{1} \end{pmatrix},$$

and we apply the contraction mapping principle to the mapping

$$\psi \mapsto \Phi(\psi) = \psi - J \cdot [T\psi - \psi]$$

which has the same fixed points as T.

A simple calculation using Estimate 1 shows that

$$\|D\Phi(\psi)\| < .9 \quad \text{for } \|\psi - \psi_0\| < .01.$$

It will then follow from the contraction mapping theorem that Φ has a fixed point in this ball provided that

$$\frac{\|\Phi(\psi_0) - \psi_0\|}{1 - .9} < .01.$$

For this we have

ESTIMATE 2.

$$\|\Phi(\psi_0) - \psi_0\| < 4 \times 10^{-6}.$$

Thus T has a fixed point in A, and DT at the fixed point, acting on A, has the hyperbolicity properties stated in Theorem 3. Domains of analyticity may be enlarged using the functional equation for g, and in this way we arrive at Theorems 1 and 3 as formulated.

Furthermore, Estimate 1 makes it possible to establish the existence of a system of expanding and contracting cones for T on $\{\psi: \|\psi - \psi_0\| < .01\}$, which in turn makes it possible to construct local stable and unstable manifolds which are not too small. This facilitates the proofs of Theorems 4 and 5.

The proofs of Estimates 1 and 2 are completely straightforward, if long. Consider, for example, Estimate 1. Since A is essentially l^1, we can think of $DT(\psi)$ as an infinite matrix. Norms of matrices acting on l^1 are easy to compute in terms of the matrix elements. Any matrix element can be expressed in terms of ψ. All but finitely many of these matrix elements are estimated analytically. For the remainder, strict upper and lower bounds are computed numerically from bounds on the Taylor coefficients for ψ. The arithmetic operations are performed in finite precision floating point arithmetic; the methods of interval arithmetic are used to control the effect of round-off error.

4. Supplementary remarks. 1. The results described here are descendants of (and improvements on) the results announced in [7]. Since that announcement, a completely different proof for the existence of g has been given by Campanino, Epstein, and Ruelle [1].

2, The approach to proving Theorem 1 outlined in the preceding section produces strict bounds on the difference between an approximate fixed point and the exact one. These estimates can be applied to higher precision calculations. Let

$$g^{(0)}(x) = 1 + \sum_{n=1}^{40} g_n^{(0)} x^{2n}$$

where the $g_n^{(0)}$ are given by Table 1.

We then have strict bounds

$$|g(z) - g^{(0)}(z)| \leqslant \begin{cases} 1.5 \times 10^{-23} & \text{for } |z|^2 \leqslant 1.5, \\ 5.5 \times 10^{-13} & \text{for } |z|^2 \leqslant 2, \\ 5 \times 10^{-7} & \text{for } |z|^2 \leqslant 6, \\ 1.7 \times 10^{-2} & \text{for } |z|^2 \leqslant 8. \end{cases}$$

These bounds are probably very conservative.

3. The domain Ω used in the statements of Proposition 2 and Theorem 3 was chosen for convenience. Many other domains, including arbitrarily small open neighborhoods of $[-1, 1]$, could have been used instead. The hyperbolicity statement of Theorem 3 is formally stronger for small domains than for larger ones. (For $\Omega_1 \subset \Omega_2$, any eigenfunction for $DT(g)$ on Ω_2 is also an eigenfunction on Ω_1). It can be shown, however, that any function analytic on a neighborhood of $[-1, 1]$ and satisfying there the formal functional equation for an eigenvector of $DT(g)$ is actually analytic and bounded on the domain Ω.

4. It follows easily from the functional equation for g that g is analytic on a neighborhood of the whole real axis. H. Epstein (private communication) has observed that a similar argument shows that it is analytic on a neighborhood

n	$g_n^{(0)}$
1	−1.52763 29970 36301 45403 58903 10240
2	0.10481 51947 87303 73321 67426 13801
3	0.02670 56705 25193 35403 26520 94944
4	−0.00352 74096 60908 70917 02341 90769
5	0.00008 16009 66547 53174 51721 90486
6	0.00002 52850 84233 96353 61762 62552
7	−2.55631 71662 78493 84635 32541 $\times 10^{-6}$
8	−9.65127 15508 91203 21637 25768 $\times 10^{-8}$
9	2.81934 63974 50409 13707 56629 $\times 10^{-8}$
10	−2.77305 11607 99011 72437 $\times 10^{-10}$
11	−3.02842 70221 30566 32983 $\times 10^{-10}$
12	2.67058 92807 48075 55396 $\times 10^{-11}$
13	9.96229 16410 28482 31059 $\times 10^{-13}$
14	−3.62420 29829 04156 08455 $\times 10^{-13}$
15	2.17965 77448 27070 47701 $\times 10^{-14}$
16	1.52923 28994 80962 60560 $\times 10^{-15}$
17	−3.18472 87899 52775 $\times 10^{-16}$
18	1.13467 21062 11871 $\times 10^{-17}$
19	1.88167 60568 25439 $\times 10^{-18}$
20	−2.27561 25646 32121 $\times 10^{-19}$
21	−9.82244 76294 21762 $\times 10^{-22}$
22	2.06412 97560 04508 $\times 10^{-21}$
23	−1.24932 00592 43689 $\times 10^{-22}$
24	−1.07706 12046 $\times 10^{-23}$
25	1.87274 68082 $\times 10^{-24}$
26	−2.57770 82101 $\times 10^{-26}$
27	−1.55419 04560 $\times 10^{-26}$
28	1.28044 34650 $\times 10^{-27}$
29	5.58505 87986 $\times 10^{-29}$
30	−1.52783 46925 $\times 10^{-29}$
31	5.04174 26639 $\times 10^{-31}$
32	1.01653 68070 $\times 10^{-31}$
33	−1.00690 $\times 10^{-32}$
34	−5.24253 $\times 10^{-34}$
35	1.72437 $\times 10^{-34}$
36	−1.31439 $\times 10^{-35}$
37	−1.85830 $\times 10^{-38}$
38	8.05506 $\times 10^{-38}$
39	−6.26717 $\times 10^{-39}$
40	1.76882 $\times 10^{-40}$

TABLE 1.

of the imaginary axis. On the other hand, it is essentially certain that g is not entire. Indeed, it appears – but has not been proved – that the singularities of g nearest to the origin occur at a set of 4 periodic points of period 2 for $z \mapsto g(-\lambda z)$, $\lambda = -g(1)$, located approximately at

$$z^2 = -3.8428 \pm i\ 9.8215.$$

5. Proposition 2 and Theorem 3 remain true if the requirement that ψ be even is dropped. In other words: No new expanding eigenvectors are introduced if we let $DT(g)$ act on functions which are not necessarily even (but which vanish *to second order* at 0).

6. Theorem 4 can be extended considerably. To formulate the extension, we need the theory of kneading sequences for unimodal mappings as developed, for example, in Chapter III.1 of Collet-Eckmann **[3]**. Let $\underline{K}$ be a finite kneading sequence. Except for the simple case $\underline{K} = RC$, there are associated with $\underline{K}$ three hypersurfaces in $\mathcal{M}$:

The set of superstable ψ's with kneading sequence $\underline{K}$.

The saddle-node or period-doubling bifurcation surface where the attracting periodic orbit passing through the critical point on the preceding surface appears.

The period-doubling bifurcation surface where that periodic orbit becomes unstable.

It can be shown that, intuitively, the unstable manifold crosses these three surfaces for each $\underline{K}$; a precise version of this statement must be formulated with the same circumspection as Theorem 4. There is no reason to doubt that these crossings are all transverse.

A simple argument using the apparatus developed in **[3]** reduces the proof of Theorem 4 and the above extension to establishing the existence, on the local unstable manifold, of one point whose kneading sequence strictly precedes, and one whose kneading sequence strictly follows, that of g (in the combinatorial ordering for kneading sequences). The proof proceeds by finding with sufficient precision two points on the unstable manifold and computing initial segments of their kneading sequences.

7. Although done by computer, the computations involved in proving the results stated are just on the boundary of what it is feasible to verify by hand. I estimate that a carefully chosen minimal set of estimates sufficient to prove Theorems 1 and 3 could be carried out, with the aid only of a nonprogrammable calculator, in a few days.

ACKNOWLEDGEMENTS. It is a pleasure for me to thank:

P. Collet, J. P. Eckmann, H. Epstein, D. Ruelle, and S. Smale for helpful discussions and encouragement.

L. Michel for his assistance in making available the computing facilities needed to carry out this work.

Director N. Kuiper for his very gracious hospitality at the IHES.

The Stiftung Volkswagenwerk for financial support during my visit to the IHES.

The National Science Foundation for continuing financial support under Grant MCS 78-06718.

REFERENCES

1. M. Campanino, H. Epstein and D. Ruelle, *On Feigenbaum's functional equation*, (IHES preprint P/80/32 (1980)) Topology (to appear).

2. M. Campanino and H. Epstein, *On the existence of Feigenbaum's fixed point*, (IHES preprint P/80/35 (1980)) Comm. Math. Phys. (1981), 261–302.

3. P. Collet and J. P. Eckmann, *Iterated maps of the interval as dynamical systems*, Birkhäuser, Boston-Basel-Stuttgart, 1980.

4. P. Collet, J. P. Eckmann and O. E. Lanford, *Universal properties of maps on an interval*, Comm. Math. Phys. **76** (1980), 211–254.

5. M. Feigenbaum, *Quantitative universality for a class of non-linear transformations*, J. Statist. Phys. **19** (1978), 25–52.

6. ———, *The universal metric properties of non-linear transformations*, J. Statist. Phys. **21** (1979), 669–706.

7. O. E. Lanford, *Remarks on the accumulation of period-doubling bifurcations*, Mathematical Problems in Theoretical Physics, Lecture Notes in Physics, vol. 116, Springer-Verlag, Berlin and New York, 1980, pp. 340–342.

8. ———, *Smooth transformations of intervals*, Séminaire Bourbaki, 1980/81, No. 563, Lecture Notes in Math., vol. 901, Springer-Verlag, Berlin, Heidelberg and New York, 1981, pp. 36–54.

INSTITUT DES HAUTES ETUDES SCIENTIFIQUES, 35, ROUTE DE CHARTRES, 91440, BURES-SUR-YVETTE, FRANCE

DEPARTMENT OF MATHEMATICS, UNIVERSITY OF CALIFORNIA, BERKELEY, CALIFORNIA 94720 (Current address)

Editor's note. There is a considerable mathematical literature proving the existence of universal functions, such as Collet, Eckmann and Lanford (1980) and Campanino and Epstein (1981). We also recommend Vul and Khanin's (1982) and Golberg, Sinai and Khanin's (1983) formulation of the universality theory.

Communications in Mathematical Physics **77** 65–86 (1980)

The Transition to Aperiodic Behavior in Turbulent Systems

Mitchell J. Feigenbaum

University of California, Los Alamos Scientific Laboratory (T-DOT MS 452), P.O. Box 1663, Los Alamos, NM 87545, USA

Abstract. Some systems achieve aperiodic temporal behavior through the production of successive half subharmonics. A recursive method is presented here that allows the explicit computation of this aperiodic behavior from the initial subharmonics. The results have a character universal over specific systems, so that all such transitions are characterized by noise of a universal internal similarity.

Introduction

A variety of numerical experiments [1–6] on systems of differential equations have demonstrated that a possible route to chaotic or turbulent behavior is a cascade of successive half harmonics of a basic mode with turbulence commencing after an infinite halving has produced non-periodic behavior. Moreover, a recent experiment [7] on Rayleigh Bénard convection has also exhibited these half harmonics as the behavior determining the onset of turbulence in the fluid. In this paper we draw upon some mathematics intimately connected with period doubling, and determine the time fluctuation spectrum of such a system at the onset of turbulence [8]. This spectrum proves to be of a *universal* construction, so that no specific formulas for the differential system are ever encountered. On the other hand, the theory presented is asymptotic and recursive, so that it requires as input the specific spectrum after several stages of period doubling. We are not concerned with this aspect here, taking this input for granted after which the entirety of the behavior through the transition is computed.

The paper is divided into three main parts. The first *assumes* the theory later to be exposited in order to present the new results with the least dedication required of the reader. The principal results are the formulae (10) and (17) which determine the subharmonic spectrum recursively. The second section reviews the universality theory for one-dimensional maps, and constructs the basic scaling function required in the first part. Finally, the last section consists of an argument establishing the relevance of one-dimensional maps to the original system of differential equations, resting upon very recent work of Collet et al. [9]. The correct formula to which (10) is a rough approximation is also determined.

Subharmonic Trajectory Scaling

We are considering a system of differential equations

$$\dot{x}_i = f_i(x_1, \ldots, x_N, \lambda) i = 1, \ldots, N \tag{1}$$

for which we know that at λ_n each $x_i(t)$ is periodic with period $T_n = 2^n T_0$ and $\lambda_n \to \lambda_\infty < \infty$. (For each T_n there is a range of suitable λ's: λ_n is chosen for each n to produce identical stabilities.) As $n \to \infty$, $x_i(t)$ is becoming aperiodic: it possesses a fundamental at $\omega_n = \omega_0/2^n$ with equally spaced harmonics, with 2^n harmonics up to ω_0, so that as $n \to \infty$, the spectrum becomes continuous. We are, in particular, interested in computing this part of the spectrum – the *subharmonics* of ω_0 – that determines the noise (or fluctuations).

Consider the system at $\lambda_n : x_i(t + T_n) = x_i(t)$, and the motion can be divided into 2^n roughly similar "cycles" of duration T_0. At λ_{n+1}, $x_i(t + T_n)$ no longer equals $x_i(t)$, but "almost" does, another T_n "cycles" required until x_i again equals $x_i(t)$. Accordingly, we focus attention on

$$\psi^{(n)}(t) \equiv x^{(n)}(t) - x^{(n)}(t + T_{n-1}) \tag{2}$$

which along the entire trajectory measures x's failure to duplicate itself after one half of its true period, T_n. Observe, by periodicity, that

$$\psi^{(n)}(t + T_n) = \psi^{(n)}(t)$$

and (3)

$$\psi^{(n)}(t + T_{n-1}) = -\psi^{(n)}(t).$$

Now imagine that the "way" in which $\psi^{(n)}$ fails to vanish is the same way in which $\psi^{(n+1)}$ will fail to vanish. More precisely, *assume* that

$$\psi^{(n+1)}(t) \cong \sigma(t/T_{n+1}) \psi^{(n)}(t) \tag{4}$$

so that $\psi^{(n+1)}$ is built of two displaced copies of $\psi^{(n)}(t)$ suitably scaled. By (3) it follows that

$$\begin{aligned} \sigma(x+1) &= \sigma(x) \\ \sigma(x+1/2) &= -\sigma(x). \end{aligned} \tag{5}$$

Also, by (4)

$$r_{n+1}(2t) \equiv \frac{\psi^{(n+1)}(2t)}{\psi^{(n)}(2t)} \cong \sigma(2t/T_{n+1}) = \sigma(t/T_n) \cong \frac{\psi^{(n)}(t)}{\psi^{(n-1)}(t)} = r_n(t) \tag{6}$$

so that with each r_n plotted against a scaled time, x, (for which $T_n = 1$), the curves $r_n(t)$ should be identical and equal to $\sigma(x)$. Figure 1 depicts the degree to which this is true for $n = 4, 5$ in the case of Duffing's equation, where λ_n has been chosen to determine that 2^n-cycle most quickly converged to from initial conditions within the basin of this cycle. In fact there is a large class of systems (1) for which Eq. (4) is correct. So, assuming that (4) is correct (for large n), let us see what form of fluctuation spectrum is implied.

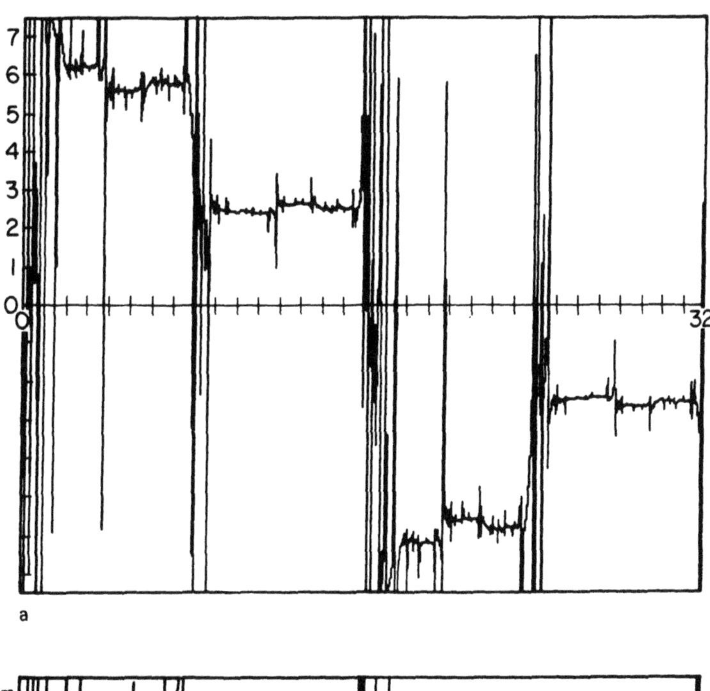

a

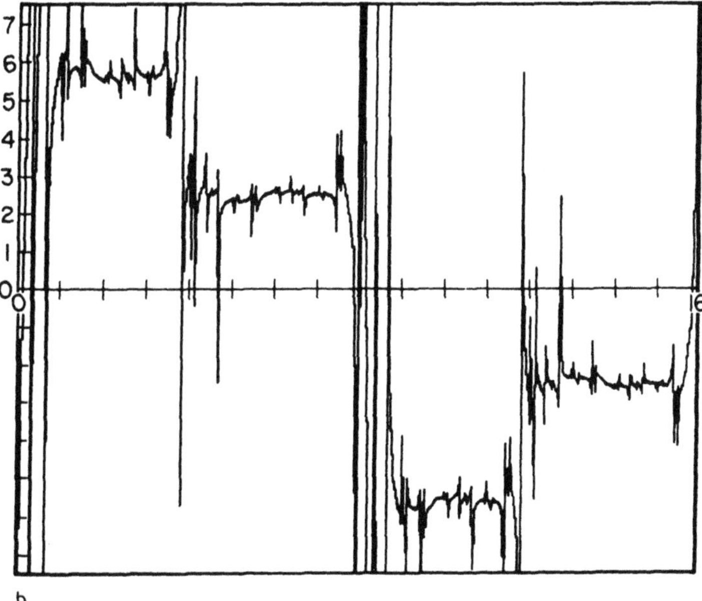

b

Fig. 1. a $r_5^{-1}(t)$; **b** $r_4^{-1}(t)$; r_n^{-1} as obtained by the direct numerical division of solutions to Duffing's equation. The divergences represent slight mismatches of zeroes, and are, for smooth integrals of (4), irrelevant

First of all, the even and odd Fourier components of $x^{(n)}$ are fundamentally of different characters. This is so simply because at λ_{n+1}, the *odd* harmonics of the fundamental, $\omega_0/2^{n+1}$, are all absent in the spectrum at λ_n, since at λ_n the fundamental is $\omega_0/2^n = 2\omega_0/2^{n+1}$. Thus at λ_{n+1}, a set of *new* components beyond those of λ_n are introduced, while those present at λ_n are located coincidentally with the *even* harmonics at λ_{n+1}. In fact, to a first approximation, the even components at λ_{n+1} are just all the old components at λ_n, so the main task in determining the spectrum at λ_{n+1} is the computation of the new components. Let us now formally render these verbal remarks, in order to see what we must compute. By definition, the kth Fourier component of $x^{(n)}(t)$ is

$$x_k^{(n)} \equiv \int_0^{T^n} \frac{dt}{T_n} x^{(n)}(t) e^{-2\pi ikt/T_n}. \tag{7}$$

For our purposes, we manipulate (7) by splitting the integral into two halves, and shifting the upper half, producing,

$$x_k^{(n)} = \int_0^{T_{n-1}} \frac{dt}{2T_{n-1}} [x^{(n)}(t) + (-1)^k x^{(n)}(t+T_{n-1})] e^{-\pi ikt/T_{n-1}}. \tag{8}$$

Consider first the even harmonics of $x^{(n+1)}$. By (8),

$$x_{2k}^{(n+1)} = \int_0^{T_n} \frac{dt}{2T_n} [x^{(n+1)}(t) + x^{(n+1)}(t+T_n)] e^{-2\pi ikt/T_n}. \tag{9}$$

However, after T_n, $x^{(n+1)}$ has almost repeated itself (σ is small), so that to first approximation,

$$x^{(n+1)}(t) \cong x^{(n+1)}(t+T_n) \cong x^{(n)}(t)$$

and by (9),

$$x_{2k}^{(n+1)} \cong \int_0^{T_n} \frac{dt}{T_n} x^{(n)}(t) e^{-2\pi ik/T_n} = x_k^{(n)}. \tag{10}$$

However, the fundamental of $x^{(n+1)}$ is $\omega_0/2^{n+1}$, so that its $2k^{\text{th}}$ harmonic is at $\omega = k\omega_0/2^n$ which is the *kth* harmonic of the fundamental $\omega_0/2^n$ of $x^{(n)}$. Thus, as verbally stated, to first approximation, the even harmonics at λ_{n+1} are just the spectrum at λ_n.

Accordingly, we now turn to the serious computation, that for the odd harmonics. By (8),

$$x_{2k+1}^{(n+1)} = \int_0^{T_n} \frac{dt}{2T_n} [x^{(n+1)}(t) - x^{(n+1)}(t+T_n)] e^{-\pi i(2k+1)t/T_n}. \tag{11}$$

The integrand of (11) is $\psi^{(n+1)}(t)$, so that by (4),

$$x_{2k+1}^{(n+1)} \cong \int_0^{T_n} \frac{dt}{2T_n} \sigma(t/2T_n)[x^{(n)}(t) - x^{(n)}(t+T_{n-1})] e^{-\pi i(2k+1)t/T_n}. \tag{12}$$

Using the inverse of (7),

$$x^{(n)}(t) = \sum_k x_k^{(n)} e^{2\pi ikt/T_n}$$

we have

$$x^{(n)}(t)-x^{(n)}(t+T_{n-1})=2\sum_{k} x_{2k+1}^{(n)} e^{2\pi i(2k'+1)t/T_n}. \tag{13}$$

Substituting (13) into (12),

$$x_{2k+1}^{(n+1)} \cong \sum_{k} x_{2k'+1}^{(n)} \int_0^{T_n} \frac{dt}{T_n} \sigma(t/2T_n) e^{2\pi i[(2k'+1)-1/2(2k+1)]t/T_n},$$

which, through the substitution $\xi = t/T_n$, becomes

$$x_{2k+1}^{(n+1)} \cong \sum_{k'} x_{2k'+1}^{(n)} \int_0^1 d\xi \sigma(\xi/2) e^{2\pi i\xi[(2k'+1)-1/2(2k+1)]}. \tag{14}$$

Thus, the scaling law (4) allows the computation of the spectrum of $x^{(n+r)}(t)$ for all $r \geqq 1$, given the spectrum of $x^{(n)}(t)$. Equation (14) has two significant properties. First, only the odd components of $x^{(n)}$ are required to determine the new (and odd) components of $x^{(n+1)}$. This is important because the "cycle" frequency, ω_0, is the 2^n harmonic of ω_n and so *even*. That is, the basic "cycle"'s spectrum is *decoupled* from (14), so that the noise spectrum introduced at each n is roughly independent of the "coherent" spectrum of the basic "cycle". Also significant is that the recursion (14) is *independent of n* (i.e. "autonomous").

It should be pointed out that if (4) is asymptotically exact as $n \to \infty$, then (14) is similarly exact. Formula (10) however is genuinely approximate, and we defer until later the correct formula, since an extra ingredient is required.

By Fig. 1, σ is marked by discontinuities (as $n \to \infty$), roughly constant at one value for $0<x<1/4$ and constant at another for $1/4<x<1/2$. In a next approximation, each of these quarters is decomposable into two halves, and so forth. Thus, σ is representable as

$$\sigma(x)=\sum \sigma_i \theta(x_{i+1}-x)\theta(x-x_i). \tag{15}$$

Let us use (15) to compute the transform integral of (14):

$$I([\ ]) \equiv \int_0^1 d\xi \sigma(\xi/2) e^{2\pi i\xi[\ ]} = \frac{1}{2\pi i[\ ]} \sigma(\xi/2) e^{2\pi i\xi[\ ]}\Big|_0^1 - \frac{1}{4\pi i[\ ]} \int_0^1 d\xi \sigma'(\xi/2) e^{2\pi i\xi[\ ]}.$$

Since $e^{2\pi i[(2k'+1)-1/2(2k+1)]} = -1$,

$$I([\ ]) = -\frac{1}{2\pi i[\ ]}\left\{(\sigma(1/2^-)+\sigma(0^+)) + 1/2\int_0^1 d\xi \sigma'(\xi/2) e^{2\pi i\xi[\ ]}\right\}.$$

By (15),

$$\sigma'(x)=\sum \sigma_i(\delta(x-x_i)-\delta(x-x_{i+1}))=\sum(\sigma_i-\sigma_{i-1})\delta(x-x_i) \equiv \sum d_i \delta(x-x_i)$$

so that

$$1/2\int_0^1 d\xi \sigma'(\xi/2) e^{2\pi i\xi[\ ]} = \sum_i d_i \int_0^1 \frac{d\xi}{2} \delta(\xi/2-x_i) e^{2\pi i\xi[\ ]} = \sum_i (d_i) e^{4\pi i x_i[\ ]}.$$

Accordingly, (14) becomes

$$x_{2k+1}^{(n+1)} \cong -\frac{1}{2\pi i}\sum_{k} x_{2k'+1}^{(n)} \frac{1}{2k'+1-1/2(2k+1)} \cdot \left\{(\sigma(1/2^-)+\sigma(0^+))+\sum_i d_i e^{4\pi i x_i[(2k'+1)-1/2(2k+1)]}\right\}. \tag{16}$$

From (16) follow various approximations by including only those discontinuities d_i in excess of a certain amount. As a first approximation, we include only the largest discontinuity at $x=1/4$, for which the exponential factor is

$$e^{\pi i[(2k'+1)-1/2(2k+1)]}=i(-1)^k,$$

so that (16) becomes

$$x_{2k+1}^{(n+1)} \cong -\frac{1}{2\pi i}[(\sigma(1/2^-)+\sigma(0^+))+i(-1)^k(\sigma(1/4^+)-\sigma(1/4^-))] \cdot \sum_{k'} \frac{x_{2k'+1}^{(n)}}{2k'+1-1/2(2k+1)} \tag{17}$$

[To gauge the accuracy of (17), given the $n=4$ spectrum for Duffing's equation, the $n=5$ spectrum is determined to within 5% of the correct values for both amplitude and phase, with the logarithm of amplitudes determined to about 0.5%.] As $T_n \to \infty$ (7) determines x_k analytically continued in k into the lower half plane, so that

$$-\frac{P}{\pi i}\int \frac{dk'}{k'-k}x(k')=x(k).$$

An integral approximation to the sum of (17) immediately produces

$$|x_k^{(n+1)}| \cong \tfrac{1}{4}\sqrt{(\sigma(1/2^-)+\sigma(0^+))^2+(\sigma(1/4^+)-\sigma(1/4^-))^2}\,|x^{(n)}k/2| \equiv \mu|x^{(n)}k/2|. \tag{18}$$

The meaning of (18) is: smoothly interpolate the *odd* components of $x^{(n)}$ and rescale by μ; the *odd* components of $x^{(n+1)}$ are "in the mean" the values of this new curve at the appropriate (odd) frequencies. Reducing by another factor of μ determines the $(n+2)$ spectrum and so forth ad infinitum. Accordingly, the fluctuation spectrum (approximately) has a simple self-similar character.

Thus, Fig. 1 completely determines the fluctuations spectrum for Duffing's equation as $n \to \infty$. What is remarkable though, is that Fig. 1 [i.e. $\sigma(x)$] is in fact *universal* over all systems (1) possessing order 2^n subharmonic production sequentially as $\lambda_n \to \lambda_\infty$! In the next section, we partially prove this result and compute $\sigma(x)$ in a universal format. Anticipating this computation, let us record here the quantities of (17) and (18) that will follow.

$$\sigma(1/2^-)=0.3995 \qquad \sigma(0^+)=[\sigma(1/2^-)]^2=0.1596$$

$$\sigma(1/4^+)=0.4191 \qquad \sigma(1/4^-)=0.1752$$

$$\mu=0.1525$$

$$-10\log_{10}\mu=8.17\,\text{db}.$$

For log-amplitudes, (18) simply means that the new components at any level n define roughly the same interpolation as the previous level's new components, but shifted 8.2 db downwards. We now proceed with the theoretical notions and the computation of $\sigma(x)$.

Universality Theory

An elementary class of systems exhibiting successive period doublings has been known to exist for some time [10, 11]. These are one-dimensional discrete systems: a one parameter family of (non-invertible) maps on an interval of the real line. For fixed parameter λ, a dynamics is determined by successive iteration of the map,

$$x_{n+1}=f(\lambda, x_n).$$

f has the crucial property of mapping an interval several-to-one onto itself, accomplished by f's attaining an extremum within the interval. For a large class of such f's there exists a monotone sequence of parameter values λ_n such that at λ_n a maximally stable 2^n-cycle exists (or more generally, for any r, an $r\cdot 2^n$ cycle for appropriate λ_n).

Several years ago this author discovered that beyond this universal doubling property, the maps of this class possess a host of universal metric properties [12–15]. More precisely, with $\hat{x}$ the location of the extremum of f, all f's with

$$|f(x)-f(\hat{x})|\propto|x-\hat{x}|^z(z>1), \quad x\sim\hat{x}$$

(for a normal *quadratic* extremum, $z=2$) for the same z share identical metric properties. Thus, with $\lambda_n\to\lambda_\infty$, it is a consequence of this theory that

$$|\lambda_n-\lambda_\infty|\propto\delta^{-n} \tag{19}$$

with δ a function only of z. (For $z=2$, $\delta=4.6692016\ldots$.) That is, as $n\to\infty$, λ_n converges to λ_∞ at a universal geometric rate independent of the global properties of f.

By way of review, recall that a map possesses an n-cycle if there are n points x_0, $x_1, \ldots, x_{n-1}$ such that

$$x_0\xrightarrow{f}x_1\xrightarrow{f}\cdots\xrightarrow{f}x_r\xrightarrow{f}x_{r+1}\longrightarrow\cdots\xrightarrow{f}x_{n-1}\xrightarrow{f}x_0.$$

Denoting the nth iterate of f by f^n:

$$f^n(x)=f(f^{n-1}(x)); \quad f^0(x)=x,$$

where the n elements of the cycle satisfy

$$f^n(x_r)=x_r \quad r=0,\ldots,n-1$$

i.e. each element is a fixed point of the nth iterate of f. Accordingly, the *stability* of a cycle is the stability of each element of the cycle viewed as a fixed point of f^n. If x^* is a fixed point of f, stability is determined by linear approximation about x^*:

writing $x_n = x^* + \xi_n$,

$$\xi_{n+1} \cong f'(x^*)\xi_n$$
$$\rightarrow \xi_n \cong \xi_n [f'(x^*)]^n .$$

x^* is stable if $x_n \rightarrow x^*$ for x_0 sufficiently close to x^*. The criterion for stability is evidently

$$|f'(x^*)| < 1$$

while $f'(x^*) = 0$ is the condition for greatest stability. For an n-cycle the condition for stability is then

$$|f^{n'}(x_r)| < 1 \text{ for each } r .$$

Indeed, by the chain rule, $f^{n'}$ is independent of r with

$$f^{n'}(x_r) = f'(x_0) f'(x_1) \cdots f'(x_{n-1}) .$$

Accordingly, a most stable n-cycle is one containing the point $\hat{x}$, and the parameter value λ_n determined as a zero of

$$F_n(\lambda_n) \equiv f^{2^n}(\lambda_n, \hat{x}) - \hat{x} . \tag{21}$$

Evidently λ_r, $r = 0, 1, \ldots, n-1$ are also zeroes of (20) together with many other values. Thus, λ_n is defined recursively: λ_0 is in general unique; λ_1 is a different zero of F_1 closest to λ_0, etc. with λ_n monotone in n. As n becomes large, F_n is increasingly time-consuming to compute while its zeroes become arbitrarily close. For large n, it is essentially impossible to locate λ_n without knowledge of (19), which is used as

$$\lambda_{n+2} - \lambda_{n+1} \cong \delta^{-1}(\lambda_{n+1} - \lambda_n)$$

to predict the next value given two previous values. (Indeed, a host of lengthy numerical studies have, in the past, observed the pattern 1, 2, aperiodic simply because $\lambda_\infty - \lambda_1 \sim \lambda_1 - \lambda_0$, while equal-parameter increment searches were performed.)

The theory actually determines (19) (and δ) secondarily, with a scaling phenomenon on the elements of a cycle primary. At λ_n a maximally stable 2^n cycle exists. As λ is increased through some interval, a stable 2^n-cycle persists until at Λ_n it loses stability "bifurcating" into a 2^{n+1} cycle. For λ slightly above Λ_n, 2^n iterations map x_0 into a point arbitrarily near to x_0 for λ arbitrarily near to Λ_n. Indeed, throughout the interval of λ up to Λ_{n+1}, $f^2(x_r)$ is that element of the 2^{n+1} cycle nearest to but distinct from x_r, and

$$\psi_r^{(n+1)} \equiv x_r - f^{2^n}(x_r) (\Lambda_n < \lambda < \Lambda_{n+1}) \tag{22}$$

is a measure of deviation from 2^n-cycle behavior throughout the cycle. [The reader will of course realize that (22) is simply the discrete, one-dimensional version of (2).] The basic result of the theory is that

$$\hat{x} - f^{2^n}(\lambda_{n+1}, \hat{x}) \cong -\alpha^{-1}(\hat{x} - f^{2^{n-1}}(\lambda_n, \hat{x})) \tag{23}$$

[compare with Eq. (4)] where α enjoys a universality identical to δ's, with value (for $z=2$)

$$\alpha = 2.502907875\ldots.$$

Moreover, in units of the distance between $\hat{x}$ and $f^{2^{n-1}}(\lambda_n, \hat{x})$ the locations of the elements of the 2^n-cycle about $\hat{x}$ are universally determined through a function $g_0(x)$ where theory determines that

$$\lim_{n\to\infty} (-\alpha)^n[f^{2^n}(\lambda_n, \hat{x}+\xi/(-\alpha)^n)-\hat{x}] = \nu g_0(\xi/\nu) \tag{24}$$

with the magnification ν the only f-dependent ingredient. [Upon remagnification, so that $\hat{x}=f^{2^{n-1}}(\lambda_n, \hat{x})\to 1$ (i.e. setting the scale as above), a universal limit is obtained.] The elements of the 2^n cycle have the property of being fixed points of g_0 at its extrema, while g_0 itself can be universally computed by the theory. Related to g_0 is a sequence of *universal* functions g_r where

$$\lim_{n\to\infty} (-\alpha)^n[f^{2^n}(\lambda_{n+r}, \hat{x}+\xi/(-\alpha)^n)-\hat{x}] = \nu g_r(\xi/\nu) \tag{25}$$

with the same ν as in (24). [The universal scale is simply $g_1(0)=1$.] These functions serve as a basis as which the operator of functional composition and rescaling becomes the shift:

$$g_{r-1}(x) = -\alpha g_r(g_r(-x/\alpha)) \equiv O[g_r]\,. \tag{26}$$

Since g_r is universal, and since for an f symmetric about $\hat{x}$, the limit in (25) is a function symmetric in ξ, each g_r is a symmetric function, so that (26) can be written as

$$g_{r-1}(x) = -\alpha g_r(g_r(x/\alpha))\,. \tag{26'}$$

In particular

$$g(x) \equiv \lim_{r\to\infty} g_r(x)$$

is the fixed-point of O:

$$g(x) = -\alpha g(g(x(\alpha)))\,. \tag{27}$$

(27) admits of a unique solution for α *and* $g(x)$ for $g(x)$ with z^{th} order extremum at 0, symmetric, and of sufficient smoothness. g_r for large r is obtained by linearizing O about g and studying the spectrum. Indeed the linearized operator has a unique eigenvalue in excess of one, this eigenvalue δ of (19). (References [13, 14] maintained the uniqueness of δ as a necessary conjecture for the validity of the entirety of this theory. Recently this conjecture has been rigorously proven [16, 17], although only for z sufficiently close to one.)

The verbal content of (23) is that points near $\hat{x}$ scale in successive bifurcations by $-\alpha$. Since these points are imaged by f to a right-most cluster of points (taking the extremum to be a maximum), and f has a zth order extremum, it follows that points about $f(\hat{x})$ must scale by α^z, i.e.

$$f(\lambda_{n+1}, \hat{x}) - f^{2^n}(\lambda_{n+1}, f(\lambda_{n+1}, \hat{x})) \cong \alpha^{-z}(f(\lambda_n, \hat{x}) - f^{2^{n-1}}(\lambda_n, f(\lambda_n, \hat{x})))\,. \tag{28}$$

The next several images of this cluster of points are through a slope of an infinitesimal linear stretch of f, and so also scale with α^z. Similarly, several *pre images* of the cluster about $\hat{x}$ scale with $-\alpha$ (several here means r, with $\frac{1}{n}\log r \ll 1$ for a 2^n-cycle). Our immediately task is to determine this scale factor along the entirety of the cycle. Towards this end write

$$\psi_r^{(n+1)} \equiv \sigma_r^{(n+1)} \psi_r^{(n)}; \tag{29}$$

setting $x_0 \equiv \hat{x}$. By (23) $\sigma_0^{(n)} = -\alpha^{-1}$, $\sigma_1^{(n)} = \alpha^{-z}$. Also, $\sigma_1^{(n)} = \sigma_2^{(n)} = \ldots$ and $\sigma_0^{(n)} = \sigma_{-1}^{(n)} = \sigma_{-2}^{(n)} = \ldots$, as $n \to \infty$. By the definitions of ψ in Eq. (22)

$$\begin{aligned}\psi_{2^n+r}^{(n+1)} &= f^{2^n}(\lambda_{n+1}, x_r) - f^{2^n}(\lambda_{n+1}, f^{2^n}(\lambda_{n+1}, x_r)) \\ &= f^{2^n}(\lambda_{n+1}, x_r) - f^{2^{n+1}}(\lambda_{n+1}, x_r) \\ &= f^{2^n}(\lambda_{n+1}, x_r) - x_r\end{aligned}$$

or,

$$\psi_{2^n+r}^{(n+1)} = -\psi_r^{(n+1)},$$

while

$$\psi_{2^{n+1}+r}^{(n+1)} = \psi_r^{(n+1)}. \tag{30}$$

Thus,

$$\sigma_{2^n+r}^{(n+1)} = -\sigma_r^{(n+1)} \tag{31}$$

and in particular,

$$\sigma_{2^n}^{(n+1)} = \alpha^{-1}. \tag{32}$$

It is clear that deviations from $-\alpha^{-1}$, α^{-z} can occur only for r such that

$$\lim_{n\to\infty} r/2^n \neq 0.$$

Accordingly, choose r as

$$r_m \equiv 2^{n-m} \tag{33}$$

so that

$$\begin{aligned}\psi_{r_m}^{(n)} &= f^{2^{n-m}}(\lambda_n, \hat{x}) - f^{2^{n-1}}(\lambda_n, f^{2^{n-m}}(\lambda_n, \hat{x})) \\ &= f^{2^{n-m}}(\lambda_n, \hat{x}) - f^{2^{n-m}}(\lambda_n, f^{2^{n-1}}(\lambda_n, \hat{x})).\end{aligned} \tag{34}$$

By (25), $f^{2^{n-1}}(\lambda_n, \hat{x}) \sim \frac{\nu}{(-\alpha)^{n-1}} g_1(0) + \hat{x}$ so that, again by (25),

$$\begin{aligned}f^{2^{n-m}}(\lambda_n, f^{2^{n-1}}(\lambda_n, \hat{x})) &\sim \frac{\nu}{(-\alpha)^{n-m}} g_m\left[\frac{(-\alpha)^{n-m}}{\nu} \frac{\nu}{(-\alpha)^{n-1}} g_1(0)\right] + \hat{x} \\ &= \frac{\nu}{(-\alpha)^{n-m}} g_m(\alpha^{1-m} g_1(0)) + \hat{x}.\end{aligned}$$

Similarly,

$$f^{2^{n-m}}(\lambda_n, \hat{x}) \sim \frac{\nu}{(-\alpha)^{n-m}} g_m(0) + \hat{x}$$

so that

$$\psi^{(n)}_{r_m} \sim \frac{\nu}{(-\alpha)^{n-m}} [g_m(0) - g_m(\alpha^{1-m} g_1(0))]. \tag{35}$$

At $n+1$, to maintain the same number of iterations into the cycle, r_m must be unchanged, so that by (33) $m \to m+1$, and (35) becomes

$$\psi^{(n+1)}_{m+1} \sim \frac{\nu}{(-\alpha)^{n-m}} [g_{m+1}(0) - g_{m+1}(\alpha^{-m} g_1(0))]. \tag{36}$$

Finally, by (29)

$$\sigma^{(n+1)}_{r_m} \sim \frac{g_{m+1}(0) - g_{m+1}(\alpha^{-m} g_1(0))}{g_m(0) - g_m(\alpha^{1-m} g_1(0))}$$

which as $n \to \infty$ is exact. Defining $x_m \equiv r_m/2^{n+1} = 2^{-m-1}$, $\sigma(x_m) \equiv \lim_{n\to\infty} \sigma^{(n)}_{r_m}$

$$\sigma(2^{-m-1}) = \frac{g_{m+1}(0) - g_{m+1}(\alpha^{-m} g_1(0))}{g_m(0) - g_m(\alpha^{1-m} g_1(0))}. \tag{37}$$

($\sigma^{(n)}_r$ is defined on integers r; we have rescaled r by 2^n so that an entire cycle, for any n occupies $x \in [0, 1]$. By (31) $\sigma(x)$ obeys (5) for, as $n \to \infty$, all $x \in [0, 1]$.) Since the g_m are all universal, it now follows that $\sigma(2^{-m})$ is universal. For $m = 0$, (37) reads

$$\sigma(1/2) = \frac{g_1(0) - g_1(g_1(0))}{g_0(0) - g_0(\alpha g_1(0))}.$$

But, by (24), $g_0(0) = 0$, while by (26),

$$0 = g_0(0) = -\alpha g_1(g_1(0)) \qquad g_1(g_1(0)) = 0$$

$$g_0(\alpha g_1(0)) = -\alpha g_1(g_1(g_1(0))) = -\alpha g_1(0)$$

so that

$$\sigma(1/2) = \alpha^{-1}$$

which is obviously correct since $\sigma(1/2) = -\sigma(0^-) = -(-\alpha^{-1})$. Also for $m \to \infty$

$$\sigma(0^+) = \lim_{m\to\infty} \frac{g(0) - g(\alpha^{-m} g_1(0))}{g(0) - g(\alpha^{1-m} g_1(0))}$$

But for x small, $g(x) = g(0) + 1/2 g''(0)x^2 + \dots$, so that $\sigma(0^+) = \alpha^{-2}$. However, $m = 1$ is non-trivial and represents a drastic change in scaling half-way around the cycle:

$$\sigma(1/4) = \frac{g_2(0) - g_2(g_1(0)/\alpha)}{g_1(0) - g_1(g_1(0))}.$$

This requires a computation using a numerically computed g_2, with result

$$\sigma(1/4) = 0.1752\ldots .$$

We now compute $\psi^{(n)}_{r_{m+1}}$, which through interchanges of numbers of iterations yields

$$\psi^{(n)}_{r_{m+1}} \sim f\left(\lambda_n, \hat{x} + \frac{\nu}{(-\alpha)^{n-m}} g_m(0)\right) - f\left(\lambda_n, \hat{x} + \frac{\nu}{(-\alpha)^{n-m}} g_m(\alpha^{1-m} g_1(0))\right)$$

$$\sim \tfrac{1}{2} f''(\lambda_n, \hat{x}) \frac{\nu^2}{(\alpha^2)^{n-m}} [g_m^2(0) - g_m^2(\alpha^{1-m} g_1(0))]$$

so that

$$\sigma^{(n+1)}_{r_{m+1}} \sim \frac{g^2_{m+1}(0) - g^2_{m+1}(\alpha^{-m} g_1(0))}{g_m^2(0) - g_m^2(\alpha^{1-m} g_1(0))}$$

which produces for $n \to \infty$

$$\sigma(2^{-m-1} + 0) = \frac{g^2_{m+1}(0) - g^2_{m+1}(\alpha^{-m} g_1(0))}{g_m^2(0) - g_m^2(\alpha^{1-m} g_1(0))}. \tag{38}$$

Special cases of (38) are

$$\sigma(1/2 + 0) = -\alpha^{-2}$$

and $\sigma(1/4 + 0) = 0.4191\ldots$.

Thus, at 1/4 an abrupt change from small to large scaling occurs: in essentially one iteration the string of images of the right most cluster suddenly goes over into the string of pre-images leading to $\hat{x}$.

So far, the largest discontinuity of $\sigma(x)$ has been computed, as well as successively smaller discontinuities at $x = 2^{-m-1}$. It is now straight forward to compute the discontinuities at *any* rational x given its binary representation. All that is necessary is to write

$$r^{(n)}_{\{m\}} = 2^{n-m_1} + 2^{n-m_2} + \ldots$$

and observe that

$$x^{(n)}_{\{m\}} = f^{2^{n-m_1}} \circ f^{2^{n-m_2}} \circ \ldots (\hat{x})$$

where $\circ$ denotes functional compositions. Repeated use of (25) then produces

$$\sigma(2^{-m_1-1} + 2^{-m_2-1} + \ldots)$$
$$= \frac{[\ldots g_{m_2+1}(\alpha^{m_1-m_2} g_{m_1+1}(0))\ldots)] - [\ldots g_{m_2+1}(\alpha^{m_1-m_2} g_{m_1+1}(\alpha^{-m_1} g_1(0))| \ldots)]}{[\ldots g_2(\alpha_m^{m_1-m_2} g_1(0))_m \ldots)] - [\ldots g_2(\alpha_m^{m_1-m_2} g_1(\alpha_m^{1-m_1} g(0)))_1 \ldots)]} \tag{39}$$

for the value of σ to the left of its discontinuity at x, while σ to the right of the discontinuity is simply (39) with each bracketed term squared. (These "composite" discontinuities rapidly decrease in amplitude.) Accordingly, $\sigma(x)$ is discontinuous on the rationals with successively milder strength in successive half fractions of intervals, but *universal*. Figure 2 depicts the universally computed $\sigma(x)$ which upon comparison with Fig. 1 is evidently the $n \to \infty$ limit of the differential equation's σ. Indeed the quoted values for μ, etc. are precisely the values of the universal σ computed on the previous few pages. At this point it is computer-experimentally plausible that $\sigma(x)$ is universal over systems (1) – Duffings's equation was drawn at random as an example. Indeed we are asserting that σ arises in Lorenz' system, in a five mode truncation of Navier-Stokes on the torus, the "three wave" problem and probably, based on the spectral test for the Bénard flow experiment, in the Navier-Stokes field equations with whatever corrections to these equations that are required to determine physical Bénard flow!

It now remains to establish why the one-dimensional universality theory presented in this section should apply to these higher dimensional flows.

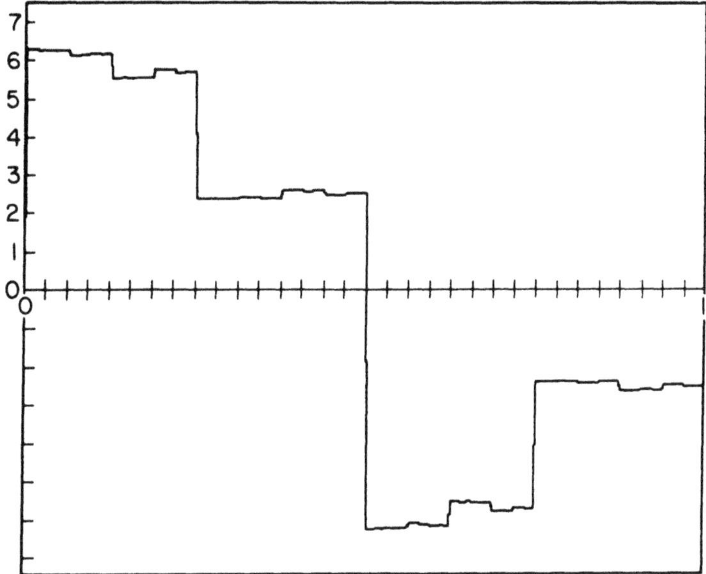

Fig. 2. σ^{-1} as universally computed

The Extension of Universality to High Dimensional Flows

Our first task is to extend the previous section to N dimensional maps [18]. Next the connection between these maps and the flow of a system of differential equations is to be drawn. Finally several details will be fixed.

The first task has been performed by Collet et al. [9]. The basic result is that if a one parameter family of maps (after a coordinate transformation if necessary) passes near to the map

$$\begin{pmatrix} x_1 \\ x_2 \\ \vdots \\ x_N \end{pmatrix} \to \begin{pmatrix} g((x_1^2 - x_2 - \ldots - x_N)^{1/2}) \\ 0 \\ \vdots \\ 0 \end{pmatrix},$$

then it will go through 2^n bifurcations ad infinitum, and exhibit essentially one dimensional behavior. That is, about a given element of a cycle, the iterated map collapses about a definite (local) *ray* in the N-dimensional space, and along this ray the mapping is one-dimensional and obeys the universality theory of the previous section. In contrast to the one-dimensional case, it is difficult in N dimensions to know if a map will be of this nature by a casual perusal of its explicit form. Numerically, however, a large variety of interesting systems exhibit this behavior.

We next turn to the connection between maps and flows. The basic idea here is the "Poincaré map". The solution to (1) is a trajectory in N dimensions which, for motions of interest here, lies in a compact region performing a "noisy" periodic motion. A map is defined by slicing transversally across the trajectory with an $N-1$ dimensional plane: each time the orbit crosses this plane in the same sense (e.g. from "below") mark its $N-1$ coordinates. The image of one such point in the plane is by definition the location of its next crossing. This map is called the Poincaré map and by elementary properties of (1) is invertible and differentiable. Should the orbit close on itself after one intersection with the plane, then the map has a fixed point; should it close after n crossings (or "basic cycles") then the map has an n-cycle. Finally, should it never close, then the map has no periodic cycle. Also, should an n-cycle of the map be stable, then by uniqueness and smoothness of the solution to (1), the orbit will converge to a definite stable orbit with identical convergence around the entire orbit. Accordingly all stability and periodicity aspects of the trajectory can be deduced from the map. It now follows by the previous paragraph that the universality theory can apply to (1). To be precise, for an appropriate λ, the trajectory returns every 2^p times successively closer to an original point along a certain ray in the $N-1$ dimensional plane. At λ_n, after 2^{n-1} crossings the trajectory has returned to the nearest adjacent point to its original crossings, and the points at 2^n, 2^{n-1}, $2^{n-2}, \ldots, 2^{n-m} m \ll n$ all lie along this special ray at spacings determined by g_0. By linearity, the same is true for any one-dimensional projection of the $N-1$ dimensional plane. Accordingly, we immediately have (19) satisfied for the parameter, and in each projection, a scaling of the form (23) from one bifurcation to the next. Now, consider two adjacent points y, z in the $N-1$ plane such that their spacing scales by σ_0. After one iteration of the map, they are adjacent points y', z' also scaling by σ_0 [i.e. they are not located at a large discontinuity of $\sigma(x)$]. Issuing out of y is a trajectory "cycle" linking it to y'; adjacent to this "cycle" is the trajectory linking z to z'. It now follows by the differential flow that these two trajectories have a spacing that also scales with σ_0 along its entire length from y to y'. Now the z adjacent to y is $f^{2^{n-1}}(y)$. But 2^{n-1} crossings later require a time $2^{n-1}T_0 \equiv T_{n-1}$ where T_0 is the time of a "basic cycle". Accordingly, $\psi^{(n)}(t)$ of Eq. (2) is precisely the spacing of the two adjacent trajectories linking y, y' and z, z'. From (29) we now immediately conclude (4) where $\sigma(x)$ *is* the universal scaling function of the previous section. Accordingly, if (1) has successive half-subharmonic behavior, then the spectrum of each projection $x_i(t)$ is determined by (16) with the universal function σ. [The map restricted to the special ray must generically be of quadratic sort, since it is differentiable, and $z \neq 2$ implies that $f''(\hat{x})=0$ which is non-generic.] These remarks, then, establish the general applicability of the one-dimensional theory.

Let us comment on some mathematical features of σ. Recall that

$$\frac{\psi^{(n+1)}(t)}{\psi^{(n)}(t)} \sim \sigma(t/T_{n+1}).$$

Denoting $x^{(n)}(0) \equiv \hat{x}_n$, $x^{(n)}(mT_0) = P^m(\hat{x})$ where P is the Poincaré map at λ_n, so that

$$\psi^{(n+1)}(mT_0) = P^m(\hat{x}_{n+1}) - P^m \circ P^{2^n}(\hat{x}_{n+1}).$$

(P is always understood to be the map at the λ value corresponding to the subscript of $\hat{x}$.) Thus,

$$\sigma(mT_0/T_{n+1}) = \sigma(m/2_{n+1}) \cong \frac{P^m(\hat{x}_{n+1}) - P^m \circ P^{2^n}(\hat{x}_{n+1})}{P^m(\hat{x}_n) - P^m \circ P^{2^{n-1}}(\hat{x}_n)}.$$

That is, σ is determined at the $n+1^{\text{st}}$ level of bifurcation at the rationals whose denominator is 2^{n+1} *purely* by the Poincaré map, and as $n \to \infty$ at the rationals generally purely by the Poincaré map with σ determined at all reals by its continuous extension. Moreover, at $m = 2^{n-r} + 2^{n-s} + \ldots$ for a *finite* such sum and $r, s \ldots \ll n$, as $n \to \infty \sigma$ converges to the universality values (and discontinuities) of (39). Also, for $m = 2^{n-r} + 2^{n-s} + \ldots + p$ for $p = 1, 2, \ldots$ while $\frac{1}{n} \ln p \ll 1$, σ is independent of p. Thus as $n \to \infty$ the continuous extension of σ has the property of $\sigma'(x) = 0$ almost everywhere with discontinuities at the rationals such that if σ has a discontinuity d at a rational r, then for rationals arbitrarily close to r, the associated discontinuities are arbitrarily smaller than d.

It is important to point out that not *all* scaling laws deduced from the associated map can extend to the trajectories. For example it is a direct consequence of the meaning of α that

$$\lim_{n \to \infty} \frac{P^{2m}(\hat{x}_{n+1}) - P^{2m}(\hat{x}_n)}{P^m(\hat{x}_n) - P^m(\hat{x}_{n-1})} = \alpha^{-1} \quad \text{for all } m.$$

However, the analogous trajectory formula

$$\frac{x^{(n+1)}(2t) - x^{(n)}(2t)}{x^{(n)}(t) - x^{(n-1)}(t)} \to \alpha^{-1} \quad \text{for all } t$$

is necessarily false. Rather, it is correct for $t_m = mT_0$, but widely oscillates over the duration of a basic cycle. Only map scaling-laws pertaining to the *same* time can be lifted by the flow to become trajectory laws. For example, let us formally deduce this property for σ. By (1),

$$\dot{x}^{(n+1)}(t) = f(x^{(n+1)}(t), \lambda_{n+1})$$

and

$$\dot{x}^{(n+1)}(t + T_n) = f(x^{(n+1)}(t + T_n), \lambda_{n+1}),$$

so that

$$\dot{\psi}^{(n+1)} \cong Df(x^{(n+1)}(t), \lambda_{n+1}) \cdot \psi^{(n+1)} \equiv M^{(n+1)}(t) \cdot \psi^{(n+1)}. \tag{40}$$

By definition,

$$\psi^{(n+1)} = \sigma_n(t)\psi^{(n)} \quad (\sigma_n \text{ is a scalar}). \tag{41}$$

Differentiating (41) and utilizing (40) and the analogous formula for $n+1 \to n$,

$$M^{(n+1)} \cdot \psi^{(n+1)} = \sigma_n M^{(n+1)} \cdot \psi^{(n)} = \dot{\sigma}_n \psi^{(n)} + \sigma_n \dot{\psi}^{(n)} = \dot{\sigma}_n \psi^{(n)} + \sigma_n M^{(n)} \cdot \psi^{(n)}$$

or,

$$\dot{\sigma}_n \psi^{(n)} \cong \sigma_n (M^{(n+1)} - M^{(n)}) \cdot \psi^{(n)}. \tag{42}$$

However, $\lambda_{n+1} - \lambda_n \sim \delta^{-n}$ and also $x^{(n+1)}(t) - x^{(n)}(t) \sim \delta^{-n}$, so that

$$\dot{\sigma}_n \psi^{(n)} \sim \delta^{-n} \sigma_n N(t) \cdot \psi^{(n)}$$

for an appropriate $N(t)$. Accordingly, as $n \to \infty$, $\dot{\sigma}_n \to 0$ for $\psi^{(n)}(t) \neq 0$. Thus, σ is preserved at the value set by the map, readjusting to its new values at those points during the basic cycle at which $\psi^{(n)}$ vanishes. (These are, of course, the discontinuities of σ occurring precisely where the ratio σ is undefined.) Had $\psi^{(n)}$ and $\psi^{(n+1)}$ been evaluated at different times – in particular at a ratio of two – then the right hand side of (42) would be oscillating at a scale of $x(t)$ and the corresponding ratio could not remain constant. It is worthy to mention at this point that the trajectory constancy of σ implies that the basic spectral recursion (16) holds for *all* spectral components, and not just below ω_0 – the purely subharmonic components would still obey (16) had σ merely averaged to the map value over each basic cycle, while the harmonics of these components at and above the basic frequency would fail to do so. *Thus, the way in which the basic cycle itself metamorphoses as $n \to \infty$ is also determined by the theory.*

We now turn to the recursion for the even spectral components. By (9)

$$x_{2k}^{(n+1)} - x_k^{(n)} = \int_0^{T_n} \frac{dt}{2T_n} [x^{(n+1)}(t) + x^{(n+1)}(t+T) - 2x^{(n)}(t)] e^{-2\pi i k t/T_n}$$

or,

$$x_{2k}^{(n+1)} - x_k^{(n)} = \int_0^{T_n} \frac{dt}{T_n} [x^{(n+1)}(t) - x^{(n)}(t)] e^{-2\pi i k t/T_n}$$

$$- \int_0^{T_n} \frac{dt}{2T_n} [x^{(n+1)}(t) - x^{(n+1)}(t+T_n)] e^{-2\pi i k t/T_n}. \tag{43}$$

The second term of (43) is, of course, determined through σ, so that the new ingredient necessary to obtain a recursion is the function

$$\frac{x^{(n+1)}(t) - x^{(n)}(t)}{x^{(n)}(t) - x^{(n-1)}(t)} \cong \varrho(t/T_{n+1}). \tag{44}$$

Formula (44) anticipates that the ratio of the left hand side has a dependence purely through the scale for t (as $n \to \infty$) and, as the reader has no doubt guessed, has a universal limit. Now, for that t one cycle, T_0, after $x^{(n)} = \hat{x}_n$, (19) implies that $\varrho = \delta^{-1}$; also ϱ will persist at δ^{-1} for some range of t above this value, lending evidence to the correctness of (44). To go further, consider once more

$$t_r = 2^{n-r} T_0$$

so that

$$\varrho(2^{-1-r}) \cong \frac{P^{2^{n-r}}(\hat{x}_{n+1}) - P^{2^{n-r}}(\hat{x}_n)}{P^{2^{n-r}}(\hat{x}_n) - P^{2^{n-r}}(\hat{x}_{n-1})}.$$

Again, employ the universality formulae (25) to obtain

$$\varrho(2^{-1-r}) \cong \frac{(\hat{x}_{n+1} - \hat{x}_n) + \dfrac{v}{(-\alpha)^{n-r}}(g_{r+1}(0) - g_r(0))}{(\hat{x}_n - \hat{x}_{n-1}) + \dfrac{v}{(-\alpha)^{n-r}}(g_r(0) - g_{r-1}(0))}. \tag{45}$$

Now the value $\hat{x}$ of the "extremal" point of the map is determined by some condition analogous to

$$D_x f(\lambda_n, \hat{x}_n) = 0$$

in the 1-dimensional case. It thus follows from (19) that

$$\hat{x}_{n+1} - \hat{x}_n \sim \eta \delta^{-n}$$

for some η. Thus, (45) becomes

$$\varrho(2^{-1-r}) \cong \frac{\eta/v\left(\dfrac{-\alpha}{\delta}\right)_n + (-\alpha)^r (g_{r+1}(0) - g_r(0))}{\delta\eta/v\left(\dfrac{-\alpha}{\delta}\right)^n + (-\alpha)^r (g_r(0) - g_{r-1}(0))}.$$

Since $\alpha = 2.5, \ldots$, $\delta = 4.6, \ldots$, $\alpha/\delta < 1$ and so, as $n \to \infty$,

$$\varrho(2^{-1-r}) = \frac{g_{r+1}(0) - g_r(0)}{g_r(0) - g_{r-1}(0)}. \tag{46}$$

By an analysis identical to that that led to (39), it is clear that ϱ is again universally determined at the rationals. However, should we attempt the deduction that led to the discontinuities of σ [e.g. Eq. (38) derivation], the δ^{-n} term remains in numerator and denominator, while instead of α^{-n}, we now have α^{-2n}:

$$\begin{aligned} P \cdot P^{2^{n-r}}(\hat{x}_{n+1}) - P \cdot P^{2^{n-r}}(\hat{x}_n) &\sim P\left(\hat{x}_{n+1} + \frac{v}{(-\alpha)^{n-r}} g_{n+1}(0)\right) \\ &\quad - P\left(\hat{x}_n + \frac{v}{(-\alpha)^{n-r}} g_r(0)\right) \\ &\cong \eta' \delta^{-n} + \tfrac{1}{2} P''(\hat{x}) \frac{v^2}{\alpha^{2(n-r)}} (g_{r+1}^2(0) - g_r^2(0)). \end{aligned}$$

Since $\alpha^2 > \delta$, we now have

$$\varrho(2^{-1-r} + 0) = \delta^{-1}.$$

Thus, the discontinuities at the rationals are now determined purely by ϱ's deviations from δ^{-1}, which are vanishing as $r\to\infty$, growing as x departs from 0. [As $r\to\infty$, (46) is simply the convergence rate of $g_r(0)$ which is δ^{-1}.] On the other hand, for several iterations after this first quadratic imagining, only a slope is involved, so that ϱ remains at δ^{-1}. Similarly several iterates prior to 2^{-1-r}, ϱ remains at $\varrho(2^{-1-r}-0)$. Next, when we consider the second generation rationals $2^{-1-r}+2^{-1-s}$, so that

$$P^{2^{n-s}}P^{2^{n-r}}(\hat{x}_n)\cong P^{2^{n-s}}\left(\hat{x}_n+\frac{v}{(-\alpha)^{n-r}}g_r(0)\right)\cong\hat{x}_n+\frac{v}{(-\alpha)^{n-s}}g_s((-\alpha)^{r-s}g_r(0))$$

then for $s\gtrsim n/2$ the g terms again dominate as $n\to\infty$, so that

$$\varrho(2^{-r-1}+2^{-s-1})=\frac{g_{s+1}(\alpha^{r-s}g_{r+1}(0))-g_s(\alpha^{r-s}g_r(0))}{g_s(\alpha^{r-s}g_r(0))-g_{s-1}(\alpha^{r-s}g_{r-1}(0))}$$

and the discontinuity is now correctly given by the same formula with each term replaced by its square. Thus, at all rationals arbitrarily close to those of large discontinuity, small discontinuities reign. However, ϱ for these large values of s is by the formula

$$g_s(x)\sim g(x)-\delta^{-s}h(x)$$

(h has quadratic extremum at $x=0$ and is universal) also just δ^{-1}. Accordingly, $\varrho'(x)=0$ almost everywhere, and up to $x=1/4$ is "fairly" close to δ^{-1}' ϱ can also be shown to have this "constancy" to δ^{-1} for $3/4<x<1$. Figure 3 depicts the ϱ of (44) for Duffings equations, while Fig. 4 depicts ϱ as determined by the quadratic recursion as the interval

$$x_{n+1}=a-x_n^2$$

for a 256-cycle. (In these figures, $x=0$ has been displaced to $x=1/4$ in order to about the "well-behaved" quarters of ϱ to form its first half-cycle.) As the cycle length is increased, convergence is of an oscillating sort as higher r values with their discontinuities are exposed. Nevertheless, since ϱ is used only as a functional (in an integral) with the large excursions occurring where the numerator and denominator are crossing zero, it is clear that for half the cycle, $\varrho=\delta^{-1}$ is a very good approximation. *Accordingly, the phase of* $x^{(n)}(t)$ *in* (43) *must be set to be* 3/4 *through the cycle in order to have* $0<t<T_n$ *as the "smooth" half cycle of* ϱ.

We now return to (43) and substitute the fourier expansions of the $x(t)$'s:

$$\begin{aligned}x_{2k}^{(n+1)}-x_k^{(n)}&=\sum_{k'}x_{k'}^{(n)}\int_0^{T_n}\frac{dt}{T}\varrho(t/2T_n)e^{2\pi i(k'-k)t/T_n}\\&\quad-\sum_{k'}x_{k'}^{(n-1)}\int_0^{T_n}\frac{dt}{T_n}\varrho(t/2T_n)e^{2\pi i(2k'-k)t/T_n}\\&\quad-\sum_{k'}x_{2k'+1}^{(n)}\int_0^{T_n}\frac{dt}{T_n}\sigma(t/2T_n)e^{2\pi i(2k'+1-k)t/T_n}\\&=\sum_{k'}(x_{2k'}^{(n)}-x_{k'}^{(n-1)})\int_0^{T_n}\frac{dt}{T_n}\varrho(t/2T_n)e^{2\pi i(2k'-k)t/T_n}\\&\quad+\sum_{k'}x_{2k'+1}^{(n)}\int_0^{T_n}\frac{dt}{T_n}[\varrho(t/2T_n)-\sigma(t/2T_n)]e^{2\pi i(2k'+1-k)t/T_n}\end{aligned}$$

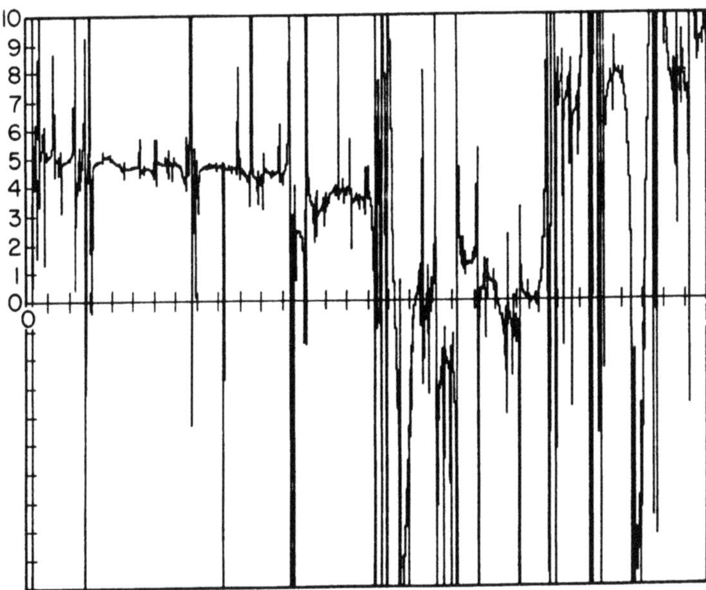

Fig. 3. ϱ^{-1} of (44) for $n=4$ by numerical division of solutions to Duffing's equation

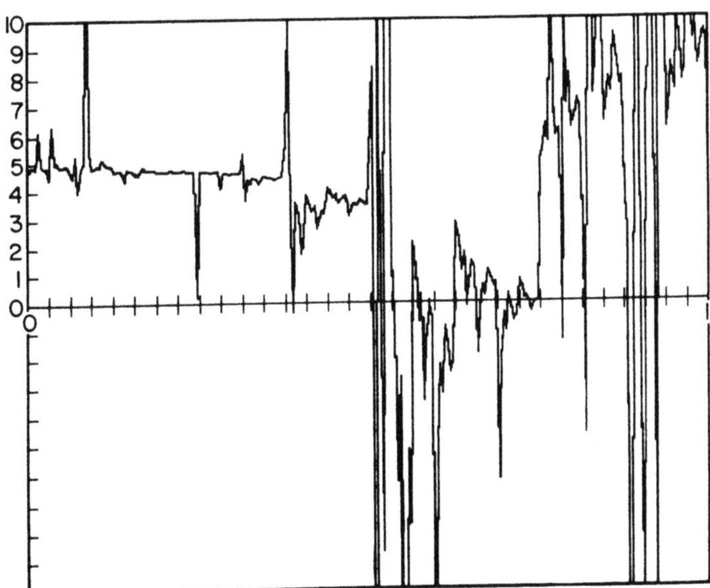

Fig. 4. ϱ^{-1} for $n=7$ by numerical division of iterates of $a-x^2$

or,

$$x_{2k}^{(n+1)}-x_k^{(n)}=\sum[x_{2k'}^{(n)}-x_{k'}^{(n-1)}]\int_0^1 d\xi\varrho(\xi/2)e^{2\pi i\xi(2k'-k)}$$
$$+\sum_{k'}x_{2k'+1}^{(n)}\int_0^1 d\xi[\varrho(\xi/2)-\sigma(\xi/2)]e^{2\pi i\xi(2k'+1-k)}. \tag{47}$$

It is immediately clear from (47) that the even spectrum is now compatible, and again through an autonomous universal recursion. However, the recurrence is now second order with superficially two levels of spectrum required to compute the next. Observe that the index in the first term (containing the second order term) is even in k', while odd in the second. With the $-1/4$ phasing, by the above discussion, ϱ can be approximated by δ^{-1} in the first integral, so that

$$\int_0^1 d\xi\varrho(\xi/2)e^{2\pi i\xi(2k'-k)}\cong\delta^{-1}\delta_{2k',k}. \tag{48}$$

Thus to good approximation [since (47) is small to begin with] the recursion is second-order only for k even. In particular, consider the "new" components, $x_{2k+1}^{(n)}$, at the nth level: at the $n+1$st level, the first term of (47) does not contribute, so that at the $n+1$st level its value is determined by a first order recurrence:

$$x_{2(2k+1)}^{(n+1)}-x_{2k+1}^{(n)}\cong\sum_{k'}x_{2k'+1}^{(n)}\int_0^1 d\xi[\varrho(\xi/2)-\sigma(\xi/2)]e^{4\pi i(k'-k)}. \tag{49}$$

It is easy to obtain a good approximation to (49), recalling, though, that the $-1/4$ phasing determines σ to be $-\alpha^{-1}$ on the first half of the integration, and $+\alpha^{-2}$ on the second:

$$\int_0^1 d\xi[\varrho(\xi/2)-\sigma(\xi/2)]e^{4\pi i\xi(k'-k)}\cong\delta^{-1}\delta_{k,k'}+\alpha^{-1}\int_0^{1/2} d\xi e^{4\pi i\xi(k'-k)}$$
$$-\alpha^{-2}\int_0^{1/2} d\xi e^{4\pi i\xi(k'-k)}$$
$$=(\delta^{-1}+\tfrac{1}{2}(\alpha^{-1}-\alpha^{-2}))\delta_{k',k}.$$

Accordingly, we have the good approximation

$$x_{2(2k+1)}^{(n+1)}\cong(1+\delta^{-1}+\tfrac{1}{2}(\alpha^{-1}-\alpha^{-2}))x_{2k+1}^{(n)}$$
$$=1.33x_{2k+1}^{(n)}. \tag{50}$$

Thus, after a given component appears at some level, it maintains its phase and grows in amplitude by a *universal* factor of

$$\mu'=1.33$$

or $10\log_{10}\mu'\cong 1.24$ db.

Next consider even k, to determine the future growth of this component:

$$x_{4k}^{(n+1)}-x_{2k}^{(n)}\cong\delta^{-1}[x_{2k}^{(n)}-x_k^{(n-1)}]$$
$$+\sum_{k'}x_{2k'+1}^{(n)}\int_0^1 d\xi[\delta^{-1}-\sigma(\xi/2)]e^{2\pi i\xi(2k'+1-2k)}. \tag{51}$$

But,

$$\int_0^1 d\xi[\delta^{-1}-\sigma(\xi/2)]e^{2\pi i\xi(2k'+1-2k)}=-\int_0^1 d\xi\sigma(\xi/2)e^{2\pi i\xi(2k'+1-2k)}$$

$$\cong\alpha^{-1}\int_0^{1/2} d\xi e^{2\pi i\xi(2k'+1-2k)}-\alpha^{-2}\int_{1/2}^1 d\xi e^{2\pi i\xi(2k'+1-2k)}$$

$$=(\alpha^{-1}+\alpha^{-2})\int_0^{1/2} d\xi e^{2\pi i\xi(2k'+1-2k)}=\frac{-(\alpha^{-1}+\alpha^{-2})}{\pi i}\frac{1}{2k'+1-2k}$$

so that

$$x_{4k}^{(n+1)}-x_{2k}^{(n)}\cong\delta^{-1}[x_{2k}^{(n)}-x_k^{(n-1)}]-\frac{1}{\pi i}(\alpha^{-1}+\alpha^{-2})\sum_{k'}\frac{1}{2k'+1-2k}x_{2k'+1}^{(n)}. \tag{52}$$

To estimate the second term, use the principal value approximation to obtain

$$+\tfrac{1}{2}(\alpha^{-1}+\alpha^{-2})x_{2k''}^{(n)}\cong\tfrac{1}{2}(\alpha^{-1}+\alpha^{-2})x_{2k+1}^{(n)}.$$

($x_{2k''}^{(n)}$ is the interpolation of the *odd* nth order components, and so of magnitude of the closest of these.) But, by (18),

$$x_{2k+1}^{(n)}\sim 0.15x_{2k}^{(n)}$$

when k is odd, and smaller otherwise. Altogether, then, the second term of (52) is $\lesssim 0.04x_{2k}^{(n)}$, and so roughly negligible compared to the first term. Thus,

$$x_{4k}^{(n+1)}-x_{2k}^{(n)}\cong\delta^{-1}[x_{2k}^{(n)}-x_k^{(n-1)}]. \tag{53}$$

Combined with (50) we now see that at the next levels the component under question grows first another $\sim 5\%$, then $\sim 1\%$ for a total further increase of several tenths of a decibel.

Altogether, we now summarize the construction of the spectrum. Start at the nth level, and smoothly interpolate the *odd* components. Shifted down by 8.2 db, this curve is the local average of the locations of the *odd* components of the $n+1$st level. Next, shift up each of the odd components of the nth level by 1.4 db to this new (and final) location. Finally, all the even components of the nth level remain in place. This (very good) approximation requires *only* the components at the nth level and now recursively determines the spectrum for $n\to\infty$. Whatever approximation is chosen, it is clear that the spectrum at the transition point is determined and our task is completed.

Conclusions

Our result is that given a system known to become erratic through a cascade of subharmonic bifurcations, after the spectrum is determined through some appropriate analysis through the first several bifurcations, the behavior down through the transition is determined. At the transition itself, the spectrum of any mode is comprised of 2^{n-1} components at the odd multiples of the 2^n subharmonic of the "original" frequency each of magnitude roughly $8.2n$ db below the basic component. The phases of these components are determined ultimately by the phases of the first bifurcations'components, although the basic formula (14) endows these high level components with rapid phase variations, so that they have a sort of random character. The total power in this spectrum (taken as the squares of the spectral amplitudes) is easily estimated at ~2.5% of the power in the basic frequency, to give a rough idea of the transition noise.

Acknowledgements. The author has greatly profited from discussions with Jean-Pierre Eckmann, and accordingly thanks D. Bessis and NATO under whose auspices, at Cargèse, this contact occurred. A seminal discussion with my colleague Harvey Rose issued directly into the research of this paper. Finally, I am grateful to P. Carruthers, E. Lieb, and A. Jaffe for their enthusiastic reception of this work and facilitation of its rapid publication.

References

1. Franceschini, V., Tebaldi, C.: Sequences of infinite bifurcation and turbulence in five-modes truncation of the Navier-Stokes equations. Istituto Matematico, Univ. di Modena preprint
2. Franceschini, V.: A Feigenbaum sequence of bifurcations in the Lorenz model, Istituto Matematico, Univ. di Modena preprint, to be published in J. Stat. Phys.
3. Computations by the author on Duffing's equation, following Ueda, Y.: J. Stat. Phys. **20** (2), 181 (1979)
4. Holmes, P.: A nonlinear oscillator with a strange attractor. Department of Theoretical and Applied Mechanics, Cornell University (preprint)
5. Huberman, B., Crutchfield, J.: Chaotic states of anharmonic systems in periodic fields. Xerox Corp., Palo Alto Research Center (preprint)
6. Marzec, C.J., Spiegel, E.A.: A strange attractor. Astronomy Department, Columbia University (preprint)
7. Libchaber, A., Maurer, J.: Une expérience de Rayleigh-Bénard de géométrie réduite. École Normale Supérieure (preprint)
8. Feigenbaum, M.J.: Phys. Lett. **74**A, 375 (1979)
9. Collet, P., Eckmann, J.-P., Koch, H.: Period doubling bifurcations for families of maps on C^n. Department of Physics, Harvard University (preprint)
10. Metropolis, N., Stein, M.L., Stein, P.R.: Combinatorial Theory **15** (1), 25 (1973)
11. May, R., Oster, G.: Amer. Naturalist **110** (974), 573 (1976). This paper independently of myself, discovers the first clue of a universal metric property
12. Feigenbaum, M.J.: Annual Report 1975–76, LA-6816-PR, Los Alamos
13. Feigenbaum, M.J.: J. Stat. Phys. **19** (1), 25 (1978)
14. Feigenbaum, M.J.: J. Stat. Phys. **21** (6) (1979)
15. Feigenbaum, M.J.: Lecture Notes in Physics **93**, 163 (1979)
16. Collet, P., Eckmann, J.-P., Lanford III, O.: Universal properties of maps on an interval (in preparation)
17. Collet, P., Eckmann, J.-P.: Bifurcations et groupe de renormalisation. IHES/P/78/250 (preprint)
18. Derrida, B., Gervois, A., Pomeau, Y.: J. Phys. A **12**, 269 (1979). This paper contains the first numerical observation of δ in a 2-dimensional map

Communicated by A. Jaffe

Received April 7, 1980; in revised form May 12, 1980

Editor's note. The 8.2 decibels quoted in the above paper and in Feigenbaum (1979b) refer to the mean amplitude drop for successive subharmonics. Libchaber and Maurer (1980, 1981) measured the power spectrum, for which Feigenbaum had predicted 16.4 decibels. The numerical calculation of Nauenberg and Rudnick (1981) (the next reprint in this selection) yields 13.5, while the Collet, Eckmann and Thomas (1981) rigorous bounds for the Fourier spectrum of the universal function exclude both estimates. Quoting a single number is rather misleading, because amplitude variations among different subharmonics at a given level are large (see figure 15 in Feigenbaum (1980a), p 49 this selection) and their envelope is not universal. A careful analysis would require that the first few subharmonics be used as input for a computation of the succeeding levels.

Physical Review B **24** 493–5 (1981)

Universality and the power spectrum at the onset of chaos

M. Nauenberg and J. Rudnick
Physics Department, University of California, Santa Cruz, California 95064
(Received 2 March 1981)

Two one-dimensional maps are iterated to evaluate the average height $\phi(k)$ of the peaks in the power spectrum corresponding to frequencies $\omega_{k,l} = (2l-1)\pi/2^k$, where $l = 1, 2, \ldots, 2^{k-1}$ and $k = 1, 2, \ldots$ at the onset of chaos. It is shown that the ratio $\phi(k)/\phi(k+1)$ is nearly constant and for large k approaches a universal limit $2\beta^{(2)} = 20.963\ldots$.

Recently there has been considerable excitement over the possibility that iterated maps of the interval might provide a mathematical model for certain physical systems that undergo a transition from periodic to chaotic behavior. One reason for this excitement is the fact that there are characteristics of the maps that behave in a universal fashion at and near the transition.[1] In fact, it has proven possible to construct an analogy with critical phenomena, derive critical exponents, and, in one case, obtain a universal scaling function.[2–5]

From a physical point of view it is clear that one should consider variables exhibiting universal behavior that are directly accessible experimentally. In this Communication we discuss just such a quantity—the autocorrelation function of points under the iterated map and the Fourier transform of this function which ought to be directly related to the power spectrum of the physical system for which the iterated map is a model.[6] This allows for direct experimental tests of the physical relevance of the recently developed models of the transition to chaos.[7,8]

Consider a one-dimensional map defined by the following recursion relation:

$$x_{k+1} = f(x_k;r) \quad . \tag{1}$$

The function $f(x;r)$ is defined on the interval $a < x < b$ which it maps into itself. It is assumed that $f(x;r)$ is a smooth function of x on the interval with a single quadratic maximum and no other extremum. The variable r controls the height of the maximum. A simple example of such a function is $rx(1-x)$ in the interval $0 \leq x \leq 1$ with $0 \leq r \leq 4$.

The autocorrelation function of the map is defined by

$$c(j) = \lim_{N\to\infty} \frac{1}{N} \sum_{k=1}^{N} x_k x_{k+j} = \langle x_0 x_j \rangle \quad , \tag{2}$$

and the Fourier transform of this autocorrelation function $C(\omega) = |x(\omega)|^2$ is

$$C(\omega) = c(0) + 2\sum_{j=1}^{\infty} c(j)\cos j\omega \quad . \tag{3}$$

The transition to chaos in the iterated map is heralded by a cascade of period-doubling bifurcations. On the periodic side of the transition at a value of r such that the stable orbit has a period 2^n (i.e., $x_{k+2^n} = x_k$), $C(\omega)$ will consist of a set of δ functions at $\omega = \pi m/2^n$, with m an integer less than 2^n. As the transition is approached so that a bifurcation takes place and the orbit has a period 2^{n+1}, new contributions to $C(\omega)$ will appear in the form of δ functions at $\omega = \pi(2m-1)/2^{n+1}$ with m an integer and $2m - 1 < 2^n$. On the chaotic side of the transition a point acted upon by the map follows a trajectory that takes it between a set of bands that merge pairwise as r is adjusted to take the map ever further into the chaotic region. This merging occurs in a sequence that is the mirror image of the bifurcation sequence on the periodic side.[9] In the chaotic regime, $C(\omega)$ will consist of the same kind of δ functions as on the periodic side with the addition of a broad-band component representing the noisy, or chaotic, aspect of trajectories under the map. In what follows we will concentrate on the δ-function contributions to $C(\omega)$. Aspects of the broad-band behavior have already been considered.[10,11]

One quantity of interest is the ratio between the coefficients of the δ functions in $C(\omega)$ which are related to the peaks in the power spectrum.

We write

$$C(\omega) = C_{0,0}\delta(\omega) + C_{0,1}\delta(\omega - \pi) + 2\sum_{k=1}^{\infty}\sum_{l=1}^{2^{k-1}} C_{k,l}\delta\left(\omega - \frac{(2l-1)\pi}{2^k}\right) \tag{4}$$

and define the average $\phi(k)$

$$\phi(k) = \frac{1}{2^{k-1}} \sum_{l=1}^{2^{k-1}} C_{k,l} \quad . \tag{5}$$

We find that the ratio $\phi(k)/\phi(k+1)$ is nearly constant (see Table I), and we show that it approaches a universal ratio for large k, provided

TABLE I. The values for $\phi(k)/\phi(k+1)$ were obtained by two methods: (a) evaluating $\phi(k)$ directly from Eq. (5), and (b) evaluating $\bar{D}(2^k)$ from Eqs. (9) and (10) and substituting into Eq. (8). Both methods gave the same result, verifying numerically our analysis.

k	$\phi(k)/\phi(k+1)$	
	Feigenbaum	Parabolic
0	21.1876	21.8911
1	20.8684	20.7707
2	20.9924	21.0453
3	20.9532	20.9383
4	20.9670	20.9744

$k << n$, independent of the map,

$$\phi(k-1)/\phi(k) = 2\beta^{(2)} \ , \tag{6}$$

where $\beta^{(2)}$ is a constant whose value[12] is 10.4817. . . . Furthermore, if we consider $\phi(n-1)$ for a value of the parameter r such that the mapping has a period 2^n just before the bifurcation to the period 2^{n+1}, and correspondingly $\phi'(n)$ for r' with period 2^{n+1} before the bifurcation to the period 2^{n+2}, we find that for large n, $\phi(n-1)/\phi'(n) = 2\beta^{(2)}$.

To arrive at the scaling relationship (6) let us see how $\phi(k)$ is calculated. Since the coefficient of the δ function $\delta(\omega = (2l-1)\pi/2^k)$ is given by

$$C_{k,l} = \lim_{N\to\infty} \frac{1}{N}\left[c(0) + 2\sum_{j=1}^{N-1} c(j)\cos\left(\frac{(2l-1)j\pi}{2^k}\right)\right] \tag{7}$$

we can obtain an expression for $\phi(k)$ by substituting Eq. (7) in Eq. (5). After some algebra we find

$$\phi(k) = \frac{1}{2^{k+2}}\bar{D}(2^k)\left(1 - \sum_{s=1}^{\infty}\frac{1}{2^s\beta_{s,k}^{(2)}}\right) \ , \tag{8}$$

where

$$\bar{D}(2^k) \equiv \lim_{M\to\infty}\frac{1}{M}\sum_{s=1}^{M} D(2^k(2s-1)) \ , \tag{9}$$

$$D(j) \equiv \lim_{N\to\infty}\frac{1}{N}\sum_{j'=0}^{N-1}[x(j+j') - x(j')]^2 = 2[c(0) - c(j)] \ , \tag{10}$$

and

$$\beta_{s,k}^{(2)} = \left(\frac{\bar{D}(2^k)}{\bar{D}(2^{k+s})}\right) \ . \tag{11}$$

From the nature of the universal map it will be shown that

$$\lim_{k\to\infty}\frac{\bar{D}(2^k)}{\bar{D}(2^{k+1})} = \beta^{(2)} = 10.4817\ldots \ . \tag{12}$$

Hence for large enough k

$$\beta_{s,k}^{(2)} = (\beta^{(2)})^s \tag{13}$$

and we obtain

$$\phi(k) = \frac{1}{2^{k+1}}\left(\frac{\beta^{(2)}-1}{2\beta^{(2)}-1}\right)\bar{D}(2^k) = \frac{0.4750}{2^k}\bar{D}(2^k) \ . \tag{14}$$

In particular

$$\frac{\phi(k)}{\phi(k+1)} = \frac{2\bar{D}(2^k)}{\bar{D}(2^{k+1})} \ ,$$

and Eq. (6) follows immediately from Eq. (12).

Some of the results of our numerical investigations are summarized in Table I. The ratio $\phi(k)/\phi(k+1)$ was calculated for both the Feigenbaum invariant map $f(x)$ and the parabolic map $rx(1-x)$. In the case of the parabolic map, r was adjusted so that there was a stable orbit of period 2^n. Our results were independent of n to the accuracy quoted here when $n \gtrsim 10$. For such an orbit we can replace N in Eq. (7) and in similar formulas by 2^n and not take the limit of an infinite number of terms. For the Feigenbaum map we adjusted the initial point so that it was mapped back into itself after 2^n iterations of the map with $n \gtrsim 10$. Again, to the accuracy quoted here, our results were independent of the precise value of n. For the initial point which mapped into itself after 2^n iterations of the map, we took $(-1/\alpha)^n x_f$, where x_f is the unstable fixed point of $f(x)$ that is closest to the origin. The period 2^n orbit executed by this point is unstable, in contrast to the orbits in the periodic regime. The results in Table I show that for both cases the universal limiting ratio $\beta^{(2)}$ is approached for sufficiently high $\phi(k)$'s.

Finally, we would like to comment briefly on our calculation of $\beta^{(2)}$ directly from the Feigenbaum invariant map $f(x)$, which is displayed in Fig. 1. This function has a single maximum at $x=0$, in the region $-1/\alpha \leqslant x \leqslant 1$, a region which is mapped into itself by $f(x)$. As shown in Fig. 1, the function $f(f(x))$, the iterate of $f(x)$, has three extrema in this region. There are two subregions, centered about two of the extrema, that are mapped into themselves by $f(f(x))$. In the subregion centered about $x=0$ the function $f(f(x))$ is exactly a scaled-down, inverted version of $f(x)$, the scale factor being α. In the other region $f(f(x))$ is approximately a scaled-down, shifted $f(x)$, that is,

$$f(f(x)) \cong (1/\alpha')f(\alpha'(x-x_0)) + x_0 \ ,$$

where $\alpha' = (1+1/\alpha)/[1-f(1/\alpha)]$. The mean-square width of these two regions is smaller than the

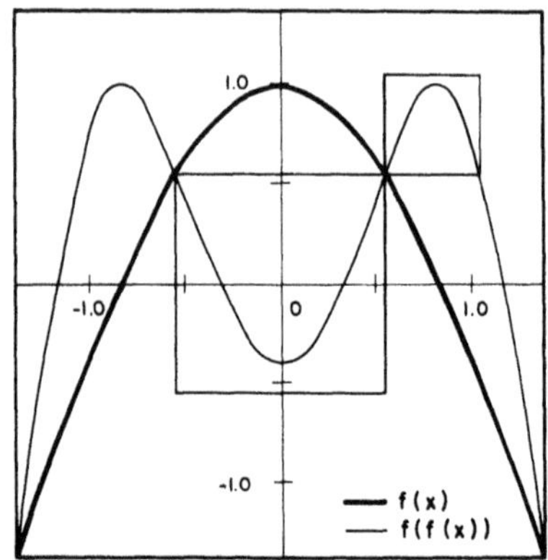

FIG. 1. Feigenbaum invariant map $f(x)$ (dark curve) and the iterated map $f(f(x))$ (light curve) are displayed. The subregions about the extrema which are mapped into themselves by $f(f(x))$ are indicated by the squares.

width of the original region, by the factor $\frac{1}{2}(1/\alpha^2 + 1/\alpha'^2)$. Likewise iterating $f(f(x))$ we find four regions mapped into themselves by

$$f^{(4)}(x) \equiv f(f(f(f(x)))) \ ,$$

with a mean-square width reduced by a factor of approximately $[\frac{1}{2}(1/\alpha^2+1/\alpha'^2)]^2$. The ratio of the mean-square width of the 2^n regions mapped into themselves by $f^{(2^n)}(x)$ to that of the $2^{(n+1)}$ regions mapped into themselves by $f^{[2^{(n+1)}]}(x)$ is just $\beta^{(2)}$ where in our approximation $\beta^{(2)} = [\frac{1}{2}(1/\alpha^2+1/\alpha'^2)]^{-1} = 10.31$. It can be readily seen that in this approximation the widths of the allowed regions are distributed according to a binomial distribution. This estimate for $\beta^{(2)}$ can be systematically improved, and $\beta^{(2)}$ can be calculated to arbitrary accuracy by considering the ratios of the mean-square widths of regions mapped into themselves by $f^{(2^n)}(x)$. The value of $\beta^{(2)}$ quoted in this paper was obtained by this procedure.

In conclusion, we note that, although it will be necessary for the experimentalists to look at orbits deep in the bifurcation scheme in order to test the limiting behavior, we predict that, with precision, it ought to be possible to observe ratios in reasonably good accord with the limit predicted by Eq. (6) early in the sequence.

ACKNOWLEDGMENTS

We thank J. D. Farmer, M. Feigenbaum, and J. Guckenheimer for useful discussions. A seminal comment by Eduardo Fradkin contributed significantly to progress on the work reported here and is gratefully acknowledged. This work was supported by a grant from the National Science Foundation.

---

[1]M. J. Feigenbaum, J. Stat. Phys. 19, 25 (1978); 21, 669 (1979).

[2]B. A. Huberman and J. Rudnick, Phys. Rev. Lett. 45, 154 (1980).

[3]J. Crutchfield and B. A. Huberman, Phys. Lett. 45, 154 (1980); J. Crutchfield, J. D. Farmer, and B. A. Huberman (unpublished).

[4]J. Crutchfield, M. Nauenberg, and J. Rudnick, Phys. Rev. Lett. 46, 933 (1981).

[5]B. Schraiman, C. E. Wayne, and P. C. Martin (unpublished).

[6]M. J. Feigenbaum, Phys. Lett. 74A, 375 (1979); Commun. Math. Phys. 77, 65 (1980).

[7]A. Libchaber and J. Maurer, J. Phys. (Paris) 41, C3-51 (1980).

[8]J. P. Gollub, S. V. Benson, and J. Steinman, Ann. N.Y. Acad. Sci. 357, 22 (1980).

[9]S. Grossman and S. Thomae, Z. Naturforsch. 32, 1353 (1977); E. N. Lorenz, Ann. N.Y. Acad. Sci. 357, 282 (1980).

[10]B. A. Huberman and A. Zisook (unpublished).

[11]J. D. Farmer (unpublished).

[12]The ratio $\phi(k)/\phi(k+1) = 20.963$ is to be compared with the ratio $\mu^2 = 43.2$ of Feigenbaum (Ref. 6). Feigenbaum's ratio describes the way in which interpolations between subharmonic peaks in the power spectrum behave. For more details the reader is referred to Refs. 6. It is noteworthy that the derivation of Feigenbaum's ratio parallels ours up to the final step, at which point he replaces a sum with an integral.

Part 4

Noise

Zeitschrift für Naturforschung a **32** 1353–63 (1977)

Invariant Distributions and Stationary Correlation Functions of One-Dimensional Discrete Processes

S. Grossmann and S. Thomae

Fachbereich Physik, Philipps-Universität, Marburg, Germany

(Z. Naturforsch. **32 a**, 1353–1363 [1977]; received September 8, 1977)

The connection between one-dimensional dynamical laws generating discrete processes and their invariant densities as well as their stationary correlaton functions is discussed. In particular the changes occuring under a special equivalence transformation are considered. Correlation functions are used to describe the gradual transition from periodic states to chaotic states via periodic motions with superimposed nonlinearity noise.

I. Introduction

There are certain processes in physics, chemistry, ecology, etc. in which the time development is described by giving the values $x(t_\tau)$ of the dynamical variables at a discrete sequence of times t_τ. Instead of a differential equation of motion or its solution one considers some function $f(x)$ which generates the process by

$$x_{\tau+1} = f(x_\tau), \quad \tau = 0, 1, 2, \ldots \,. \tag{1}$$

In what follows we denote the τth iterate of $f(x)$ by

$$f^{[\tau]}(x) := f[f^{[\tau-1]}(x)], \quad f^{[0]}(x) := x. \tag{2}$$

If there are properties of the sequence $\{f^{[\tau]}(x_0)\}$ which do not depend on the initial conditions x_0, these invariants reflect characteristics of $f(x)$ alone.

In this paper we study for some dynamical laws $f(x)$ not only the asymptotic density functions they generate,

$$\varrho_f^*(x) = \lim_{\substack{m(I)\to 0\\ I\to x}} \lim_{N\to\infty} \frac{1}{m(I)} \frac{1}{N} \sum_{\tau=0}^{N-1} \chi_I[f^{[\tau]}(x_0)] \tag{3}$$

but also the time correlation of the process,

$$C_f(\tau) = \langle \delta x\, \delta x_\tau \rangle := \langle x f^{[\tau]}(x) \rangle - \langle x \rangle^2 . \tag{4}$$

This latter is a useful means to understand why apparently stochastic processes may arise from simple deterministic dynamical laws. The mean $\langle\ldots\rangle$ is either a time mean or, if f turns out to be ergodic, the ensemble mean with ϱ_f^*. $m(I)$ denotes the size of the interval I (Lebesgue measure) and χ_I its characteristic function (1 if x inside I, 0 elsewhere).

In Sect. II we cite two theorems of basic importance for our subject and give an ergodicity criterion for dynamical laws f based on a theorem by Li and Yorke. In Sect. III we discuss a method to compute ϱ_f^* and in IV we show how the set of accessible dynamical laws can be extended by introducing an equivalence relation. The computation of correlation functions in V leads us in VI to a further discussion of the previously mentioned equivalence relation. In VII we show that under certain well defined conditions the processes generated by nonlinear dynamical laws may be described as periodic motions in state-space superimposed by noise.

There is great recent interest in such "nonlinearity noise" in connection with experimental [1, 2] and theoretical [3, 4] studies of hydrodynamic turbulence, chemical turbulence [5], etc.

Reprint requests to Prof. Dr. S. Grossmann, Fachbereich Physik, Philipps-Universität, Renthof 6, *D-3550 Marburg*, Germany.

II. Two Theorems on Asymptotic Distributions

In statistical physics dynamical laws f which are ergodic or even strongly mixing are of major importance. The existence of a unique L^1-integrable invariant density ϱ_f^* generating an absolutely continuous measure

$$\mu_f^*(I) = \int_I \varrho_f^*(x)\, dx \tag{5}$$

is equivalent to the ergodicity of f [6, 7]. Lasota, Li and Yorke proved the following two theorems on the existence and uniqueness of such an ϱ_f^* [8].

T1: *Existence* [9]

Let f: $[0, 1] \to [0, 1]$ be a piecewise C^2-function such that

$$\inf_x \left| \frac{df^{[n_0]}(x)}{dx} \right| > 1 \tag{6}$$

for one positive integer n_0. Then there is a function $\varrho_f^*(x)$ with the following properties:

(a) $\varrho_f^* \geqq 0$,

(b) $\int_0^1 \varrho_f^*(x)\,dx = 1$,

(c) ϱ_f^* is invariant under f, i. e.

$$\varrho_f^*(x) = \frac{d}{dx} \int_{f^{-1}([0,x])} \varrho_f^*(x')\,dx' .$$

The main purpose of the infimum-condition (6) is to exclude the stability of possibly existing periodic points

$$x^* = f^{[\tau]}(x^*), \quad \tau = 1, 2, 3, \ldots .$$

It implies a severe restriction on the applicability of the theorem since e. g. all continuous differentiable functions on $[0, 1]$ are excluded. In IV we will see that this restriction can, at least partially, be overcome.

The next theorem gives some insight into the structure of the set R_f of invariant densities of a dynamical law f from which the uniqueness of ϱ_f^* can be inferred:

T2: *Structure of* R_f [10]

Let f: $[0, 1] \to [0, 1]$ be a piecewise C^2-function such that

$$\inf_x \left| \frac{df(x)}{dx} \right| > 1 .$$

The set $J := \{y_0, y_1, \ldots, y_k\}$ of points, where $df(y)/dy$ does not exist, be finite [11].

Then there is a finite collection of sets

$$M_1, M_2, \ldots, M_n$$

and a set R_f of L^1-functions of bounded variation, which are invariant under f, with a subset

$$\{\varrho_1, \varrho_2, \ldots, \varrho_n\}$$

such that

(a) each $M_i\,(1 \leqq i \leqq n)$ is a finite union of closed intervals;

(b) $M_i \cap M_j$ contains at most a finite number of points if $i \neq j$;

(c) each $M_i\,(1 \leqq i \leqq n)$ contains at least one point $y_j\,(1 \leqq j \leqq k)$ in its interior; hence $n \leqq k$;

(d) $\varrho_i(x) = 0$ for $x \notin M_i\,(1 \leqq i \leqq n)$ and $\varrho_i(x) > 0$ for almost all $x \in M_i$.

T2 immediately yields a first criterion of uniqueness, already formulated by Li and Yorke [10]:

C1: The invariant density ϱ_f^* of f is uniquely determined if f fulfills the propositions of T2 and has exactly one point of discontinuity in $(0, 1)$.

This criterion implicitly contains a restriction to f with one hump because any further hump would imply a further point of discontinuity within $(0, 1)$. Therefore we give a second criterion starting from the sets M_i of T2: Since M_i constitutes something like the closure of the support of ϱ_i, it is invariant under f up to a finite subset. So two different sets $M_i, M_j, i \neq j$ cannot be mapped on the same interval by any iterate of f. On the other hand, according to T2 (c) each M_i contains a complete ε-neighbourhood of at least one point of discontinuity. This leads us to another uniqueness criterion:

C2: Let f satisfy the propositions of T2.

$$U_\varepsilon(y_j) := (y_j - \varepsilon, y_j + \varepsilon) \cap (0, 1), \quad \varepsilon > 0$$

be the ε-neighbourhood of any point of discontinuity y_j. The invariant density ϱ_f^* is unique if there are positive integers n_1, n_2 for each pair (y_i, y_j) such that for arbitrarily small $\varepsilon > 0$

$$m\left(f^{[n_1]}[U_\varepsilon(y_i)] \cap f^{[n_2]}[U_\varepsilon(y_j)]\right) \neq 0 .$$

III. Computation of the Invariant Density ϱ_f^*

The computation of an invariant density ϱ_f^* for a given dynamical law f constitutes a problem which to the best of our knowledge has not yet been solved in complete generality. Usually one falls back to numerical methods. But even these, if the sequence $\{x_\tau\}$ is evaluated directly, are often extremely susceptible to round-off errors [12].

A numerically stable procedure yielding an approximation by step functions was proposed by Ulam [13]. The fundamental idea starts from the Frobenius-Perron operator $\hat{P}_f$, which describes the "time-evolution" of an arbitrary density ϱ_τ under f:

$$\varrho_{\tau+1}(x) = \hat{P}_f\,\varrho_\tau(x) := \frac{d}{dx} \int_{f^{-1}([0,x])} \varrho_\tau(x')\,dx', \qquad \varrho_\tau \in L^1 . \tag{7}$$

The desired invariant density ϱ_f^* obviously is a fixed point of $\hat{P}_f$. Ulam conjectured that $\hat{P}_f$ can be approximated by a matrix $\mathbb{P}$ with elements

$$p_{ij} = \frac{m[I_j \cap f^{-1}(I_i)]}{m(I_j)} . \tag{8}$$

$I_i\,(1 \leqq i \leqq n)$ are the elements of an arbitrary decomposition of $[0, 1]$. p_{ij} is that fraction of I_j

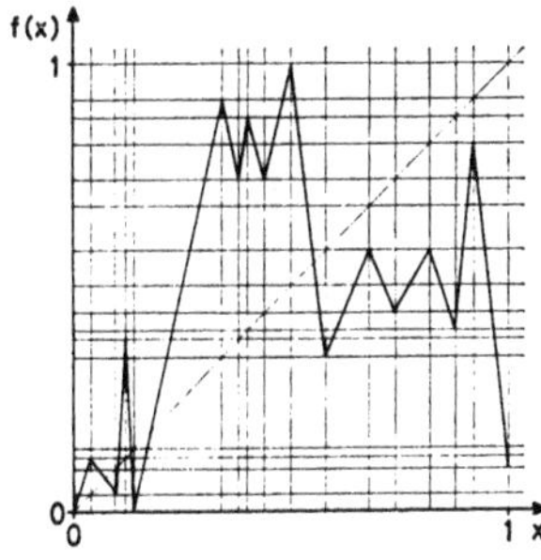

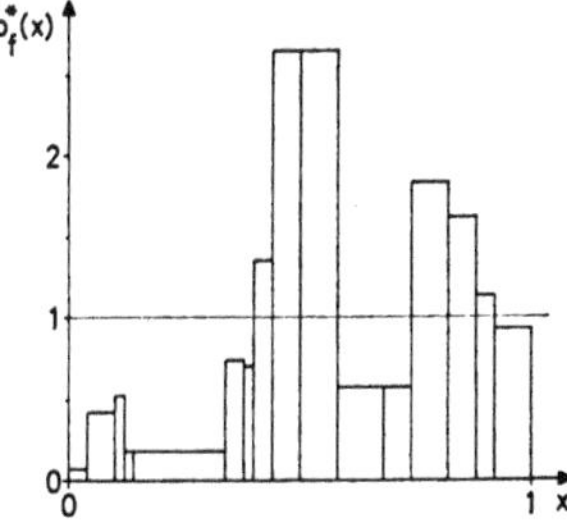

Fig. 1. The exact asymptotic distribution $\varrho_f^*(x)$ for a dynamical law f with finite K. If the sizes of the intervals I_i do not vary too much, the relative magnitude of ϱ_f^* in I_i can roughly be estimated by the number of branches of f mapping into I_i.

which is mapped into the interval I_i by $f(x)$. Assuming that $\hat{P}_f$ has exactly one fixed point ϱ_f^*, Ulam's conjecture was proved by Li[14], who showed that with increasing n the fixed point of $\mathbb{P}$ converges to ϱ_f^*.

It may be worthwhile to point out that for special classes of dynamical laws f one may even obtain an exact ϱ_f^* already for finite n. This holds for all broken linear transformations f satisfying T1 and C2 with a finite set $K := \bigcup_{\tau=0}^{\infty} f^{[\tau]}(J)$. One gets an analytic expression for ϱ_f^* from that $\mathbb{P}$ which particularly uses the decomposition of $[0,1]$ generated by K. An example is given in Figure 1. Proof: If K is finite, each interval I_i of the decomposition is mapped linearly onto a union of such intervals. So all step functions being constant in each I_i are transformed by $\hat{P}_f$ into functions of the same type.

In these finite dimensional cases $\varrho(x)$ may be represented by a vector $\boldsymbol{r}$ with n components r_i defined by

$$r_i := \varrho(x \in I_i) \cdot m(I_i)\,; \quad 1 \leqq i \leqq n\,, \tag{9}$$

the probability of I_i. ϱ_f^* corresponds to the eigenvector $\mathbb{P}\,\boldsymbol{r}_f^* = \boldsymbol{r}_f^*$ with eigenvalue 1. Since $\mathbb{P}$ is a stochastic, indecomposable matrix, the largest eigenvalues are of modulus 1 and the largest real eigenvalue is 1[15]. So $\boldsymbol{r}_f^*$ can also be found by repeated multiplication of an arbitrary normalized vector $\boldsymbol{r}$ by $\mathbb{P}$:

$$\lim_{\tau\to\infty} \mathbb{P}^\tau \boldsymbol{r}_0 = \boldsymbol{r}_f^*$$

normalization condition: $\sum_{i=1}^{n} r_{0i} = 1$.

Finally we give some results concerning a special family of broken linear transformations. We consider

$$N_p(x) := (-1)^{[px]}\, p\, x \,(\mathrm{mod}\ 1)\,, \quad x \in [0,1]\,, \quad p \in \mathbb{R}\,. \tag{10}$$

$[p\,x]$ is the integer part of $p\,x$. $N_p(x)$ is ergodic for $p > 1$. The reader may easily prove this statement by using T1 and C1 if $1 < p \leqq 2$ and applying C_2 if $2 < p$. Moreover the relations

$$N_p^{[\tau]}(x) = N_{p^\tau}(x)\,, \quad p, \tau = 1, 2, 3, \ldots \tag{11}$$

and

$$\varrho^*_{N_p}(x) = 1\,, \quad p = 2, 3, 4, \ldots \tag{12}$$

hold, i.e., the sequence $\{x_\tau\}$ in the course of the physical process covers the interval $(0,1)$ everywhere with equal weight.

IV. Dynamical Laws Related by Conjugation

As we already pointed out the propositions of T1 and T2 severely restrict the admitted dynamical laws f. To extend the range of validity of these theorems we make use of an equivalence relation described by Halmos[16] and by Ulam[17].

Two transformations $f\colon I \to I$ and $g\colon J \to J$ on intervals I and J are called conjugate if there exists a one-to-one map $h\colon I \xrightarrow{\text{onto}} J$ such that

$$g(x) = h\big(f[h^{-1}(x)]\big)\,. \tag{13}$$

In what follows the conjugating function h will always be assumed to be continuous and sufficiently smooth.

h establishes a one-to-one correspondence between the number sequences $\{x_\tau\}$ and $\{\Theta_\tau\}$ generated by

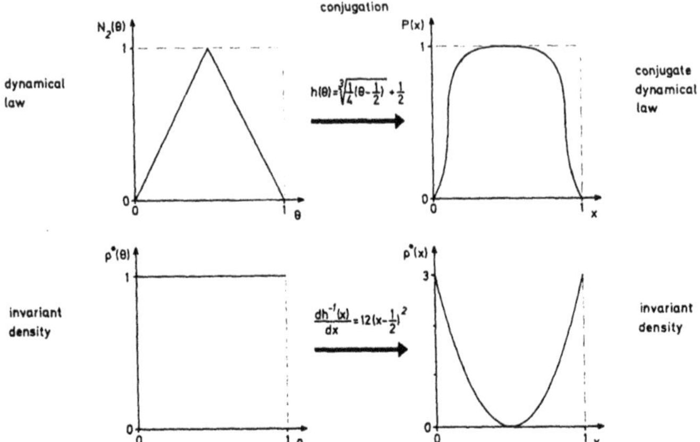

Fig. 2. Two conjugate dynamical laws and their corresponding invariant densities. $N_2(\Theta)$ satisfies the propositions of T1 and T2 but its derivate by conjugation, $P(x) = \sqrt{0.125-2\,|x-0.5|^3}+0.5$, obviously does not. Nevertheless T1 and T2 can be applied to $P(x)$ due to the conjugation. The corresponding transformation of the density is controlled by the first derivative of $h^{-1}(x)$.

$x_{\tau+1} = g(x_\tau)$ and $\Theta_{\tau+1} = f(\Theta_\tau)$, respectively. Since the invariant density is determined by the totality of generated sequences, h must also yield a relation between ϱ_f^* and ϱ_g^*. Using conservation of probability one may easily prove the relation

$$\varrho_g^*(x) = \varrho_f^*[h^{-1}(x)]\left|\frac{\mathrm{d}h^{-1}(x)}{\mathrm{d}x}\right|. \qquad (14)$$

Because of the one-to-one correspondence between sequences under g and f, ϱ_g^* is the asymptotic density for almost all $\{x_\tau\}$ if ϱ_f^* has this property with regard to $\{\Theta_\tau\}$. So the mere existence of a conjugating function h suffices to guarantee the existence and uniqueness of ϱ_g^* provided ϱ_f^* exists and is unique. h need not be a linear map. Thus g may be a function not satisfying the infimum condition (6) whereas f does so (Figure 2). Consequently the range of validity of T1 and T2 is extended to all dynamical laws which turn out to be conjugated to any f satisfying the propositions of T1 and T2 [17a].

Using the invariance of ϱ_g^* as well as the equivalence of ergodicity and the existence of a unique, invariant density which generates an absolutely continuous invariant measure, one may easily prove the following statement:

T3: Let f and g be dynamical laws conjugate by a one-to-one piecewise differentiable h. Then,

(a) if f is ergodic, g is so, too;

(b) if f is strongly mixing, so is g.

That means ergodicity and strong mixing are not characteristics of single dynamical laws, but properties of complete equivalence classes generated by conjugation.

Further we remark that Eq. (14) provides us with a possibility to construct dynamical laws g generating a required density ϱ_g^* by choosing an appropriate h to a given f with already known density ϱ_f^*. In practice one must, of course, restrict oneself to dynamical laws f whose densities have a sufficiently simple structure because of the implicit dependence of ϱ_f^* on x via h^{-1}. Figure 3 shows some examples of dynamical laws and their corresponding invariant densities found by conjugation.

Adler and Rivlin have proven the strongly mixing property of Tchebychev-polynomials [18]

$$T_p(x) = \cos(p \arccos x), \quad p = 2, 3, 4, \ldots . \qquad (15)$$

These are conjugate to $N_p(\Theta)$ via $h(\Theta) = \cos(\pi\Theta)$. Therefore, according to T3, $N_p(\Theta)$, $p = 2, 3, 4, \ldots$ is strongly mixing, too.

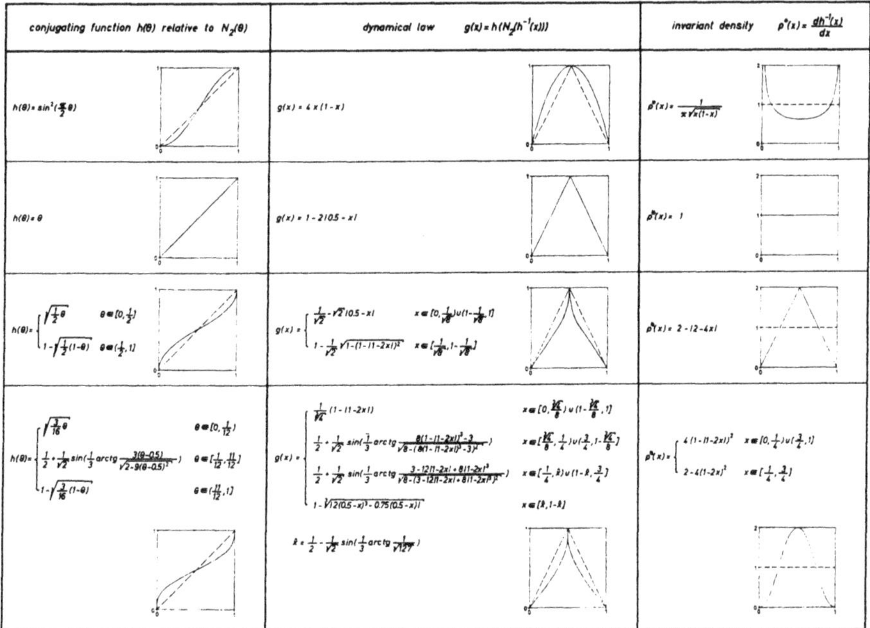

Fig. 3. Some dynamical laws, their invariant densities, and their conjugating functions relative to $N_2(\Theta)$. The second derivative of the dynamical law and of the respective invariant density have opposite signs. Slight changes in g may result in appreciable changes of $\varrho_g{}^*$.

V. Correlation Functions

We now study the time development of processes $\{x_\tau\}$ generated by a dynamical law f using the stationary correlation function (4). In this section we temporarily confine ourselves to ergodic laws f.

A process will be called δ-correlated if its correlation function has the structure

$$C_f(\tau) = C_f(0)\cdot\delta_{0\tau}\,. \tag{16}$$

All dynamical laws g: $[0,1]\to[0,1]$ which are conjugate to $N_{2n}(\Theta)$ with integer n and which have even symmetry in $[0,1]$ generate such processes. Proof: The conservation of symmetry requires the conjugating function $h(\Theta)$ to have odd symmetry relative to the point $[\frac{1}{2},\frac{1}{2}]$; for examples see Figure 3. That means the derivative of its inverse is of even symmetry in $[0,1]$ and so is the transformed density $\varrho_g{}^*(x)$ according to (14). Using these properties of g and $\varrho_g{}^*$ the correlation function $C_g(\tau)$ may be written in the form

$$C_g(\tau) = \int_{-\frac{1}{2}}^{\frac{1}{2}} x\, g^{[\tau]}(x+\tfrac{1}{2})\,\varrho_g{}^*(x+\tfrac{1}{2})\,\mathrm{d}x\,. \tag{17}$$

This integral vanishes for all $\tau = 1, 2, 3, \ldots$, and therefore the process generated by g must be δ-correlated. Another way to state the relation proven above is: The δ-correlation property of a process generated by a dynamical law with even symmetry is preserved under conjugation with odd symmetry. Notice: That does not mean δ-correlated processes can only be generated by symmetrical laws. The Tchebychev-polynomials $T_{2n+1}(x)$, $n = 1, 2, 3, \ldots$ are examples for antisymmetric laws generating δ-correlated processes. But this property in general is not preserved under conjugation with odd symmetry.

We now turn our attention to dynamical laws with less symmetry. To get a "feeling" for the changes occuring in the correlation function if there are deviations from even symmetry we consider a generalized version $\tilde{N}_p(x)$ of the broken linear transformation $N_p(x)$ of Section III. Let $\{y_\mu(\tau)\}$ with $y_{\mu-1}(\tau)<y_\mu(\tau)$ denote the set of points of discontinuity of $\tilde{N}_p^{[\tau]}$ and $[y_0(\tau), y_{p^\tau}(\tau)]=[0,1]$. Then

$$\tilde{N}_p[x;\{y_\mu(1)\}]: \tag{18}$$

$$:=\tfrac{1}{2}\left(1-(-1)^\mu\,\frac{2x-y_{\mu-1}(1)-y_\mu(1)}{y_\mu(1)-y_{\mu-1}(1)}\right)$$

if $x\in[y_{\mu-1}(1), y_\mu(1)]$; $\mu=1,2,\ldots,p$; $p\geqq 2$.

To calculate $C_{\tilde{N}_p}(\tau)$ we decompose the integration from 0 to 1 into subintegrals over $[y_{\mu-1}(\tau), y_\mu(\tau)]$, $\mu=1,2,\ldots,p$. Furthermore we use $\varrho^*_{\tilde{N}_p}(x)=1$ which is easily concluded from the structure of the Frobenius-Perron operator.

$$\int_0^1 x\tilde{N}_p^{[\tau]}(x)\,\mathrm{d}x$$

$$=\sum_{\mu=1}^{p^\tau}\int_{\frac{y_{\mu-1}(\tau)-y_\mu(\tau)}{2}}^{\frac{y_\mu(\tau)-y_{\mu-1}(\tau)}{2}}\left\{x+\frac{1}{2}\,[y_{\mu-1}(\tau)+y_\mu(\tau)]\right\}$$

$$\cdot\left\{\frac{1}{2}-(-1)^\mu\,\frac{x}{y_\mu(\tau)-y_{\mu-1}(\tau)}\right\}\mathrm{d}x\,.$$

Thus

$$C_{\tilde{N}_p}(\tau)=\tfrac{1}{6}\sum_{\mu=1}^{p^\tau}(-1)^\mu y_\mu(\tau)\,y_{\mu-1}(\tau)-\tfrac{1}{12}(-1)^{(p^\tau)}\,. \tag{19}$$

An immediate consequence of (19) is

$$C_{\tilde{N}_p}(0)=1/12\,. \tag{20}$$

For $\tau=1,2,3,\ldots$ a relation can be derived using the following equations which can be understood in terms of the geometric nature of the iteration process:

$$y_0(\tau)=0,\quad y_{p^\tau}(\tau)=1,\quad p=2,3,4,\ldots,\quad \tau=0,1,2,\ldots; \tag{21'}$$

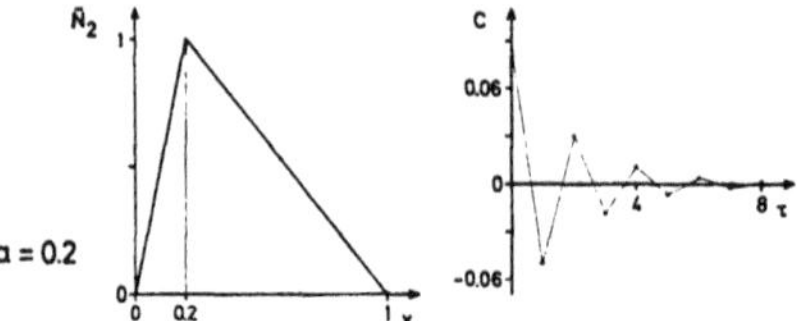

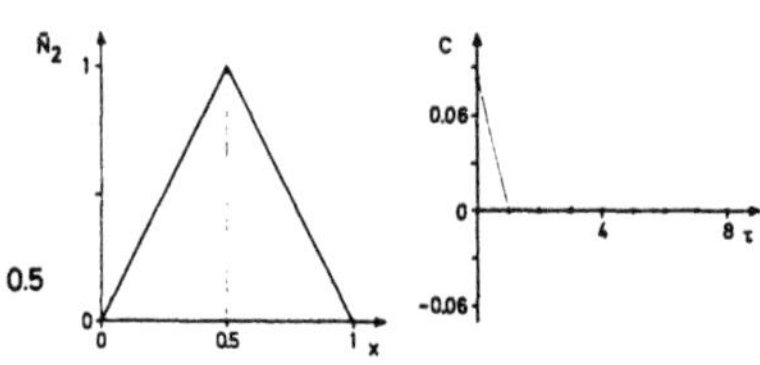

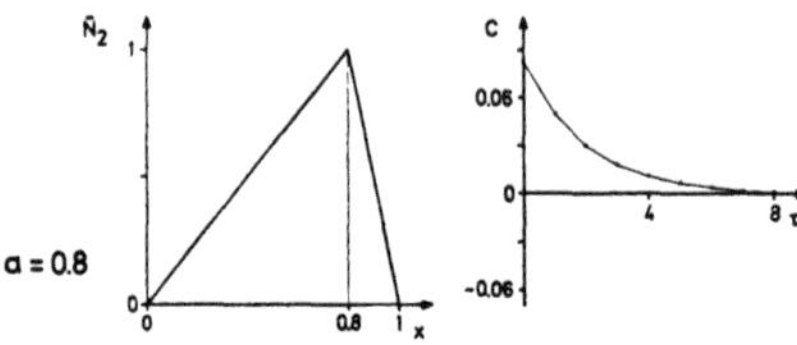

Fig. 4. **$\tilde{N}_2(x;a)$ and the corresponding correlation functions for three different values of a. The skewness of $\tilde{N}_2$ destroys the δ-correlation and results in exponentially decaying monotone or oscillating correlations.**

In general $\tilde{N}_p(x)$ yields an exponential decay of the time correlation with a characteristic time

$$\tau_c=-1/\ln\left|\sum_{a=1}^{p}(-1)^{a-1}\delta_a^2\right|. \tag{24}$$

According to the sign of $\Sigma\ldots C_{\tilde{N}_p}(\tau)$ may be monotone or oscillating. If $\Sigma\ldots=0$, $\tilde{N}_p$ generates a δ-correlated process. Figure 4 shows these three charac-

$$q:=p^{\tau-1},\quad \delta_a:=y_a(1)-y_{a-1}(1)\,,$$

$$\left.\begin{aligned} y_{(a-1)q+\mu}(\tau)&=\delta_a y_\mu(\tau-1)+y_{a-1}(1), & a&=1,3,5,\ldots\\ y_{(a-1)q+\mu}(\tau)&=-\delta_a y_{q-\mu}(\tau-1)+y_a(1), & a&=2,4,6,\ldots\end{aligned}\right\}\begin{aligned}&\tau=1,2,\ldots,\\ &\mu=0,1,2,\ldots,q\,.\end{aligned} \tag{21''}$$

If (21″) is introduced in (19) a recursion formula for $C_{\tilde{N}_p}(\tau)$ results:

$$C_{\tilde{N}_p}(\tau)=\left(\sum_{a=1}^{p}(-1)^{a-1}\delta_a^2\right)C_{\tilde{N}_p}(\tau-1)\;\begin{cases}p=2,3,4,\ldots,\\ \tau=1,2,3,\ldots,\end{cases} \tag{22}$$

$$C_{\tilde{N}_p}(\tau)=\frac{1}{12}\left(\sum_{a=1}^{p}(-1)^{a-1}\delta_a^2\right)^\tau\quad\begin{cases}p=2,3,4,\ldots,\\ \tau=0,1,2,\ldots\,.\end{cases} \tag{23}$$

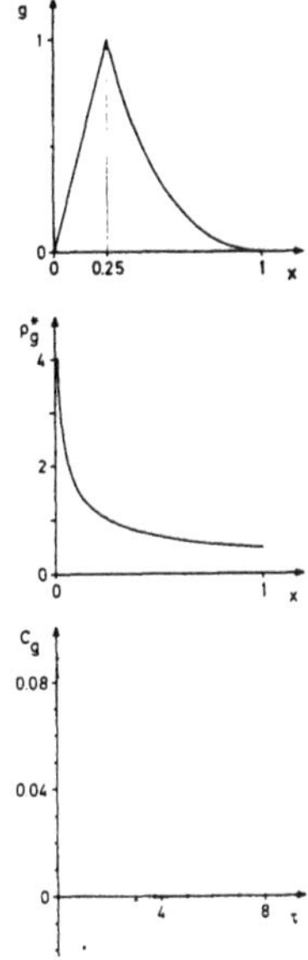

Fig. 5. The asymmetric dynamical law $g(x) = (1 - 2\,|\,0.5 - \sqrt{x}\,|)^2$, its invariant density, and its correlation function.

teristic cases for $\tilde{N}_2(x; a)$. In particular we get:

$$\left.\begin{aligned} C_{N_{2p}}\;(\tau) &= \tfrac{1}{12}\,\delta_{0\tau} \\ C_{N_{2p-1}}(\tau) &= \tfrac{1}{12}(2\,p+1)^{-2\tau} \end{aligned}\right\} p = 1, 2, 3, \ldots \quad (25)$$

The generic feature revealed by this calculation is that any skewness in the dynamical law tends to destroy the δ-correlated character of the process. The characteristic time τ_c increases with the skewness. But, of course, there are exceptions, e. g. the already mentioned Tchebychev-polynomials whose special curvature just compensates the effect of skewness.

If we want to generate non-δ-correlated processes, we might do so by making use of our above observations and applying a non-antisymmetric conjugating function on a symmetric dynamical law:

$$f(\Theta) = N_2(\Theta)\,, \quad h(\Theta) = \Theta^2\,.$$

The conjugation yields (see also Fig. 5):

$$g(x) = (1 - 2\,|\,0.5 - \sqrt{x}\,|)^2\,, \qquad (26)$$

$$\varrho_g{}^*(x) = 1/\sqrt{4\,x}\,, \qquad (27')$$

$$C_g(\tau) = \begin{cases} \frac{4}{45} & \tau = 0\,, \\ -\frac{7}{90}\left(\frac{1}{4}\right)^\tau & \tau = 1, 2, \ldots\,. \end{cases} \qquad (27'')$$

VI. A Misbehaved Conjugating Function

In the course of a synthesis of a dynamical law which generates processes with prescribed invariant density as well as prescribed correlation it might be valuable to have methods at hand that admit an independent adjustment of these two characteristics. We have already seen that conjugation with odd symmetry exclusively effects the density of symmetric laws but leaves the δ-correlation unchanged. Further we saw how the correlation functions of $\tilde{N}_p$ could be influenced by shifting their points of discontinuity whereas the density remained unchanged. Is this shifting procedure equivalent to a conjugation? We do not know the answer to this general problem. But the investigation of

$$\tilde{N}_2(x; a) = \begin{cases} x/a & x \in [0, a]\,, \\ (1-x)/(1-a) & x \in [a, 1]\,, \end{cases} \qquad (28)$$

reveals an obviously generic feature of the conjugating function

$$h_a(\Theta): \;\; \tilde{N}_2(x; a) = h_a(N_2[h_a{}^{-1}(x)])\,. \qquad (29)$$

There must be an h_a for all $a \in (0, 1)$ since $\tilde{N}_2(x; a)$ can be generated by a series of infinitesimal shift operations which obviously cannot alter the totality of generated processes in its structure. So a one-to-one correspondence can be established.

We might start trying to find an h_a by mapping the left branch of $N_2(\Theta)$ on the left branch of $\tilde{N}_2(x; a)$. This is accomplished by

$$h_a(\Theta) = \Theta^{-\ln a/\ln 2}\,. \qquad (30)$$

But this cannot be the correct h_a because

(i) it does not take into account the right branches of both laws; and

(ii) the first derivative of h_a must, according to (14), be constant and equal to unity, since both laws generate a constant density.

So we have to try another method using intrinsic properties of N_2 as well as $\tilde{N}_2$. Both laws assign a unique predecessor to the point 1. Under $N_2(\Theta)$ the predecessor is $\frac{1}{2}$, under $\tilde{N}_2(x; a)$ it is a. We now generate sequences of predecessors for both laws and construct h_a by relating corresponding points to each other (see Figure 6). (It should be noticed that this method is of rather general character and in many cases yields a numerical access to the conjugating function.) If we write L for choosing the

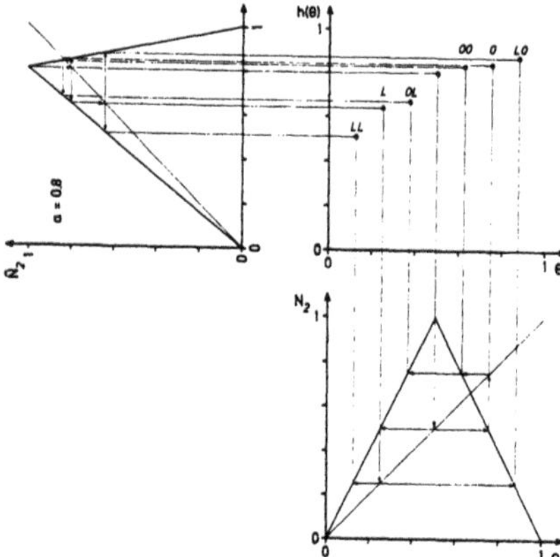

Fig. 6. Construction of a conjugating function $h(\Theta)$ by relating corresponding points of two dynamical laws to each other.

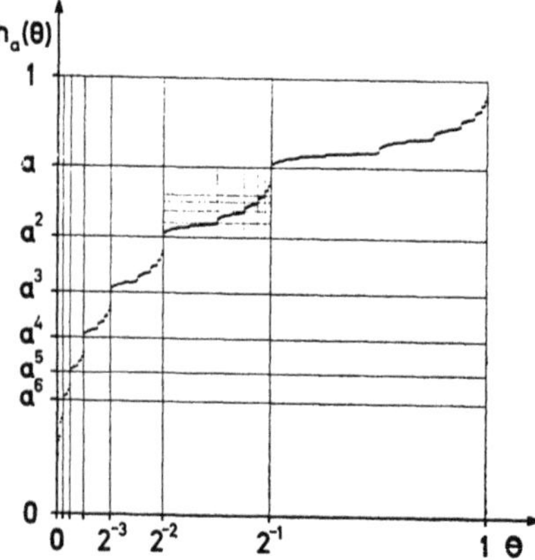

Fig. 7. The misbehaved conjugating function $h_a(\Theta)$ relating $N_2(\Theta)$ to $\tilde{N}_2(x; a)$ for $a=0.8$. The grid reveales the Cantor set structure of h_a.

predecessor on the left branch and 0 for choosing it on the right branch, each point is uniquely determined by its L0-pattern, which might be interpreted as a binary number. Figure 7 shows a result obtained this way by taking into account all points corresponding to binary numbers up to eight bits. Is this solution compatible with the condition on the derivative of h_a? A closer inspection of this problem shows that all points constructed above are points of discontinuity where h_a is not differentiable. The set of these is not countable since the corresponding binary numbers take on all values in $(0, 1)$. h_a is an everywhere continuous and nowhere differentiable, strictly monotone function with Cantor set structure.

VII. Pure Chaos and Mixed States

Up to now we have considered ergodic dynamical laws. All iterates of these were ergodic, too. Physical systems described by this kind of laws are comparatively easy to deal with. Since ergodicity implies the instability of all periodic processes, which

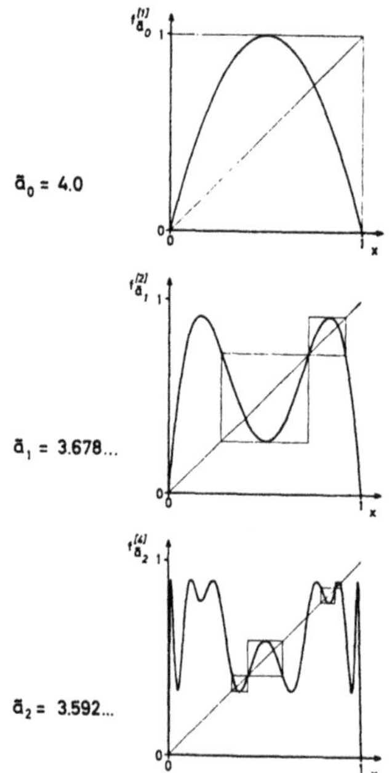

Fig. 8. Iterates of $f_a(x) = a\,x(1-x)$. If the parameter a takes on the critical value $\tilde{a}_\mu$, the iterate $f_a{}^{[2^\mu]}$ maps certain subintervals of $[0, 1]$ (indicated by squares) onto themselves.

$\tilde{a}_0 = 4.0$

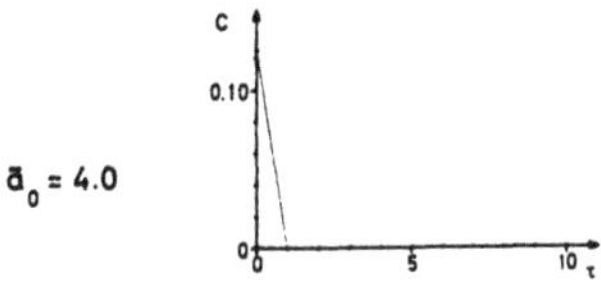

$\tilde{a}_1 = 3.678...$

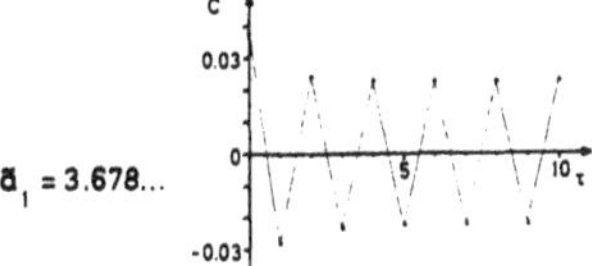

$\tilde{a}_2 = 3.592...$

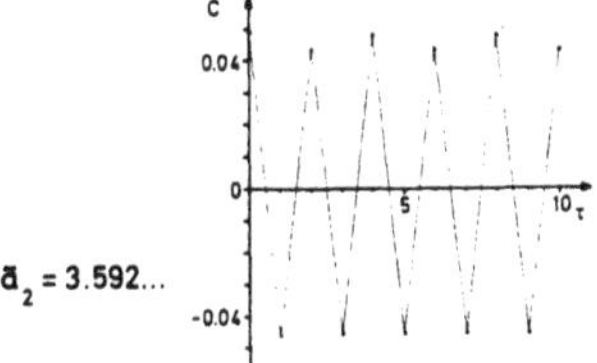

Fig. 9. Correlation functions of $f_a(x) = a\,x(1-x)$ for the critical values $\tilde{a}_\mu$, $\mu = 0, 1, 2$.

therefore become completely unimportant from the physical point of view, aperiodic behaviour prevails. If not only the dynamical law itself but also all of its iterates are ergodic, we call the system purely chaotic.

But there are also mixed states where periodic and chaotic time development are mingled with each other. In what follows we discuss these generic mixed states in the logistic equation [19]

$$f_a(x) = a\,x(1-x)\,, \quad a \in [1, 4]\,. \tag{31}$$

If a is sufficiently small, f_a generates periodic processes which, as a is increased, change their period by repeated bifurcation from 1 to $2, 4, 8, \ldots, 2^n, \ldots$ There is a threshold $a_\infty \cong 3.5699$ of the parameter a where the limit 2^n, $n \to \infty$ for the periodicity is reached. F or $a = 4$ f_a and all its iterates are ergodic, i. e. f_4 generates pure chaos. Within $(a_\infty, 4)$ there are values of a where some iterates $f_a^{[\tau]}$ are decomposable despite the ergodicity of f_a itself. Figure 8 shows some examples of iterates at such values. Under these iterates the complete interval $[0, 1]$ obviously breaks up into several independent subintervals on which the respective iterate is ergodic. The dynamical law f_a periodically maps these subintervals onto each other and thus establishes a connection between them. Therefore the complete process can be understood as a superposition of a periodic motion and a chaotic motion in state space, which becomes particularly obvious by means of correlation functions. Figure 9 shows some of these corresponding to the cases depicted in Figure 8. The periodic part of the motion yields the oscillations whereas the chaotic part produces a decrease of the amplitudes to constant values. Since the subintervals mentioned above are comparatively small with respect to their relative distances this decrease is small in comparison to the oscillations.

To survey the solution set of the logistic equation we indicate these states of mixed and pure chaos by their invariant densities in a generalized bifurcation diagram (Figure 10). The gaps between the critical values $\tilde{a}_\mu$ might be filled by considering other periodic processes, e. g. periods three or five, and the corresponding states of mixed chaos.

A numerical analysis shows that the sequences $\{a_\mu\}$ and $\{\tilde{a}_\mu\}$ of bifurcation points approximately satisfy exponential laws:

$$\begin{aligned} &a_\mu \cong c_0 - c_1 \cdot e^{-c_2\mu}\,; \\ &c_0 = 3.56994567 \pm 1.3 \cdot 10^{-7}\,, \\ &c_1 = 2.628 \pm 0.13\,, \\ &c_2 = 1.543 \pm 0.02\,, \end{aligned} \tag{32}$$

$$\begin{aligned} &\tilde{a}_\mu \cong \tilde{c}_0 + \tilde{c}_1 \cdot e^{-\tilde{c}_2\mu}\,; \\ &\tilde{c}_0 = 3.56994565 \pm 1.3 \cdot 10^{-7}\,, \\ &\tilde{c}_1 = 2.152 \pm 0.06\,, \\ &\tilde{c}_2 = 1.530 \pm 0.03\,. \end{aligned} \tag{33}$$

Each sequence numerically converges to

$$a_\infty = 3.5699456\ldots \tag{34}$$

which therefore is the threshold of chaos.

VIII. Summarizing Discussion

We have studied the statistics arising from discrete, one-dimensional processes in terms of the invariant asymptotic distributions and of the stationary correlations between successive events during

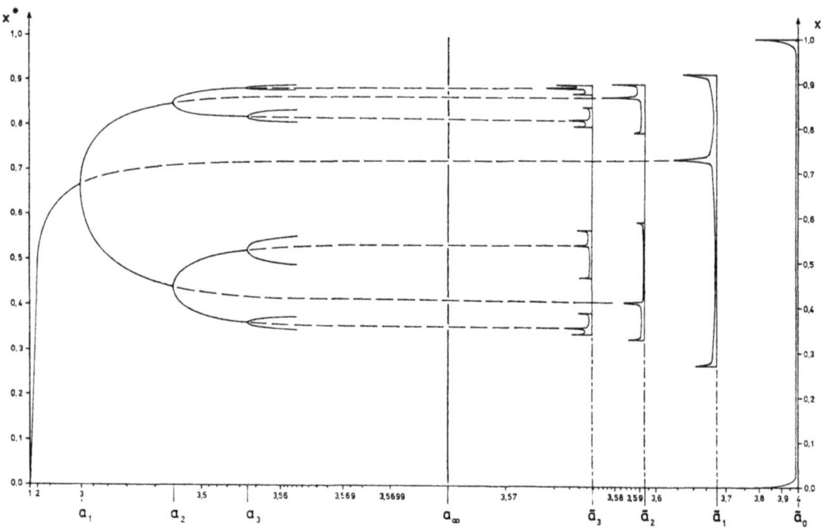

Fig. 10. Generalized bifurcation diagram for $f_a(x) = a\,x(1-x)$. The chaos states at $\tilde{a}_\mu$ are indicated by their invariant densities. [To achieve a better graphical representation a nonlinear scale was used on the abscissa: $a' = \sqrt[5]{\tanh(a-a_\infty)}$.]

the process. Following Rössler's [20] considerations of chaotic degrees of freedom one might relate this kind of process to multi-dimensional continuous systems. The results reported may shed some light on the remarkable fact that simple deterministic processes may physically appear as being chaotic. Chaos needs a probabilistic description. It then is necessary to derive the connection between the dynamical process and its probability distribution or correlations. In a conservative $N \sim 10^{23}$ particle system this connection is: Hamilton's equations of motion $\rightarrow$ canonical distribution. In a $N=1$ degree of freedom system, being discrete in time, the form of the dynamical law determines the probabilistic properties much more detailed, although general concepts like ergodicity apply as well. It is this implication of a given nonlinear deterministic process on its corresponding probability distribution function which lies at the basis of many classical problems as for example the statistical theory of turbulence. The occurence of mixed states reveals a kind of steady transition from purely periodic to purely chaotic behaviour. It will probably be very difficult to get experimentally hold of these mixed states since they require a very precise setting of test parameters.

1 G. Ahlers, Phys. Rev. Lett. **33**, 1185 [1974].
2 J. P. Gollub and H. L. Swinney, Phys. Rev. Lett. **35**, 927 [1975].
3 E. N. Lorenz, J. Atmos. Sci. **20**, 130 [1963].
4 J. B. McLaughlin and P. C. Martin, Phys. Rev. Lett. **33**, 1189 [1974].
5 O. E. Rössler, Z. Naturforsch. **31 a**, 1168 [1976].
6 P. R. Halmos, Lectures on Ergodic Theory, Chelsea Publ. Comp., New York 1956, p. 25.
7 V. I. Arnold and A. Avez, Ergodic Problems of Classical Mechanics, W. A. Benjamin Inc., 1968, p. 17.
8 We confine us to those parts of their statements which we will use later on.
9 A. Lasota and J. A. Yorke, Trans. Am. Math. Soc. **186**, 481 [1973].
10 T.-Y. Li and J. A. Yorke, Ergodic Transformations from an Interval into Itself, Preprint 1975.
11 The elements of J will be referred to as points of discontinuity. In what follows it will be useful to include 0 and 1 in J.
12 See example 2.1 in [14].

[13] S. M. Ulam, A Collection of Mathematical Problems, Interscience Tracts in Pure and Applied Mathematics **8**, pp. 73 [1960].
[14] T.-Y. Li, Journal of Approximation Theory **17**, 177 [1976].
[15] F. R. Gantmacher, Matrizenrechnung I, II, VEB Deutscher Verlag der Wissenschaften, Berlin 1958.
[16] same as [6], p. 44.
[17] same as [13], pp. 69.
[17a] As we have been informed by O. E. Rössler, conjugation relations are also considered by S. Smale and R. F. Williams, J. Math. Biol. **3**, 1 [1976].
[18] R. L. Adler and T. J. Rivlin, Proc. Am. Math. Soc. **15**, 794 [1964].
[19] F. C. Hoppensteadt and J. M. Hyman, Periodic Solutions of a Logistic Difference Equation (Preprint 1976), submitted for publication to SIAM J. Appl. Math.
A general parabolic transformation is discussed by P. J. Myrberg, Ann. Acad. Sci. Fennicae, series A, 256 [1958]; 268 [1959]; 336/3 [1963].
[20] O. E. Rössler, Z. Naturforsch. **31 a**, 1664 [1976].

Annals of the New York Academy of Sciences **357** 282–91 (1980)

NOISY PERIODICITY AND REVERSE BIFURCATION*

Edward N. Lorenz

Department of Meteorology
Massachusetts Institute of Technology
Cambridge, Massachusetts 02139

In studying apparently periodic phenomena modeled in the laboratory or simulated on the computer, we often find, upon close examination, that the periodicity is noisy. FIGURE 1a is a laboratory example. It is a trace of the temperature of a fluid at a fixed location in a rotating, differentially heated vessel recorded by Hide et al. during a 20-minute interval as a chain of waves passed by.[1] Since no fluid experiment can be perfectly controlled, one might assume that the failure of each peak in the curve to duplicate the third peak preceding it represents the experimental uncertainty. The authors have established, however, that the differences between the peaks are a real feature of the process they are investigating.

FIGURE 1b is a computer example. It shows the variations of one of a set of three variables governed by a system of ordinary differential equations during 930 iterations, after transient effects have died out. Here, likewise, the failure of each peak to duplicate the second peak preceding it does not represent computational uncertainty; it is the behavior to be expected.

We shall first examine noisy periodicity as a phenomenon that might be produced by iteration of a differentiable mapping

$$x_{n+1} = f(x_n). \tag{1}$$

FIGURE 1c is an example; the 16 line segments connecting successive iterates of x_0 are included only to make the chronological order stand out. Again, every peak fails to duplicate the second peak preceding it.

We shall denote by $\{x_0\}$ and $\{x_0\}_N$ the sequences $\{x_0, x_1, x_2, \ldots\}$ and $\{x_0, x_N, x_{2N}, \ldots\}$ generated by f and its Nth iterate, f^N. A sequence $\{x_0\}$ is *periodic* of period N if $x_N = x_0$ and $x_m \neq x_0$ when $0 < m < N$. It is *eventually periodic* if $x_{k+N} = x_k$ for some $k > 0$ and *asymptotically periodic* if $x_k - y_k \rightarrow 0$ as $k \rightarrow \infty$ for some periodic sequence $\{y_0\}$. Otherwise, it is *aperiodic*. It is *steady* if it is periodic of period 1.

We shall call an aperiodic sequence $\{x_0\}$ *semiperiodic* of period N if the ranges of $\{x_k\}_N$ are disjoint for $0 \leq k < N$ and the ranges of $\{x_k\}_m$ overlap for $0 \leq k < m$ when $m > N$. An aperiodic sequence that is not otherwise semiperiodic is semiperiodic of period 1. A variance spectrum of a semiperiodic sequence with $N > 1$ contains lines superposed on a continuum.

If the ranges of $\{x_k\}_N$ are very narrow, $\{x_0\}$ may be mistaken for a periodic sequence from which transient effects have not yet disappeared, but, if sufficient precision is used, the periodicity will be seen to be noisy. The sequence in FIGURE 1c is semiperiodic of period 4. The curve in FIGURE 1b, and possibly that in FIGURE 1a, is like FIGURE 1c in that the sequences of successive maxima are semiperiodic.

*This research was supported by the Climate Dynamics Program of the National Science Foundation, grant no. NSF-g 77 10093 ATM.

A sequence $\{x_0\}$ is stable if $y_k - x_k \to 0$ as $k \to \infty$ for every sequence $\{y_0\}$ where $y_0 - x_0$ is sufficiently small. Otherwise, it is unstable. A periodic sequence $\{x_0\}$ is stable if $|\Lambda| < 1$ and unstable if $|\Lambda| > 1$, where Λ is the product of the N values of the derivative $f'(x)$.

We shall now restrict our attention to the quadratic mapping

$$x_{n+1} = \tfrac{1}{2} x_n^2 - a. \tag{2}$$

For suitable values of a, (**2**) is equivalent to the familiar quadratic mapping of the unit interval. We have chosen the form (**2**) to make $f'(x) = x$. Many of the statements that follow apply to more general mappings.

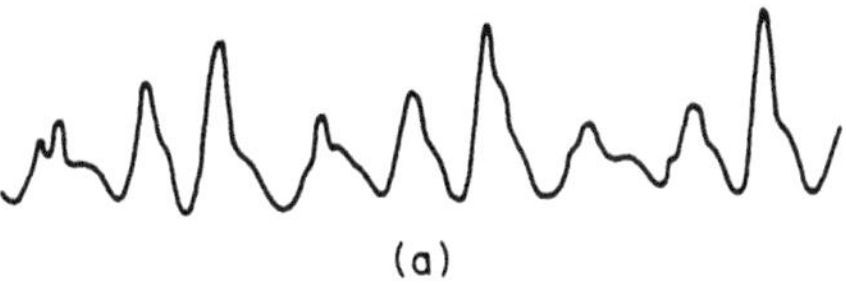

(a)

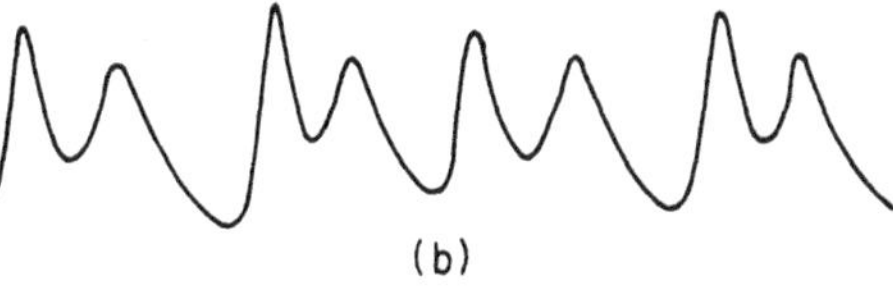

(b)

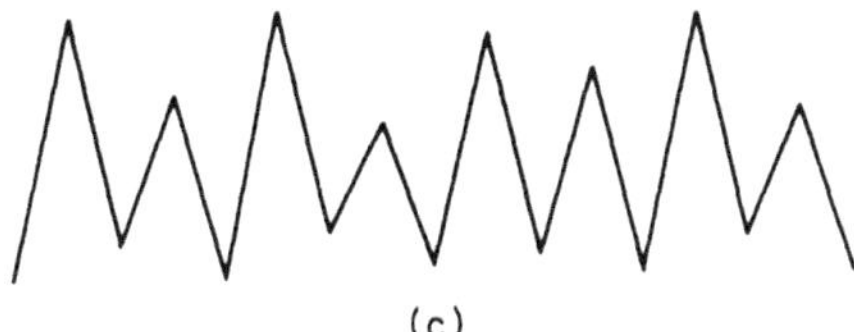

(c)

FIGURE 1. Examples of semiperiodic or apparently semiperiodic variables (a) from the laboratory, (b) from a system of differential equations, and (c) from a mapping.

A value w_0 of x for which $f'(w_0) = 0$ is a *singularity* of f. For (**2**), the lone singularity is at $w_0 = 0$. We shall call the sequence $\{w_0\}$ the singular sequence. A frequently cited theorem states that if a stable periodic sequence $\{x_0\}$ exists, then $w_k - x_{k+m} \to 0$ for some m as $k \to \infty$.[2-4] A corollary is that there is at most one stable periodic sequence. We shall say that a is periodic if there is a stable periodic sequence and aperiodic otherwise.

For $a > -\tfrac{1}{2}$, the mapping (**2**) generates two steady sequences, $\{s_0\}$ and $\{u_0\}$, where $s_0 = 1 - (1 + 2a)^{1/2}$ and $u_0 = 2 - s_0$. For a in $(-\tfrac{1}{2}, 4)$, the interval $(-a, u_0)$ of x is mapped into itself. We shall call the interval $(-\tfrac{1}{2}, 4)$ of a the *principal band*.

In the principal band, $\{u_0\}$ is always unstable, but, for $-\tfrac{1}{2} < a < \tfrac{3}{2}$, $\{s_0\}$ is stable, so a is periodic of period 1. For some values of a in $(-\tfrac{3}{2}, 4)$, including those for which the singular sequence $\{w_0\}$ is eventually, but not immediately, steady, a is aperiodic. Outside the principal band, if $a < -\tfrac{1}{2}$, all sequences go to infinity as $n \to \infty$. If $a > 4$, $\{s_0\}$ and $\{u_0\}$ are still steady and unstable, but some sequences $\{x_0\}$ with x_0 in $(-a, u_0)$, including $\{w_0\}$, go to infinity.

We shall call a value of a_0 of a for which the singular sequence $\{w_0\}$ is periodic a singular value of a. A singular value is stable because $\Lambda = 0$. For period 1, the singular value is $a_0 = 0$.

For variations of a and x_0 about a_0 and w_0,

$$dx_1 = x_0 dx_0 - da, \tag{3}$$

whence, by iteration,

$$dx_N = Px_0 dx_0 - Q da. \tag{4}$$

where

$$P = \prod_{m=1}^{N-1} x_m,$$

$$Q = 1 + \sum_{k=1}^{N-1} \prod_{m=1}^{k} x_m.$$

For sufficiently small variations, even though x_N may change sign, x_m will remain relatively close to w_m for $0 < m < N$, and P and Q will not vary greatly.

In general, P and Q are large. If P and Q were true constants, then **(4)** could be integrated, yielding

$$Px_n = \tfrac{1}{2}(Px_0)^2 - PQ(a - a_0). \tag{5}$$

The Nth iterate of **(2)**, i.e., **(5)**, would then be identical in form with **(2)**. Thus, for $a > a'$, where $PQ(a' - a_0) = -\tfrac{1}{2}$, the mapping would generate steady sequences $\{t_0\}_N$ and $\{v_0\}_N$, and, hence, periodic sequences $\{t_0\}$ and $\{v_0\}$, where $Pt_0 = 1 - (1 + 2PQ(a - a_0))^{1/2}$ and $Pv_0 = 2 - Pt_0$. For a in (a', a''), where $PQ(a'' - a_0) = 4$, the interval $(-b, v_0)$ of x, where $b = Q(a - a_0)$, would be mapped into itself by **(5)**. Moreover, **(2)** would map $(-b, v_0)$ into an interval near $-a$ that would be disjoint from $(-b, v_0)$. Hence, for a in (a', a'') and x in $(-b, v_0)$, aperiodic sequences would be semiperiodic, and all periods would be multiples of N.

The actual variation of P and Q alters the values of a' and a'', but does not appear to invalidate the qualitative conclusions. For example, the lone singular value a_0 for period 3 is 3.50976, so $P = -9.299$ and $Q = -5.649$. The estimated values of a' and a'' would, therefore, be 3.50024 and 3.58590. The true values, which may be found by rapidly converging algorithms using the estimated values as initial approximations, are 3.5 (exactly) and 3.58066. The algorithms are based on equation 5 and make use of the existence of sequences $\{x_0\}$ with $x_N = x_0$ and $Px_0 = +1$, if $a = a'$, and $x_0 = 0$ and $x_{2N} = -x_N$, if $a = a''$.

We shall call a true interval (a', a'') a *semiperiodic band* of a. We shall say that a is semiperiodic if it is aperiodic and lies in a semiperiodic band. The principal band is semiperiodic of period 1. Different bands of the same period, N, are distinguished by the sequence of plus and minus signs in $\{w_0\}$.

In a semiperiodic band, $\{v_0\}$ is always unstable, but $\{t_0\}$ is stable for the smaller values of a, and a is periodic. Some of the larger values of a are semiperiodic. Outside the band, if $a < a'$, all sequences $\{x_0\}_N$ leave $(-b, v_0)$ and no periodic sequence of period N with the appropriate sequence of signs exists. If $a > a''$, then $\{t_0\}$ and $\{v_0\}$ are still periodic and unstable, but $\{w_0\}_N$ leaves $(-b, v_0)$, and a is no longer semiperiodic.

Since a semiperiodic band is topologically a small copy of the principal band, which contains semiperiodic bands, each semiperiodic band must contain semiperiodic bands, which in turn contain more semiperiodic bands, etc. A band that is contained in no other band except the principal band will be called a *prime* band; other bands will be called *composite*. The period of a composite band is obviously a composite number, but the converse does not hold. The sequence in FIGURE 1c satisfies (**2**) for $a = 2.85$, which is in a composite band of period 4. TABLE 1 gives (a', a'') for all prime bands of period ≤ 6.

We can now describe a routine that will yield (with an infinite amount of work) the complete structure of the principal band, i.e., the arrangement of the steady, periodic, semiperiodic, and aperiodic values of a. We first find all prime semiperiodic bands and place them with their proper periods in their proper locations in the band; these are countable in number and do not overlap. To the left of the first prime band, a is steady; between prime bands, a is aperiodic. We then place composite bands in the prime bands, and more composite bands in the composite bands, until each prime band, and hence each composite band, has attained the structure of the principal band.

TABLE 1

LOWER AND UPPER LIMITS a', a'' AND WIDTHS $a'' - a'$ OF PRIME SEMIPERIODIC BANDS OF PERIOD $N \leq 6$ FOR EQUATION 2

N	a'	a''	$a'' - a'$
2	1.50000	3.08738	1.58738
5	3.24879	3.26672	0.01793
3	3.50000	3.58066	0.08066
5	3.72117	3.72466	0.00349
6	3.81450	3.81503	0.00053
4	3.88110	3.88552	0.00442
6	3.93353	3.93369	0.00016
5	3.97082	3.97108	0.00026
6	3.99275	3.99277	0.00002

We are not aware of a proof that the Lebesgue measure of the set of aperiodic values of a exceeds zero, but we have previously offered considerable evidence favoring a positive measure,[5] while Collet and Eckmann have given a proof for a somewhat similar mapping.[6] Let r_1, r_2, and r_3 denote the fractions of the principal band for which a is periodic of period 1, in a prime semiperiodic band, and between prime bands, respectively, and let s_1 and s_3 denote the fractions for which a is periodic and aperiodic. Then $r_1 + r_2 + r_3 = 1$ and $s_1 + s_3 = 1$; $r_1 = 0.44$ and, from a crude extrapolation of TABLE 1, we estimate that $r_2 = 0.39$ and $r_3 = 0.17$. With the approximation that each semiperiodic band is a small but otherwise exact copy of the principal band, $s_3/s_1 = r_3/r_1$, so $s_1 = 0.70$ and $s_3 = 0.30$, and the fraction of the principal band for which a is semiperiodic of period >1 is $s_3 - r_3 = 0.13$.

Perhaps the most interesting prime band is the single band of period 2, which begins at $a' = 1.5$ with a bifurcation of period 1 and terminates at $a'' = 3.0874$, where the condition $w_4 = -w_2$ can be satisfied. We shall determine what the structure of the principal band would be if there were no other prime bands, so that the periods of all composite bands would be powers of 2 and all values of $a > a''$ would be aperiodic. FIGURE 2a shows schematically, for each $a < a'$, the value s_0 that w_k approaches as $k \rightarrow \infty$, and, for $a > a''$, the continuum of values that would form the range of $\{w_0\}$. A gap has been left between a' and a''.

Since the band of period 2 belonging in the gap must be a small copy of the principal band with all periods doubled and with the number of admissible values of x doubled where it is finite, we must fill the gap with two reduced copies of FIGURE 2a, one upside down. FIGURE 2b shows the result of doing this. A band of period 4 now belongs in the remaining gap, so we must fill it with four copies of FIGURE 2b. This gives us FIGURE 2c, which still contains a gap. We see eight narrow continua to the right of the gap reaching out like fingers to meet the eight curves to the left. The gap becomes filled by a nested sequence of bands when the process is continued to infinity.

On the left in FIGURE 2c, we see the familiar bifurcations to periods of successively higher powers of 2.[7,8] The successive values of a', which appear in TABLE 2 for periods up to 2^{12}, converge to $a_2 = 2.80231$, and the ratios of successive differences, $a_2 - a'$,

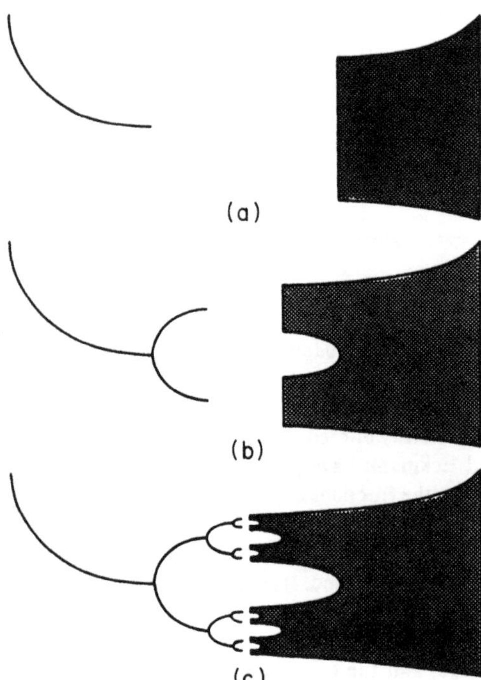

FIGURE 2. A schematic illustration of the procedure for constructing a nested sequence of semiperiodic bands. See text for details.

TABLE 2

LOWER AND UPPER LIMITS a', a'' AND WIDTHS $a'' - a'$ OF NESTED SEMIPERIODIC BANDS OF PERIOD $N = 2^M$, $M \leq 12$ FOR EQUATION 2

M	N	a'	a''	$a'' - a'$
0	1	−0.50000000	4.00000000	4.50000000
1	2	1.50000000	3.08737803	1.58737803
2	4	2.50000000	2.86071526	0.36071526
3	8	2.73619788	2.81481024	0.07861236
4	16	2.78809231	2.80498435	0.01689204
5	32	2.79926248	2.80288299	0.00362051
6	64	2.80165748	2.80243330	0.00077582
7	128	2.80217054	2.80233664	0.00016610
8	256	2.80228043	2.80231600	0.00003557
9	512	2.80230396	2.80231158	0.00000762
10	1024	2.80230900	2.80231064	0.00000164
11	2048	2.80231008	2.80231043	0.00000035
12	4096	2.80231032	2.80231039	0.00000007

converge rapidly to 0.21417, the reciprocal of a ratio found by Feigenbaum to be characteristic of a wide class of mappings.[8]

On the right in FIGURE 2c, we see transitions to semiperiodicities of successively higher powers of 2 as a decreases. The successive values of a'', which also appear in TABLE 2, also converge to a_2, and the ratios of successive values of $a'' - a_2$ converge equally rapidly to 0.21417. The ratio of $a'' - a_2$ to $a_2 - a'$ converges to 0.18781, which appears to be another universal value.

We shall call the process of transition to successively lower semiperiodicities *reverse bifurcation.* We feel that the reverse bifurcation of semiperiodicities to successive powers of 2 as a decreases is as significant a feature of the structure of the principal band as the more familiar bifurcation of periodicities to successive powers of 2 as a increases. FIGURE 3 shows the same nested sequence of bands, drawn to scale for equation 2.

We must now consider the effect of the remaining prime bands, which we neglected in constructing FIGURES 2 and 3. Since these bands occur only where $a > a_2$, their effect on the left portion of FIGURE 2a and, hence, of FIGURES 2b and 2c, is nil. In the right portion of FIGURE 2a, however, the solid shading must now be interpreted as meaning that, for some of the included values of a, the range of $\{w_0\}$ is as shown. Other values of a lie in semiperiodic bands of period ≥ 3. Similarly, just to the right of the gap in FIGURE 2c, the indicated ranges of $\{w_k\}_8$ for $0 \leq k < 8$ are valid only for some values of a; other values lie in composite bands. Needless to say, within every semiperiodic band of period N, prime or composite, there is a nested sequence of composite bands whose periods are the products of N with successive powers of 2.

To extend the concept of semiperiodicity to solutions of differential equations, we might define a function of time to be semiperiodic if its spectrum contains lines and a continuum. If the function possesses a clearly defined succession of maxima and minima, we might instead define it to be semiperiodic if its sequence of maximum or minimum values is semiperiodic. The two definitions are not equivalent, since, in the

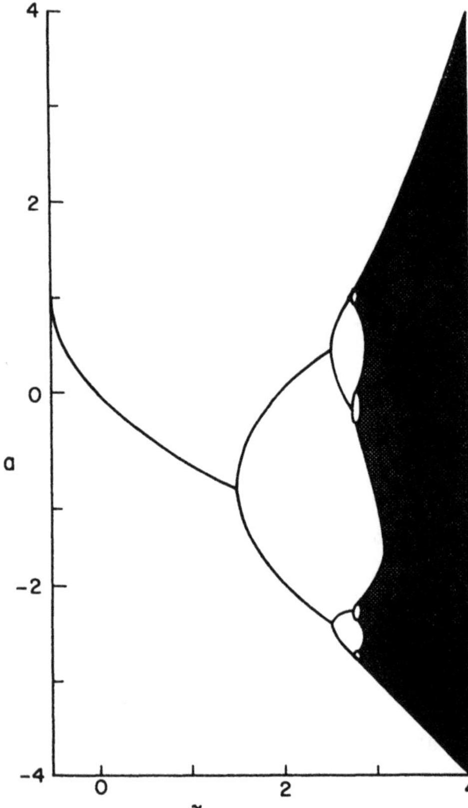

FIGURE 3. Bifurcations to periodicities and reverse bifurcations to semiperiodicities of successive powers of 2 for equation 2.

latter case, the time intervals between successive maxima are, in general, neither uniform nor periodic. Hence, the function may pass in and out of phase with a sine curve of any chosen frequency, and no line need appear in the spectrum.

We shall investigate the semiperiodicity of solutions of the system

$$\frac{dx}{dt} = -\sigma x + \sigma y,$$

$$\frac{dy}{dt} = -xz + rx - y, \tag{6}$$

$$\frac{dz}{dt} = xy - bz.$$

For $b = 8/3$, $\sigma = 10$, and $r = 28$, we have found that the general solution is aperiodic.[9] Robbins has found that the solution is periodic for much higher values of r.[10] Its projection on the xz or yz plane resembles a figure eight with equal maxima of z, while the values of x or y at successive maxima of z differ only in sign. She has also observed bifurcations to periods of successively higher powers of 2 as r decreases.

By examining extended numerical solutions of (**6**) using a fourth-order Taylor series procedure with a time step, $\Delta t = (256b(r - 1))^{-1/2}$, we have found that, for $b = 8/3$ and $\sigma = 10$, the sequence of successive maxima of z bifurcates to period 2 when r is decreased to 312.98. As r is further decreased, a succession of bifurcations to higher powers of 2 culminates at $r_2 = 215.364$. This is followed by reverse bifurcations to semiperiodicities of successively lower powers of 2 until the sequence becomes completely aperiodic when r passes 203.04. In a sense, the total solution is still semiperiodic, since x and y continue to alternate in sign at successive maxima of z. This alternation is replaced by aperiodic behavior when r passes 197.6, which is the highest value of r for which the unstable fixed point (0, 0, 0) is in the attractor.

TABLE 3 gives the limiting values r'' and r' for the nested semiperiodic bands. To the precision of the computations, the lower part of TABLE 3, where r is near r_2, is a linear transformation of the lower part of TABLE 2, where a is near a_2. The same limiting ratios, 0.21417 and 0.18781, are present.

FIGURE 4 shows a spectrum of z for $r = 205$, where the sequence of maxima is semiperiodic of period 2. To perform the analysis, we made Fourier analyses of 16 runs of 2^{15} time steps or 87.81 time units each, and then averaged the squares of the real and imaginary parts of each Fourier component. Each run spans 344 maxima of z. The figure shows only the first 800 of the $2^{14} + 1$ spectral amplitudes, and these have been smoothed by averaging in groups of 5. The remainder of the spectrum tapers off to zero.

There is a continuum with several prominent wide bands, but superposed on this are three apparent lines at 172, 344, and 516 waves per run; these are completely unresolvable as bands even with the high resolution used. The corresponding wavelengths are 2, 1, and 2/3 maxima of z. The lines account for 25, 30, and 7 per cent of the variance of z, respectively, the remaining 38 percent being contained in the continuum. Evidently, z is also semiperiodic according to the spectral definition.

FIGURE 5 is a similarly obtained spectrum for $r = 200$, where the sequence of the maxima of z is aperiodic. The lines at 172 and 516 waves per run have been replaced by strong bands, but the line at 344, which accounts for 26 percent of the total variance, is as unresolvable as before. Again, z is spectrally semiperiodic.

Since the time intervals between successive maxima vary with an aperiodic

TABLE 3

LOWER AND UPPER LIMITS r'', r' AND WIDTHS $r' - r''$ OF NESTED SEMIPERIODIC BANDS OF PERIOD $N = 2^M$, $M \leq 6$ FOR EQUATION 6

M	N	r''	r'	$r' - r''$
1	2	203.04	312.98	109.94
2	4	212.94	229.40	16.46
3	8	214.82	218.21	3.39
4	16	215.252	215.967	0.715
5	32	215.340	215.492	0.152
6	64	215.359	215.393	0.034

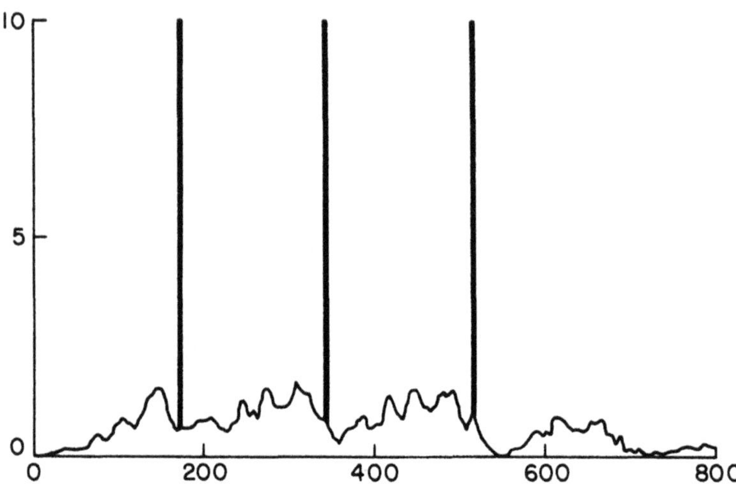

FIGURE 4. The spectrum of variable z satisfying equation 6 for $b = 8/3$, $\sigma = 10$, and $r = 205$. The abscissa is the frequency in waves per run of 2^{15} time steps. The ordinate is the variance contained in given frequency. The vertical lines at frequencies 172, 344, and 516 represent delta functions containing 25, 30, and 7 percent of the total variance.

component, the lines in the spectrum can occur only if the long intervals are compensated almost immediately by short intervals. Accordingly, in a run of 2^{20} time steps spanning $M = 11\,008$ maxima of z, we have determined the time, t_m, of occurrence of each maximum for $m \leq$ M. For an optimally chosen time interval, $\tau = 0.255575$, the range of $t_m - m\tau$ is only 0.16513. Thus, the maxima of a suitably chosen

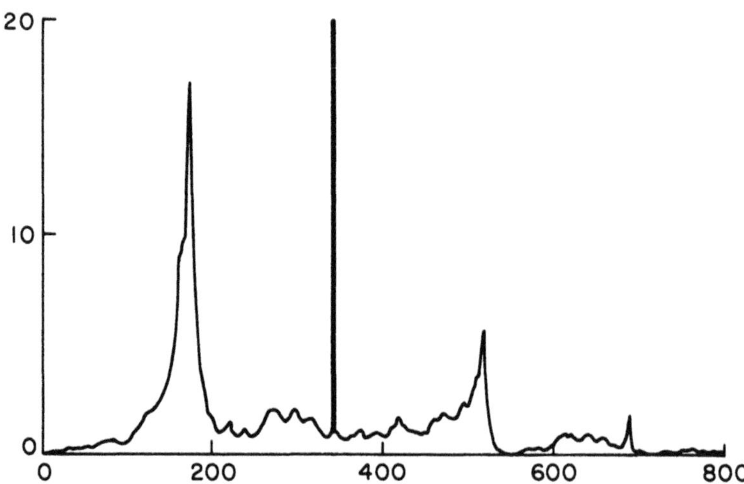

FIGURE 5. Same as in FIGURE 4, but for $r = 200$. The vertical line at frequency 344 represents a delta function containing 26 percent of the total variance.

sine curve never become completely out of phase with the maxima of z. If the apparent lines are actually narrow bands, it will require a much higher-resolution spectrum to demonstrate this.

We had not anticipated this feature of the spectrum from other known properties of the equations. The periodic solutions that are stable when $r > r_2$ still exist as unstable solutions when $r < r_2$, but the failure of the resulting fully developed aperiodic disturbances to destroy the periodicity of the solution was not expected.

The semiperiodic variable in FIGURE 1b is the variable z in equation 6 for $r = 210$. Whether or not the experimentally recorded variable in FIGURE 1a is semiperiodic we cannot say without further analysis, but we do know that systems of equations used to simulate rotating-fluid experiments possess semiperiodic solutions. More generally, semiperiodicity seems to be a normal phenomenon in mathematical and physical systems.

ACKNOWLEDGEMENT

The author has benefited from many discussions of the present subject with Dr. James Curry.

REFERENCES

1. HIDE, R., P. J. MASON & R. A. PLUMB. 1977. Thermal convection in a rotating fluid subject to a horizontal temperature gradient: Spatial and temporal characteristics of fully developed baroclinic waves. J. Atmos. Sci. **34:** 930–50.
2. JULIA, G. 1918. Mémoire sur l'itération des fonctions rationelles. J. Math. Pures. Appl. **4:** 47–245.
3. LORENZ, E. N. 1964. The problem of deducing the climate from the governing equations. Tellus **16:** 1–11.
4. SINGER, D. 1978. Stable orbits and bifurcation of maps of the interval. SIAM J. Appl. Math. **35:** 260–67.
5. LORENZ, E. N. 1979. On the prevalence of aperiodicity in simple systems. *In* Global Analysis. M. Grmela and J. E. Marsden, Eds.: 53–75. Springer-Verlag. New York.
6. COLLET, P. & J.-P ECKMANN. 1979. On the abundance of ergodic behavior for maps on the interval. Preprint.
7. MAY, R. M. 1976. Simple mathematical models with very complicated dynamics. Nature **261:** 459–67.
8. FEIGENBAUM, M. J. 1978. Quantitative universality for a class of nonlinear transformations. J. Stat. Phys. **19:** 25–52.
9. LORENZ, E. N. 1963. Deterministic nonperiodic flow. J. Atmos. Sci. **20:** 130–41.
10. ROBBINS, K. A. 1979. Periodic solutions and bifurcation structure at high R in the Lorenz model. SIAM J. Appl. Math. **36:** 457–72.

Physical Review Letters **45** 154–6 (1980)

Scaling Behavior of Chaotic Flows

B. A. Huberman
Xerox Palo Alto Research Center, Palo Alto, California 94304

and

J. Rudnick
Physics Department, University of California, Santa Cruz, California 95064
(Received 29 April 1980)

It is shown that in the turbulent regime of systems with period-doubling subharmonic bifurcations, the maximum Lyapunov characteristic exponent behaves like $\bar{\lambda} = \bar{\lambda}_0(r - r_c)^t$, with t a universal exponent which is calculated to be $t = 0.449\,806\,9...$. This result is in agreement with the available data on $\bar{\lambda}$ for a number of dynamical systems.

PACS numbers: 05.20.Dd, 05.40.+j, 47.25.-c

There exist a large number of physical systems for which the nonlinear equations describing their dynamics display transitions into a chaotic regime in the absence of external noise sources. This regime, which is characterized by broadband noise in the power spectral densities, has been extensively studied in simple fluids, plasmas, chemical reactions, and various mathematical models.[1] In the case of dissipative systems, a pervasive pathway to turbulent behavior appears to be made of a cascade of period-doubling subharmonic bifurcations into a strange attractor in phase space. This has been observed in some experiments on the onset of fluid turbulence,[2-4] studies of driven anharmonic oscillators,[5] nonlinear saturation of unstable plasma modes,[6] and several other mathematical models.[7-9] This cascading behavior in the periodic regime has been shown to display universal features, independent of the detailed nature of the governing equations.[10] More recently, it has been established[11-13] that beyond the onset of chaos another set of fiburcations takes place whereby 2^n bands of the attractor successively merge in a mirror sequence of the cascading bifurcations found in the periodic regime.

A hallmark of ergodic and mixing behavior for nonlinear dynamical systems is their sensitive dependence on initial conditions. Two trajectories in phase space that initially differ by a small amount will separate exponentially in time, with the divergence rate measured by a positive value of the maximum Lyapunov characteristic exponent, $\bar{\lambda}$, associated with the flow.[14,15] For systems that display period-doubling subharmonic bifurcations, the emergence of a positive value for the envelope of $\bar{\lambda}$ as the control parameter r exceeds the onset value r_c, takes place in a steep and continuous fashion, a behavior reminiscent of critical-point phenomena in phase transitions. As illustrated in Fig. 1, which has been computed for a one-dimensional map, as $r \to r_c^+$ the envelope of $\bar{\lambda}$ seems to approach its zero value with power-law behavior, the sharp dips corresponding to stable orbits in the chaotic regime.

In this paper we show that the power-law behavior for the envelope of $\bar{\lambda}$ suggested by Fig. 1 is indeed universal for dynamical systems exhibiting period-doubling subharmonic bifurcations; the characteristic exponent behaves as

$$\bar{\lambda} = \bar{\lambda}_0(r - r_c)^t \tag{1}$$

with $\bar{\lambda}_0$ a constant of order unity and t an expo-

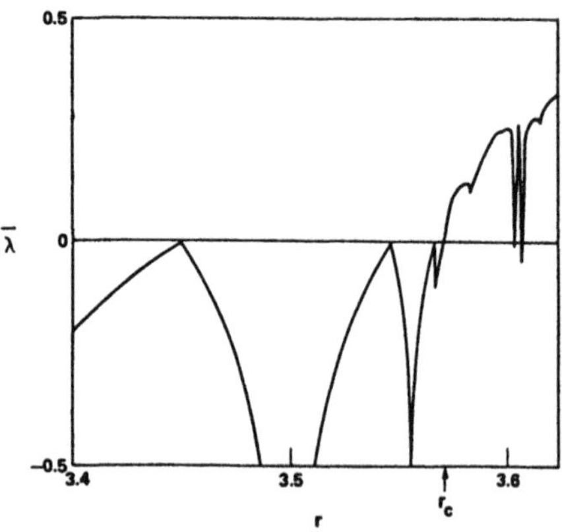

FIG. 1. The Lyapunov characteristic exponent for the one-dimensional map $x_{k+1} = rx_k(1 - x_k)$, with $0 \leqslant x_k \leqslant 1$ and $r_c = 3.57$. The sharp dips in the region $r > r_c$ correspond to periodic orbits. (See Ref. 18.)

nent which is given by

$$t = \frac{\ln 2}{\ln \delta} = 0.449\,806\,9\ldots, \qquad (2)$$

where δ is a constant which has been determined to be given by $\delta = 4.669\,261\,6\ldots$ (Ref. 10). This result is in agreement with all present computations and measurements of $\bar{\lambda}$ (albeit not extremely accurate) available for these systems.

Consider a one-dimensional map,[16] defined by $x_{k+1} = f(x_k, r)$ with $f(x, r)$ a continuous, single-hump function with a parabolic maximum in the interval $0 \leqslant x \leqslant 1$, and r a variable that controls its steepness. If $f^{(n)}(x, r)$ denotes the nth iterate of the map, the Lyapunov characteristic exponent, $\bar{\lambda}(r)$, can be written as

$$\bar{\lambda}(r) = \lim_{N\to\infty} \frac{1}{N} \sum_{k=1}^{N} 2^{-n} \ln \left| \frac{df(x_k, r)^{(2^n)}}{dx_k} \right|. \qquad (3)$$

If we assume that there exists a probability distribution, $p_r(x)$, that is invariant under the operations of the map [i.e., $p_r(x_i) = \sum_k p_r(x_{i-1}{}^k) dx_{i-1}{}^k / dx_i$, where the sum is over the x_{i-1}'s that are mapped into x_i], we can express Eq. (3) as

$$\bar{\lambda}(r) = 2^{-n} \int p_r(x) \ln \left| \frac{df(x, r)^{(2^n)}}{dx} \right| dx. \qquad (4)$$

In the chaotic regime (i.e., $r > r_c$) the existence of the reverse set of bifurcations that we described above implies that up to the $(n+1)$ bifurcation, $f^{(2^n)}(x_i, r)$ maps points within a given band of the attractor into the same band. Moreover, since the distance between bands scales like α^{-n} with α a universal constant,[10] the probability distribution $p(x, r)$ will consist of a set of 2^n narrow strips of width α^{-n} and height h, separated by regions in which it is identically zero. This in turn implies that Eq. (4) can be broken into 2^n integrals over each band evaluated with $f^{(2^n)}(x, r)$. If in the spirit of the Feigenbaum scaling study of the periodic regime[10] we assume that $f^{2^n}(x, r) - x = \alpha^{-n}\varphi[\alpha^n x, \delta^n(r - r_c)]$ with φ a universal function of x, we can write Eq. (4) as

$$\bar{\lambda}(r) = \alpha^{-n} \int_{\substack{\text{single}\\ \text{band}}} p(y/\alpha^n) \times \ln|\varphi'[y, \delta^n(r - r_c)] + 1|\, dy, \qquad (5)$$

where $y = \alpha^n x$ and $\delta^n(r - r_c)$ is of order unity. In order to evaluate this integral, first note that the normalization condition for the probability function written as

$$2^n \int_{\substack{\text{single}\\ \text{band}}} p(x)\, dx \simeq 2^n \alpha^{-n} h = 1 \qquad (6)$$

enables one to write $p(y/\alpha^n)$ as $hR(y)$, where $R(y)$ is a uniform function of y over the width of the band, and $h = \alpha^n 2^{-n}$. Equation (5) then becomes

$$\bar{\lambda}(r) = 2^{-n} \bar{\lambda}_0{}'[\delta^n(r - r_c)] \qquad (7)$$

where the functional dependence of $\bar{\lambda}_0{}'$ on $\delta^n(r - r_c)$ describes the structure beneath the envelope. Since we are not dealing with that structure we replace it by a constant $\bar{\lambda}_0{}''$. We therefore see that $\bar{\lambda}$ increases as the number of bands within the strange attractor merge pairwise into a single one.

In order to obtain a scaling relation that involves the control parameter r, we note that in the highly bifurcated regime the value of r for which the nth bifurcation takes places behaves like[10]

$$r - r_c = c\delta^{-n} \qquad (8)$$

with c a constant, so that $n = c' + \ln(r - r_c)/(-\ln\delta)$. Use of this equality in Eq. (7) results in

$$\bar{\lambda}(r) = \bar{\lambda}_0 2 \ln(r - r_c)/\ln\delta, \qquad (9)$$

where $\bar{\lambda}_0 = 2^{-c'} \bar{\lambda}_0{}''$ or, equivalently

$$\bar{\lambda}(r) = \bar{\lambda}_0 (r - r_c)^t \qquad (10)$$

with the universal exponent t, given by $t = \ln 2/\ln\delta = 0.449\,806\,9\ldots$. Since points in phase space separated by an initial distance d will separate after m iterations of the map like $Me^{\bar{\lambda} m}$ with M a constant, Eq. (10) expresses the fact that the rate of divergence will grow like a power law as one enters the turbulent regime.

There exists at the present time a number of calculations[8,11,17,18] and measurements of $\bar{\lambda}$ as a function of r for many different systems displaying cascades of period-doubling bifurcations into a chaotic state. To within the accuracy with which we can compare them with the predictions of our scaling theory, they are all in good agreement with Eq. (10). It is clear, however, that more accurate calculations and measurements will have to be made in order to have a precise test of the theory. Furthermore, since the effect of external noise is to produce a bifurcation gap in the sequence of available states[13] the scaling behavior of $\bar{\lambda}$ can only be checked in the limit of small fluctuations or truncation errors. This appears quite feasible.

In concluding, we point out that the theory we have presented applies only to a region near the onset value r_c for which the mirror sequence of period-doubling bifurcation takes place. Although in some dynamical systems[11,13] this sequence exhausts most of the turbulent regime, it is well known that there are other nonlinear problems for which there exist beyond the single-band attractor a rich structure of orbits and bands, and where our arguments may no longer apply. In that region the observed growth of the envelope of $\bar{\lambda}$ as a function of r, although it still reflects the effects of larger bandwidths, will require a different theoretical approach from the one presented here. Nevertheless it is rewarding to have a measure of chaos whose universal behavior near onset is exactly calculable.

We have benefitted from conversations with J. Crutchfield. This work was supported in part by the National Science Foundation Contract No. PHY 79-29545.

---

[1]A fairly complete set of references can be found in the Proceedings of the 1978 Oji seminar at Kyoto, Prog. Theor. Phys. Suppl. 64 (1978).

[2]Yu. N. Belyaev, A. A. Monakhov, S. A. Scherbakov, and I. M. Yavorskaya, Pis'ma Zh. Eksp. Teor. Fiz. 29, 329 (1979) [JETP Lett. 29, 295 (1979)].

[3]A. Libchaber and J. Maurer, to be published.

[4]J. P. Gollub, S. V. Benson, and J. Steinman, to be published.

[5]B. A. Huberman and J. P. Crutchfield, Phys. Rev. Lett. 43, 1743 (1979).

[6]J. M. Wersinger, J. M. Finn, and E. Ott, Phys. Rev. Lett. 44, 453 (1980).

[7]N. Metropolis, M. L. Stein, and P. R. Stein, J. Combinatorial Theory 15, 25 (1973); R. May, Nature (London) 261, 459 (1976).

[8]T. Nagashima, Ref. 1, p. 368.

[9]C. Boldrighini and V. Franceschini, Commun. Math. Phys. 64, 159 (1979).

[10]M. J. Feigenbaum, J. Statist. Phys. 19, 25 (1978).

[11]J. Crutchfield, D. Farmer, N. Packard, R. Shaw, G. Jones, and R. J. Donnelly, Phys. Lett. 76A, 1 (1980).

[12]E. Lorenz, to be published; P. Collet, J. P. Eckmann, and O. Lanford, to be published.

[13]J. P. Crutchfield and B. A. Huberman, to be published.

[14]V. I. Oseledec, Trans. Moscow Math. Soc. 19, 197 (1968).

[15]G. Benettin, L. Galgani, and J. M. Strelcyn, Phys. Rev. A 14, 2338 (1976).

[16]For the sake of simplicity, we will deal with a one-dimensional map, which possess only one Lyapunov characteristic exponent. In the case of many-dimensional dynamical systems, the corresponding quantity is the maximum characteristic exponent, as defined in Ref. 15.

[17]S. D. Feit, Commun. Math Phys. 61, 249 (1978).

[18]R. Shaw, unpublished.

Physics Letters **83A** 184–7 (1981)

UNIVERSAL POWER SPECTRA FOR THE REVERSE BIFURCATION SEQUENCE

Alan WOLF and Jack SWIFT
Department of Physics, University of Texas at Austin, Austin, TX 78712, USA

Received 26 February 1981

Many dynamical systems exhibit forward and reverse period-doubling bifurcation sequences, the latter being intrinsically noisy. Feigenbaum has predicted the amplitude of sharp spectral components in the forward sequence from universality arguments. In the same spirit we derive the approximate form of the broad band features in the reverse sequence. Our results give a power-law behavior of the integrated noise spectrum similar to that recently reported by Huberman and Zisook.

Some dynamical systems give rise to a strange attractor in phase space that can be conveniently (though not completely) described by a one-dimensional (1-D) discrete mapping. This function indicates the manner in which trajectories are reinjected into the flow after one traversal of the attractor. The data string generated by iterating the map will first exhibit transient behavior whose nature and duration depend on the choice of initial conditon. Post transient behavior is periodic or "chaotic" depending on the value of a Reynolds number-like "bifurcation parameter", L.

There has been much recent interest in 1-D maps in which the chaotic state is reached by a forward sequence of period-doubling bifurcations [1–3] followed by a reverse sequence of noisy limit cycles [4–7]. A noisy limit cycle consists of 2^n bands of finite measure as opposed to 2^n points in a forward sequence. Work by Feigenbaum [1–3] and others [8] suggests that many dynamical features are independent of the exact form of the 1-D map ("universality"). Thus simple quadratic forms can be studied without regard to an underlying physical system. There is ample [1,2] numerical evidence for universality among the commonly used 1-D maps.

Power spectra are often obtained in laboratory experiments on dynamical systems, while properties such as Lyapunov exponents, which can be readily calculated in numerical studies, are difficult or impossible to determine for real physical systems. Therefore, it is important to understand the scaling properties of power spectra if universality concepts are to be tested in the laboratory.

Exploiting universality, Feigenbaum has predicted [3] the amplitude of sharp components in the power spectra for the forward sequence. Here we predict the functional form of broad band spectral features in a similar fashion for the reverse sequence. After outlining the calculation we discuss verification of these universal spectra, and the connection between our work and the recent work of Huberman and Zisook [5].

Our numerical studies of iterates of the initial point for a noisy 2^n cycle suggest that the time series generated is decomposable into 2^n functions; one for each band and identical save for the following points: (1) The functions differ by various dc shifts having to do with the center of the 2^n bands. (2) The functions are scaled [1,2] by various powers of the universal number α ($\alpha = 2.5029$). These scalings are related to the widths of the bands (see below). (3) Factors of minus one, having to do with inversions about the centers of the band, occur in the functions. (4) The functions are shifted in time by one increment in the discrete time variable.

This suggests that we can write the time series for a noisy 2^n cycle in the following form:

$$\begin{aligned} X_m &= a_1 + b_1 f(t), \qquad m = 2^n t + 1\,, \\ &\ \ \vdots \\ &= a_{2n} + b_{2n} f(t), \qquad m = 2^n t + 2^n\,, \end{aligned} \tag{1}$$

where X_m is the mth iterate of an initial point X_1, t is a discrete time variable, the a_i refer to the band centers discussed in point (1) above, the b_i are factors discussed below, and $f(t)$ describes the stochastic nature of the 2^0 cycle.

The implication is that noise is introduced into the time series only on every 2^nth iteration of the map. This becomes clear when one studies the 2^n submaps sampled by the time series. Of these, only the one containing the extremum of the map is mixing, the others are (asymptotically) linear transformations. The time series is then decomposed into a strictly periodic 2^n-cycle and a noise term that acts once per limit cycle. We have confirmed this point directly from the time series and less directly from probability distributions. Applying the definition of fourier transform we find

$$X_k^n = \frac{1}{2^n T}\sum_{m=1}^{2^n T} X_m \mathrm{e}^{-\mathrm{i}\omega m} = \frac{1}{2^n T}\sum_{t=0}^{T-1}\sum_{j=1}^{2^n}[a_j + b_j f(t)]\,\mathrm{e}^{-\mathrm{i}\omega(2^n t+j)}$$

$$= \frac{1}{2^n T}\left[\left(\sum_{j=1}^{2^n} a_j \mathrm{e}^{-\mathrm{i}\omega j}\right)\left(\sum_{t=0}^{T-1}\mathrm{e}^{-\mathrm{i}\omega 2^n t}\right) + \left(\sum_{j=1}^{2^n} b_j \mathrm{e}^{-\mathrm{i}\omega j}\right)\left(\sum_{t=0}^{T-1} f(t)\mathrm{e}^{-\mathrm{i}\omega 2^n t}\right)\right]$$

$$= (1/2^n T)\,[g_a(\omega) T\delta(2^n\omega/2\pi, p) + g_b^n(\omega)\tilde{f}(2^n\omega)]\ , \qquad (2)$$

where T is the formal period of $f(t)$, $\delta(x, y)$ is the Kronecker delta function, p is an integer,

$$g_a(\omega) = \sum_{j=1}^{2^n} a_j \mathrm{e}^{-\mathrm{i}\omega j}\ , \quad g_b^n(\omega) = \sum_{j=1}^{2^n} b_j \mathrm{e}^{-\mathrm{i}\omega j}\ , \quad \tilde{f}(\omega) = \sum_{t=0}^{T-1} f(t)\mathrm{e}^{-\mathrm{i}\omega t}\ ,$$

and the transform is evaluated at the frequencies $\omega = 2k\pi/2^n T$, for $-(2^n T/2 - 1) \leqslant k \leqslant 2^n T/2$, which become the interval $(-\pi, \pi)$ in the limit $T \to \infty$.

As we are only interested in the broad band features, we ignore the first term. Calculation of the remaining term requires knowledge of $\tilde{f}$ and the b_i coefficient ("bandwidths"). The former is easily disposed of: $\tilde{f}(\omega)$ is approximately white noise (but not exactly, see ref. [5]) and so $\tilde{f}(2^n\omega)$ is approximately flat.

The band widths can be generated to first order by extending certain forward sequence results to the reverse sequence. One of these concerns the relative location of limit-cycle elements and remains exact for the reverse sequence, due to simple properties of the map. The sign of each bandwidth is determined uniquely by these properties. Another result is that the width of the *chaotic* submap shrinks by the universal factor $\alpha = 2.5029$. At each bifurcation this has been confirmed numerically to the required precision. We now claim that to first order all other band widths scale by factors of α as well. The arguments for this are approximate and rather involved. This is not an unexpected result and is confirmed numerically to be approximately true. Individual band widths do differ from our predicted values by as much as 30%, but the mean error is typically ≈5% for any specified limit cycle.

Arguments based on the single extrema nature of the map (monotone-up to monotone-down or vice versa) give us the order in which the bands are visited in the time series. Then the properly ordered b_i coefficients are known and $g_b^n(\omega)$ may be computed recursively. Choosing $g_b^0 = 1$ as a normalization condition we find

$$g_b^{n+1}(\omega) = 2^{-2} g_b^n(2\omega) \times (1 - \mathrm{e}^{\mathrm{i}\omega}/\alpha)/\alpha\ . \qquad (3a)$$

Power spectra then follow:

$$|X_k^0|^2 = 1 \times \text{noise}\ ,$$

$$|X_k^1|^2 = 1/2^2\,|(1 - \mathrm{e}^{\mathrm{i}\omega}/\alpha)|^2/\alpha^2 \times \text{noise}\ ,$$

$$\vdots$$

$$|X_k^n|^2 = 1/2^{2n}\,|(1 - \mathrm{e}^{\mathrm{i}\omega}/\alpha)\ldots(1 - \mathrm{e}^{2^{n-1}\mathrm{i}\omega}/\alpha)|^2/\alpha^{2n} \times \text{noise}\ , \qquad (3b)$$

where "noise" is approximately flat. These forms are compared with numerical spectra for the 1-D Lorenz [4] quadratic, $X_{n+1} = \frac{1}{2}x_n^2 - L$ in fig. 1. They have also been checked to be in good agreement with spectra for the 2-D

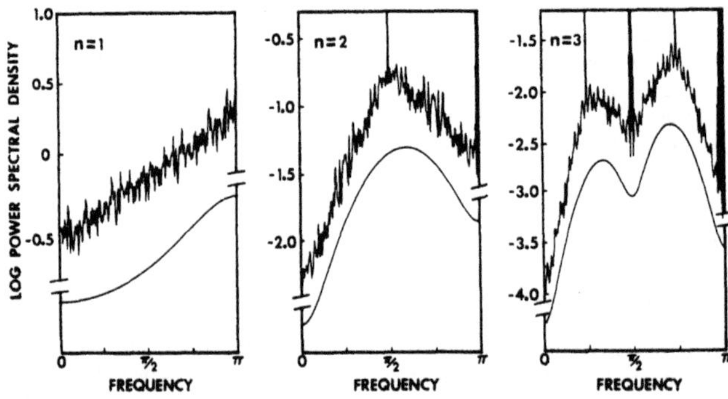

Fig. 1. Numerically computed noisy power spectra for the Lorenz map are compared with universal power spectra [Eq. (3)]. The "white noise" factor and sharp spectral components are not shown in the universal spectra which have been shifted down for clarity. The numerical spectra have been computed for values of the bifurcation parameter L = 3.087 378 03 (n = 1), 2.860 715 26 (n = 2), and 2.814 810 24 (n = 3), respectively.

Henon [9] mapping

$$\begin{pmatrix} X_{n+1} \\ Y_{n+1} \end{pmatrix} = \begin{pmatrix} 1 + Y_n - LX_n^2 \\ bX_n \end{pmatrix} ,$$

with $b = 0.3$. In the latter case a more sophisticated choice of time series (reflecting the locally 1-D universal character of the 2-D system) would likely improve the result.

In all cases bifurcation parameter values are chosen to generate formally equally stable limit cycles, which are the only ones suitable for comparison. This is done to mimic forward sequence analysis with the aid of Lorenz' universal number [4] 0.1871, although the definition of stability in the chaotic regime is far from obvious.

Our calculation is related to the work of Huberman and Zisook [7] on the intergrated noise spectrum near L_c. They find power-law behavior,

$$N(L) = N_0(L - L_c)^{2 \ln \beta / \ln \delta}$$

for $\beta = 3.257$, a universal constant and $\delta = 4.669$ another universal number [1,2]. They suggest that β, which describes the scaling of the rms band width at successive bifurcations, is independent of all previously discovered universal numbers.

Our spectra are consistent with this power-law behavior and our first-order arguments [9] predict an approximate value for $\beta = 3.28 = \alpha^2\sqrt{2}/(\alpha^2 + 1)^{1/2}$. Again this assumes exact $1/\alpha$ bandwidth scaling. Writing the integrated spectrum for a noisy 2^n-cycle as

$$P_n = \frac{1}{2^n \alpha^{2n}} \int_0^\pi \left| (1 - e^{i\omega}/\alpha)(1 - e^{2i\omega}/\alpha) \ldots (1 - e^{2^{n-1} i\omega}/\alpha) \right|^2 \frac{d\omega}{\pi} \quad (n > 0) , \tag{4a}$$

we find

$$P_{n+1} = \tfrac{1}{2}[(\alpha^2 + 1)/\alpha^4] P_n . \tag{4b}$$

The n dependence may be replaced with the desired L dependence with the aid of the definition of the bifurcation velocity and the result obtains.

In summary, using ideas based on universality, we have predicted the approximate form of the noisy power spectra in the reverse bifurcation region and have found excellent agreement between our predictions and numerically generated spectra for the 1-D Lorenz quadratic map.

We thank Drs. W. McGormick and H. Swinney for many helpful suggestions and conversations. We thank Dr. Jack Turner for generating the figures. This research was supported in part by the National Science Foundation under Grant No. DMR 76-11426 and by the Robert A. Welch Foundation under Grant No. F-767.

References

[1] M.J. Feigenbaum, J. Stat. Phys. 19 (1978) 25.
[2] M.J. Feigenbaum, J. Stat. Phys. 21 (1979) 669.
[3] M.J. Feigenbaum, Phys. Lett. 74A (1979) 374.
[4] E.N. Lorenz, Noisy periodicity and reverse bifurcation, Ann. NY Acad. Sci. (1981), to be published.
[5] S. Grossman and S. Thomae, Z. Naturforsch. 32A (1977) 1353.
[6] B.A. Huberman and J. Rudnick, Phys. Rev. Lett. 45 (1980) 154.
[7] B.A. Huberman and A.B. Zisook, Power spectra of strange attractors, preprint.
[8] N. Metropolis, M.L. Stein and P.R. Stein, J. Combinatorial Theory 15 (1973) 25.
[9] M. Henon, Commun. Math. Phys. 50 (1976) 6977.

Physical Review Letters **46** 626–8 (1981)

Power Spectra of Strange Attractors

B. A. Huberman and Albert B. Zisook[a]
Xerox Palo Alto Research Center, Palo Alto, California 94304
(Received 29 December 1980)

It is shown that, for systems which enter chaos through period doubling bifurcations, the integrated noise power spectrum in the chaotic regime behaves as $N(r)=N_0(r-r_c)^\sigma$, with $\sigma=1.5247...$. Furthermore, the existence of a new universal constant which describes the scaling behavior of the average bandwidth in the strange attractor is reported. These results are directly applicable to experiments probing the onset of turbulence in physical systems.

PACS numbers: 05.20.Dd, 05.40.+j, 47.25.-c

A number of physical systems, such as stressed fluids, high-temperature plasmas, and Josephson junctions have been observed to undergo a transition into a turbulent regime characterized by broadband noise in the power spectra. A possible explanation for these phenomena is that the phase trajectories for the complete nonlinear many-body problem enter a low-dimensional region of phase space containing a strange attractor. A strange attractor is a region in phase space such that nearby trajectories must enter it but once inside they diverge from each other. Hence we arrive at a description of turbulence involving only very few degrees of freedom. The effectively stochastic motion which these few degrees of freedom undergo gives rise to the observed noise in the power spectra. One common route into this turbulent regime is a universal cascade of period doubling bifurcations which occur as some control parameter is varied.[1-3] This cascade can be easily understood when, through the construction of return maps associated with the Poincaré maps, the dynamical system is mapped onto one-dimensional (1D) recursion relations which possess the same bifurcation structure.[4]

Recently, it has been shown that once in the chaotic regime, the Lyapunov exponent, which measures the rate of divergence of nearby trajectories, behaves very much like the order parameter of a phase transition near the critical point, i.e., it obeys a universal scaling law.[5] This development allows, in principle, for the application of techniques developed in the study of critical phenomena to the onset of turbulence in these nonlinear systems.

[a]Permanent address: James Frank Institute, The University of Chicago, 5640 S. Ellis Ave., Chicago, Ill. 60637.

Appealing as these ideas might be, they suffer from the fact that one cannot directly measure Lyapunov exponents or discern the topology of attractors in experiments dealing with physical systems. We therefore believe that what is needed is a theory of the power spectra associated with strange attractors above the chaotic threshold.

This paper presents the main results of such a theory. In particular, we show that, for systems which enter the chaotic regime through a sequence of period doubling bifurcations, the integrated noise power spectrum in the chaotic regime, $N(r)$, behaves as

$$N(r)=N_0(r-r_c)^\sigma, \tag{1}$$

where $\sigma=1.5247...$. Furthermore, we report the existence of a new universal constant, β, associated with 1D maps displaying bifurcation cascades, which describes the scaling behavior of the average bandwidth in the strange attractor. Our results, which are in very good agreement with numerical simulations, should be directly applicable to experiments probing the onset of turbulence in a variety of physical systems.

It has been shown that the return map for dynamical systems displaying period doubling bifurcations, such as anharmonic systems,[4] corresponds to simple recursion relations of the form $x_{n+1}=f(x_n)$, where $f(x)$ has a single parabolic maximum. Because of the universality of the bifurcation structure of these 1D maps, we may consider the simplest one defined on the inverval [0, 1] (with $0\le r\le 4$), i.e.,

$$f(x)=rx(1-x). \tag{2}$$

For such a map it is well known[6] that, as the control parameter r is increased, a cascade of period doubling bifurcations takes place for $r = r_n$, $n = 1,2,3,\ldots$, until a value $r = r_c$ is reached beyond which chaotic behavior ensues. Beyond r_c, a reverse set of bifurcations, or band mergings. occurs for the values $r = \tilde{r}_n$, $n = 1,2,3,\ldots$.[7] Moreover, at $r = \tilde{r}_n$ the action of $f^{(2^n)}(x)$ on any fixed band is completely chaotic, its invariant measure being a scaled down version of that of the map $f(x)$ with $r = 4$.

If the iterates generated by the map are denoted by $x_k = f^{(k)}(x_0)$, we can represent the sequence x_k (for $r = \tilde{r}_n$) by

$$x_k = \sum_j A_j \exp[i(\omega_j)k] + n(k) \tag{3}$$

with

$$A_j = \lim_{N\to\infty} (1/N) \sum_{n=1}^{N} \exp[-i(\omega_j)n] x_n, \tag{4}$$

where the frequencies ω_j are integral multiples of $2\pi/2^n$ and $n(k)$ is the noise term generated by the deterministic map. Since $n(k)$ is a scaled down version of the noise obtained from the map $f(x)$ with $r = 4$ we can write[7]

$$\lim_{T\to\infty}\left[T^{-1} \sum_{t=1}^{T} n(t+k)n(t)\right] = W^2 \delta(k) \tag{5}$$

with W^2 a constant proportional to the average of the square of the width of one of the 2^n bands. The power spectrum $G(\omega)$ is simply calculated from Eq. (3). We thus obtain

$$G(\omega) = |x(\omega)|^2 = |\sum_j A_j \delta(\omega - \omega_j) + n(\omega)|^2. \tag{6}$$

Therefore, the power spectrum in the chaotic regime will consist of a set of instrumentally narrow peaks sitting on top of a broadband noise background. The existence of these δ-function peaks in the chaotic regime are a reflection of the time-translation–invariant property of the driving term in the original dynamical system.[8]

In order to proceed further in our analysis of the broadband noise we first establish a new universal property of 1D recursion relations with period doubling bifurcations. We have found that for highly bifurcated orbits in the chaotic regime (or $r = \tilde{r}_n$ with n large) the root-mean-square bandwidth obeys the scaling relation

$$W_n = W_0 \beta^{-n} \tag{7}$$

with $\beta = 3.2375\ldots$.[9] We believe that this new universal constant is independent of the constants $\alpha = 2.5029\ldots$ and $\delta = 4.6692\ldots$ introduced by Feigenbaum.[10] It should be noted that Eq. (7) also describes the scaling of the average spacing of the most highly bifurcated pairs in the periodic regime with $r = r_n$.

The experimental quantity of interest in the chaotic regime is the integrated power density $N(r)$, which is given by

$$N(r) = \int d\omega\, |n(\omega)|^2, \tag{8}$$

where we have subtracted the periodic structure given by the δ functions. The range of integration is $0 \leq \omega \leq \omega_d$, where ω_d is the driving frequency in the original system. For discrete maps, the driving frequency is 2π. To determine the scaling behavior, we use an argument similar to that used by Huberman and Rudnick in their study of Lyapunov exponents.[5] First we notice that, as r goes from $\tilde{r}_n$ to $\tilde{r}_{n+1}$ in the reverse bifurcation sequence, Eqs. (5), (7), and (8) imply that

$$N(\tilde{r}_{n+1}) = N(\tilde{r}_n)\beta^{-2}. \tag{9}$$

Furthermore, since for large n, the value of $r = \tilde{r}_n$ for which the nth reverse bifurcation, or band merging, takes place behaves as[7,10]

$$\tilde{r}_n = r_c + \text{const}\,\delta^{-n}, \tag{10}$$

we obtain, from Eqs. (9) and (10),

$$N(r) = \text{const}(r - r_c)^{\sigma}, \tag{11}$$

with σ a universal exponent which is given by $\sigma = 2\ln(\beta)/\ln(\delta) = 1.5247\ldots$. Therefore, the total noise power, which in turn defines an effective noise temperature, T_{eff}, obeys a scaling law near the chaotic threshold.

In order to test these predictions we have measured the power spectrum associated with the 1D map of Eq. (2). To within the accuracy of our calculations we obtained $\sigma = 1.527 \pm 0.005$, in excellent agreement with the theoretical value. We thus believe that experiments in systems displaying period doubling bifurcations will observe the scaling behavior given by Eq. (11).

Our arguments rest on the existence of the special points $\tilde{r}_n$ at which the bands are completely mixing. Since one may question the validity of our prediction for $r \neq \tilde{r}_n$ because of the existence of highly bifurcated shallow periodic attractors, it is important to consider the effects of external noise. As has been recently shown[11] the addition of small amounts of external noise to the dynamical system results in the disappearance of these shallow attractors. Therefore, experimental determinations of our scaling predictions will unavoidably interpolate smoothly between the special reverse bifurcation points, leading to a simple determination of the power-law behavior predicted by Eq. (11).

In summary, we have shown that (a) the power spectrum of the chaotic phase of period doubling systems consists of δ functions and broadband noise, (b) the average noise scales near onset with a universal power law, and (c) our theoretically determined exponent agrees with our numerical experiments.

A. B. Z. wishes to acknowledge his Robert R. McCormick and National Science Foundation Fellowships, and gives special thanks to Xerox Palo Alto Research Center for their support and hospitality.

---

[1]Yu. N. Bekyaev, A. A. Monakhov, S. A. Scherbakov, and I. M. Yavorskaya, Pis'ma Zh. Eksp. Teor. Fiz. 29, 329 (1979) [JETP Lett. 29, 295 (1979)].

[2]A. Libchaber and J. Maurer, J. Phys. (Paris), Colloq. 41, C3-51 (1980).

[3]J. P. Gollub, S. V. Benson, and J. Steinman, to be published.

[4]B. A. Huberman and J. P. Crutchfield, Phys. Rev. Lett. 43, 1743 (1979).

[5]B. A. Huberman and J. Rudnick, Phys. Rev. Lett. 45, 154 (1980).

[6]See, for example, R. M. May, Nature (London) 261, 459 (1976).

[7]S. Grossmann and S. Thomae, Z. Naturforsch. A 32, 1353 (1977).

[8]One exception, which is unimportant to our situation, is discussed by S. J. Shenker and L. P. Kadanoff, to be published. We should also point out that for systems other than period-doubling ones, the existence of sharp peaks is by no means generic.

[9]The first estimate of β was given by M. Nauenberg and J. Rudnick (private communication), who obtained $\beta = 3.2$. We thank them for discussing this result with us.

[10]M. J. Feigenbaum, J. Stat. Phys. 19, 25 (1978).

[11]J. P. Crutchfield and B. A. Huberman, Phys. Lett. 77A, 407 (1980); J. P. Crutchfield, D. Farmer, and B. A. Huberman, to be published.

Physical Review Letters **47** 179–82 (1981)

Spectral Broadening of Period-Doubling Bifurcation Sequences

J. Doyne Farmer
Dynamical Systems Group, Physics Board, University of California at Santa Cruz, Santa Cruz, California 95064
(Received 19 January 1981)

A perturbation calculation shows that the power spectrum of strange attractor near the accumulation parameter of a period-doubling bifurcation sequence consists of peaks broadened by a phase modulation, with broad skirts created by an amplitude modulation. Moving toward the accumulation parameter, at each bifurcation the total noise power decreases by a factor of 10.48, the average peak width decreases by a factor of 20.96, and the spectral bandwidth of the skirts decreases by a factor of 2.

PACS numbers: 47.10.+g, 05.40.+j

This paper discusses properties of the power spectrum of a continuous dynamical system in the chaotic regime of period-doubling sequences. The universal properties of power spectra on the periodic side of doubling sequences were originally discussed by Feigenbaum,[1] and his predictions are in qualitative agreement with convection experiments by Libchaber and Maurer,[2] and Gollub, Benson, and Steinman.[3] Since there is numerical evidence that mathematical models with period-doubling sequences contain strange attractors, this agreement supports the hope that chaotic fluid flow can be modeled by strange attractors. The results presented here provide a more severe test of this theory. They support previous[4,5] and concurrent[6] work, and, in addition, treat the general case of dynamical systems that are not periodically driven.

A dynamical system with a period-doubling sequence that accumulates at parameter r_c has on one side a sequence of limit cycles whose period repeatedly doubles as r_c is approached. On the other side of r_c, numerical evidence[7-9] indicates that there is a sequence of strange attractors as shown in Fig. 1. To a good approximation, a strange attractor near r_c is a thin two-dimensional ribbon that makes 2^n loops and then closes onto itself. Aspects of this behavior can be summarized with use of a return map, constructed as follows: The intersection of the attractor and a transverse surface is approximately a curve. When this curve is parametrized by a variable y, successive crossings at times t_i yield a sequence y_i given by a recursion relation $y_{i+1}=F(y_i)$, where F is a continuous function (see Shaw[10]).

On the chaotic side, near r_c the probability density of y_i is nonzero on 2^n bands, corresponding to the 2^n loops of the continuous attractor. Motion between bands is periodic with period 2^n, but motion within each band is chaotic. This chaotic motion, which introduces broad components into a power spectrum, can be thought of as an amplitude modulation of an otherwise periodic orbit.

For a limit cycle, for example, the sequence of return times $T_i=t_{i+1}-t_i$ is constant. The re-

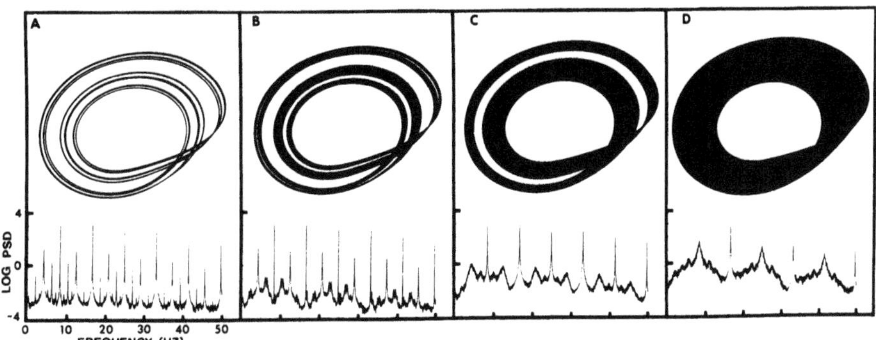

FIG. 1. Four simulations of strange attractors of the Rössler dynamical system, taken from Ref. 9. Case (a) is closest to r_c, and is a period-8 attractor. A power spectrum is shown below each frame.

turn times T_i are also constant for a strange attractor of a periodically driven system, as long as the surface of section used to construct the return map is taken at a constant phase of the driving force. The power spectrum in this case contains δ-function peaks superimposed on the broad background created by the amplitude modulation.

For the more general case, the return times T_i are *not* periodic. Nevertheless, numerical evidence indicates that the chaotic sequence T_i can be approximated as a continuous function of y_i, i.e., $T_i = T(y_i)$. Thus, orbits can gain or lose phase due to the chaotic behavior of $T(y_i)$. Letting $T_0 = \langle T_i \rangle$ (time average), and $\omega_0 = 2\pi f_0 = 2\pi/T_0$, the net phase fluctuation in completing a cycle is $\delta\theta_i = \omega_0(T_i - T_0)$. The chaotic return times effectively create a random "phase modulation" that broadens the peaks of the power spectrum.

When the central-limit theorem holds for $\delta\theta_i$, it ensures that the cumulative phase drift $\theta_k = \theta(t_k) = \sum_{i=1}^{k} \delta\theta_i$ has a Gaussian probability density for large k. Ratner[11] has shown that the central-limit theorem holds for dynamical systems that satisfy Axiom A.[12] Unfortunately, there are no known dynamical systems that satisfy Axiom A and also have a period-doubling sequence. Fortunately, behavior qualitatively similar to that in which we are interested can be simulated by "arbitrarily" choosing a return map F with a period-doubling sequence, "arbitrarily" choosing a continuous time transformation T, and using this pair to generate sequences $\delta\theta_i$. In every case studied, the coarse-grained probability density $\chi(\theta_k)$ approached a Gaussian. (See Fig. 2.)

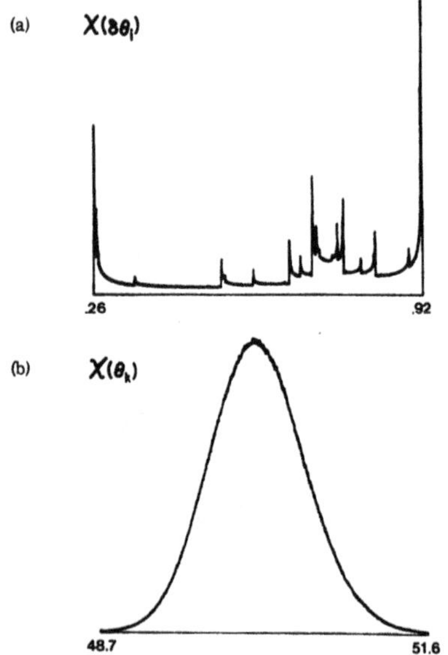

Fig. 2. A coarse-grained probability density of $\delta\theta_i = T(y_i)$, obtained by iterating, by 10^6 times, the one-dimensional map $y_{i+1} = 3.7 y_i(1-y_i)$ and sorting the result into 1000 bins in order to estimate the frequency of occurrence over each bin. For this case $T(y) = y$. (b) Similar to (a), except the probability density is constructed for $\theta_k = \sum_{i=1}^{k} \delta\theta_i$, with $k = 75$. Several different choices of smooth time transformations T all show $\chi(\theta_{75})$ approximately a Gaussian.

Naturally, as time increases the spread in the cumulative phase fluctuations θ_k gets larger. It can be shown that the variance σ^2 of $\chi(\theta_k)$ asymptotically grows linearly in time at a rate c. (This proof assumes that the sum of the autocorrelation function of T_i is finite.) For a limit cycle, or a periodically driven system, $\delta\theta_i = 0$, which implies that $c = 0$. In general, however, $c \neq 0$.

We are now ready to compute the form of the power spectrum. As a first approximation, the attractor is a limit cycle $p(\omega_0 t)$, with period $2^n(2\pi)$. To take the chaotic motion into account, write the transverse displacement from the limit cycle p as $w(\omega_0 t)R(\omega_0 t)$. $w(\omega_0 t)$ is the width of the attractor at phase $\omega_0 t$, and is periodic with period $2^n(2\pi)$. Thus, all of the chaotic behavior of the amplitude is contained in R. To take into account the chaotic phase drifting, write the phase at time t as $\varphi(t) = \omega_0 t + \theta(t)$. A trajectory on the attractor can be written as

$$x(t) = p(\varphi(t)) + w(\varphi(t))R(\varphi(t)). \quad (1)$$

$p(\varphi(t))$ is constructed so that $\langle x(t)\rangle = p(\omega_0 t)$, and R is constructed so that $\langle R\rangle = 0$.

A complication in the application of these ideas is that experiments are normally conducted with use of a projection onto a single coordinate. All of the following remarks remain true, however, if x, R, p, w, and y_i are consistently considered to be the projected values.

The autocorrelation of x can be computed from Eq. (1) by assuming that x is uncorrelated with p and w (see Thomae and Grossman[4]):

$$Q_x(t) = Q_p(t) + Q_w(t)Q_R(t). \quad (2)$$

Q_x, Q_p, Q_w, and Q_R are the autocorrelation functions of x, p, w, and R, respectively. In the absence of phase fluctuations, $c = 0$, $\varphi(t) = \omega_0 t$, and therefore $Q_p(t)$ and $Q_w(t)$ are periodic. Including phase fluctuations has the effect of multiplying Q_p and Q_w by a damping factor $e^{-ct/2}$. [To do this calculation it is necessary to assume that $\theta(t)$ is ergodic with a Gaussian probability density, and convert time averages to ensemble averages.] Letting P_k and W_k be the complex Fourier coefficients of p and w, and $f_k = (k/2^n)f_0$, Fourier transforming Eq. (2) gives the power spectrum of x,

$$S_x(f)$$
$$= 2 \sum_{k=1}^{\infty} [\,|P_k|^2 L_c(f-f_k) + |W_k|^2 S_R(f-f_k)]. \quad (3)$$

S_R is the power spectrum of R, and $L_c(f-f_k)$ is a Lorentzian peak of half power width $c/4\pi$ centered at f_k, i.e.,

$$L_c(f-f_k) = 2c/\{c^2 + [4\pi(f-f_k)]^2\}. \quad (4)$$

The effect of phase modulations has been neglected in the second term of Eq. (3), since for small values of c this is a second-order effect. In the first term, however, the phase modulations are responsible for the broadening of the δ-function peaks into Lorentzians. We will refer to the terms $|W_k|^2 S_R(f-f_k)$ as "skirts" because they are convolved about each peak. Taken together, they form a broad background caused by the amplitude modulation.

As a parameter r is varied, the power spectrum changes in a manner that becomes universal[1] as r approaches r_c. Let r_n be the parameter value where the number of distinct bands in the return map goes from 2^n to 2^{n-1}. At any given parameter value r_n, the width $w_i(r_n)$ of each band is not constant, and varies considerably in completing a cycle. Nevertheless, our numerical investigations show that

$$\lim_{n\to\infty} [\langle w_i^2(r_n)\rangle / \langle w_i^2(r_{n+1})\rangle] \to \gamma, \quad (5)$$

where $\gamma \cong 10.48$ is a universal number (see also Ref. 6). Parseval's theorem implies that the total noise power $\sum_{k=1}^{\infty} W_k^2$ also decreases by a factor of γ at each bifurcation. In addition, the ratio of the square of the separation of the adjacent bands at r_n to that at r_{n+1} is given by γ. This implies that the total power added to the periodic part of the spectrum in going from r_n to r_{n+1} is a factor of γ smaller than that added in going from r_{n-1} to r_n. A more detailed prediction of the behavior of the Fourier coefficients P_k (which behave just as they do on the periodic side of the bifurcation sequence) has been made by Feigenbaum,[1] and, with somewhat different results, by Nauenberg and Rudnick.[13] At $r = r_n$, if the 2^n iterate of F is restricted to a given band and rescaled appropriately, a universal function is approached. In passing to $r = r_{n+1}$ the 2^{n+1} iterate must be used; the number of iterations needed to construct a universal function consequently doubles. As a result, in passing from r_n to r_{n+1}, the frequency of S_R must be rescaled by a factor of 2, i.e., $2S_R(r_{n+1}, 2f) = S_R(r_n, f)$. (The factor of 2 in amplitude is necessary to maintain the integral of S_R constant.) Thus, the characteristic frequency of S_R at $r = r_{n+1}$ is half that at $r = r_n$.

The half power width of the Lorentzian peaks in the spectrum depends on c, the rate of growth of the variance of the cumulative phase fluctuations. At each bifurcation, the decrease by γ in the mean-square width of the bands causes a corresponding decrease in the mean-square value of the phase fluctuations $\delta\theta_i$. In addition, twice as many iterations are needed to complete a cycle and return to a universal function; the rate c must decrease altogether by a factor of $2\gamma \cong 20.96$ at each bifurcation. In passing through successive bifurcations c decreases rapidly, justifying the assumptions used to compute Eq. (4). After only a few bifurcations the peaks become experimentally indistinguishable from δ functions. This explains the sharpness of the peaks seen by Gollub, Benson, and Steinman.[3] It does not explain, however, why these sharp peaks frequently persist long after all the bands merge, far away from r_c.[8,14]

I would like to thank Bill Burke, Jim Crutchfield, Jim Curry, Patrick Huerre, Ed Lorenz, Mitchell Feigenbaum, Michael Nauenberg, Norman Packard, David Rand, Joe Rudnick, and Rob Shaw for valuable conversations. I would also like to thank Richard Kaplan and the Aerospace Engineering Department of University of Southern California for their gracious hospitality and the use of their Digital Equipment Corp. PDP-11/55 minicomputer. This work was funded by the Hertz Foundation.

[1]M. J. Feigenbaum, J. Stat. Phys. 19, 25 (1978), and 21, 669 (1979).

[2]A. Libchaber and J. Maurer, J. Phys. (Paris), Colloq., 41, C3-51 (1980).

[3]J. Gollub, S. Benson, and J. Steinman, Ann. N.Y. Acad. Sci. 357, 22 (1980).

[4]S. Thomae and S. Grossman, "Correlations and Spectra of Periodic Chaos Generated by the Logistic Parabola" (to be published), and "A Scaling Property in Critical Spectra of Discrete Systems" (to be published).

[5]A. Wolf and J. Swift, "Universal Power Spectra of Strange Attractors" (to be published).

[6]B. Huberman and A. Zisook, Phys. Rev. Lett. 46, 626 (1981).

[7]S. Grossman and S. Thomae, Z. Naturforsch. 32A, 1353 (1977).

[8]E. N. Lorenz, Ann. N.Y. Acad. Sci. 357, 282 (1980).

[9]J. Crutchfield, D. Farmer, N. Packard, R. Shaw, G. Jones, and R. Donnelly, Phys. Lett. 76A, 1 (1980).

[10]R. Shaw, Z. Naturforsch. 36A, 80 (1981).

[11]M. Ratner, Isr. J. Math. 16, 181 (1973).

[12]S. Smale, Bull. Amer. Math. Soc. 13, 747 (1967).

[13]M. Nauenberg and J. Rudnick, "Universality and the Power Spectrum at the Onset of Chaos" (to be published).

[14]D. Farmer, J. Crutchfield, H. Froehling, N. Packard, and R. Shaw, Ann. N.Y. Acad. Sci. 357, 453 (1980).

Physics Letters **77A** 407–10 (1980)

FLUCTUATIONS AND THE ONSET OF CHAOS

J.P. CRUTCHFIELD [1] and B.A. HUBERMAN
Xerox Palo Alto Research Center, Palo Alto, CA 94304, USA

Received 3 April 1980

We consider the role of fluctuations on the onset and characteristics of chaotic behavior associated with period doubling subharmonic bifurcations. By studying the problem of forced dissipative motion of an anharmonic oscillator we show that the effect of noise is to produce a bifurcation gap in the set of available states. We discuss the possible experimental observation of this gap in many systems which display turbulent behavior.

It has been recently shown that the deterministic motion of a particle in a one-dimensional anharmonic potential, in the presence of damping and a periodic driving force, can become chaotic [1]. This behavior, which appears after an infinite sequence of subharmonic bifurcations as the driving frequency is lowered, is characterized by the existence of a strange attractor in phase space and broad band noise in the power spectral density. Furthermore, it was predicted that under suitable conditions such turbulent behavior may be found in strongly anharmonic solids [2]. Since condensed matter is characterized by many-body interactions, one may ask about the effects that random fluctuating forces have on both the nature of the chaotic regime and the sequence of states that lead to it. This problem is also of relevance to the behavior of stressed fluids, where it has been suggested that strange attractors play an essential role in the onset of the turbulent regime [3]. Although there are experimental results supporting this conjecture [4–6], other investigations have emphasized the possible role of thermodynamic fluctuations directly determining the chaotic behavior [7].

With these questions in mind, we study the role of fluctuations on the onset and characteristics of chaotic behavior associated with period doubling subharmonic bifurcations. We do so by solving the problem of forced dissipative motion in an anharmonic potential with the aid of an analog computer and a white-noise generator. As we show, although the structure of the strange attractor is very stable even under the influence of large fluctuating forces, their effect on the set of available states is to produce a symmetric gap in the deterministic bifurcation sequence. The magnitude of this bifurcation gap is shown to increase with noise level. By keeping the driving frequency fixed we are also able to determine that increasing the random fluctuations induces further bifurcations, thereby lowering the threshold value for the onset of chaos. Finally, the universality of these results is tested by observing the effect of random errors on a one-dimensional map, and suggestions are made concerning the possible role of temperature in experiments that study the onset of turbulence.

Consider a particle of mass m, moving in a one-dimensional potential $V = a\eta^2/2 - b\eta^4/4$, with η the displacement from equilibrium and a and b positive constants. If the particle is acted upon by a periodic force of frequency ω_d and amplitude F, and a fluctuating force $f(t)$, with its coupling to all other degrees of freedom represented by a damping coefficient γ, its equation of motion in dimensionless units reads

$$\frac{d^2\psi}{dt^2} + \alpha\frac{d\psi}{dt} + \psi - 4\psi^3 = \Gamma\cos\left(\frac{\omega_d}{\omega_0}\right)t + f(t) \qquad (1)$$

with $\psi = \eta/2\eta_0$, the particle displacement normalized to the distance between maxima in the potential ($\eta_0 = (a/b)^{1/2}$), $\alpha = \gamma/(ma)^{1/2}$, $\Gamma = Fb^{1/2}/2a^{3/2}$, $\omega_0 = (a/m)^{1/2}$ and $f(t)$ a random fluctuating force such that

$$\langle f(t)\rangle = 0 \qquad (2a)$$

and

$$\langle f(0)f(t)\rangle = 2A\,\delta(t) \qquad (2b)$$

with A a constant proportional to the noise temperature of the system.

The range of solutions of eq. (1), in the case where $f(t) = 0$ (the deterministic limit) has been investigated earlier [1]. For values of Γ and ω_d such that the particle can go over the potential maxima, as the driving frequency is lowered, a set of bifurcations takes place in which orbits in phase space acquire periods of 2^n times the driving period, T_d. At a threshold frequency ω_{th}, a chaotic regime sets in, characterized by a

[1] Permanent address: Physics Department, University of California, Santa Cruz, CA 95064, USA.

strange attractor with "periodic" bands. Within this chaotic regime, as the frequency is decreased even further, another set of bifurcations takes place whereby 2^m bands of the attractor successively merge in a mirror sequence of the 2^n periodic sequence that one finds for $\omega \to \omega_{th}^+$. The final chaotic state corresponds to a single band strange attractor, beyond which there occurs an irreversible jump into a periodic regime of lower amplitude.

In order to study the effects of random fluctuations on the solutions we have just described, we solved eq. (1) using an analog computer in conjunction with a white-noise generator having a constant power spectral density over a dynamical range two orders of magnitude larger than that of the computer. Time series and power spectral densities were then obtained for different values of Γ, A and ω_d. While we found that the folding structure of the strange attractor is very stable under the effect of random forces, the bifurcation sequence that is obtained in the presence of noise differs from the one encountered in the deterministic limit.

Our results can be best summarized in the phase diagram of fig. 1, where we plot the observed set of bifurcations (or limiting set) as a function of the noise level, N, normalized to the rms amplitude of the driving term, Γ. The vertical axis denotes the possible states of the system, labeled by their periodicity $P = 2^n$, which is defined as the observed period normalized to the driving period, T_d. As can be seen, with increasing noise level a symmetric bifurcation gap appears, depleting states both in the chaotic and periodic phases. This set of inaccessible states is characterized by the fact that the longest periodicity which is observed before a strange attractor appears is a decreasing function of N, with the maximum number of bands which appear in the strange attractor behaving in exactly the same fashion. This gap extends over a large range of noise levels (up to $N = 1.5$), beyond which the motion either becomes unstable (i.e., $|\psi| \to \infty$; lower dashed line) or an amplitude jump takes place from the chaotic regime to a limit cycle of period 1 (upper dashed line).

We can illustrate this behavior by looking at the power spectral densities, $S(\omega)$ at fixed values of the driving frequency while increasing N. Fig. 2 shows such a sequence for $\omega_d/\omega_0 = 0.6339$, $\Gamma = 0.1175$, and $\alpha = 0.4$. Fig. 2(a) corresponds to $S(\omega)$ near the deterministic limit which, for the parameter values used, displays a limit cycle of period four. As N is increased, a transition takes place into a chaotic regime characterized by broad band noise with subharmonic content of periodicity $P = 4$ (fig. 2(b)) [‡1]. As the noise is increased even further, a new bifurcation occurs from which a new chaotic state with $P = 2$ emerges. Physically, this sequence reflects the fact that a larger effective noise temperature (and hence a larger fluctuating force) makes the particle gain enough energy so as to sample increasing nonlinearities of the potential, with a resulting motion which in the absence of noise could only occur for longer driving periods [‡2].

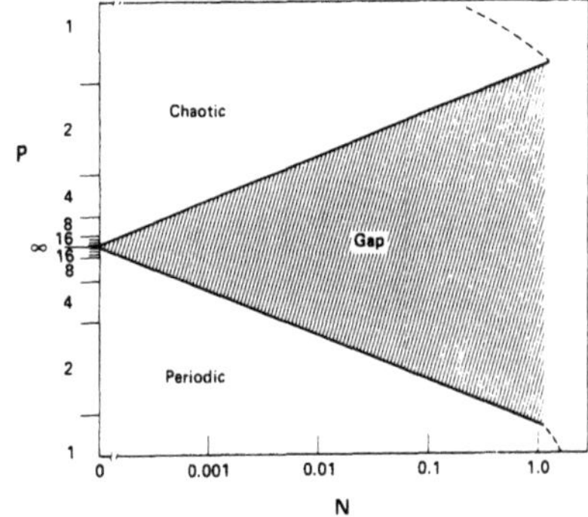

Fig. 1. The set of available states of a forced dissipative anharmonic oscillator as a function of the noise level. The vertical axis denotes the periodicity of a given state with $P = T/T_d$. The noise level is given by $N = A/\Gamma_{rms}$. The shaded area corresponds to inaccessible states.

A different set of states appears if the noise level is kept fixed while changing the driving frequency. In this case the observed states of the system correspond to vertical transitions in the phase diagram of fig. 1, with the threshold value of the driving period, T_{th}^n, at which one can no longer observe periodicities $P \geqslant 2^n$, behaving like

$$T_{th}^n = T_{th}^\infty (1 - N_n^\gamma) \tag{3}$$

for $0 \leqslant N_n \leqslant 1$, with N_n the corresponding noise level, T_{th}^∞ the value of the driving period for which the deterministic equation undergoes a transition into the chaotic state, and γ a constant which we determined to be $\gamma = \sim 1$ for $P \geqslant 2$ [‡3].

In order to test the universality of the bifurcation gap we have just described, we have also studied the bifurcation structure of the one-dimensional map described by

$$x_{L+1} = \lambda x_L (1 - x_L) + n_L(0, \sigma^2) \tag{4}$$

where $0 \leqslant x_L \leqslant 1$, $0 \leqslant \lambda \leqslant 4$, and n_L is a gaussian random number of zero mean and standard deviation σ. For $n_L = 0$, eq. (4) displays a set of 2^n periodic states universal to all single hump maps [8–9], with a chaotic regime characterized by 2^m bands that merge pairwise with increasing λ [10]. For $n_L \neq 0$ and a given value of σ, the effect of random errors on the stability of the limiting set is to produce a bifurcation gap analogous to the one shown in fig. 1.

The above results are of relevance to experimental studies of turbulence in condensed matter, for they

‡1 We should mention that the Poincare map corresponding to this state clearly shows a four-band strange attractor with a single fold.

‡2 In the regime of subharmonic bifurcation the dependence of response amplitude on driving frequency is almost linear.

‡3 Using the scaling relation $(T_{th} - T_n)/(T_{th} - T_{n+1}) = \delta$ [1] this implies that the threshold noise level scales like $N_n/N_{n+1} = \delta$, with $\delta = 4.669201609$ ....

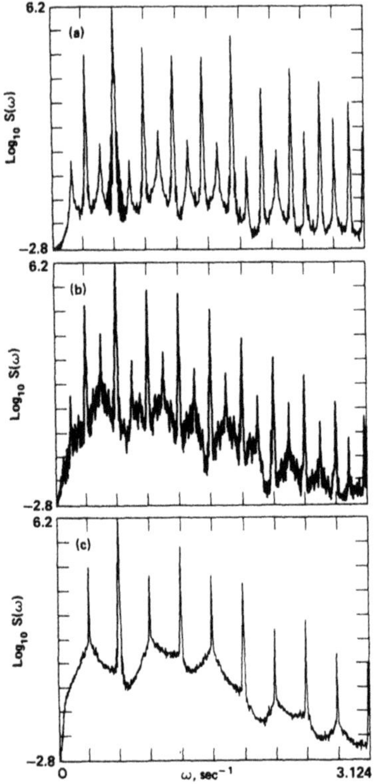

Fig. 2. Power spectral densities at increasing values of the effective noise temperature, for $\Gamma = 0.1175$, $\alpha = 0.4$, and $\omega_d = 0.6339\ \omega_0$. Fig. 2(a): $N = 10^{-4}$. Fig. 2(b): $N = 0.005$. Fig. 2(c): $N = 0.357$.

show that temperature plays an important role in the observed behavior of systems belonging to this same universality class. In particular, Belyaev et al. [11], Libchaber and Maurer [12] and Gollub et al. [13] have reported that under certain conditions the transition to turbulence is preceded and followed by different finite sets of 2^n subharmonic bifurcations. It would therefore be interesting to see if temperature changes or external sources of noise in the fluids can either reduce or increase the set of observed frequencies, thus providing for a test of these ideas. In the case of solids such as superionic conductors, the exponential dependence on temperature of their large diffusion coefficients might provide for an easily tunable system with which to study the existence of bifurcation gaps. Last, but not least, these studies can serve as useful calibrations on the relative noise temperature of digital and analog simulations.

In concluding we would like to emphasize the wide applicability of the effects that we have reported. Beyond the experimental studies of turbulence, there exist other systems which belong to the same universality class as the anharmonic oscillator and one-dimensional maps. These systems range from the ordinary differential equations studied by Lorenz [10], Robbins [14], and Rossler [15] to partial differential equations describing chemical instabilities [16]. Since period doubling subharmonic bifurcation is a universal feature of all these models, our results provide a quantitative measure of the effect of noise on their non-linear solutions.

The authors wish to thank D. Farmer, N. Packard, and R. Shaw for helpful discussions and the use of their simulation system.

References

[1] B.A. Huberman and J.P. Crutchfield, Phys. Rev. Lett. 43 (1979) 1743.
[2] See also, C. Herring and B.A. Huberman, Appl. Phys. Lett. 36 (1980) 976.
[3] See, D. Ruelle, in Lecture notes in physics, eds. G. Dell'Antonio, S. Doplicher and G. Jona-Lasinio (Springer-Verlag, New York, 1978), Vol. 80, p. 341.
[4] G. Ahlers, Phys. Rev. Lett. 33 (1975) 1185; G. Ahlers and R.P. Behringer, Prog. Theor. Phys. (Japan) Suppl. 64 (1978) 186.
[5] J.P. Gollub and H.L. Swinney, Phys. Rev. Lett. 35 (1975) 927; P.R. Fenstermacher, H.L. Swinney, S.V. Benson and J.P. Gollub, in: Bifurcation theory in scientific disciplines, eds. D.G. Gorel and D.E. Rossler (New York Academy of Sciences, 1978).
[6] A. Libchaber and J. Maurer, J. Physique Lett. 39 (1978) L-369.
[7] G. Ahlers and R.W. Walden, preprint (1980).
[8] T. Li and J. Yorke, in: Dynamical systems, an International Symposium, ed. L. Cesari (Academic Press, New York, 1972), Vol. 2, 203.
[9] M. Feigenbaum, J. Stat. Phys. 19 (1978) 25.
[10] E.N. Lorenz, preprint (1980).
[11] Yu.N. Belyaev, A.A. Monakhov, S.A. Scherbakov and I.M. Yavorshaya, JETP Lett. 29 (1979) 295.
[12] A. Libchaber and J. Maurer, preprint (1979).
[13] J.P. Gollub, S.V. Benson and J. Steinman, preprint (1980).
[14] K.A. Robbins, SIAM J. Appl. Math. 36 (1979) 451.
[15] O.E. Rossler, Phys. Lett. 57A (1976) 397; J.P. Crutchfield, D. Farmer, N. Packard, R. Shaw, G. Jones and R.J. Donnelly, to appear in Phys. Lett.
[16] Y. Kuramoto, preprint (1980).

Physical Review Letters **46** 935–9 (1981)

Scaling Theory for Noisy Period-Doubling Transitions to Chaos

Boris Shraiman, C. Eugene Wayne, and Paul C. Martin

Department of Physics and Division of Applied Sciences, Harvard University, Cambridge, Massachusetts 02138

(Received 19 December 1980)

The effect of noise on systems which undergo period-doubling transitions to chaos is studied. With the aid of nonequilibrium field-theoretic techniques, a correlation-function expression for the Lyapunov parameter (which describes the sensitivity of the system to initial conditions) is derived and shown to satisfy a *scaling theory*. Since these transitions have previously been shown to exhibit *universal behavior*, this theory predicts *universal effects* for the noise. These predictions are in good agreement with numerical experiments.

PACS numbers: 64.60.Fr, 02.90.+p, 47.25.Mr

During the past few years, the onset of chaotic behavior, after a sequence of period-doubling transitions, has been extensively studied. Feigenbaum[1] has observed that these transition sequences exhibit "universal" features akin to those of phase transitions; Collet and Eckmann[2] have noted that these universal features are shared by differential equations and multidimensional maps in which chaos is preceded by a sequence of period doublings; and Libchaber and Maurer[3] have observed this phenomenon in a convective cell with small aspect ratio. Recently, Huberman and Rudnick[4] have related one of the pretransitional parameters identified by Feigenbaum with the growth of disorder (i.e., the Lyapunov parameter) in the chaotic regime, and Huberman and Crutchfield[5] have examined numerically the effect of external noise on the onset of chaos. Nevertheless, many connections between period-doubling chaotic transitional phenomena and the critical phenomena at second-order phase transitions remain unclear.

The purpose of this Letter is the following: (1) to present a scaling theory (in which "noise" and "stress" play the role of external field and temperature) for systems that become chaotic via the period-doubling mechanism, and (2) to compare the dependence on noise and stress of the Lyapunov exponent predicted by this theory with numerical experiments performed by Huberman and collaborators.[6] Our results, found by field-theoretic methods designed for nonequilibrium systems, clarify some connections between phase transitions and the onset of chaos. These methods identify the Lyapunov exponent with the "long-time" limit of the nonequilibrium response function introduced by Martin, Siggia, and Rose.[7]

The Lyapunov exponent, λ, describes how solutions that were initially close to one another evolve after a long time (or many steps). Its sign and magnitude provide a measure of the "sensitivity of the system to initial conditions"; a large negative value implies great insensitivity to initial differences, a vanishing value implies that initial differences neither grow nor decay, and a large positive value implies rapid separation and great sensitivity.

Our principal quantitative result is that, at corresponding stress points between successive period doubling transitions, as a function of the noise amplitude, σ, of the noise and the magnitude of the difference between the stress, r, and the stress, r_∞, at which the onset to chaos occurs without noise, the Lyapunov parameter satisfies

$$\lambda(r_\infty - r;\, \sigma) = (r_\infty - r)^t\, \Phi((r_\infty - r)^{-t}\sigma^u); \tag{1}$$

with

$$t = (\ln 2)/\ln\delta = 0.4498\ldots \tag{2a}$$

and

$$u = (\ln 2)/\ln\beta \approx 0.34\ldots. \tag{2b}$$

The quantity, $\delta = 4.669\ldots$, is Feigenbaum's universal scaling parameter for functions $f(x)$ with quadratic maxima and β is a scaling parameter, associated with the noise, whose value we have calculated in a second-order approximation to be $\beta \approx 7.7\ldots$.

Let us consider the one-dimensional difference equation

$$x_{m+1} = f_r(x_m) + \xi_m. \tag{3}$$

The dynamical variable x_m ranges over the interval $[-1, 1]$, the function $f_r(x)$ has its maximum value, $f_r(0) = 1$, at $x = 0$, and ξ_m is a Gaussian random variable with $\langle \xi_m \rangle = 0$ and $\langle \xi_m \xi_{m'} \rangle = \sigma^2 \delta_{mm'}$. We analyze this stochastic difference equation, using a discrete version of the path-integral formulation[8] developed for stochastic nonlinear Langevin equations. The average of the functional $F[\{x\}]$ over sequences $\{x\}$ which obey Eq. (3)

is given by

$$\langle F[\{x\}]\rangle = Z^{-1}\langle \pi_m \int dy_m\, \delta(y_m - x_m) F[\{y\}]\rangle = Z^{-1}\int [\mathfrak{D}y][\mathfrak{D}s] F[\{y\}]\langle \exp\{i\sum_m s_m[y_{m+1} - f_r(y_m) - \xi_m]\}\rangle$$
$$= Z^{-1}\int [\mathfrak{D}x][\mathfrak{D}s] F[\{y\}] \exp\Omega_r(x, s, \sigma)$$

with

$$\Omega_r(x, s, \sigma) \equiv \sum_m \{is_m[x_{m+1} - f_r(x_m)] - \tfrac{1}{2}\sigma^2 s_m^2\}. \tag{4}$$

Specifically, the correlation function is given by

$$\langle x_m x_{m'}\rangle = Z^{-1}\int [\mathfrak{D}x][\mathfrak{D}s] x_m x_{m'} \exp\Omega_r(x, s, \sigma). \tag{5}$$

We also introduce the Martin-Siggia-Rose response function, $R(r, \sigma; m - m') \equiv i\langle x_m s_{m'}\rangle$,

$$R(r, \sigma; m - m') = Z^{-1}\int [\mathfrak{D}x][\mathfrak{D}s]\, ix_m s_{m'} \exp\Omega_r(x, s, \sigma) \tag{6}$$

which depends on both x and s, the variable "conjugate" to x, to define the Lyapunov parameter in the presence of noise. Let $X_N(y)$ be the expectation value of x_N in the ensemble with the initial condition $x_0 = y$. We then have for large N

$$\exp(\lambda N) \equiv \lim_{N\to\infty,\, \epsilon\to 0} [X_N(y+\epsilon) - X_N(y)]/\epsilon = Z^{-1}\epsilon^{-1}\int [\mathfrak{D}x][\mathfrak{D}s] x_N \exp\Omega_r(x, s, \sigma)\{\exp[i\epsilon s_0 f_r'(x_0)] - 1\}$$
$$\approx if_r'(y)\langle x_N s_0\rangle = f_r'(y) R(r, \sigma; N). \tag{7}$$

As observed originally by Feigenbaum, it is sometimes preferable to study the iterated equation which, for weak noise, takes the form

$$x_{m+1} = f_r(x_m; n) + \xi_m' g_r(x_m; n)$$

in terms of the 2^n-th iterate of f_r

$$f_r(x; n) = \underbrace{f_r \circ f_r \circ f_r \circ \cdots \circ f_r}_{2^n \text{ terms}}(x). \tag{8}$$

We call this iterative process "coarse-graining." If $f_r(x) = f_r(x; 0)$ has a period-2 limit cycle, then $f_r(x; 1)$ has a pair of isolated fixed points; similarly, a four-cycle becomes a pair of two-cycles, etc. As a result of coarse graining, the function $g_r(x; n)$ appears, which, along with $f_r(x; n)$, is assumed to approach a fixed point through the coarse graining and rescaling transformations. Approximate expressions for the scaling variables are obtained by explicitly integrating over every other x_m variable in the functional integral.

With these ideas in mind, let us introduce $\langle\,\rangle_n$ to describe expectations computed at the nth coarse-graining level, i.e.,

$$R(r, \sigma; N; n) \equiv i\langle x_N s_0\rangle_n = Z^{-1}\int [\mathfrak{D}x][\mathfrak{D}s]\, ix_N s_0 \exp\Omega_r(x, s, \sigma; n)$$

with

$$\Omega_r(x, s, \sigma; n) \equiv \sum_m \{is_m[x_{m+1} - f_r(x_m; n)] - \tfrac{1}{2}\sigma^2 s_m^2 g_r^2(x; n)\}. \tag{9}$$

We can also express the Lyapunov parameter in terms of the coarse-grained response function,

$$\exp(\lambda N) = f_r'(y; n) R(r, \sigma; N; n). \tag{10}$$

Note that the quantity $f_r'(y; n)$ is the "stability parameter" for a 2^n-cycle with given r.

To find the desired recursion relations for $f_r(x)$ and $g_r(x)$ we first integrate over every even s_m in the "partition function"

$$Z \equiv N\int [\mathfrak{D}x][\mathfrak{D}s] \exp\Omega_r(x, s, \sigma; n) = N\int [\mathfrak{D}x][\mathfrak{D}s]_{\text{odd}} \exp\bar{\Omega}_r(x, s, \sigma; n)$$

with

$$\bar{\Omega}_r(x, s, \sigma; n) \equiv \sum_{\text{odd } m} \{is_m[x_{m+1} - f_r(x_m; n)] - \tfrac{1}{2}\sigma^2 s_m^2 g_r^2(x_m; n)\} \tag{11}$$
$$- \tfrac{1}{2}\sigma^{-2} \sum_{\text{even } m} \{g_r^{-2}(x_m; n)[x_{m+1} - f_r(x_m; n)]^2\}.$$

We next calculate the x_m integrals for odd values of m in the saddle-point approximation, obtaining

$$Z = N \int_{\text{even}}[\mathfrak{D}x]\,[\mathfrak{D}s]_{\text{odd}} \exp\Omega_r'(x, s, \sigma; n)$$

with

$$\Omega_r'(x, s, \sigma; n) \equiv \sum_{\text{odd } m} \{ is_m [x_{m+1} - f_r(f_r(x_{m-1}; n); n)] - \tfrac{1}{2}\sigma^2 s_m^2 [g_r^2(f_r(x_{m-1}; n); n) + f_r'^2(f_r(x_{m-1}; n); n) g_r^2(x_{m-1}; n)] \}. \tag{12}$$

Equation (12) leads to the coarse-graining recursion relations,

$$f_r(x; n+1) = f_r(f_r(x; n); n), \quad g_r^2(x; n+1) = f_r'^2(f_r(x; n); n) g_r^2(x; n) + g_r^2(f_r(x; n); n). \tag{13}$$

Rescaling Eq. (13) by

$$x_m' = -\alpha^{-1} x_{2m} \text{ and } s_m' = -\alpha s_{2m+1},$$

using Feigenbaum's result,

$$f_{r_{n+1}}(x; n+1) = -\alpha^{-1} f_{r_n}(-\alpha x; n),$$

and assuming the existence of a fixed point for

$$g_{r_{n+1}}(x; n+1) = \beta\alpha^{-1} g_{r_n}(\alpha x; n)$$

(where β is a multiplicative renormalization constant for the noise amplitude, and r_n and r_{n+1} are, respectively, points with corresponding stability for the 2^n- and 2^{n+1}-cycle), we find

$$\Omega_{r_{n+1}}(x, s, \sigma; n+1) = \Omega_{r_n}(x', s', \beta\sigma; n). \tag{14}$$

This leads to a scaling form for the response function

$$R(r_n, \sigma; N; n) = Z^{-1} \int [\mathfrak{D}x][\mathfrak{D}s]\, ix_N s_0 \exp\Omega_{r_n}(x, s, \sigma; n)$$
$$= Z^{-1} \int [\mathfrak{D}x'][\mathfrak{D}s']\, ix_{N/2}' s_0' \exp\Omega_{r_{n-1}}(x', s', \beta\sigma; n-1) = R(r_{n-1}, \beta\sigma; \tfrac{1}{2}N; n-1). \tag{15}$$

The factor $\frac{1}{2}$ multiplies N since coarse graining doubles the length of the unit iteration step.

We have searched for the approximate fixed points of Eq. (13) by using linear and quadratic approximations for $g_r(x)$. The linear approximation leads to an expression for $\beta^2 = \alpha^2 + \delta^2$, entirely in terms of Feigenbaum's universal constants. In the quadratic approximation we obtain the value[9] $\beta \approx 7.7$. For large enough n, Eqs. (10) and (15) imply that $\lambda(r_n; \sigma) = 2\lambda(r_{n+1}; \beta^{-1}\sigma)$ which, upon iteration, yields

$$\lambda(r_n; \sigma) = 2^m \lambda(r_{n+m}; \beta^{-m}\sigma). \tag{16}$$

Let us first examine the behavior of λ as a function of σ at r_∞. From Eq. (16), we see that

$$\lambda(r_\infty; \sigma) = 2^m \lambda(r_\infty; \beta^{-m}\sigma);$$

whence, assuming that $\lambda(r_\infty; \sigma)$ is proportional to σ^u, we find

$$u = (\ln 2)/\ln\beta \approx 0.34\ldots. \tag{2b$'$}$$

Since $r_k - r_\infty$ is proportional to δ^{-k}, we can rewrite Eq. (16) as

$$\lambda(r_\infty - r_n; \sigma) = 2^m \lambda(\delta^{-m}(r_\infty - r_n); \beta^{-m}\sigma). \tag{17}$$

Introducing $t = (\ln 2)/\ln\delta = 0.4498\ldots$, and fixing the first argument on the right-hand side of Eq. (17) at some small constant value, we are led to

$$\lambda(r_\infty - r_n; \sigma) = (r_\infty - r_n)^t \Phi((r_\infty - r_n)^{-t}\sigma^u), \tag{1$'$}$$

and the zero-noise scaling relation,

$$\lambda(r_\infty - r_n; 0) \sim (r_\infty - r_n)^t. \tag{18}$$

Equation (18) gives the same scaling exponent for the Lyapunov parameter below threshold that Huberman and Rudnick previously obtained for this parameter beyond threshold in the chaotic regime. Our relation, which complements theirs, holds only when the Lyapunov exponent is calculated for corresponding values of the stability parameter, i.e., it describes the curve connecting the points in each 2^n-cycle which have equal stability parameters. With this understanding, we see that the same power law describes the Lyapunov exponent above and below threshold.

From Eq. (1), we can calculate how, as the external noise is varied, the point r_c, at which the Lyapunov parameter changes sign, is shifted. Since $(r_\infty - r_c)^{-t}\sigma^u$ must be constant for $\sigma \neq 0$, we see that γ, defined by $(r_\infty - r_c) \sim \sigma^\gamma$, satisfies

$$\gamma = u/t \approx 0.75\ldots. \tag{19}$$

In careful numerical experiments, Huberman and collaborators[6] have found

$$u = 0.37 \pm 0.02 \text{ and } \gamma = 0.82 \pm 0.02$$

which agree satisfactorily with our simple approximate values, 0.34 and 0.75.

This work is supported in part by the National Science Foundation under Grant No. DMR-77-10210. One of us (C. E. W.) is a National Science Foundation Predoctoral Fellow.

---

[1]M. Feigenbaum, J. Statist. Phys. 19, 25 (1978), and 21, 669 (1979).

[2]P. Collet and J.-P. Eckmann, *Iterated Maps on the Interval as Dynamical Systems* (Birkhäuser, Boston, 1980), and references therein.

[3]A. Libchaber and J. Maurer, J. Phys. (Paris), Colloq. 41, C3-51 (1980).

[4]B. A. Huberman and J. Rudnick, Phys. Rev. Lett. 45, 154 (1980).

[5]J. P. Crutchfield and B. A. Huberman, Phys. Lett. 77A, 407 (1980).

[6]We are grateful to Dr. B. A. Huberman (private communication) for carrying out more detailed and accurate calculations akin to those in Ref. 4, for informing us of the results prior to publication and permitting us to quote them, and for several helpful comments.

[7]P. C. Martin, E. D. Siggia, and H. A. Rose, Phys. Rev. A 8, 423 (1973).

[8]R. Graham, *Quantum Statistics in Optics and Solid-State Physics*, Springer Tracts in Modern Physics Vol. 66 (Springer-Verlag, New York, 1973), p. 1; C. DeDominicis, J. Phys. C 1, 247 (1976); R. Phythian, J. Phys. A 9, 269 (1976), and J. Phys. A 10, 777 (1977); H. K. Janssen, Z. Phys. B 23, 377 (1977); B. Jouvet and R. Phythian, Phys. Rev. A 19, 1350 (1979).

[9]An additional interesting property of β has been brought to our attention: β is approximately equal to μ, the ratio of the intensity of successive subharmonics in the power spectrum studied by M. Feigenbaum [Phys. Lett. 74A, 375 (1979)], who found $\mu \simeq 4\alpha[2(1+\alpha^{-2})]^{1/2}$. The current best values of β and μ are, respectively, 6.618 and 6.557. That β and the "noise-free" parameter μ are related can be made plausible by requiring the ratio of the intensities of the noise-induced power spectra for chaotic transitions at r_n and r_{n+1} to coincide with the ratio of the spectral peaks at corresponding values of the control parameter, and performing a field-theoretic calculation which identifies the former ratio with the ratio of noise levels causing the transition.

Editor's note. We recommend also Feigenbaum and Hasslacher (1982) for their discussion of how the noise problem can be treated in terms of path integrals.

Physical Review Letters **46** 933–5 (1981)

Scaling for External Noise at the Onset of Chaos

J. Crutchfield, M. Nauenberg, and J. Rudnick
Physics Department, University of California, Santa Cruz, California 95064
(Received 8 December 1980)

The effect of external noise on the transition to chaos for maps of the interval which exhibit period-doubling bifurcations are considered. It is shown that the Liapunov characteristic exponent satisfies scaling in the vicinity of the transition. The critical exponent for noise is calculated with the use of Feigenbaum's renormalization group approach, and the scaling function for the Liapunov characteristic exponent is obtained numerically by iterating a map with additive noise.

PACS numbers: 64.60.Fr, 02.90.+p, 47.25.Mr

The notion that the transition to turbulence in fluids has universality properties similar to those of critical phenomena has been suggested by Feigenbaum[1] on the basis of the scaling behavior of mathematical models near the onset of chaos.[2] A further impetus for an analogy between the transition to chaos and critical point phase transitions was given[3] by the observation that as a control parameter r in these models increases past a critical value r_c into the chaotic regime the measure-theoretic entropy—the Liapunov characteristic exponent $\bar{\lambda}$—has an envelope curve of the form $(r-r_c)^\tau$. The universal exponent τ is given by $\tau = \ln 2/\ln\delta = 0.449\,806\,9\ldots$, where δ is the maximum eigenvalue associated with perturbations about the invariant map[1] of the interval. The transition to chaos in these models is heralded by a cascade of period-doubling bifurcations,[2] which is also of interest to an understanding of the onset of turbulence in physical systems.[4]

Motivated by the interpretation of experiments in fluids[5] and solids and by some recent numerical calculations,[6,7] we have considered theoretically the effect of added external noise on the transition to chaos in maps of the interval. The main result to be reported here is that the noise amplitude behaves as a *scaling variable* and that the dependence of the Liapunov characteristic exponent $\bar{\lambda}$ on the noise amplitude σ and $\bar{r} = (r-r_c)/r_c$ is of the scaling form

$$\bar{\lambda}(r,\sigma) = \sigma^\theta L(\bar{r}/\sigma^\gamma) \tag{1}$$

with $L(y)$ a universal function, and θ and γ universal exponents. In the limit of vanishing noise $\sigma \to 0$ we have $\bar{\lambda} \propto \bar{r}^\tau$ which implies that as $y\to\infty$, $L(y) \propto y^\tau$, and leads to the exponent relation $\theta = \gamma\tau$.

The idea that the noise plays a role parallel to that of the ordering field in a ferromagnetic transition was conjectured previously in Ref. 7. The noise exponent θ is a new critical exponent which we evaluate from an extension of Feigenbaum's scaling theory. Our result agrees with the recently observed value[7] of θ to within the limits of accuracy of the measurement. We also report on the measured form of the scaling function $L(y)$.

We start out by specifying the form of the one-dimensional map with additive noise. It is defined by the stochastic recursion relation

$$x_{\kappa+1} = f(x_\kappa; r) + \xi_\kappa \sigma \tag{2}$$

with $f(x; r)$ a continuous function of x in a finite interval having a parabolic maximum, and r a parameter that controls the shape of the function.[2] A common example is the function $rx(1-x)$ with $0 \le r \le 4$, and $0 \le x \le 1$. The quantity ξ_κ is a random variable controlled by an even distribution of unit width, and σ is a variable that controls the width (or amplitude) of the noise. Note that when $\sigma = 0$ the map is perfectly deterministic.

We consider successive iterations of the stochastic map, Eq. (2) with r at the critical value r_c, following techniques introduced by Feigenbaum. Setting the origin of coordinates to the x for which the function $f(x; r)$ is a maximum and rescaling this maximum to 1, the 2^nth iterate of $f(x; r_c)$ converges to $(-\alpha)^{-n} g(\alpha^n x)$, where $g(x)$ is a universal map satisfying the equation

$$g(g(x)) = -\alpha^{-1} g(\alpha x) \tag{3}$$

with $\alpha = -1/g(1)$. Adding a small amount of noise $\xi\sigma$, we assume that the corresponding 2^nth iterate of the map converges to $(-\alpha)^{-n}[g(\alpha^n x) + \xi\sigma\kappa^n D(\alpha^n x)]$ with $D(x)$ a universal x-dependent noise amplitude function and κ a constant. When σ is small enough, we have

$$g(g(x)+\xi\sigma D(x)) + \xi'\sigma D(g(x)+\xi\sigma D(x)) \approx g(g(x)) + \xi\sigma g'(g(x))D(x) + \xi'\sigma D(g(x)) + O(\sigma^2)$$
$$= g(g(x)) + \xi''\sigma\{[g'(g(x))D(x)]^2 + [D(g(x))]^2\}^{1/2}. \tag{4}$$

In going to the last line we used the fact that ξ and ξ' are independent random variables, and that ξ'' is also a random variable. This and our above assumption implies that $D(x)$ must satisfy the eigenvalue equation

$$KD(\alpha x) = \alpha\{[g'(g(x))D(x)]^2 + [D(g(x))]^2\}^{1/2}. \quad (5)$$

We have solved Eq. (5) for the eigenvalue κ and the corresponding eigenfunction $D(x)$ using the known results[1] for α and $g(x)$. Carrying out a calculation involving a polynomial interpolation for $D(x)$ we have found $\kappa = 6.61903\ldots$.

In the immediate vicinity above the transition to chaos the invariant probability distribution associated with the stochastic map will consist of 2^n bands, where n is an integer that grows in the case of the deterministic map by unit steps to infinity as the transition is approached.[8,9] In the case of the stochastic map, n grows to a finite value—and then decreases by unit steps as one passes to the other side of the transition. This modification of the deterministic bifurcation sequence is called a bifurcation gap.[6]

We now extend to the present case the previous discussion in Ref. 2 of the scaling behavior of the Liapunov characteristic exponent $\bar{\lambda}$, given by

$$\bar{\lambda} = \lim_{N\to\infty} \frac{1}{N} \sum_{k=1}^{N} \ln|f'(x_k; r)|, \quad (6)$$

or alternatively

$$\bar{\lambda} = \int p(x) \ln|f'(x; r)|\, dx, \quad (7)$$

where $p(x)$ is the invariant probability distribution associated with the map. Applying the above-mentioned considerations we obtain[10]

$$\bar{\lambda} = 2^{-n} L(\delta^n \bar{r}, \kappa^n \sigma). \quad (8)$$

Now, we assume that there will be 2^n bands in the chaotic regime when $\kappa^n\sigma$ is of order unity so that $n = -\ln\sigma/\ln\kappa$. Substituting this result into Eq. (9) we obtain Eq. (1) for $\bar{\lambda}$ with the two exponents θ and γ given in terms of Feigenbaum's eigenvalue δ and the new eigenvalue κ by $\theta = \ln 2/\ln\kappa = 0.366754\ldots$ and $\gamma = \ln\delta/\ln\kappa = 0.815359\ldots$. The appearance of a bifurcation gap implies that $L(y)$ vanish at some $y = y_0$ which in turn implies that the maximum number n of bifurcations is determined by the relation $\bar{r}_{n\,\max} = y_0\sigma^\gamma$. This behavior has been observed numerically.[6]

Measurements of the behavior of $\bar{\lambda}$ as a function of σ at $\bar{r} = 0$ have already been made by numerically calculating $\bar{\lambda}$ according to Eq. (6) with varying amounts of noise.[7] The measured value for θ is 0.37 ± 0.01. This agrees with our theoretical value for θ to within the experimental error.

To verify the existence of the scaling function $L(y)$ of Eq. (1) we used our values of θ and γ to plot $\bar{\lambda}\sigma^{-\theta}$, with $\bar{\lambda}$ the result of numerical calculations of Eq. (6), as a function of $\bar{r}\sigma^{-\gamma}$. The results are shown in Figs. 1 and 2 for three different noise levels: $\sigma = 10^{-6}$, 10^{-8}, and 10^{-10}. The results for those three different noise levels all fall on a universal curve in the chaotic regime, and in its immediate vicinity, Fig. 1, and fit the asymptotic behavior $L(y) \sim y^\tau$ for large y. The results do *not* coincide in the periodic regime, Fig. 2, but they could have been made to agree if we had chosen noise amplitudes differing by factors of κ, instead of factors of 100. This more restricted scaling follows from considerations of the type enunciated above.

These results appear to us to be both exciting and highly provocative. A theoretical picture of the transition to turbulence is just beginning to emerge; the analogy to critical phenomena should lead to new and important insights into the nature and characteristics of this transition.

The authors have benefited from conversations with B. A. Huberman and wish to thank him for

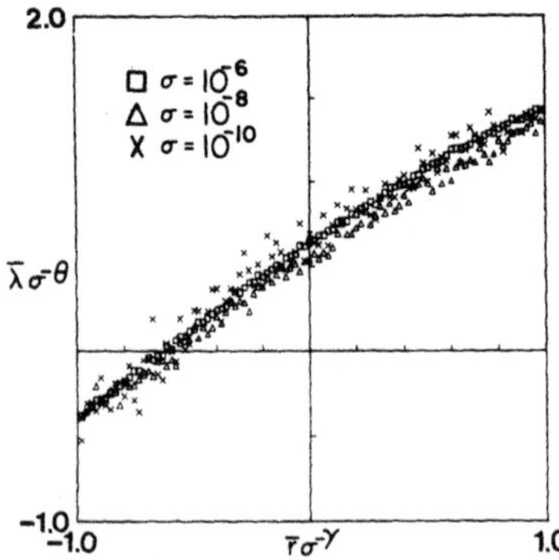

FIG. 1. Numerical determination of the scaling function $L(y)$, Eq. (1). The quantity $\bar{\lambda}\sigma^{-\theta}$ is plotted against 100 values of $y = \bar{r}\sigma^{-\gamma}$ at each of three noise levels: $\sigma = 10^{-6}$, 10^{-8}, and 10^{-10}. $\bar{\lambda}$ was calculated with use of Eq. (6), with $N = 10^6$ and with ξ_K a uniformly distributed random number of standard deviation σ.

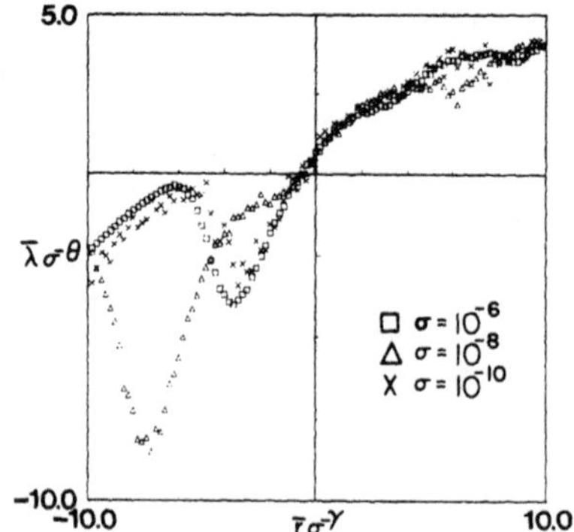

FIG. 2. $\bar{\lambda}\sigma^{-\theta}$ is plotted again, but over a wider range of $y = \bar{r}\sigma^{-\gamma}$ to illustrate the scaling regime. See text for discussion of various features. The details are the same as in Fig. 1, expect that $\bar{\lambda}$ was calculated with $N = 10^5$ in Eq. (6).

the use of computing facilities at Xerox Palo Alto Research Center. One of us (J.C.) would also like to acknowledge useful discussions with N. Packard and the receipt of a University of California Regents Fellowship. This work is supported by the National Science Foundation.

[1] M. J. Feigenbaum, J. Statist. Phys. 19, 25 (1978).

[2] For a recent monograph on this subject, see P. Collet and J. P. Eckmann, *Iterated Maps of the Interval as Dynamical Systems* (Birkhäuser, Boston, 1980).

[3] B. A. Huberman and J. Rudnick, Phys. Rev. Lett. 45, 154 (1980). The exponent τ appears as t in this reference. We have replaced the latin by a greek letter for consistency with other critical exponents.

[4] A. Libchaber and J. Maurer, J. Phys. (Paris), Colloq. 41, C3-51 (1980); M. J. Feigenbaum, Phys. Lett. 74A, 375 (1979); J. P. Gollub, S. V. Benson, and J. Steinman, Ann. N.Y. Acad. Sci. (to be published); B. A. Huberman and J. P. Crutchfield, Phys. Rev. Lett. 43, 1743 (1979).

[5] G. Ahlers, private communication.

[6] J. P. Crutchfield and B. A. Huberman, Phys. Lett. 77A, 407 (1980).

[7] J. P. Crutchfield, J. D. Farmer, and B. A. Huberman, to be published.

[8] J. P. Crutchfield, J. D. Farmer, N. Packard, R. Shaw, G. Jones, and R. J. Donnelly, Phys. Lett. 76A, 1 (1980).

[9] E. N. Lorentz, Ann. N.Y. Acad. Sci. (to be published).

[10] The details of the derivation of the result (8), which involves a careful consideration of the structure of the bands, will be presented in a future paper.

Part 5

Intermittency

Communications in Mathematical Physics **74** 189–97 (1980)

Intermittent Transition to Turbulence in Dissipative Dynamical Systems

Yves Pomeau and Paul Manneville*

Commissariat à l'Énergie Atomique, Division de la Physique, Service de Physique Théorique, F-91190 Gif-sur-Yvette, France

Abstract. We study some simple dissipative dynamical systems exhibiting a transition from a stable periodic behavior to a chaotic one. At that transition, the inverse coherence time grows continuously from zero due to the random occurrence of widely separated bursts in the time record.

Introduction

A number of investigators [1] have observed in convective fluids an intermittent transition to turbulence. In these experiments the external control parameter, say r, is the vertical temperature difference across a Rayleigh-Bénard cell. Below a critical value r_T of this parameter, measurements show well behaved and regular periodic oscillations. As r becomes slightly larger than r_T the fluctuations remain apparently periodic during long time intervals (which we shall call "laminar phases") but this regular behavior seems to be randomly and abruptly disrupted by a "burst" on the time record. This "burst" has a finite duration, it stops and a new laminar phase starts and so on. Close to r_T, the time lag between two bursts is seemingly at random and much larger than – and not correlated to – the period of the underlying oscillations. As r increases more and more beyond r_T it becomes more and more difficult and finally quite impossible to recognize the regular oscillations (see Fig. 1).

This sort of transition to turbulence is also present in simple dissipative dynamical systems [2] such as the Lorenz model [2a]. We present here the results of some numerical experiments on this problem.

When a burst starts at the end of a laminar phase this denotes an instability of the periodic motion due to the fact that the modulus of at least one Floquet multiplier [3] is larger than one. This may occur in three different ways: a real Floquet multiplier crosses the unit circle at $(+1)$ or at (-1) or two complex conjugate multipliers cross simultaneously. To each of these three typical crossings we may associate one type of intermittency that we shall call for convenience type

* DPh. G. PSRM, Cen Saclay, Boîte Postale 2, F-91190 Gif-sur-Yvette, France

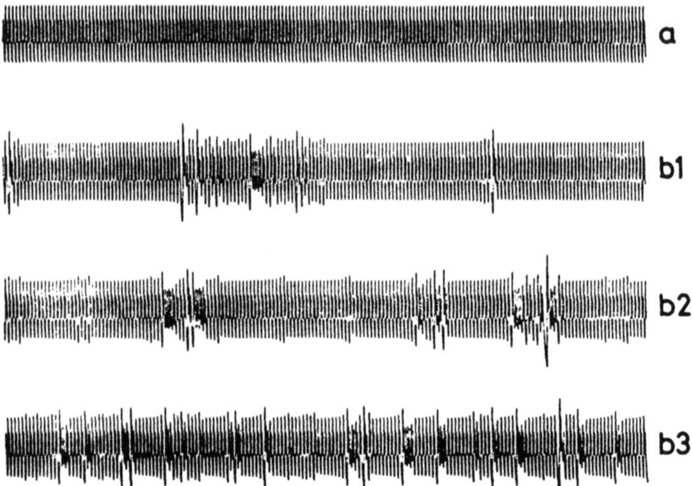

Fig. 1a and b. Time record of one coordinate (z) in the Lorenz model. **a** Stable periodic motion for $r=166$. **b** Above the threshold the oscillations are interrupted by bursts which become more frequent as r is increased

1: crossing at $(+1)$; type 2: complex crossing; and type 3: crossing at (-1) respectively. In all these three cases our numerical studies show that the Lyapunov number grows continuously from zero beyond the onset of turbulence. In what follows we shall present some simple estimates for the "critical behavior" of this Lyapunov number in the vicinity of the turbulence threshold and compare them with the results of numerical experiments.

Type 1. Intermittency in the Lorenz Model

The Lorenz system reads [4]:

$$\frac{dx}{dt}=\sigma(y-z)\,; \quad \frac{dy}{dt}=-xz+rx-y\,; \quad \frac{dz}{dt}=xy-bz\,, \tag{1}$$

where σ, b, and r are parameters. We have kept b and σ fixed at their original values ($\sigma=10$, $b=8/3$). Integrating system (1) around $r=166$ one finds for r slightly less than $r_T(\simeq 166.06)$ regular and stable oscillations for a random choice of initial condition (Fig. 1a). For r slightly larger than r_T these oscillations are interrupted by bursts (Fig. 1b). This can be explained quite simply by studying the Poincaré map (restricted here to be 1-dimensional without loss of significance). Let f be the function such that $y_{n+1}=f(y_n, r)$ where y_n is the y-coordinate of the nth crossing of the plane $x=0$. Near $r=r_T$ the curve of equation $y'=f(y,r)$ is nearly tangent to the first bissectrix (Fig. 2). For r slightly below r_T, this curve has two intersections with the bisectrix, they collapse into a single point at $r=r_T$ while for $r>r_T$ the curve is lifted up and no longer crosses the first bisectrix so that a "channel" appears between them (Fig. 3). Hence the successive iterates generated by the map

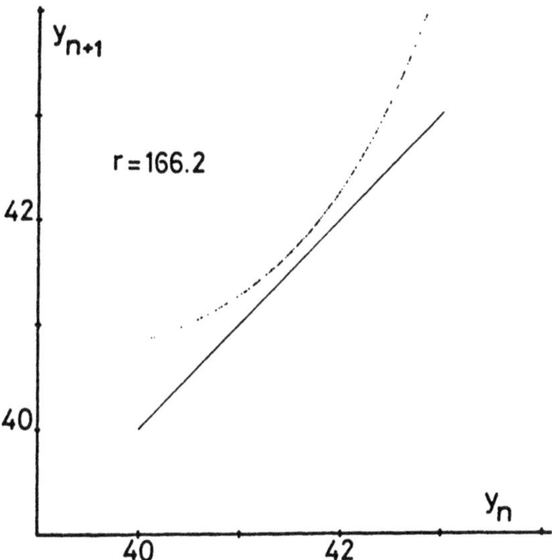

Fig. 2. A part of the Poincaré map along the y-coordinate for $r = 166.2$ slightly beyond the intermittency threshold ($r_T \simeq 166.06$)

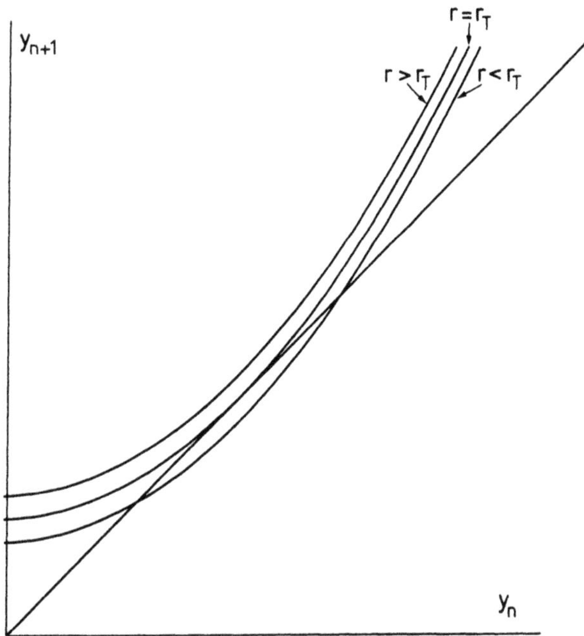

Fig. 3. Idealized picture of the deformation of $y_{n+1}(y_n)$ explaining the transition via intermittency. For $r < r_t$ two fixed points coexist one stable the other unstable. They collapse at $r = r_T$ and then disappear leaving a channel between the curve and the first bisectrix

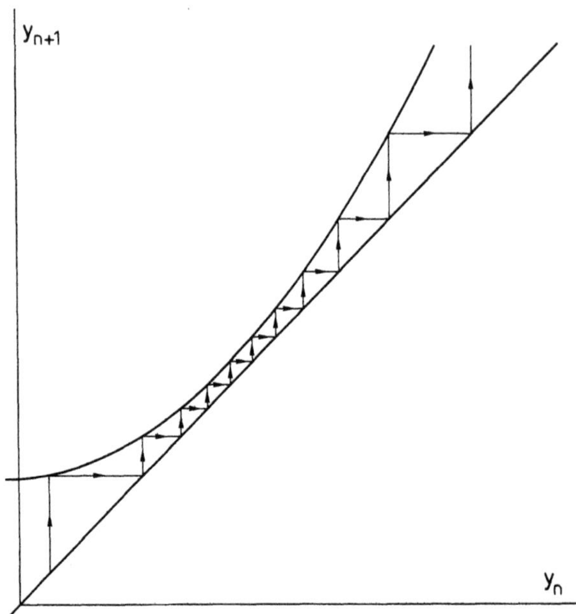

Fig. 4. The motion through the channel corresponds to the laminar phase of the movement. The slow drift is quite imperceptible on the time record of Fig. 1b

$y \to f(y, r)$ travel along this channel, which requires a large number of iterations (Fig. 4). To estimate this number let us consider a "generic form" for $f(y, r)$ in the region considered:

$$f(y, r) = y + \varepsilon + y^2 \text{ (+ higher order terms } - \text{H.O.T.)},$$

where $\varepsilon = (r - r_T)/r_T$. Near $\varepsilon = 0_+$ the difference equation

$$y_{n+1} = y_n + \varepsilon + y_n^2 \text{ (+ H.O.T.)}$$

can be approximated by a differential equation over n and an elementary estimate shows that a number of iteration of the order of $\varepsilon^{-1/2}$ is needed to cross the channel. This is in nice agreement with our numerical simulation of system (1) (Fig. 5). After each transfer the burst destroys the coherence of the motion. This leads one to conclude that near $\varepsilon = 0_+$ the Lyapunov number varies as $\varepsilon^{1/2}$. Though this is consistent with our first numerical estimates, close to the intermittency threshold the Lyapunov number converges so slowly that it is difficult to get with precision, so we have preferred to turn to a modelling of the Poincaré map. We have got a qualitatively similar behavior for the following map of $S^1 = [0, 1[$

$$\theta \to 2\theta + r \sin 2\pi\theta + 0.1 \sin 4\pi\theta \,(\text{mod}\, 1). \tag{2}$$

As shown in Fig. 6 this applies S^1 twice on itself and it is intermittent at $r_T \simeq -0.24706$. In this model, as well as those we shall consider later, the

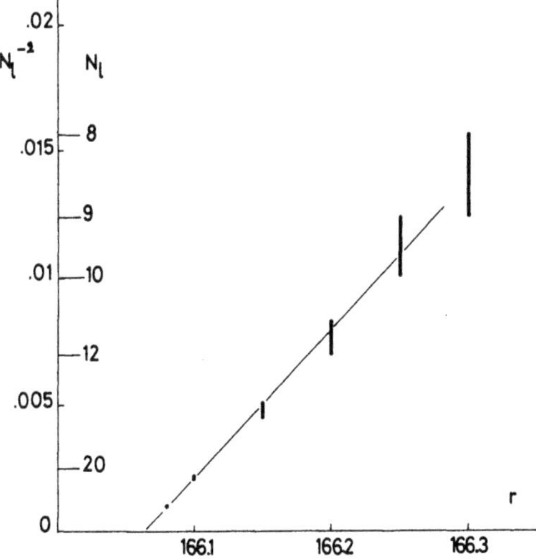

Fig. 5. The square of N_l the largest number of cycles during a laminar period is inversely proportional to the distance from the threshold $r - r_T$. N_l is given within 1 cycle to account for the uncertainty in the definition of the beginning/end of a laminar phase

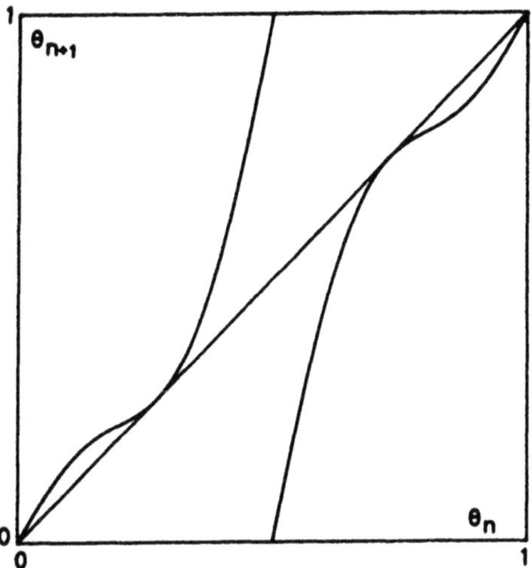

Fig. 6. Model mapping displaying qualitatively the same behavior as the Lorenz model around $r = 166$

possibility of starting a laminar phase after a burst comes from the fact that the map is not invertible. In diffeomorphisms the "relaminarization" cannot occur in this way due to the uniqueness of preimages. However dynamical systems for which the "reduced" Poincaré map takes a form similar to (2) [and later to (3) or (5)] can be constructed simply by adding other dimensions along which fluc-

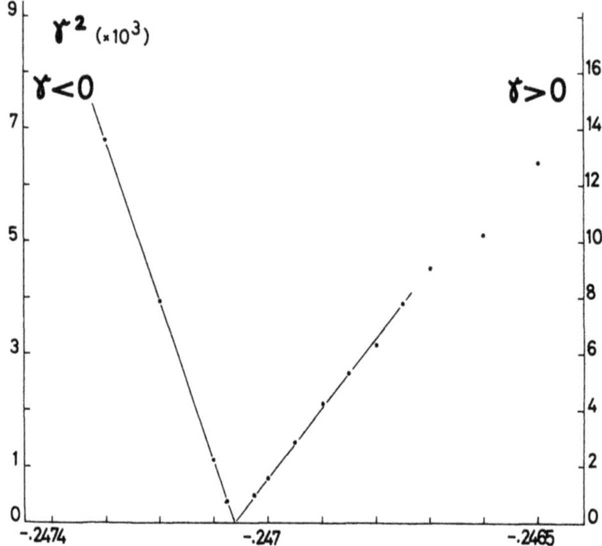

Fig. 7. For the model mapping $r_T \simeq -0.24706$. For $r<r_T$ the Lyapunov number γ is negative and varies as $-\sqrt{r_T-r}$ while for $r>r_T$ it is positive and grows like $\sqrt{r-r_T}$

tuations are stable [5]. Numerical simulation of the model defined by (2) can easily be performed using a desk-top computer. As expected the Lyapunov number grows with the 1/2-power near the threshold (Fig. 7).

Type 2. Intermittency

In order to study numerically this case we have considered a map that applies the torus $T^2 = [0,1[\times [0,1[$ four times onto itself: in complex notations $z = x + iy$

$$z' = \lambda z + \mu |z|^2 z \text{ close to the origin } (\lambda \text{ complex and } \mu \text{ real}), \tag{3a}$$

$$z' = 2z \text{ far from the origin} \tag{3b}$$

with a smooth interpolation inbetween. The fixed point $z=0$ of this map looses its stability when the complex parameter λ crosses the unit circle, the coefficient μ of the cubic term being so chosen as to avoid the appearance of a stable limit cycle or equivalently to make the bifurcation subcritical i.e. $\mu = \mu$ Re $\{\lambda\} > 0$. Iterations of the above map show intermittency when $|\lambda| = 1 + \varepsilon$ and $\varepsilon \to O_+$. Once an iterate falls near $z=0$, it enters a laminar phase and a large number of further iterations are needed to expell it towards the "bursting region" (where correlations are broken) defined by $|z| > \varrho^*$, ϱ^* being fixed and ε-independent, roughly in the interpolating region. To find how the Lyapunov number grows near the intermittency threshold one may reason as follows: Let $\varrho_j = |z_j|$ be the distance of the jth iterate to the fixed point. The iterates rotate around the fixed point due to the complex nature of λ but we shall neglect the angular variation and only consider the growth of the modulus

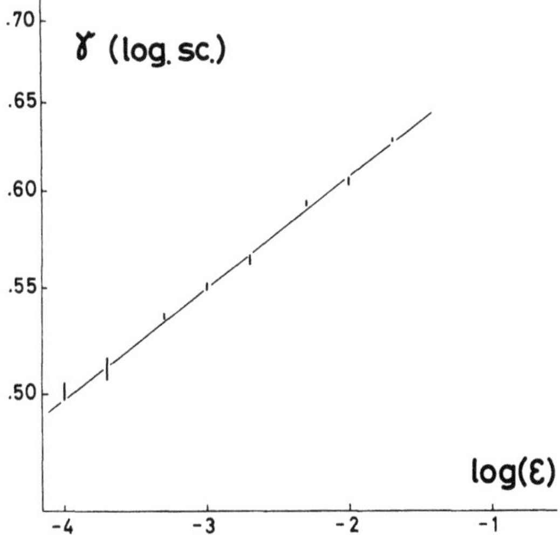

Fig. 8. On the torus T_2 for $\lambda=(1+\varepsilon)\exp i\varphi$, $\varphi=0.05$ rd and $\mu=20$ "mean field" theory predicts $\gamma \sim \ln(1/\varepsilon)$ while the numerical simulation gives $\gamma \sim \varepsilon^{\alpha}$ with $\alpha \sim 0.04$

ϱ. Near $\varrho=0$ it is approximatively given by

$$\varrho_{n+1}=(1+\varepsilon)\varrho_n+\bar{\mu}\varrho_n^3+\text{H.O.T.} \tag{4}$$

Now let us examine a laminar cycle with starting point at $\varrho=\tilde{\varrho}\ll\varrho^*$. If $\tilde{\varrho}\gg\varepsilon^{1/2}$ one easily sees that a number of iteration of order $1/\tilde{\varrho}^2$ are needed to reach ϱ^* and enter a turbulent burst. On the other hand if $\tilde{\varrho}\ll\varepsilon^{1/2}$ the laminar cycle ends after $\varepsilon^{-1}\ln\varrho$ iterations approximately. Assuming then that $\tilde{\varrho}$ is at random with probability $\tilde{\varrho}\,d\tilde{\varrho}$ in the circle of center 0 and radius ϱ^* the estimates given above yield $\ln(1/\varepsilon)$ as an order of magnitude for both the mean duration of a laminar period and the inverse Lyapunov number near $\varepsilon=0_+$. This is in slight disagreement with our computer experiments which seem to indicate rather a power-like growth of the Lyapunov number $\gamma\sim\varepsilon^{\alpha}$ α small and positive (Fig. 8). This descrepancy between the naive theory presented above and computer results may come from the neglect of fluctuations about the mean length of the laminar cycles, which makes the procedure used sound much like a "mean field theory" in the usual jargon of phase transitions (it may also come from the neglect of the rotation of iterates affecting the statistics in an unknown way).

Type 3. Intermittency

The last type of intermittency we shall examine may occur when the Floquet multiplier is real and crosses the unit circle at (-1). Although a differential system has been found which displays this kind of behavior [2b] we shall report here on the simulation of the following mapping of the circle S_1 onto itself:

$$\theta\to 1-2\theta-\frac{1}{2\pi}(1-\varepsilon)\cos\left[2\pi\left(\theta-\frac{1}{12}\right)\right](\mathrm{mod}\,1). \tag{5}$$

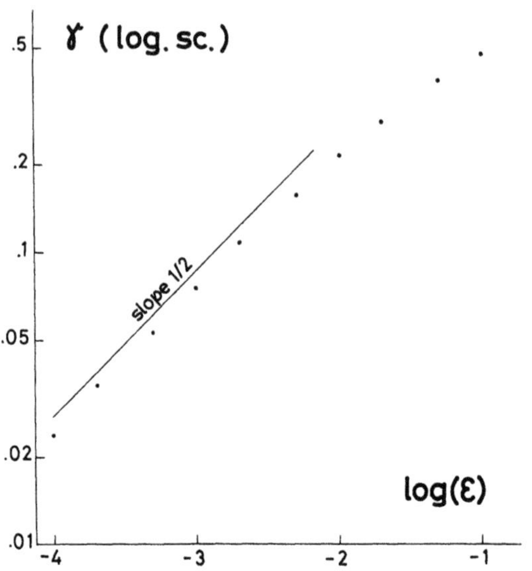

Fig. 9. On the torus T_1 for type 3 intermittency the Lyapunov number γ grows as $\varepsilon^{1/2}$

This map applies S^1 twice onto itself and it reverse the orientation so that the eigenvalue of the map linearized near the fixed point $\theta_F = 1/3$ can easily be made negative. Near the fixed point the map expands as

$$\bar{\theta}_{n+1} = -(1+\varepsilon)\bar{\theta}_n - \frac{(2\pi)^2}{6}\bar{\theta}_n^3 + \text{H.O.T.}\,(\bar{\theta} = \theta - \theta_F)\,. \tag{6}$$

The most general form would be

$$\bar{\theta}_{n+1} = -(1+\varepsilon)\bar{\theta}_n + a\theta_n^2 + b\theta_n^3\,, \tag{7}$$

a and b being constant. If the r.h.s. of (7) has a positive Schwarzian derivative that is here $b + a^2 < 0$ then the bifurcation at $\varepsilon = 0$ is subcritical and type 3 intermittency can occur. This is precisely the case with (5) since $a = 0$ and $b < 0$. To estimate the mean length of a laminar phase one considers instead of (6) or (7) the equation giving $\bar{\theta}_{n+2}$ in function of $\bar{\theta}_n$. This relation is basically of the same form as Eq. (4) (quadratic terms vanish at $\varepsilon = 0$ and are in inessential for ε small enough). Thus one reasons as for type 2 intermittency with the difference that now the problem is strictly unidimensional so that the probability measure for for the starting point of a laminar cycle is now the usual Lebesgue measure instead of $\tilde{\varrho}\, d\tilde{\varrho}$ previously. An elementary calculation shows that the Lyapunov number should grow like $\varepsilon^{1/2}$ near threshold, this time in agreement with the computer experiment (Fig. 9).

Conclusion

Intermittency is a quite common phenomenon in experimental turbulence. The theory sketched in this paper is more especially related with the case of convection in confined geometries [1] but intermittency is also well known in boundary layers and pipe flows [6] and even in $1/f$ – noise theory [7]. Despite the different meanings of the term "intermittency", the possibility remains that the kind of dynamics described by the models we have studied could afford a qualitative understanding of all these phenomena.

References

1. Maurer, J., Libchaber, A., Bergé, P., Dubois, M.: Personal communications
2. (a) Manneville, P., Pomeau, Y.: Phys. Lett. **75**A, 1 (1979)
   (b) Arneodo, A., Coullet, P., Tresser, C.: Private communication
3. Iooss, G.: Bifurcation of maps and applications. In: North-Holland Math. Studies 36. Amsterdam, New York: North-Holland 1979
4. Lorenz, E.N.: J. Atmos. Science **20**, 130 (1963)
5. Pomeau, Y.: Intrinsic stochasticity in plasmas, Cargèse 1979. (eds. G. Laval, D. Gresillon). Orsay: Editions de Physique 1979
6. Tritton, D.J.: Physical fluids dynamics. New York: Van Nostrand-Reinhold 1977
7. Mandelbrot, B.: Fractals form chance and dimension. San Francisco: Freeman 1977

Communicated by D. Ruelle

Received January 17, 1980

Physics Letters **87A** 391–3 (1982)

INTERMITTENCY IN THE PRESENCE OF NOISE: A RENORMALIZATION GROUP FORMULATION

J.E. HIRSCH and M. NAUENBERG [1]
Institute for Theoretical Physics, University of California, Santa Barbara, CA 93106, USA

and

D.J. SCALAPINO
Institute for Theoretical Physics and Department of Physics, University of California, Santa Barbara, CA 93106, USA

Received 14 November 1981

A renormalization group (RG) formulation of the transition to chaotic behavior via intermittency in one-dimensional maps is presented. The known scaling behavior of the length of the laminar regions in the presence of external noise is obtained from the leading relevant eigenvalues of the RG transformation. In addition, the complete spectrum of eigenvalues and corresponding eigenfunctions is found.

The renormalization group equations describing a phase transition are regular functions of the parameters such as temperature and external field which determine the state of the system. As these parameters are continuously varied through a phase transition, the singular behavior of the system arises from an infinite iteration of the regular renormalization group equations. Recent work suggests that a similar point of view provides a useful framework for understanding the onset of irregular or chaotic behavior of dynamical systems as a parameter is continuously varied. In particular, iterates of the one-dimensional logistic map

$$x_{n+1} = Rx_n(1 - x_n) , \tag{1}$$

can change from a regular to an irregular pattern as R is varied. The logistic map exhibits two types of such transitions. One of these involves an infinite cascade of period-doubling or pitchfork bifurcations, while the other arises from a saddle or tangent bifurcation leading to intermittency. Feigenbaum [1] and others [2] have developed a renormalization approach to describe the scaling and universal properties of the transition to chaos through period-doubling for the class of one-dimensional maps $x_{n+1} = f(x_n)$, with

$$f(x) = 1 - a|x|^z , \tag{2}$$

where $z = 2$ for the logistic map. A scaling theory describing the effect of external noise on the period-doubling cascade has also been developed [3,4].

Here we are interested in the second type of transition exhibited by the logistic map. Following the initial ideas of Pomeau and Manneville [5], the onset of chaotic behavior characterized by the occurrence of regular or "laminar" sequences of x_n values separated by intermittent bursts has also been shown to scale [6,7]. For the class of saddle point maps

$$f(x) = x + a|x|^z + \epsilon , \tag{3}$$

with $z > 1$, the length of the laminar regions l varies for small ϵ as $\epsilon^{-(1-1/z)}$. In the presence of a stochastic noise source of amplitude g, l satisfies the scaling equation

$$l(\epsilon, g) = \epsilon^{-(1-1/z)} f(g/\epsilon^{(z+1)/2z}) . \tag{4}$$

These relations were established by considering a Langevin equation describing the map near the saddle point, and using Fokker–Planck techniques to determine the time of passage in the presence of noise.

Here we develop a renormalization approach for saddle point maps which puts the known scaling results for intermittency in the same framework as Feigenbaum's treatment of the period-doubling cascade. We consider the class of maps given by eq. (3) in the presence of external noise,

$$x' = f(x) + g\xi , \tag{5}$$

where ξ is a random variable of unit standard deviation. The idea of the renormalization approach is to evaluate the map $x \to x''$ associated with two consecutive iterations of eq. (5) and by rescaling cast it back into the original form. This requires new parameters, ϵ', g' which in the limit $\epsilon, g \to 0$ satisfy the relation

$$\epsilon' = \lambda_\epsilon \epsilon, \quad g' = \lambda_g g , \tag{6}$$

where λ_ϵ and λ_g are the largest relevant eigenvalues of the linearized renormalization group transformation. Then the length $l(\epsilon, g)$ satisfies the homogeneity relation

$$l(\epsilon, g) = 2l(\epsilon', g') , \tag{7}$$

[1] Permanent address: Natural Science, University of California, Santa Cruz, CA 95060, USA.

which leads in the usual way to the scaling relation

$$l(\epsilon,g) = \epsilon^{-\nu} f(g/\epsilon^{\mu}) , \quad (8)$$

with exponents

$$\nu = \log 2/\log \lambda_\epsilon , \quad \mu = \log \lambda_g/\log \lambda_\epsilon . \quad (9)$$

The functional recursion relation we use to define our renormalization procedure is the same as in Feigenbaum's case:

$$T\{f(x)\} = \alpha f(f(x/\alpha)) , \quad (10)$$

where α is a rescaling factor, but with boundary conditions

$$f(0) = 0 , \quad f'(0) = 1 , \quad (11)$$

appropriate to a saddle point bifurcation at $x = 0$. It can be readily verified that

$$f^*(x) = x/(1 - ax) , \quad (12)$$

is a fixed point of the transformation (10) with $\alpha = 2$ and a an arbitrary constant. For small x this solution corresponds to eq. (3) for $\epsilon = 0$ and $z = 2$. For $z \neq 2$, we can find the fixed point of (10) by series expansion, and to third non-vanishing order obtain

$$f^*(x) = x + a|x|^z + \tfrac{1}{2} z a^2 |x|^{2z-1} + ... , \quad (13)$$

with the scale factor $\alpha = 2^{1/(z-1)}$.

The next step is to consider the effect of small perturbations around the fixed point. We write

$$f(x) = f^*(x) + \epsilon h(x) , \quad (14)$$

where ϵ is a small parameter, and determine the eigenfunction from the usual condition of form invariance after rescaling:

$$f^{*\prime}(f^*(x))h_n(x) + h_n(f^*(x)) = (\lambda_n/\alpha)h_n(\alpha x) , \quad (15)$$

where λ_n is the nth eigenvalue. For the case $z = 2$ we find $\lambda_n = 4/2^n$ and obtain the eigenfunctions h_n by series expansions. The relevant eigenfunction with eigenvalue $\lambda_\epsilon = 4$ is, to second order in x,

$$h_\epsilon(x) = 1 + ax + \tfrac{4}{3}a^2x^2 + ... \quad (16)$$

The other relevant eigenfunction, with eigenvalue $\lambda_1 = 2$, does not correspond to the physical situation of interest here and will not be considered [‡1]. The marginal eigenfunction, with $\lambda_2 = 1$, is associated with the arbitrary constant a in the map eq. (12) and can be found in closed form:

$$h_2(x) = x^2/(1 - ax)^2 . \quad (17)$$

Finally, the irrelevant eigenfunctions with eigenvalue λ_n, $n > 2$, have the leading behavior $h_n(x) = x^n + ...$

In the general case $z > 1$, the form of the eigenvalues is $\lambda_n = 2^{(z-n)/(z-1)}$, and the leading behavior of the eigenfunctions is x^n. The largest relevant eigenvalue is $\lambda_\epsilon = 2^{z/(z-1)}$. The case $n = 1$ is again not of interest here. For $1 < n < z$ the perturbation is still relevant and gives a crossover to a behavior described by the map eq. (3) with z replaced by n. The marginal eigenvalue, for $n = z$, is again associated with the arbitrary constant a, and the eigenfunctions with $n > z$ are irrelevant.

We consider now the effect of adding a small amount of external noise to the fixed point function:

$$f(x) = f^*(x) + g(x)\xi . \quad (18)$$

Under iteration, this leads to the eigenvalue equation [4]

$$f^{*\prime}(f^*(x))g^2(x) + g^2(f^*(x)) = (\lambda_g^2/\alpha^2)g^2(\alpha x) . \quad (19)$$

For the leading eigenvalue λ_g one obtains the exact result

$$\lambda_g = 2^{(z+1)/2(z-1)} , \quad (20)$$

and the corresponding eigenfunction is

$$g(x) \propto 1 + \tfrac{1}{2} z a |x|^z/x + ... \quad (21)$$

Using the above results for λ_ϵ and λ_g we obtain for the exponents defined in eq. (9),

$$\nu = (z-1)/z , \quad \mu = (z+1)/2z , \quad (22)$$

and from eq. (8) the scaling behavior eq. (4) follows.

In conclusion, we have shown that the scaling behavior for the average length of the laminar regions in the transition to chaos via intermittency can be easily derived from a renormalization group formulation of the problem. We have obtained exact results for the complete spectrum of eigenvalues and the leading terms of the corresponding eigenfunctions of the renormalization group transformation.

Two of us (J.E.H. and M.N.) would like to acknowledge support by the NSF under PHY77-27084 and PHY78-22253. D.J.S. would like to acknowledge the support of the ONR under N0014-79-C-0707.

References

[1] M. Feigenbaum, J. Stat. Phys. 19 (1978) 25; 21 (1979) 669.

[2] P. Collet and J.-P. Eckmann, in: Iterated maps on the interval as dynamical systems (Birkhäuser, Boston, 1980).

[3] J.P. Crutchfield, M. Nauenberg and J. Rudnick, Phys. Rev. Lett. 46 (1981) 933.

[4] B. Shraiman, C.E. Wayne and P.C. Martin, Phys. Rev. Lett. 46 (1981) 935.

[5] P. Manneville and Y. Pomeau, Phys. Lett. 75A (1979) 1; Commun. Math. Phys. 74 (1980) 189.

[6] J.-P. Eckmann, L. Thomas and P. Wittwer, to be published.

[7] J.E. Hirsch, B.A. Huberman and D.J. Scalapino, to be published.

[‡1] The eigenfunction corresponding to $\lambda = 2$ has the leading behavior $h_1(x) = x$ which changes the character of eq. (3), eliminating the intermittent behavior.

Editor's note. The long version of this work is given in Hirsch, Huberman and Scalapino (1982). The same solution to universality equations was independently obtained by Cosnard (1981). Hu and Rudnick (1982) have generalised these solutions to non-quadratic maps.

Part 6

Period-doubling in Higher Dimensions

Communications in Mathematical Physics **50** 69–77 (1976)

A Two-dimensional Mapping with a Strange Attractor

M. Hénon
Observatoire de Nice, F-06300 Nice, France

Abstract. Lorenz (1963) has investigated a system of three first-order differential equations, whose solutions tend toward a "strange attractor". We show that the same properties can be observed in a simple mapping of the plane defined by: $x_{i+1} = y_i + 1 - ax_i^2$, $y_{i+1} = bx_i$. Numerical experiments are carried out for $a = 1.4$, $b = 0.3$. Depending on the initial point (x_0, y_0), the sequence of points obtained by iteration of the mapping either diverges to infinity or tends to a strange attractor, which appears to be the product of a one-dimensional manifold by a Cantor set.

1. Introduction

Lorenz (1963) proposed and studied a remarkable system of three coupled first-order differential equations, representing a flow in three-dimensional space. The divergence of the flow has a constant negative value, so that any volume shrinks exponentially with time. Moreover, there exists a bounded region R into which every trajectory becomes eventually trapped. Therefore, all trajectories tend to a set of measure zero, called *attractor*. In some cases the attractor is simply a point (which is then a stable equilibrium point) or a closed curve (known as a limit cycle). But in other cases the attractor has a much more complex structure; it appears to be locally the product of a two-dimensional manifold by a Cantor set. This is known as a *strange attractor*. Inside the attractor, trajectories wander in an apparently erratic manner. Moreover, they are highly sensitive to initial conditions. These phenomena are of interest for weather prediction (Lorenz, 1963) and more generally for turbulence theory (Ruelle and Takens, 1971; Ruelle, 1975). Further numerical explorations of the Lorenz system have been made by Lanford (1975) and Pomeau (1976).

We present her a "reductionist" approach in which we try to find a model problem which is as simple as possible, yet exhibits the same essential properties as the Lorenz system. Our aim is (i) to make the numerical exploration faster and more accurate, so that solutions can be followed for a longer time, more

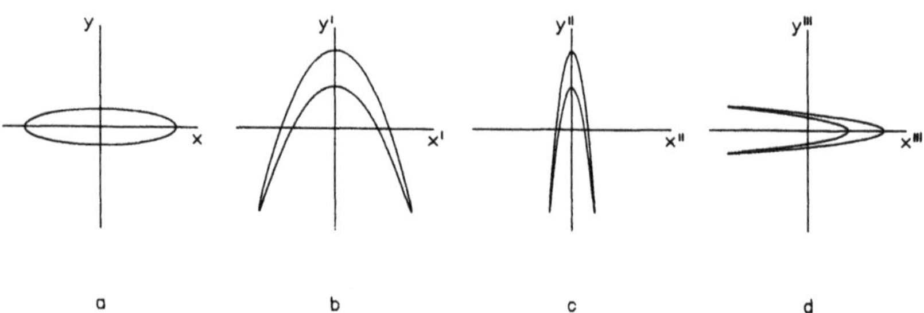

Fig. 1. The initial area a is mapped by T' into b, then by T'' into c, and finally by T''' into d

detailed explorations can be conducted, etc.; (ii) to provide a model which might lend itself more easily to mathematical analysis.

2. The Model

Our first step is classical (Birkhoff, 1917) and consists in considering not the whole trajectories in the three-dimensional space, but only their successive intersections with a two-dimensional *surface of section* S. We define a mapping T of S into itself as follows: given a point A of S, we follow the trajectory which originates from A until it intersects S again; this new point is $T(A)$. This mapping is sometimes called a *Poincaré map*. A trajectory is thus replaced by an infinite set of points in S, obtained by repeated application of the mapping T. The essential properties of the trajectory are reflected into corresponding properties of the set of points. We have thus formally reduced the problem to the study of a two-dimensional mapping.

At this point, however, the only advantage really gained is in clarity of presentation of the results; the actual computation of the mapping still requires the numerical integration of the differential equations. Now comes the second and decisive step: we forget about the differential system, and we define a mapping T by explicit equations, giving directly $T(A)$ when A is known. This of course simplifies the computation drastically. The new mapping T does not any more correspond to the Lorenz system; however, by choosing it carefully we may hope to retain the essential properties which we wish to study. Past experience in the measure-preserving case (see Hénon, 1969, and references therein) has shown indeed that the same features are found in dynamical systems defined by differential equations and in mappings defined as such.

The third step consists in specifying T. Here we have been inspired by the numerical results of Pomeau (1976) on the Lorenz system, which show clearly how a volume is stretched in one direction, and at the same time folded over itself, in the course of one revolution. This folding effect has been also described by Ruelle (1975, Fig. 5 and 6). We simulate it by the following chain of three mappings of the (x, y) plane onto itself. Consider a region elongated along the x axis (Fig. 1a). We begin the folding by

$$T' : x' = x, \quad y' = y + 1 - ax^2, \tag{1}$$

which produces Figure 1b; a is an adjustable parameter. We complete the folding by a contraction along the x axis:

$$T'': x''=bx', \quad y''=y', \tag{2}$$

which produces Figure 1c; b is another parameter, which should be less than 1 in absolute value. Finally we come back to the orientation along the x axis by

$$T'': x'''=y'', \quad y'''=x'', \tag{3}$$

which results in Figure 1d.

Our mapping will be defined as the product $T=T'''T''T'$. We write now (x_i, y_i) for (x, y) and (x_{i+1}, y_{i+1}) for (x''', y''') (as a reminder that the mapping will be iterated) and we have

$$T: x_{i+1}=y_i+1-ax_i^2, \quad y_{i+1}=bx_i. \tag{4}$$

This mapping has some interesting properties. Its Jacobian is a constant:

$$\frac{\partial(x_{i+1}, y_{i+1})}{\partial(x_i, y_i)}=-b. \tag{5}$$

The geometrical interpretation is quite simple: T' preserves areas; T''' also preserves areas but reverses the sign; and T'' contracts areas, multiplying them by the constant factor b. The property (5) is welcome because it is the natural counterpart of the constant negative divergence in the Lorenz system.

A polynomial mapping satisfying (5) is known as an *entire Cremona transformation*, and the inverse mapping is also given by polynomials (Engel, 1955, 1958). Indeed we have here

$$T^{-1}: x_i=b^{-1}y_{i+1}, \quad y_i=x_{i+1}-1+ab^{-2}y_{i+1}^2. \tag{6}$$

Thus T is a one-to-one mapping of the plane onto itself. This is also a welcome property, because it is the natural counterpart of the fact that in the Lorenz system there is a unique trajectory through any given point.

The selection of T could have been approached in a different way, by looking for the "simplest" non-trivial mapping. It is natural then to consider polynomial mappings of progressively increasing order. Linear mappings are trivial, so the polynomials must be at least of degree 2. The most general quadratic mapping is

$$\begin{aligned} x_{i+1}&=f+ax_i+by_i+cx_i^2+dx_iy_i+ey_i^2, \\ y_{i+1}&=f'+a'x_i+b'y_i+c'x_i^2+d'x_iy_i+e'y_i^2 \end{aligned} \tag{7}$$

and depends on 12 parameters. But if we impose the condition that the Jacobian is a constant, some relations must be satisfied by these parameters. We can further reduce the number of parameters by an appropriate linear change of coordinates in the plane. In this way, by a slight extension of the results of Engel (1958), it can be shown that the general form (7) is reducible to a "canonical form" depending on two parameters only. This is a generalization of our earlier result (Hénon, 1969) that a quadratic *area-preserving* mapping can be brought into a form depending on one parameter only. The canonical form can be written in several different ways; and one of them turns out to be identical with (4), which is thus reached by an entirely different road! The mapping (4), which was initially constructed in empirical fashion, is in fact the most general quadratic mapping with constant Jacobian.

One difference with the Lorenz problem is that the successive points obtained by repeated application of T do not always converge towards an attractor; sometimes they "escape" to infinity. This is because the quadratic term in (4) dominates when the distance from the origin becomes large. However, for particular values of a and b it is still possible to prove the existence of a bounded "trapping region" R, from which the points can never escape once they have entered it (see below Section 5).

T has two invariant points, given by

$$x=(2a)^{-1}[-(1-b)\pm\sqrt{(1-b)^2+4a}], \qquad y=bx. \tag{8}$$

These points are real for

$$a>a_0=(1-b)^2/4. \tag{9}$$

When this is the case, one of the points is always linearly unstable, while the other is unstable for

$$a>a_1=3(1-b)^2/4. \tag{10}$$

3. Choice of Parameters

We select now particular values of a and b for a numerical study. b should be small enough for the folding described by Figure 1 to occur really, yet not too small if one wishes to observe the fine structure of the attractor. The value $b=0.3$ was found to be adequate. A good value of a was found only after some experimenting. For $a<a_0$ or $a>a_3$, where a_0 is given by (9) and a_3 is of the order of 1.55 for $b=0.3$, the points always escape to infinity: apparently there exists no attractor in these cases. For $a_0<a<a_3$, depending on the initial values (x_0, y_0), either the points escape to infinity or they converge towards an attractor, which appears to be unique for a given value of a. We concentrate now on this attractor. For $a_0<a<a_1$, where a_1 is given by (10), the attractor is the stable invariant point. When a is increased over a_1, at first the attractor is still simple and consists of a periodic set of p points. (An equivalent attractor in the Lorenz problem would be a limit cycle intersecting the surface of section p times). The value of p increases through successive "bifurcations" as a increases, and appears to tend to infinity as a approaches a critical value a_2, of the order of 1.06 for $b=0.3$. For $a_2<a<a_3$, the attractor is no more simple, and the behaviour of the points becomes erratic. This is the case in which we are interested. We adopt the following values:

$$a=1.4, \qquad b=0.3. \tag{11}$$

4. Numerical Results

Figure 2 shows the result of plotting 10000 successive points, obtained by iteration of T, starting from the arbitrarily chosen initial point $x_0=0$, $y_0=0$; the vertical scale is enlarged to give a better picture. Figure 3 shows the result of 10000

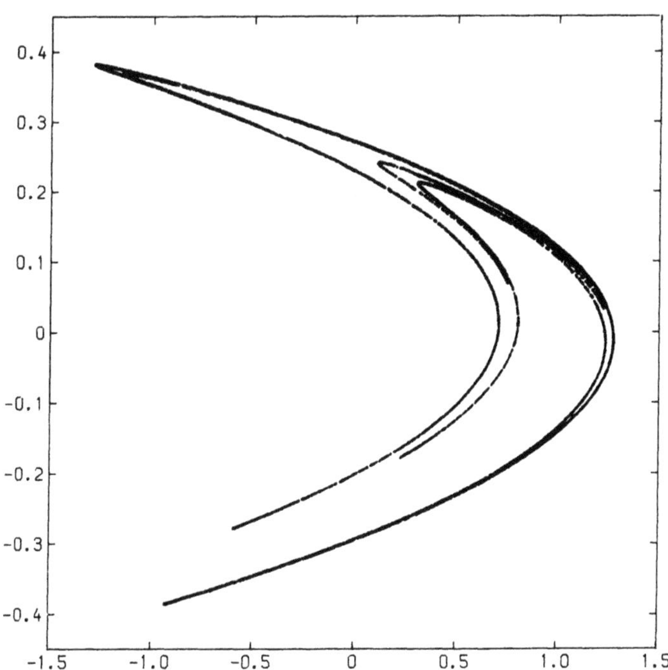

Fig. 2. 10000 successive points obtained by iteration of the mapping T starting from $x_0=0$, $y_0=0$

iterations of T again, starting from a different point: $x_0=0.63135448$, $y_0=0.18940634$ (this choice will be explained below). The two figures are seen to be almost identical. This suggests strongly that what we see in both figures is essentially the attractor itself: the successive points quickly approach the attractor and soon become undistinguishable from it at the scale of the figure. This is confirmed if one looks at the first few points on Figure 2. The initial point at $x_0=0$, $y_0=0$ and the first iterate at $x_1=1$, $y_1=0$ are clearly visible; the second iterate is still visible at $x_2=-0.4$, $y_2=0.3$; the third iterate can barely be distinguished at $x_3=1.076$, $y_3=-0.12$; and the fourth iterate at $x_4=-0.7408864$, $y_4=0.3228$ is already lost inside the attractor at the resolution of Figure 2. The following points then wander over the attractor in an apparently erratic manner.

One of the two unstable invariant points has the coordinates, given by (8):

$$x=0.63135448\ldots, \quad y=0.18940634\ldots . \tag{12}$$

This point appears to belong to the attractor. The two eigenvalues λ_1, λ_2 and the slopes p_1, p_2 of the corresponding eigenvectors are

$$\begin{aligned} &\lambda_1=0.15594632\ldots, \quad p_1=1.92373886\ldots, \\ &\lambda_2=-1.92373886\ldots, \ p_2=-0.15594632\ldots . \end{aligned} \tag{13}$$

The instability is due to λ_2. The corresponding slope p_2 appears to be tangent to the "curves" in Figure 2.

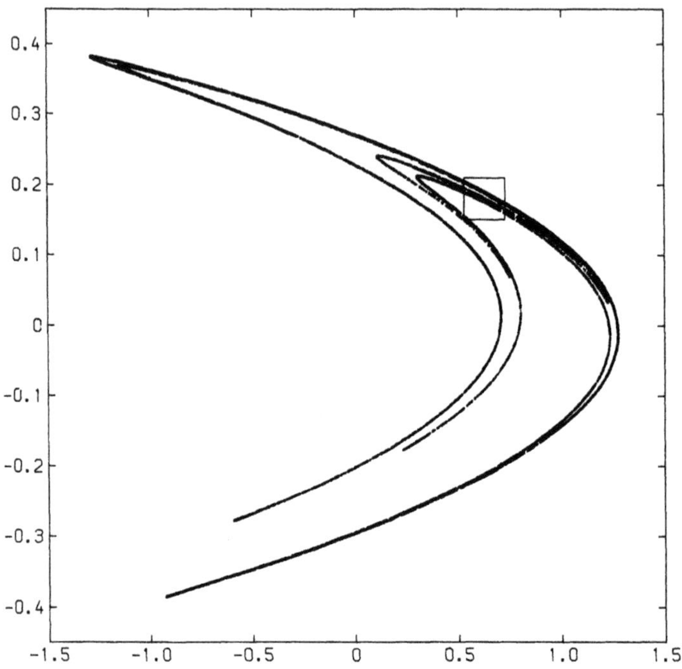

Fig. 3. Same as Figure 2, but starting from $x_0 = 0.63135448$, $y_0 = 0.18940634$

These properties allow us to eliminate the "transient regime" in which the points approach the attractor, and which is not of much interest: we simply start from the close vicinity of the unstable point (12), by rounding off its coordinates to 8 digits. This is done in Figure 3 and in the following figures. The points quickly move away along the line of slope p_2 since $|\lambda_2|$ is appreciably larger than 1.

The attractor appears to consist of a number of more or less parallel "curves"; the points tend to distribute themselves densely over these curves. The few gaps that can still be seen on Figures 2 and 3 have probably no particular significance. Their locations are not the same on the two figures. They are simply due to statistical fluctuations in the quasi-random distribution of points, and they would disappear if more moints were plotted. Thus, the *longitudinal structure* of the attractor (along the curves) appears to be simple, each curve being essentially a one-dimensional manifold.

The *transversal structure* (across the curves) appears to be entirely different, and much more complex. Already on Figures 2 and 3 a number of curves can be seen, and the visible thickness of some of them suggests that they have in fact an underlying structure. Figure 4 is a magnified view of the small square of Figure 3: some of the previous "curves" are indeed resolved now into two or more components. The number n of iterations has been increased to 10^5, in order to have a sufficient number of points in the small region examined. The small square in Figure 4 is again magnified to produce Figure 5, with n increased to 10^6: again the

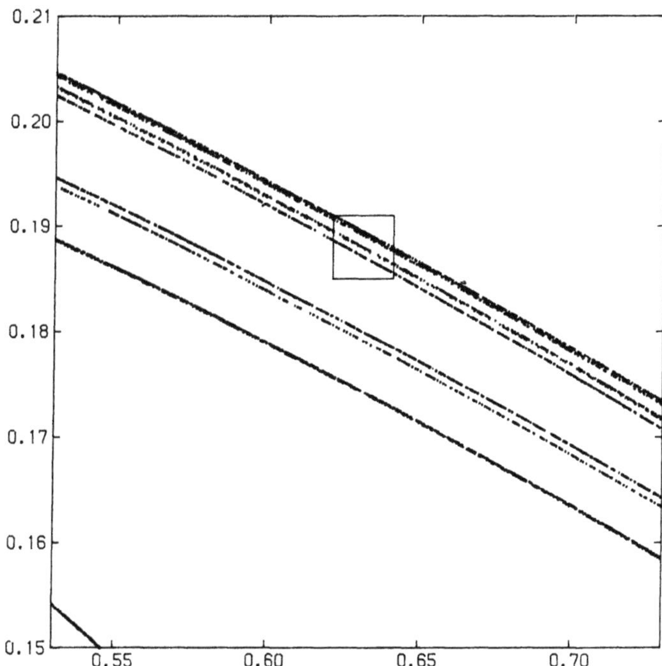

Fig. 4. Enlargement of the squared region of Figure 3. The number of computed points is increased to $n = 10^5$

number of visible "curves" increases. One more enlargement results in Fig. 6, with $n = 5 \times 10^6$: the points become sparse but new curves can still easily be traced.

These figures strongly suggest that the process of multiplication of "curves" will continue indefinitely, and that each apparent "curve" is in fact made of an infinity of quasi-parallel curves. Moreover, Figures 4 to 6 indicate the existence of a hierarchical sequence of "levels", the structure being practically identical at each level save for a scale factor. This is exactly the structure of a Cantor set.

The frames of Figures 4 to 6 have been chosen so as to contain the invariant point (12). This point appears to lie on the upper boundary of the attractor. Surprisingly, its presence is completely invisible on the figures; this contrasts with the area-preserving case, were stable and unstable invariant points play a very conspicuous role (see for instance Hénon, 1969). On the other hand, the presence of the invariant point explains, locally at least, the hierarchy of similar structures: at each application of the mapping, the scale of the transversal structure is multiplied by λ_1 given by (13). At the same time, the points spread out along the curves, as dictated by the value of λ_2.

5. A Trapping Region

The fact that even after 5×10^6 iterations the points have not diverged to infinity suggests that there is a region of the plane from which the points cannot escape.

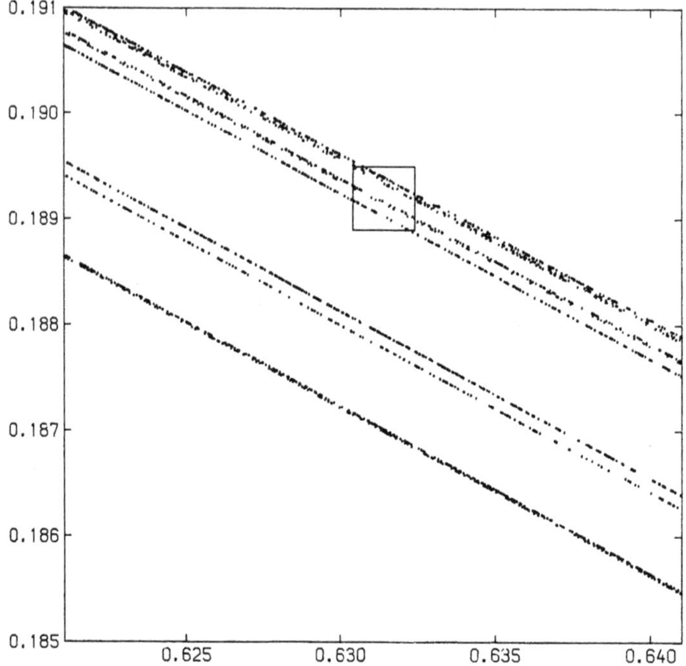

Fig. 5. Enlargement of the squared region of Figure 4; $n=10^6$

This can be actually proved by finding a region R which is mapped inside itself. An example of such a region is the quadrilateral $ABCD$ defined by

$$x_A=-1.33, \quad y_A=0.42, \qquad x_B=1.32, \qquad y_B=0.133,$$

$$x_C=1.245, \qquad y_C=-0.14, \quad x_D=-1.06, \quad y_D=-0.5. \tag{14}$$

The image of $ABCD$ is a region bounded by four arcs of parabola, and it can be shown by elementary algebra that this image lies inside $ABCD$. Plotting the quadrilateral on Figure 2 or 3, one can verify that it encloses the observed attractor.

6. Conclusions

The simple mapping (4) appears to have the same basic properties as the Lorenz system. Its numerical exploration is much simpler: in fact most of the exploratory work for the present paper was carried out with a programmable pocket computer (HP-65). For the more extensive computations of Figures 2 to 6, we used a IBM 7040 computer, with 16-digit accuracy. The solutions can be followed over a much longer time than in the case of a system of differential equations. The accuracy is also increased since there are no integration errors.

Lorenz (1963) inferred the Cantor-set structure of the attractor from reasoning, but could not observe it directly because the contracting ratio after one "circuit"

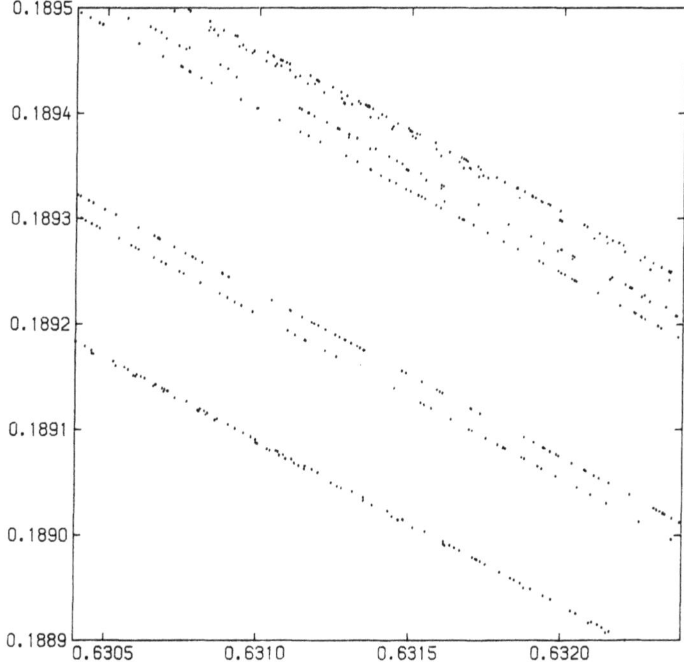

Fig. 6. Enlargement of the squared region of Figure 5; $n = 5 \times 10^6$

was too small: 7×10^{-5}. A similar experience was reported by Pomeau (1976). In the present mapping, the contracting ratio after one iteration is 0.3, and one can easily observe a number of successive levels in the hierarchy. This is also facilitated by the larger number of points.

Finally, for mathematical studies the mapping (4) might also be easier to handle than a system of differential equations.

References

Birkhoff, G. D.: Trans. Amer. Math. Soc. **18**, 199 (1917)
Engel, W.: Math. Annalen **130**, 11 (1955)
Engel, W.: Math. Annalen **136**, 319 (1958)
Hénon, M.: Quart. Appl. Math. **27**, 291 (1969)
Lanford, O.: Work cited by Ruelle, 1975
Lorenz, E. N.: J. atmos. Sci. **20**, 130 (1963)
Pomeau, Y.: to appear (1976)
Ruelle, D., Takens, F.: Comm. math. Phys. **20**, 167; **23**, 343 (1971)
Ruelle, D.: Report at the Conference on "Quantum Dynamics Models and Mathematics" in Bielefeld. September 1975

Communicated by K. Hepp

Received March 25, 1976

Editor's note. Period doublings and the Feigenbaum numbers for Hénon mapping were found and measured by Derrida, Gervois and Pomeau (1979).

Physical Review A **24** 1640–2 (1981)

Universal effects of dissipation in two-dimensional mappings

Albert B. Zisook
The James Franck Institute and The University of Chicago, 5640 South Ellis Avenue, Chicago, Illinois 60637
(Received 4 June 1981)

One can define an effective dissipation for highly iterated two-dimensional maps or recursion relations. The convergence rate $\delta_{\rm eff}$ of the sequence of period-doubling bifurcations is shown to depend in a universal way on this effective dissipation in the limit of long 2^n cycles. These conclusions are based on both a renormalization argument and numerical calculations. The paper concludes with a brief discussion of the implications for physical systems.

I. INTRODUCTION

At present there is much interest in using one- and two-dimensional recursion relations to represent the dynamics of both dissipative and nondissipative systems which undergo a transition to chaotic behavior. One typically describes dissipative systems, such as turbulent fluids[1,2] and Josephson junctions,[3,4] with a one-dimensional recursion relation. In order to describe conservative systems, such as particle storage rings[5] or magnetically confined plasmas,[6] one chooses an area-preserving map. This area-preserving property represents the conservation of volume under flows in phase space. Both the one- and two-dimensional recursion relations are known to undergo period-doubling bifurcations.[7–12]

The convergence rate in the one-dimensional case, δ_1, which tells how quickly successive bifurcations occur as some control parameter is changed, is equal to 4.6692 .... In the two-dimensional case the convergence rate $\delta_2 = 8.7210$ .... This paper examines the behavior of the effective convergence rate $\delta_{\rm eff}$ for two-dimensional maps with a small amount of dissipation. A universal crossover is found whereby $\delta_{\rm eff}$ goes from δ_2 to δ_1 in a predictable way.

Using the Helleman renormalization approach[13] we prove that an effective dissipation, $B_{\rm eff}$, can be defined and that $\delta_{\rm eff}$ depends only on $B_{\rm eff}$ (not on n or the details of the mapping) for large n. Since this renormalization approach is only approximate we verify our results with a numerical calculation for a specific two-dimensional recursion relation. The qualitative behavior is exactly that predicted by the renormalization approach.

II. RENORMALIZATION APPROACH

The recursion relation used in this paper is Helleman's "standard form"[13]:

$$x_{n+1} = 2Cx_n + 2Cx_n^2 - y_n \ , \qquad y_{n+1} = Bx_n \ . \tag{1}$$

The Jacobian determinant of this mapping is constant and equal to B (we take $0 \leq B \leq 1$). When $B = 1$ the mapping is area preserving and models conservative systems. For various other values of the parameters B and C, the mapping (1) is equivalent to many other mappings studied, such as Henon's mapping[14,15] or the standard logistic equation ($B = 0$).

For $|2C| < 1 + B$ the fixed point $x_n = y_n = 0$ is stable. When $2C = -1 - B$ this fixed point bifurcates into a stable 2 cycle. If one iterates the recursion relation (1) twice, one gets a new recursion relation for x_{n+2} and y_{n+2} in terms of x_n and y_n. The essence of the renormalization scheme is to rescale the variables x and y and throw away higher-order terms to put this new recursion relation into the standard form (1). The result is Helleman's renormalization formula:

$$C' = f(B,C) = -2C^2 + 2(1+B)C + 2B^2 + 3B + 2 \ , \qquad B' = B^2 \ . \tag{2}$$

When $|2C'| < 1 + B'$ the 2 cycle will be stable. When $2C' = -1 - B'$ the 2 cycle will bifurcate into a stable 4 cycle. By iterating Eq. (2) n times one obtains a recursion relation which describes the 2^n cycle.

Using the renormalization equations one can calculate how δ_{eff} depends on B and n, where

$$\delta_{\text{eff}}(B,n) = \frac{\Lambda_{n-1}(B) - \Lambda_n(B)}{\Lambda_n(B) - \Lambda_{n+1}(B)} \quad . \tag{3}$$

Here $\Lambda_n(B)$ is the value of the parameter C for which the 2^n cycle is maximally stable. By maximally stable we mean that the real parts of the two eigenvalues of the linear stability matrix vanish. When $B=0$ this corresponds to what is usually called superstability. The fixed point $x_n = y_n = 0$ is maximally stable for $C=0$; so $\Lambda_0(B)=0$. The value of C for which the 2 cycle is maximally stable is easy to obtain. It must be the solution of the equation $C'=0=f(B,C)$. By solving $f(B,C)=0$ one finds

$$\Lambda_1(B) = C = \tfrac{1}{2}[1+B-(5B^2+8B+5)^{1/2}] \quad . \tag{4}$$

This result, obtained using the renormalization Eq. (2), happens to be exact, but for $n>1$ one obtains only approximate values for $\Lambda_n(B)$ with this method. It is convenient to define the inverse of $f(B,C)$ by $g(B,C)$ so that

$$g(B,f(B,C)) = C \quad . \tag{5}$$

The appropriate choice of signs in the quadratic formula gives

$$g(B,C) = \tfrac{1}{2}[1+B-(5B^2+8B+5-2C)^{1/2}] \quad . \tag{6}$$

Now one may calculate $\Lambda_2(B)$. The condition for maximal stability of the $2^2=4$ cycle is that $C'' = f(B',C') = 0$. So one must have $C'=g(B',0) = g(B^2,0)$. But $C'=f(B,C)$ so $C=g(B,C') = g(B,g(B^2,0))$. The result is

$$\Lambda_2(B) = g(B,g(B^2,0)) \quad . \tag{7}$$

One may continue this process to obtain an explicit formula for $\Lambda_n(B)$,

$$\Lambda_n(B) = g(B,g(B^2,g(B^{2^2}, \ldots g(B^{2^{n-1}},0)))) \quad . \tag{8}$$

From the formula (8) one can conclude immediately that

$$\Lambda_n(B) = g(B, \Lambda_{n-1}(B^2)) \quad . \tag{9}$$

Using this equation one can derive an identity for $\delta_{\text{eff}}(B,n)$. Using the variables defined by $\Lambda_n(B) = \Lambda_\infty(B) + D_n(B)$ Eq. (9) may be rewritten as

$$D_n(B) = g(B, \Lambda_\infty(B^2) + D_{n-1}(B^2)) - \Lambda_\infty(B) \quad . \tag{10}$$

When n is large, regardless of the value of B, $D_n(B)$ will be very small. From Eq. (9) one may conclude that $\Lambda_\infty(B) = g(B, \Lambda_\infty(B^2))$. So, using the smallness of $D_n(B)$, one may approximate the right-hand side of Eq. (10)

$$D_n(B) = \frac{\partial g(B, \Lambda_\infty(B^2))}{\partial C} D_{n-1}(B^2) \quad . \tag{11}$$

It follows from formula (3) that

$$\delta_{\text{eff}}(B,n) = \frac{D_{n-1}(B) - D_n(B)}{D_n(B) - D_{n+1}(B)} \quad . \tag{12}$$

Combining Eqs. (11) and (12) one obtains an identity for $\delta_{\text{eff}}(B,n)$

$$\delta_{\text{eff}}(B,n) = \delta_{\text{eff}}(B^2,n-1) \quad . \tag{13}$$

From this identity comes the final theoretical result, that δ_{eff} depends on B and n only through the combination $B_{\text{eff}} \equiv B^{2^n}$:

$$\delta_{\text{eff}}(B,n) \equiv \delta_{\text{eff}}(B^{2^n}) \equiv \delta_{\text{eff}}(B_{\text{eff}}) \quad . \tag{14}$$

When $B=B_{\text{eff}}=0$ the map is one dimensional; so it must be that

$$\delta_{\text{eff}}(0) = \delta_1 = 4.6692 \ldots \quad . \tag{15}$$

When $B=B_{\text{eff}}=1$ the map is area preserving, so

$$\delta_{\text{eff}}(1) = \delta_2 = 8.7210 \ldots \quad . \tag{16}$$

For $0 \leq B_{\text{eff}} \leq 1$, $\delta_{\text{eff}}(B_{\text{eff}})$ should be a smoothly varying, universal function.

III. NUMERICAL CALCULATION OF $\delta_{\text{eff}}(B_{\text{eff}})$

Because the renormalization approach is only approximate it is possible that the conclusions of the last section could be qualitatively wrong. These conclusions are certainly quantitatively wrong (by about 5–10%) as the renormalization approach yields the values $\delta_1 \approx 5.12$ and $\delta_2 \approx 9.06$. In order to test the validity of the results, we have performed numerical

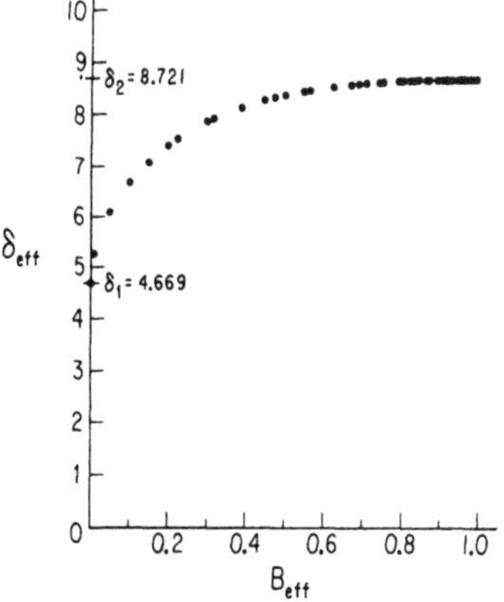

FIG. 1. Data from eight different values of B and $n=4,5,6,7,8$ all lie on the same smooth curve.

calculations of the exact maximally stable C values, $\Lambda_n(B)$, for eight different values of B and $n = 4, 5, 6, 7, 8$. This calculation was carried out using the recursion relation (1). A plot of $\delta_{\text{eff}}(B_{\text{eff}})$ based on these computations is shown in Fig. 1. As one can see, δ_{eff} appears to be a well-defined (single-valued) function of $B_{\text{eff}} = B^{2^n}$. $\delta_{\text{eff}}(0) = \delta_1$ and $\delta_{\text{eff}}(1) = \delta_2$, as they must. On the basis of this evidence we conclude that we have indeed found a new universal behavior in two-dimensional recursion relations with dissipation.

IV. DISCUSSION

The motivation for this work is the search for robust behavior in two-dimensional maps. For conservative systems, with $B = B_{\text{eff}} = 1$, there is no truly stable cyclic behavior in the sense that the cycles persist under small perturbations. Rather, the two eigenvalues of the linear stability matrix lie on the unit circle in the complex plane. In real physical systems there is always a small amount of dissipation ($B = 1 - \epsilon$). This can be thought of as due to a weak coupling to a heat bath. In this case the 2^n cycles will be stable (eigenvalues inside the unit circle) and hence potentially observable in real systems. Since ϵ will be small in such cases, B_{eff} will not go to 0 until after many, many bifurcations. In this case the convergence rate δ_{eff} will take on the new universal values between δ_2 and δ_1.

ACKNOWLEDGMENTS

The author wishes to acknowledge his National Science Foundation and Robert R. McCormick Fellowships. He also thanks L. P. Kadanoff, S. J. Shenker, and M. Widom for helpful discussions. This research was supported in part by the NSF.

---

[1]A. Libchaber and J. Maurer, J. Phys. (Paris) 41, C3-51 (1980).

[2]J. P. Gollub, S. V. Benson, and J. Steinman (unpublished).

[3]B. A. Huberman, J. P. Crutchfield, and N. H. Packard, Appl. Phys. Lett. 37, 750 (1980).

[4]Y. Taur and P. L. Richards, J. Appl. Phys. 48, 1321 (1977).

[5]*Nonlinear Dynamics and the Beam-Beam Interaction*, edited by M. Month and J. C. Herrera, AIP Conf. Proc. No. 57 (AIP, New York, 1979).

[6]B. V. Chirikov, Phys. Rep. 52, 263 (1979).

[7]R. M. May, Nature 261, 459 (1976).

[8]M. J. Feigenbaum, J. Stat. Phys. 19, 25 (1978).

[9]S. Grossmann and S. Thomae, Z. Naturforsch. A 32, 1353 (1977).

[10]G. Benettin, C. Cercignani, L. Galgani, and A. Giorgilli, Lett. Nuovo Cimento 28, 1 (1980).

[11]T. C. Bountis, Physica D (in press).

[12]J. M. Greene, R. S. MacKay, F. Vivaldi, and M. J. Feigenbaum (unpublished).

[13]R. H. G. Helleman, in *Fundamental Problems in Statistical Mechanics*, edited by E. G. D. Cohen (Elsevier, New York, 1980), Vol. 5, p. 165.

[14]M. Henon, Commun. Math. Phys. 50, 69 (1976).

[15]Period doubling bifurcations in the Henon mapping were observed by B. Derrida, A. Gervois, and Y. Pomeau, J. Phys. A 12, 269 (1979).

Editor's note. The relationship between the universal numbers for conservative and dissipative two-dimensional systems was also discussed by Helleman (the last reprint in this volume) and Hu (1981).

Journal of Statistical Physics **25** 1–14 (1981)

Period Doubling Bifurcations for Families of Maps on $\mathbb{R}^n$

P. Collet,[1] J.-P. Eckmann,[2] and H. Koch[2]

Received February 20, 1980

Infinite sequences of period doubling bifurcations in one-parameter families (1-pf) of maps enjoy very strong universality properties: This is known numerically in a multitude of cases and has been shown rigorously for certain 1-pf of maps on the interval. These bifurcations occur in 1-pf of analytic maps at values of the parameter tending to a limit with the asymptotically geometric ratio 1/4.6692 . . . . In this paper we indicate the main steps of a proof that the same is true for 1-pf of analytic maps from $\mathbb{C}^n$ to $\mathbb{C}^n$, whose restriction to $\mathbb{R}^n$ is real.

KEY WORDS: Subharmonic bifurcations; hydrodynamics; analytic maps.

1. MOTIVATION

Physical models of hydrodynamics or of other dissipative systems tend to be very complicated. In addition, the laws describing such systems are only known approximately. One is thus faced with the problem of isolating and if possible answering new types of questions which are more or less independent of a detailed knowledge of the dynamics of any given physical system. Such questions then have answers which are *universal*. A well-known field where universal answers have been obtained is the renormalization group analysis of critical phenomena in statistical mechanics.

A new universal property has been discovered by Feigenbaum[(4)] for families of maps of the interval to itself, which depend on a parameter. It states that if such families present subharmonic bifurcations, then one can

Work supported by the Fonds National Suisse, and by the National Science Foundation under Grant PHY-79-16812.

[1] Physics Department, Harvard University, Cambridge, Massachusetts 02138.
[2] Département de Physique Théorique, Université de Genève, 1211 Genève 4, Switzerland.

expect an infinite cascade of such bifurcations, as the parameter is varied, and they accumulate in a fashion independent of the detailed structure of the one-parameter family of maps. Soon afterwards, several authors[3,6,7] noted in numerical experiments that this phenomenon is not restricted to maps of the interval, but that it also occurs in discrete or continuous time approximations to the kind of equations one encounters in hydrodynamics. The purpose of this paper is to show why this is true in the discrete time case.

The continuous time case is indirectly also covered by our results because there are many situations in which the continuous evolution can be described in terms of a discrete map by use of a Poincaré map. Thus our theorems below are directly relevant to hydrodynamical equations in finitely many variables, and they show that for many of these one should observe the universally scaled accumulation of period doubling bifurcations. For the Bénard flow in liquid helium, Libchaber and Maurer[9] have measured a power spectrum for such cascades which (at least qualitatively) looks similar to the power spectrum which can be predicted from the general theory developed in this paper. Feigenbaum[5] gives a heuristic derivation for this spectrum, which is based on our results. Thus we have a first indication of a possible link between the abstract theory presented here and experiments.

2. INTRODUCTION

Universal properties of one-parameter families of maps on an interval were discovered numerically by Feigenbaum[4] and investigated from a rigorous point of view in Ref. 2. In that paper the authors considered one-parameter families of transformations of the interval $[-1, 1]$ into itself of the form

$$x \mapsto g_\mu(|x|^{1+\epsilon})$$

Here, μ is the parameter, $g_\mu(0) = 1$, $\epsilon > 0$ is small, and $g_\mu(\cdot)$ is analytic in some neighborhood of $[-1, 1]$ and satisfies some other technical conditions on the μ dependence. Assuming moreover that g_μ is near to a specific function f, the following result was found.

Theorem I.[2] There is a manifold $\mathcal{W}_s$ (of codimension one in a space of analytic functions) such that the following is true. If the family $\mu \mapsto g_\mu$ is transverse to $\mathcal{W}_s$ and if ϵ is sufficiently small, then

(1) The family $\mu \mapsto g_\mu$ has infinitely many bifurcation points. These points correspond to successive bifurcations from a stable period 2^n to a stable period 2^{n+1}.

(2) If $\{\mu_i\}_{i\in\mathbb{N}}$ is the sequence of values of μ for which a period described in (1) appears, then

$$\lim_{n\to\infty} \frac{\log|\mu_\infty - \mu_n|}{n} = -\log\delta$$

where δ is a universal number, which depends only on ϵ, but not on $\mu \mapsto g_\mu$.

The most interesting case $\epsilon = 1$ is not covered by this theorem but has been numerically studied by Feigenbaum.[4] In particular, he computed the value of the universal number: $\delta = 4.669\ldots$.

Infinite sequences of period doubling bifurcations have also been observed in higher-dimensional systems. One of them is the Hénon map in $\mathbb{R}^2$:

$$\begin{pmatrix} x \\ y \end{pmatrix} \mapsto \begin{pmatrix} 1 - \mu x^2 + y \\ bx \end{pmatrix}$$

For $b = 0.3$, the first 11 values of μ for which a doubling bifurcation occurs were computed with good accuracy in Ref. 3. They appear to satisfy the universal behavior described in Theorem I with the same number δ. Other examples of this behavior in higher-dimensional flows were described in Refs. 6 and 7; cf. also Ref. 1.

The aim of the present paper is to outline a proof of the universal behavior for maps in finite-dimensional spaces. We shall assume that the results proven in Ref. 2 for small ϵ extend to $\epsilon = 1$. Lanford reported recently on some decisive progress in this direction,[8] and our hypotheses are inspired by his results. See also Ref. 10.

Our argument is organized as follows. We first state some hypotheses on the one-dimensional case for $\epsilon = 1$. We then explain how the renormalization group program can be realized for certain maps on $\mathbb{C}^n$, $n > 1$. This includes the search for a fixed point of a nonlinear transformation and the study of its linearization at this fixed point. A second part of the argument should include a more detailed description of the stable and unstable manifolds $\mathcal{W}_s$ and $\mathcal{W}_u$ and of their intersection with certain submanifolds of codimension one. We have not worked out the tedious functional analytic details associated with this part of the argument. They should be similar to those of Ref. 2, and the reader is referred to that paper for an explanation of the geometrical ideas of the method. *Throughout this paper maps $\mathbb{C}^n \to \mathbb{C}^n$ are implicitly considered to be real on $\mathbb{R}^n$.*

The main result of the present paper is the following theorem.

Theorem II. There is a map Φ from $\mathbb{C}^n$ to $\mathbb{C}^n$, and a submanifold $\mathcal{W}_s$ in the space of analytic functions on $\mathbb{C}^n$ to $\mathbb{C}^n$ (of codimension one and passing through Φ) such that the following is true:

(1) Every once continuously differentiable one-parameter family $\mu \mapsto G_\mu$ of analytic maps from $\mathbf{C}^n$ to $\mathbf{C}^n$ which crosses transversally through $\mathcal{W}_s$ near Φ has infinitely many bifurcations from a stable period 2^m to a stable period 2^{m+1}.

(2) If $\{\mu_m\}_{m\in\mathbf{N}}$ is the sequence of values of μ for which a period 2^m described in (1) appears, then

$$\lim_{m\to\infty} \frac{\log|\mu_\infty - \mu_m|}{m} = -\log\delta$$

where $\delta = 4.669\ldots$ does not depend on the family G_μ.

3. THE ONE-DIMENSIONAL CASE. ASSUMPTIONS

Before stating our hypotheses we recall some definitions associated with the one-dimensional problem. Let $\mathfrak{M}$ be the set of functions g which map $[-1,1]$ into itself and for which $-1 < g(1) < 0$. For $g \in \mathfrak{M}$ we define $\mathfrak{T}$ by

$$\mathfrak{T}g(x) = g(1)^{-1} g \circ g(g(1)x)$$

Our first set of assumptions is the following.

(M1) The equation $\mathfrak{T}g = g$ has a solution $\phi \in \mathfrak{M}$ which is analytic in some neighborhood D_1 of $[-1,1]$.

(M2) ϕ is symmetric, $\phi(x) = f(x^2)$, and $f'(t) \neq 0$ for $t \in [0,1]$.

(M3) Define $\lambda = \phi(1)$.[3] For some positive γ,

$$2\lambda^2 \sup\{|f'(z^2)| : z \in D_1\} < 1 - \gamma < 1.$$

(M4) f has exactly one zero in $[0,1]$.

From (M1) we can investigate the derivative of $\mathfrak{T}$ at the fixed point ϕ. We obtain

$$(D\mathfrak{T}\phi h)(z) = \lambda^{-1} h(\phi(\lambda z)) + \lambda^{-1}\phi'(\phi(\lambda z))h(\lambda z)$$

Lemma 1.[(4)] If $\sigma(y) = y^n$ for some integer $n \geqslant 0$ then

$$\psi_\sigma(x) = -\sigma(\phi(x)) + \phi'(x)\sigma(x)$$

is an eigenvector of $D\mathfrak{T}_\phi$ with eigenvalue λ^{n-1}.

Proof. We consider the following family $S(t)$ of maps of the complex plane,

$$z \mapsto S(t)z = z + t\sigma(z)$$

[3] For the solution found by Feigenbaum and Lanford, $\phi(x) = 1 - 1.401\ldots x^2$, $\lambda = \phi(1) = -0.3995\ldots$.

which is well defined and invertible on a neighborhood of $[-1,1]$ for small t.

If M_λ denotes the operator of multiplication by λ in $\mathbb{C}$, we have

$$\begin{aligned}\mathfrak{T}\big(S(t)^{-1}\circ\phi\circ S(t)\big) &= M_\lambda^{-1}\circ S(t)^{-1}\circ M_\lambda\circ\big(M_\lambda^{-1}\circ\phi\circ\phi\circ M_\lambda\big)\\ &\qquad\circ M_\lambda^{-1}\circ S(t)\circ M_\lambda\\ &= \big(M_\lambda^{-1}\circ S(t)^{-1}\circ M_\lambda\big)\circ\phi\circ\big(M_\lambda^{-1}\circ S(t)\circ M_\lambda\big)\end{aligned}$$

Differentiating with respect to t and setting $t=0$ we obtain

$$D\mathfrak{T}_\phi(\psi_\sigma) = \psi_{M_\lambda^{-1}\circ\sigma\circ M_\lambda}$$

by using that $\psi_\sigma = \partial_t[S(t)^{-1}\circ\phi\circ S(t)]|_{t=0}$. From this the result follows. ■

Our last set of hypotheses is the following.

(M5) The operator $D\mathfrak{T}_\phi$ has a simple eigenvalue $\delta > 1$ which is different from λ^{-1}, λ^{-2}.[4] The corresponding eigenvector ρ is even. We define $r(x^2) = \rho(x)$.

(M6) The eigenvalues δ, λ^{-1}, 1 are the only eigenvalues of modulus $\geqslant 1$. Their corresponding spectral projections are one dimensional.

Lanford has essentially completed the proof of (M1), . . . , (M4).[(8)] His method can be extended to prove (M5) and (M6). See also Ref. 10.

Remark. In view of Lemma 1 it could seem reasonable to assume that one has found a total set of eigenvectors for $D\mathfrak{T}_\phi$. This is not the case and in fact the family $\{\psi_\sigma : \sigma \text{ analytic}\}$ has infinite codimension. This can be seen as follows: For all σ, one can easily check that ψ_σ satisfies the relations

$$\begin{aligned}&\phi''(x_0)\Big[\lambda^{-j}(\phi+\epsilon\psi_\sigma)^{2^j}(\lambda^j x_0) - x_0\Big]\\ &\qquad -(\phi'(x_0)-1)\Big[\frac{d}{dx}(\phi+\epsilon\psi_\sigma)^{2^j}(x)|_{x=\lambda^j x_0} - \phi'(x_0)\Big] = 0\\ &\qquad\qquad \operatorname{mod} O(\epsilon^2),\quad j=0,1,2,\ldots\end{aligned}$$

where x_0 is defined by $\phi(x_0) = x_0$. This is true because $\epsilon\psi_\sigma$ is generated through a (nonlinear) coordinate transformation, and this leaves the derivatives at the periodic points invariant. It is also easy to see that the relations for different j are independent.

[4] In fact, $\delta = 4.6692\ldots$.

4. MAPS ON $\mathbb{C}^n$

We introduce some notations for the n-dimensional problem. We will use a fixed decomposition of $\mathbb{C}^n$ into a direct sum $\mathbb{C}^n = \mathbb{C} \oplus \mathbb{C}^{n-1}$. If z is a vector of $\mathbb{C}^n$, its components will be written $(z_0, \mathbf{z})$. $\|\cdot\|$ will be the norm of $\mathbb{C}^n$ given by

$$\|z\| = \|\mathbf{z}\|_{\mathbb{C}^{n-1}} + |z_0|$$

where $\|\mathbf{z}\|_{\mathbb{C}^{n-1}} = (\mathbf{z} \cdot \bar{\mathbf{z}})^{1/2}$ is the usual norm in $\mathbb{C}^{n-1}$. For an open subset D of $\mathbb{C}^n$ let $\mathcal{H}(D)$ denote the space of analytic and bounded maps from D to $\mathbb{C}^n$. Equipped with the norm

$$\|h\| = \sup\{\|h(z)\| : z \in D\}$$

this space is a Banach space. We shall mostly consider $\mathcal{H}_\Delta = \mathcal{H}(D(\Delta))$, where $D(\Delta)$ is the convex set

$$D(\Delta) = \{z \in \mathbb{C}^n : \|z - (y_0, \mathbf{0})\| < \Delta \quad \text{for some } y_0 \in [-1,1]\}$$

We now fix a nonzero vector $\boldsymbol{\alpha}$ in $\mathbb{C}^{n-1}$ whose norm is bounded by two. Φ will always denote the map

$$z \mapsto \Phi(z) = (f(\zeta(z)), \mathbf{0}) \tag{1}$$

where $\zeta(z) = z_0^2 - \boldsymbol{\alpha} \cdot \mathbf{z}$ and where f is the function described in (M2). Note that if Δ is sufficiently small, so that $\{\zeta^{1/2}(z) : z \in D(\Delta)\}$ is contained in D_1, then Φ belongs to $\mathcal{H}_\Delta$. The universal behavior asserted in Theorem II will be proven in the sequel for one-parameter families of maps which are near to Φ.

It might seem that this is an undue restriction on the one-parameter family. Note, however, that our problem is invariant under C^1 coordinate transformations. This means that given a one-parameter family $\tilde{G}_\mu$ of maps one might find transformations τ_μ such that $G_\mu = \tau_\mu^{-1} \circ \tilde{G}_\mu \circ \tau_\mu$ satisfies the conditions of Theorem II. In particular, τ_μ might be constant and change the direction or length of $\boldsymbol{\alpha}$. The conclusions of Theorem II are then valid for G_μ as well as for $\tilde{G}_\mu$.

We next define the renormalization transformation. Let Λ be the diagonal $n \times n$ matrix given by

$$\Lambda z = (\lambda z_0, \lambda^2 \mathbf{z})$$

where $\lambda = \phi(1) = -0.3995\ldots$.

Lemma 2. If Δ is sufficiently small and if G belongs to $\mathcal{H}_\Delta$, and $\|G - \Phi\| < \gamma\Delta$,[5] then $\Lambda^{-1} \circ G \circ G \circ \Lambda$ also belongs to $\mathcal{H}_\Delta$.

[5] Cf. (M3) for the definition of γ.

Proof. From the hypotheses on f and on $u = G - \Phi$ it is easy to verify, using (M3), that for $z \in D(\Delta)$,

$$G(\Lambda z) = \big(f(\lambda^2\zeta(z)) + u_0(\Lambda z), \mathbf{u}(\Lambda z)\big)$$

is again in $D(\Delta)$, and therefore $\Lambda^{-1} \circ G \circ G(\Lambda z)$ is well defined. The analyticity follows from that of G and of $G(\Lambda\,\cdot\,)$. ■

Definition. The transformation $\mathfrak{N}$ given by

$$\mathfrak{N} : G \mapsto \Lambda^{-1} \circ G \circ G \circ \Lambda$$

will be called the renormalization transformation.

Owing to Lemma 2, this transformation maps the ball in $\mathcal{H}_\Delta$ centered at Φ and of radius $\gamma\Delta$ into $\mathcal{H}_\Delta$.

Lemma 3. Φ is a fixed point of $\mathfrak{N}$.

Proof. This is an easy consequence of (M1), (M2), and of the definition of $\mathfrak{N}$ and Φ. ■

According to the general philosophy of the renormalization group analysis, we shall now investigate the spectrum of the derivative of $\mathfrak{N}$ at Φ. One can easily derive the following expression for the derivative $D\mathfrak{N}_G$, where $G \in \mathcal{H}_\Delta$ and $\|G - \Phi\| < \Delta\gamma/2$:

$$(D\mathfrak{N}_G h)(z) = \Lambda^{-1}\big[h(G(\Lambda z)) + DG_{G(\Lambda z)}h(\Lambda z)\big]$$

Proposition 4. For sufficiently small Δ, the following assertions hold.

(1) $\mathfrak{N}$ is a C^2 transformation on $\mathcal{H}_\Delta$ defined in a ball centered at Φ and with radius $\gamma\Delta/2$. $\|D^2\mathfrak{N}_{\Phi+u}(h,k)\| \leqslant O(1)\|h\|\,\|k\|$ provided $\|u\| < \gamma\Delta/2$.

(2) $D\mathfrak{N}_\Phi$ is a compact operator from $\mathcal{H}_\Delta$ into itself.

Proof. We remark that for $\|u\| < \gamma\Delta/2$ and sufficiently small Δ, $(\Phi + u)$ maps the closure of $\Lambda D(\Delta)$ into $D(\Delta)$ by (M1) and (M3). The assertion (1) is now verified by a direct computation. The compactness of $D\mathfrak{N}_\Phi$ is a consequence of Montel's theorem. ■

We start now an analysis of the spectrum of $D\mathfrak{N}_\Phi$. Our result is

Theorem 5

(1) If $\sigma = (\sigma_0, \boldsymbol{\sigma})$ is an analytic map from $\mathbb{C}^n$ to $\mathbb{C}^n$, and if $\Lambda^{-1} \circ \sigma \circ \Lambda = \lambda^m\sigma$ for some integer $m \geqslant -2$, then the map Ψ_σ defined by

$$\Psi_\sigma(z) = -\big(\sigma_0(f(\zeta(z)),\mathbf{0}), \boldsymbol{\sigma}(f(\zeta(z)),\mathbf{0})\big) + \big(2z_0 f'(\zeta(z))\sigma_0(z) - f'(\zeta(z))\boldsymbol{\alpha}\cdot\boldsymbol{\sigma}(z), \mathbf{0}\big)$$

is an eigenvector of $D\mathfrak{N}_\Phi$ with eigenvalue λ^m.

(2) δ, λ^{-2}, λ^{-1}, and 1 are the only eigenvalues of $D\mathfrak{N}_\Phi$ of modulus greater or equal to 1. The corresponding spectral subspaces are spanned (i) for δ by $P = (r \circ \zeta, \mathbf{0})$, where r is the function defined in (M5), and (ii) for λ^{-2}, λ^{-1}, 1 by the vectors Ψ_σ with $\Lambda^{-1} \circ \sigma \circ \Lambda$ equal to $\lambda^{-2}\sigma$, $\lambda^{-1}\sigma$, and σ, respectively. They have the dimensions $1, n-1, n$, and $n^2 - n + 1$.

Proof

(1) This can be shown by a method analogous to that of Lemma 1.

(2) It is sufficient to prove the assertions for the restrictions of $D\mathfrak{N}_\Phi$ to a closed linear subspace which contains $D\mathfrak{N}_\Phi\mathcal{H}_\Delta$. The direct sum $\hat{\mathcal{H}}(\Delta) = \mathcal{H}_0(\Delta) \oplus \mathcal{H}(\Delta)$ has this property, where

$$\mathcal{H}_0(\Delta) = \{h : h = (h_0, \mathbf{0}) \in \mathcal{H}_\Delta\}$$

$$\mathbf{H}(\Delta) = \{h : h(z) = (0, \mathbf{h}_1(\zeta(z))) \text{ with } \mathbf{h}_1 \text{ analytic and bounded on } D_0(\Delta)\}$$

where

$$D_0(\Delta) = \{\zeta(z) : z \in D(\Delta)\}$$

The restriction of $D\mathfrak{N}_\Phi$ to $\hat{\mathcal{H}}(\Delta)$ will be denoted by A. Let $\mathcal{H}' = \mathcal{H}(\Phi(D(\Delta)) \cup D(\Delta))$. The following two lemmas will be proven below.

Lemma 6. Let $(A-\mu)^\nu\Psi_\tau = 0$ for some $\tau \in \mathcal{H}'$ and some $\nu \in \mathbb{N}$. If $\mu \in \{\lambda^k : k = -2, -1, 0, \ldots\}$ then $\Psi_\tau = \Psi_\sigma$ for some σ which is analytic in $\mathbb{C}^n$ and for which $\Lambda^{-1} \circ \sigma \circ \Lambda = \mu\sigma$. Otherwise $\Psi_\tau = 0$.

Lemma 7. Let $u \in \hat{\mathcal{H}}(\Delta)$ and let $(A-\mu)^\nu u = 0$ for some μ with $|\mu| \geqslant 1$. Then $u = cP + \Psi_\tau$ for some $c \in \mathbb{C}$ and some $\tau \in \mathcal{H}'$.

We complete the proof of Theorem 5, part (2). Let $u \in \hat{\mathcal{H}}(\Delta)$ and in a spectral subspace of $D\mathfrak{N}_\Phi$ with eigenvalue μ, $|\mu| \geqslant 1$. Since $D\mathfrak{N}_\Phi$ is compact, and $\mu \neq 0$, this means that for some $\nu \in \mathbb{N}$, $(A-\mu)^\nu u = 0$. By Lemma 7, we conclude that for some c, $u = cP + \Psi_\tau$ for some $\tau \in \mathcal{H}'$. Suppose $\mu = \delta$. Then $(A-\mu)^\nu cP = 0$ and hence $(A-\mu)^\nu\Psi_\tau = 0$, so that by Lemma 6, $\Psi_\tau = 0$. If $\mu \neq \delta$ then $-cP = (\delta-\mu)^{-\nu}(A-\mu)^\nu\Psi_\tau = \Psi_\kappa$ for some $\kappa \in \mathcal{H}'$. From $(A-\delta)^\nu cP = 0$ we have thus $(A-\delta)^\nu\Psi_\tau = 0$ and applying Lemma 6 again we must either have $\Psi_\tau = 0$ or $\mu = \lambda^k$, $k \in \{-2, -1, 0\}$ and $u = \Psi_\tau = \Psi_\sigma$ for a σ with $\Lambda^{-1} \circ \sigma \circ \Lambda = \lambda^k$. This completes the proof of Theorem 5. ■

Proof of Lemma 6. For $\epsilon \neq 0$ let $N(\epsilon) = \{z \in \mathbb{C}^n : \|z\| < |\epsilon|\}$. We define the bounded operators[6]

$$\hat{A} : \mathcal{H}(N(\epsilon)) \to \mathcal{H}(N(\lambda^{-1}\epsilon)) \qquad \text{by } \hat{A}\sigma = \Lambda^{-1} \circ \sigma \circ \Lambda$$

[6] To shorten the notation we omit the ϵ dependence of $\hat{A}$. This can be rendered completely rigorous by writing the proper injections, etc.

and for $|\lambda\eta| < 1$,

$$\hat{U}(\eta) : \mathcal{H}(N(\epsilon)) \to \mathcal{H}(N(\lambda^2\epsilon))$$

$$\text{by } (\hat{U}(\eta)\sigma)(z) = (\sigma_0(\eta^2 z_0, \eta\mathbf{z}), \eta\boldsymbol{\sigma}(\eta^2 z_0, \eta\mathbf{z}))$$

One can easily check that

$$\hat{U}(\eta)\hat{A} = \hat{A}\hat{U}(\eta) = \lambda^{-2}\hat{U}(\lambda\eta)$$

By defining $U(\eta)\Psi_\kappa = \Psi_{\hat{U}(\eta)\kappa}$ for $\kappa \in \mathcal{H}(N(\lambda^{-3}))$, we obtain the corresponding relation

$$U(\eta)A\Psi_\kappa = AU(\eta)\Psi_\kappa = \lambda^{-2}U(\lambda\eta)\Psi_\kappa$$

where we have used that $A\Psi_\kappa = \Psi_{A\kappa}$.

Let us now assume that $(A-\mu)^\nu\Psi_\tau = 0$ for some integer $\nu \geqslant 1$ and some $\tau \in \mathcal{H}'$. By choosing $M \in \mathbb{N}$ sufficiently large, we can achieve that $\sigma = \hat{A}^M\hat{D}^M\tau$, where $\hat{D} = -\sum_{j=1}^{\nu}\binom{\nu}{j}\hat{A}^{j-1}(-\mu)^{-j}$, belongs to $\mathcal{H}(N(\lambda^{-3}))$. Moreover, $\Psi_\tau = \Psi_\sigma$, and

$$\eta \mapsto U(\eta)\Psi_\sigma$$

defines an analytic family of maps in $\mathcal{H}_\Delta$ for $|\lambda\eta| < 1$. Thus in a disc of radius larger than one the following series converges:

$$U(\eta)\Psi_\sigma = \sum_{k=0}^{\infty} \eta^k \Psi_{\sigma_{0,k}}$$

where $\sigma_{\eta,k} = (1/k!)\partial_\eta^k\hat{U}(\eta)\sigma$. This is an expansion into eigenvectors of $D\mathfrak{N}_\Phi$ since

$$\hat{A}\sigma_{0,k} = \frac{1}{k!}\partial_\eta^k\hat{A}\hat{U}(\eta)\sigma|_{\eta=0} = \lambda^{-2}\frac{1}{k!}\partial_\eta^k\hat{U}(\lambda\eta)\sigma|_{\eta=0}$$

$$= \frac{\lambda^{k-2}}{k!}\partial_\eta^k\hat{U}(\eta)\sigma = \lambda^{k-2}\sigma_{0,k}$$

For all η with $|\eta\lambda| < 1$ one has

$$0 = U(\eta)(A-\mu)^\nu\Psi_\sigma = (A-\mu)^\nu U(\eta)\Psi_\sigma$$

This implies that for $k = 0, 1, 2, \ldots$

$$(A-\mu)^\nu\Psi_{\sigma_{0,k}} = 0$$

But $(A-\mu)^\nu\Psi_{\sigma_{0,k}} = (\lambda^{k-2}-\mu)^\nu\Psi_{\sigma_{0,k}}$, which leads to the conclusion that either $\Psi_{\sigma_{0,k}} = 0$ or $\mu = \lambda^{k-2}$. The assertion follows by using that

$$\Psi_\tau = U(1)\Psi_\sigma = \sum_{k=0}^{\infty}\Psi_{\sigma_{0,k}} \qquad \blacksquare$$

Proof of Lemma 7. It will be useful to construct a partition of the operator 1 on $\hat{\mathcal{H}}(\Delta)$ which commutes with A. For a map $h = (h_0, \mathbf{h}_1 \circ \zeta)$ in

$\hat{\mathcal{H}}(\Delta)$ we define

$$\begin{aligned} \boldsymbol{\nu} &= \mathbf{h}_1(f^{-1}(0)) \\ \mathbf{h}_2(\zeta) &= (\mathbf{h}_1(\zeta) - \boldsymbol{\nu})/f(\zeta) \end{aligned} \tag{2}$$

and

$$\begin{aligned} (\Pi_0 h)(z) &= \left(h_0(z) - f'(\zeta(z))\boldsymbol{\alpha}\cdot\boldsymbol{\nu} - \tfrac{1}{2}\boldsymbol{\alpha}\cdot\mathbf{h}_2(\zeta(z)), \mathbf{0}\right) \\ (\Pi_1 h)(z) &= \left(f'(\zeta(z))\boldsymbol{\alpha}\cdot\boldsymbol{\nu}, \boldsymbol{\nu}\right) \qquad \{=\Psi_{(0,-\boldsymbol{\nu})}\} \\ (\Pi_2 h)(z) &= \left(\tfrac{1}{2}\boldsymbol{\alpha}\cdot\mathbf{h}_2(\zeta(z)), f(\zeta(z))\mathbf{h}_2(\zeta(z))\right) \end{aligned}$$

This defines linear operators Π_0, Π_1, Π_2 on $\hat{\mathcal{H}}(\Delta)$, which by our hypotheses on f are bounded by $O(\Delta^{-1})$. From the explicit form of A and the fact that $f(\lambda^2 f^{-1}(0))^2 = f^{-1}(0)$, one easily obtains the relations

$$\begin{aligned} \Pi_0 + \Pi_1 + \Pi_2 &= 1 \\ \Pi_i\Pi_j &= \delta_{ij}\Pi_j, \qquad i,j = 0,1,2 \\ \Pi_i A &= A\Pi_i, \qquad i = 0,1,2 \end{aligned} \tag{3}$$

In view of (3) it is sufficient to show that the hypotheses of the lemma imply $\Pi_i u \sim cP$ for some $c \in \mathbb{C}$, where $h \sim k$ means $h - k \in \{\Psi_\sigma : \sigma \in \mathcal{H}'\}$.

The case $i = 1$ follows since for all $h \in \hat{\mathcal{H}}(\Delta)$, $\Pi_1 h$ is an eigenvector of A of the form Ψ_σ with eigenvalue λ^{-2}, and thus $\Pi_1 u \sim 0$.

Next we will show that $\Pi_2 u \sim 0$. For maps h in $\mathcal{H}_2(\Delta) = \Pi_2\hat{\mathcal{H}}(\Delta)$ we define

$$\mathbf{B}(h) = \mathbf{h}_2 \circ f^{-1}$$

where $\mathbf{h}_2$ is defined in Eq. (2).

Since, by (M2), $m^{-1} < |f'(\zeta)| < m$ for some $m \in \mathbb{N}$ and for all $\zeta \in D_0(\Delta)$, the components of $\mathbf{B}(h)$ are well defined as analytic and bounded functions on $f(D_0(\Delta))$, which contains an open neighborhood of zero. It is easily seen that the action of A on $\mathcal{H}_2(\Delta)$ induces

$$\mathbf{B}(Ah)(z) = \lambda^{-1}\mathbf{B}(h)(\lambda z)$$

i.e., it enlarges the domain of analyticity. Since Π_i commutes with A, we find from $(A - \mu)^\nu u = 0$ that

$$\Pi_i u = -\sum_{k=1}^{\nu}\binom{\nu}{k}(-\mu)^{-k}A^k\Pi_i u, \qquad i = 0,1,2 \tag{4}$$

This implies that $\mathbf{B}(\Pi_2 u)$ is analytic on $\mathbb{C}$. Furthermore $\Pi_2 u = \Psi_\sigma$ with

$$\sigma(z) = -(2\boldsymbol{\alpha}\cdot\mathbf{B}(\Pi_2 u)(z_0), z_0\mathbf{B}(\Pi_2 u)(z_0))$$

and thus $\Pi_2 u \sim 0$.

Finally we will show that $\Pi_0 u \sim cP$ for some $c \in \mathbb{C}$. Define the bounded operators $S : \mathcal{H}_0(\Delta) \to \mathcal{H}(D_0(\Delta))$ and $S^* : \mathcal{H}(D_0(\Delta)) \to \mathcal{H}_0(\Delta)$ by

$$(Sh)(z_0) = h_0(z_0, \mathbf{0})$$
$$(S^*k)(z_0, \mathbf{z}) = (k(z_0), \mathbf{0})$$

and set $R = S^*S$. These operators have the following properties:

$$SS^* = Id_{\mathcal{H}_0(\Delta)}$$
$$RA(1 - R)|_{\mathcal{H}_0(\Delta)} = 0$$
$$SAS^* = D\mathcal{T}_\phi \qquad \text{(the linearized operator for one dimension)}$$

Now let $C = RAR$. Since $CS^* = S^*D\mathcal{T}_\phi$ and $C(1 - R) = 0$, the spectral subspace $\mathcal{H}$ corresponding to eigenvalues of modulus greater than or equal to 1 of C is spanned by the eigenvectors $S^*\rho$, $S^*\psi_{S\kappa}$, and $S^*\psi_{S\tau}$, where $\kappa(z) = (1, \mathbf{0})$ and $\tau(z) = (z_0, \mathbf{0})$. They differ from P, Ψ_κ, and Ψ_τ (these are eigenvectors of A with corresponding eigenvalues) only by elements in $\mathcal{H}_0(\Delta)$ which map vectors $(z_0, \mathbf{0})$ to zero. Since for Δ sufficiently small the functions f, f', and r are analytic and bounded on $\zeta(\Phi(D(\Delta)) \cup D(\Delta))$, these differences can be written in the form

$$\Psi_\sigma(z) = -(f'(\zeta(z))\boldsymbol{\alpha} \cdot \boldsymbol{\sigma}(z), \mathbf{0}) \tag{5}$$

with $\sigma = (0, \boldsymbol{\sigma})$, $\boldsymbol{\sigma}(z_0, \mathbf{0}) = 0$, and $\Psi_\sigma \sim 0$. In other words,

$$S^*\rho \sim P$$
$$S^*\psi_{S\kappa} \sim \Psi_\kappa \sim 0$$
$$S^*\psi_{S\tau} \sim \Psi_\tau \sim 0$$

The assertion $\Pi_0 u \sim cP$ will now be proven by showing that $\Pi_0 u \sim k$ for some $k \in \mathcal{H}$.

By using that $RA(1 - R) = 0$ we obtain

$$\begin{aligned} 0 &= (A - \mu)^\nu \Pi_0 u \\ &= (C - \mu)^\nu \Pi_0 u + (-\mu)^\nu v \\ &= (C - \mu)^\nu (\Pi_0 u + v) \end{aligned}$$

where

$$v = (1 - R)A \sum_{k=1}^{\nu} (-\mu)^{-k}(A - \mu)^{k-1}\Pi_0 u$$

This implies that $\Pi_0 u + v \in \mathcal{H}$. From (4) it follows that for every $k \in \mathbb{N}$ there is a $w_k \in \mathcal{H}_0(\Delta)$ such that $v = (1 - R)A^k w_k$. The operator A substitutes $z \overset{X_1}{\mapsto} f(\lambda^2\zeta(z))$ or $z \overset{X_2}{\mapsto} \Lambda z$ in the argument of the function on which it acts. By (M3), for sufficiently large k, $\prod_{i=1}^k X_{j_i}$ and $\prod_{i=1}^k X_{j_i}\Phi$ are contractions on $D(\Delta)$. It follows that v is analytic and bounded on

$\Phi(D(\Delta)) \cup D(\Delta)$, and since $v(z_0, \mathbf{0}) = 0$, it is of the form (5), and therefore we have $v \sim 0$. This concludes the proof of Lemma 7. ■

5. THE DEFINITION OF A RENORMALIZATION TRANSFORMATION

In the usual renormalization group analysis, the derivative of the renormalization transformation at the fixed point has only one eigenvalue greater than or equal to 1. We shall now show that the eigenvalues λ^{-2}, λ^{-1}, and 1 of $D\mathfrak{N}_\Phi$ can be removed by the appropriate choice of a new renormalization transformation T.

From the definition of Ψ_σ,

$$\Psi_\sigma = \partial_t (I + t\sigma)^{-1} \circ \Phi \circ (I + t\sigma)\big|_{t=0}$$

it can be seen that the eigenvalues λ^n, $n = -2, -1, 0$, correspond to degrees of freedom associated to some change of coordinates (in particular the eigenvalue 1 corresponds to transformations which are compatible with our choice of the z_0 axis, i.e., which commute with Λ). Since we intend to describe only coordinate-independent properties, the eigenvalues λ^n can be eliminated and ultimately play no role in the universal behavior. We shall now work towards the construction of a new renormalization transformation whose derivative at the fixed point has spectrum inside the unit circle except for δ.

Let E denote the spectral projection of $D\mathfrak{N}_\Phi$ associated to the eigenvalues λ^{-2}, λ^{-1}, 1. The first step is the definition of a map $h \mapsto \sigma[h]$ which satisfies

$$\Psi_{\sigma[h]} = Eh$$

Proposition 8. Define $D' = \Phi(D(\Delta)) \cup D(\Delta)$. For any h in $\mathcal{H}_\Delta$, the equation

$$\Psi_{\sigma[h]} = Eh$$

has a unique solution $\sigma[h]$ in $\mathcal{H}(D')$. The map $h \mapsto \sigma[h]$ is linear and bounded.

Proof. Let $\mathcal{K}$ be the following finite-dimensional subspace of $\mathcal{H}(D')$:

$$\mathcal{K} = \Big\{ \sigma : \sigma(z) = \nu + z_0 \nu' + \big(0, z_0^2 \mu + \mu'(\mathbf{z})\big) \text{ with } \nu, \nu' \in \mathbb{C}^n, \mu \in \mathbb{C}^{n-1}$$
$$\text{and } \mu' \text{ a linear operator from } \mathbb{C}^{n-1} \text{ into itself} \Big\}$$

It is easy to verify that $\sigma \mapsto Q\sigma = \Psi_\sigma$ is a bounded linear operator from $\mathcal{K}$ to $E\mathcal{H}_\Delta$. By Theorem 5, part 2, we have $\dim Q\mathcal{K} = \dim E\mathcal{H}_\Delta$. Therefore Q has an inverse Q^{-1} and we can define $\sigma[h] = Q^{-1}Eh$. ■

We are now able to define our final renormalization transformation T. The explicit expression is

$$T : h \mapsto$$
$$(I + \sigma[D\mathfrak{N}_\Phi h]) \circ \Lambda^{-1} \circ (\Phi + h) \circ (\Phi + h) \circ \Lambda \circ (I + \sigma[D\mathfrak{N}_\Phi h])^{-1} - \Phi$$

Since for sufficiently small $\|h\|$ the transformation $z \mapsto z + \sigma[D\mathfrak{N}_\Phi h](z)$ maps $D(\Delta)$ analytically and one-to-one onto some neighborhood of $D(\Delta/2)$, this transformation T is well defined in some neighborhood of zero in $\mathcal{H}_\Delta$.

The properties of T are summarized in the following theorem.

Theorem 9. If Δ is sufficiently small, then

(1) T is a C^2 transformation from a neighborhood of zero in $\mathcal{H}_\Delta$ to $\mathcal{H}_\Delta$.
(2) $DT_0 = (1 - E)D\mathfrak{N}_\Phi$, where $DT_0 = DT_G$ at $G = 0$.
(3) DT_0 is compact and its spectrum consists of the simple eigenvalue δ, and a remainder strictly inside the unit disk. The eigenvector corresponding to δ is

$$P(z) = (r(\zeta(z)), \mathbf{0})$$

Proof. The proof is an immediate consequence of our previous results. ■

The theorem above is the main ingredient for the analysis of universal behavior of maps, which now follows very closely the one given in Ref. 2. We have not, however, worked out all the details of the proofs of the steps of this construction, but we believe they should not be very different from the one-dimensional case. From the existence of the fixed point 0 for the map T and from its spectral properties one deduces the existence of stable and unstable manifolds $\mathcal{W}_s$ and $\mathcal{W}_u$ for T in a neighborhood of 0. The main point of our preceding analysis is that $\mathcal{W}_s$ will have codimension one and $\mathcal{W}_u$ will have dimension one; furthermore T is expanding by a factor δ on $\mathcal{W}_u$. Now let $\Sigma_0 = \{G - \Phi : G$ has one fixed point in $D(\Delta)$ and DG has one eigenvalue -1 at this fixed point$\}$. Let $\Sigma_m = T^{-m}(\Sigma_0)$. These manifolds are, for sufficiently large m, transversal to $\mathcal{W}_u$, and the intersection of a curve $\mu \to G_\mu$ (which is near $\mathcal{W}_u$) with Σ_m corresponds to the point of bifurcation from a stable period 2^m to a stable period 2^{m+1} for G_μ. Owing to the differentiability of T near 0, the distance of Σ_m from 0 goes as constant δ^{-m} and this is the main ingredient of the proof of Theorem II.

REFERENCES

1. P. Collet and J.-P. Eckmann, Universal Properties of Continuous Maps of the Interval to Itself, in *Lecture Notes in Physics*, Vol. 74 (Springer, New York, 1979).
2. P. Collet, J.-P. Eckmann, and O. E. Lanford III, Universal Properties of Maps on an Interval, *Commun. Math. Phys.* **76**:211–254 (1980).
3. B. Derrida, A. Gervois, and Y. Pomeau, *J. Phys.* **A12**:269 (1979).
4. M. Feigenbaum, *J. Stat. Phys.* **19**:25 (1978); *J. Stat. Phys.* **21**:6 (1979).
5. M. Feigenbaum, *Phys. Lett.* **74A**:375 (1979), and The Transition to Aperiodic Behaviour in Turbulent Systems, *Commun. Math. Phys.* **77**:65–86 (1980).
6. V. Franceschini, A Feigenbaum Sequence of Bifurcations in the Lorenz Model, *J. Stat. Phys.* **22**:397–406 (1980).
7. V. Franceschini and C. Tebaldi, Sequences of Infinite Bifurcations and Turbulence in a 5-Modes Truncation of the Navier-Stokes Equations, *J. Stat. Phys.* **21**:707–726 (1979).
8. O. E. Lanford, III, Remarks on the Accumulation of Period-Doubling Bifurcations, in *Lecture Notes in Physics*, Vol. 74 (Springer, New York, 1979).
9. A. Libchaber and J. Maurer, Une expérience de Rayleigh–Bénard de géométrie réduite, *J. de Physique* **41**, Colloque C3:51–56 (1980).
10. M. Campanino, H. Epstein, and D. Ruelle, On Feigenbaum's Functional Equation, preprint IHES; M. Campanino and H. Epstein, On the Existence of Feigenbaum's Fixed Point, *Commun. Math. Phys.*, to appear.

Editor's note. A very readable account of this work is given in Collet and Eckmann's (1980a) monograph.

Journal of the Atmospheric Sciences **20** 130–41 (1963)

Deterministic Nonperiodic Flow[1]

EDWARD N. LORENZ

Massachusetts Institute of Technology

(Manuscript received 18 November 1962, in revised form 7 January 1963)

ABSTRACT

Finite systems of deterministic ordinary nonlinear differential equations may be designed to represent forced dissipative hydrodynamic flow. Solutions of these equations can be identified with trajectories in phase space. For those systems with bounded solutions, it is found that nonperiodic solutions are ordinarily unstable with respect to small modifications, so that slightly differing initial states can evolve into considerably different states. Systems with bounded solutions are shown to possess bounded numerical solutions.

A simple system representing cellular convection is solved numerically. All of the solutions are found to be unstable, and almost all of them are nonperiodic.

The feasibility of very-long-range weather prediction is examined in the light of these results.

1. Introduction

Certain hydrodynamical systems exhibit steady-state flow patterns, while others oscillate in a regular periodic fashion. Still others vary in an irregular, seemingly haphazard manner, and, even when observed for long periods of time, do not appear to repeat their previous history.

These modes of behavior may all be observed in the familiar rotating-basin experiments, described by Fultz, *et al.* (1959) and Hide (1958). In these experiments, a cylindrical vessel containing water is rotated about its axis, and is heated near its rim and cooled near its center in a steady symmetrical fashion. Under certain conditions the resulting flow is as symmetric and steady as the heating which gives rise to it. Under different conditions a system of regularly spaced waves develops, and progresses at a uniform speed without changing its shape. Under still different conditions an irregular flow pattern forms, and moves and changes its shape in an irregular nonperiodic manner.

Lack of periodicity is very common in natural systems, and is one of the distinguishing features of turbulent flow. Because instantaneous turbulent flow patterns are so irregular, attention is often confined to the statistics of turbulence, which, in contrast to the details of turbulence, often behave in a regular well-organized manner. The short-range weather forecaster, however, is forced willy-nilly to predict the details of the large-scale turbulent eddies—the cyclones and anticyclones—which continually arrange themselves into new patterns. Thus there are occasions when more than the statistics of irregular flow are of very real concern.

In this study we shall work with systems of deterministic equations which are idealizations of hydrodynamical systems. We shall be interested principally in nonperiodic solutions, i.e., solutions which never repeat their past history exactly, and where all approximate repetitions are of finite duration. Thus we shall be involved with the ultimate behavior of the solutions, as opposed to the transient behavior associated with arbitrary initial conditions.

A closed hydrodynamical system of finite mass may ostensibly be treated mathematically as a finite collection of molecules—usually a very large finite collection—in which case the governing laws are expressible as a finite set of ordinary differential equations. These equations are generally highly intractable, and the set of molecules is usually approximated by a continuous distribution of mass. The governing laws are then expressed as a set of partial differential equations, containing such quantities as velocity, density, and pressure as dependent variables.

It is sometimes possible to obtain particular solutions of these equations analytically, especially when the solutions are periodic or invariant with time, and, indeed, much work has been devoted to obtaining such solutions by one scheme or another. Ordinarily, however, nonperiodic solutions cannot readily be determined except by numerical procedures. Such procedures involve replacing the continuous variables by a new finite set of functions of time, which may perhaps be the values of the continuous variables at a chosen grid of points, or the coefficients in the expansions of these variables in series of orthogonal functions. The governing laws then become a finite set of ordinary differential

[1] The research reported in this work has been sponsored by the Geophysics Research Directorate of the Air Force Cambridge Research Center, under Contract No. AF 19(604)-4969.

equations again, although a far simpler set than the one which governs individual molecular motions.

In any real hydrodynamical system, viscous dissipation is always occurring, unless the system is moving as a solid, and thermal dissipation is always occurring, unless the system is at constant temperature. For certain purposes many systems may be treated as conservative systems, in which the total energy, or some other quantity, does not vary with time. In seeking the ultimate behavior of a system, the use of conservative equations is unsatisfactory, since the ultimate value of any conservative quantity would then have to equal the arbitrarily chosen initial value. This difficulty may be obviated by including the dissipative processes, thereby making the equations nonconservative, and also including external mechanical or thermal forcing, thus preventing the system from ultimately reaching a state of rest. If the system is to be deterministic, the forcing functions, if not constant with time, must themselves vary according to some deterministic rule.

In this work, then, we shall deal specifically with finite systems of deterministic ordinary differential equations, designed to represent forced dissipative hydrodynamical systems. We shall study the properties of nonperiodic solutions of these equations.

It is not obvious that such solutions can exist at all. Indeed, in dissipative systems governed by finite sets of *linear* equations, a constant forcing leads ultimately to a constant response, while a periodic forcing leads to a periodic response. Hence, nonperiodic flow has sometimes been regarded as the result of nonperiodic or random forcing.

The reasoning leading to these concludions is not applicable when the governing equations are nonlinear. If the equations contain terms representing advection—the transport of some property of a fluid by the motion of the fluid itself—a constant forcing can lead to a variable response. In the rotating-basin experiments already mentioned, both periodic and nonperiodic flow result from thermal forcing which, within the limits of experimental control, is constant. Exact periodic solutions of simplified systems of equations, representing dissipative flow with constant thermal forcing, have been obtained analytically by the writer (1962a). The writer (1962b) has also found nonperiodic solutions of similar systems of equations by numerical means.

2. Phase space

Consider a system whose state may be described by M variables $X_1, \cdots, X_M$. Let the system be governed by the set of equations

$$dX_i/dt=F_i(X_1,\cdots X_M), \quad i=1,\cdots,M, \tag{1}$$

where time t is the single independent variable, and the functions F_i possess continuous first partial derivatives. Such a system may be studied by means of *phase space*—an M-dimensional Euclidean space Γ whose coordinates are $X_1, \cdots, X_M$. Each *point* in phase space represents a possible instantaneous state of the system. A state which is varying in accordance with (1) is represented by a moving *particle* in phase space, traveling along a *trajectory* in phase space. For completeness, the position of a stationary particle, representing a steady state, is included as a trajectory.

Phase space has been a useful concept in treating finite systems, and has been used by such mathematicians as Gibbs (1902) in his development of statistical mechanics, Poincaré (1881) in his treatment of the solutions of differential equations, and Birkhoff (1927) in his treatise on dynamical systems.

From the theory of differential equations (e,g., Ford 1933, ch. 6), it follows, since the partial derivatives $\partial F_i/\partial X_j$ are continuous, that if t_0 is any time, and if $X_{10}, \cdots X_{M0}$ is any point in Γ, equations (1) possess a unique solution

$$X_i=f_i(X_{10},\cdots,X_{M0},t), \quad i=1,\cdots,M, \tag{2}$$

valid throughout some time interval containing t_0, and satisfying the condition

$$f_i(X_{10},\cdots,X_{M0},t_0)=X_{i0}, \quad i=1,\cdots,M. \tag{3}$$

The functions f_i are continuous in $X_{10}, \cdots, X_{M0}$ and t. Hence there is a unique trajectory through each point of Γ. Two or more trajectories may, however, approach the same point or the same curve asymptotically as $t\rightarrow\infty$ or as $t\rightarrow-\infty$. Moreover, since the functions f_i are continuous, the passage of time defines a continuous deformation of any region of Γ into another region.

In the familiar case of a conservative system, where some positive definite quantity Q, which may represent some form of energy, is invariant with time, each trajectory is confined to one or another of the surfaces of constant Q. These surfaces may take the form of closed concentric shells.

If, on the other hand, there is dissipation and forcing, and if, whenever Q equals or exceeds some fixed value Q_1, the dissipation acts to diminish Q more rapidly then the forcing can increase Q, then $(-dQ/dt)$ has a positive lower bound where $Q\geqq Q_1$, and each trajectory must ultimately become trapped in the region where $Q<Q_1$. Trajectories representing forced dissipative flow may therefore differ considerably from those representing conservative flow.

Forced dissipative systems of this sort are typified by the system

$$dX_i/dt=\sum_{j,k} a_{ijk}X_jX_k-\sum_j b_{ij}X_j+c_i, \tag{4}$$

where $\sum a_{ijk}X_iX_jX_k$ vanishes identically, $\sum b_{ij}X_iX_j$ is positive definite, and $c_1, \cdots, c_M$ are constants. If

$$Q=\tfrac{1}{2}\sum_i X_i^2, \tag{5}$$

and if $e_1, \cdots, e_M$ are the roots of the equations

$$\sum_j (b_{ij}+b_{ji})e_j = c_i, \tag{6}$$

it follows from (4) that

$$dQ/dt = \sum_{i,j} b_{ij}e_ie_j - \sum_{i,j} b_{ij}(X_i-e_i)(X_j-e_j). \tag{7}$$

The right side of (7) vanishes only on the surface of an ellipsoid E, and is positive only in the interior of E. The surfaces of constant Q are concentric spheres. If S denotes a particular one of these spheres whose interior R contains the ellipsoid E, it is evident that each trajectory eventually becomes trapped within R.

3. The instability of nonperiodic flow

In this section we shall establish one of the most important properties of deterministic nonperiodic flow, namely, its instability with respect to modifications of small amplitude. We shall find it convenient to do this by identifying the solutions of the governing equations with trajectories in phase space. We shall use such symbols as $P(t)$ (variable argument) to denote trajectories, and such symbols as P or $P(t_0)$ (no argument or constant argument) to denote points, the latter symbol denoting the specific point through which $P(t)$ passes at time t_0.

We shall deal with a phase space Γ in which a unique trajectory passes through each point, and where the passage of time defines a continuous deformation of any region of Γ into another region, so that if the points $P_1(t_0)$, $P_2(t_0)$, $\cdots$ approach $P_0(t_0)$ as a limit, the points $P_1(t_0+\tau)$, $P_2(t_0+\tau)$, $\cdots$ must approach $P_0(t_0+\tau)$ as a limit. We shall furthermore require that the trajectories be uniformly bounded as $t\rightarrow\infty$; that is, there must be a bounded region R, such that every trajectory ultimately remains with R. Our procedure is influenced by the work of Birkhoff (1927) on dynamical systems, but differs in that Birkhoff was concerned mainly with conservative systems. A rather detailed treatment of dynamical systems has been given by Nemytskii and Stepanov (1960), and rigorous proofs of some of the theorems which we shall present are to be found in that source.

We shall first classify the trajectories in three different manners, namely, according to the absence or presence of transient properties, according to the stability or instability of the trajectories with respect to small modifications, and according to the presence or absence of periodic behavior.

Since any trajectory $P(t)$ is bounded, it must possess at least one *limit point* P_0, a point which it approaches arbitrarily closely arbitrarily often. More precisely, P_0 is a limit point of $P(t)$ if for any $\epsilon>0$ and any time t_1 there exists a time $t_2(\epsilon,t_1)>t_1$ such that $|P(t_2)-P_0|<\epsilon$. Here absolute-value signs denote distance in phase space. Because Γ is continuously deformed as t varies, every point on the trajectory through P_0 is also a limit point of $P(t)$, and the set of limit points of $P(t)$ forms a trajectory, or a set of trajectories, called the *limiting trajectories* of $P(t)$. A limiting trajectory is obviously contained within R in its entirety.

If a trajectory is contained among its own limiting trajectories, it will be called *central*; otherwise it will be called *noncentral*. A central trajectory passes arbitrarily closely arbitrarily often to any point through which it has previously passed, and, in this sense at least, separate sufficiently long segments of a central trajectory are statistically similar. A noncentral trajectory remains a certain distance away from any point through which it has previously passed. It must approach its entire set of limit points asymptotically, although it need not approach any particular limiting trajectory asymptotically. Its instantaneous distance from its closest limit point is therefore a transient quantity, which becomes arbitrarily small as $t\rightarrow\infty$.

A trajectory $P(t)$ will be called *stable at a point* $P(t_1)$ if any other trajectory passing sufficiently close to $P(t_1)$ at time t_1 remains close to $P(t)$ as $t\rightarrow\infty$; i.e., $P(t)$ is stable at $P(t_1)$ if for any $\epsilon>0$ there exists a $\delta(\epsilon,t_1)>0$ such that if $|P_1(t_1)-P(t_1)|<\delta$ and $t_2>t_1$, $|P_1(t_2)-P(t_2)|<\epsilon$. Otherwise $P(t)$ will be called *unstable* at $P(t_1)$. Because Γ is continuously deformed as t varies, a trajectory which is stable at one point is stable at every point, and will be called a *stable* trajectory. A trajectory unstable at one point is unstable at every point, and will be called an *unstable* trajectory. In the special case that $P(t)$ is confined to one point, this definition of stability coincides with the familiar concept of stability of steady flow.

A stable trajectory $P(t)$ will be called uniformly stable if the distance within which a neighboring trajectory must approach a point $P(t_1)$, in order to be certain of remaining close to $P(t)$ as $t\rightarrow\infty$, itself possesses a positive lower bound as $t_1\rightarrow\infty$; i.e., $P(t)$ is uniformly stable if for any $\epsilon>0$ there exists a $\delta(\epsilon)>0$ and a time $t_0(\epsilon)$ such that if $t_1>t_0$ and $|P_1(t_1)-P(t_1)|<\delta$ and $t_2>t_1$, $|P_1(t_2)-P(t_2)|<\epsilon$. A limiting trajectory $P_0(t)$ of a uniformly stable trajectory $P(t)$ must be uniformly stable itself, since all trajectories passing sufficiently close to $P_0(t)$ must pass arbitrarily close to some point of $P(t)$ and so must remain close to $P(t)$, and hence to $P_0(t)$, as $t\rightarrow\infty$.

Since each point lies on a unique trajectory, any trajectory passing through a point through which it has previously passed must continue to repeat its past behavior, and so must be *periodic*. A trajectory $P(t)$ will be called *quasi-periodic* if for some arbitrarily large time interval τ, $P(t+\tau)$ ultimately remains arbitrarily close to $P(t)$, i.e., $P(t)$ is quasi-periodic if for any $\epsilon>0$ and for any time interval τ_0, there exists a $\tau(\epsilon,\tau_0)>\tau_0$ and a time $t_1(\epsilon,\tau_0)$ such that if $t_2>t_1$, $|P(t_2+\tau)-P(t_2)|$

$<\epsilon$. Periodic trajectories are special cases of quasi-periodic trajectories.

A trajectory which is not quasi-periodic will be called *nonperiodic*. If $P(t)$ is nonperiodic, $P(t_1+\tau)$ may be arbitrarily close to $P(t_1)$ for some time t_1 and some arbitrarily large time interval τ, but, if this is so, $P(t+\tau)$ cannot remain arbitrarily close to $P(t)$ as $t \rightarrow \infty$. Nonperiodic trajectories are of course representations of deterministic nonperiodic flow, and form the principal subject of this paper.

Periodic trajectories are obviously central. Quasi-periodic central trajectories include multiple periodic trajectories with incommensurable periods, while quasi-periodic noncentral trajectories include those which approach periodic trajectories asymptotically. Nonperiodic trajectories may be central or noncentral.

We can now establish the theorem that a trajectory with a stable limiting trajectory is quasi-periodic. For if $P_0(t)$ is a limiting trajectory of $P(t)$, two distinct points $P(t_1)$ and $P(t_1+\tau)$, with τ arbitrarily large, may be found arbitrary close to any point $P_0(t_0)$. Since $P_0(t)$ is stable, $P(t)$ and $P(t+\tau)$ must remain arbitrarily close to $P_0(t+t_0-t_1)$, and hence to each other, as $t \rightarrow \infty$, and $P(t)$ is quasi-periodic.

It follows immediately that a stable central trajectory is quasi-periodic, or, equivalently, that a nonperiodic central trajectory is unstable.

The result has far-reaching consequences when the system being considered is an observable nonperiodic system whose future state we may desire to predict. It implies that two states differing by imperceptible amounts may eventually evolve into two considerably different states. If, then, there is any error whatever in observing the present state—and in any real system such errors seem inevitable—an acceptable prediction of an instantaneous state in the distant future may well be impossible.

As for noncentral trajectories, it follows that a uniformly stable noncentral trajectory is quasi-periodic, or, equivalently, a nonperiodic noncentral trajectory is not uniformly stable. The possibility of a nonperiodic noncentral trajectory which is stable but not uniformly stable still exists. To the writer, at least, such trajectories, although possible on paper, do not seem characteristic of real hydrodynamical phenomena. Any claim that atmospheric flow, for example, is represented by a trajectory of this sort would lead to the improbable conclusion that we ought to master long-range forecasting as soon as possible, because, the longer we wait, the more difficult our task will become.

In summary, we have shown that, subject to the conditions of uniqueness, continuity, and boundedness prescribed at the beginning of this section, a central trajectory, which in a certain sense is free of transient properties, is unstable if it is nonperiodic. A noncentral trajectory, which is characterized by transient properties, is not uniformly stable if it is nonperiodic, and, if it is stable at all, its very stability is one of its transient properties, which tends to die out as time progresses. In view of the impossibility of measuring initial conditions precisely, and thereby distinguishing between a central trajectory and a nearby noncentral trajectory, all nonperiodic trajectories are effectively unstable from the point of view of practical prediction.

4. Numerical integration of nonconservative systems

The theorems of the last section can be of importance only if nonperiodic solutions of equations of the type considered actually exist. Since statistically stationary nonperiodic functions of time are not easily described analytically, particular nonperiodic solutions can probably be found most readily by numerical procedures. In this section we shall examine a numerical-integration procedure which is especially applicable to systems of equations of the form (4). In a later section we shall use this procedure to determine a nonperiodic solution of a simple set of equations.

To solve (1) numerically we may choose an initial time t_0 and a time increment Δt, and let

$$X_{i,n}=X_i(t_0+n\Delta t). \tag{8}$$

We then introduce the auxiliary approximations

$$X_{i(n+1)}=X_{i,n}+F_i(P_n)\Delta t, \tag{9}$$

$$X_{i((n+2))}=X_{i(n+1)}+F_i(P_{(n+1)})\Delta t, \tag{10}$$

where P_n and $P_{(n+1)}$ are the points whose coordinates are

$$(X_{1,n},\cdots,X_{M,n}) \quad \text{and} \quad (X_{1(n+1)},\cdots,X_{M(n+1)}).$$

The simplest numerical procedure for obtaining approximate solutions of (1) is the forward-difference procedure,

$$X_{i,n+1}=X_{i(n+1)}. \tag{11}$$

In many instances better approximations to the solutions of (1) may be obtained by a centered-difference procedure

$$X_{i,n+1}=X_{i,n-1}+2F_i(P_n)\Delta t. \tag{12}$$

This procedure is unsuitable, however, when the deterministic nature of (1) is a matter of concern, since the values of $X_{1,n}, \cdots, X_{M,n}$ do not uniquely determine the values of $X_{1,n+1}, \cdots, X_{M,n+1}$.

A procedure which largely overcomes the disadvantages of both the forward-difference and centered-difference procedures is the double-approximation procedure, defined by the relation

$$X_{i,n+1}=X_{i,n}+\tfrac{1}{2}[F_i(P_n)+F_i(P_{(n+1)})]\Delta t. \tag{13}$$

Here the coefficient of Δt is an approximation to the time derivative of X_i at time $t_0+(n+\frac{1}{2})\Delta t$. From (9) and (10), it follows that (13) may be rewritten

$$X_{i,n+1}=\tfrac{1}{2}(X_{i,n}+X_{i((n+2))}). \tag{14}$$

A convenient scheme for automatic computation is the successive evaluation of $X_{i(n+1)}$, $X_{i((n+2))}$, and $X_{i,n+1}$ according to (9), (10) and (14). We have used this procedure in all the computations described in this study.

In phase space a numerical solution of (1) must be represented by a jumping particle rather than a continuously moving particle. Moreover, if a digital computer is instructed to represent each number in its memory by a preassigned fixed number of bits, only certain discrete points in phase space will ever be occupied. If the numerical solution is bounded, repetitions must eventually occur, so that, strictly speaking, every numerical solution is periodic. In practice this consideration may be disregarded, if the number of different possible states is far greater than the number of iterations ever likely to be performed. The necessity for repetition could be avoided altogether by the somewhat uneconomical procedure of letting the precision of computation increase as n increases.

Consider now numerical solutions of equations (4), obtained by the forward-difference procedure (11). For such solutions,

$$Q_{n+1}=Q_n+(dQ/dt)_n\Delta t+\tfrac{1}{2}\sum_i F_i^2(P_n)\Delta t^2. \qquad (15)$$

Let S' be any surface of constant Q whose interior R' contains the ellipsoid E where dQ/dt vanishes, and let S be any surface of constant Q whose interior R contains S'.

Since $\sum F_i^2$ and dQ/dt both possess upper bounds in R', we may choose Δt so small that P_{n+1} lies in R if P_n lies in R'. Likewise, since $\sum F_i^2$ possesses an upper bound and dQ/dt possesses a *negative* upper bound in $R-R'$, we may choose Δt so small that $Q_{n+1}<Q_n$ if P_n lies in $R-R'$. Hence Δt may be chosen so small that any jumping particle which has entered R remains trapped within R, and the numerical solution does not blow up. A blow-up may still occur, however, if initially the particle is exterior to R.

Consider now the double-approximation procedure (14). The previous arguments imply not only that $P_{(n+1)}$ lies within R if P_n lies within R, but also that $P_{((n+2))}$ lies within R if $P_{(n+1)}$ lies within R. Since the region R is convex, it follows that P_{n+1}, as given by (14), lies within R if P_n lies within R. Hence if Δt is chosen so small that the forward-difference procedure does not blow up, the double-approximation procedure also does not blow up.

We note in passing that if we apply the forward-difference procedure to a conservative system where $dQ/dt=0$ everywhere,

$$Q_{n+1}=Q_n+\tfrac{1}{2}\sum_i F_i^2(P_n)\Delta t^2. \qquad (16)$$

In this case, for any fixed choice of Δt the numerical solution ultimately goes to infinity, unless it is asymptotically approaching a steady state. A similar result holds when the double-approximation procedure (14) is applied to a conservative system.

5. The convection equations of Saltzman

In this section we shall introduce a system of three ordinary differential equations whose solutions afford the simplest example of deterministic nonperiodic flow of which the writer is aware. The system is a simplification of one derived by Saltzman (1962) to study finite-amplitude convection. Although our present interest is in the nonperiodic nature of its solutions, rather than in its contributions to the convection problem, we shall describe its physical background briefly.

Rayleigh (1916) studied the flow occurring in a layer of fluid of uniform depth H, when the temperature difference between the upper and lower surfaces is maintained at a constant value ΔT. Such a system possesses a steady-state solution in which there is no motion, and the temperature varies linearly with depth, If this solution is unstable, convection should develop.

In the case where all motions are parallel to the $x-z$-plane, and no variations in the direction of the y-axis occur, the governing equations may be written (see Saltzman, 1962)

$$\frac{\partial}{\partial t}\nabla^2\psi=-\frac{\partial(\psi,\nabla^2\psi)}{\partial(x,z)}+\nu\nabla^4\psi+g\alpha\frac{\partial\theta}{\partial x}, \qquad (17)$$

$$\frac{\partial}{\partial t}\theta=-\frac{\partial(\psi,\theta)}{\partial(x,z)}+\frac{\Delta T}{H}\frac{\partial\psi}{\partial x}+\kappa\nabla^2\theta. \qquad (18)$$

Here ψ is a stream function for the two-dimensional motion, θ is the departure of temperature from that occurring in the state of no convection, and the constants g, α, ν, and κ denote, respectively, the acceleration of gravity, the coefficient of thermal expansion, the kinematic viscosity, and the thermal conductivity. The problem is most tractable when both the upper and lower boundaries are taken to be free, in which case ψ and $\nabla^2\psi$ vanish at both boundaries.

Rayleigh found that fields of motion of the form

$$\psi=\psi_0 \sin(\pi aH^{-1}x)\sin(\pi H^{-1}z), \qquad (19)$$

$$\theta=\theta_0 \cos(\pi aH^{-1}x)\sin(\pi H^{-1}z), \qquad (20)$$

would develop if the quantity

$$R_a=g\alpha H^3\Delta T\nu^{-1}\kappa^{-1}, \qquad (21)$$

now called the ***Rayleigh number***, exceeded a critical value

$$R_c=\pi^4a^{-2}(1+a^2)^3. \qquad (22)$$

The minimum value of R_c, namely $27\pi^4/4$, occurs when $a^2=\tfrac{1}{2}$.

Saltzman (1962) derived a set of ordinary differential equations by expanding ψ and θ in double Fourier series in x and z, with functions of t alone for coefficients, and substituting these series into (17) and (18). He arranged the right-hand sides of the resulting equations in double-Fourier-series form, by replacing products of trigonometric functions of x (or z) by sums of trigonometric functions, and then equated coefficients of similar functions of x and z. He then reduced the resulting infinite system to a finite system by omitting reference to all but a specified finite set of functions of t, in the manner proposed by the writer (1960).

He then obtained time-dependent solutions by numerical integration. In certain cases all except three of the dependent variables eventually tended to zero, and these three variables underwent irregular, apparently nonperiodic fluctuations.

These same solutions would have been obtained if the series had at the start been truncated to include a total of three terms. Accordingly, in this study we shall let

$$a(1+a^2)^{-1}\kappa^{-1}\psi = X\sqrt{2}\sin(\pi a H^{-1}x)\sin(\pi H^{-1}z), \tag{23}$$

$$\pi R_c^{-1}R_a\Delta T^{-1}\theta = Y\sqrt{2}\cos(\pi a H^{-1}x)\sin(\pi H^{-1}z) - Z\sin(2\pi H^{-1}z), \tag{24}$$

where X, Y, and Z are functions of time alone. When expressions (23) and (24) are substituted into (17) and (18), and trigonometric terms other than those occurring in (23) and (24) are omitted, we obtain the equations

$$X^{\cdot} = -\sigma X + \sigma Y, \tag{25}$$

$$Y^{\cdot} = -XZ + rX - Y, \tag{26}$$

$$Z^{\cdot} = XY - bZ. \tag{27}$$

Here a dot denotes a derivative with respect to the dimensionless time $\tau = \pi^2 H^{-2}(1+a^2)\kappa t$, while $\sigma = \kappa^{-1}\nu$ is the *Prandtl number*, $r = R_c^{-1}R_a$, and $b = 4(1+a^2)^{-1}$. Except for multiplicative constants, our variables X, Y, and Z are the same as Saltzman's variables A, D, and G. Equations (25), (26), and (27) are the convection equations whose solutions we shall study.

In these equations X is proportional to the intensity of the convective motion, while Y is proportional to the temperature difference between the ascending and descending currents, similar signs of X and Y denoting that warm fluid is rising and cold fluid is descending. The variable Z is proportional to the distortion of the vertical temperature profile from linearity, a positive value indicating that the strongest gradients occur near the boundaries.

Equations (25)–(27) may give realistic results when the Rayleigh number is slightly supercritical, but their solutions cannot be expected to resemble those of (17) and (18) when strong convection occurs, in view of the extreme truncation.

6. Applications of linear theory

Although equations (25)–(27), as they stand, do not have the form of (4), a number of linear transformations will convert them to this form. One of the simplest of these is the transformation

$$X' = X, \quad Y' = Y, \quad Z' = Z - r - \sigma. \tag{28}$$

Solutions of (25)–(27) therefore remain bounded within a region R as $\tau \to \infty$, and the general results of Sections 2, 3 and 4 apply to these equations.

The stability of a solution $X(\tau)$, $Y(\tau)$, $Z(\tau)$ may be formally investigated by considering the behavior of small superposed perturbations $x_0(\tau)$, $y_0(\tau)$, $z_0(\tau)$. Such perturbations are temporarily governed by the linearized equations

$$\begin{bmatrix} x_0 \\ y_0 \\ z_0 \end{bmatrix}^{\cdot} = \begin{bmatrix} -\sigma & \sigma & 0 \\ (r-Z) & -1 & -X \\ Y & X & -b \end{bmatrix} \begin{bmatrix} x_0 \\ y_0 \\ z_0 \end{bmatrix}. \tag{29}$$

Since the coefficients in (29) vary with time, unless the basic state X, Y, Z is a steady-state solution of (25)–(27), a general solution of (29) is not feasible. However, the variation of the volume V_0 of a small region in phase space, as each point in the region is displaced in accordance with (25)–(27), is determined by the diagonal sum of the matrix of coefficients; specifically

$$V_0^{\cdot} = -(\sigma+b+1)V_0. \tag{30}$$

This is perhaps most readily seen by visualizing the motion in phase space as the flow of a fluid, whose divergence is

$$\frac{\partial X^{\cdot}}{\partial X} + \frac{\partial Y^{\cdot}}{\partial Y} + \frac{\partial Z^{\cdot}}{\partial Z} = -(\sigma+b+1). \tag{31}$$

Hence each small volume shrinks to zero as $\tau \to \infty$, at a rate independent of X, Y, and Z. This does not imply that each small volume shrinks to a point; it may simply become flattened into a surface. It follows that the volume of the region initially enclosed by the surface S shrinks to zero at this same rate, so that all trajectories ultimately become confined to a specific subspace having zero volume. This subspace contains all those trajectories which lie entirely within R, and so contains all central trajectories.

Equations (25)–(27) possess the steady-state solution $X = Y = Z = 0$, representing the state of no convection. With this basic solution, the characteristic equation of the matrix in (29) is

$$[\lambda+b][\lambda^2+(\sigma+1)\lambda+\sigma(1-r)] = 0. \tag{32}$$

This equation has three real roots when $r>0$; all are negative when $r<1$, but one is positive when $r>1$. The criterion for the onset of convection is therefore $r=1$, or $R_a = R_c$, in agreement with Rayleigh's result.

When $r>1$, equations (25)–(27) possess two additional steady-state solutions $X = Y = \pm\sqrt{b(r-1)}$, $Z = r-1$.

For either of these solutions, the characteristic equation of the matrix in (29) is

$$\lambda^3+(\sigma+b+1)\lambda^2+(r+\sigma)b\lambda+2\sigma b(r-1)=0. \quad (33)$$

This equation possesses one real negative root and two complex conjugate roots when $r>1$; the complex conjugate roots are pure imaginary if the product of the coefficients of λ^2 and λ equals the constant term, or

$$r=\sigma(\sigma+b+3)(\sigma-b-1)^{-1}. \quad (34)$$

This is the critical value of r for the instability of steady convection. Thus if $\sigma<b+1$, no positive value of r satisfies (34), and steady convection is always stable, but if $\sigma>b+1$, steady convection is unstable for sufficiently high Rayleigh numbers. This result of course applies only to idealized convection governed by (25)–(27), and not to the solutions of the partial differential equations (17) and (18).

The presence of complex roots of (34) shows that if unstable steady convection is disturbed, the motion will oscillate in intensity. What happens when the disturbances become large is not revealed by linear theory. To investigate finite-amplitude convection, and to study the subspace to which trajectories are ultimately confined, we turn to numerical integration.

TABLE 1. Numerical solution of the convection equations. Values of X, Y, Z are given at every fifth iteration N, for the first 160 iterations.

N	X	Y	Z
0000	0000	0010	0000
0005	0004	0012	0000
0010	0009	0020	0000
0015	0016	0036	0002
0020	0030	0066	0007
0025	0054	0115	0024
0030	0093	0192	0074
0035	0150	0268	0201
0040	0195	0234	0397
0045	0174	0055	0483
0050	0097	−0067	0415
0055	0025	−0093	0340
0060	−0020	−0089	0298
0065	−0046	−0084	0275
0070	−0061	−0083	0262
0075	−0070	−0086	0256
0080	−0077	−0091	0255
0085	−0084	−0095	0258
0090	−0089	−0098	0266
0095	−0093	−0098	0275
0100	−0094	−0093	0283
0105	−0092	−0086	0297
0110	−0088	−0079	0286
0115	−0083	−0073	0281
0120	−0078	−0070	0273
0125	−0075	−0071	0264
0130	−0074	−0075	0257
0135	−0076	−0080	0252
0140	−0079	−0087	0251
0145	−0083	−0093	0254
0150	−0088	−0098	0262
0155	−0092	−0099	0271
0160	−0094	−0096	0281

7. Numerical integration of the convection equations

To obtain numerical solutions of the convection equations, we must choose numerical values for the constants. Following Saltzman (1962), we shall let $\sigma=10$ and $a^2=\frac{1}{2}$, so that $b=8/3$. The critical Rayleigh number for instability of steady convection then occurs when $r=470/19=24.74$.

We shall choose the slightly supercritical value $r=28$. The states of steady convection are then represented by the points $(6\sqrt{2}, 6\sqrt{2}, 27)$ and $(-6\sqrt{2}, -6\sqrt{2}, 27)$ in phase space, while the state of no convection corresponds to the origin (0,0,0).

We have used the double-approximation procedure for numerical integration, defined by (9), (10), and (14). The value $\Delta\tau=0.01$ has been chosen for the dimensionless time increment. The computations have been performed on a Royal McBee LGP-30 electronic com-

TABLE 2. Numerical solution of the convection equations Values of X, Y, Z are given at every iteration N for which Z possesses a relative maximum, for the first 6000 iterations.

N	X	Y	Z	N	X	Y	Z
0045	0174	0055	0483	3029	0117	0075	0352
0107	−0091	−0083	0287	3098	0123	0076	0365
0168	−0092	−0084	0288	3171	0134	0082	0383
0230	−0092	−0084	0289	3268	0155	0069	0435
0292	−0092	−0083	0290	3333	−0114	−0079	0342
0354	−0093	−0083	0292	3400	−0117	−0077	0350
0416	−0093	−0083	0293	3468	−0125	−0083	0361
0478	−0094	−0082	0295	3541	−0129	−0073	0378
0540	−0094	−0082	0296	3625	−0146	−0074	0413
0602	−0095	−0082	0298	3695	0127	0079	0370
0664	−0096	−0083	0300	3772	0136	0072	0394
0726	−0097	−0083	0302	3853	−0144	−0077	0407
0789	−0097	−0081	0304	3926	0129	0072	0380
0851	−0099	−0083	0307	4014	0148	0068	0421
0914	−0100	−0081	0309	4082	−0120	−0074	0359
0977	−0100	−0080	0312	4153	−0129	−0078	0375
1040	−0102	−0080	0315	4233	−0144	−0082	0404
1103	−0104	−0081	0319	4307	0135	0081	0385
1167	−0105	−0079	0323	4417	−0162	−0069	0450
1231	−0107	−0079	0328	4480	0106	0081	0324
1295	−0111	−0082	0333	4544	0109	0082	0329
1361	−0111	−0077	0339	4609	0110	0080	0334
1427	−0116	−0079	0347	4675	0112	0076	0341
1495	−0120	−0077	0357	4741	0118	0081	0349
1566	−0125	−0072	0371	4810	0120	0074	0360
1643	−0139	−0077	0396	4881	0130	0081	0376
1722	0140	0075	0401	4963	0141	0068	0406
1798	−0135	−0072	0391	5035	−0133	−0081	0381
1882	0146	0074	0413	5124	−0151	−0076	0422
1952	−0127	−0078	0370	5192	0119	0075	0358
2029	−0135	−0070	0393	5262	0129	0083	0372
2110	0146	0083	0408	5340	0140	0079	0397
2183	−0128	−0070	0379	5419	−0137	−0067	0399
2268	−0144	−0066	0415	5495	0140	0081	0394
2337	0126	0079	0368	5576	−0141	−0072	0405
2412	0137	0081	0389	5649	0135	0082	0384
2501	−0153	−0080	0423	5752	0160	0074	0443
2569	0119	0076	0357	5816	−0110	−0081	0332
2639	0129	0082	0371	5881	−0113	−0082	0339
2717	0136	0070	0395	5948	−0114	−0075	0346
2796	−0143	−0079	0402				
2871	0134	0076	0388				
2962	−0152	−0072	0426				

puting machine. Approximately one second per iteration, aside from output time, is required.

For initial conditions we have chosen a slight departure from the state of no convection, namely (0,1,0). Table 1 has been prepared by the computer. It gives the values of N (the number of iterations), X, Y, and Z at every fifth iteration for the first 160 iterations. In the printed output (but not in the computations) the values of X, Y, and Z are multiplied by ten, and then only those figures to the left of the decimal point are printed. Thus the states of steady convection would appear as 0084, 0084, 0270 and −0084, −0084, 0270, while the state of no convection would appear as 0000, 0000, 0000.

The initial instability of the state of rest is evident. All three variables grow rapidly, as the sinking cold fluid is replaced by even colder fluid from above, and the rising warm fluid by warmer fluid from below, so that by step 35 the strength of the convection far exceeds that of steady convection. Then Y diminishes as the warm fluid is carried over the top of the convective cells, so that by step 50, when X and Y have opposite signs, warm fluid is descending and cold fluid is ascending. The motion thereupon ceases and reverses its direction, as indicated by the negative values of X following step 60. By step 85 the system has reached a state not far from that of steady convection. Between steps 85 and 150 it executes a complete oscillation in its intensity, the slight amplification being almost indetectable.

The subsequent behavior of the system is illustrated in Fig. 1, which shows the behavior of Y for the first 3000 iterations. After reaching its early peak near step 35 and then approaching equilibrium near step 85, it undergoes systematic amplified oscillations until near step 1650. At this point a critical state is reached, and thereafter Y changes sign at seemingly irregular intervals, reaching sometimes one, sometimes two, and sometimes three or more extremes of one sign before changing sign again.

Fig. 2 shows the projections on the X-Y- and Y-Z-planes in phase space of the portion of the trajectory corresponding to iterations 1400–1900. The states of steady convection are denoted by C and C'. The first portion of the trajectory spirals outward from the vicinity of C', as the oscillations about the state of steady convection, which have been occurring since step 85, continue to grow. Eventually, near step 1650, it crosses the X-Z-plane, and is then deflected toward the neighborhood of C. It temporarily spirals about C, but crosses the X-Z-plane after one circuit, and returns to the neighborhood of C', where it soon joins the spiral over which it has previously traveled. Thereafter it crosses from one spiral to the other at irregular intervals.

Fig. 3, in which the coordinates are Y and Z, is based upon the printed values of X, Y, and Z at every fifth iteration for the first 6000 iterations. These values determine X as a smooth single-valued function of Y and Z over much of the range of Y and Z; they determine X

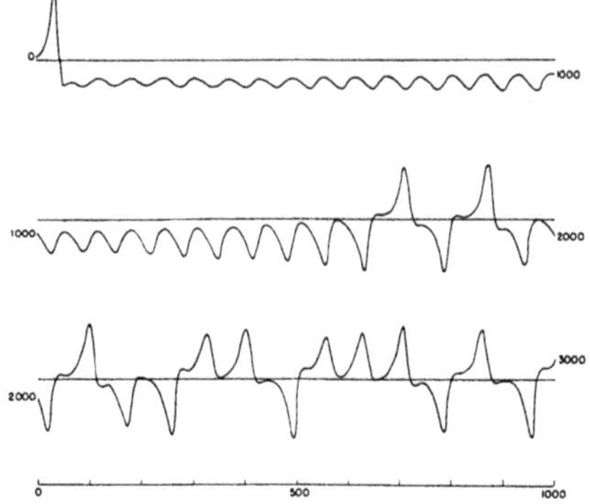

Fig. 1. Numerical solution of the convection equations. Graph of Y as a function of time for the first 1000 iterations (upper curve), second 1000 iterations (middle curve), and third 1000 iterations (lower curve).

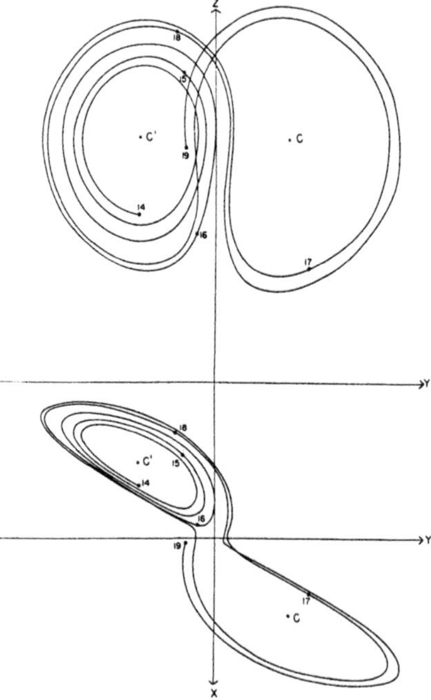

Fig. 2. Numerical solution of the convection equations. Projections on the X-Y-plane and the Y-Z-plane in phase space of the segment of the trajectory extending from iteration 1400 to iteration 1900. Numerals "14," "15," etc., denote positions at iterations 1400, 1500, etc. States of steady convection are denoted by C and C'.

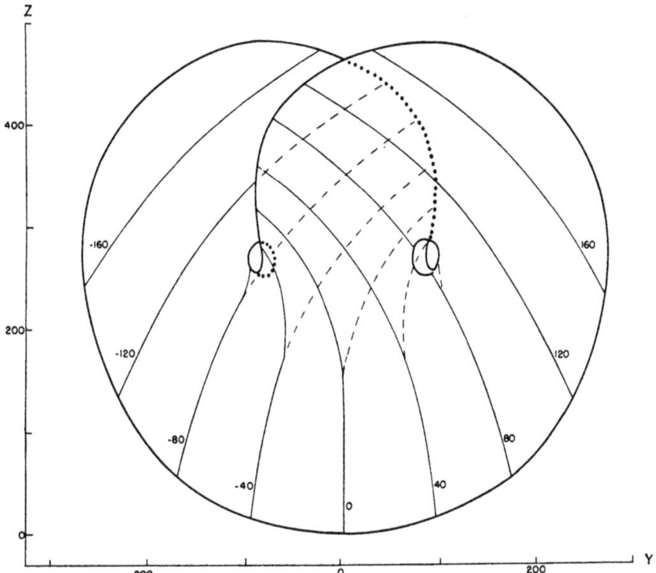

FIG. 3. Isopleths of X as a function of Y and Z (thin solid curves), and isopleths of the lower of two values of X, where two values occur (dashed curves), for approximate surfaces formed by all points on limiting trajectories. Heavy solid curve, and extensions as dotted curves, indicate natural boundaries of surfaces.

as one of two smooth single-valued functions over the remainder of the range. In Fig. 3 the thin solid lines are isopleths of X, and where two values of X exist, the dashed lines are isopleths of the lower value. Thus, within the limits of accuracy of the printed values, the trajectory is confined to a pair of surfaces which appear to merge in the lower portion of Fig. 3. The spiral about C lies in the upper surface, while the spiral about C' lies in the lower surface. Thus it is possible for the trajectory to pass back and forth from one spiral to the other without intersecting itself.

Additional numerical solutions indicate that other trajectories, originating at points well removed from these surfaces, soon meet these surfaces. The surfaces therefore appear to be composed of all points lying on limiting trajectories.

Because the origin represents a steady state, no trajectory can pass through it. However, two trajectories emanate from it, i.e., approach it asymptotically as $\tau \rightarrow -\infty$. The heavy solid curve in Fig. 3, and its extensions as dotted curves, are formed by these two trajectories. Trajectories passing close to the origin will tend to follow the heavy curve, but will not cross it, so that the heavy curve forms a natural boundary to the region which a trajectory can ultimately occupy. The holes near C and C' also represent regions which cannot be occupied after they have once been abandoned.

Returning to Fig. 2, we find that the trajectory apparently leaves one spiral only after exceeding some critical distance from the center. Moreover, the extent to which this distance is exceeded appears to determine the point at which the next spiral is entered; this in turn seems to determine the number of circuits to be executed before changing spirals again.

It therefore seems that some single feature of a given circuit should predict the same feature of the following circuit. A suitable feature of this sort is the maximum value of Z, which occurs when a circuit is nearly completed. Table 2 has again been prepared by the computer, and shows the values of X, Y, and Z at only those iterations N for which Z has a relative maximum. The succession of circuits about C and C' is indicated by the succession of positive and negative values of X and Y. Evidently X and Y change signs following a maximum which exceeds some critical value printed as about 385.

Fig. 4 has been prepared from Table 2. The abscissa is M_n, the value of the nth maximum of Z, while the ordinate is M_{n+1}, the value of the following maximum. Each point represents a pair of successive values of Z taken from Table 2. Within the limits of the round-off in tabulating Z, there is a precise two-to-one relation between M_n and M_{n+1}. The initial maximum $M_1=483$ is shown as if it had followed a maximum $M_0=385$, since maxima near 385 are followed by close approaches to the origin, and then by exceptionally large maxima.

It follows that an investigator, unaware of the nature of the governing equations, could formulate an empirical prediction scheme from the "data" pictured in Figs. 2 and 4. From the value of the most recent maximum of Z, values at future maxima may be obtained by repeated applications of Fig. 4. Values of X, Y, and Z between maxima of Z may be found from Fig. 2, by interpolating between neighboring curves. Of course, the accuracy of predictions made by this method is limited by the exactness of Figs. 2 and 4, and, as we shall see, by the accuracy with which the initial values of X, Y, and Z are observed.

Some of the implications of Fig. 4 are revealed by considering an idealized two-to-one correspondence between successive members of sequences $M_0, M_1, \cdots$, consisting of numbers between zero and one. These sequences satisfy the relations

$$\begin{aligned} M_{n+1}&=2M_n && \text{if } M_n<\tfrac{1}{2}\\ M_{n+1}&\text{ is undefined} && \text{if } M_n=\tfrac{1}{2}\\ M_{n+1}&=2-2M_n && \text{if } M_n>\tfrac{1}{2}. \end{aligned} \tag{35}$$

The correspondence defined by (35) is shown in Fig. 5, which is an idealization of Fig. 4. It follows from repeated applications of (35) that in any particular sequence,

$$M_n=m_n\pm 2^n M_0, \tag{36}$$

where m_n is an even integer.

Consider first a sequence where $M_0=u/2^p$, where u is odd. In this case $M_{p-1}=\frac{1}{2}$, and the sequence terminates. These sequences form a denumerable set, and correspond to the trajectories which score direct hits upon the state of no convection.

Next consider a sequence where $M_0=u/2^p v$, where u and v are relatively prime odd numbers. Then if $k>0$, $M_{p+1+k}=u_k/v$, where u_k and v are relatively prime and u_k is even. Since for any v the number of proper fractions u_k/v is finite, repetitions must occur, and the sequence is periodic. These sequences also form a denumerable set, and correspond to periodic trajectories.

The periodic sequences having a given number of distinct values, or phases, are readily tabulated. In particular there are a single one-phase, a single two-phase, and two three-phase sequences, namely,

$$\begin{gathered} 2/3, \cdots,\\ 2/5, 4/5, \cdots,\\ 2/7, 4/7, 6/7, \cdots,\\ 2/9, 4/9, 8/9, \cdots. \end{gathered}$$

The two three-phase sequences differ qualitatively in that the former possesses two numbers, and the latter only one number, exceeding $\frac{1}{2}$. Thus the trajectory corresponding to the former makes two circuits about C, followed by one about C' (or vice versa). The trajectory corresponding to the latter makes three circuits about C, followed by three about C', so that actually only Z varies in three phases, while X and Y vary in six.

Now consider a sequence where M_0 is not a rational fraction. In this case (36) shows that M_{n+k} cannot equal

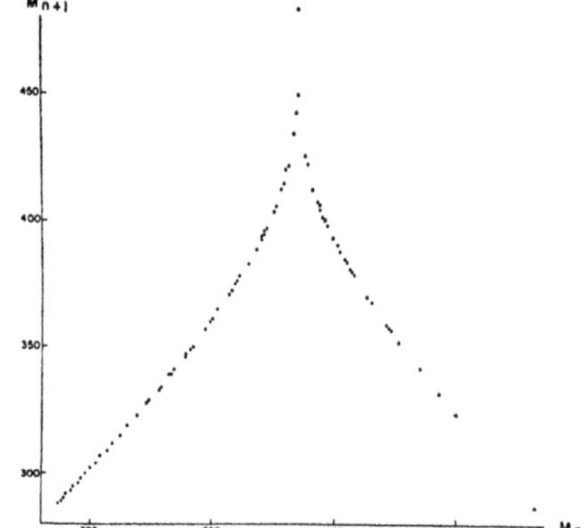

FIG. 4. Corresponding values of relative maximum of Z (abscissa) and subsequent relative maximum of Z (ordinate) occurring during the first 6000 iterations.

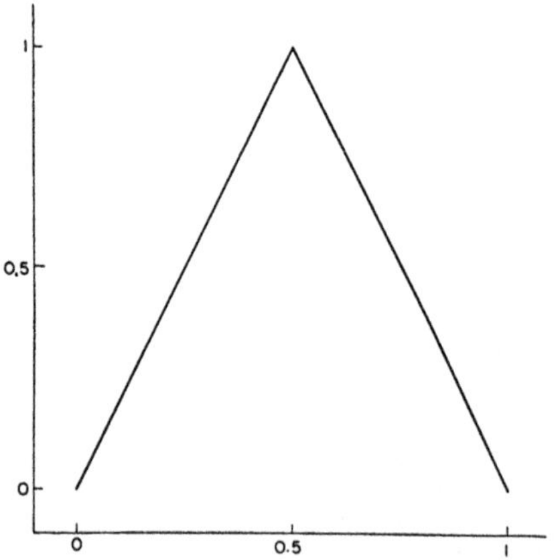

FIG. 5. The function $M_{n+1}=2M_n$ if $M_n<\frac{1}{2}$, $M_{n+1}=2-2M_n$ if $M_n>\frac{1}{2}$, serving as an idealization of the locus of points in Fig. 4.

M_n if $k>0$, so that no repetitions occur. These sequences, which form a nondenumerable set, may conceivably approach periodic sequences asymptotically and be quasi-periodic, or they may be nonperiodic.

Finally, consider two sequences M_0, M_1, $\cdots$ and M_0', M_1', $\cdots$, where $M_0'=M_0+\epsilon$. Then for a given k, if ϵ is sufficiently small, $M_k'=M_k\pm 2^k\epsilon$. All sequences are therefore unstable with respect to small modifications. In particular, all periodic sequences are unstable, and no other sequences can approach them asymptotically. All sequences except a set of measure zero are therefore nonperiodic, and correspond to nonperiodic trajectories.

Returning to Fig. 4, we see that periodic sequences analogous to those tabulated above can be found. They are given approximately by

$$\begin{aligned}&398, \cdots,\\ &377, 410, \cdots,\\ &369, 391, 414, \cdots,\\ &362, 380, 419, \cdots.\end{aligned}$$

The trajectories possessing these or other periodic sequences of maxima are presumably periodic or quasi-periodic themselves.

The above sequences are temporarily approached in the numerical solution by sequences beginning at iterations 5340, 4881, 3625, and 3926. Since the numerical solution eventually departs from each of these sequences, each is presumably unstable.

More generally, if $M_n'=M_n+\epsilon$, and if ϵ is sufficiently small, $M_{n+k}'=M_{n+k}+\Lambda\epsilon$, where Λ is the product of the slopes of the curve in Fig. 4 at the points whose abscissas are M_n, $\cdots$, M_{n+k-1}. Since the curve apparently has a slope whose magnitude exceeds unity everywhere, all sequences of maxima, and hence all trajectories, are unstable. In particular, the periodic trajectories, whose sequences of maxima form a denumerable set, are unstable, and only exceptional trajectories, having the same sequences of maxima, can approach them asymptotically. The remaining trajectories, whose sequences of maxima form a nondenumerable set, therefore represent deterministic nonperiodic flow.

These conclusions have been based upon a finite segment of a numerically determined solution. They cannot be regarded as mathematically proven, even though the evidence for them is strong. One apparent contradiction requires further examination.

It is difficult to reconcile the merging of two surfaces, one containing each spiral, with the inability of two trajectories to merge. It is not difficult, however, to explain the *apparent* merging of the surfaces. At two times τ_0 and τ_1, the volumes occupied by a specified set of particles satisfy the relation

$$V_0(\tau_1)=e^{-(\sigma+b+1)(\tau_1-\tau_0)}V_0(\tau_0), \tag{37}$$

according to (30). A typical circuit about C or C' requires about 70 iterations, so that, for such a circuit, $\tau_2=\tau_1+0.7$, and, since $\sigma+b+1=41/3$,

$$V_0(\tau_1)=0.00007\,V_0(\tau_0). \tag{38}$$

Two particles separated from each other in a suitable direction can therefore come together very rapidly, and appear to merge.

It would seem, then, that the two surfaces merely appear to merge, and remain distinct surfaces. Following these surfaces along a path parallel to a trajectory, and circling C or C', we see that each surface is really a pair of surfaces, so that, where they appear to merge, there are really four surfaces. Continuing this process for another circuit, we see that there are really eight surfaces, etc., and we finally conclude that there is an infinite complex of surfaces, each extremely close to one or the other of two merging surfaces.

The infinite set of values at which a line parallel to the X-axis intersects these surfaces may be likened to the set of all numbers between zero and one whose decimal expansions (or some other expansions besides binary) contain only zeros and ones. This set is plainly nondenumerable, in view of its correspondence to the set of all numbers between zero and one, expressed in binary. Nevertheless it forms a set of measure zero. The sequence of ones and zeros corresponding to a particular surface contains a history of the trajectories lying in that surface, a one or zero immediately to the right of the decimal point indicating that the last circuit was about C or C', respectively, a one or zero in second place giving the same information about the next to the last circuit, etc. Repeating decimal expansions represent periodic or quasi-periodic trajectories, and, since they define rational fractions, they form a denumerable set.

If one first visualizes this infinite complex of surfaces, it should not be difficult to picture nonperiodic deterministic trajectories embedded in these surfaces.

8. Conclusion

Certain mechanically or thermally forced nonconservative hydrodynamical systems may exhibit either periodic or irregular behavior when there is no obviously related periodicity or irregularity in the forcing process. Both periodic and nonperiodic flow are observed in some experimental models when the forcing process is held constant, within the limits of experimental control. Some finite systems of ordinary differential equations designed to represent these hydrodynamical systems possess periodic analytic solutions when the forcing is strictly constant. Other such systems have yielded nonperiodic numerical solutions.

A finite system of ordinary differential equations representing forced dissipative flow often has the property that all of its solutions are ultimately confined within the same bounds. We have studied in detail the properties of solutions of systems of this sort. Our principal results concern the instability of nonperiodic solutions. A nonperiodic solution with no transient component must be unstable, in the sense that solutions temporarily approximating it do not continue to do so. A nonperiodic solution with a transient component is sometimes stable, but in this case its stability is one of its transient properties, which tends to die out.

To verify the existence of deterministic nonperiodic flow, we have obtained numerical solutions of a system of three ordinary differential equations designed to represent a convective process. These equations possess three steady-state solutions and a denumerably infinite set of periodic solutions. All solutions, and in particular the periodic solutions, are found to be unstable. The remaining solutions therefore cannot in general approach the periodic solutions asymptotically, and so are nonperiodic.

When our results concerning the instability of nonperiodic flow are applied to the atmosphere, which is ostensibly nonperiodic, they indicate that prediction of the sufficiently distant future is impossible by any method, unless the present conditions are known exactly. In view of the inevitable inaccuracy and incompleteness of weather observations, precise very-long-range forecasting would seem to be non-existent.

There remains the question as to whether our results really apply to the atmosphere. One does not usually regard the atmosphere as either deterministic or finite, and the lack of periodicity is not a mathematical certainty, since the atmosphere has not been observed forever.

The foundation of our principal result is the eventual necessity for any bounded system of finite dimensionality to come arbitrarily close to acquiring a state which it has previously assumed. If the system is stable, its future development will then remain arbitrarily close to its past history, and it will be quasi-periodic.

In the case of the atmosphere, the crucial point is then whether analogues must have occurred since the state of the atmosphere was first observed. By analogues, we mean specifically two or more states of the atmosphere, together with its environment, which resemble each other so closely that the differences may be ascribed to errors in observation. Thus, to be analogues, two states must be closely alike in regions where observations are accurate and plentiful, while they need not be at all alike in regions where there are no observations at all, whether these be regions of the atmosphere or the environment. If, however, some unobserved features are implicit in a succession of observed states, two successions of states must be nearly alike in order to be analogues.

If it is true that two analogues have occurred since atmospheric observation first began, it follows, since the atmosphere has not been observed to be periodic, that the successions of states following these analogues must eventually have differed, and no forecasting scheme could have given correct results both times. If, instead, analogues have not occurred during this period, some accurate very-long-range prediction scheme, using observations at present available, may exist. But, if it does exist, the atmosphere will acquire a quasi-periodic behavior, never to be lost, once an analogue occurs. This quasi-periodic behavior need not be established, though, even if very-long-range forecasting is feasible, if the variety of possible atmospheric states is so immense that analogues need never occur. It should be noted that these conclusions do not depend upon whether or not the atmosphere is deterministic.

There remains the very important question as to how long is "very-long-range." Our results do not give the answer for the atmosphere; conceivably it could be a few days or a few centuries. In an idealized system, whether it be the simple convective model described here, or a complicated system designed to resemble the atmosphere as closely as possible, the answer may be obtained by comparing pairs of numerical solutions having nearly identical initial conditions. In the case of the real atmosphere, if all other methods fail, we can wait for an analogue.

Acknowledgments. The writer is indebted to Dr. Barry Saltzman for bringing to his attention the existence of nonperiodic solutions of the convection equations. Special thanks are due to Miss Ellen Fetter for handling the many numerical computations and preparing the graphical presentations of the numerical material.

REFERENCES

Birkhoff, G. O., 1927: *Dynamical systems.* New York, Amer. Math. Soc., Colloq. Publ., 295 pp.

Ford, L. R., 1933: *Differential equations.* New York, McGraw-Hill, 264 pp.

Fultz, D., R. R. Long, G. V. Owens, W. Bohan, R. Kaylor and J. Weil, 1959: Studies of thermal convection in a rotating cylinder with some implications for large-scale atmospheric motions. *Meteor. Monog*, 4(21), Amer. Meteor. Soc., 104 pp.

Gibbs, J. W., 1902: *Elementary principles in statistical mechanics.* New York, Scribner, 207 pp.

Hide, R., 1958: An experimental study of thermal convection in a rotating liquid. *Phil. Trans. Roy. Soc. London,* (A), **250**, 441–478.

Lorenz, E. N., 1960: Maximum simplification of the dynamic equations. *Tellus,* **12**, 243–254.

——, 1962a: Simplified dynamic equations applied to the rotating-basin experiments. *J. atmos. Sci.*, **19**, 39–51.

——, 1962b: The statistical prediction of solutions of dynamic equations. *Proc. Internat. Symposium Numerical Weather Prediction,* Tokyo, 629–635.

Nemytskii, V. V., and V. V. Stepanov, 1960: *Qualitative theory of differential equations.* Princeton, Princeton Univ. Press, 523 pp.

Poincaré, H., 1881: Mémoire sur les courbes définies par une équation différentielle. *J. de Math.*, **7**, 375–442.

Rayleigh, Lord, 1916: On convective currents in a horizontal layer of fluid when the higher temperature is on the under side. *Phil. Mag.*, **32**, 529–546.

Saltzman, B., 1962: Finite amplitude free convection as an initial value problem—I. *J. atmos. Sci.*, **19**, 329–341.

Editor's note. Another paper which had played a key role in shaping the modern theory of the onset of turbulence was Ruelle and Takens (1971). We have unfortunately not been able to include it in this selection due to space limitations. A short account of the Ruelle–Takens 'scenario' is given by Eckmann (1981), p 94 this selection.

Journal of Statistical Physics **21** 707–26 (1979)

Sequences of Infinite Bifurcations and Turbulence in a Five-Mode Truncation of the Navier–Stokes Equations

Valter Franceschini[1] **and Claudio Tebaldi**[2]

Received May 21, 1979

Two infinite sequences of orbits leading to turbulence in a five-mode truncation of the Navier–Stokes equations for a 2-dimensional incompressible fluid on a torus are studied in detail. Their compatibility with Feigenbaum's theory of universality in certain infinite sequences of bifurcations is verified and some considerations on their asymptotic behavior are inferred. An analysis of the Poincaré map is performed, showing how the turbulent behavior is approached gradually when, with increasing Reynolds number, no stable fixed point or periodic orbit is present and all the unstable ones become more and more unstable, in close analogy with the Lorenz model.

KEY WORDS: Navier–Stokes equations; turbulence; strange attractors; Poincarè map; infinite sequences of periodic orbits; stable and hyperbolic orbits collapse; universal properties in infinite sequences of bifurcations.

1. INTRODUCTION

A model obtained by a suitable five-mode truncation of the Navier–Stokes equations for a two-dimensional incompressible fluid on a torus has been presented in Ref. 1.

The system of nonlinear ordinary differential equations resulting from such a truncation is

$$\begin{aligned}
\dot{x}_1 &= -2x_1 + 4x_2x_3 + 4x_4x_5 \\
\dot{x}_2 &= -9x_2 + 3x_1x_3 \\
\dot{x}_3 &= -5x_3 - 7x_1x_2 + r \\
\dot{x}_4 &= -5x_4 - x_1x_5 \\
\dot{x}_5 &= -x_5 - 3x_1x_4
\end{aligned}$$

[1] Istituto Matematico, Università di Modena, Modena, Italy.

[2] Dipartimento di Matematica, Università di Ancona, Ancona, Italy and Istituto di Fisica, Università di Bologna, Bologna, Italy.

(where r is the Reynolds number), and exhibits an interesting variety of different behaviors for different ranges of r. Keeping the same symbols as in Ref. 1 for the critical values of r, the most interesting feature is the stochastic behavior observed when $R_{12} < r < R_{13}$, with $28.73 < R_{12} < 29.0$ and $R_{13} \approx 33.43$.

In recent years much attention has been devoted to the study of models exhibiting such a feature when one or more parameters increase beyond certain critical values. The best known models of this kind are certainly the ones by Lorenz[(2–4)] and Hénon.[(5,6)] Ruelle and Takens[(7)] explain this stochastic behavior as a consequence of the appearance of an attractor with a complicated nature ("strange attractor"), on which the motion seems completely chaotic ("turbulence"). In addition to detailed studies on the nature of these attractors (see, for example, Lanford,[(8)] and Hénon and Pomeau[(6)]), strong interest has been focused upon the study of the mechanism of their generation.

In Ref. 1 it is shown that turbulence is reached through a long and rather complicated sequence of bifurcations related to two sequences of orbits: the former consists of four orbits with periods T, $2T$, $4T$, and $8T$, respectively, and the latter of five orbits of a different type with periods T^*, $2T^*$, $4T^*$, $8T^*$, and $16T^*$. In the following, $\mathscr{C}_i$, $i = 0, 1, 2, 3$, will refer to the orbits of the former sequence and $\mathscr{C}_i^*$, $i = 0, 1, \ldots, 4$, to the orbits of the latter.[3] The orbit $\mathscr{C}_0^*$ is found for a value of r larger than but very close to the largest value for which $\mathscr{C}_3$ is still found.[4] It is then suggested that the sequence $\mathscr{C}_i$ is finite, different from $\mathscr{C}_i^*$, and $\mathscr{C}_3$ bifurcates in $\mathscr{C}_0^*$. Since this transition remains an obscure point, because it does not fit very well with the ideas of bifurcation theory, it seems interesting to us to investigate more deeply the two sequences of orbits.

A further reason for this investigation is to verify if the behavior of the sequence $\mathscr{C}_i^*$ is compatible with the strongly suggestive idea of universality in certain infinite sequences of bifurcations developed by Feigenbaum.[(9)] This exhaustive study has been possible because we have been able to apply numerical schemes that are more efficient for studying the stability of orbits and especially in searching for new orbits, even unstable ones. With these new techniques, for a better understanding of the generation of

[3] More precisely, $\mathscr{C}_i$ must be regarded as one of four symmetrically placed, identical orbits going through identical behavior, and the same for $\mathscr{C}_i^*$. There are then four sequences $\mathscr{C}_i$ and four $\mathscr{C}_i^*$, although we will refer simply to "the" sequence $\mathscr{C}_i$ or $\mathscr{C}_i^*$, when not otherwise required. The presence of quadruples of periodic solutions is accounted for by the symmetries $(x_1, x_2, x_3, -x_4, -x_5) \leftrightarrow (x_1, x_2, x_3, x_4, x_5)$, $(-x_1, -x_2, x_3, -x_4, x_5) \leftrightarrow (x_1, x_2, x_3, x_4, x_5)$, $(-x_1, -x_2, x_3, x_4, -x_5) \leftrightarrow (x_1, x_2, x_3, x_4, x_5)$.

[4] $\mathscr{C}_3$ is observed up to $r = 28.6660$, while $\mathscr{C}_0^*$ is first found for $r = 28.6662$.

the "strange attractor" we also have been able to reconsider the transitions through R_{12}, for r increasing, and R_{13}, for r decreasing, with which the system goes over to turbulent behavior. The results of these investigations are described in the following.

In Section 2 a detailed analysis of the two sequences of orbits $\mathscr{C}_i$ and $\mathscr{C}_i^*$ is given, showing that both of them are very likely to be infinite, with a phenomenon of hysteresis because of the simultaneous presence of the orbits $\mathscr{C}_i$, $i \geqslant 3$, with $\mathscr{C}_0^*$.

In Section 3 it is shown that the appearance of the "strange attractor" for r decreasing to R_{13} follows the collapse of the stable orbit present for the high-r regime[5] with an unstable hyperbolic one and that analogous phenomenology is present in the appearance of $\mathscr{C}_0^*$. An interpretation using the theory of the bifurcation of periodic orbits in generic conditions[(7)] is tried.

In Section 4 the compatibility of the, now two, infinite sequences of bifurcations with the universality theory developed by Feigenbaum is verified and some considerations on their asymptotic behavior are inferred.

In Section 5 a detailed analysis of the Poincaré map shows how the turbulent behavior is approached gradually when the previously stable periodic orbits, now all unstable, become more and more unstable for r increasing.

Finally, a schematic picture of the features exhibited by the system is presented in Section 6, together with some concluding remarks.

2. TWO INFINITE SEQUENCES OF BIFURCATIONS

In Ref. 1 it is shown that for a certain value of the Reynolds number r ($r = R_3 = 22.85370163 \cdots$) four previously stable fixed points become unstable and four stable periodic orbits, referred to as $\mathscr{C}_0$ in Section 1, arise via a Hopf bifurcation[6] around each fixed point. With increasing r, the periodic orbits are shown to go through a number of bifurcations, doubling in period and winding up twice as many times around the fixed points from which they are generated. This is shown to happen up to $r = 28.6660$, when three successive bifurcations have taken place, giving rise to the orbits $\mathscr{C}_1$, $\mathscr{C}_2$, $\mathscr{C}_3$. For $r = 28.6662$ four new stable orbits $\mathscr{C}_0^*$ are found, with structure and period different from the previous ones of $\mathscr{C}_3$, each of them winding up around two of the fixed points. It is stressed that no definite statement can be made about the fact that no further similar bifurcation

[5] We recall that for $r \to R_{13}$ from above, this stable periodic orbit bifurcates with a real eigenvalue crossing the unit circle at $+1$.

[6] For a detailed theory concerning the Hopf bifurcation and its applications see Ref. 11.

takes place in the sequence $\mathscr{C}_i$; all the phenomenology observed, however, leads to the conjecture that $\mathscr{C}_3$ bifurcates in $\mathscr{C}_0^*$.

The difficulty in interpreting this point according to the bifurcation theory motivated us to reconsider it, applying Newton's method to obtain and analyze periodic solutions. Once the approximate initial point is close enough, the convergence of the method is fast, both for stable and unstable periodic orbits. The main purpose of such a method is in fact to be able to find unstable periodic solutions too.

Going back to the study of the first sequence of orbits, we have been able to determine the bifurcation points with a very good accuracy and, much more important, to find two more orbits in the sequence $\mathscr{C}_i$, i.e., $\mathscr{C}_4$ and $\mathscr{C}_5$ (see Table I). We have also verified that each of the orbits in the sequence $\mathscr{C}_i$ becomes unstable when an eigenvalue of the Liapunov matrix of the Poincaré map crosses the unit circle at the point -1.[7] The agreement with what is predicted by the bifurcation theory in this case is now complete, since upon bifurcation the previously stable orbit becomes unstable and a new stable orbit appears, with the period doubled (see Ref. 7). At this point it is reasonable to infer that the sequence $\mathscr{C}_i$ is infinite too. We have not carried out a further investigation for higher bifurcated orbits because of the high amount of computational time required even to consider only the next one.

Also for the sequence of orbits $\mathscr{C}_i^*$ the bifurcation points have been determined very accurately using the Newton method, up to the orbit $\mathscr{C}_4^*$ (see Table II). It has been verified that $\mathscr{C}_i^*$ is generated by a sequence of bifurcations of the same kind as those for $\mathscr{C}_i$, since also each orbit $\mathscr{C}_i^*$ becomes unstable with an eigenvalue of -1 in the Liapunov matrix of the Poincaré map.

[7] This was already seen in Ref. 1 for the orbit $\mathscr{C}_0$.

Table I. Bifurcation points ρ_i of the Periodic Orbits in the Sequence $\mathscr{C}_i$ and Relative Periods $T(\rho_i)$

i	ρ_i	$T(\rho_i)$
0	28.4105	0.81621
1	28.6399	1.64567
2	28.6641	3.29334
3	28.66776	6.58741
4	28.668463	13.17507
5	28.668611	26.35026

Table II. Bifurcation points ρ_i^* of the Periodic Orbits in the Sequence $\mathscr{C}_i^*$ and Relative Periods $T^*(\rho_i^*)$

i	ρ_i^*	$T^*(\rho_i^*)$
0	28.7013	3.80928
1	28.71606	7.62056
2	28.71926	15.24271
3	28.719947	30.48597
4	28.720103	60.97222

A question left open is how the orbit $\mathscr{C}_0^*$ appears. The answer is that an unstable orbit exists simultaneously with $\mathscr{C}_0^*$, very close to it, and the two collapse upon bifurcation for r decreasing. All the details will be given in the next section.

An important feature must be emphasized concerning the two sequences of orbits $\mathscr{C}_i$ and $\mathscr{C}_i^*$, i.e., the simultaneous presence of different stable orbits for the same value of the Reynolds number r in a certain range of r. For $r \geqslant 28.663$ in fact the stable orbit $\mathscr{C}_0^*$ is present together with one of the sequence $\mathscr{C}_i$. It is clear that $\mathscr{C}_0^*$ appears in the beginning with a very small basin of attraction; with increasing r, this becomes larger and larger, while that of the simultaneous orbit $\mathscr{C}_i$ gets smaller and smaller. $\mathscr{C}_4$ already appears with a very small basin of attraction and $\mathscr{C}_5$ even more so. The simultaneous presence of more than one attracting orbit is termed hysteresis and has the effect of causing a rather sensitive dependence of the asymptotic solution on the initial conditions. In this kind of model this was found by Curry[(12)] in a generalized Lorenz system and by us[(13)] as a very strong feature in a seven-mode truncation model of the two-dimensional Navier–Stokes equations.

3. COLLAPSE OF A STABLE ORBIT WITH AN UNSTABLE ONE

One of the interesting results of Ref. 1 is the observation that, after the second infinite sequence of orbits, the system shows the presence of two symmetric attractors, with all the characteristics of a "strange attractor," on which a random motion takes place. With increasing Reynolds number r, each attractor seems to shrink to a stable periodic orbit, present for all the high-r regime considered. Analysis of the stability of this orbit shows that, with decreasing r toward $R_{13} = 33.43$, an eigenvalue of the Poincaré map approaches the unit circle at the point $+1$. It seems of interest to us to attempt to obtain a better understanding of the way this bifurcation takes place, as it is connected with the transition to turbulence.

We have verified that after the bifurcation, i.e., for $r < R_{13}$, the orbit is no longer present. This fact suggests the hypothesis that the stable orbit could disappear by collapse with an unstable one present at the same time (see, for example, Brunowsky[(14)]). We have looked at how the fixed point of the Poincaré map for the stable orbit, using a fixed hyperplane, would move when r decreases toward R_{13}. Keeping in mind the hypothesis of collapse, by extrapolation we have been able to find the fixed point for an unstable orbit at a value of r close to the critical one. We have then followed the unstable orbit present together with the stable one, quite close to it, and disappearing for $r < R_{13}$. It has been verified that, with decreasing r toward R_{13}, the two orbits become closer and closer (see Fig. 1) and so do their periods, that for the stable orbit increasing, that of the unstable one decreasing. Figure 2, where the fixed points of the Poincaré map for the stable and unstable orbits are represented for different values of r approaching R_{13} from above, shows the phenomenon of collapse quite clearly.

The same detailed analysis performed on the stable orbit present in the high-r regime has been carried on for $\mathscr{C}_0{}^*$, since a study of its stability shows an eigenvalue of the Poincaré map approaching the unit circle at $+1$ for r

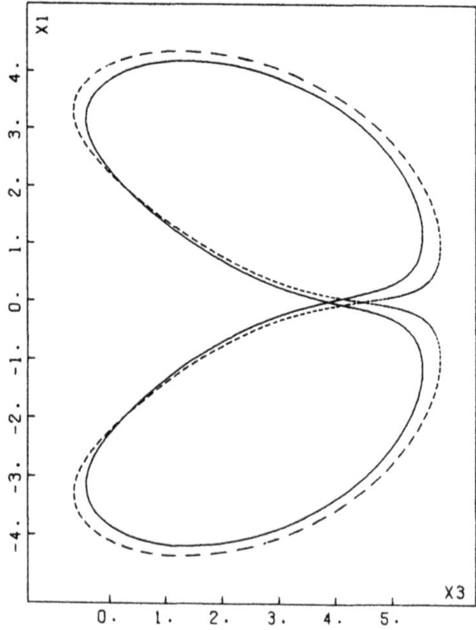

Fig. 1. Stable (——) and hyperbolic (- - -) orbits for $r = 33.60$.

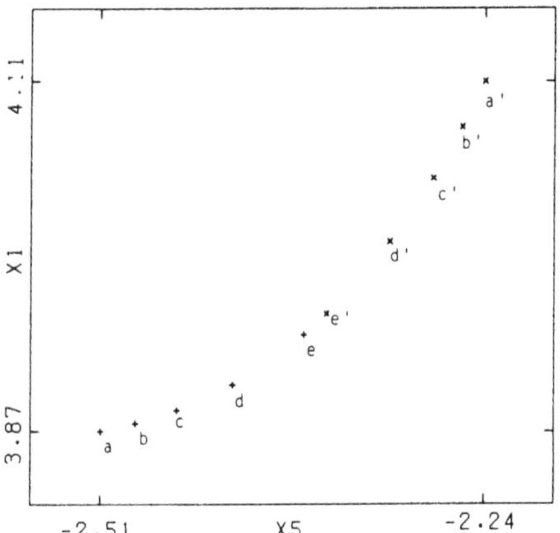

Fig. 2. Fixed points of the Poincaré map for the stable (+) and hyperbolic (×) orbits for r approaching R_{13} from above: (a, a') $r = 33.80$; (b, b') $r = 33.70$; (c, c') $r = 33.60$; (d, d') $r = 33.50$; (e, e') $r = 33.44$.

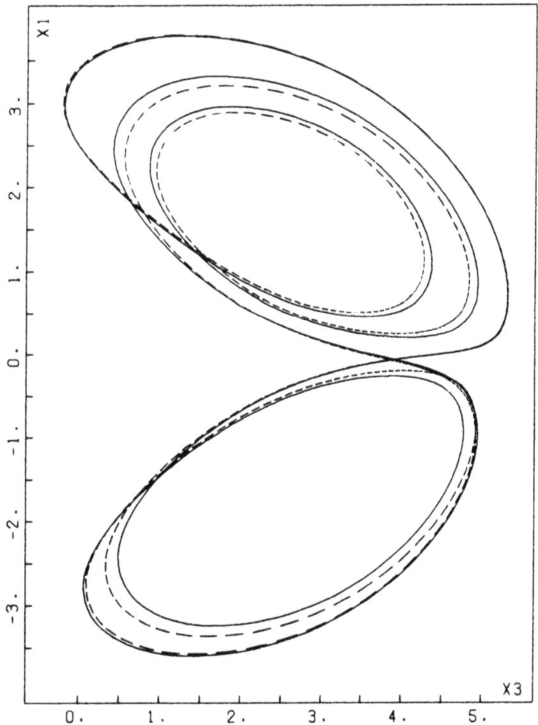

Fig. 3. Stable (——) and hyperbolic (- - -) orbits for $r = 28.695$.

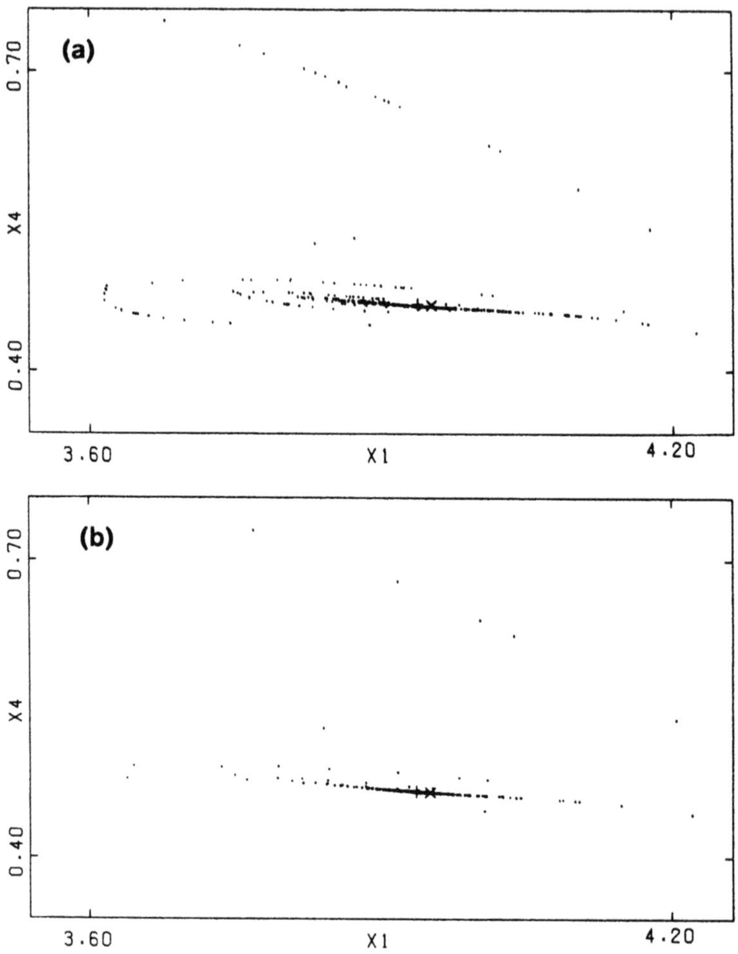

Fig. 4. Projections of the Poincaré map on the plane (x_1, x_4) for $r =$ (a) 33.300; (b) 33.430; (c) 33.4385; (d) 33.440. The symbol $+$ $(\times)$ represents the fixed point of the stable (hyperbolic) orbit for $r = 33.440$.

decreasing toward 28.663, the orbit disappearing below that value. For $\mathscr{C}_0{}^*$ too we have verified the identical phenomenon of collapse with an unstable orbit $\bar{\mathscr{C}}_0{}^*$ (see Fig. 3); the difference from the previous case is that no attractor close to the orbit is present after the bifurcation. The presence of the orbit $\bar{\mathscr{C}}_0{}^*$, besides explaining the bifurcation for $r = 28.663$, plays a role in the explanation of how the model goes over to "turbulent" behavior, as will be seen in Section 5.

In their fundamental paper, among other considerations, Ruelle and

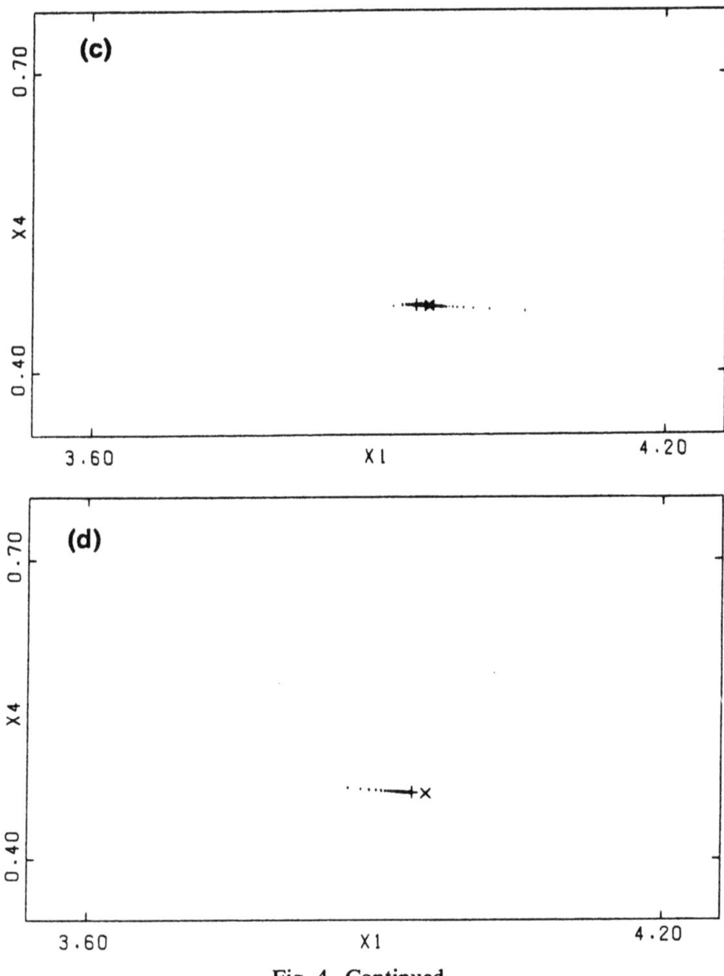

Fig. 4. Continued.

Takens[7] analyze the bifurcations of a stable periodic orbit in generic systems when the Poincaré map has only a finite number of isolated eigenvalues with modulus 1 and the others inside the open unit circle. They find that when only one eigenvalue crosses the unit circle at the point +1, one should expect the attracting closed orbit to disappear together with a hyperbolic closed one and no attractor close to the orbit to appear after the bifurcation has taken place. This is the exact phenomenology found for $\mathscr{C}_0^*$, but at first sight not that for the orbit in the high-r regime. In this last case in fact at the bifurcation point we observe the collapse of the two orbits, but after the bifurcation a strange attractor is present, apparently close to the orbits.

This phenomenology does not seem to be in agreement with Ref. 7, so we have tried to study in more detail the transition from the strange attractor to the periodic orbits.

For different values of r approaching R_{13} from below, we have considered the Poincaré map on the hyperplane $x_3 = 3.0$, $x_1 \geqslant 0$, plotting its projection on (x_1, x_4) together with, as a reference, the fixed points of the stable and hyperbolic orbits for r slightly greater than R_{13} ($r = 33.44$).

Figures 4a–4c clearly show how, as r approaches R_{13}, the points tend to dispose themselves along a line, getting denser and denser on a segment containing the fixed points of the two orbits. Looking at the projection of the intersection points on the plane (x_1, x_4), this segment is described from left to right, with a return mechanism which redescribes it always progressing in the same direction. Moreover, numerical evidence has been found for the two following facts: for $r \to R_{13}$, $33.4385 < R_{13} < 33.4390$, the length of the segment does not seem to tend to zero; at $r = R_{13}$ a stable fixed point for the Poincaré map appears on the segment.

A possible qualitative explanation for the phenomenology is the existence of some attracting variety, containing the "strange attractor" for $r < R_{13}$, and a stable periodic orbit, together with an unstable one, for $r > R_{13}$. The presence of such a manifold, attracting also for $r > R_{13}$, appears to be confirmed in Fig. 4d, where the approach to the fixed point of the stable orbit on the Poincaré map is shown for $r = 33.44$. Looking at the phenomenon for r increasing, we have then that the strange attractor does not "shrink" to the orbit, in agreement with the bifurcation theory.

For r decreasing, the stable orbit and the hyperbolic one collapse and, disappearing, are replaced by a "larger" attractor which occupies a portion of a manifold to which the orbits are always attracted for r near R_{13}, no matter whether larger or smaller; the diameter of the attractor does not tend to zero as $r \to R_{13}$.

4. COMPATIBILITY WITH A CONJECTURE OF UNIVERSALITY IN INFINITE SEQUENCES

In a recent paper, Feigenbaum[9] develops a very interesting theory concerning a large class of recursion relations $x_{n+1} = \lambda f(x_n)$ exhibiting infinite bifurcations, varying the parameter λ in the open interval (0, 1). They are shown to possess a structure essentially independent of the recursion function that, among other properties, is supposed to map the closed interval [0, 1] on itself and have a unique, twice differentiable maximum $\bar{x}$. For such a class of f, a λ_n exists such that a stable 2^n-point limit cycle including $\bar{x}$ exists. It is shown as numerical evidence that

$$\lim_{n\to\infty} \delta_n = \lim_{n\to\infty} \frac{\lambda_n - \lambda_{n-1}}{\lambda_{n+1} - \lambda_n} = \delta$$

i.e., the λ_n geometrically converge to a certain λ_∞ at the rate δ, independent of the specific function f.

The same asymptotic behavior is shown to take place for

$$\frac{\Lambda_n - \Lambda_{n-1}}{\Lambda_{n+1} - \Lambda_n}$$

where Λ_n is the nth bifurcation point. Moreover, when λ is increased in order to obtain the transition from a stable 2^n-point to a stable 2^{n+1}-point limit cycle, the local structure about $\bar{x}$ reproduces itself on a scale α_n times smaller. It is shown that

$$\lim_{n\to\infty} \alpha_n = \alpha$$

with α also f-independent. Both the numbers α and δ depend only on the order of the maximum $\bar{x}$ of f; for a normal (i.e., quadratic) maximum, it is found that $\delta = 4.6692 \cdots$ and $\alpha = 2.5029 \cdots$.

In a previous section we have described two sequences of bifurcations that are very likely to be infinite, even if for the reasons exposed there we could obtain only a limited number of terms. We have tried to verify numerically if the universal metric properties pointed out by Feigenbaum for one-dimensional mappings could hold in our dynamical system too, hoping for a convergence of α_n and δ_n as fast as that of one of the examples in Ref. 9. We have computed the ratios $\delta_i = (\rho_i - \rho_{i-1})/(\rho_{i+1} - \rho_i)$, $i = 1,\ldots, 5$, for the sequence $\mathscr{C}_i$ and $\delta_i^* = (\rho_i^* - \rho_{i-1}^*)/(\rho_{i+1}^* - \rho_i^*)$ for $\mathscr{C}_i^*$, where ρ_i and ρ_i^* are the bifurcation points given in Tables I and II. Both sequences δ_i and δ_i^*, listed in Table III, seem to indicate a convergence, more rapid for δ_i^*, to numbers quite compatible with the one found by Feigenbaum. A comment is required by the last term δ_4^*. This has been computed knowing quite well that it might not have been completely reliable. In fact the numerical errors due to the large value of the period of $\mathscr{C}_4^*$ now become relevant compared with the very small variation of the parameter r. We have computed the term anyway because of the small number of terms available otherwise, to verify at least a persistence of the sequence around the value of δ.

Table III

i	δ_i	δ_i^*
1	24.22	2.57
2	9.48	4.63
3	6.54	4.64
4	5.29	4.42
5	4.73	—

An even more striking instance of the compatibility of our sequences with the one considered by Feigenbaum is found by looking at how the fixed points of the Poincaré map reproduce themselves upon transition from each orbit to the next one in the sequence. Calling Φ_r the Poincaré map on a hyperplane transverse to each orbit $\mathscr{C}_i$, we can write a recursion relation

$$x_{n+1} = \Phi_r(x_n)$$

When, for $\rho_{i-1} < r < \rho_i$, we consider the stable orbit $\mathscr{C}_i$, Φ_r has a stable 2^i-point limit cycle. In this way we have a recursion function similar to the one in Ref. 9 since in the ith bifurcation point ρ_i a 2^i-point limit cycle becomes unstable and a stable 2^{i+1}-point limit cycle appears, with the stable orbit $\mathscr{C}_{i+1}$ appearing.

A comparison with Feigenbaum's scale factors α_n can be attempted once something corresponding to $\bar{x}$ is found. Considering our limit cycles at the bifurcation points, we have observed the following. Calling $P_j^{(i)}$, $i = 0, 1,\dots, 5$, $j = 1,\dots, 2^i$, the points of the 2^i-point cycle, and $P_{2j-1}^{(i+1)}$ and $P_{2j}^{(i+1)}$ the two points bifurcating from $P_j^{(i)}$ in the 2^{i+1}-point cycle, we let

$$Q_1^{(i)} = P_{2k_0-1}^{(i)}, \qquad Q_2^{(i)} = P_{2k_0}^{(i)}$$

where k_0 is the index for which $d(P_{2k-1}^{(i)}, P_{2k}^{(i)})$ is maximum,[8] $k = 1,\dots, 2^{i-1}$. Then, for $i \geqslant 2$, we have that $Q_1^{(i)}$ and $Q_2^{(i)}$ correspond either to $Q_1^{(i-1)}$ or to $Q_2^{(i-1)}$. This is equivalent to saying that if we consider the binary tree with the points $P_j^{(i)}$ as nodes of level $i + 1$, a path from $P_1^{(0)}$ to a certain $P_j^{(5)}$ exists, along which the points $P_j^{(i)}$ reproduce themselves with a scale factor that is maximum.

In Ref. 9 the scale factor, by definition $1/\alpha_i$, by which a cluster about a point of a 2^i-cycle reproduces itself is maximum if the point is $\bar{x}$. For this reason it seems relevant to compute the α_i along the path we have specified before, proposing in this way a correspondence between $Q^{(i)}$ and $\bar{x}$. In Table IV we list the values of the α_i for the Poincaré map on the hyperplane $x_3 = 1.0$ for each coordinate and the Euclidean distance.

The same procedure has been followed for the sequence $\mathscr{C}_i^*$, the corre-

[8] d is the usual Euclidean distance in R^5.

Table IV

	x_1	x_2	x_4	x_5	d
α_1	4.20	4.21	5.05	5.09	4.51
α_2	3.43	3.43	3.29	3.30	3.39
α_3	2.90	2.90	2.94	2.94	2.91
α_4	2.53	2.53	2.52	2.52	2.53

Table V

	x_1	x_2	x_4	x_5	d
$\alpha_1{}^*$	2.31	2.31	2.38	2.38	2.34
$\alpha_2{}^*$	2.65	2.65	2.60	2.60	2.63
$\alpha_3{}^*$	2.43	2.43	2.45	2.45	2.44

sponding limit cycle being $m \cdot 2^i$-point, where m is dependent on the plane chosen for the Poincaré map. The scale factors $\alpha_i{}^*$ for this case are given in Table V, also for the Poincaré map on $x_3 = 1.0$, limited to $x_1 > 0$, with $m = 3$.

For different choices of the hyperplane for the Poincaré map the results are essentially unchanged.

The compatibility of our numerical values for α_i and $\alpha_i{}^*$ with the asymptotic value of α computed by Feigenbaum seems evident, even if we could compute only a few terms.

Our results make possible the hypothesis that universal metric properties of one-dimensional mappings also hold in dynamical systems with infinite sequences of bifurcations. The fact that complicated n-dimensional phenomena possess characteristics in some sense "one-dimensional" appears significant and very suggestive.[9]

These arguments give more support to our hypothesis of Section 2 on the two sequences being infinite and allow us now to estimate the asymptotic values for the critical Reynolds number ρ_i and $\rho_i{}^*$. We obtain for them

$$\rho_\infty = 28.668652, \qquad \rho_\infty{}^* = 28.720135$$

A complete numerical definition of these values from the model appears impossible, however.

5. ONSET OF TURBULENCE

In Ref. 1 it is shown that for $R_{12} < r < R_{13}$, randomly chosen initial data lead to two attractors (see Fig. 5), on which the motion appears to be completely random, the trajectories looking exactly like the ones found by Lorenz in his model.[(2)] These two "strange attractors" are localized in two

[9] A detailed study generalizing Feigenbaum's results has been carried out by Derrida *et al.*[(10)] Moreover, they have pointed out that also in the Hénon two-dimensional mapping[(5)] the bifurcation rate δ for the sequence of stable periods 2^n is the same as in Ref. 9. An intuitive explanation for this is indicated by the contracting nature of the transformation.

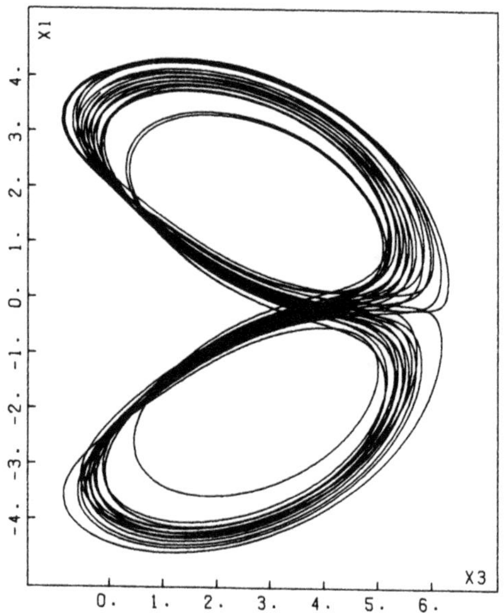

Fig. 5. One of the two "strange attractors" for $r = 33.0$.

symmetric regions, each of them surrounding two fixed points, two sequences of orbits $\mathscr{C}_i$, two sequences $\mathscr{C}_i{}^*$,[10] and two orbits $\bar{\mathscr{C}}_0{}^*$, all unstable in this range of the Reynolds number r.

In the following we give a detailed analysis of the transition from the periodic behavior of the sequences $\mathscr{C}_i{}^*$ to the turbulent behavior on the two attractors.

We consider the motion with random initial data for different values of r, starting from $r = 28.72$, studying the Poincaré map on the hyperplane $x_3 = 1.4$, limited to the region $x_1 \geqslant 0$, $x_4 \geqslant 0$ for simplicity.[11] The orbits of only one of the sequences $\mathscr{C}_i$, the orbits of two of the sequences $\mathscr{C}_i{}^*$, and two orbits $\bar{\mathscr{C}}_0{}^*$, all intersecting the hyperplane $x_3 = 1.4$, are present in the region considered. These intersections are the elements of n-point limit cycles for the Poincaré map. Denoting by c_i a 2^i-point limit cycle related to an orbit $\mathscr{C}_i$, and by $c^*_{m,i}$ an $m \cdot 2^i$-point cycle related to an orbit $\mathscr{C}_i{}^*$, we find the Poincaré map then has one sequence of cycles c_i, one $c^*_{2,i}$, one $c^*_{3,i}$, one cycle $\bar{c}^*_{2,0}$, and one $\bar{c}^*_{3,0}$. The complexity of the situation is evident from Fig. 6a, where for $r = 28.72$ we have represented only the cycle $c^*_{3,4}$, stable for this value of r, and the unstable cycles c_0, c_1, $c^*_{2,0}$, $c^*_{3,0}$, $\bar{c}^*_{2,0}$, $\bar{c}^*_{3,0}$. The

[10] For a better understanding see Ref. 1, especially Figs. 1a, 7a–7d, and 9–11.

[11] Suitable changes in sign, allowed for by the symmetries present in the model, make it possible to study any orbit in this region.

points of the stable cycle $c_{3,4}^*$ accumulate in three groups in a neighborhood of the three points of the cycle $c_{3,0}^*$ from which they have bifurcated. Because of the scale factor chosen in order to represent a complete picture of the Poincaré map, the points in each group appear to be practically indistinguishable, but show in the plane (x_1, x_4) a line along which they duplicate at each bifurcation. This fact is clearly evident in Fig. 6a′, where on an enlarged scale the central group of the points of the stable cycle $c_{3,4}^*$ is represented together with the points of the cycles $c_{3,0}^*$ and $c_{3,1}^*$, now unstable, from which they have bifurcated: they also appear on the line of the points of $c_{3,4}^*$. Even if we give no evidence for this, because of the high computational time required, it is reasonable to think that the points of $c_{3,2}^*$ and $c_{3,3}^*$ also stay on the same line.

A natural extension of this argument is the hypothesis that also for $r > \rho_\infty{}^*$ all the points of the full sequence $c_{3,i}^*$ are disposed in an analogous way along the same line. The same study carried on for the sequence c_i shows an analogous phenomenology.

Let us examine now the behavior of the flow for r slightly larger than 28.72. Figures 6*b*–6d show the projections on the plane (x_1, x_4) of the hyperplane chosen for the Poincaré map for $r = 28.721$, $r = 28.723$, and $r = 28.730$, respectively. In all three figures we see the results of 400 intersections of the solution curve with our codimension-one section. It is observed that the behavior of the flow is still very much analogous to that observed for $r = 28.72$ when the stable orbit $\mathscr{C}_4{}^*$ is present. The numerical data obtained do not allow us at all to state whether the observed motion is periodic, possibly with a very long period, or not. The figures show, however, that with increasing r the intersection points keep disposing themselves only on arcs along the direction identified by the points of the cycle $c_{3,4}^*$, now unstable, but become more and more spread, and they tend to approach the points of the cycle $\bar{c}_{3,0}^*$ (see Figs. 6a–6d). An examination of the values of the coordinates of all these points seems to show more and more randomness with increasing r, confirmed by Fig. 6e, where we see that the behavior of the flow for $r = 28.732$ is definitely changed. In fact it is possible to observe that now also the points of the cycles $c_{2,i}^*$ due to the second sequence $\mathscr{C}_i{}^*$ and of the cycle $\bar{c}_{2,0}^*$ due to the second orbit $\overline{\mathscr{C}}_0{}^*$ contribute to the behavior of the solution curve. Moreover, some intersection points now seem to be arranged rather randomly and not to be connected with any one of the unstable cycles of the Poincaré map. With continued increasing r, we observe a more and more chaotic behavior, due to a gradual involvement of the orbits $\mathscr{C}_i$ also, since the intersection points are now also close to the fixed points of the orbits $\mathscr{C}_0$ and $\mathscr{C}_1$. Figure 6f shows the motion for $r = 28.80$, appearing fully random around the points of the unstable n-cycles present in the region.

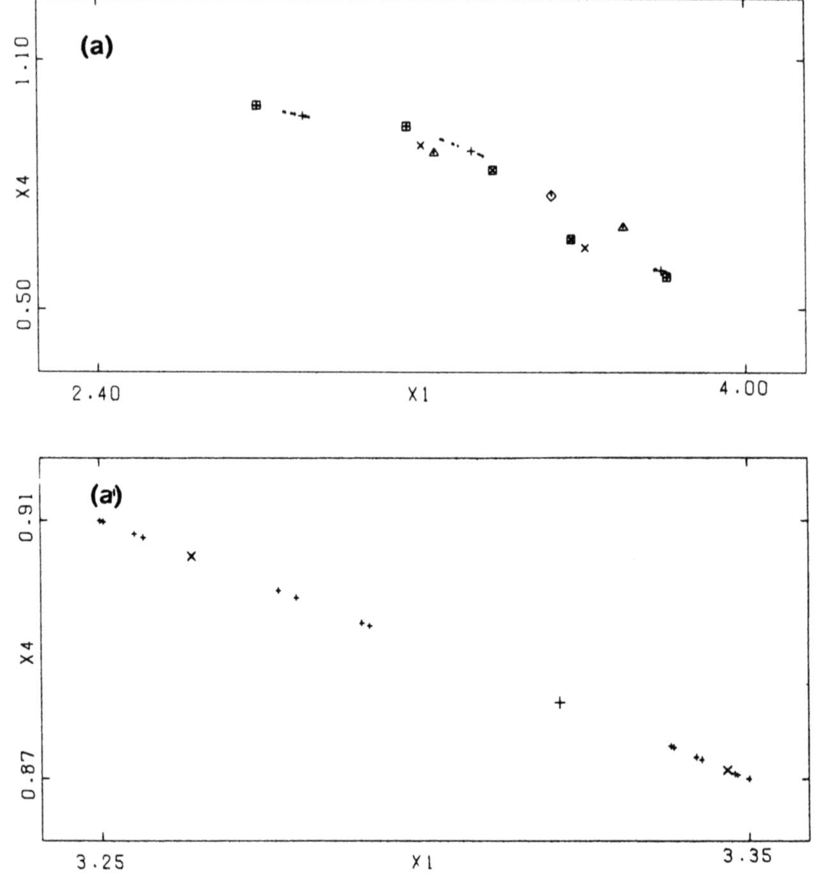

Fig. 6. Projections of the Poincaré map on the plane (x_1, x_4) for r = (a) 28.720; (b) 28.721; (c) 28.723; (d) 28.730; (e) 28.732; (f) 28.800. (+) the three points of the cycle $c^*_{3,0}$; (×) the two points of the cycle $c^*_{2,0}$; (⊞) the three points of the cycle $\bar{c}^*_{3,0}$; (⊠) the two points of the cycle $\bar{c}^*_{2,0}$; (◇) the fixed point of c_0; (△) the two points of the cycle c_1. (a′) Points of the central group [part (a)] of the stable cycle $c^*_{3,4}$ (+) on an enlarged scale, with the points of $c^*_{3,0}$ (+) and $c^*_{3,1}$ (×) from which they have bifurcated.

The phenomenology described above does not allow a rigorous definition of the mechanism of the onset of turbulence. In fact we are unable to evaluate $\rho_\infty{}^*$ exactly, i.e., the critical value of the Reynolds number r for which the sequence $\mathscr{C}_i{}^*$ exhausts itself, and we cannot state definitely what happens for r slightly greater than $\rho_\infty{}^*$. The existence of more infinite sequences of stable periodic orbits in very small ranges of r or with very long periods then cannot be rejected.

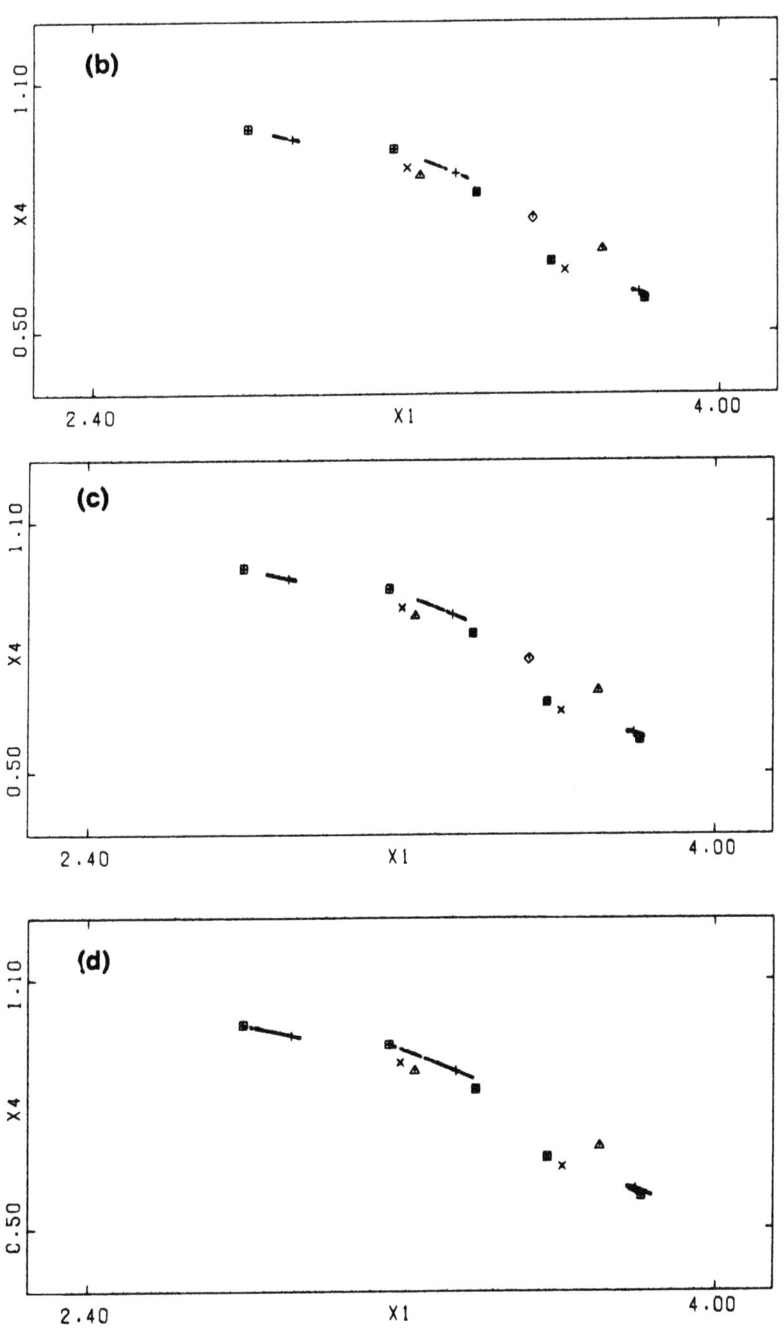

Fig. 6. Continued.

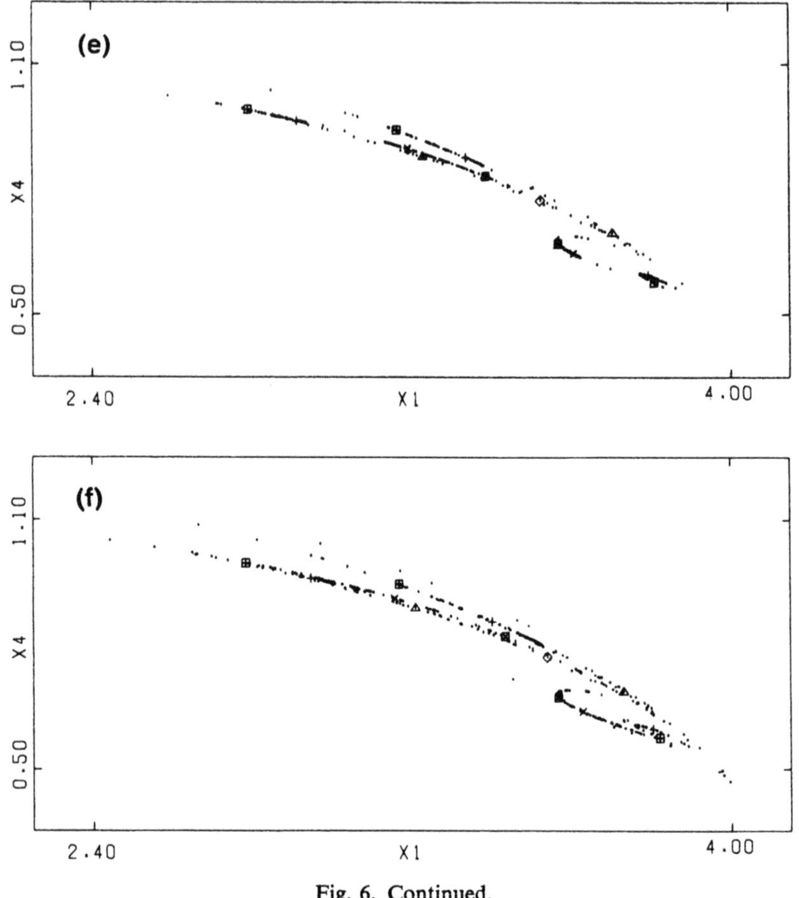

Fig. 6. Continued.

We think, however, that we can interpret the numerical results in the following way. The characteristics of the sequence $\mathscr{C}_i^*$ indicate that it tends rapidly to exhaust itself when r approaches a value very likely to be quite close to ρ_∞^*, computed in the previous section according to Feigenbaum's theory. For $r > \rho_\infty^*$, when all the orbits have become unstable, the system possesses only unstable fixed points and periodic orbits: the seven fixed points (see Ref. 1), the four sequences of orbits $\mathscr{C}_i$, the four sequences $\mathscr{C}_i^*$, and the four orbits $\overline{\mathscr{C}}_0^*$. Up to $r \cong 28.73$ any random initial value is attracted by the orbits of one of the four sequences $\mathscr{C}_i^*$, which are definitely unstable, but being less unstable than the others, succeed in "catching" the point and keep it trapped in their neighborhood, at least for the long time intervals observed. With increasing r, the orbits $\mathscr{C}_i^*$ become more unstable and gradually lose the ability to keep the point trapped, allowing it also to approach the other unstable periodic orbits. For r further increased all the orbits have lost more and more stability and the point "jumps" more easily from the neighborhood of an orbit to that of another. The motion then becomes more and more chaotic, even remaining confined in one of the two distinct symmetric regions where the unstable orbits are localized.

The two strange attractors in our system then appear as a consequence of the instability of all the orbits present, i.e., with a mechanism perfectly analogous to the one in the "standard" Lorenz attractor,[3] even if much more complicated.

6. CONCLUSION

Because of the complicated phenomenology present in the considered model of the five-mode truncated Navier–Stokes equations, it seems useful to present a schematic picture of the features found. Redefining the sequence of the critical values of r with $R_1' = R_3$, $R_2' = 28.663$, $R_3' = \rho_\infty$, $R_4' = \rho_\infty^*$, and $R_5' = R_{13}$, we have:

(a) For $0 < r \leqslant R_1'$ the model exhibits only stationary solutions (see Ref. 1 for details).

(b) For $R_1' < r < R_3'$ the system, through an infinite sequence of bifurcations, gives rise to four infinite sequences of symmetric orbits $\mathscr{C}_i$, each one with a period double that of the previous one.

(c) For $R_2' \leqslant r < R_4'$ a further sequence of infinite bifurcations gives rise to four more infinite sequences of orbits $\mathscr{C}_i^*$, also symmetrically placed and with doubled period, but with a more complicated spatial structure.

(d) For $R_4' \leqslant r < R_5'$ all the periodic orbits present in the system are unstable and an erratic, chaotic motion takes place on two symmetric "strange" attractors, analogous to the Lorenz model ("turbulence").

(e) For $r \geqslant R_5'$ two stable periodic orbits are present.

At this point a detailed knowledge of the phenomenology of the model seems to have been reached. In particular we remark the fact that the turbulent behavior is reached gradually when no stable fixed point or periodic orbit is present and all the unstable ones keep losing stability with increasing r. Also a relevant feature is the fact that the two infinite sequences of bifurcations present seem to possess certain characteristics or universality analogous to the ones found by Feigenbaum in nonlinear transformations of an interval in itself.

We conclude by proposing two basic questions: How does a different choice of the five modes for the truncated Navier–Stokes equations effect the behavior of the model, and how does an increase in the number of modes in the truncation affect the model? Concerning the last question, an ongoing study of a seven-mode truncation obtained by adding two more modes to the five used in this study seems to show a rather different phenomenology, with much more variety and strong features of hysteresis.

ACKNOWLEDGMENTS

We are deeply indebted to G. Gallavotti for his interest in this work and his continuous help and to P. Collet and J. P. Eckmann for informing us of the Feigenbaum conjecture and for suggesting its test. We are also grateful to V. Grecchi and F. Marchetti for many useful conversations.

REFERENCES

1. C. Boldrighini and V. Franceschini, *Comm. Math. Phys.* **64**:159 (1979).
2. E. N. Lorenz, *J. Atmos. Sci.* **20**:130 (1963).
3. J. Marsden, in *Lecture Notes in Mathematics*, No. 615 (1976).
4. D. Ruelle, in *Lecture Notes in Mathematics*, No. 565 (1976).
5. M. Hénon, *Comm. Math. Phys.* **50**:69 (1976).
6. M. Hénon and Y. Pomeau, in *Lecture Notes in Mathematics*, No. 565 (1976).
7. D. Ruelle and F. Takens, *Comm. Math. Phys.* **20**:167 (1977).
8. O. E. Lanford, in *Proceedings of Corso CIME held in Bressanone* (June 1976).
9. M. J. Feigenbaum, Quantitative Universality for a Class of Nonlinear Transformations, Preprint, Los Alamos (1977).
10. B. Derrida, A. Gervois, and Y. Pomeau, Universal Metric Properties of Bifurcations of Endomorphisms, Preprint.
11. J. E. Marsden and M. McCracken, *The Hopf Bifurcation and its Applications* (Applied Mathematical Sciences, No. 19; 1976).
12. J. H. Curry, *Comm. Math. Phys.* **60**:193 (1978).
13. V. Franceschini and C. Tebaldi, in preparation.
14. P. Brunowsky, in *Lecture Notes in Mathematics*, No. 206 (1971).

Editor's note. There is a very large literature on the numerical simulations of dynamical systems and observations of period doubling—it is of varying quality and none of it gives instructions as to how the calculations are actually done. Some of the articles worth looking at are McLaughlin (1981), Ueda (1979), Rossler (1976), Tomita and Kai (1978), Tomita (1982), Kai (1982), Tedeschini-Lalli (1982), etc.

Physics Letters **76A** 1–4 (1980)

POWER SPECTRAL ANALYSIS OF A DYNAMICAL SYSTEM ☆

J. CRUTCHFIELD, D. FARMER, N. PACKARD and R. SHAW
Department of Physics, University of California, Santa Cruz, CA 95064, USA

and

G. JONES and R.J. DONNELLY
Department of Physics, University of Oregon, Eugene, OR 97403, USA

Received 27 November 1979
Revised manuscript received 7 January 1980

Power spectra for chaotic transitions in three dimensions are presented for a dynamical system first proposed by Rössler. Relations between the spectra and the topology of the corresponding strange attractor are discussed.

Modern experiments in Couette and Bénard flows often use power spectral analysis as a measure of the temporal behavior of fluid motions, and the transition to turbulence [1]. At the same time there is interest in the study of simple dynamical systems which may exhibit chaotic behavior owing to the existence of a "strange attractor" [2]. The most familiar example is the highly idealized model for Bénard convection of Lorenz [3]. It is natural to study the power spectra of simple dynamical systems, in the hope of gaining insight for the interpretation of the spectra of real fluids.

There has been some work on the power spectra of strange attractors, by Ueda [4], Holmes [5] and others, but there remains a need for systematic studies of the changes in power spectra of dynamical systems as they bifurcate from one attractor topology to another. We present here results for a particular bifurcation sequence, one which occurs widely in the transition to chaotic behavior of vector fields in three dimensions.

The construction of digital computer solutions to dynamical systems followed by power spectral analysis is time-consuming. We have developed a hybrid computer system by solving the equations on an analog computer and performing power spectral analysis on a digital computer. The analog computer is only accurate to a percent or so, but high accuracy is not needed for these largely qualitative studies.

Fig. 1 displays the transition in a system originally studied by Rössler [6]:

$$\dot{x} = -(y + z), \quad \dot{y} = x + 0.2y, \qquad \dot{z} = 0.2 + xz - Cz, \tag{1}$$

as the parameter C is varied. Projections onto the x–y plane of these equations after transients have been allowed to die out are shown. We chose this particular system for study because the "branched manifold" [7,8], enclosing its attractor has the simplest topology which will still produce a strange attractor.

To obtain these power spectra we solved the equations on an analog computer, with the natural time unit taken to be 0.01 s, sampled the solutions until 4096 points were accumulated, and used an FFT to compute the discrete spectra. For each parameter value this process was repeated 10 times and the resulting spectra averaged. Some of the results for time series obtained from $z(t)$ are shown in fig. 1. The time series derived from $x(t)$ and $y(t)$ gave similar results.

The sequence of bifurcations is schematically represented in fig. 2. The bifurcation sequence for $C < C_\infty \approx 4.20$, consisting of the successive appearance of subharmonics, is familiar from the work of May on one-dimensional maps [9], was mentioned by Brunovsky [10], was observed by Feit in two-dimensional maps [11] and has been reported in a driven oscillator system by Coullet et al. [12]. At each step the limit cycle "unwinds", roughly doubling the period of a complete orbit. This bifurcation was known to Poincaré [13].

The doubling process reaches an accumulation point at C_∞, however, and is succeeded by a qualitatively different behavior. The largest non-zero Liapunov characteristic exponent $\bar{\lambda}$ (see fig. 3) becomes positive, reflecting the exponential divergence of trajectories [14–16]. Families of orbits remain confined to thin bands, which rejoin in a pairwise manner, as illustrated. A complete bifurcation sequence has been described by Shimada and Nagashima [17] and is visible in a figure in a paper by Li and Yorke [18]. This sequence occurs quite generally in maps or flows con-

☆ This research was supported at U of O by an NSF Grant, ENG 78-18405.

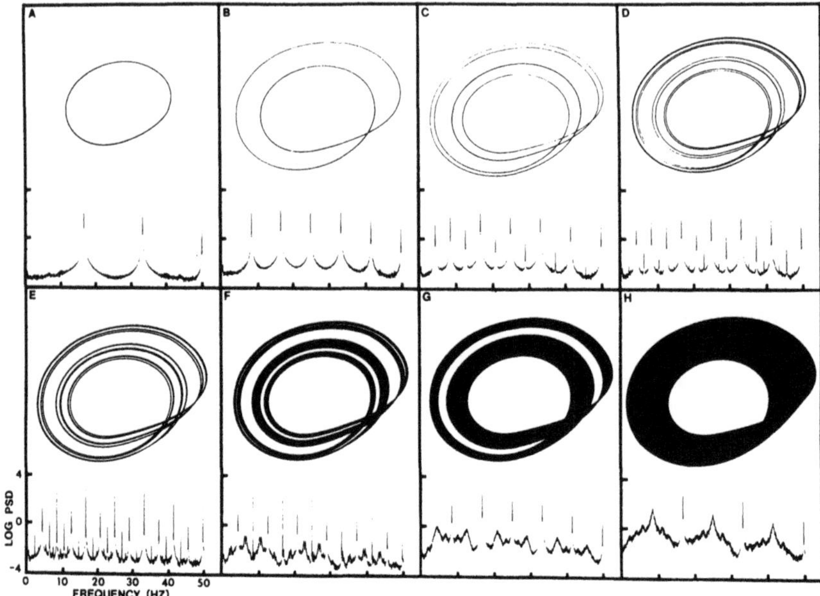

Fig. 1. Plots of the solutions of eq. (1) on the $x-y$ plane after transients have died out. Directly below each phase trajectory is the corresponding power spectral density (PSD) as a function of frequency. Details of these plots are contained in table 1.

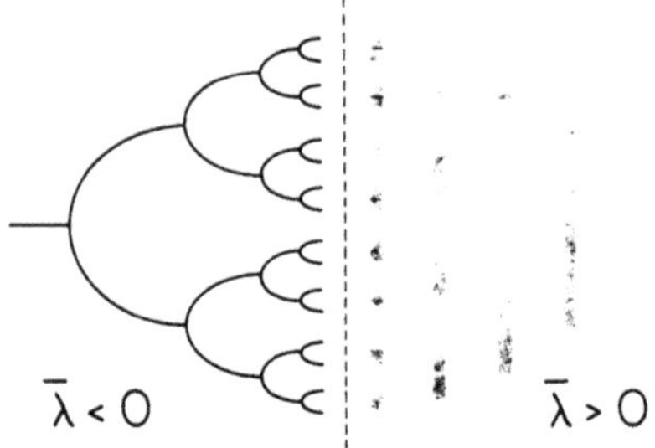

Fig. 2. The bifurcation sequence obtained from the data of fig. 1. $\bar{\lambda}$ increases from left to right. The shaded regions denote strange bands.

taining simple folds. It has been observed in the Rössler system, the Lorenz system for high values of the R parameter [17], and we have observed it in the driven Van der Pol equations, the driven Duffing equations [19], and several other sets of equations, and one-dimensional maps. Periodic orbits, or other behavior, may be interspersed in the sequence, however.

Fig. 1 displays the corresponding power spectra. Each subharmonic bifurcation doubles the number of sharp frequency components, and each pairwise rejoining broadens every other sharp spike. The fundamental, at about 16 Hz, moves little during the series. Typical parameters are given in table 1.

A remarkable feature of this transition is the presence of sharp frequency components in a chaotic attractor. Even at the final stage, when the trajectories fill out the complete Rössler attractor, the spectrum retains a peak which appears to be instrumentally sharp. We believe this to be a feature of some attractors whose branched manifolds are simply connected, in the sense that all trajectories are constrained to revolve about a single hole [20]. Attractors containing fixed points, such as the familiar Lorenz, do not have this property, and do not have sharp spikes in their power spectra.

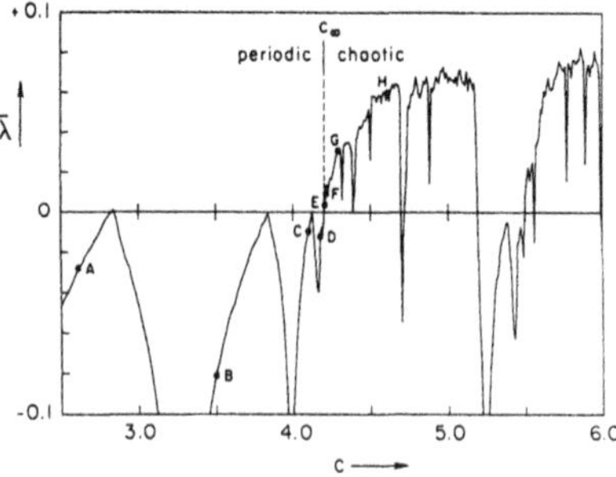

Fig. 3. Plot of the largest non-zero characteristic exponent $\bar{\lambda}$ as a function of C. Letters refer to the plots of fig. 1. A line has been drawn through around 300 calculated points.

Table 1
Data for the bifurcation sequence shown in figs. 1 and 2. f and T refer to the natural frequency (~16 Hz) and period of the system. Subharmonic bifurcations occur where the curve of fig. 3 has a point of tangency with the line $\bar{\lambda} = 0$.

Fig.	C	Phase trajectory	Spectral lines	$10^3 N$
1A	2.6	period 1T limit cycle	f, and harmonics	1.17
1B	3.5	period 2T limit cycle	$\frac{1}{2}f$, f and harmonics	2.40
1C	4.1	period 4T limit cycle	$\frac{1}{4}f$, $\frac{1}{2}f$, f and harmonics	2.19
1D	4.18	period 8T limit cycle	$\frac{1}{8}f$, $\frac{1}{4}f$, $\frac{1}{2}f$, f and harmonics	1.99
a)				
1E	4.21	broadening of band of period ~ 8T	$\frac{1}{8}f$, $\frac{1}{4}f$, $\frac{1}{2}f$, f and harmonics (weak $\frac{1}{16}f$ present)	1.86
1F	4.23	broadening of band of period ~ 4T	$\frac{1}{4}f$, $\frac{1}{2}f$, f and harmonics	1.84
1G	4.30	broadening of band of period ~ 2T	$\frac{1}{2}f$, f and harmonics	2.73
1H	4.60	broadening of band of period ~ T	f and harmonics	5.97

a) The accumulation point C_∞ occurs near here. We are able to resolve a period 16T limit cycle and band of period ~ 16T, not illustrated.

It would be convenient to have a simple measure over the power spectrum of "chaos". Several measures are capable of distinguishing sharp spikes from broad features in power spectra, but it is unclear to the writers which, if any, might have a close relation to the topology or the characteristic exponents of the underlying attractor.

One such quantity is tabulated in table 1. The number of "degrees of freedom" of a discrete spectrum of n frequencies is given by:

$$N = \Big(\sum_{i=1}^{n} P_i\Big)^2 \Big/ n \sum_{i=1}^{n} P_i^2 , \qquad (2)$$

where P_i is the power at the ith frequency. For a sine wave, $N = 1/n$, for white noise, $N = 1$. One can see from table 1 that changes in N reflect bifurcations in the system, but are decisive only when the single strange band appears as in fig. 1H. Another measure can be obtained by normalizing the spectral power density and summing $P_i \log P_i$. This measure tends toward $\log n$ for a sine wave, and is equal to zero for white noise. Still another measure of the $P \log P$ type, due to Aikake [21], has been used by Yahata [22] to characterize theoretically derived power spectra relating to Couette flow. We have used the measures N and $P_i \log P_i$ in some unpublished experimental research on transitions in Couette flow. We are not convinced that they are experimentally useful measures of the power spectra, and expect further work is required on the characterization of power spectra.

The transition from a simple to a chaotic attractor occurs, in three dimensions, in one of a few distinct ways. Each of these has a characteristic signature in the power spectra domain. We have presented one in this letter, the others, as well as some higher dimensional results, will be detailed in a forthcoming paper.

We would like to acknowledge, with thanks, discussions with Professors Ralph Abraham, W.L. Burke and Michael Tabor, and the assistance of Harold Froehling.

References

[1] J.P. Gollub and H.L. Swinney, Phys. Rev. Lett. 35 (1975) 927;
G. Ahlers and R.P. Behringer, Phys. Rev. Lett. 40 (1978) 712;
R.W. Walden and R.J. Donnelly, Phys. Rev. Lett. 42 (1979) 301.
[2] D. Ruelle and F. Takens, Commun. Math. Phys. 20 (1971) 167.
[3] E.N. Lorenz, J. Atmos. Sci. 20 (1963) 130.
[4] Y. Ueda, J. Stat. Phys. 20 (1979) 181.
[5] P. Holmes, Appl. Math. Modeling 1 (1977) 362.
[6] O.E. Rössler, Phys. Lett. 57A (1976) 397.
[7] R.F. Williams, Berkeley Turbulence Seminar 1976–77, eds. P. Bernard and T. Ratiu, Lecture Notes in Mathematics, Vol. 615 (Springer, 1977).
[8] R. Shaw, UCSC preprint (1978).
[9] R.M. May, Nature 261 (1976) 459.
[10] P. Brunovsky, Symp. on Diff. eqs. and dynamical systems (Warwick, 1968).
[11] S. Feit, Commun. Math. Phys. 61 (1978) 249.
[12] P. Coullet, C. Tresser and A. Arneodo, Phys. Lett. 72A (1979) 268.
[13] R. Abraham and J.E. Marsden, Foundations of mechanics (Benjamin/Cummings, 1978).
[14] G. Benettin, L. Galgani, A. Giorgilli and J.M. Strelcyn, C.R. Acad. Sci. Paris 286 (1979).
[15] I. Shimada and T. Nagashima, Prog. Theor. Phys. 61 (1979) 1605.
[16] J. Crutchfield, UCSC Senior Thesis (1979).
[17] I. Shimada and T. Nagashima, Prog. Theor. Phys. 59 (1978) 1033.
[18] T. Li and J. Yorke, in: Dynamical systems, an Intern. Symp. (Academic Press, 1978).
[19] B. Huberman and J. Crutchfield, to be published (1979).
[20] D. Farmer and R. Shaw, to be published (1979).
[21] H. Aikake, in: System identification: advances in case studies, eds. R.K. Mehra and D.G. Lainiotis (1979) p. 27.
[22] H. Yahata, Prog. Theor. Phys. 57 (1977) 1490.

Part 7

Beyond the One-dimensional Theory

Physica **5D** 405–11 (1982)

SCALING BEHAVIOR IN A MAP OF A CIRCLE ONTO ITSELF: EMPIRICAL RESULTS*

Scott J. SHENKER†
The James Franck Institute and The University of Chicago, 5640 South Ellis Avenue, Chicago, Illinois 60637, USA

Received 22 December 1981
Revised 28 January 1982

Motion on a torus is studied via a return map of a circle onto itself. The map $T(\theta) = \theta + \Omega - (K/2\pi)\sin 2\pi\theta$ exhibits unusual scaling behavior for $K = 1$. Some evidence is presented which suggests this scaling behavior is universal.

1. Introduction

One common scenario for the appearance of turbulent motion in dissipative dynamical systems involves a transition from quasiperiodic motion lying on an attracting invariant torus in phase space to chaotic motion on a strange attractor [1]. To elucidate the nature of this transition, one can study the dynamics on the torus by means of a return map of the circle onto itself. Parameterizing the circle by θ and identifying the points θ and $\theta + 1$, the return map T must be invertible and satisfy

$$T(\theta + 1) = T(\theta) + 1. \qquad (1)$$

The map considered here is

$$T(\theta) = \theta + \Omega - \frac{K}{2\pi}\sin 2\pi\theta, \qquad (2)$$

with $K > 0$ [2]. (The map with negative K is equivalent to the map with positive K under the substitution $\theta \to \theta + \frac{1}{2}$). For $0 \leq K < 1$, T and its inverse are analytic. At $K = 1$, the inverse still exists but is not differentiable at $\theta = 0$ where it has a cube root singularity. T^{-1} does not exist for $K > 1$.

When T is invertible, the winding number W

$$W(K, \Omega) = \lim_{n\to\infty} \frac{1}{n}(T^n(\theta) - \theta) \qquad (3)$$

is well defined and independent of θ. The winding number is useful for classifying the nature of trajectories on the torus, as it represents the ratio of frequencies in the quasiperiodic regime. It is the purpose of this paper to investigate the nature of trajectories having irrational winding numbers. It is found that these trajectories exhibit scaling behavior in the limit $K \to 1^-$. Evidence is presented which suggests that this scaling behavior is universal.

This work is essentially a one-dimensional version of the calculation L.P. Kadanoff and I did for KAM surfaces [3]. As in the earlier paper, I have relied heavily on the physical insights and numerical methodology of J. Greene [4, 5].

2. Methodology

Diffeomorphisms of the circle have been studied in great depth by Denjoy [6], Arnol'd [7], and Herman [8] among others, and they have provided us with a set of powerful results which can be applied to our particular map for $0 \leq K < 1$. The function $W(K, \Omega)$ is continuous and monotonic in Ω. For a given nonzero K, the function exhibits plateaus at every rational

* Support in part by the Materials Research Laboratory Program of the National Science Foundation at the University of Chicago under Grant 79-24007.
† Robert R. McCormick and National Science Foundation Fellow.

value of W; the set of Ω values that produces a given winding number W is an interval if W is rational and a single point if W is irrational. Furthermore $W(K, \Omega)$ is rational, say $W = P/Q$ for some relatively prime integers P, Q, if and only if there is some θ_0 such that

$$T^Q(\theta_0) = \theta_0 + P. \tag{4}$$

There are no periodic points when W is irrational. Denjoy's theorem [6] states that if $W(K, \Omega)$ is any irrational winding number, T is topologically equivalent to a simple rotation by W; i.e., there exists a continuous invertible function $\theta(t)$ such that

$$\theta^{-1} T \theta = R \tag{5}$$

and

$$\theta(t+1) = \theta(t) + 1, \tag{6}$$

$$\theta(0) = 0, \tag{7}$$

where $R(t) = t + W$. Herman [8] has shown that for almost all irrational winding numbers, the function $\theta(t)$ is analytic.

As in the earlier work [3], attention is focused primarily on one particular winding number

$$\bar{W} = \frac{\sqrt{5} - 1}{2}, \tag{8}$$

the reciprocal of the Golden Mean. This number is convenient because of its extremely simple continued fraction representation

$$\bar{W} = \langle 1111\ldots\rangle = \cfrac{1}{1 + \cfrac{1}{1 + \cfrac{1}{1 + \cdots}}} \tag{9}$$

Direct determination of the value of Ω that produces $\bar{W}$ as a winding number is clearly impossible given the infinite limit in definition (3). To avoid a similar problem in Hamiltonian systems Greene proposed a method which includes studying a sequence of rational winding numbers $W_i = P_i/Q_i$ that converge to $\bar{W}$

$$\lim_{i\to\infty} W_i = \bar{W}. \tag{10}$$

Greene points out that the optimal way to construct the sequence W_i is to consider successive truncations of the continued fraction representation of the irrational winding number. For $\bar{W}$, this produces the sequence

$$\begin{aligned} W_1 &= \langle 1\rangle = 1/1, \\ W_2 &= \langle 11\rangle = 1/2, \\ W_3 &= \langle 111\rangle = 2/3, \\ &\vdots \\ W_i &= \langle 111\ldots 1\rangle = F_i/F_{i+1}, \end{aligned} \tag{11}$$

where F_i is the ith Fibonacci number. We define $-l^2$ as the convergence rate of this sequence

$$\lim_{i\to\infty} \frac{W_{i+1} - W_i}{W_i - W_{i-1}} \equiv -l^2. \tag{12}$$

For the case of our particular winding number, $l = \bar{W}$.

Given a value of K, one can easily define a series $\Omega_i(K)$ of Ω values where

$$T^{Q_i}(0) = P_i, \tag{13}$$

so there is a Q_i cycle passing through $\theta = 0$ with winding number W_i. The crucial assumption is then that the $\Omega_i(K)$ will converge to the single value $\bar{\Omega}(K)$ that produces $\bar{W}$ as winding number

$$\lim_{i\to\infty} \Omega_i(K) = \bar{\Omega}(K), \tag{14}$$

$$W(K, \bar{\Omega}(K)) = \bar{W}. \tag{15}$$

3. Representations of cycles and rotations

As in our earlier work [3], we shall study the function $\theta(t)$ defined for irrational winding numbers in a representation that is equally applicable to rational winding numbers. For irrational winding numbers, we write

$$u(t) = \theta(t) - t \tag{16}$$

and note that, from eq. (6), $u(t)$ is periodic. For rational winding numbers $W_i = P_i/Q_i$ we use a similar representation with a discrete "time" variable

$$t_j = j(P_i/Q_i), \quad j = 0, 1, \ldots, Q_i, \tag{17}$$

and construct a function $\theta^i(t_j)$ defined on this discrete set which satisfies

$$\theta^i(t_{j+1}) = T(\theta^i(t_j)) \tag{18}$$

and

$$\theta^i(0) = 0, \tag{19}$$

$$\theta^i(t_j + 1) = \theta^i(t_j) + 1. \tag{20}$$

From this, we can define the periodic quantity $u^i(t_j)$,

$$u^i(t_j) \equiv \theta^i(t_j) - t_j. \tag{21}$$

The Fourier transform $A^i(\omega)$ of $u^i(t)$ is defined in the standard form

$$A^i(\omega) = \frac{1}{Q_i} \sum_{j=0}^{Q_i-1} u^i(t_j)\, e^{i2\pi\omega t_j}, \tag{22}$$

from $\omega = 0, \ldots, Q_i$. The assumption here is that any properties of $\theta^i(t)$, $u^i(t)$, and $A^i(\omega)$ that persist in the limit of large i reflect the properties of the continuous functions $\theta(t)$ and $u(t)$, along with the Fourier transform $A(\omega)$.

Another quantity of interest is the residue. Let $D_i(K)$ denote the stability parameter of the Q_i cycle;

$$D_i(K) = \frac{d}{d\theta} T^{Q_i}(\theta)\bigg|_{\theta=0}, \tag{23}$$

for $\Omega = \Omega_i(K)$. The residue R is then given by

$$R_i(K) = \tfrac{1}{4}(1 - D_i(K)). \tag{24}$$

The cycle is stable when $0 < R < 1$ and unstable for $R > 1$ and $R < 0$. When $R = 0$, the cycle has stability parameter 1 and is on the verge of a tangent bifurcation. Period doubling bifurcations occur when $R = 1$.

4. Results

The results fall into three categories; (i) global properties of the map as described by $u(t)$, (ii) convergence and scaling properties in the limit $i \to \infty$, and (iii) evidence that the scaling behavior is universal.

4.1. *Global results*

Herman's work [8] shows that for $0 \leq K < 1$, the function $u(t)$ is analytic. Figs. 1a and 2a display $U^{17}(t)$ and $\omega|A^{17}(\omega)|$ for $K = 0.5$, revealing a smooth curve and an exponential decay in the Fourier transform. The quantity $\omega|A(\omega)|$ has been shown so as to exhibit the high ω end of the spectrum more clearly. For $K = 1$, the function $u(t)$ exhibits extremely fine scale structure (fig. 1b). The peaks in Fourier transform now fall off algebraically, as $1/\omega$ (fig. 2b). The peaks occur at Fibonacci numbers, reflecting the fact that the motion is almost periodic after F_i iterations; subpeaks occur at sums and differences of non-adjacent Fibonacci numbers. The function $u(t)$ appears to have a self-similarity property in that the major structure in $\omega|A(\omega)|$ between any two adjacent peaks is essentially the same. This can be seen most clearly in fig. 3,

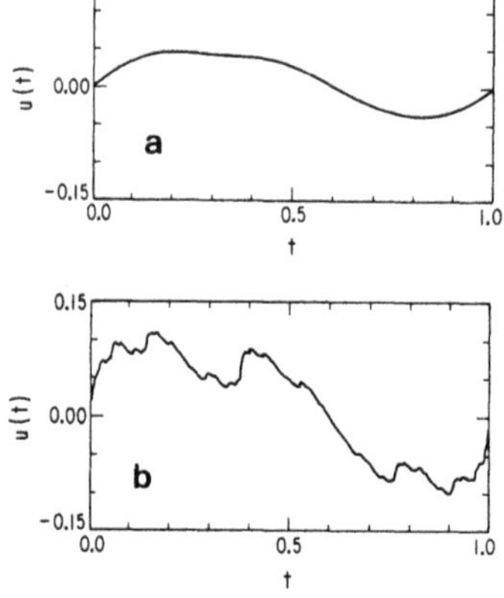

Fig. 1. The function $U^{17}(t)$ for (a) $K = 0.5$ and (b) $K = 1.0$ ($W_{17} = 1597/2584$).

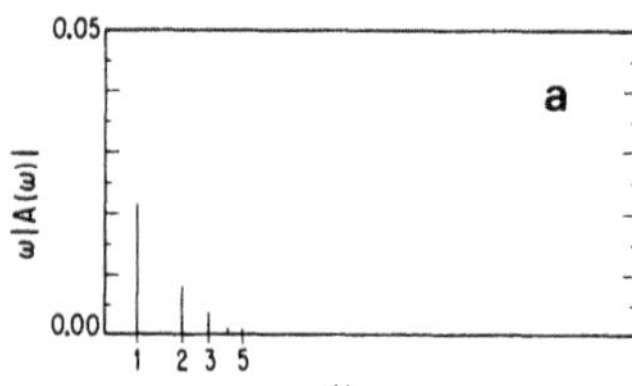

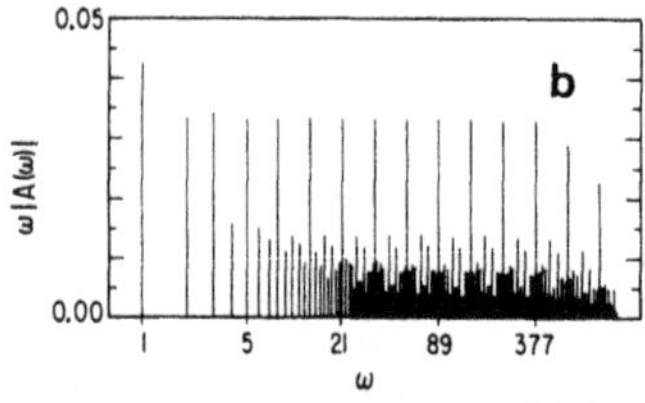

Fig. 2. The Fourier coefficients $A^{17}(\omega)$ for (a) $K = 0.5$ and (b) $K = 1.0$. The coefficients are shown in the form $\omega|A(\omega)|$ vs. ω with the abscissa having a log scale. Since $A^i(Q_i - \omega) = (A^i(\omega))^*$, only the first half of the spectrum is displayed.

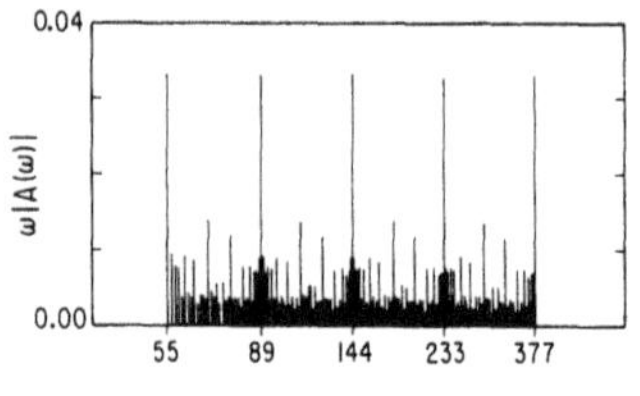

Fig. 3. The Fourier coefficients $A^{17}(\omega)$ for $K = 1.0$ and $55 \leqslant \omega \leqslant 377$.

where $\omega|A^{17}(\omega)|$ has been shown for $55 \leqslant \omega \leqslant 377$.

4.2. *Scaling results*

Followng Feigenbaum [9], we can consider the convergence behavior of the sequence $\Omega_i(K)$. Define

$$\delta_i(K) = \frac{\Omega_{i-1}(K) - \Omega_i(K)}{\Omega_i(K) - \Omega_{i+1}(K)}. \tag{25}$$

This sequence displays two distinct convergence behaviors;

$$\lim_{i\to\infty} \delta_i(K) = -l^{-2}, \quad 0 \leqslant K < 1, \tag{26}$$

$$\lim_{i\to\infty} \delta_i(K) = -l^{-y} \equiv \delta, \quad K = 1, \tag{27}$$

where $y = 2.16443 \pm 0.00002$.

Continuing in Feigenbaum's footsteps, it is possible to find a rescaling ratio α by letting d_i denote the value of the element on the Q_i cycle closest to $\theta = 0$ (modulo 1). It turns out that

$$d_i = T^{Q_{i-1}}(0) - P_{i-1}. \tag{28}$$

If we then define

$$\alpha_i(K) = d_{i-1}/d_i, \tag{29}$$

we again find two convergence behaviors,

$$\lim_{i\to\infty} \alpha_i(K) = -l^{-1}, \quad 0 \le K < 1, \tag{30}$$

$$\lim_{i\to\infty} \alpha_i(K) = -l^{-x} \equiv \alpha, \quad K = 1, \tag{31}$$

with $x = 0.52687 \pm 0.00002$. In both cases, the convergence behavior for $0 \le K < 1$ can easily be understood as resulting from the underlying convergence rate of the W_i. The $K = 1$ convergence rates are more puzzling.

The three quantities $R_i(K)$, $\alpha_i(K)$, and $\delta_i(K)$ all depend on i and K independently. However, in the limit of large i and small $\epsilon \equiv 1 - K$, all three appear to be functions of the single quantity ϵQ_i^ν with $\nu = 1.053744 \pm 0.00015$. The value of ν was determined in two ways. First, data for several values of Q_i and ϵ was plotted versus ϵQ_i^ν for various values of ν (see figs. 4a, b, c). The data appeared to fall on a single curve for ν values in the range $1.0536 < \nu < 1.0539$. We then fit the data to a polynomial in ϵQ_i^ν and summed the total squared deviations from this curve. The value of ν which minimized these deviations was $\nu = 1.053744 \pm 0.000002$. The quoted uncertainty in ν, as in x and y, is merely an estimate of the internal consistency of the data, and does not in any way reflect an analysis of the error from first principles.

The existence of the scaling parameter α makes plausible the conjecture that for small θ

$$f^{Q_i}(\theta) - P_i \approx \alpha^{-i}\bar{f}(\alpha^i\theta), \tag{32}$$

for $K = 1$, $\Omega = \bar{\Omega}(1)$, and $\bar{f}$ some universal function. One can define approximates $\bar{f}_i$ to $\bar{f}$

$$\bar{f}_i(\theta) = \alpha^i(f^{Q_i}(\alpha^{-i}\theta) - P_i), \tag{33}$$

for $K = 1$, $\Omega = \bar{\Omega}(1)$. For $|\theta| < 1.0$, the maximum difference between $f_i(x)$ and $f_{i+1}(x)$ decreases by roughly a factor of 1.7 for each unit increase in i, and for $i = 17$ the difference is approximately

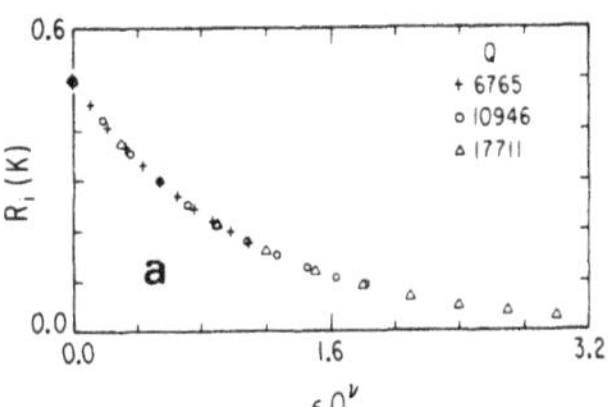

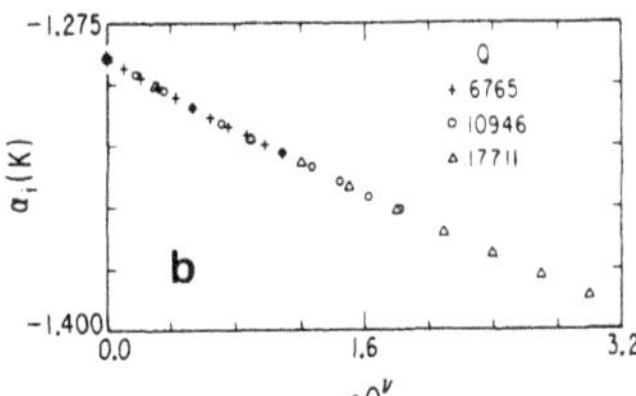

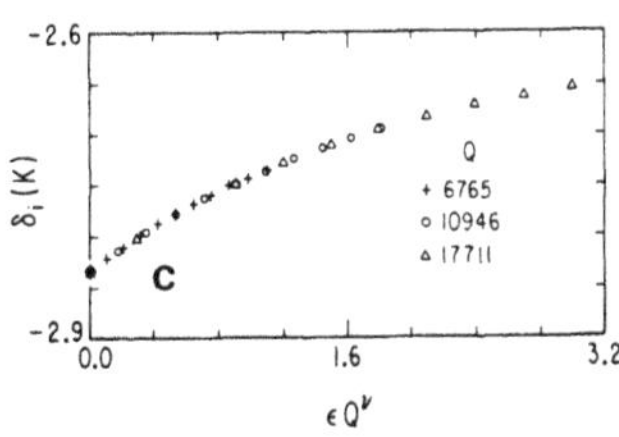

Fig. 4. The quantities (a) $R_i(K)$, (b) $\alpha_i(K)$, and (c) $\delta_i(K)$ are plotted versus ϵQ_i^ν with $\epsilon \equiv 1 - K$ and $\nu = 1.053744$. The data points represent three values of Q and eleven values of ϵ.

10^{-4}. Thus, it does appear that the $\bar{f}_i$ are converging onto a single function $\bar{f}$ (see fig. 5).

Using the recursion relation for Fibonacci numbers

$$F_{i+1} = F_i + F_{i-1} \tag{34}$$

it is possible to define two fixed point equations

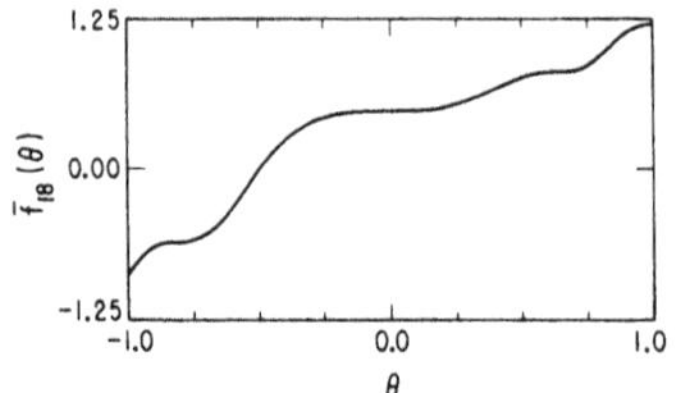

Fig. 5. The quantity $\bar{f}_{18}(\theta)$ is shown for $|\theta| < 1.0$.

for $\bar{f}(\theta)$

$$\bar{f}(\theta) = \alpha\bar{f}(\alpha\bar{f}(\alpha^{-2}\theta)), \tag{35}$$

$$\bar{f}(\theta) = \alpha^2\bar{f}(\alpha^{-1}\bar{f}(\alpha^{-1}\theta)). \tag{36}$$

Attempts to verify that equations (35) and (36) have a single analytic solution have been inconclusive, save for the trivial solution $f(\theta) = -1 + \theta$ which represents the case $|K| < 1$. Tsuda [10] has discussed piecewise linear solutions to these equations.

4.3. *Universality*

The scaling behavior described in the previous section dealt with one particular map and one particular winding number. To investigate how universal this scaling behavior is, three similar problems were analyzed.

(a) $T(\theta) = \theta + \Omega - \dfrac{K}{2\pi}\sin 2\pi\theta,$

$W = \langle 3141111\ldots\rangle;$

(b) $T(\theta) = \theta + \Omega - \dfrac{K}{2\pi}(\sin 2\pi\theta + 0.2\sin 6\pi\theta),$

$W = \langle 11111\ldots\rangle;$

(c) $T(\theta) + \theta + \Omega - \dfrac{K}{2\pi}\sin 2\pi\theta,$

$W = \langle 2222\ldots\rangle.$

In each case, scaling behavior described by three exponents x, y, and $\epsilon\nu$ was observed (in base b, the critical value of K is 0.625). The results are compiled in table I. The exponents for cases a and b were the same as our original problem to within numerical uncertainty, indicating that the scaling behavior does not depend on early digits in the continued fraction representation of the winding number, nor on the details of the map. I do expect however, to find different universality classes of maps. Case c yielded exponents which were significantly different from those found in our original problem. This suggests that the asymptotic nature of the continued fraction representation is important in determining the scaling behavior.

Table I
Compilation of scaling exponents

Problem	x	y	ν
$T(\theta) = \theta + \Omega - \frac{K}{2\pi}\sin 2\pi\theta$ $W = \langle 11111\ldots\rangle$	0.52687 ± 0.00002	2.16443 ± 0.00002	1.053744 ± 0.00015
$T(\theta) = \theta + \Omega - \frac{K}{2\pi}\sin 2\pi\theta$ $W = \langle 3141111\ldots\rangle$	0.527 ± 0.001	2.164 ± 0.001	1.0537 ± 0.0010
$T(\theta) = \theta + \Omega - \frac{K}{2\pi}(\sin 2\pi\theta + \sin 6\pi\theta)$ $W = \langle 11111\ldots\rangle$	0.527 ± 0.001	2.164 ± 0.001	1.0537 ± 0.0010
$T(\theta) = \theta + \dot{\Omega} - \frac{K}{2\pi}\sin 2\pi\theta$ $W = \langle 2222\ldots\rangle$	0.5239 ± 0.0002	2.1748 ± 0.0002	1.04776 ± 0.001

Acknowledgements

I am extremely grateful to L.P. Kadanoff for initially suggesting this problem, and for his astute advice and patient guidance during the course of the research. I also wish to thank Stephen H. Shenker, D. Huse, M. Widom, A. Zisook, and E. Siggia for helpful discussions.

References

[1] E. Ott, Reviews of Modern Physics 53 (1981) 655; D. Ruelle and F. Takens, Commun. Math. Phys. 20 (1971) 167.

[2] Similar maps have been discussed in P. Coullet, C. Tresser, and A. Arneodo, Phys. Lett. 77A (1980) 327; G. Iooss and W.F. Langford Nonlinear Dynamics, Vol. 357, R.H.G. Helleman, ed. (New York Academy of Sciences, New York, 1981) p. 489.

[3] S.J. Shenker and L.P. Kadanoff, J. Stat. Phys. 27 (1982) 631.

[4] J.M. Greene, J. Math. Phys. 9 (1968) 760.

[5] J.M. Greene, J. Math. Phys. 20 (1979) 1183.

[6] A. Denjoy, C.R. Acad. Sci. 195 (1932) 478.

[7] V.I. Arnol'd, Transl. Amer. Soc. 2nd Series 46 (1965) 213.

[8] M.R. Herman, Lecture Notes in Mathematics, vol. 597 (Springer, Berlin, 1977).

[9] M. Feigenbaum, J. Stat. Phys. 21 (1979) 669.

[10] I. Tsuda, Progress Theor. Phys. 66 (1981).

Editor's note. The above work arose from the Shenker and Kadanoff (1982) study of the dissolution of KAM tori, and was followed up by Feigenbaum, Kadanof and Shenker's (1982), Rand, Ostlund, Sethna and Siggia's (1982) and Ostlund, Rand, Sethna and Siggia's (1983) formulation of the universal equations for the golden mean windings. Feigenbaum and Hasslacher (1982) have extended this universality to noisy circle maps.

Long-Time Prediction in Dynamics (ed C W Horton Jr, L E Reichl and A G Szebehely) 127–34 (1983)

PERIOD DOUBLING AS A UNIVERSAL ROUTE TO STOCHASTICITY

ROBERT S. MACKAY
Plasma Physics Laboratory
Princeton University, Princeton, New Jersey

Abstract. *In dissipative systems, period doubling is known to provide a route, with universal properties, from coherent to chaotic behavior.*[1] *This chapter describes an analogous transition in area preserving maps, which represent the simplest conservative systems. This transition almost completely eliminates closed invariant curves from the vicinity of an originally stable periodic orbit.*

1. THE UNIVERSAL PERIOD DOUBLING SEQUENCE

As an example, consider the DeVogelaere form of the area preserving quadratic map:

$$T\colon x' = -y + f(x) \qquad (1)$$

$$y' = x - f(x')$$

with

$$f(x) = px - (1-p)x^2 \qquad (2)$$

where p is a parameter. For $|p| < 1$ the fixed point at the origin is elliptic. As p decreases through -1 it loses its stability and gives birth to a stable two-cycle (Fig. 1). As p decreases further this two-cycle itself goes unstable and gives birth to a stable four-cycle (Fig. 2).

This period doubling process repeats infinitely often, but in a finite range of parameter, accumulating at some value $p^* = -1.26631127692$. The parameter values p_n, at which the nth period doubling occurs, are found to converge

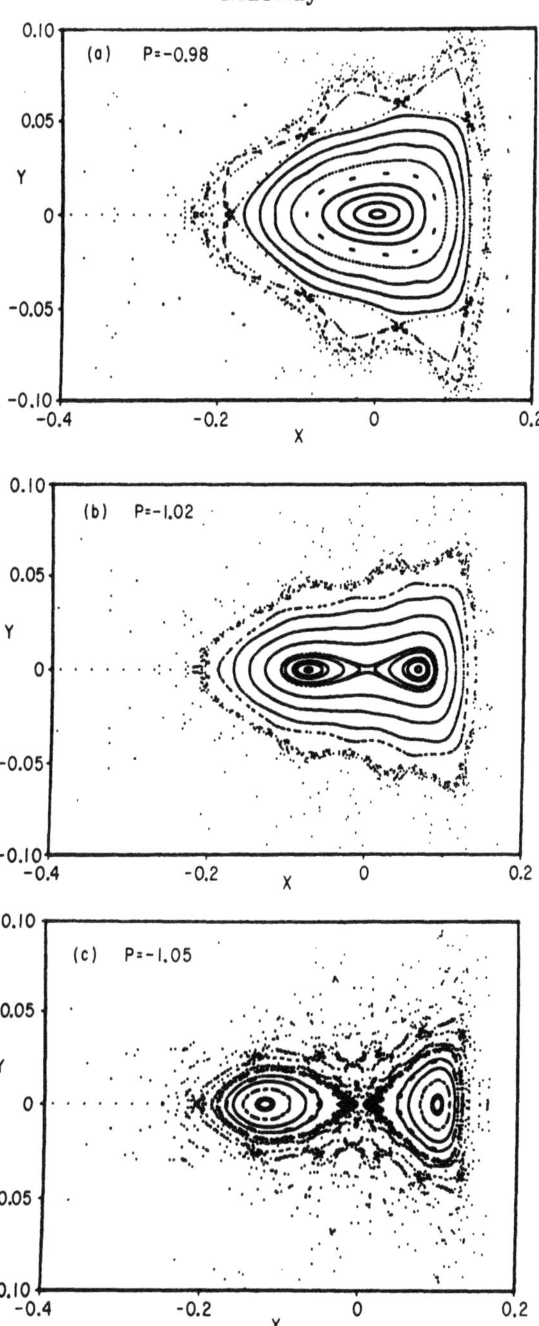

Figure 1. Some orbits of the map (1, 2) for three parameter values, showing the first-period doubling.

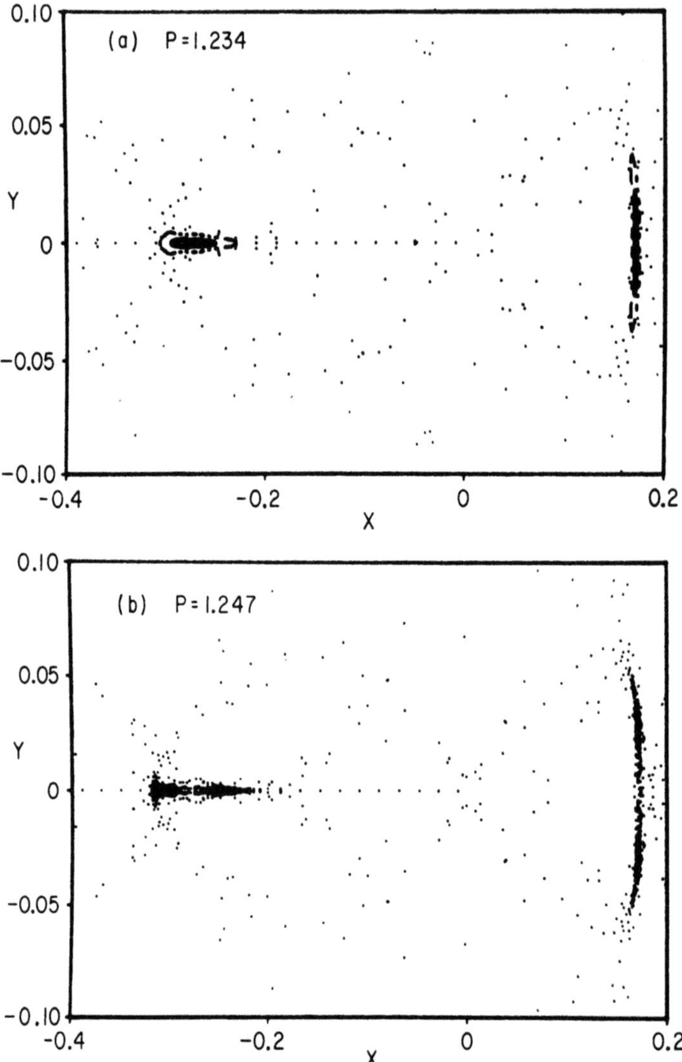

Figure 2. Some orbits of the map (1, 2) for two parameter values, showing the second-period doubling.

asymptotically geometrically to p^*, with ratio $1/\delta$, where:

$$\delta = 8.721097200\ldots \tag{3}$$

The period doubling process possesses further self-similarity, easiest revealed by exploiting the symmetry evident in Figs. 1 and 2. A map S is said to be a *symmetry* of a map T if:

$$S^2 = \text{identity} \tag{4}$$

S reverses orientation, and

$$S^{-1}TS = T^{-1}$$

The last condition implies that S transforms T into its inverse. Thus maps with symmetry are said to be *reversible*.[2] For example, DeVogelaere maps [Eq. (1)] are reversible with the following symmetry:

$$S\colon x' = x \qquad (5)$$

$$y' = -y$$

The fixed points of a symmetry often form lines, called its *symmetry lines*. It can be shown[3] that given a fixed point on a symmetry line, there is a *dominant symmetry* such that precisely two points of each periodic orbit of its period doubling sequence lie on the fixed line of the dominant symmetry. Thus the period doubling sequence can be followed by looking for periodic points on this *dominant line* only, rather than in the whole plane. Incidentally, the DeVogelaere form of the quadratic map is used to give the dominant symmetry the simple form, Eq. (5).

Plotting the positions on the dominant line of such periodic points against parameter yields a slice of the *period doubling tree* (Fig. 3). It can be seen to repeat itself on smaller and smaller scales, accumulating at a point (x^*, p^*),

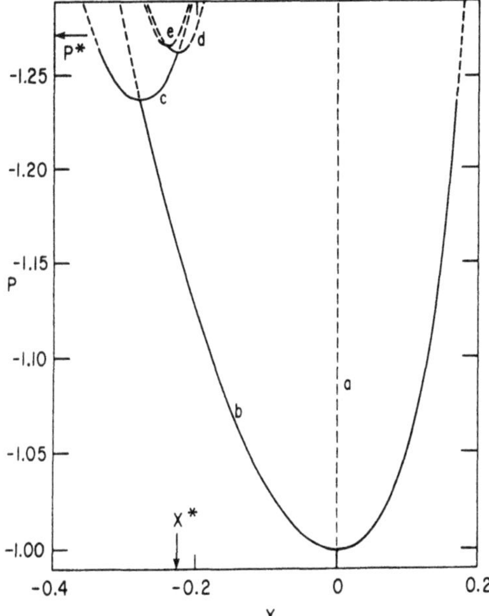

Figure 3. The section of the period doubling tree lying on the dominant symmetry.

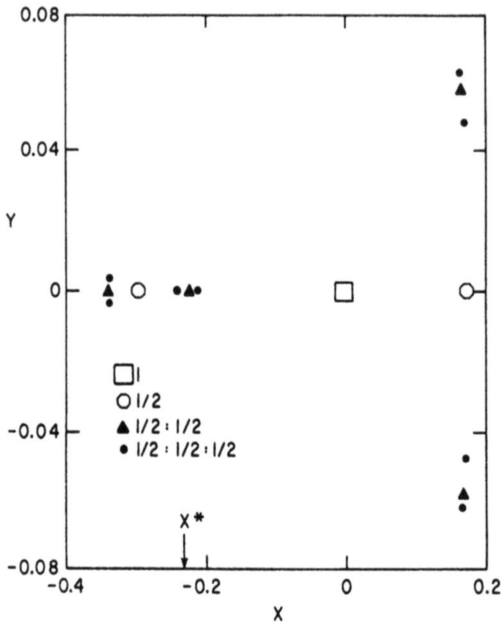

Figure 4. The first few periodic orbits of the period doubling sequence at the accumulation parameter value p^*.

with rescaling factors δ in parameter, and α along the dominant line:

$$\alpha = -4.018076704\ldots \tag{6}$$

The rescaling factors δ and α were already known.[4-6] We found[7] that this self-similarity extends off the dominant line too (Fig. 4), with a rescaling factor β across the line:

$$\beta = 16.363896879\ldots \tag{7}$$

In fact, the whole map shares properties similar to those of its period doubling sequence, as follows. At the accumulation parameter value p^*, T converges under the operation A of squaring (i.e., composing with itself) and rescaling by α, β in x, y about $x^*, 0$. The limit T^* is a map invariant under A. Furthermore the whole one-parameter family converges under the operation D of squaring and rescaling in parameter as well as in x, y, to a one-parameter family T_p^* (T^* is the special case of T_p^* when $p = p^*$). For more details see Ref. 7.

This behavior is all the more remarkable in that it appears to be universal; that is, almost all one-parameter families of reversible area preserving maps with period doubling sequences have the same δ, α, β, T^*, and T_p^* (up to simple changes of coordinates and parametrization).[4-9]

2. COMPLETE STOCHASTICITY

One of the most striking consequences of period doubling, as Figs. 1 and 2 indicate, is the destruction of closed invariant curves, and the corresponding reduction in the area they trap. Figure 5 shows, for a range of parameter values, the points of the dominant line whose orbits are bounded. Also for some parameter values beyond p^* we found that all points of a large grid in the plane escaped. Thus, as conjectured by J. B. McLaughlin[10] and R. H. G. Helleman,[1] for example, it might appear that beyond p^* there are no closed invariant curves in the vicinity. We would call this a transition to *complete stochasticity*, but the stochasticity turns out to be not quite complete. There are small universal islets of stability beyond p^*.[20]

In general, of course, there can also be large, nonuniversal regions of stability, since T_p^* exerts its influence only in a neighborhood of the original periodic orbit. Thus, for instance, the period doubling sequence of the stable five-cycle of the island chain born at $p = \cos(2\pi/5) = 0.30901699\ldots$, accumulates at $p = 0.119353761903\ldots$, leaving plenty of invariant curves around the fixed point, but destroying completely the fifth-order island chain. As another example, the period doubling sequence of the fixed point of the standard map accumulates at $k = 2\pi \times 1.05597806\ldots$,[4] but at $k = 2\pi$ an islet of stability pops out by tangent bifurcation, whose fixed point does not lose stability until $k^2 = (2\pi)^2 + 16$,[11] that is, $k = 2\pi \times 1.1854471\ldots$.

This transition should be distinguished from that to "connected stochasticity,"[12] when the last of the original invariant curves of an originally integrable map is destroyed, permitting orbits unconfined in action. That occurs at $k = 0.971635\ldots$ in the standard map, well before the fixed point has doubled at all.

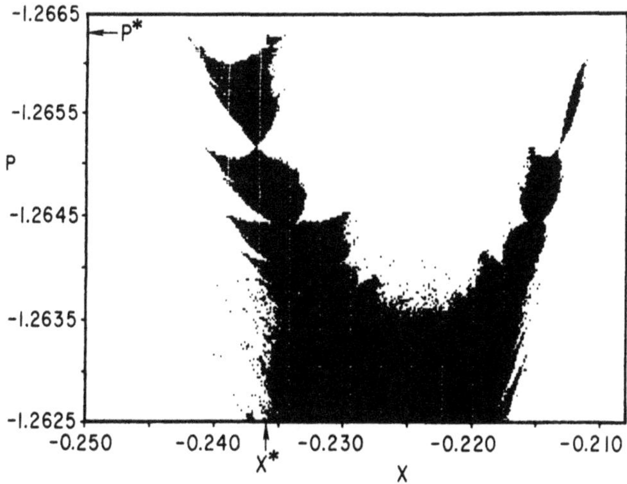

Figure 5. Points of the dominant line, for a range of parameter values, that remain bounded under T_p^*.

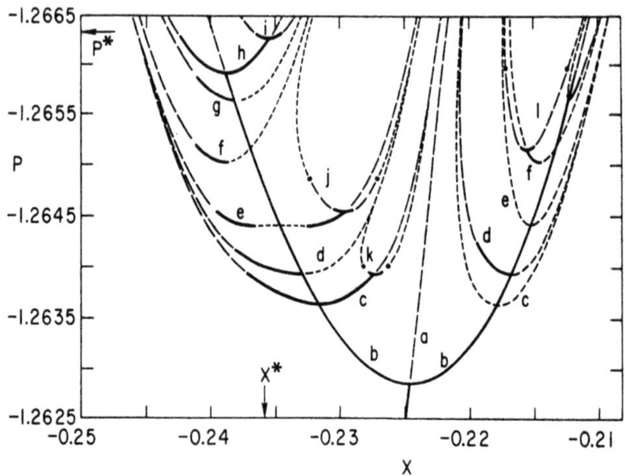

Figure 6. Some of the branches of the bifurcation tree of T_p^* with periods as follows: a, 1; b, 2; c, 12; d, 10; e, 8; f, 6; g, 10; h, 4; i, 8; j, 16; k, 24; l, 12. Thick line indicates elliptic; long dashes, inversion hyperbolic; short dashes, ordinary hyperbolic.

3. CONCLUDING REMARKS

We conclude with three questions:

1. What about maps with no symmetry? These are rare in practice. For example, all maps derived from a Hamiltonian even in the momenta are reversible.[13] We know only one example that can be shown to have no symmetry,[14] although numerical evidence indicates that Rannou's map[15] has no symmetry either.[16] Reversibility is not generic, however, in one-parameter families of area preserving maps. This follows from the work of Rimmer.[17] Nevertheless, we expect that nonreversible maps will exhibit the same universal behavior. This is supported by finding that Rannou's map has period doubling with the same δ,[16] and by analogy with the one-dimensional case where the universal one-parameter family of maps has a symmetry property. It is even, and yet it attracts both even and noneven maps.

2. What about bifurcations other than period doubling? A period n-tupling bifurcation of a periodic orbit requires a multiplier* to pass through an nth root of unity. The (only) multiplier λ of a periodic orbit of a one-dimensional map is always real, so the only bifurcations are tangent ($\lambda = +1$) and period doubling ($\lambda = -1$). In two-dimensional area preserving maps the two multipliers are reciprocal reals or conjugate points on the unit circle. Thus in the latter case all bifurcations are possible (Fig. 6). Note the connection between Figs. 5 and 6, in particular the temporary destruction of closed invariant curves associated with the period tripling bifurcation. One can follow

*The multipliers of a periodic orbit are the eigenvalues of the linearization of the map around the orbit.

infinite bifurcation sequences other than period doubling, such as period tripling, and find that they have self-similarity too.[7] Period doubling, however, is the most significant, since it is the last bifurcation before a stable periodic orbit loses its stability, and it leads to complete stochasticity.

3. What about higher dimensional systems? The results for one-dimensional maps extend to higher dimensional dissipative systems.[18] In contrast, it appears that the two-dimensional area preserving results do not extend to higher dimensional symplectic maps.[19] It is an open question, however, whether they might have their own universal period doubling sequences.

ACKNOWLEDGMENTS

I thank John Greene for his inspiration, guidance, and encouragement, Jon Schonfeld for arranging for me to speak, and H. Greenwood of Keele University, England, for the use of computing facilities. This work was supported by the U.S. Department of Energy contract DE-AC02-76-CHO-3073, and the U.K. Science Research Council grant B/80/3001.

REFERENCES

1. M. J. Feigenbaum, this volume; R. H. G. Helleman, and R. S. MacKay, "One Mechanism for the Onset of Large-Scale Chaos in Conservative and Dissipative Systems," this volume.
2. R. Devaney, *Trans. Am. Math. Soc.*, **218**, 89 (1976).
3. R. S. MacKay, "The Dominant Symmetry of Reversible Maps," notes (1981).
4. G. Benettin, C. Cercignani, L. Galgani, and A. Giorgilli, *Lett. Nuovo Cimento* **28**, 1 (1980).
5. T. C. Bountis, *Physica 3D*, 577 (1981).
6. F. Vivaldi, and J. Ford, private communication on period doubling in a forced Duffing oscillator.
7. J. M. Greene, R. S. MacKay, F. Vivaldi, and M. J. Feigenbaum, *Physica 3D*, 468 (1981).
8. G. Benettin, L. Galgani, and A. Giorgilli, Lett. Nuovo Cimento *29*, 163 (1980).
9. P. Collet, J.-P. Eckmann, and H. Koch, *Physica 3D*, 457 (1981).
10. J. B. McLaughlin, *J. Stat. Phys.* **24**, 375 (1981).
11. B. V. Chirikov, *Phys. Rep.* **52**, 263 (1979).
12. J. M. Greene, *J. Math. Phys.* **20**, 1183 (1979).
13. J. M. Greene, in *Nonlinear Orbit Dynamics and the Beam–Beam Interaction*, M. Month and J. C. Herrera, Eds., American Institute of Physics Conference Proceedings Series, Vol. 57 (AIP, New York, 1979), p. 257.
14. J. N. Mather, private communication.
15. F. Rannou, *Astron. Astrophys.* **31**, 289 (1974).
16. J. M. Greene, private communication.
17. R. Rimmer, *J. Differ. Equations* **29**, 329 (1978); "Generic Bifurcations from Fixed Points of Involutory Area Preserving Maps," P. Math. Res. Paper 79-9, La Trobe University, Melbourne, Australia (1979).
18. P. Collet, J.-P. Eckmann, and H. Koch, *J. Stat. Phys.* **25**, 1 (1981).
19. R. S. MacKay and T. C. Bountis, "Period Doubling in 4-D Symplectic Maps," notes (1981).
20. R. S. MacKay, "Islets of Stability Beyond Period Doubling," Physics Letters *87A*, 321 (1982).

Fundamental Problems in Statistical Mechanics vol **5** (ed E G D Cohen) pp 165–233 (1980)

SELF-GENERATED CHAOTIC BEHAVIOR IN NONLINEAR MECHANICS §

Robert H.G. Helleman

Theoretical Physics Group
Twente University of Technology
P.O. Box 217, Enschede
The Netherlands

Several of the exciting developments in the *Classical* Mechanics of Conservative and Dissipative Systems are briefly reviewed here, at this summerschool on *Statistical* Mechanics, assuming only a rudimentary knowledge of some graduate course in Mechanics: It turns out that the phase space of most Hamiltonian- and many Dissipative-systems is dotted with 'Chaotic' Regions in which some properties of many orbits are *as random as coin tosses*, even though the system is deterministic. While statistical methods appear to be applicable to those chaotic regions they are incompatible with the very smooth regular, 'quasi-periodic', behavior in other regions, of which there is an abundance as well. Hence *'Ergodicity' and the 'Approach to Thermal Equilibrium' do not hold for most Hamiltonian systems.* Parts of those chaotic regions do survive under some *dissipative* perturbations but the regular regions do not: Instead of being 'attracted' to the origin (damping) or to a periodic orbit ('limit cycle'), the orbits -from the previously regular region- may now be attracted to the remainder of a chaotic region. The motion *on* these so called 'Strange Attractors' can be chaotic, ergodic and even 'mixing'. The transition from simple limit cycles to very complicated ones, to a regime with Strange Attractors has recently been derived. This 'Period-Doubling' Transition provides a model for the Onset of Turbulent Behavior.

CONTENTS

§ Reprinted here with a few literature updates (> Sept. 1980) and corrections.

1. INTRODUCTION

"In spite of their great efforts over a long period Poincaré, Birkhoff, Siegel and others have not succeeded in making use of perturbation theory to obtain precise qualitative (!) conclusions about the behavior of the majority of the trajectories, as $t \to \infty$.

The non-integrable problems of dynamics appeared inaccessible to the tools of modern mathematics.

Essential progress was made in 1954 when A.N. Kolmogorov proved his theorem.... [22]. *Apart from this, little is known about the behavior of the trajectories".*

V.I. ARNOLD,

ref. 22, p. 12 (1963).

Contrary to the impression created, when we studied 'Classical Mechanics' in our (under-) graduate years [7-9], Mechanics is not in good shape; even though exciting progress is being made. A number of review articles have recently appeared to describe these new developments [1-6, 90-94, 215, 226, 250, 251, 301, 17].

The above quoted inability to approximate the long-time behavior of most orbits stems not so much from a lack of ingenuity (what with Newton, Poincaré, Birkhoff, Siegel, Kolmogorov, Arnold and Moser in the field) as it does from the incredibly complicated behavior of the actual orbits in most systems. In fact, the $\vec{q},\vec{p}$ phase space of *most systems* is dotted with regions in which some properties of many orbits *are as random as* the outcome of *coin tosses* or of *a throw of the dice* (moving dice and coins are mechanical systems...). To people in *Statistical* Mechanics it may come as a surprise that these 'quasi-random' orbits and this chaotic behavior appear already in systems with only 2 or 3 degrees of freedom (used as we are to things going right only in the thermodynamic limit, $N \to \infty$). Moreover these are not pathological, exceptional, systems: *Most* Hamiltonian systems exhibit this chaotic behavior. But then, how could we possibly have overlooked such obvious effects in our usual graduate mechanics courses? The answer is that, without saying so ..., we select the examples and exercises from a (small) subclass, the *'Integrable'* (separable) systems, in which there is none of this chaos. It contains the familiar N harmonic oscillators, the (celestial) 2-body problem, etc. However, already the (celestial) 3-body problem is *'Nonintegrable'* ($\approx$ non-separable) and exhibits chaotic behavior. It is only this 'Three-Body Problem' which has acquired enough notoriety over the last 100 years to be mentioned in some of our courses, unfortunately most often as somewhat of a curiosity rather than as being truly representative of *most* Hamiltonian systems. The reader may get a taste of what is meant by a chaotic region from Figure 1, where the intersection points are plotted of a few orbits with a 2 dimensional plane in phase space.

The existence of chaotic regions was known already to Poincaré [18], Einstein [115], and to a few others. Apparently its consequences failed to sink in in the physics community which was more excited about *quantum* mechanics (for integrable systems ...) during the ensuing decades. It is very surprising that even today no authoritative graduate-physics textbook on Classical Mechanics has appeared which does show pictures like Figure 1, let alone warn the student that *most* systems exhibit such behavior or that all solution methods, derived in its text, *fail* (diverge) for most Hamiltonian systems. As far as I knew, the only such textbook, in physics, which began to mention these problems was one on *Statistical* Mechanics [94]; now see refs. [17]. As the title of this paper implies, chaotic behavior appears already without using additional noise sources, stochastic processes, ensembles or 'molecular chaos' assumptions ('stoszzahlansatz'). Hence we merely consider single orbits in deterministic mechanical systems.

While the existence of chaotic regions will cheer up those who seek to derive and explain statistical mechanics from classical mechanics ('ergodic theory', etc.

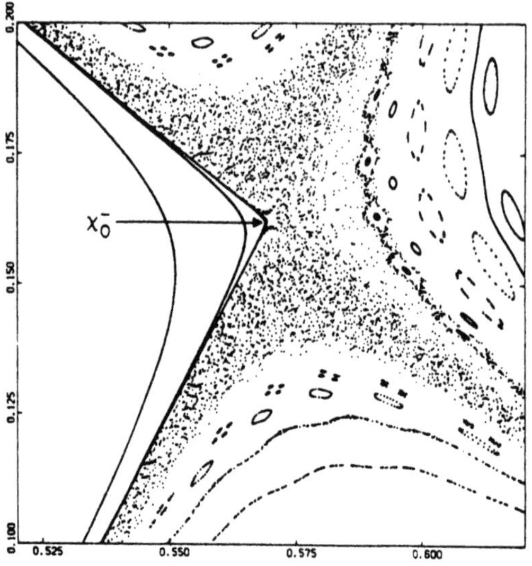

Figure 1
Intersection points of some (≈ 15) orbits in Phase Space with a 2 dimensional plane. Note that all chaotic dots in the central region are created by *one* orbit (cf. eqs. (2.16), (2.12) and (2.20) with $F(y) = y^2$; taken from refs. 35 and 48).

[90-98]) it sows terror in the hearts of those designing and building extensive (expensive) mechanical systems, e.g. the multi (-hundred) million dollar intersecting storage rings of high energy physics [200-211].

Notwithstanding the abundance of chaotic regions, there is also an abundance of 'Regular' regions in phase space. This follows from the so-called K.A.M. Theorem [1-6, 12-17, 19-23] which shows that many of the simple 'quasi-periodic' solutions of an Integrable system will still be present, virtually unchanged, in a Nonintegrable system, in general. Moreover a *finite* fraction ('measure') of all orbits are of this regular type [22]. An 'invariant torus' (car-tire), on which such a regular orbit lies, does not come arbitrarily close to every point on the 'surface-of-constant-energy', in general. Hence Ergodicity and the 'Approach to Thermal Equilibrium' do not hold for most Hamiltonian systems [3,13,90-94]. These results differ substantially from earlier ones by Fermi [135,134] and Peierls [136].

Yet, in the chaotic regions, there is evidence for the onset of some diffusive motion, called 'Arnold Diffusion' [80-87]. Thus one is eager to investigate for which class of nonintegrable Hamiltonians it might be true that the measure of the regular K.A.M. regions approaches zero in the 'Thermodynamic Limit' ($E, N \to \infty$, while $\varepsilon \equiv E/N$ is kept constant). In that case there might still be some approach to equilibrium or ergodicity, in the thermodynamic limit (and/or at sufficiently high ε ?) [79,139,140].

To complete this confusing picture it should be noted that there are Hamiltonian systems for which the K.A.M. theorem does not hold, e.g. when $H(\vec{q},\vec{p})$ is not smooth enough. An example is the hard sphere Boltzmann gas, whose Hamiltonian is not [156] differentiable. For this system one (Sinai) has been able to prove ergodicity and an approach to equilibrium, already for *two* particles (hard-core) or more [90-98], without any need to go to the thermodynamic limit.

At the same time all this also gave rise to a number of interesting results for *Integrable* systems, as we heard during this school [34]. As a matter of fact it was the surprising failure to find chaos (numerically) in some nonlinear solid-state models, e.g. in those of Fermi, Pasta and Ulam [33], that led to the discovery of a number of Integrable systems [32-35] and the complete 'Soliton' solutions of the 'KdeV' equation as well as the 'Toda Chain' [33,34], cf. (2.5-.6)

We have the astronomers and mathematicians to thank for passing the field on to us, physicists, in a much better shape [12-22] than we left it to them, 70 years ago. As a result of the exciting developments during 1960-1980 [1-6, 12-43] it is rapidly becoming popular in many branches of physics, engineering, chemistry, hydrodynamics and even biology [44-302].

Even more surprising, at least to me, are the very recent developments in *Dissipative* Systems: one might think that adding a friction term to the (conservative) equations of motion would damp out all motion, i.e. all orbits would eventually be *'attracted'* to the "origin", i.e. the point of lowest energy of the conservative system. This may happen indeed. Yet, it has been proven that the quasi-random orbits *can persist* under some dissipative perturbations [15,23,274-279] while the regular quasi-periodic orbits will disappear [13], in general. Such chaotic objects, in dissipative systems, are particularly easy to find when *they* (instead of the origin) 'attract' the orbits from the other, previously regular K.A.M., regions. A number of these so called *'Strange Attractors'* have now been exhibited in very simple systems, numerically and analytically [250-272]. It is the motion *along* the attractor which can be chaotic, ergodic and even 'mixing'. The rapid uniform spreading, along a Strange Attractor, of a set of initially close points can be seen in Figure 2:

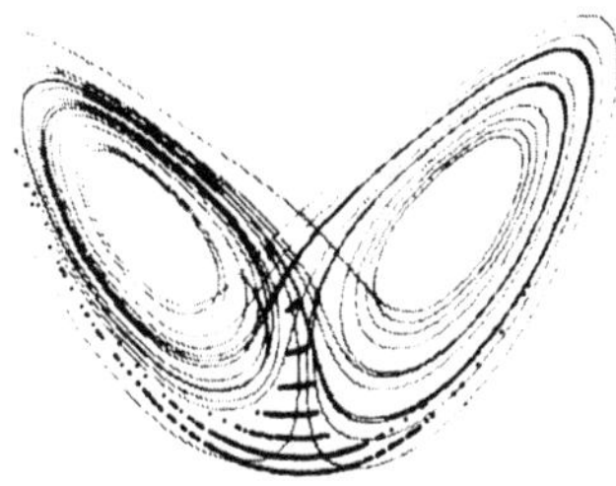

Figure 2
Projection of a Strange Attractor in 3-d. space onto a 2-d. plane. The thin lines belong to the (Lorenz-) Attractor (3.35-.37). At $t=0$ the set of different starting points cannot be distinguished from a single point at this resolution. The first picture shows the break-up, after some time, of the set into filaments (thick lines); the second picture, a little later, shows the eventual distribution of this set, without the attractor super-imposed (taken from [250/301]), cf. Fig. 25.

To the naked eye the shape of the attractor may seem simple, yet one usually reserves the name Strange Attractor for the (frequent) case when the intersection of the object, e.g. in Figure 2, with a straight line is a 'Cantor Set', i.e. an infinite number of points which are not dense on any interval. This infinitely nested structure can be seen in each repeated minor magnification, e.g. Figure 22. Hence the attractor is not (merely) a complicated *periodic* orbit. Loosely speaking, we can in the Fourier spectrum of an orbit on the attractor see many periods come and go "in the course of time". This is reminiscent of the decay of the vortices behind an object in a water flow, into smaller and smaller vortices, of increasing wave number. Thus the Strange Attractors are thought to model some aspects of Turbulence. They were in fact discovered in (truncated) mode equations for the Navier-Stokes equation of hydrodynamics [250-257,234,235].

However, as one changes the value of one of the parameters, μ, which plays a role similar to that of the "Reynolds" number, *simple periodic attractors* can exist in such equations as well. An example of one equation with one simple periodic attractor, or 'Limit Cycle', is the 'van der Pol Equation' [282-289], describing a triode oscillator in electronics. There exists an amazingly successful theory about the transition from simple periodic attractors to complicated ones, to a regime with Strange Attractors. As one changes this "Reynolds" Number, μ, the simple periodic attractor of period T changes, at μ_1, into a double-loop attractor of period 2T. At another value, μ_2, it changes into a four-loop attractor of period 4T, etc. Eventually the attractor can become very complicated and resemble the one in Figure 2. Amazingly enough, the resulting 'Feigenbaum Sequence' $\{\mu_k\}$ *converges* geometrically, at a rate $\delta = 4.669201...$ which seems to be a universal constant for many models [215-239]. Somewhere above the finite critical value, μ_∞, Strange Attractors arise. Hence the onset of Turbulence is well understood for some model systems. Most of these results were first found numerically. Similar 'period-doubling' has now been observed in real turbulence experiments [241-243,256]. Analogous Feigenbaum sequences, with a different δ value, have just been found in conservative systems as well [63-68], see section 2.6. Since the original periodic orbits become unstable after each period-doubling the values of μ_∞ might provide estimates for the appearance of large scale chaotic regions.

The frequent appearance of 'structurally stable' *ergodic* attractors has led Smale to speculate that we might have to resort to slightly dissipative systems if we want to revive (read: rescue) the Ergodic Hypothesis of Boltzmann and Birkhoff [273] Finally, it might be pointed out that the above mentioned picture of turbulence differs substantially from that of Landau and Hopf [298,299].

The selection of topics to be reviewed heuristically, in the next sections, is displayed in the table of contents, on the title page.

2. CONSERVATIVE SYSTEMS

These are systems with at least one constant of the motion which is a well behaved, e.g. continuous, function of the variables. While there exist other conservative systems, e.g. the Volterra equations of population dynamics [288], we shall only consider the Hamiltonian systems of physics. The volume of a small volume element, in the (phase) space of all variables, is 'conserved' forever (or bounded, away from 0 and ∞) during its subsequent motion [7-11]: In the following sections we can often also allow Hamiltonians which are explicit *periodic* functions of the time [146]. This contrasts with the behavior of a Dissipative system, in which the volume of an element of phase space shrinks to zero, as $t \to \infty$.

2.1 Non-Integrable versus Integrable Hamiltonian Mechanics.

The equations of motion for an N-'degrees of freedom' system are obtained from a Hamiltonian, H, as:

$$\dot{q}_k = \partial H(\vec{q},\vec{p})/\partial p_k, \qquad \dot{p}_k = -\partial H(\vec{q},\vec{p})/\partial q_k, \tag{2.1}$$

with $k = 1,\ldots,N$ and $(\vec{q})_k \equiv q_k$. These are 2N coupled nonlinear first order differential equations, in general. A change of variables, from $\vec{q},\vec{p}$ to $\vec{Q},\vec{P}$, is called a 'canonical transformation' if the transformed equations of motion (2.1) for $\vec{Q},\vec{P}$ have the same appearance as (2.1), using the transformed $H(\vec{Q};\vec{P})$. A particularly popular canonical transformation is the one to *'Action- and Angle Variables'* $\vec{I},\vec{\Theta}$. These new variables are *defined* as variables which will transform the given $H(\vec{q},\vec{p})$ into an $H(\vec{\Theta};\vec{I})$, i.e. one which does not depend on half the new variables, $\vec{\Theta}$ [7-9]. *If* we succeed in doing this ..., the rewards are great since the new equations of motion yield:

$$\dot{I}_k = -\partial H(\vec{\Theta};\vec{I})/\partial\Theta_k = 0, \qquad \textit{whence } \vec{I}(t) = \vec{I}(0), \textit{ constant,} \tag{2.2}$$

$$\dot{\Theta}_k = \partial H(\vec{\Theta};\vec{I})/\partial I_k \equiv \omega_k(\vec{I}), \qquad \textit{whence } \vec{\omega} \textit{ is constant,} \tag{2.3}$$

$$\textit{thus } \Theta_k(t) = \omega_k t + \Theta_k(0), \tag{2.4}$$

i.e. we have explicitly, and trivially, integrated the equations of motion and obtained 2N constants of the motion $\vec{I},\vec{\Theta}(0)$. Hence *a Hamiltonian system is called 'Integrable' if we can obtain (the transformation to) these Action- and Angle Variables as a function of* $\vec{q},\vec{p}$. As functions of $\vec{q},\vec{p}$ the 2N constants of the motion are called 'Integrals'. Clearly the last N integrals $\vec{\Theta}(0)$ are easily obtained via (2.3), once we have the first N integrals $\vec{I}(\vec{q},\vec{p})$.
Alternatively, if we know N other independent, single valued analytic, integrals $\vec{J}(\vec{q},\vec{p})$ the system is again explicitly Integrable, by 'quadratures' [14], under certain mild conditions on the $\vec{J}$ and if $\{J_k,J_m\} = 0$.
An example similar to (2.2) is the transformation to polar coordinates in the celestial 2-body problem. Since the resulting H does not depend on the ('cyclic') angle ϕ itself the corresponding (angular-) momentum is a constant of the motion.

Thus we have shifted the problem, from solving the equations of motion - trivially accomplished by (2.2-4) - to finding a transformation to (or from) Action & Angle variables. The latter is the subject of 'Canonical Perturbation Theory' [8-10,12] where, *in effect*, one writes each of the old variables $q_1,..,q_N,p_1,..,p_N$ as a power series in all the new variables $I_1,..,I_N,\Theta_1,..,\Theta_N$. The as yet unknown coefficients in the series are derived by substituting the series into $H(\vec{q},\vec{p})$ and the equations of motion (2.1). Requiring those expressions to become identical to (2.2-3), up to the lowest order terms in $\vec{I},\vec{\Theta}$, we can solve for the lowest-order coefficients. Having obtained these we start all over and obtain the next higher-order coefficients, etc. Hence, *whenever* the series obtained do *converge* we have an *Integrable* system by definition, and we have the $\vec{q},\vec{p}$ as explicit functions of the constants $\vec{I}$ and as (quasi-/) periodic functions of the $\vec{\Theta}$ [12-15]. Also, in that case (finite region of convergence) the $\vec{q},\vec{p}$ are nice 'real-analytic' (infinitely differentiable) functions of $\vec{I},\vec{\Theta}$, which may be inverted to yield the *Integrals* $\vec{I}(\vec{q},\vec{p})$ as nice *analytic* functions, in some $\vec{q},\vec{p}$ region.

Such series, and some variants [12], are named after Birkhoff who investigated their *formal* existence [12,19,20,132,133], i.e. can the above iterative procedure indeed be carried out to infinite order? Often the procedure breaks off due to one coefficient acquiring a zero denominator [20,46]. This signals that we either need another variant of the Birkhoff-series or that *no* truly *convergent* series exists *at all*, i.e. that one of the Integrals is a singular function of $\vec{q},\vec{p}$. In the latter case the system is called *'Non-Integrable'*. Whenever we have *infinite* Birkhoff series the Non-Integrability or Integrability of a given system *can* be immediately decided analytically (Rüssmann has shown [12] (for N = 2, but cf. [28]) that in this case the Birkhoff series must converge if $\{J_1,J_2\} = 0$). [12,19]. Hence, if they diverge we have a Non-Integrable system [28] and if they converge an Integrable system. Practical tests are discussed in section 2.4.

Examples of Integrable Systems

a) *All 1-degree of freedom problems* (with H(q,p) analytic). In that case we solve for $\dot{q}$, from H = E, and obtain a first order differential equation, whose solution is a 'hyperelliptic function', e.g. the pendulum in section 18 of ref. 9.
b) All systems with *linear equations* of motion. A 'normal-mode' transformation

[7-9] reduces the system to uncoupled 1-degree of freedom equations. In particular: N coupled harmonic oscillators and the celestial 2-body problem (linear in new coordinates, $s \equiv 1/r$, $t \to \phi$ [9]).

c) All *nonlinear* systems which one can *separate* into uncoupled 1-degree of freedom systems [32-34], hopefully by a simpler transformation of variables than the Birkhoff series itself. In particular: the one-dimensional (-in space) chains with the following interparticle potentials, $\phi(x)$,

$$\phi(x) = a\exp(-bx) + cx + d, \qquad \textit{the Toda Chain,} \tag{2.5}$$

[34] which in one limit (Flaschka), as $N \to \infty$, yields the Korteweg-de-Vries partial differential equation [33,34],

$$u_t - 6uu_x + u_{xxx} = 0, \qquad \textit{the 'KdeV' eq.,} \tag{2.6}$$

which turns out to have an infinite number of integrals [34]. An example which was first solved quantum mechanically is [32]:

$$\phi(x) = ax^2 + bx^{-2}, \qquad \textit{the Calogero "Fluid",} \tag{2.7}$$

between any 2 particles, and many higher order partial differential equations [33].

d) All textbook-exercises and -examples students are supposed to solve.

Properties of Integrable Systems

Since the H and ω_k are analytic functions of the $\vec{I}$, the equations of motion (2.3) can be further transformed into uncoupled, *separated*, 1-degree of freedom equations [28,17,11,18]. Hence *all integrable systems* with N-degrees of freedom can be nonlinearly transformed into each other and are in this sense *equivalent to N pendula or even N harmonic oscillators*. Hence we expect their phase planes to look like nonlinearly deformed versions of the phase plane for the pendulum, Figure 3.

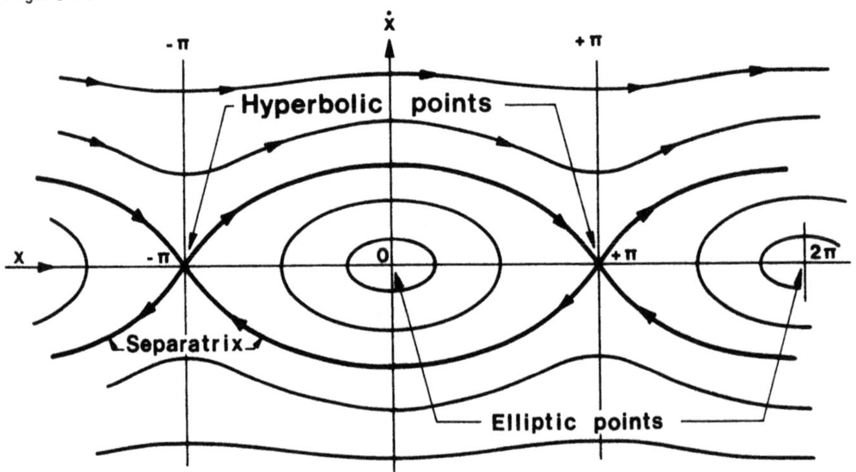

Figure 3
Phase plane plot for the pendulum, $\ddot{x} = -\sin x$. Note that it takes an infinite time to get from one saddle- ('hyperbolic') point (at rest at $x = -\pi$) to the next (at rest at $x = \pi$), along the 'separatrix', here separating bounded (oscillation) and unbounded motion (rotation). This plot is typical for most integrable systems ('elliptic': linearly stable).

Also, as we can guess from Figure 3 and eqs. (2.2-4), each bounded orbit lies on a (deformed) torus with an N dimensional surface [2,15,14], in 2N-dimensional phase space. They are called *'Invariant Tori'* since *any* orbit starting on one must stay on that one forever. The actions $\vec{I}$ "measure" the N radii of the torus and $\vec{\Theta}$ the N angles of a point on the torus.

It might be mentioned that nearly all of the known iterative-, perturbative-, series- and variational schemes [283,284], plus the new ones that continue to be invented, work fine for integrable systems. In particular, the Hamilton-Jacobi partial differential equation [7-11] can, in principle, be solved in terms of nice single-valued, functions [13,15,88].

Since each bounded orbit is constrained to a smooth N dimensional torus, it cannot come arbitrarily close to every point on the (2N-1) dimensional surface of constant energy, when $N \geq 2$. Hence, *an Integrable system cannot be Ergodic* ($N \geq 2$), let alone approach thermal equilibrium. In order to be ergodic an orbit should, in general, cover the surface of constant energy both densely and uniformly (some individual variables though, but not all, might still behave 'ergodically' [30], e.g. their time averages might equal their ensemble averages; also, a single harmonic oscillator is ergodic).

Since an integrable system can be nonlinearly transformed into harmonic oscillators there exist, in general, nonlinear versions of the (harmonic) 'normal modes', e.g. the 'Soliton' solutions of (2.5-7), which run all through the chain "without losing their shape, or energy", in general. Hence *integrable systems are bad candidates to try to establish 'Transport Laws' for*, e.g. Fick's law of diffusion or Fourier's law of heat conduction [29]. Again however, one may establish transport laws for some individual variables, in the thermodynamic limit of the system [30], but not for all. It is this ideal conduction by Solitons, etc. without any scattering or dispersion, which makes it difficult to even create the density- and/or temperature-*gradients* required in the above transport laws. In lattice-systems the calculated heat current varies with the lengths of the lattices in various ways but rarely inversely proportional to the length [29] as it should, according to Fourier's 'Law' of heat conduction.

One might speculate that integrable systems (and the invariant tori of nonintegrable systems, cf. section 2.5) could profitably be employed in superconductivity and superfluidity, i.e. their quantum versions, cf. section 2.8.

In the next section, 2.2, the chaotic properties of a Nonintegrable system are discussed, its most striking difference from an Integrable system. In section 2.5 the existence of regular 'invariant tori' in a Nonintegrable system is discussed, the most striking feature it has in common with an Integrable system.

2.2 Sensitive Dependence on Initial Conditions, and Random Behavior

We saw that for a Nonintegrable system the Canonical Perturbation series, or Birkhoff series, *diverge*, in all likelihood due to (or causing..) a singularity in one of the integrals $I_k(\vec{q},\vec{p})$. Moreover there appears to be an infinite set of such singularities (which may even be dense) in phase space, judging by the theorems quoted in section 2.4. This makes it hard to explicitly display such integrals [26,12,2], witness the efforts to calculate the, formally existing [31] and adiabatically invariant, Action of plasma physics [31] (cf. below (2.63)).

Thus, two orbits with their initial conditions arbitrarily close together, but "on different sides of such a singularity", can have vastly different behavior, i.e. the orbits 'depend sensitively on the initial conditions'. This dependence often is so sensitive that we cannot calculate the orbit for any time interval of interest, due to the experimental- or numerical uncertainty in the initial conditions: It is quite common to have effects of order 1 from causes of order 10^{-16} after a short time [90-91]. In fact, we shall see that there are regions in phase space where one has such effects even from an *arbitrarily small* change in the initial conditions of certain orbits. Therefore the results of experiments or numerical calculations can be very unpredictable and chaotic in such regions.

Some familiar examples of this unpredictable, chaotic, behavior are:

a) Throwing Dice, tossing coins or playing roulette. In these mechanical systems we have the initial conditions in our own hands, literally. Yet, the outcome is so sensitively dependent on the initial conditions that we use them as models of random behavior in conversation (Dutch weather is.., "as random as a throw of the dice") and in probability theory [69,6,79].

b) Pin-Ball Machines, i.e. oldfashioned ones with real pins and without flippers (Galton Boards). The balls move unpredictably and have a Binomial probability distribution at the bottom [69,6,79].

c) A Random-Number Generator in a digital computer. This is a deterministic algorithm (program), i.e. it will always give the same numbers from the same initial conditions. Yet, even after it has generated many numbers we still cannot estimate the value of the next number. Depending on the size of the computer we can make the sequence as random as desired [79].

This sensitive dependence on initial conditions is generated, in the first two examples, by singularities in the Hamiltonian itself, i.e. H is not differentiable at the edges of the dice, coins, balls and pins. It shows that even a few singularities in an Integral can cause chaotic behavior. Since I want to stress nonintegrable systems with a nice *analytic* Hamiltonian one might suppose that the above examples would be atypical. However, remember that an integrable system can be separated into N uncoupled equations, each with its own Action-Integral which can be thought of as the separate "Hamiltonian" for that equation, cf. N harmonic oscillators.

Hence the other Integrals are not unlike Hamiltonians and all it takes to make the system nonintegrable is to put some singularities in one of them. The resulting behavior is as chaotic as in the above examples or worse since we usually have an *infinite* number of singularities.

Examples of Non-Integrable Systems

The canonical example is the Hénon-Heiles system [25],

$$\ddot{x} = -x - 2xy \tag{2.8}$$

$$\ddot{y} = -y + y^2 - x^2 \quad , \quad \textit{or} \tag{2.9}$$

$$V(x,y) = \tfrac{1}{2}(x^2 + y^2) + x^2y - y^3/3 \quad , \tag{2.10}$$

cf. [24-26,2,12,15,90,91,41,51,61,62,49,45,81,202,100,112,105], about the simplest nonintegrable system one could imagine, according to examples a) and b) of section 2.1. This system has not been solved (globally, see refs. [100]) using any analytic method. Numerical solutions can however be obtained over a time interval of physical interest when sufficiently many numerical precautions are taken. How do we look for possible chaotic behavior in the 4-dimensional phase space of $x,\dot{x},y,\dot{y}$? If we restrict the initial conditions to one value of the energy, E, there are only 3 independent coordinates left over, due to the constraint $H(x,\dot{x},y,\dot{y}) = E$. Since 3-d. space is still difficult to draw we only plot the intersection points of the orbit $x(t)$, $y(t)$, $\dot{y}(t)$ with a 2-d. plane, y versus $\dot{y}$ (at $x = 0$). If the system were integrable this (Poincaré-) 'Surface of Section' would look like a dotted version of Figure 3, for the pendulum, i.e. "dotted ellipses" traced out by disjunct dots rather than by one continuous curve, dotted separatrices, etc. Surfaces of section for the Hénon-Heiles system (2.8-9) are shown in Figure 4. Inspecting its left column we see that at some low energy, $E = 1/12$ (2.10), the phase plot does look like a dotted deformed version of the one for an integrable system, cf. Figure 3 and its caption. Note the three most visible hyperbolic points and the separatrices "connecting" them. However at $E = 1/8$ these seem to have disappeared, replaced by a chaotic collection of dots (from *one* orbit) through which it would be difficult to draw a nice simple curve. At a higher energy, $E = 1/6$, nearly all of the available phase plane seems chaotically "filled" with the intersection dots of just *one* orbit. Yet some minuscule 'islands', with an '*elliptic*', i.e. linearly stable point at their center, can still be seen. Those orbits do not wander. A '*hyperbolic*' orbit is one that is unstable under linear perturbation.

If this Hénon-Heiles system had a second analytic integral $I(x,\dot{x},y,\dot{y})$, besides

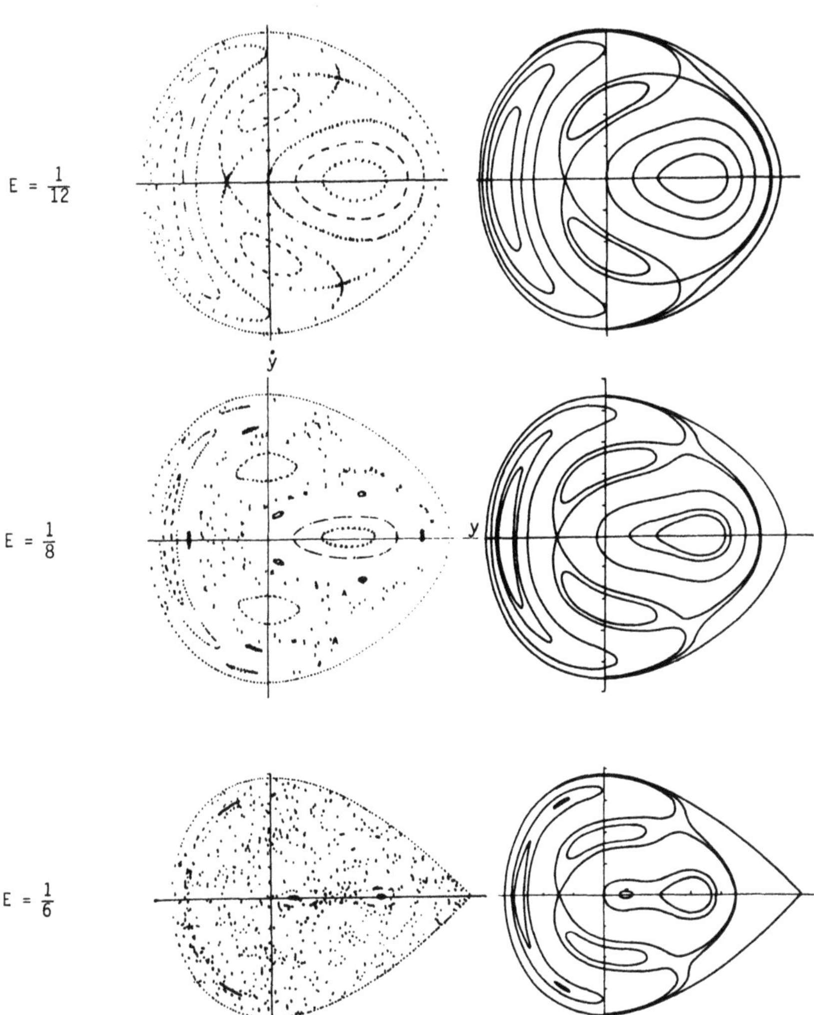

Figure 4
Surfaces of Section, in the y (horizontal) versus $\dot{y}$ (vertical) plane at $x = 0$, for the Hénon-Heiles system (2.8-9). The left hand column is generated by numerical integration of individual orbits and plotting the intersection points (dots) with this plane. At energy $E = 1/12$ several orbits are used. Note that all chaotic dots, at $E = 1/6$ and the ones between the 'islands' at $E = 1/8$, are created by *one* orbit (taken from [25,12]). The right hand column is generated by calculating a formal integral, with a variant of the Birkhoff series, and plotting its intersection curves at different values of this integral (taken from [26,12]). N.B. Only the first 80 dots, or so, of the chaotic orbits, can be computed reliably ($ ... !).

its Hamiltonian (2.10), it would be integrable according to section 2.1. Gustavson tried to calculate such an integral through 7th. order, in a variant of the Birkhoff canonical perturbation series [26,12]. This yields another surface, than the surface of constant energy, to which the orbit should be confined as well. Plotting the intersection curve of this new surface with the $y,\dot{y}$ plane he obtained the right column of Figure 4. Hence any single orbit starting on any such ('invariant') curve should return through that curve. Several curves are drawn at values of I, corresponding to the initial conditions in the left column. The curves agree quite well with the visible behavior of the orbits on the left, at $E = 1/12$, and several orbits at $E = 1/8$. However the chaotic "space filling" orbits, visible at $E = 1/6$ and 1/8, demonstrate that the curves on the right, corresponding to the same initial conditions, are incorrect and even misleading as to the qualitative behavior of the orbits. This provides a strong indication that the Birkhoff series is divergent here, i.e. that the Hénon-Heiles system is nonintegrable [100]. When one finds such chaotic dots, through which it is difficult to draw a nice curve, it is usually assumed that the system is nonintegrable. While one might hope that this chaotic (often called 'stochastic') behavior will justify the use of statistical methods, e.g. at $E = 1/6$, it is clear that statistical methods should not be used at $E = 1/12$.

The Hénon-Heiles system describes the motion of a test-star in the effective potential due to the other stars in a galaxy [25,26] and a number of other phenomena in physics and chemistry [100].

A second example, to which we return often, is the motion of protons in the *intersecting storage rings* of high energy physics where one would like to *avoid* chaotic behavior at all cost ($\leq$ 300 M\$). The ('betatron') oscillations of the protons about their ideal circular orbit within one ring are well described by the simple harmonic oscillator

$$d^2y/d\phi^2 = -Q^2y \quad , \tag{2.11}$$

where y is the proton's displacement perpendicular to the ring and ϕ its azimuthal angle, proportional to the time [204-208]. The 'tune' Q is determined by the gradient of the containing magnetic field. The effect of a second, colliding, beam is modelled by [146]:

$$d^2y/d\phi^2 = -Q^2y + BF(y)\left[\sum_{t=0,1,2,}^{\infty} \delta(\phi - t\,2\pi)\right] \quad , \tag{2.12}$$

cf. [201],(3.2). The second beam, from another ring, intersects the first ring only over a very small ϕ interval, whence the δ-functions. Due to their small size very few protons actually collide. Yet, they do feel a strong nonlinear electromagnetic force, BF(y), as they briefly pass through the second beam [200,201,203,80]. Thus, in order to get enough true collisions, the protons are required (/requested) to participate in $\approx 6 \times 10^{11}$ such near-collisions without exhibiting large oscillations, exceeding the width of the vacuum chamber, and be lost. The integer t in (2.12) counts this number of passages through an intersection region. This required number of revolutions is about the same as the earth and moon have ever made about the sun. Numerical solutions cannot be done over such cosmic times and one has to resort to other methods [201,200-211,245].

Here, we note that *between* the δ-pulses eq. (2.12) is linear (2.11) and can be trivially solved. *During* a pulse the momentum, $p \equiv dy/d\phi$, is changed by an amount $BF(y(t2\pi))$. Hence eq. (2.12) can be translated exactly [245] into the difference equations,

$$y_{t+1} = Cy_t + (S/Q)p_t \quad , \tag{2.13}$$

$$p_{t+1} = -SQy_t + Cp_t + BF(y_{t+1}), \tag{2.14}$$

$$\textit{where } C \equiv \cos(2\pi Q) \quad , \quad S \equiv \sin(2\pi Q), \tag{2.15}$$

with $y_t \equiv y(t2\pi+), p_t \equiv p(t2\pi+)$. This is *an algebraic mapping of the y,p phase plane on itself, exactly equivalent to* (2.12). In the previous example Hénon and Heiles were forced to numerically integrate out their equations of motion (2.8-9) just to get from one point in the plane to the next. In this example we have such a mapping explicitly and, given some F(y), e.g. $=y^2$ or $=y^3$, we can use a pocket calculator to get hundreds of points [146]. The mapping is further simplified by eliminating p_t between (2.13-14) (with $t+1 \to t$):

$$y_{t+1} + y_{t-1} = 2Cy_t + (BS/Q)F(y_t) \qquad (2.16)$$

The y,p phase plane is a linear transformation (2.13) of the y_t, y_{t+1} plane, which we plot instead, in Figure 5. We choose initial conditions y_0, y_1, plot this point, calculate y_2 from (2.16) at $t=1$ and plot y_1, y_2, etc. In Figure 5 only separate dots are printed. However so many points were calculated and plotted that several seem to merge into continuous curves. The centers of the 4 large islands are one orbit of period 4. In between, where the islands touch, are 4 'hyperbolic' points, i.e. an orbit which is linearly unstable under perturbation [201,48,40,43,35]. Magnifying the region about one of those hyperbolic points we obtain the lower picture [209]. Note that all chaotic dots are points on *one single* orbit. We could take our initial point nearly anywhere in this region and still obtain an orbit which appears to "fill" the region chaotically. It might be mentioned again that Figure 5 is the 'Surface of Section' for the differential equation (2.12), as Figure 4 is for the Hénon-Heiles system (2.8-9). Figure 5 was obtained with $F(y) = 2\,[1-\exp(-y^2/2)]/y$ [209] but similar pictures are easily obtained for many other choices of F(y) [1-6,35-40,43,48,60-62,80-82,90-93,100,200-202,209-211]. Fig. 1, in the Introduction, is a similar magnification about a hyperbolic point, of period 5, obtained with $F(y) = y^2$ [35,48]. In ref. 36 a comparison with the results of the Birkhoff series is made. We confine ourselves here to analytic nonlinear F(y)'s.

Looking at Figs. 5,4,1 one wonders how random, or even ergodic, the behavior in those chaotic regions truly is. I am often asked whether these could not be due to computer (round-off) error or, worse, programmer error. While there are numerical examples inherently without round-off error [37,4] we turn to the oldest known nonintegrable system for some analytical results which are typical for most non-integrable systems [12-21,1-6,70]:

A final analytical example is the *celestial Three-Body System*. Take the usual celestial 2-body system, with two equal heavy masses M, and let loose a small third mass m above their center of mass, its initial velocity perpendicular to the x,y plane of revolution of the 2-body system. All masses attract each other according to Newton's Law of gravitation. Thus m moves "along" the z-axis, z(t), crossing the x,y plane at $z=0$, once, repeatedly or never, depending on its initial conditions. The time intervals between subsequent crossings, i.e. zero's, are $\tau_1, \tau_2, \tau_3 \ldots$ If the orbit is periodic all τ's are forever equal. If m escapes, the last τ is infinite. Loosely speaking, the analytic result is that, given *any* such sequence of τ's, there are initial conditions for a 3-body system such that z(t) does have those given intervals. In particular, if we tear a page out of a book of *random* numbers there is an orbit with those τ's. If we then tear out a second page we can still find another orbit which "follows" both pages. If, on the other hand, we did not get a second page the last τ becomes infinite. Again, there is another orbit such that m follows the first page but then 'escapes' and never crosses again. Thus, no matter how many τ's we observe we cannot say anything about the next τ.

The actual technical results, due to Sitnikov and Alexeev [70-75,13,2], employ *integer* sequences $\{n_k\}$, with: $n_k \equiv$ Integer Part (τ_k/T) where T is the period of the 2-body system. Under some minor conditions on the 2-body system it is shown that, given 'almost' *any* integer sequence $\{n_k\}$, there are initial conditions $z(0), \dot{z}(0)$ (only) such that z(t) will yield this integer sequence. If we also vary the initial conditions for the 2-body sytem we can change T as well, whence the "loose" version given earlier.

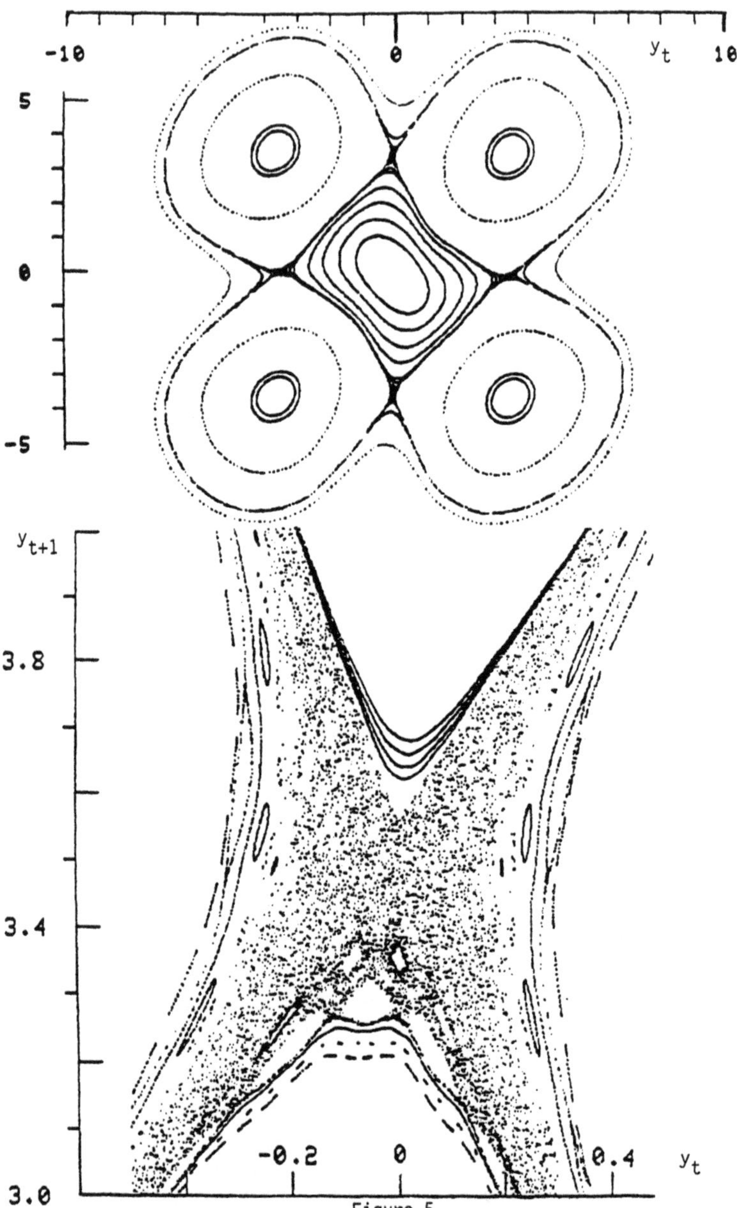

Figure 5
Phase Plot for Intersecting Storage Rings (2.16-12), i.e. vertically y_{t+1} versus y_t, horizontally. The lower picture is a magnification around a hyperbolic point in the top picture. All chaotic dots are created by *one* orbit [209]. N.B. see caution below Fig. 4.

The various initial conditions mentioned above can all be found within one chaotic region of the 3-body system. Similar results apply to the chaotic regions of most nonintegrable systems [13,76] but are more easily visualized in this 3-body system. We therefore conclude that *some properties of many orbits are as random as coin tosses*, yet we shall see in sections 2.5 and 2.8 that a nonintegrable system is not ergodic in general. There is an infinite number of periodic orbits interspersed as well. Some have cosmic periods and may, in practice, be indistinguishable from the above 'quasi random' orbits.

You may agree by now that simple nonintegrable systems exist, with some random properties, but still wonder how exceptional such systems are and also wonder how prevalent chaotic regions are within the phase space of one system. The first question is addressed below; the second in section 2.4.

2.3 Most Hamiltonians are Non-Integrable (Siegel, Poincaré)

If we write down randomly chosen analytic Hamiltonians ($N \geq 2$) most of them will be nonintegrable and only rarely will we encounter an integrable one. These sweeping results were obtained by Siegel [78,77,12,13], after earlier results for the 3-body system by Poincaré (1890) and Bruns (1884) [10,18,13,88,19].

For example, a 2-degree of freedom Hamiltonian $H(x,y,\dot{x},\dot{y})$ which is analytic has a convergent Taylor expansion,

$$H(x,y,\dot{x},\dot{y}) = \sum_{j,k,m,n=2}^{\infty} c_{j,k,m,n} \; x^j y^k \dot{x}^m \dot{y}^n \tag{2.17}$$

The c's are taken as components of an infinite vector $\vec{c}$ [12,78]. Given some $\vec{c}$ the theorem states that in any neighborhood of $\vec{c}$ there is a (convergent) $\vec{c}\,'$ which yields another Hamiltonian H', via (2.17), with an infinite Birkhoff series which is divergent, i.e. H' is nonintegrable. A neighborhood of $\vec{c}$ consists here of an infinite set of positive intervals,

$$|c'_{j,k,m,n} - c_{j,k,m,n}| < \varepsilon_{j,k,m,n} \quad , \tag{2.18}$$

for all j,k,m,n (≥ 2), where the ε's may be any set of positive numbers, including ones with ε smallest for small j,k,m,n.

Thus *the Nonintegrable Hamiltonians are dense* in this sense *among all analytic Hamiltonians* [78]. Moreover, Siegel also showed [77] that *the Integrable Hamiltonians* - the ones with a convergent Birkhoff series - *are not dense* in this sense, but cf. [89], i.e. not dense-everywhere since there clearly exist 1-parameter families of integrable systems somewhere in $\vec{c}$ space, e.g. the Toda chain (2.5). It may also interest you that the nonintegrable Hénon-Heiles Hamiltonian [100] (2.10) becomes integrable if we change the sign of its y^3 term [131]. Still, many nonintegrable systems cannot be well approximated therefore by integrable systems, including series methods, cf. Fig. 4. It also shows that truncated Hamiltonians may not yield good approximations (no matter how high the order from where on we set $c = 0$ in (2.17)). For instance 3-particle Hamiltonians of the type,

$$H = (\dot{x}_1^2 + \dot{x}_2^2 + \dot{x}_3^2) / 2m + \phi(x_2 - x_1) + \phi(x_3 - x_2) + \phi(x_1 - x_3), \tag{2.19}$$

include the integrable Toda chain (for some a,b,c in (2.5) $\phi(x)$ has a minimum at $x = 0$). Yet, their Taylor expansion up to, and including, 3rd. order terms in $(x_2 - x_1)$, etc. (if $\neq 0$) can always be reduced and transformed [61,62] into the non-integrable Hénon-Heiles Hamiltonian (2.10). The regular Soliton solutions of the Toda chain at all energies [62,61,2] are however in marked contrast with the chaotic solutions of the Hénon-Heiles system, cf. Fig. 4.

In the next section we discuss the mechanism responsible for this chaotic behavior and find it to be wide-spread.

2.4 Chaotic Regions are Abundant in Phase Space

In this section chaotic regions are related to the wild behavior of the 'separatrices' emanating from hyperbolic points, of which there are infinite numbers in most nonintegrable systems. Inspecting Fig. 4 ($E = 1/12$ vs. $1/8$), Fig. 5 (bottom vs. top) and Fig. 1 it *seems* that some separatrices have changed into chaotic dots whereas they remain visible at all energies in an integrable system, simply connecting two hyperbolic points with a finite curve, cf. Fig. 3. In a nonintegrable system a separatrix does so with a curve of *infinite* length. Its infinitely many loops, plies and pleats are miraculously folded away in the chaotic regions. The same aplies to the second separatrix "connecting" the two hyperbolic points. Thus an arbitrarily small error, in this region, may move the orbit from one separatrix to the other, etc. to larger and larger errors [90,91].

For example, consider a dynamical mapping of the type (2.16) [35,43,48],

$$y_{t+1} + y_{t-1} = 2Cy_t + 2y_t^2 \ , \tag{2.20}$$

$t = 0,1,2,\ldots$ Its solutions of period 1, $\hat{y}_0 = \hat{y}_1 = \hat{y}_2$ etc., are obviously the origin and,

$$\hat{y}_1 = \hat{y}_0 = 1 - C \tag{2.21}$$

which is *a 'hyperbolic' point* as we shall see, linearizing (2.20) about it:

$$\Delta y_{t+1} + \Delta y_{t-1} = (2C + 4\hat{y}_t)\ \Delta y_t, \tag{2.22}$$

$$\textit{whence}\ \Delta y_{t+1} + \Delta y_{t-1} = (4-2C)\ \Delta y_t, \tag{2.23}$$

with $y_t \equiv \hat{y}_t + \Delta y_t$. Its solutions are

$$\lambda^t \ \textit{and}\ \lambda^{-t}\ \textit{with}\ \ \lambda + \lambda^{-1} \equiv 4 - 2C, \tag{2.24}$$

as checked by substitution. Thus we have (hyperbolic) real eigenvalues [49(1981)],

$$\lambda_1 > 1 \quad \textit{and} \quad \lambda_2 = 1/\lambda_1 < 1 \ , \tag{2.25}$$

when $C < 1$. The corresponding eigenvectors in the phase plane (y_t vs. y_{t+1}) are

$$\vec{e}_1 = (1,\lambda_1)\ \textit{and}\ \vec{e}_2 = (1,1/\lambda_1), \tag{2.26}$$

cf. (2.23). We plot a Separatrix by taking thousands of points in some small interval $\hat{y}z$ along $\vec{e}_1$, see Figure 6, and repeatedly mapping each point. In an integrable system the mapped points return to the hyperbolic point $\hat{y}$ *along* the second eigenvector $\vec{e}_2$ via a separatrix of *finite* length, cf. the hand-drawing in Fig. 6a (after Fig. 19.11b [11] and Fig. 7 of [56]). However, using the Hénon mapping (2.20), at $C = 0$, we see in Fig. 6b that this 'unstable separatrix' gyrates, showing no inclination to join $\vec{e}_2$ [202]. To test this we take a similar interval along $\vec{e}_2$ already and map it backwards ($y_0, y_1 \nrightarrow y_{-1}$ with $t = 0$ in (2.20), etc.). *This 'stable separatrix' intersects the unstable one*, in ..A,B,C,D,E,F,G,.. of Fig. 6b; note the analytic argument in appendix A. From this we deduce, below, that there must be an *infinite* number of intersection points and loops between A and $\hat{y}$. Continuing the unstable separatrix we find longer and longer loops between intersection points, coming closer and closer to each other, cf. Fig. 6c [56,202]. Some loops follow the separatrix all the way back to near z, before returning and intersecting A-$\hat{y}$ again. The stable separatrix is mapped backwards for only one additional loop, so as not to clutter up Fig. 6c. However the picture should be symmetric about the line " $y = x$ " and one can imagine how cluttered it does become, below $\hat{y}$ and near the so called *'homoclinic' points* ..A,B,C,.. [13,14], as we let $t \to \infty$. In those chaotic regions we have, arbitrarily close together and moving in opposite directions, loops from two different separatrices as well as from one and

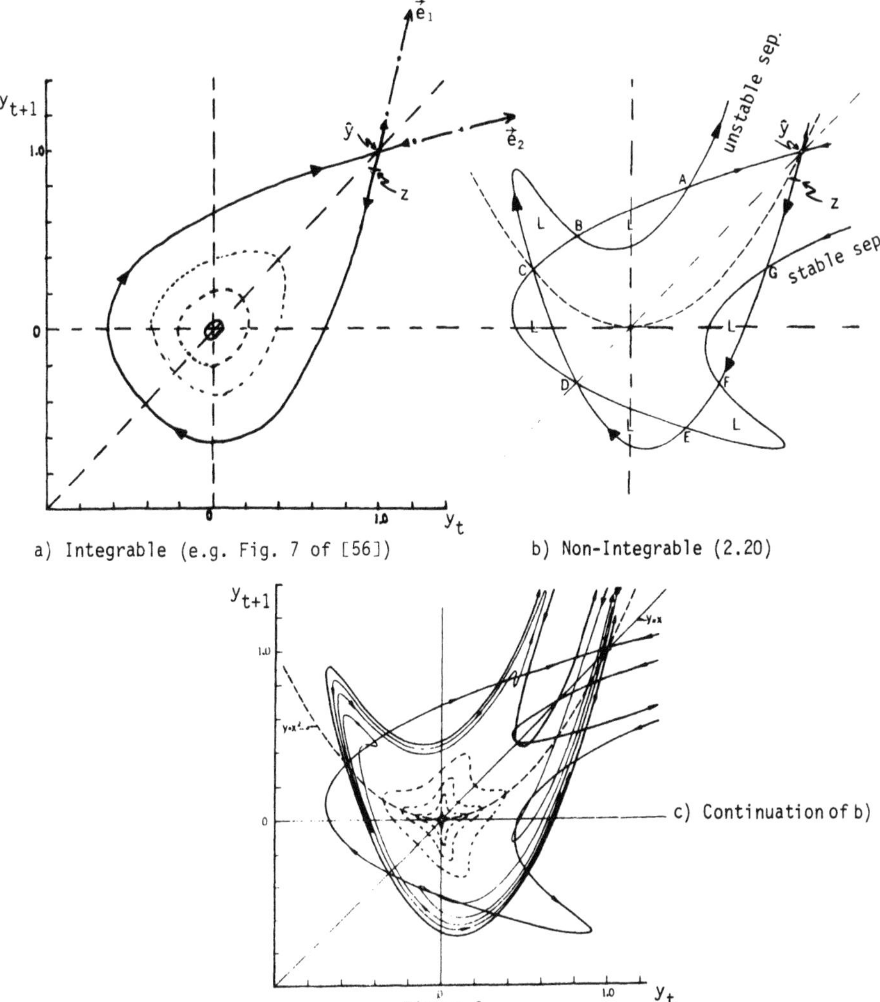

Figure 6

Intersecting-wild-Separatrices in the phase plot of the nonintegrable mapping (2.20) at C = 0, [56,202]; but only one *single finite* separatrix in the integrable case a). Points in some small interval $\hat{y}z$, along the eigenvector $\vec{e}_1$ of the hyperbolic point $\hat{y}$ of period 1, are mapped repeatedly (horizontally y_t, vertically y_{t+1}). In a) they return to $\hat{y}$ along $\vec{e}_2$. However, if we map an interval along $\vec{e}_2$ backwards in b) and c) we get a second separatrix *intersecting* the first one in infinitely many points, ..A,B,C,D,E,F,G,.. The area of each loop remains a constant, L in b). Since there are an infinite number of intersection points between $\hat{y}$ and A (and G,$\hat{y}$) the length of the loops approaches infinity. Note the dissipative loops in Figs. 22 and 23.

the same separatrix, cf. Fig. 6c. One can even prove [302] that near every such homoclinic point there is a region in which the homoclinic points, of these same two separatrices, are *dense* [13]. Hence the orbits there have a very 'sensitive dependence' on the initial conditions and 'quasi-random' orbits, mentioned in section 2.2 (3-body system), can be shown to exist in these regions [13,14,76]. For some C^∞ mappings we have analytic arguments for the splitting of separatrices, see Appendix A, i.e. those near piecewise-linear ones. Homoclinic points can be found numerically. This also serves as a practical test for the (non)integrability of a system since Moser proves that an analytic integral cannot (even 'regionally') exist if there are homoclinic points [13] (for other numerical tests see refs. 90,91).

Finally, the postponed technical argument: The homoclinic point E belongs to both separatrices, cf. Figure 6. So, when we map it we get another homoclinic point, C, by definition. Mapping it again we obtain yet another homoclinic point A, etc. Since the stable separatrix approaches $\hat{y}$ along $\vec{e}_2$ as λ_2^t, and since $\lambda_2 < 1$ (2.25), there is an infinite number of homoclinic points between A and $\hat{y}$. The lengths of the loops between them must approach infinity then, due to the fact that the area enclosed by each loop is a constant, L in Fig. 6b. Actually the area of *any* closed loop remains constant under the mapping:

$$\oiint_{loop} dy_{t-1}dy_t = \oiint_{\substack{mapped\\loop}} J^{-1}dy_t dy_{t+1} = \oiint_{\substack{mapped\\loop}} dy_t dy_{t+1} , \tag{2.27}$$

since the Jacobian, J, the determinant of the 2×2 matrix in:

$$\begin{pmatrix} dy_{t+1} \\ dy_t \end{pmatrix} = \begin{pmatrix} 4y_t + 2C & -1 \\ 1 & 0 \end{pmatrix} \begin{pmatrix} dy_t \\ dy_{t-1} \end{pmatrix} , \tag{2.28}$$

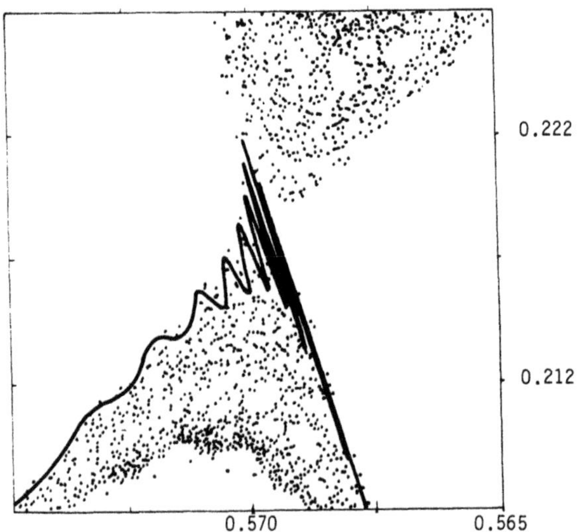

Figure 7
First few loops of *one* of the intersecting Separatrices, emanating from a hyperbolic point of period 6, using the Hénon mapping (2.20) at $C = 0.4$ (axes rotated compared to Fig. 6; taken from [2]. Chaotic dots created by one orbit).

cf. (2.22), equals 1 everywhere. For the relation of this 'area-preserving' property to Liouville's Theorem, and Poincaré invariants in general [7], see refs. 14-12, 255. We shall frequently employ this property of a conservative system.

Another example is shown in Figure 7, for one of the smaller chaotic regions of the Hénon map (2.20), much like the one displayed in Fig. 1 (period 5). A second (intersecting) separatrix can be obtained but even the first few loops in this smaller region required a substantial numerical effort (Cuthill, in [2]).

We saw earlier that there is a demonstrable abundance of homoclinic points and chaotic regions as soon as a first homoclinic point is obtained. The latter is accomplished by a theorem of Zehnder [59,13] which leads to the conclusion that for *'most'* mappings (2.16) the homoclinic points are dense near an elliptic orbit. Those, in turn, are easier to find [40-55], e.g. the origin 0,0 is an elliptic orbit of (2.20) and (2.16) ($|C| < 1$). The word 'most' is used here in the same sense as earlier, in section 2.3.

In the next section we find an infinite number of elliptic orbits and about them an abundance of very smooth 'quasi-periodic' orbits, in addition to the above 'quasi-random' orbits...

2.5 Smooth Regions are Abundant (K.A.M. Theorem)

By a 'smooth' or 'regular' region is meant here a region of phase space in which the majority of orbits are confined to *'Invariant Tori'*. Any orbit starting on an invariant torus remains confined to it, by definition. If the system were integrable all bounded orbits would lie on invariant tori, cf. section 2.1. In this section we encounter the celebrated Kolmogorov-Arnold-Moser Theorem which proves the existence of such regular regions about most elliptic orbits [22,12-15,1-6]. But we know from the previous section that chaotic regions are dense there in general [59]. So, even within the regular regions there still are small chaotic regions, in the gaps between the 'K.A.M. tori', as can be seen in magnifications of the regular regions of Figs. 4($E = 1/12$) or 5 [202]. Recalling that in a chaotic region there are periodic orbits also, some of which are elliptic in general, we conclude that *inside the regular (smooth) regions are chaotic regions, inside which are regular regions, etc.* This fantastic hierarchy is partly visible between the higher order islands in Figs. 1, 5 (bottom) and in Figs. 13-15 of ref. 202. It brings us to the Poincaré-Birkhoff theorem which shows that sufficiently close to an elliptic orbit exist 2k other periodic orbits ($k \geq 1$), half of which are hyperbolic and half of which are elliptic, in general [133,154,15,14,12]. Close enough to the latter we apply the theorem again, etc. Thus there exist regions with an infinity of elliptic orbits and an abundance of regular as well as chaotic regions therefore.

The fact that the invariant tori have finite measure may be frustrating in statistical mechanics (but can be comforting in mechanics [201]): An invariant torus has an N dimensional surface which in general cannot come arbitrarily close to every point of the 2N-1 dimensional surface of constant energy. This destroys earlier beliefs that most nonlinear Hamiltonian systems would be ergodic [135, 134, 3], would have decent transport laws and approach thermal equilibrium [136,134] and the hope that most dynamical mappings would be ergodic [134,137,138,13]. However, a system can still be ergodic when the equations of motion are singular enough [156] for the K.A.M. theorem not to hold [95,96,137,138,153,90-94] or perhaps in some thermodynamic limit [139,140,79], even if K.A.M. does hold, cf. section I.

For example, consider again a dynamical mapping of the type (2.16), used for intersecting storage rings [47,201,25],

$$y_{t+1} + y_{t-1} = 2Cy_t + y_t^3 , \qquad t = 0,1,2,.., \qquad (2.29)$$

with a cubic nonlinearity, for which the next argument is simpler than for the y^2 used earlier (2.20), cf. [48]. The origin is an *elliptic* orbit ($|C| < 1$) as we see, remembering $C \equiv \cos(2\pi Q)$ (2.15) and linearizing about the origin:

$$y_{t+1} + y_{t-1} = 2\cos(2\pi Q)y_t \ , \tag{2.30}$$

$$\therefore \quad y_t = r[\exp(it2\pi Q + \Theta_0) + c.c.]/2 \ , \tag{2.31}$$

as checked by substitution [149]. Guided by this we consider some asymptotic expression of the same type for the nonlinear (2.29)

$$y_t = r[\exp(it2\pi\sigma(r) + \Theta_0) + c.c.]/2 + \textit{higher harmonics of } \sigma, \tag{2.32}$$

$$\textit{with} \quad \sigma(0) = Q \ . \tag{2.33}$$

Substituting this in (2.29) and equating the lowest Fourier coefficients we immediately derive

$$2\cos(2\pi\sigma(r)) = 2\cos(2\pi Q) + 3r^2/4 \tag{2.34}$$

($Q \neq n/4$ or $n/3$ [150]) and obtain a 3rd. harmonic proportional to $r^3 (r \to 0)$. Iterating (2.29) one could find higher order (in r) contributions to (2.34) and to the other harmonics (2.32) [151], which all vanish faster than r, as $r \to 0$. Hence, *for small enough* y, *the nonlinear mapping* (2.29) *can be approximated by:*

$$y_{t+1} + y_{t-1} = 2\cos(2\pi\sigma(r_t))y_t \ , \tag{2.35}$$

where $\sigma(r)$ is determined by (2.34) ($r \to 0$) and r_t is the constant amplitude of y_t, cf. [151]. Changing to polar coordinates, in the original phase plane,

$$r_t^2 \equiv x_t^2 + y_t^2 \ \textit{and} \ \tan\Theta_t \equiv x_t/y_t \ , \textit{with} \ x_t \equiv (Cy_t - y_{t+1})/S, \tag{2.36}$$

cf. (2.15), the mapping (2.35) becomes even simpler,

$$r_{t+1} = r_t \ \textit{and} \ \Theta_{t+1} = \Theta_t + 2\pi\sigma(r_t), \tag{2.37}$$

cf. (2.32), a so called '*Twist Mapping*' [151]. It can trivially be integrated producing the simple phase plot sketched in Figure 8: Note that both its 'rational' and 'irrational tori' are invariant tori, in this integrable approximation (2.37-35). In order to get back to our nonintegrable mapping (2.29) we must add to (2.35) the function

$$f(y_t, r_t) \equiv 2(C - \cos 2\pi\sigma(r_t))y_t + y_t^3 \ . \tag{2.38}$$

One easily sees from (2.34-36) that f vanishes as r^3. This addition changes the twist mapping (2.37) into

$$r_{t+1} = r_t + f(r_t,\Theta_t) \ \textit{and} \ \Theta_{t+1} = \Theta_t + 2\pi\sigma(r_t) + g(r_t,\Theta_t), \tag{2.39}$$

with functions f,g which could be obtained by transforming the addition f (2.38) to polar coordinates (2.36). The f,g "perturbations" can again be seen to vanish faster than r, close enough to the origin.

The K.A.M. theorem can be invoked now, having transformed (2.29) into (2.39)[156] It states that in some neighborhood of the origin *most of the invariant irrational tori*, we found in Fig. 8 for (2.37), *still exist for the nonintegrable* (2.39), *i.e.* (2.29). All orbits on such an invariant 'K.A.M. Torus' are quasi-periodic with the same irrational winding number σ (2.34) [150]. Each orbit covers the torus densely. Essential to the proofs [22,148,12-15] is the '*Twist Condition*':

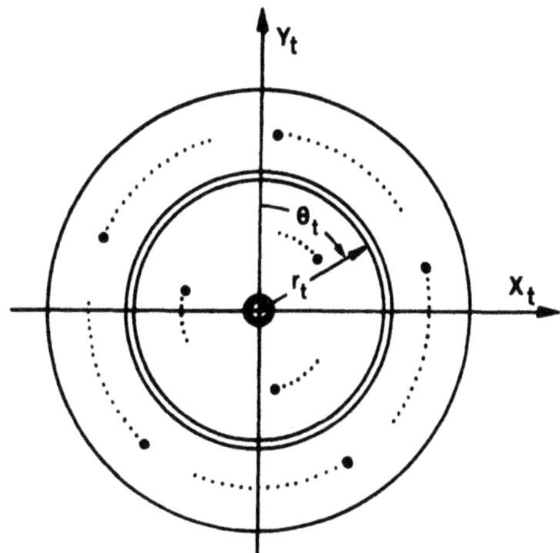

Figure 8
Phase Plot of the integrable twist mapping (2.37), an approximation of the nonintegrable (2.29) at small y values. At some r values the 'winding number' $\sigma(r)$ is an *irrational* number and we obtain invariant "tori" (continuous circles) which are densely covered by even a single orbit. When $\sigma(r) = m_1/m_2$, some *rational* number, we obtain invariant tori (dotted circles) which are not densily covered by a single orbit since each orbit is m_2- periodic (several orbits are plotted at $\sigma = 1/5$ and 1/3). The orbits on the 'irrational tori' are all 'quasi-periodic'.

$$d\sigma / dr^2 \neq 0, \qquad \textit{the Twist Condition} \qquad (2.40)$$

at the origin, which is satisfied by (2.34). The shape of the resulting KAM tori differs very little from the continuous circles of Fig. 8. So, one might innocently hope that the proofs would be simple, at least when $r \to 0$. However we shall see that the interspersed *rational* circles do not survive (2.39) and that some of the orbits on them become the quasi-random orbits of earlier sections. Hence the long-time behavior of the orbits, and the K.A.M. proofs, depend on "how irrational" σ is! This runs counter to many people's physical intuition but happens to be the truth in these systems. The winding number $\sigma(r)$ is '*sufficiently irrational*' for the K.A.M. proofs to hold [156] when it satisfies:

$$|\sigma - n/m| \geq c/|m|^{2+\beta}, \quad \textit{for all integers } n,m, \textit{ some } \beta: \ 0 < \beta < \tfrac{1}{2} \qquad (2.41)$$

and a $c > 0$. It means that *those* tori survive (2.39) whose σ cannot be well approximated by rationals. Since the majority of the irrational numbers satisfy (2.41) [40] the majority of the invariant tori will survive (2.39) (in a small neighborhood of an elliptic orbit [149]). This already shows that the majority of all orbits in that neighborhood will be confined to it *for all time*. The nonlinear stability of the remainder is discussed in section 2.7. Results valid for all time had been sought for over a century [13,10], when the K.A.M. theorem was announced [22]. Recent work: see [40] and (1982-) refs. to Mather and to Katok, there.

While the K.A.M. proofs are constructive no practical -and general- scheme has yet been demonstrated for the numerical construction of a K.A.M. torus, given some σ. Techniques for particular classes of invariant tori are found in refs. 40-42, 56-58,201, 202. However most people believe that they can be seen in numerical phase plots, e.g. in Figs. 5 and 4.

The fate of the *rational* invariant "tori" of Fig. 8 is the subject of the Poincaré-Birkhoff Theorem: Take a point r_0, Θ_0 on a torus with $\sigma(r_0) = m_1/m_2$ and map it m_2-times. In the integrable case (2.37/34) we get:

$$\Theta_{m_2} = \Theta_0 + m_2\, 2\pi\sigma(r_0) = \Theta_0, \qquad \textit{modulo}\ 2\pi, \tag{2.42}$$

with $r_{m_2} = r_0$, i.e. a periodic orbit for any value of Θ_0. In the nonintegrable case (2.39) we get:

$$\Theta_{m_2} = \Theta_0 + m_2(2\pi\sigma(r_0) + h(r_0,\Theta_0)), \tag{2.43}$$

with $r_{m_2} \neq r_0$ in general, where $h(r,\Theta_0)$ is some well defined function that could be obtained by substituting f,g (2.39), and the expansion of (2.34) about r_0, m_2-times into each other. It will suffice to check that h vanishes faster than r, as f,g did. Therefore we can always find, near r_0, some ρ_0 with

$$2\pi\sigma(\rho_0) + h(\rho_0,\Theta_0) = 2\pi m_1/m_2\ , \tag{2.44}$$

since σ changes linearly about r_0, cf. (2.40), (2.34) (r_0 sufficiently close to the origin). Hence, mapping this point we do get $\Theta_{m_2} = \Theta_0$, but $\rho_{m_2} \neq \rho_0$ in general. Repeating this for all Θ_0 values one would obtain some curve $\rho_0(\Theta)$ whose points do not rotate but would be mapped radially only, as sketched in Fig. 9, onto

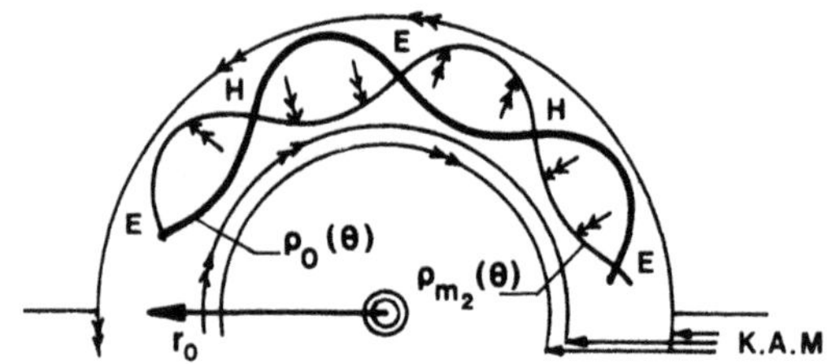

Figure 9
Phase Plane for (the intersecting storage ring) equation (2.39/29) after m_2 revolutions, near a rational torus of Fig. 8 at $\sigma(r_0) = m_1/m_2$ (sketch, after [15]). In the previous integrable approximation (2.37) each point of that invariant torus in Fig. 8 would be m_2-periodic. For the present - nonintegrable - mapping (double arrows) we obtain a curve $\rho_0(\Theta)$ whose points are mapped radially only. Thus the intersections E,H,E,H,... with its own image $\rho_{m_2}(\Theta)$ yield the only m_2-periodic orbits to survive from the rational torus of Fig. 8.

the curve $\rho_{m_2}(\Theta)$. The two curves must intersect each other in some $2k(\geq 2)$ points, since the area enclosed by the first curve is the same as by the mapped curve, cf. (2.27-28) with (2.29/39). Hence the intersection points E,H,E,H,... do yield m_2-periodic orbits and we obtain the (Poincaré-Birkhoff) theorem that the invariant rational tori of (2.37) 'break up' into 2k periodic orbits under the nonintegrable perturbation (2.39/29). Inspecting the directions of the m_2-mapping about those points in Fig. 9 we see that k orbits are elliptic ($r_0 \to 0$) and k are hyperbolic. Remember that the latter were tied up with the chaotic regions by their wild separatrices. A composite picture is sketched in Fig. 10. Shifting the variables by a constant so that another elliptic point becomes the new origin we are back to where we started in (2.29): *Thus Fig. 10 should be repeated within itself* [154], *about the new elliptic orbits, and ad infinitum about subsequent elliptic orbits!* This demonstrates at once an abundance of chaotic as well as regular regions.

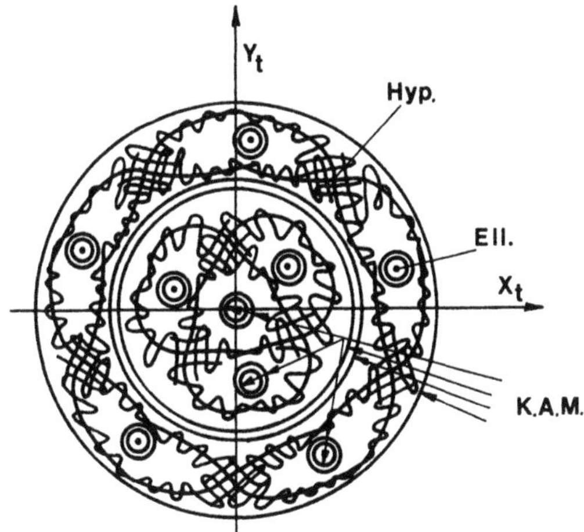

Figure 10
Composite Phase Plot near the elliptic origin, for the intersecting storage rings (2.39/29) (after [22,15,2]). The rational tori of Fig. 8 have broken up into disjunct elliptic and hyperbolic orbits, cf. Fig. 9. From the latter now emanate the wild separatrices (sketch) and chaotic regions discussed in section 2.4. Concentric with the origin are irrational K.A.M. tori which do survive, containing the majority of all orbits in this picture. Magnifications about the elliptic orbits yield pictures similar to this Fig. 10, etc.! Of that *infinite hierarchy* only the next K.A.M. tori are sketched.

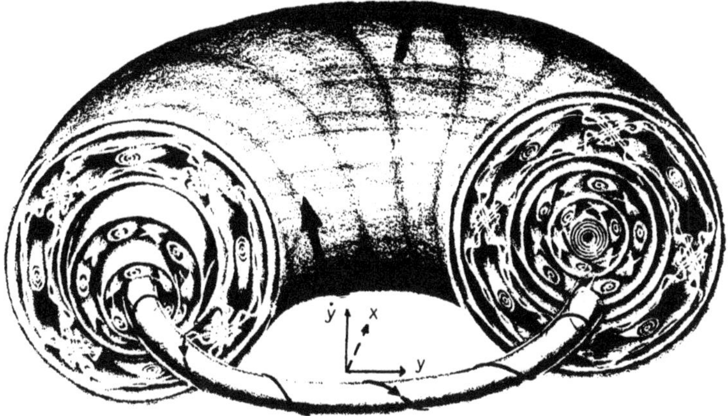

Figure 11
Nested K.A.M. tori in the 3-dimensional $x,y,\dot{y}$ surface of constant energy of a Hamiltonian system like the Hénon-Heiles system (2.8-10); see figs. 10 and 4 (taken from [16]). Note the smooth regular K.A.M. tori about the center and about the elliptic orbits which, together with the chaotic regions, populate the gaps between the K.A.M. tori A magnification about such an elliptic orbit yields the same picture all over, *etc. ad infinitum*.

In the full phase space of the Hamiltonian system, for which our mapping is a 'surface of section', these results are more difficult to sketch. For a 2-degree of freedom system the phase space, reduced via $H(x,y,\dot{x},\dot{y}) = E$, is 3-dimensional. Hence Figure 11 puts Fig. 10 in true perspective. The above theorems, and their conditions, can also be expressed directly in terms of the Hamiltonian of the system and the frequencies of the orbits. Several variants applicable to different situations are listed by Arnold in Appendices 8 (and 9) of ref. 14. One obtains an integrable approximation, like our twist mapping (2.37/35), by constructing only the first few orders of the (divergent) Birkhoff series, discussed in section 2.1. This yields an approximate-Hamiltonian $\bar{H}$ and -equations of motion (2.2-4) which are integrable. Our '*Twist Condition*' (2.40) on σ, is replaced by one on the frequencies ω_k of that integrable approximation:

$$\det |\partial\omega_j(\vec{I})/\partial I_k| \neq 0 \quad , \textit{ i.e. } \quad \det|\partial^2\bar{H}(\vec{I})/\partial I_j\partial I_k| \neq 0, \tag{2.45}$$

according to (2.3). The origin of our mapping can be replaced by an elliptic orbit or equilibrium position of H. Resonances of order 4 or less are again [150] not allowed between the components of $\vec{\omega}(\vec{0})$, for the K.A.M. Theorem to hold [13,14]. The frequencies $\omega(\vec{I})$ are '*sufficiently irrational*', as in our (2.41), if

$$|\vec{\omega}(\vec{I})\cdot\vec{m}| \geq c/|\vec{m}|^{N+1} \quad \textit{for all integer vectors } \vec{m} \text{ (N-diml.)} \tag{2.46}$$

and a $c > 0$. [12-15,152]. Hence those tori of $\bar{H}$ which satisfy (2.46) will survive and become the K.A.M. tori of H. In some neighborhood of the equilibrium position these contain again the majority of all orbits [156]. The K.A.M. theorem might be considered a modern version of Ehrenfest's 'Adiabatic Theorem' [134,125-127,94].

One of the outstanding problems is to *quantify* "how chaotic" a region is or "how smooth" [81,80], in view of the fact that both types are intertwined. Yet it is clear that Figure 4 looks "more chaotic" at $E = 1/6$ than at $E = 1/12$ [40]. Recent developments which may help us to answer such questions, in the future, are mentioned below. The long-time (in)stability of orbits not on K.A.M. tori is discussed in section 2.7.

2.6 Bifurcating Periodic Orbits

Absolutely crucial to the regular behavior and other phenomena of the last section were the linearly *stable* periodic orbits. Here we shall see that such an 'elliptic' orbit ($\sigma = m_1/m_2$) becomes *unstable* as one increases the magnitude μ of the perturbation, e.g. of some coefficient in the *f*,f,g,h or H-$\bar{H}$ of the last section. At that point μ_1, a new elliptic orbit ($\sigma = 2m_1/2m_2$) splits off the old one which continues as a 'hyperbolic' orbit. At some higher point μ_2 the new orbit turns hyperbolic itself and yet another elliptic orbit ($\sigma = 4m_1/4m_2$) splits off, etc. Using a simple scaling argument we find that the resulting 'Feigenbaum Sequence' $\{\mu_k\}$ *converges* geometrically and rapidly! Thus, near μ_1, there is a point μ_∞ at which an originally elliptic orbit has branched into an *infinite* number of *hyperbolic* orbits ($\sigma = 2^k m_1/2^k m_2$, $k\to\infty$). The hyperbolic branches of this (fig-)tree converge geometrically also, to some limit-orbit (period: $2^k m_2$ and $k\to\infty$). Since the wild separatrices and chaotic regions of section 2.4 emanate from those hyperbolic orbits we may see, near the limit-orbit, large scale chaos, as those regions interconnect So, as we change from μ_1 to μ_∞ the single regular island about an elliptic orbit breaks up into smaller and smaller islands to end up as a very chaotic region. The scaling argument would allow a calculation of the distances from elliptic to [49(1981)] hyperbolic branches (period: $2^k m_2$) and thus yield the maximum size of the island of order k ($\mu < \mu_\infty$). Figures 1 and 4, at one μ value.., show some of the higher order islands belonging to different basic periods m_2. If this size estimate could be repeated for all islands - up to order k - in a region, we might in some future perhaps be able to quantify "how regular" (or irregular) a system is, to order k or to the minimum island area that can still be distinguished experimentally (e.g. to Planck's constant). It would allow us to decide for which class of non-integrable Hamiltonians it might be true that the measure of the regular regions approaches zero in the Thermodynamic Limit. Under those circumstances there might in practice be some *Approach to Equilibrium and Ergodicity* (at sufficiently high E/N?) even when $H(\vec{q},\vec{p})$ is smooth enough for the K.A.M. theorem to hold [156,139,140].

This splitting of orbits is easily understood from the earlier Figure 9. If the perturbation strength (2.38-39) were reduced the picture would more resemble the integrable mapping in Fig. 8, i.e. the rotation about E in Fig. 9 would be reduced. If on the other hand the perturbation strength is increased the rotation about E increases also [154]. Thus the curve $\rho_0(\Theta)$ (2.44) and its radial image $\rho_{m_2}(\Theta)$ separate even further and must be rotated in opposite directions. At some perturbation strength (μ_1) they must *intersect* in two additional points E_1, E_2, as sketched in Figure 12 (old E→H.R.).

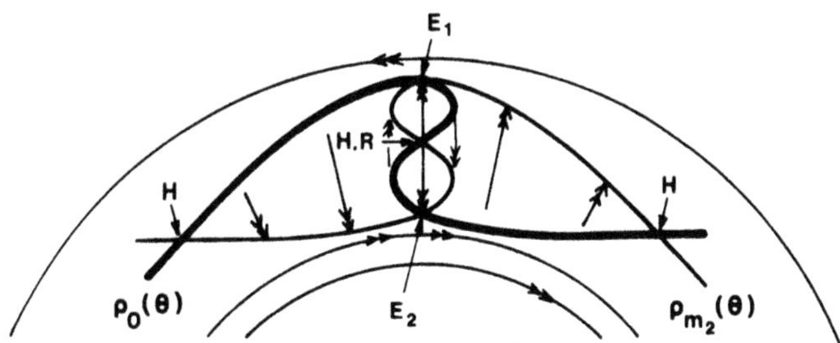

Figure 12
Same Phase Plot as in Figure 9, but at a larger perturbation. The rotation about H.R. (= E of Fig. 9) is larger also and the two ρ-curves intersect in E_1, E_2. The mapping (double arrows) "shows" that these points are part of an elliptic orbit of *double-period:* $2m_2$ while H.R. is now hyperbolic (with reflection [90], [202]: Fig. 12).

Since E_1 and E_2 each belong to both curves they are mapped onto each other, by definition, creating one orbit of period: $2m_2$. Inspecting the mapping directions we "see" that this new orbit is elliptic while the originally elliptic orbit has turned hyperbolic (with reflection), at H.R. Taking E_1 as the origin and mapping a region about it $2m_2$ times we could start all over and find another such '*Period-Doubling Bifurcation*', etc. as shown below, algebraically.

For example, consider again a dynamical mapping of the type (2.20/16/29):

$$y_{t+1} + y_{t-1} = 2Cy_t + 2y_t^2, \qquad t = 0,1,2.., \tag{2.47}$$

cf. [35,48]. In a sense, *this is the most general mapping since locally every nonlinear mapping is quadratic* in general, as we see from a Taylor expansion about an orbit (e.g. E_1 in Fig. 12). We show that (2.47) exhibits a period-doubling bifurcation and that, after scaling, the local mapping about the new elliptic orbit is again of the form (2.47) (+ higher corrections). Starting all over we obtain a '*renormalization scheme*' which explains, and closely approximates [245], numerical data for the present - recently discovered - Feigenbaum Sequence [63-68]. As the parameter μ we choose: 2-C, cf. (2.47) with (2.38/39) [157].

The origin is a periodic orbit of period 1, which is elliptic when $|C| < 1$ and hyperbolic with reflection ($\lambda_1 < -1$) when $C < -1$, cf. (2.25) and (2.30). As expected from the earlier discussion an elliptic orbit of period 2 appears when $C < -1$,

$$\hat{y}_t = a + |b|(-1)^t, \quad \textit{with} \quad 2a \equiv -1 - C \textit{ and } 4b^2 \equiv (C+1)(C-3), \tag{2.48}$$

as checked by substitution. Hence,

$$C_0 = +1, \qquad C_1 = -1, \qquad \mu \equiv 2-C, \tag{2.49}$$

defining C_k as the point of "creation" of a period-2^k orbit. Substituting $y \equiv \hat{y}+\Delta y$ the original mapping becomes

$$\Delta y_{t+1} + \Delta y_{t-1} = (2C + 4\hat{y}_t)\Delta y_t + 2\Delta y_t^2 , \tag{2.50}$$

due to (2.48). Adding (2.50) at $t = 2\tau + 1$, $2\tau - 1$ and substituting it at 2τ we get

$$\Delta y_{2\tau+2} + \Delta y_{2\tau-2} = 2C'\Delta y_{2\tau} + 2\gamma\Delta y_{2\tau}^2 \quad + \quad 2\,[\Delta y_{2\tau-1}^2 + \Delta y_{2\tau+1}^2] , \tag{2.51}$$

$$\textit{where } C' = \tfrac{1}{2}\beta\gamma - 1 = -2C^2 + 4C + 7, \textit{ with} \tag{2.52}$$

$$\beta \equiv 2C + 4\hat{y}_0 , \quad \gamma \equiv 2C + 4\hat{y}_1 \textit{ and } \alpha \equiv \gamma + \tfrac{1}{2}\beta^2 . \tag{2.53}$$

The square-bracket term can also be expressed as $\frac{1}{2}\beta^2\Delta y_{2\tau}^2$ + h.o., as you can see in notes [155] and [164]. Rescaling (2.51) we finally arrive at:

$$y'_{\tau+1} + y'_{\tau-1} = 2C'y'_\tau + 2y'^2_\tau \quad + \quad \textit{higher orders} , \tag{2.54}$$

$$\textit{with } y'_\tau \equiv \alpha\,\Delta\,y_{2\tau} , \tag{2.55}$$

using the *scaling factor* α *(2.53)*. The difference with our original mapping (2.47) is arbitrarily small, close enough to the origin. Hence there also is a period-doubling bifurcation off the origin at $C' = -1$. Repeating the above cycle we obtain the subsequent bifurcation points C_k as follows: Setting $C' = -1$ $(= C_1)$ in (2.52) and solving for C we find $C_2 = 1 - \sqrt{5}$; Setting $C' = C_2$ we find $C_3 = 1-\sqrt{4 + \sqrt{5/4}}$, etc. This Feigenbaum Sequence $\{C_k\}$ converges to a limit C_∞ as we shall see. Since C' must equal C at any limit we obtain from our simple *renormalization (2.52)*:

$$C_\infty = (3 - \sqrt{65})/4 \quad = \quad -1.2656.. \quad , \tag{2.56}$$

while $C_\infty = -1.266311276922099..$, numerically [65].
Substituting $C_k = C_\infty + A\delta^{-k} + \ldots$ in (2.52) we find that the sequence ends up *converging geometrically at a rate:*

$$\delta = -4C_\infty + 4 = 1 + \sqrt{65} \quad = \quad 9.06.. \quad , \tag{2.57}$$

while $\delta = 8.721097200$, numerically [64,65,63] (a renormalization scheme over 2 cycles, using the exact period-4 solution [245] yields $\delta = 8.87..$). From (2.55) we see that the orbits of period 2^k end up converging geometrically as α^{-k}, with the *scaling factor* α (2.53):

$$\alpha = \gamma + \tfrac{1}{2}\beta^2 = -4.0955.. , \tag{2.58}$$

while $|\alpha| = 4.018076704..$, numerically [65,63].

The situation was nicely condensed by van Zeyts in a doubly logarithmic plot, Figure 13, of the present Feigenbaum (sequence) which he obtained numerically over 12 bifurcations [65]. There also is a 'mirror-Feigenbaum' $\{C_{-k}\}$, with $C_{-k} = 2 - C_k$, see [157] and cf. (3.21).

Since every mapping is locally quadratic in general, we expect that the above rates δ, α are *'universal constants'*. They have indeed been found to be the same - numerically - for a few different conservative systems and orbits [63-68]. All this was triggered off by the discovery of the Feigenbaum Sequences in *dissipative* systems, which I shall discuss in section 3.2. To such systems some renormalization techniques had already been applied by the discoverers [223,238,218]. So, if we know just 2 bifurcation points - e.g. μ_0, μ_1 - we 'universally' expect the infinite number of bifurcated hyperbolic orbits at about 13% above this interval already, since:

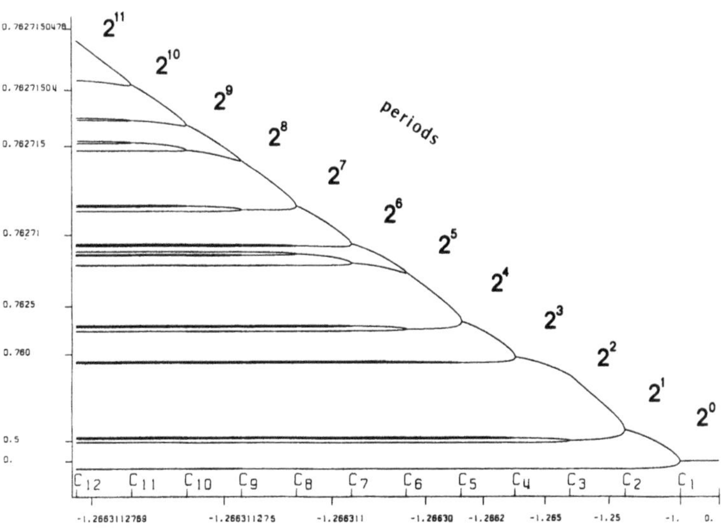

Figure 13
Period-Doubling Feigenbaum Sequence $\{C_k\}$ for a conservative system (2.47) [65]. Vertically plotted are the y_t of the elliptic orbit of period 2^k, splitting off at C_k from the elliptic orbit of period 2^{k-1} which continues as a hyperbolic (with reflection) orbit, not plotted here. Note the constant rates δ (2.57) - at which the C_k converge - and α (2.58) at which the orbits converge, in this doubly-logarithmic plot (due to a symmetry [155] in the orbits there are only $\frac{1}{2}\,2^k+1$ different y_t-values at period 2^k; $k \geq 1$). To the left of C_∞ (2.56) an infinite number of hyperbolic (w.r.) orbits remain. Note the dissipative "tree" in Fig. 19.

$$\mu_\infty \simeq \mu_0 + (\mu_1 - \mu_0)\,\delta/(\delta - 1) \simeq \mu_0 + 1.1295\,(\mu_1 - \mu_0), \tag{2.59}$$

cf. (2.49), (2.56-57) and the geometric convergence. Actually we may think of a period-doubling bifurcation as the *last* of the Poincaré-Birkhoff bifurcations: We saw from (2.34) that, as C (or $-\mu$ or C') decreases below $C_1(m_1,m_2) \equiv \cos(2\pi m_1/m_2)$, pairs of hyperbolic and elliptic orbits of period m_2 bifurcate of the elliptic origin in general [154,150,132,133], cf. Fig. 9. Also, the phenomena of this section, as well as the two previous ones, repeat themselves in general [150,154] about each new elliptic orbit, i.e. also about the ones of Fig. 13 itself etc. ad infinitum. None of the K.A.M. tori about elliptic Feigenbaum branches survive in general beyond μ_∞ for that tree. However at μ_∞ of a *high*-order tree, rooted for instance in an E of Figs. 9,10 , the K.A.M. tori of a *low*-order tree can still exist, e.g. the tori concentric with the origin in Figs. 9,10 [56,202]. In that case the μ_∞ of the lowest-order tree may be taken as a *Threshold for wide-spread Stochasticity* in the KAM gaps [57,40-42,58,81], as done in (2.56) for dynamical mappings of the type (2.16/47). The disappearing regular islands and appearing chaotic regions, as $C \to C_\infty$, are visible in Figure 14. It still remains to be seen (1983) how useful (/less) *conservative* Feigenbäume truly are.

The gaps between K.A.M. tori thus acquire a lot and lots of quasi-random orbits

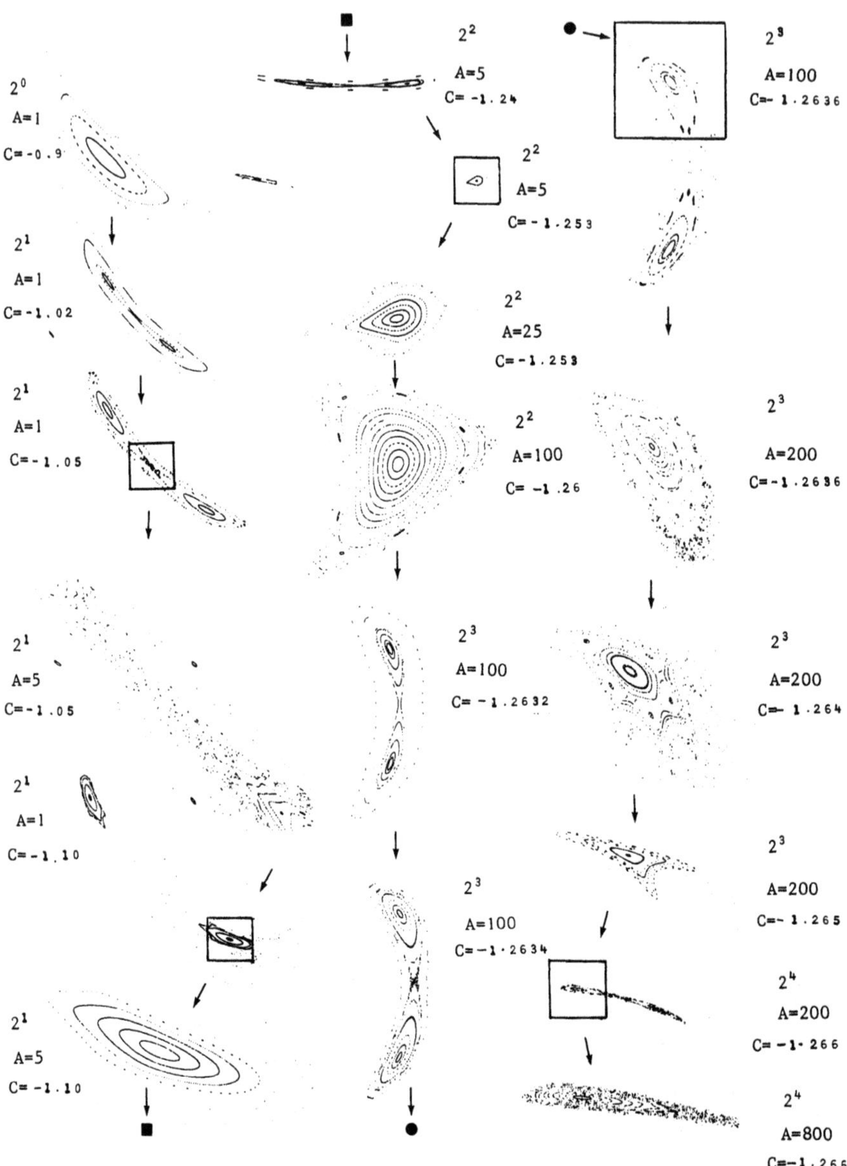

Figure 14
Phase Plots of the Period-Doubling Feigenbaum Sequence (2.48-2.58) [65], i.e. y_t horizontally vs. y_{t+1} vertically. A Sequence of magnifications: A of the regions about the orbits of period: 2^k, starts at the top, on the left, going downwards. The C is the parameter in (2.47).

as μ is increased; e.g. with $\mu \equiv H - \bar{H}$. Non-periodic orbits there also include the limit orbits of Feigenbaum sequences and likely some perturbed quasi-periodic orbits not 'sufficiently irrational' for (2.41/46). In the next section we see that gaps between unrelated (non-concentric) K.A.M. tori can be interconnected ($N \geq 3$) allowing those gap orbits to 'diffuse' through all available phase space.

2.7 Arnold Diffusion, Lack of Nonlinear Stability

Here we consider the stability of a nonintegrable system, i.e. we ask whether there are regions of phase space which will contain and confine *all* orbits $\vec{q}(t),\vec{p}(t)$ starting in it, for *all time*. The answer might be of interest also to designers of storage rings, satellites or moving parts in cars, trains and planes, not to mention their customers. There is some evidence that *a nonintegrable system may not be stable in general* ($N \geq 3$): A number of orbits can "drift" from one gap, between K.A.M.tori, to another unrelated gap, etc. and eventually get very far ($N \geq 3$). This process, called 'Arnold Diffusion', can be astronomically slow. So, even if our solar system is unstable [5,4] there need not be immediate cause for panic. Yet, even astronomically slow Arnold Diffusion can become noticeable in proton storage rings where 10^{11} revolutions do not take 10^{11} years but less than a day, cf. section 2.1 [200], or in molecular vibrations/dissociation where they take less than a millisecond [100,105]. Statistical Mechanicians may be interested in numerical evidence, below, that the quasi-random orbits involved can exhibit a form of Brownian motion and yield local diffusion coefficients [82,83].

An orbit $\vec{q}(t),\vec{p}(t)$ cannot intersect an invariant torus, by definition. As a result K.A.M. tori close off many directions in which orbits might escape ($N \geq 3$), but not all: They are but N dimensional surfaces within a total surface of constant energy which is 2N-1 dimensional. Orbits escaping from a gap can apparently be "obstructed" by various K.A.M. tori in other directions and "repelled" by hyperbolic orbits often enough for the motion to resemble a stochastic process. So it is only when $N = 2$ that the K.A.M. tori lock up *all* enclosed orbits for *all time*, providing at once the only general stability result for nonintegrable systems. Since a dynamical mapping of the plane is a 'surface of section' for a two degree of freedom system, cf. (2.13-.16) [146], its invariant 'circles' also lock up forever all orbits starting inside [201].

Thus the simplest systems on which to study Arnold Diffusion have 3 degrees of freedom or '2+1' [146], e.g. coupled mappings of 2 planes. Consider for example two identical, but coupled, mappings of an x,α plane and a y,β plane:

$$\alpha_{t+1} = \alpha_t - 2ak \sin kx_t - 2\varepsilon k \sin(kx_t + ky_t) \quad \text{and} \quad x_{t+1} = x_t + 2h \tan\alpha_{t+1}, \qquad (2.60)$$

$$\beta_{t+1} = \beta_t - 2ak \sin ky_t - 2\varepsilon k \sin(kx_t + ky_t) \quad \text{and} \quad y_{t+1} = y_t + 2h \tan\beta_{t+1}, \qquad (2.61)$$

with periodic boundary conditions. These 4 first-difference equations which model a 3 dimensional billiards problem [82,85] look simpler than their corresponding 2 second-difference equations, cf. (2.13-14) vs. (2.16). When $\varepsilon = 0$ the mappings (2.60) and (2.61) are uncoupled and one finds for each the familiar phase plot of a nonintegrable system in Figure 15a: Note the thin 'stochastic' layer about the separatrix and the thick one near the α (or β) boundaries. In all coupled [17] experiments ($\varepsilon \neq 0$) Tennyson, Lieberman and Lichtenberg [82,85] take $x_0 \simeq 0 \simeq \alpha_0$, i.e. near the origin. When y_0,β_0 are chosen in the *thick* 'stochastic' layer they find the relatively fast diffusion of Figure 15b (left column), i.e. x and α diffuse away from the origin "filling" the plane after 30000 mappings. When y_0,β_0 were chosen in the *thin* 'stochastic' layer the x,α diffused much slower and had not even filled the central region after 10000000 mappings, cf. Figure 15c (right column). Finally, when they put y_0,β_0 near the origin *no* diffusion could be discerned in the given(/paid) computer time and the resulting 'Lissajous figures' seemed identical to those of an integrable system. One might still see diffusion after some astronomical number of mappings if the numerical precision (and budget) were increased correspondingly. When the motion in Figs. 15b,c truly is diffusion

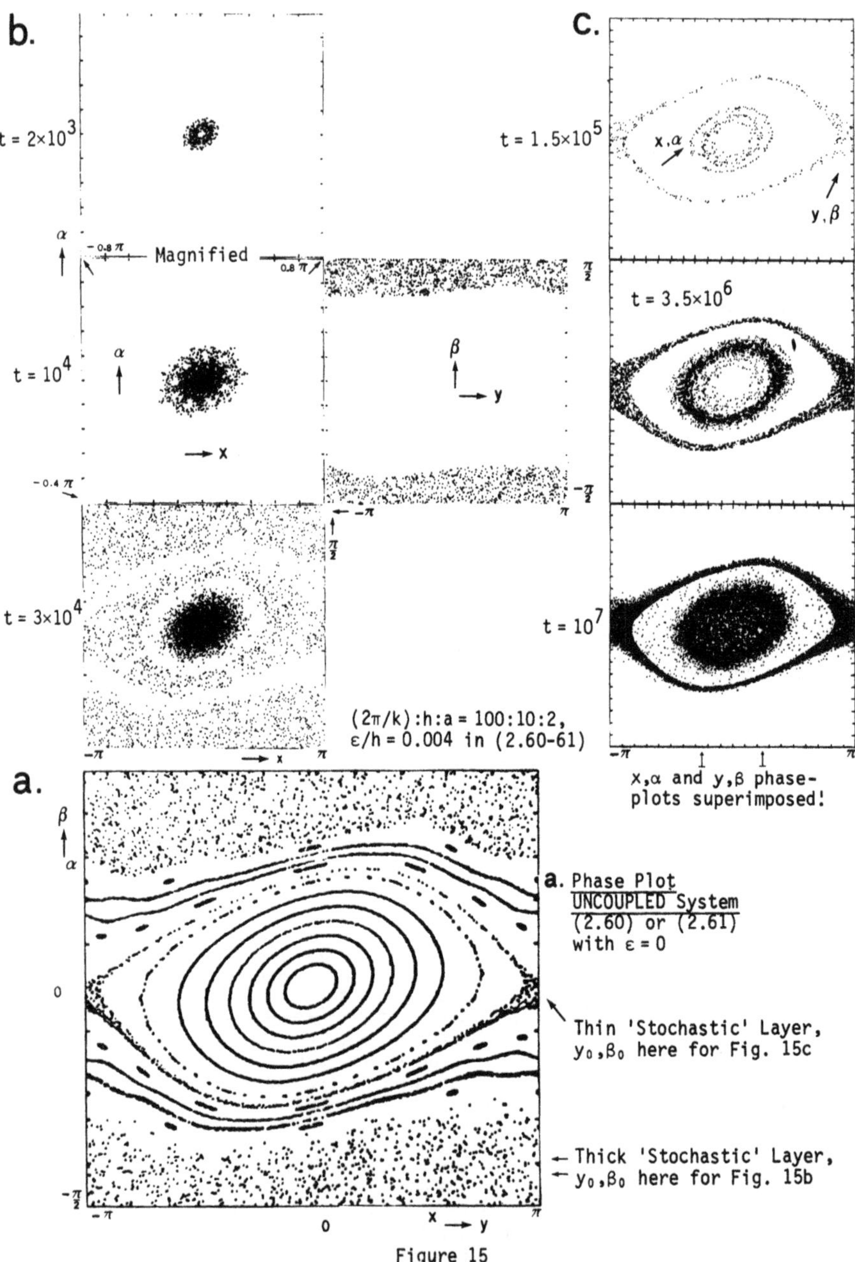

Figure 15
Arnold Diffusion in a 3 degree of freedom system (2.60-61) [82,85], <u>see text</u>.

the x,α should exhibit *Brownian Motion*, i.e. the mean square displacement should be proportional to t: $<\alpha_t^2> = 2Dt$, where D is a diffusion coefficient [94]. Taking an ensemble average, over 100 random values of y_0,β_0 within a stochastic layer, they found this to be true, as shown in Figure 16.

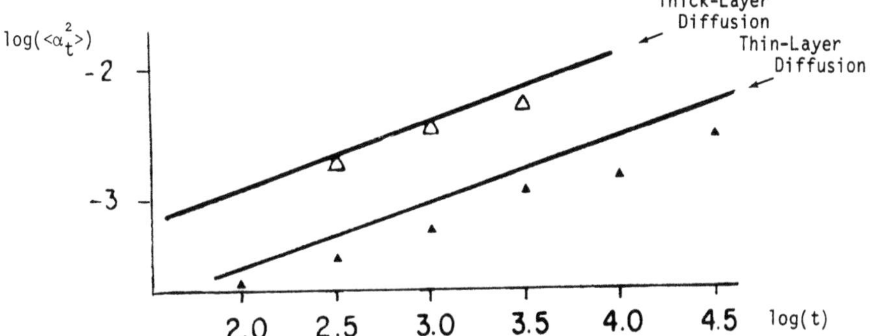

Figure 16
Brownian motion and Arnold Diffusion in the system (2.60-61) of Fig. 15 [82,85]. Note the reasonably linear dependence - of the mean square displacement - on t. Each triangle gives the spread from 100 β_0,y_0 initial conditions, randomly chosen within the Thick (/Thin) 'stochastic' Layer of Fig. 15a. The straight lines result from analytical estimates of the diffusion coefficient [82,85]. Unlike in Fig. 15, some parameters of (2.60) differ from those of (2.61), $\varepsilon/h = 0.0001$, [Thick Layer:] $(2\pi/k_x):h:a_x =$ 10:10:1 and $(2\pi/k_y):h:a_y = 100:10:1.7$, [Thin Layer:] 100:10:1 and 100:10:1.8. The Thick-Layer diffusion corresponds to the left column in Fig. 15b. Also see chapt. 6 of ref. 17 (1983).

In the 'stochastic' layers one might assume the average y,β to be random and take (2.60) only, with a 'noise' term instead of y. Hence, using different calculations for the 2 layers they approximated the values of the local diffusion coefficient D. These analytical estimates [82,85] agree quite well with their numerical results, cf. Fig. 16. At a summerschool on statistical mechanics I should stress that D is at best a *highly local* variable which changes abruptly all during the diffusion so that we measure some average which must depend on the size of the regions and on the scale of the experiment: Earlier we noted that near the y,β origin the diffusion, if any, is astronomically slow. However, even there, minute chaotic regions can exist, cf. sections 2.4-2.6. So it seems likely that we could find a much faster rate of diffusion there if y_0,β_0 were placed in such a region and if we were interested in minute changes of x, α only. Thus again, the intertwining chaotic and regular regions present problems of scale.

Arnold Diffusion has been demonstrated *analytically* for the '2+1' degree of freedom differential system:

$$\ddot{\phi} = -a \sin(\phi) [1 + \varepsilon(\cos\Theta + \cos t)], \tag{2.62}$$

$$\ddot{\Theta} = -a\varepsilon \sin(\Theta) (1 + \cos\phi) , \tag{2.63}$$

near $\phi(0) \simeq \pi$, $\dot{\phi}(0) \simeq 0$, $\Theta(0) \simeq 0 \simeq |\dot{\Theta}(0)| << 1$, (equivalent to the case) studied by Arnold [87] and Nekhoroshev [86]. When $\varepsilon = 0$ these equations are uncoupled and we find for ϕ the familiar phase plot of the integrable pendulum, cf. Figure 3: Hence, the initial conditions (for ϕ) lie again near a separatrix, where we expect a 'stochastic' layer when $\varepsilon \neq 0$. They showed that $|\dot{\Theta}(t)|$ will reach *finite* values, $\simeq 1$, as $t \to \infty$, *no matter how small* a, ε and $|\dot{\Theta}(0)|$ are chosen. Note that $\dot{\Theta}$ would be a constant of the motion in the uncoupled case. The *"diffusion velocity"* - or average $\ddot{\Theta}$ - was shown to be *of the order of*: $\exp(-1/\sqrt{\varepsilon})$ [86,87], which has an essential singularity and becomes astronomically small as $\varepsilon \to 0$, *smaller than any power of* ε! Thus, this diffusion cannot be obtained from perturbation series in ε - no matter how high the order at which we truncate - and such power series usually diverge therefore, as we saw in sections 2.1-2.4 (the angle between 2 intersecting separatrices also is of the same small order in ε [14]). Equivalent estimates hold for $\dot{\Theta}(0) \to 0$, at fixed ε.

Arnold Diffusion may lead to loss of protons from the colliding beams of intersecting storage rings and pose an ultimate limit to the performance of such machines [200-210,4]. Note that our previous beam-beam equations (2.12-16) had only '1+1' degrees of freedom, i.e. no Arnold Diffusion. Invariant tori can then be used to obtain absolute stability results [201]. In general however, there is yet a second set of equations like (2.12-16), e.g. for the radial x-oscillations, *coupled* to those y-equations (2.12-16) [209,210,200,203], i.e. a '2+1' degree of freedom system. You will have noticed that Arnold Diffusion is very much a subject of research, e.g. the extensive numerical investigation of 2 coupled cubic mappings by Chirikov, Ford and Vivaldi [83,84]. Few analytical results are available, except the ones for (2.62-63), nor has the existence of Arnold Diffusion been rigorously proven for many systems [85]. The eventual probability distribution in phase space resulting from Arnold Diffusion or large scale 'stochasticity' may be characterized by asymptotic measures, (Liapunov-)'characteristic exponents' and the (Kolmogorov-Sinai-)'entropy' [27,265,252,97,197].

The emerging picture of phase space ($N \geq 3$) is that of a *sponge* with plastic coated holes representing K.A.M. tori [Moser]; the fibres represent interconnected gap orbits (the 'Arnold Web'): When we put a drop of water anywhere on the sponge it will eventually get *everywhere*, except in the holes. If we magnify a portion of this sponge we see smaller sponges with smaller holes, etc. ad infinitum. Only in quantum mechanics will this hierarchy be truncated when the volumes of sponges and holes become less than Planck's constant, to the N-th power. This also is a current research subject, as we see below.

2.8 Quantization of Non-Integrable Systems

Since the Hamiltonian of a *Non*-Integrable system cannot be separated its Schrödinger Equation cannot be solved by a similar separation of variables, cf. section 2.1. In an Integrable system it is this separation which leads to N separate 1-dim. Schrödinger Equations and N separate good quantum numbers n_k, one for each of the Action Integrals I_k (2.2). Using the Bohr-Wilson-Sommerfeld quantization rule one would simply set them equal to $n_k\hbar$ or $(n_k + \frac{1}{2})\hbar$ [118]. This approximation is justified in modern *'semiclassical'* quantum mechanics [111,110, 116,117,115] where the energy levels of an *integrable* system are given by

$$E_{\vec{n}} = H(\vec{I} = (\vec{n} + \vec{\alpha}/4)\hbar), \tag{2.64}$$

where $\vec{\alpha}$ has integer components, determined from the classical orbits [116] (often equal to 2), and $H(\vec{I})$ is to be obtained via the methods of section 2.1. In a *Non*-Integrable system one tries to quantize the - highly local - constant of the motion associated with the orbits on one K.A.M. torus by methods similar to the above B.W.S. rule. A number of interesting results have been obtained for nonintegrable systems using such semiclassical methods [108-122]. However it is not clear what to do at energies where the invariant tori have largely disappeared and large scale 'stochasticity' has set in, cf. section 2.7. In the extreme case of an ergodic system no K.A.M. tori are present at all and few methods for solving its Schrödinger equation remain [103-105]. While ergodic systems are exceptional, as we saw in section 2.5, we want to compare the energy spectrum of such a completely chaotic system to that of a completely regular, integrable, system in order to single out those effects, which are due to chaotic behavior. The main conclusion is that energy levels can coincide and even cross each other, as we vary one parameter of an *integrable* system, whereas they *narrowly avoid such level crossings in an ergodic system*.

We have a 'degeneracy' if 2 different choices of quantum numbers, $\vec{n}$ (or n), yield the same energy. If no degeneracy is present, it is usually easy to create one -in an *integrable* system- by varying a parameter ε in *one* of the action integrals $\vec{I}$, i.e. there is a *'level crossing'* at some ε_0. In a *Non*-Integrable system the same will hold if it still has *a few* ($\leq$ N - 1) nice integrals. We can then use those few integrals to separate off a few variables [17, 11, 18] (and get a few good quantum numbers). Thus we expect that the remaining *'completely nonintegrable' system*, of N' ($\leq$ N) degrees of freedom, *has only one good quantum number*, n (whereas N' $\geq$ 2), quantizing the only remaining nice integral H into energy levels E_n. Since its eigenfunctions depend on only one quantum number, most of the above level crossings or degeneracies are eliminated. Yet, some global constants of the motion can remain which are not nice integrals, as defined in section 2.1. One example is the 'parity' P, with eigenvalues $\pm$ 1 in a symmetric system, e.g. one with H(-x) = = H(x): the symmetric and antisymmetric eigenfunctions span up 2 *invariant* subspaces, between which H has vanishing matrix elements [118]. Energy levels in those invariant subspaces can be different functions of ε, for instance when H depends on εP itself, and can therefore cross at some ε_0. Such *symmetry-induced crossings* [103 - 105, 119] are eliminated -'desymmetrized'- when one confines his calculations to one of those invariant subspaces, another type of separation. Already in 1929 Hund postulated that a 'sufficiently general' system would not exhibit 1-parameter crossings except those due to integrals or symmetries [123]. However, even in a system as general as a desymmetrized completely nonintegrable system, cf. section 2.3, a third type, the *'accidental' crossing*, can exist [125, 126]. It seems natural to suggest that a number of these are due to the highly local constants of the motion associated with the invariant K.A.M. tori. In order to introduce the K.A.M. theorem in section 2.5 we considered an H which near its minimum, 0, is 'very close' to some integrable $\bar{H}(\vec{I})$. Using those action variables the *non*-integrable H would of course depend also on corresponding angle variables (2.4): $H(\vec{\theta}, \vec{I})$. As we let $\vec{I} \to \vec{0}$ in section 2.5, their difference H - $\bar{H}$ has to vanish *faster* than H and $\bar{H}$ vanish themselves. For example for 2 degrees of freedom it will suffice when they *locally* behave like

$$\bar{H}(\vec{I}) = \omega I_1 + \varepsilon I_2, \tag{2.65}$$

$$H(\vec{\theta},\vec{I}) = \omega I_1 + \varepsilon I_2 + f(I_1, I_2; \cos\theta_1, \cos\theta_2), \tag{2.66}$$

e.g. after separating $\bar{H}$ (only) into 2 uncoupled harmonic oscillators of frequencies ω and ε, where f is some nonlinear term which vanishes *faster* than $|\vec{I}|$, cf. (2.38 - 39). The integrable $\bar{H}$ of (2.65) is easily quantized, e.g. via (2.64), and it is obvious that it has level crossings as we vary ε. Also it is clear that the "nonintegrable levels", of H (2.66), thus approach the "integrable levels", of $\bar{H}$ (2.65), *faster* than both approach their ground states, as we let $\vec{I} \to \vec{0}$ (or $\vec{n} \to \vec{0}$ in (2.64)). If the crossing of the integrable levels is at a sufficiently low energy, for f to be exceedingly small, it would seem possible to have a level crossing in the nonintegrable spectrum as well, when we vary ε. The "correct" choice of parameter would appear to be important. Also, if the classical tori about the equilibrium point of H vanish already below the crossing point in $\bar{H}$ - i.e. if f is not small enough - we expect to see an *'avoided-crossing'* [105] in the spectrum of H. If the classical tori happen to vanish well before even the first semiclassical level, i.e. well before $|\vec{I}| \simeq \hbar/2$ (2.64), no crossing is to be expected there at all, whence the name 'accidental' crossing. One would still expect to see the classical hierarchy of sections 2.6 - 7 and Arnold Diffusion, in the limit of large quantum numbers. If there is any analogue of these effects at small quantum numbers, I would expect there to be a truncation of that hierarchy and an elimination of the - slowest ? - Arnold Diffusion between the smallest gaps (of cross-section $\leq \hbar^N$). Analogous (avoided-) crossings are to be expected near the energies of the invariant K.A.M. tori about (other) elliptic orbits. Since these come at various energies and in various sizes, the *accidental* nature of these crossings becomes clear. In section 2.7 we have associated the onset of 'stochasticity' with the disappearing of K.A.M. tori. This suggests that *an ergodic system can exhibit 'avoided crossings'* [105, 103], *but no actual crossings* (note that a system is sometimes loosely called ergodic if it only is so after removal of a trivial symmetry or integral, e.g. for its center-of-mass motion). The converse is discussed below.

Accidental crossings can also appear in discrete (spin-) systems for which no analogue of the K.A.M. theorem has been suggested, as yet. Nevertheless, the pres-

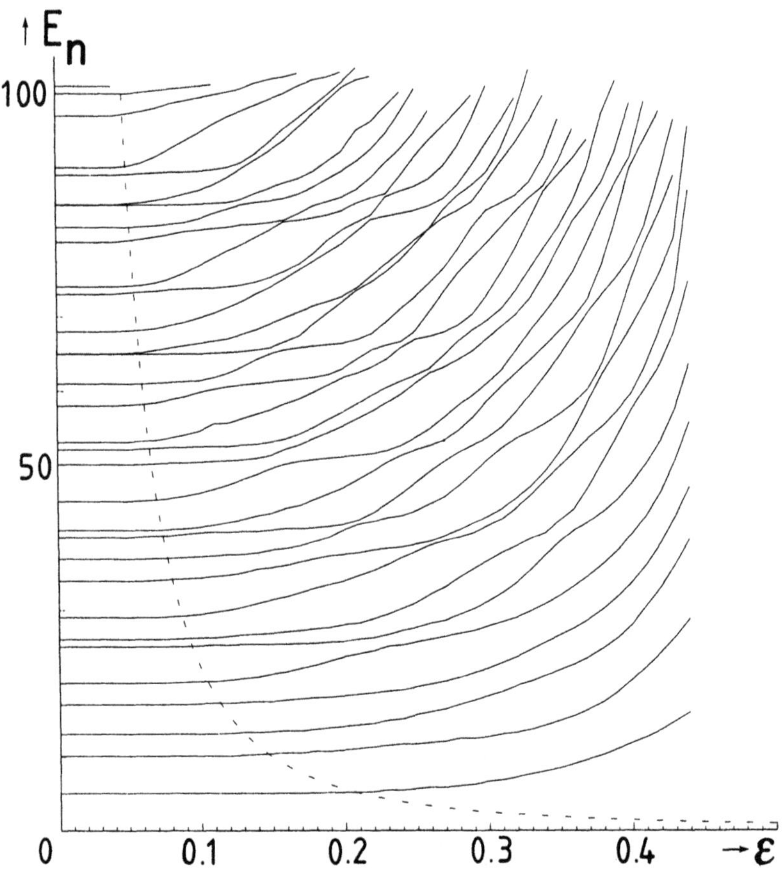

Figure 17
Energy levels of the ergodic Lorentz gas (taken from [103]). Along the horizontal axis: the radius ε of a fixed disk, a 'scatterer' (at $\varepsilon = 0$ the system is integrable, and degenerate). Note the large number of 'avoided crossings' and some avoided multiple-crossings. Magnification, as well as analytic results [103], show that any apparent crossings, above, are actually narrowly avoided crossings.

Text: see next page.

ence of an accidental crossing is known to generate a new, singular, constant of the motion here as well [126, 125] (the action of a discrete "torus"?) which might prevent *some* phase function from being 'ergodic' (as do the K.A.M. tori): a single variable is called *'ergodic'* if its time average equals its (microcanonical ensemble-) average over the *whole* surface of constant energy. An individual variable can thus be 'ergodic' even when the system is not, cf. section 2.1 [30, 125 - 130]; the converse being obviously false. Combining results by Caspers, Mazur, and Valkering, we find that if there are no level crossings under variation of some particular ε (in a desymmetrized completely nonintegrable system), the corresponding "magnetization/polarization", $dH/d\varepsilon$, is 'ergodic'. A complete absence of crossings under variations of any and all parameters in such a system thus implies the 'ergodicity' of *many* but not necessarily *all* variables. Hence, such a system without any crossings is not necessarily ergodic, while an ergodic system appears to have no crossings. This is compatible with a result by Von Neumann and Wigner (1929), Teller, and Arnold, - in different settings [124] - that *'most'* Hamiltonians do not exhibit level crossing if we vary *only one* parameter ε. Since ergodic systems are exceptional, 'most' (desymmetrized, completely) nonintegrable systems apparently have no level crossings, even though many of those have (small ?) K.A.M. tori: you may want to compare the above to Siegel's classical results, in section 2.3. The previous arguments serve to relate an abundance of avoided crossings (scarcity of crossings) to an abundance of ergodic or chaotic behavior, in general Yet, integrable systems with a number of avoided crossings can be constructed.

Consider for example the ideal Lorentz gas [95/301], in which noninteracting particles move in a two-dimensional plane, bouncing off fixed disks of radius ε, the so-called 'scatterers'. For $\varepsilon > 0$ this is an ergodic system [95-98], otherwise known as Sinai's Billiard. Consider its simplest form, a unit square with only one scatterer ($\varepsilon < \frac{1}{2}$), but with periodic boundary conditions, i.e. a unit torus. Confining the calculations to the antisymmetric wave functions renders *all* phase functions 'ergodic'. The energy levels of this desymmetrized billiard were obtained by Berry [103], after adapting the K.K.R. determinant method of solid-state physics. Some 32 levels are plotted as a function of ε, in Figure 17. A spectrum like Fig. 17, with its many avoided crossings [105, 103], may in the future be an indication of chaotic or ergodic behavior.

Another ergodic system is a particle in a 'stadium', an infinite potential well in the x,y plane bordered by 2 circle halves (radius R), connected by straight lines of length ε (> 0). Energy levels and eigenfunctions for the desymmetrized stadium were obtained by McDonald and Kaufman [104], verifying some predictions in the literature [107, 109 - 110], about the most probable spacing of eigenvalues (: positive) as opposed to that of a typical integrable system (: $\to$ 0). Also predicted was a 'random' orientation of the wave vector $\vec{k}$ at different positions. Some confirmation of these predictions is also found in the nodal lines of the eigenfunctions, plotted in Figure 18 for a typical eigenfunction.

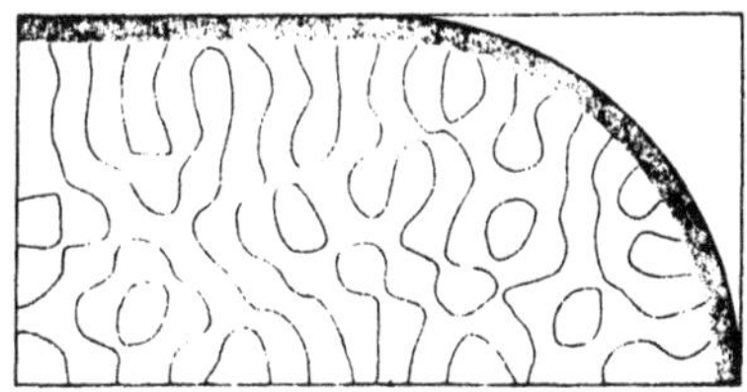

Figure 18
Nodal curves, $\psi(x,y) = 0$, for (one quadrant of) an eigenfunction of the ergodic 'stadium' (taken from [104]). Note the 'random' direction of the curves and their constant average separation. Parameters: $\hbar^2/2m = 1$, $E = k^2$, $k = 50.158$, $\frac{1}{2}\varepsilon = R = 0.665$.

Thirdly, results by Pomphrey and Percival on the Hênon-Heiles system may show that the second difference of the levels, with respect to variation of a coefficient ε of y^3 in (2.10), increases steeply in the chaotic regions [112, 111]. Results connecting the value of $d^2 E_n/d\varepsilon^2$ - a "susceptibility" - to the ergodicity of $dH/d\varepsilon$ ($= y^3$ for (2.10)) [126, 125 - 127] might be related to this.

I assume you are convinced by now that there are many outstanding problems in the quantum- and classical mechanics of chaotic conservative systems. I refer you to the literature [1 - 156] for better answers to some of your questions and invite you to investigate the remainder.

Related, very recent, results on chaotic behavior in the presence of damping, or 'dissipation', are mentioned below.

3. DISSIPATIVE SYSTEMS

These are systems in which the volume of an element of phase space shrinks to zero as $t \to \infty$. Hence the long-time behavior of a dissipative system can be [145,128] easier to predict than that of a conservative system, cf. [282-289] and their refs. However, note that if we consider as one system: a conservative system and a truly 'damped' system - in which *each* variable decays -, not coupled to each other, the total system is dissipative since some sides of a volume element in this larger phase space will decay to zero. Thus some of the conservative problems persist in lower dimensional subspaces [145,276]. This simple example also suggests a number of structures whose persistence we would like to investigate under a slight coupling of those 2 systems:

1. A Simple '*Periodic Attractor*': A periodic orbit (formerly elliptic) with variational equations whose solutions all decay to zero, i.e. the orbits near the periodic orbit are '*attracted*' to it. In a 'damped' system there is only one - trivial - attractor: the point of rest (at the origin). The origin might sometimes be '*repelling*' however, if its variational solutions become unbounded. If positive damping prevails further out in a phase plane we can have a periodic orbit in between, approached by all orbits as $t \to \infty$. Such '*Limit Cycles*' or Simple Periodic Attractors will be obtained in section 3.1. One example is the van der Pol equation: $\ddot{y} - (1-y^2)\dot{y} + y = 0$ [282-289], and limit cycle.
2. '*Period-Doubling Attractors*': In section 2.7 we discussed 'Feigenbaum Sequences' of elliptic periodic orbits bifurcating into pairs of one elliptic orbit - of double-period - and one hyperbolic orbit. From the above 1. we expect here Feigenbaum sequences of periodic *attractors* bifurcating into pairs, one stable orbit - of double-period - and one unstable. These will be found in section 3.2. Actually Feigenbaum sequences were earlier discovered in dissipative systems [223] than in conservative systems [63-68].
3. '*Strange Attractors*': Some of the quasi-random orbits discussed in section 2.4 can persist under dissipative coupling [15,23,274-279] and might thus become attractors. Such non-periodic attractors can be very complicated and produce 'Cantor sets' in their surfaces of section (as can the original quasi-random orbits). *Along* the attractor we would still have the sensitive dependence on initial conditions, characteristic of the chaotic regions in sections 2.2-4, cf. Fig. 2, i.e. *along* the attractor points can be repelled by each other. In section 2.7 we argued that large-scale 'stochasticity' might appear after all Feigenbaum orbits have turned hyperbolic. Here we might not expect Strange Attractors until all Feigenbaum attractors have turned hyperbolic. However, note that the regular K.A.M. tori in those arguments do not persist under a dissipative coupling [13], in general. We encounter strange attractors in section 3.3 [145,128].

The above phenomena have also been found solving the equations of motion for the (truncated) Fourier-modes of the Navier-Stokes equation and have even been recognized in some of the Fourier spectra of experimental Turbulence, as mentioned in section 3.4

For example, consider an *electron storage ring*, rather than the proton storage ring of section 2.1. Since electrons oscillate much faster, at the same energies, we must now include radiation-*damping* [200]. We might classically approximate the electron ('betatron'-) oscillations by means of a damped harmonic oscillator:

$$d^2y/d\phi^2 = -Q^2y - Rp \quad , \textit{with} \quad p \equiv dy/d\phi, \tag{3.1}$$

cf. (2.11). The effect of a second, colliding, beam is modelled by

$$d^2y/d\phi^2 = -Q^2y - Rp + \mathcal{B}F(y)\left[\sum_{t=0,1,2,}^{\infty} \delta(\phi - t\,2\pi)\right], \tag{3.2}$$

cf. (2.12), where F is some nonlinear function. *Between* the δ-pulses (3.2) is linear (3.1) and trivially solved in terms of: $\exp(q_{\pm}\phi)$, with

$$q^2 + Rq + Q^2 = 0 \tag{3.3}$$

During a pulse the momentum, p, is changed by an amount $\mathcal{B}F(y(t2\pi))$. Hence (3.2) can be translated *exactly* [245] into 2 first-difference equations for $y_t \equiv y(t2\pi+)$ and $p_t \equiv p(t2\pi+)$, cf. (2.13-14). Eliminating p_t between those 2 equations we arrive at the single second-difference equation:

$$y_{t+1} + B\,y_{t-1} = 2Cy_t + 2F(y_t), \tag{3.4}$$

$$\textit{with} \quad B \equiv \exp(-2\pi R), \tag{3.5}$$

$$\textit{and} \quad 2C \equiv \exp(2\pi q_+) + \exp(2\pi q_-), \tag{3.6}$$

where $\mathcal{B}$ in (3.2) is chosen suitably [166]. The present y_t, y_{t+1} phase plane is a linear transformation of the more usual y,p phase plane. Applying (2.27-28) to our present mapping (3.4) we see that the Jacobian J for this mapping equals B (3.5). Hence the area of any closed loop is multiplied by B, each time we map $y_{t-1}, y_t \to$ $\to y_t, y_{t+1}$ (3.4). When $|B| < 1$ we have a *dissipative, i.e. 'area contracting', mapping*. Note that when the damping R in (3.2) is positive we do have $B < 1$ (3.5) If $|B| > 1$ (e.g. if $R < 0$) we would have an 'area expanding' mapping. In that case the mapping becomes dissipative if we consider $t \to -\infty$ [163]. When $|B| = 1$ the mapping is conservative and, at $B = 1$, the same as (2.16). Negative B values could arise when R is 'capacitive' or 'inductive' (e.g. $= i/2$).

All quadratic mappings (with a fixed point) can be brought into our *standard form*,(3.4) with $F(y) = y^2$ [159-161,245],

$$y_{t+1} + By_{t-1} = 2Cy_t + 2y_t^2 \ , \textit{with} \quad |B| \le 1 \tag{3.7}$$

At $B = 1$ this obviously *includes Hénon's conservative mapping* (2.47). At $B = -b$ it *includes Hénon's dissipative mapping* [260,261-264]:

$$x_{t+1} - bx_{t-1} = 1 - ax_t^2 \ , \tag{3.8}$$

as we see in [160]. This mapping has a type of strange attractor at $b = 0.3$ and $a = 1.4$ [260]. At $B = 0$ (3.7) *includes the 'Logistic Equation'* [215,216-240,297]:

$$x_{t+1} = ax_t(1-x_t), \tag{3.9}$$

as we see in [161,158],note Lorenz (1964) [239], and mappings of the type [226]:

$$x_{t+1} = 1 - \mu x_t^2 \ , \tag{3.10}$$

as we see in [162,158]. In many ways the quadratic mapping (3.7) is the most general one [159] since locally every nonlinear mapping is quadratic, in general. In the next section we find some of its limit cycles and in section 3.2 some of its Feigenbaum sequences [226,245,213-249].

3.1 Limit Cycles, Simple Periodic Attractors

In this section we calculate simple periodic attractors of period-1 and period-2 for the dissipative mapping (3.7) which includes Hénon's mappings and the logistic equation as we saw before. We also find a hyperbolic orbit. These will be used in section 3.2.

An obvious periodic orbit of (3.7) is the origin, $\hat{y}_t = 0$, which has period-1. Its variational - equation and - solutions are [149,49(1981)]:

$$y_{t+1} + By_{t-1} = 2Cy_t, \qquad |B| \leq 1\,, \tag{3.11}$$

$$\textit{whence } y_t = a\lambda_1^{\,t} + b\lambda_2^{\,t}\,,\ \textit{with } \lambda^2 - 2C\lambda + B = 0, \tag{3.12}$$

cf. (2.30). From this we can check that $|\lambda_{1,2}| < 1$ when $|2C| < 1+B$. Hence the origin is an *attractor* in that range of C values. Another periodic orbit of (3.7) with period-1 is $\hat{y}_t = (1+B)/2-C$. Its variational-equation is:

$$\Delta y_{t+1} + B\Delta y_{t-1} = 2(1+B-C)\Delta y_t, \qquad |B| \leq 1, \tag{3.13}$$

with $y \equiv \hat{y} + \Delta y$, cf. (2.22). One similarly checks that in the same C range: $|\lambda_1| > 1$, for only *one* of the λ's in the solution, cf. (3.12). This orbit is a period-1 hyperbolic orbit [158,49(1981)].

A period-2 orbit of (3.7) is:

$$\hat{y}_t = a + |b|(-1)^t,\ \textit{with } 2a \equiv -(1+B)/2-C \textit{ and } 4b^2 \equiv (C+[1+B]/2)(C-3[1+B]/2), \tag{3.14}$$

cf. (2.48), as checked by substitution. From the third part we see that it exists, real, when $2C \leq -(1+B)$ (or above $3(1+B)$ [158]), i.e. it starts from $y = 0$ at the C value where the origin becomes hyperbolic, cf. (3.12). Denoting the starting point of a period-2^k attractor by C_k we just found

$$C_0 = (1+B)/2 \quad \textit{and} \quad C_1 = -(1+B)/2 \tag{3.15}$$

Substituting $y \equiv \hat{y} + \Delta y$ the mapping (3.7) becomes

$$\Delta y_{t+1} + B\Delta y_{t-1} = (2C + 4\hat{y}_t)\Delta y_t \quad + \quad 2\Delta y_t^2\,, \tag{3.16}$$

cf. (2.50). The variational equation of $\hat{y}$ (3.14) is the *linear* part of (3.16). We eliminate Δy_{t+1} and Δy_{t-1} between the 3 variational equations, at t-1,t,t+1, to obtain [245]:

$$\Delta y_{t+2} + B'\Delta y_{t-2} = 2C'\Delta y_t\,, \tag{3.17}$$

$$\textit{with } B' \equiv B^2 \tag{3.18}$$

$$\textit{and } C' \equiv \beta\gamma/2-B = -2C^2 + 2(1+B)C + 2B^2 + 3B + 2, \tag{3.19}$$

$$\textit{where } \beta \equiv 2C + 4\hat{y}_0 \textit{ and } \gamma \equiv 2C + 4\hat{y}_1, \qquad \textit{cf. (2.53)}, \tag{3.20}$$

Equation (3.17) - with t even - is the variational equation about $\hat{y}_0$ and - with t odd - about $\hat{y}_1$. Note that it is the same in both cases and that it has the same form as our original eq. (3.11). Similarly, the orbit (3.14) is an *attractor of period-2* when $|2C'| < 1 + B'$. Substituting (3.18-19) we obtain a quadratic equation in C. Hence (3.14) is an attractor for all C values between its two roots, the highest of which equals C_1 of course. The lowest root must be C_2, the starting point of a period-4 attractor, just as C_1 is the lowest C until which the origin is a period-1 attractor. In section 3.2 we continue this iterative procedure to find subsequent C_k's.

I remind you that the above discrete attractors of periods 1 and 2 correspond exactly to *continous* orbits of the storage rings (3.2) which therefore exhibit corresponding *Limit Cycles* of periods 2π and 4π (in ϕ). There exists an extensive classic literature on Limit Cycles [282-289]. The recently discovered sequences of more and more complicated attractors are discussed below.

3.2 Bifurcating Periodic Attractors (Feigenbaum), Onset of Turbulent Behavior

Here we find that one of the variational solutions of a simple attractor (/ limit cycle) of period m_2 becomes unbounded, as one increases the magnitude of some [248] parameter μ which plays a role similar to that of the 'Reynolds Number' in fluid mechanics. At that point μ_1 a new attractor, with 2 loops - whence of period $2m_2$ splits off the old one which continues hyperbolic. At some higher point μ_2 the new attractor becomes hyperbolic itself and yet another attractor, of period $4m_2$ splits off, etc. Using a renormalization argument similar to that of section 2.6 we find that the resulting 'Feigenbaum Sequence' $\{\mu_k\}$ converges geometrically and rapidly, albeit at a different rate from the conservative sequences. Thus, near μ_1 there is a point μ_∞ at which the original simple attractor has bifurcated into an infinite number of branches. The hyperbolic branches of this tree converge to some limit orbit, of "infinite period", again at a rate different from the one for a conservative Feigenbaum. Our renormalization scheme shows why the dissipative rates are the same for all values of the damping $|B|$ ($\neq 1$) in (3.7). In particular it shows that near the limit of each tree the local mapping converges to one described by a *first*-difference equation. Hence Hénon's mapping (3.8) and others all have the same Feigenbaum rates as the logistic equation (3.9) [219]. It also demonstrates why $B = \pm 1$ (conservative system) yields different rates. In order to relate the various parameter values for different equations to each other the reader should first apply the transformation-to-standard-form (3.7), discussed earlier, cf. (3.7-10) [158-163]. Secondly it should be pointed out that for each tree $\{C_k\}$ of (3.7) (or other phenomenon) there is a '*mirror-tree*' [158] (/ mirror-phenomenon) at:

$$C_{-k} = 1 + B - C_k, \tag{3.21}$$

$-\infty < k < \infty$, see [245]. The literature on dissipative systems usually discusses the increasing sequence [215-270,161] while the decreasing (mirror-)sequence is normally used in conservative systems, cf. section 2.6 [63-68]. The period-1 and -2 attractors are slightly simpler for the decreasing sequence [158] whence our choice of standard form (3.7) with $C_1 \le C_0$ (3.15). As the parameter μ we choose: $1 + B - C$, cf. (2.49) [245]. Finally, when $|B| > 1$ we divide (3.7) by B and obtain a similar equation [163] which has period-doubling *attractors* if we let $t \to -\infty$ [163]. Hence, as we let $t \to +\infty$, there are trees of '*Period-Doubling Repellors*', for $|B| > 1$. Before bifurcation each such repellor has a variational equation with *two* unbounded basis solutions ($B \neq 0$). The existence of a 'Strange Attractor', cf. section 3.3, at some value of B ($\neq \pm 1, 0$) therefore implies the existence of a '*Strange Repellor*' at 1/B. *Repelling* quasi-random orbits and Cantor sets have been discovered [234-235,251,261-264].

In section 3.1 we constructed one such period-doubling bifurcation, at C_1 (3.15). Had we carried the quadratic term as well, going from (3.16) to the modulo-2 equation (3.17) about the period-2 solution, we would have obtained *exactly*:

$$\Delta y_{2\tau+2} + B^2 \Delta y_{2\tau-2} = 2C'\Delta y_{2\tau} + 2\gamma \Delta y_{2\tau}^2 + 2\,[\Delta y_{2\tau+1}^2 + B\Delta y_{2\tau-1}^2] \tag{3.22}$$

with C' (3.19) and γ (3.20). This can be checked by adding (3.16) at $t = 2\tau + 1$ to B times (3.16) at $t = 2\tau - 1$ and substituting (3.16) at $t = 2\tau$. The square bracket term does contribute a term $\propto \Delta y_{2\tau}^2$ this time, cf. (2.51-54) [155], as can be seen in [164]. Rescaling (3.22) we finally obtain the *renormalized mapping*:

$$y'_{\tau+1} + B'y'_{\tau-1} = 2C'y'_\tau + 2y'^2_\tau \quad + \quad \textit{higher orders}, \tag{3.23}$$

$$\textit{with } y'_\tau \equiv \alpha \Delta y_{2\tau} \quad , \quad \alpha \equiv \gamma + \beta^2/(1+B), \qquad \textit{and:} \tag{3.24}$$

$$B' \equiv B^2, \tag{3.25}$$

cf. [164] and (2.54-55). This is again of the same form as our original quadratic mapping (3.7). Hence, there also is a period-doubling bifurcation when $2C' = -(1+B')$ as we saw in the previous section. Substituting (3.18-19) we obtain a quadratic equation which can be solved for C_2, e.g. $2C_2 = 1 - \sqrt{6}$ at $B = 0$. Starting with (3.23), instead of (3.7), we could repeat the whole calculation to obtain a C" which is the same function of C', B' as C' is of C, B: (3.19). At $B = 0$ this yields $2C_3 \simeq 1 - \sqrt{4 + \sqrt{6}}$. Iterating this simple renormalization (3.19 + 25) produces a Feigenbaum sequence C_k which converges, as we shall see. It is clear from (3.25) that B' vanishes in the renormalization limit for a dissipative system, since $|B| < 1$. Hence, *every dissipative mapping* (3.7) *becomes a first-difference mapping in the renormalization limit*. Our calculations would have been much simpler had I confined myself to $B = 0$, e.g. to (3.9). The present approach however explains why Hénon's mapping (3.16) has the same asymptotic rates of convergence δ (3.27) and α (3.28) as the logistic equation (3.9). It also shows why a conservative system, with $|B| = 1$, has different δ, α values (2.57-58).

At the limit C_∞ we must have $C' = C$, and $B \to 0$. From (3.19) we thus obtain,

$$2C_\infty = (1 - \sqrt{17})/2 = -1.56155.., \qquad |B| \to 0, \tag{3.26}$$

while $2C_\infty = -1.569945671870945..$, numerically [65] *at* $B = 0$! For the logistic equation (3.9) the latter result implies: $a_\infty = 2 - 2C_\infty = 3.5699...$, as observed [161,158,223]. For Hénon's mapping (3.8), at $b = 0.3$, it suggests $a_\infty \simeq 1.044..$[245]. It was observed , directly , to be of the order of 1.058 [260]; It has been noted that the amplitudes of the Fourier components of y_t increase by an order of magnitude, when going from $a \simeq 1.15$ to 1.16 (at some frequencies [165,263]). Substituting $C_k = C_\infty + A\delta^{-k}$, and $B \to 0$, in (3.19) we see that the sequence ends up *converging geometrically at the rate:*

$$\delta = -4C_\infty + 2 = 5.12.., \qquad |B| < 1, \tag{3.27}$$

while $\delta = 4.66920160910299055.$, numerically [65,220] (a second order approximation [225] yields $\delta = 2+2\sqrt{2} = 4.83..$). The scaling equation (3.24) shows that the orbits of period 2^k end up converging geometrically as α^{-k}, with the *scaling factor* α (3.24), evaluated at C_∞ (3.26) and $B \to 0$:

$$\alpha = \gamma + \beta^2 = -2.2399.., \qquad |B| < 1, \tag{3.28}$$

($\gamma = -2.60..$) while $|\alpha| = 2.50290787509589284..$, numerically [220,65]. The present Feigenbaum, obtained numerically at $B = 0$ [65], is displayed in Figure 19. The present tree of period-doubling discrete orbits corresponds *exactly* (3.2-3.7) to a Feigenbaum of period-doubling continuous orbits of the storage rings (3.2), with periods $2^{k+1}\pi$ (in ϕ), $k \to \infty$ [245].

Similar period-doubling has been observed in several other differential systems, for instance in the '*Rössler Attractor*' [254(1976)]:

$$\dot{x} = -y - z , \tag{3.29}$$

$$\dot{y} = x + ey , \tag{3.30}$$

$$\dot{z} = f + xz - \mu z , \tag{3.31}$$

with parameters e, f, μ [254,232,251(1980)]. The divergence of this flow in the 3-dimensional phase space is $e + x(t) - \mu$. Hence for $x < \mu - e$ any volume-element shrinks continually (by Gauss'theorem) as it flows through space, i.e. we have a dissipative system. Such rate equations are widely used throughout physics, chemistry, hydrodynamics, meteorology, population dynamics and biology [239-302,170-190]. The Rössler eqs. have a limit cycle for some choice of the parameters, as shown in Figure 20A. Its first few period-doubling bifurcations are

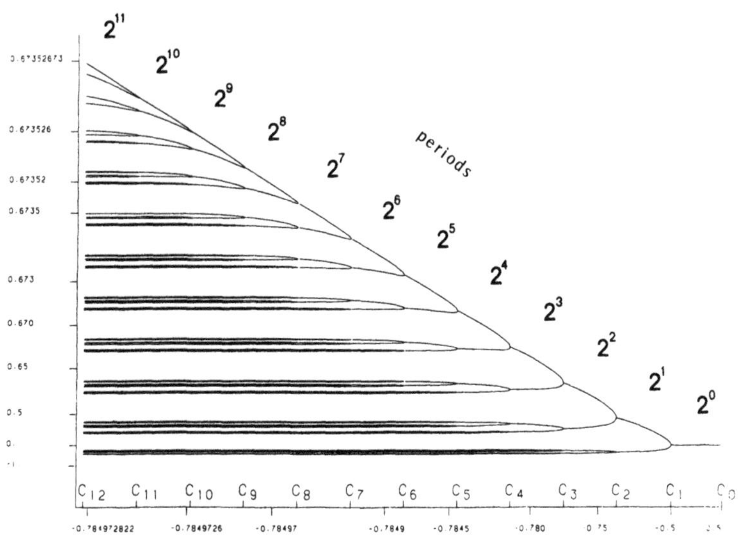

Figure 19
Period Doubling Feigenbaum Sequence $\{C_k\}$ [65] for a dissipative system (3.7) with $B=0$, equivalent to the logistic equation (3.9) with $a = 2 - 2C$ [161,158]. Vertically plotted are the y_t of the attractor of period 2^k, splitting off - at C_k - from an attractor of period 2^{k-1} which continues hyperbolic, not plotted here. Note the constant rates δ (3.27) - at which the C_k converge - and α (3.28) at which the orbits converge, in this doubly-logarithmic plot. To the left of C_∞: an infinite number of hyperbolic orbits; cf. the conservative Feigenbaum in Figure 13.

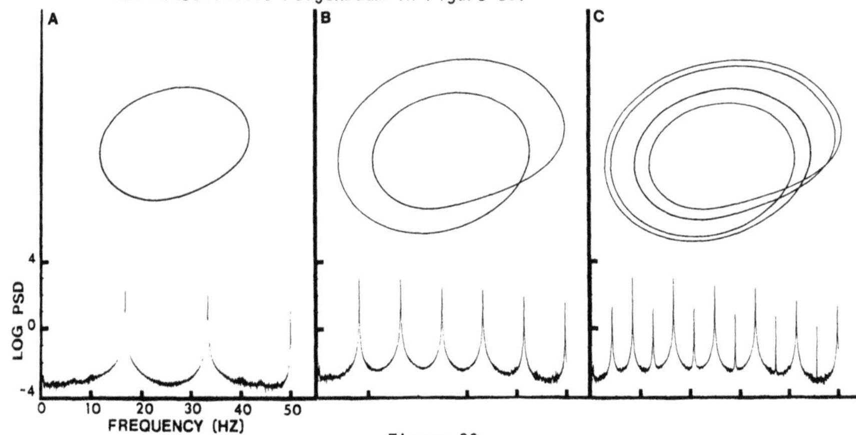

Figure 20
Period-Doubling Bifurcations of the Rössler Attractor (3.29-31), projected on the x,y plane (taken from [232]). Parameters: $e = f = 0.2$; $\mu = 2.6$(A), $= 3.5$(B), 4.1(C). Note the sub-harmonics in the power spectra of $z(t)$.

shown in Figures 20B, and 20C (at $\mu = 4.1$). Its μ_∞ is observed near $\mu \simeq 4.20$ [232]. Below the x,y projections of the orbits the corresponding power spectra of z are plotted, i.e. the squares of the amplitudes of the Fourier components of z(t). Note that each period-doubling introduces a half-frequency *sub-harmonic* (and many higher harmonics). Similar Feigenbaum sequences have also been found in the Lorenz system (3.35-37) and systems of 5 and 7 rate equations for the (spatial-) Fourier modes of the Navier-Stokes Equation in hydrodynamics [234-235,219(1980)], a periodically driven nonlinear oscillator in the conservative [64,2nd] and dissipative cases [167,249,277-280], and other systems [215-248].

In *conservative* systems we expect chaotic behavior on a Larger scale when $\mu \gtrsim \mu_\infty$, cf. section 2.6, due to the absence of K.A.M. tori then. In a *dissipative* system one would expect comparable chaotic behavior when $\mu \gtrsim \mu_\infty$ due to the infinity of hyperbolic orbits near the limit orbit. When these are *globally* hyperbolic the orbits will quickly move out to other regions and no chaos will be visible (similar "escapes" are responsible for the absence of chaotic points in many of the blank regions of Fig. 14). Yet, in several cases they seem to be *local* - or regional - objects which still attract orbits that are further away, towards this hyperbolic region. These *strange* kinds of attractors are largely unexplained. Some very strange ones will be discussed in the next section.

If we know just 2 bifurcation points we expect the infinite number of repellors at about 27% above this interval, since

$$\mu_\infty \simeq \mu_0 + (\mu_1 - \mu_0)\delta/(\delta-1) \simeq \mu_0 + 1.2725\ (\mu_1 - \mu_0), \qquad (3.32)$$

cf. (3.27). Applying it to (3.15) yields $2C_\infty \simeq -1.545$ at $B = 0$, cf. (3.26). As μ_0, μ_1 are functions of B, while δ is not, this expression is not as useful as the one in the conservative case (2.59). Only if one knows μ_0, μ_1 as a function of B can we take $B \to 0$ and use (3.32). The μ values in Fig. 20 are not the bifurcation points. Applying (3.32) nevertheless we guesstimate $\mu_\infty \simeq 4.26$. It is numerically observed at $\mu_\infty \simeq 4.20$ [232]. The dependence on B is derived in [245].

When the μ of the Rössler attractor (3.29-31) is increased beyond μ_∞ '*strange bands*' of chaotic hyperbolic behavior appear, while the total object remains an attractor, see Figure 21, and Figure 4 of [245].

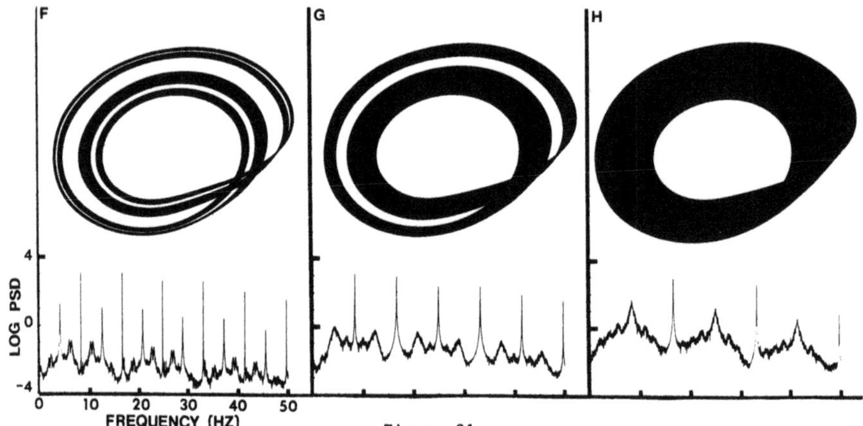

Figure 21
Strange bands of chaotic behavior appear in the Rössler Attractor (3.29-31) when $\mu > \mu_\infty$ ($\simeq$ 4.20); same e = f as in Fig. 20 (taken from [232]). The bands appear at $\mu \simeq 4.23$ (F) about the previous orbit of "period-4", at $\mu \simeq 4.30$ (G) about period-2, and at $\mu \simeq 4.60$ (H) about period-1.

Note that the "continuous" part of the Fourier power spectrum increases by orders of magnitude, especially near zero-frequency, i.e. for very long periods. The same holds for experimental Turbulence spectra. Hence, we speak of the '*Onset of Turbulence*'. Yet, sharp peaks at the basic frequency remain in Fig. 21H. When the object develops more than just one circular band, at other parameter values, those peaks may disappear as well, cf. Fig. 24. These pictures of the Rössler attractor were obtained by the Santa-Cruz (/Eugene, OR) group on an *analogue*-computer [232, 251] which facilitates parameter searches for interesting attractors. The resolution is not high enough therefore to resolve the behavior inside the strange bands. For the Hénon attractor (3.8) this will be done, with the aid of a digital computer [260], in the next section.

In a *conservative* system we have many Feigenbäume, of different basic periods m_2, simultaneously present at a given value of μ. There are many interweaving branches of different trees coming very close together in infinite hierarchies, cf. sections 2.4-2.7. For the *attracting* (repelling) branches of these trees in a dissipative system such infinite hierarchies of different trees would be difficult to construct since, close enough to an attractor, *all* orbits move towards it forever. In a dissipative system we find Feigenbäume of different basic periods m_2 (in $m_2 2^k$) within different μ-regions, one after the other [215,240,235;234], sometimes with μ-gaps between them. This makes it even more difficult to predict where the "last" tree ends [240,215] than in a conservative system, cf. section 2.6, and where large scale chaos might set in [245].

Feigenbaum sequences have been established *rigorously* for mappings of the interval [-1,1] into itself which are of the form [226],

$$y_{t+1} = f(|y_t|^{1+\varepsilon},\mu), \qquad 0 < \varepsilon << 1, \qquad (3.33)$$

where $y_t \to y_{t+1}$ is smooth, except at f's maximum at $y = 0$ [226], under some mild technical conditions on f [237]. These results involve perturbation expansions in ε and are restricted to small ε values, whereas the interesting case is $\varepsilon = 1$ (3.9). These analytical results have been extrapolated and extended to N-dimensional mappings [227,226,268,269]. Proofs exist now (1983) for $\varepsilon = 1$ as well.

Period-doubling bifurcations are observed in real turbulence experiments as we shall see in section 3.4. Below we discuss chaotic, non-periodic, attractors.

3.3 Strange Attractors

A more descriptive, but less attractive, name for these objects might be 'Stochastic Attractors', since the *motion along the attractor should be 'ergodic'*, cf. chapter 2, *and 'mixing'* [250-256,272,270,92,90-98], i.e. t-dependent correlation functions should vanish as $t \to \infty$. The latter precludes periodic attractors. A *non-periodic attractor* has infinitely many intersection points with a transverse ("perpendicular") surface. Yet, there cannot be any continuous 'curve' in this 'surface of section' with the attractor passing near *every* point of an 'interval' on that curve; otherwise it would not be an attractor *along* that 'interval' [270]. One possibility remaining is for the Strange Attractor to pass through a '*Cantor set*' an infinite number of points which are not dense on any 'interval' [168]. Examples of such attractors have been constructed [250-272,169]. None of the above properties of a Strange Attractor are easy to test, for a given system of equations and different opinions can be heard on the strangeness of particular attractors [272, 260-264]. It is apparent however that there is a bewildering variety of types of attractors, *in between* Strange Attractors and simple Limit Cycles, cf. [250-279]. I remind you that there also is a great variety of conservative systems - in between ergodic- and integrable systems - as we saw in chapter 2. Some attractors [250-302] are stranger than others. An attractor is often called 'Strange' already if it doesnot consist of a single *periodic* orbit [272]. In that case no chaotic behavior is required along the attractor and 'Aperiodic Attractor' might be a more descriptive name [224,239,257,259,262].

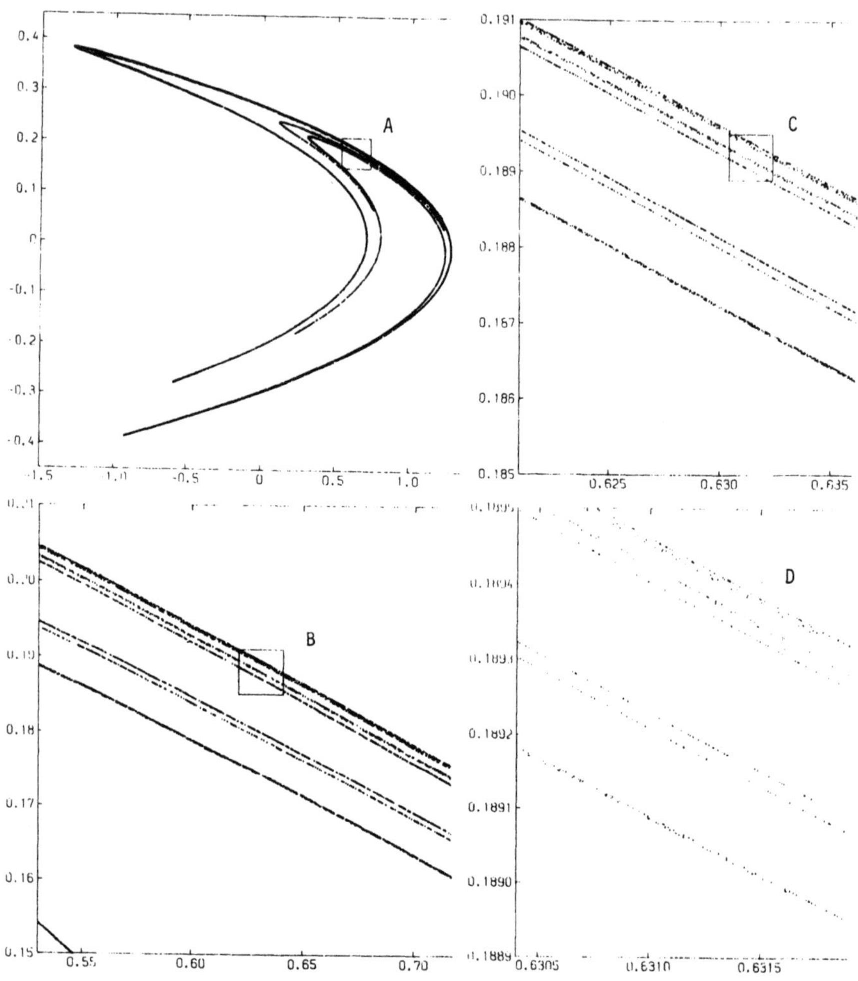

Figure 22
The Hénon-Attractor, at a = 1.4 and b = 0.3 in (3.8). Horizontally plotted is x_t, vertically: bx_{t-1} (taken from [260]). The complete attractor (A) seems simple, yet a 15 fold magnification of the little "box", in A, shows more "curves" (B). A further magnification (x10), of the small box in B, shows several more (C) "curves". A final magnification (x10), of the box in C, again shows new ones (D). This suggests a 'Cantor Set' cross-section. Along the curves points are repelled. Points are "transversally" attracted to the curves. Note the conservative curves in Figs. 6b,c. Text: see next page.

Consider for example Hénon's mapping [260,165]

$$x_{t+1} - bx_{t-1} = 1 - ax_t^2 \, , \qquad ,\text{cf. } (3.8),$$

and [260-264]. We have seen that its (2^k-) Feigenbaum sequence ends near $a_\infty \simeq 1.044..$. This may not be the "last" Feigenbaum or periodic attractor in it [263]: e.g. there exists a period-7 attractor at $a = 1.3$. At $a = 1.4$ Hénon numerically found what appears to be a Strange Attractor. Starting with some x_1, x_0, in (3.8), the subsequent orbit $\{x_{t+1}, x_t\}$ is rapidly attracted to the object plotted in Figure 22A. If we take 2 - initially close - points *on* this attractor we find that the 2 subsequent orbits - along the attractor - separate at an exponential rate, whence a 'sensitive dependence on initial conditions' and chaotic behavior, cf. sections 2.2-4. The attractor looks deceptively simple. Yet, repeated magnification in Figures 22B,C,D, shows that there is a nested set of attracting "curves". Each picture seems to reveal nests within nests, etc. up to the maximum feasible magnification, cf. sections 2.5-6. This suggests that its cross-section might indeed be a Cantor set [261-264,272]. Hence, the dark band that appeared in the Rössler attractor, cf. Fig. 21, has in Hénon's attractor been resolved into nests of attracting curves. Figure 22 *is also* the surface of section for the storage ring eq. (3.2) with a capacitive (/inductive) load (3.5).

The mechanism responsible for this chaotic behavior is similar to that in a conservative system, discussed in section 2.4. Again we find wild "separatrices" emanating from a hyperbolic point, *intersecting each other in 'homoclinic' points* [261-264,270,272-275,76,13]. Again, the "separatrices" intersect in infinitely many points while miraculously folding away their loops and plies within the strange bands of the attractor. Once homoclinic points have been established 'quasi-random' orbits and Cantor-set cross-sections can be shown to exist *nearby*, by the same theorems mentioned in section 2.4 [76,13,14]. Many of these may not be attractors but - interspersed - strange repellors [128].

At small enough a-values Hénon's mapping (3.8) has a period-1 attractor which - using our y variables (3.7) - lies at the origin, see (3.11). At $a = 1.4$ it has become a period-1 repellor with $\lambda_1 < -1$ and $0 < \lambda_2 < 1$ ($\lambda_1\lambda_2 = -b$). According to sections 2.2-4 this point $\hat{y}$ is hyperbolic (with reflection) therefore. We take (thousands of points in) some small interval $\hat{y}z$ along $\vec{e}_1$, see Figure 23, and repeatedly map with (3.8). The resulting *"unstable separatrix" coincides with Hénon's strange attractor* of Figure 22A. The hyperbolic point $\hat{y}$ is not a vertex, as in the earlier conservative case (of Figure 6), since our present $\hat{y}$ is hyperbolic *with reflection*, i.e. we sample $-\vec{e}_1$ as often as $\vec{e}_1$. So the points move on a line *through* $\hat{y}$, rather than ending *at* $\hat{y}$. Taking a similar interval along $\vec{e}_2$ and mapping it *backwards* in time ($x_1, x_0 \leftrightarrow x_{-1}$ at $t = 0$ in (3.8), etc.) we obtain the *"stable separatrix"* which intersects the previous separatrix in the homoclinic points ..A,B,C.. of Figure 23a. Mapping a homoclinic point (forwards), e.g. A in Figure 23a, we find another homoclinic point (between A an $\hat{y}$) by definition, etc. Subsequent homoclinic points approach $\hat{y}$, with their distance to $\hat{y}$ vanishing like λ_2^t, as $t \to +\infty$. Since $0 < \lambda_2 < 1$, there is an infinite number of homoclinic points between A and $\hat{y}$. Following (3.6) we saw that the area of the loops, enclosed by the separatrices between 2 homoclinic points, vanishes like $|b|^t$, i.e. not as fast as λ_2^t ($= |b/\lambda_1|^t$). Hence *the lengths of the loops of the unstable separatrix*, between 2 homoclinic points on A,$\hat{y}$, *must approach infinity as* $t \to +\infty$. This suggests that the "curves" plotted in Figure 22B-D are parts of an infinitely long unstable separatrix which would forever reveal new "curves" upon further magnification. Mapping the homoclinic points backwards along the stable separatrix (A,B,C,D,...) the area of the loops approaches infinity like $|b|^t$, with $t \to -\infty$. Doing this, the distances between any 2 homoclinic points cannot become unbounded because the other, unstable, separatrix can be contained within one finite square, for all t [260,264]. Thus *the lengths of the loops of the stable separatrix*, between 2 homoclinic points, *must approach infinity as* $t \to -\infty$. The two types of loops apparently exhibit a behavior similar to that in the conservative case ($|b| = |B| = 1$), cf. Figure 6. The loops of the stable separatrix were obtained by Franceschini and Russo, see Figure 23b (their original picture [261] is much sharper). So, *near* $\hat{y}$, we have arbitrarily close together, and moving in opposite directions, infinitely many loops of the unstable separatrix as well as the stable one, cf. Figs. 22,23. Hence the orbits there have a very 'sensitive dependence on initial conditions' [252] and 'quasi-random' orbits exist near $\hat{y}$ [13,128], as mentioned in section 2.4. The existence of homoclinic points for (3.8) at $a = 1.4$ was derived recently

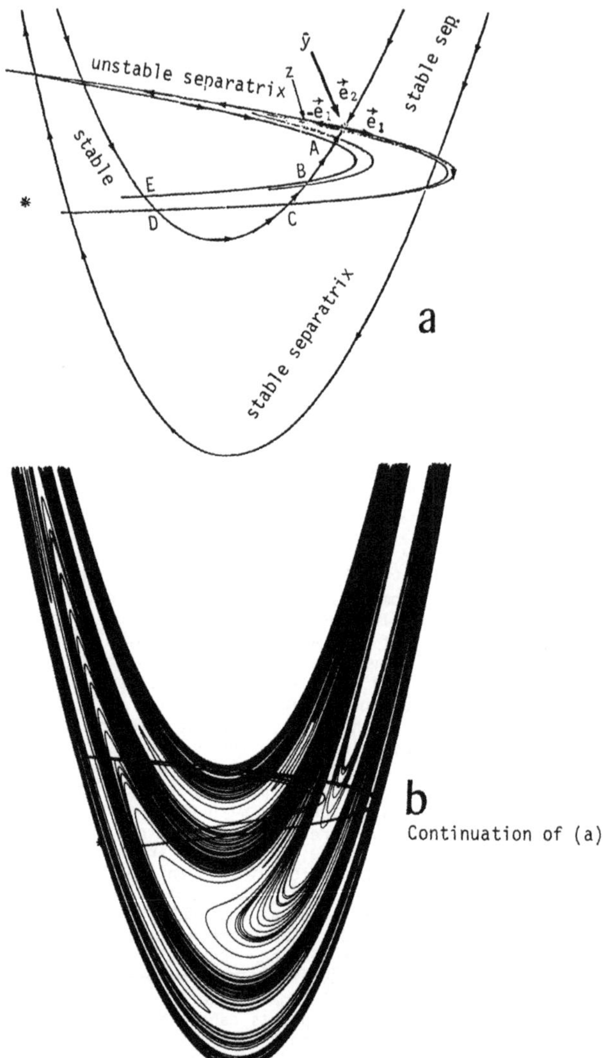

Figure 23
(a) Stable and Unstable Separatrices of the period-1 orbit, $\hat{y}$, of the Hénon mapping (3.8) intersect each other at ..A,B,C,.. (taken from [261(2nd)]). The unstable separatrix, along $\vec{e}_1$, is the strange attractor of Fig. 22A (reduced scales, x_t versus x_{t-1}).
(b) The Stable Separatrix, along $-\vec{e}_2$, is plotted backwards for a longer time [261(2nd)]; also see Fig. 44 of [251 (1981)] and cf. the conservative curves of Figs. 6b-c.

[261-262] but made plausible earlier by numerical plotting and by computing Lyapunov's 'characteristic exponents' (and/or the Kolmogorov 'entropy') [263-265, 252(2nd),97].

These homoclinic points have been found numerically from $a \simeq 1.16$ ($\gtrsim a_\infty$, as discussed earlier [165,263]) through $a \simeq 1.42$ [263]. Thus there is a chaotic region (certainly near $\tilde{y}$) throughout this a-range. If and when the homoclinic points are evenly spread out over the unstable separatrix we expect to have a 'chaotic' or strange attractor. Inspecting Figure 23b, to the extent that the loops are visible or plotted, this appears to be the case. Repeating the calculations at $a = 1.3$ however, one has found 7 *attracting* points [263-264] *on* the unstable separatrix and apparent gaps in the distribution of homoclinic points near those points [261(2nd)]. On the other hand, for $1.427 \lesssim a \lesssim 2.65$ points appear *on* the unstable separatrix which *repell* (transversally) [263-264]. *If* motion along the remainder of the separatrix *were* (to remain) ergodic in both cases, such points would be "reached" eventually. In the first case the orbit ends up on a period-7 attractor, in the second case the slightest round-off error moves it out to anywhere, as observed [264,263]. In the conservative case ($|b|=1$) no direct analogue of these 2 cases is known. Yet, two separatrices from *different* periodic orbits can [60,15] intersect each other, in 'heteroclinic' points [12-18]. In view of the above, some doubts remain [272] that, even at $a = 1.4$, the apparent strange attractor might ("merely") be a periodic attractor with an astronomic period, cf. [261-264,165].

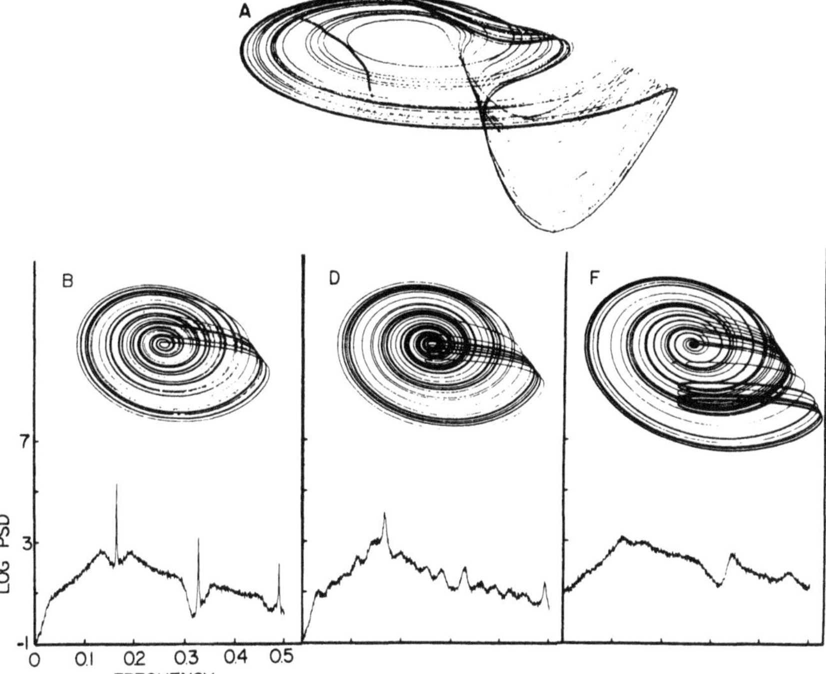

Figure 24
The sharp peaks in the power spectrum of the Rössler Attractor (3.29-31) disappear as one changes the parameter e (taken from [251(1980)]). The attractor develops a fold ('funnel'), cf. the perspective view in (A) and the x,y projections above (B) at $e = 0.17$, (D) at $e = 0.19$, and (F) at $e = 0.3$. Parameters: $f = 0.4$ and $\mu = 8.5$; cf. Figs. 20,21. Text: see next page.

The behavior of the Hénon attractor, at $a = 1.4$, has some similarity with *turbulent behavior* [261-264,165], in particular: time dependent correlation functions which decay to zero, as $t \to \infty$, and frequency spectra which seem to be continuous, to the best numerical resulution [165,263]. Such numerically *continuous power spectra* can also be obtained for the previous Rössler Attractor (3.29-31), at parameter values different from the ones used for Figs. 20,21, see Figure 24 [251(1980)]. This seems due to the appearance of a fold (or 'funnel') in the attractor, cf. the perspective view in Figure 24A. Note the increase in the spectral density at zero frequency as the orbits come closer to the center and develop a larger fold. A similar transition from a δ-peak spectrum, as in Fig. 20, to a "continuous" spectrum as in Fig. 24F is exhibited by Duffing's equation,

$$\ddot{x} = -x - a\dot{x} + 4x^3 - b\exp(i\omega t) \tag{3.34}$$

as one increases ω [249(1979)], cf. [224]. Strange Attractors [277-279] and period-doubling bifurcations [64(2nd),244] have been found for it as well. There is an extensive classic literature on periodically driven nonlinear oscillators [280-289]. Another attractor with a "continuous" spectrum is the famous *Lorenz Attractor*:

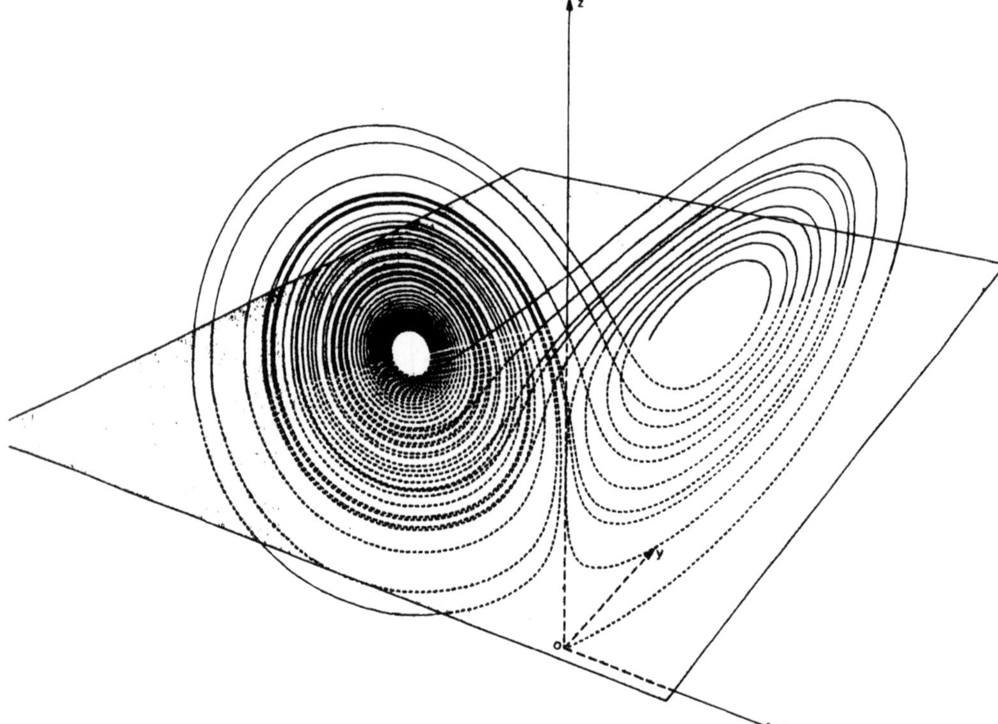

Figure 25
The Lorenz Attractor (3.35-37) in perspective (taken from [250]). Compare this against the projection in Fig. 2 [251(301)]. A particular solution, starting "off" the origin, is plotted. The horizontal plane is at $z = 27$. The "Reynolds Number", μ, is 28. Transients die rapidly.

$$\dot{x} = 10y - 10x, \tag{3.35}$$

$$\dot{y} = \mu x - y - xz, \tag{3.36}$$

$$\dot{z} = xy - 8z/3, \tag{3.37}$$

a strange attractor at $\mu = 28$ [257,258,250-253,271-272,170-188,290]. A perspective view, obtained by Lanford [222], is shown in Figure 25. The orbit shown makes a number of revolutions, in each of the 2 loops (or folds). The precise number in each is difficult to predict since it depends very sensitively on the initial conditions, i.e. we have chaotic behavior on the attractor. Its 'mixing' properties and ergodicity were earlier displayed in Fig. 2; its "continuous" spectrum is shown in Figure 26.

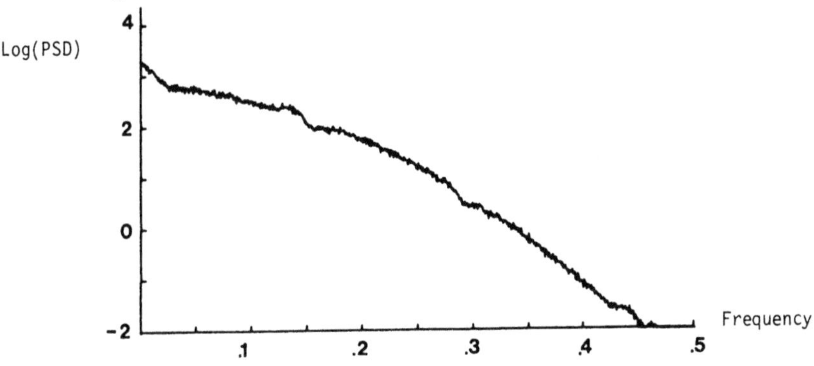

Figure 26
Power Spectrum of x(t) in the Lorenz Attractor (3.35-37) at $\mu = 28$ (taken from [251(1980)]). Here the lowest frequencies have the highest power. Such a "continuous" spectrum reflects the non-periodic chaotic behavior along the attractor.

If the motion on these attractors were '*quasi-periodic*' their spectra would consist of δ-peaks (at frequencies which are not rationally related, cf. (2.41/46)). Yet, strictly speaking, the motion would not be periodic. Thus, it is the *chaotic* behavior on the non-periodic attractor which is responsible for the "continuous" spectrum, e.g. completely random motion has a continuous 'white noise' spectrum [94,134,30].

The Lorenz system is dissipative everywhere in space since its divergence is a negative constant (-13.666..). Three stationary points are easily found from (3.35-37). One is the origin, O, and Figure 25 displays (1 branch of) its unstable separatrix. Note that it does not return to the origin, at this μ value. There are 2 other stationary points O_1 and O_2, at the centers of the 2 loops in Figure 25, within the plane depicted at $z = \mu-1$. The unstable separatrix of O can be attracted back to O (: a homoclinic orbit) at different μ values or can be attracted to O_1 (or O_2) at other μ values. At some μ value a "circular" periodic orbit bifurcates off O_1 (and a second off O_2) which, in turn, can attract the unstable separatrix of O, over yet another range of μ values [253,296,271,170-179,257,290]. If μ is increased slightly beyond the last range we get a strange attractor, about these two "circles", with the unstable separatrix of O switching unpredictably between the two loops (shown in Fig. 25). For $24.06 \leq \mu \leq 24.74$ there are 3 attractors: the strange one, about the "circles", plus O_1 and O_2 [253,271]. Above this range (but <50) the strange one is the only attractor. Its chaotic behavior may be studied from the (Poincaré) mapping of the plane $z = \mu-1$ on itself (via the 3-d. orbits). A model of this mapping can be analytically shown to have quasi-random orbits, and to be 'mixing', ergodic, etc. [271]. The mechanism for this chaotic behavior is similar to that discussed for Hénon's attractor, with the aid of Figure 23 [271,187-188].

The system exhibits various *decreasing* Feigenbaum sequences *above* the previous chaotic-attractor range [235,170,175-178]. In addition there are 'inverse' bifurcations [214,290,170-179] which are known in conservative systems as well, e.g. the period-3 orbits in [35,48], cf. [50,150].

Systematic extensions of the Lorenz equations to 5 and 7 variables, cf. Fig. 29 [234,176,257(2nd)], and 14 coupled rate equations [177-178,165(2nd)], exhibit similar phenomena, embedded in a more complicated structure of strange phenomena. The Lorenz equations and their extensions [170-188.234,257] are derived from the Navier-Stokes partial differential equation of fluid mechanics [298], cf. section 3.4, confirming again the richness of the N.S.' solutions. The Lorenz system models several other phenomena as well [170(1977),194,183,171-188,290-302]. Although its numerical investigation had been published by 1963, in the meteorological literature, by a mathematician [257(1st)], the Lorenz attractor did not start to attract wide-spread attention until a decade later [290,259].

Much of the current interest was stimulated by the 1971 article of Ruelle and Takens [259] who coined the name 'Strange Attractor', at least in print [250], and suggested a new mechanism for the onset of turbulence [19,145]. Earlier we have employed bifurcations of periodic orbits into more and more periodic orbits [214-217] as a mechanism for obtaining non-periodic (and possibly chaotic) behavior in the Feigenbaum limit, and beyond. Ruelle and Takens investigate the "bifurcation" of a periodic orbit into an *n-dimensional torus, with* $n \geq 3$ [259,252 (1978),255,186-188], cf. [248], which can be an attracting torus. The motion on the torus is periodic if its n fundamental frequencies ω_k are rationally dependent and 'quasi-periodic' if they are not, e.g. (2.46). In the latter case we would get a non-periodic attractor with a δ-peak spectrum. It was shown that slight perturbations can then make the motion chaotic [252(1978),276,169,15] and no longer confined to a torus (but still an attractor). This may sound familiar since we encountered comparable behavior upon perturbation of (/away from) K.A.M. tori in a conservative system, cf. Figs. 11,10. In both cases have we seen that systems become "more chaotic" the more periodic orbits turn *hyperbolic*. Anosov [274,15,145] and Smale [76,169,252(1978)] considered the extreme case and defined classes of [128] abstract "hyperbolic systems" in which *all* (recurrent) orbits are hyperbolic. This may be formal, yet Sinai proved that the hard-sphere Boltzmann gas belongs to such a class [95-98] and is 'mixing' and ergodic, therefore [15]. Proofs of the chaotic behavior near homoclinic points of a hyperbolic orbit, which we employed in both chapters, require these concepts as well [76,13,15,128,302]. The theory of Strange Attractors is in good shape for such hyperbolic systems due to the work of Bowen, Ruelle and Sinai, cf. the overviews [252(1978),169] and their references. Such hyperbolic systems have the pleasant property of being 'structurally stable', i.e. their strange attractors do not disappear under small perturbations [169,15]. Since these systems are 'mixing' and ergodic one has established a *statistical mechanics* for them. In the statistical mechanics of a conservative system one employs a stationary probability distribution which is invariant under the motion and therefore a function of the only nice integral, H. The familiar Gibbs distribution is $\propto \exp(-H/kT)$ in the 'Canonical Ensemble' Theory [134,94] of thermal equilibrium. Similar stationary probability (Gibbs-)distributions, equilibrium states and even an entropy and pressure have been defined for hyperbolic systems, in particular for their strange attractors [169,252,98]. All this has led Smale to speculate that we might have to resort to ergodic attractors 'close to' a Hamiltonian system if we want to rescue the ergodic hypothesis for (approximately) Hamiltonian systems [273]. A stationary probability distribution is known explicitly for the logistic equation (3.9) at $a = 4$ [252(1978)], where (3.9) is exactly solvable [297], and known to *exist* at a_∞ and other a values [252(1978)]; cf. [217(1981)].None of the concrete systems of equations, we discussed before, belong to such a class of hyperbolic systems. Yet, the systems we did discuss do have properties in common with them. So, those hyperbolic systems also serve as models of our models. Below, we discuss some real turbulence experiments.

3.4 Turbulent Models, Relation to the Navier Stokes Equation

Here we relate the above models, and the 'Onset of Turbulence' they exhibit, to turbulence experiments and to the Navier-Stokes equation, commonly used in fluid mechanics [298]. The onset of *experimental* turbulence is usually studied on circular Couette flow or Rayleigh-Bénard convection [256]. In the latter experiment a fluid layer is heated from below. The warm fluid rises to the top, cools and sinks again, etc. This "onset of boiling" produces a roll pattern whose velocity field can be measured from the Doppler shift of laser light scattered by the fluid. The experimental Rayleigh number, R, is proportional to the temperature difference between top and bottom, in this case. Experimental period-doubling, discussed in section 3.2, is found and some measure of evidence for the Ruelle-Takens picture, mentioned in section 3.3, is reported for Rayleigh-Bénard experiments.

Both routes to non-periodic behavior, discussed in section 3.3, require very few independent Fourier frequencies: only *one* in the case of Feigenbaum's period-doubling sequences and only *three or four* in the case of Ruelle-Takens' strange attractors and other chaotic attractors of section 3.3. This is in marked contrast to the picture of turbulence suggested by Landau [298] and Hopf [299]. They observe that periodic solutions of frequency ω_j may bifurcate [214] off another periodic solution, laminar flow or steady state solution as one increases our "Reynolds number", μ. When this μ is large they assumed that a large number of (rationally-) independent frequencies ω_j have appeared, in the Fourier spectrum of the motion (/velocity field). It is known from results by Weyl that the mean motion is chaotic then (and ergodic if all Fourier amplitudes are the same), cf. app. 13 of ref. 15. Only after an infinite number of (rationally-)independent frequencies ω_j have appeared would the spectrum appear continuous and would the ergodicity of the mean motion produce chaotic effects noticeable after a short time. One of the main objections, among others [187,188], against this picture is that in actuality the chaos in experimental flows does not gradually increase with the Reynolds number R, but sharply and suddenly at a particular R value. The Landau-Hopf model may yield ergodic motion but its spectra cannot become continuous (i.e. no 'mixing') after only a few bifurcations. This seems to be achieved however by the models discussed before and is also observed experimentally, as reported below.

High resolution power spectra of the velocity field in a Rayleigh-Bénard experiment on water, recently obtained by Gollub, Benson and Steinman [243(1980)], do exhibit a number of period-doubling bifurcations, see Figure 27. The direct transition from subharmonics of one frequency to a "continuous" spectrum, in Figs. 27c-e seems incompatible with the Landau-Hopf picture [298-299]. Actually more can be derived from Feigenbaum's sequences: knowing that the scale of motion is reduced by a factor α (3.28) after each (period-doubling) bifurcation Feigenbaum has predicted [218,224] that the new subharmonics will have Fourier amplitudes which eventually will be smaller than those of the last harmonics by a '*universal factor*': $4|\alpha|/\sqrt{2+2/\alpha^2} \simeq 6.6$ [218]. The agreement with experiment is quite good as we can see from the two straight lines in Fig. 27d: the upper one is drawn through the $f_2/2$ harmonics, the lower one is the same but divided by this 6.6. Note that most of the (new) $f_2/4$ harmonics peak near this line. (In Fig. 27c the bifurcation has just occurred, the new peaks are small and the theory of [218, 224] does not yet apply.) Another confirmation of these predictions can be found in the temperature spectra of liquid Helium recently obtained by Libchaber and Maurer [242(2nd)], see Figure 28. Also note the experiments in [241] and [189].

The relation of the turbulent models of sections 3.2-3 to the above Rayleigh-Bénard experiments and to the Navier-Stokes partial differential equation of fluid mechanics appears when one adds the relevant heat-conduction equations and takes the (Boussinesq) approximation of this system of equations, valid for incompressible fluids [290,257,177-178,165(2nd),186-192]. Substituting spatial Fourier modes (of the experimental container) an infinite set of coupled nonlinear equations of motion is obtained for the mode-amplitudes. To first order in the

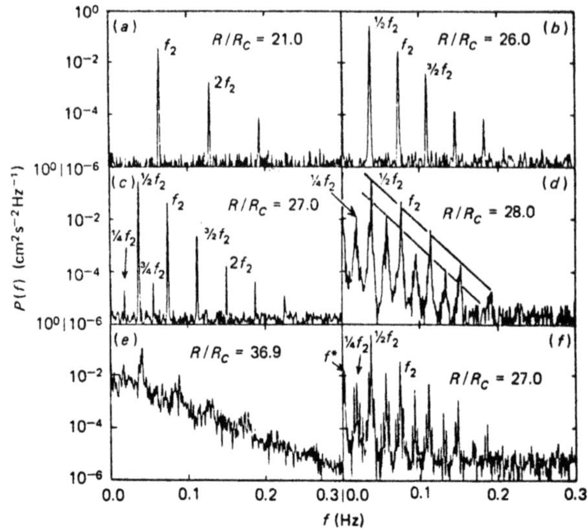

Figure 27
Period-Doubling sequences in the velocity Power Spectra of water in a Rayleigh-Bénard experiment (taken from [243(1980)]). Note the creation of sub-harmonics at increasing Reynolds numbers, R, the creation of "bands" in (d) and a "continuous non-periodic" spectrum in (e) (upon return to the R of (c) a second frequency, f*, appears in (f)). Compare these spectra against those of Figs. 20,21,24,25 and [249(1979)].

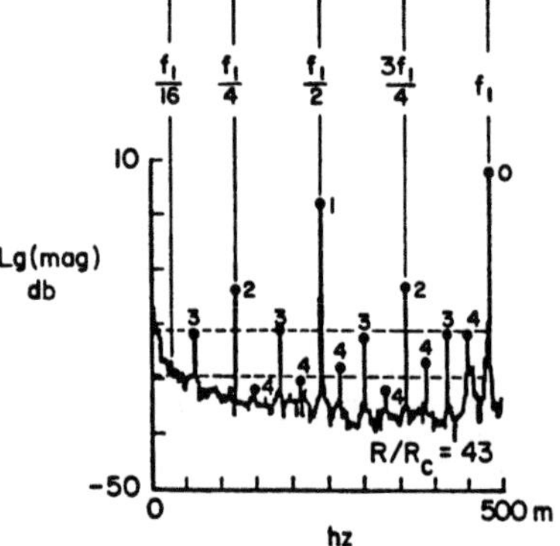

Figure 28
Period-Doubling sub-harmonics in the temperature Power Spectrum of liquid Helium in a Rayleigh-Bénard experiment (taken from [242(2nd), 218]). The dashed lines are at the predicted [218] relative magnitudes of the $f_1/2^3$ and $f_1/2^4$ components. Noting that 6.6 ≈ 8.2 db, we see that the agreement is good, cf. also Fig. 27d.

amplitudes these are linear and exhibit no chaotic behavior. To second order only 3 nonvanishing amplitudes remain [290,257]. Their equations of motion, at this order, are precisely the Lorenz equations (3.35-37) of the previous section [257-258]. Extensions to systems of 5, 7 and 14 modes [234,176,257(2nd),177-178,165 (2nd),170-194] naturally arose from the above (Boussinesq) system. Since all these equations numerically exhibit Feigenbaum sequences of period-doubling bifurcations, cf. section 3.3, it is gratifying to find them again in the spectra of the underlying experiments, in Figs. 27 and 28.

The route to turbulence via bifurcations into strange attractors with 2 or 3 *independent* frequencies [248,290,186-189], as in the Ruelle-Takens picture of section 3.3 [259], has received some experimental confirmation [256,250,189-192,290] and some critical evaluation [186,253,187-193]. One might attribute the second frequency, f*, in Fig. 27 (f) to the appearance of such an attractor. Similar independent frequencies were observed in other (and earlier) Rayleigh-Bénard and Couette experiments [190-192]. Yet, the ratio of these frequencies can be continuously varied [191(2nd.)], which seems difficult to combine with their being rationally independent after a 'Ruelle-Takens' bifurcation, mentioned in section 3.3.

Similar turbulence phenomena are found in several fields [290-295,216,300-302]. I remind you however that we only considered the onset of turbulence and not the 'fully-developed' turbulence encountered in high velocity flow through pipes, in boiling, etc. Many additional problems of spatially inhomogeneous motion and sudden ('intermittent') phenomena are encountered there [291-295,298] and remain to be solved [268,269].

While the progress discussed in this paper may still better explain turbulence and mixing among theoretical physicists and applied mathematicians than in fluids and gases, it *is* exciting to detect any progress at all on the venerable problems of chaotic behavior in deterministic systems [302] and see some agreement with experiments.

4. ACKNOWLEDGEMENTS

I wish to compliment the organizers for keeping the great tradition of this summerschool alive, and well. Their continued attention to nonlinear dynamics, cf. [90], is greatly appreciated.

Some of the work discussed here has not appeared in print yet and I thank many of you for communicating your results by letter or early-preprint, especially those who sent original drawings. I apologize to those whose work I may have overlooked, and refer them to the previous sentence. Much credit goes to Joe Ford who keeps up an inspiring (/cajoling) flow of information, in addition to Nonlinear Science Abstracts [300]. I thank him, Joel Lebowitz and Elliot Montroll for encouraging my forays into their fields.

Analytical and numerical help in the preparation of this paper was extended by Tassos Bountis, Charles Eminhizer and Johan van Zeyts. Much constructive criticism from Jouke Heringa is gratefully acknowledged, and incorporated. Several lengthy discussions with Wim Caspers and Theo Valkering facilitated the writing of section 2.8. For the sections on storage rings I owe much to the enthusiasm and support of Mel Month and David Sutter.

I thank Elly Reimerink for equanimously suffering my handwriting and converting it into excellent typewriting. Figures 3, 8, 9 and 12 were made by mr. P. Kops. Parts of this study were supported by D.O.E. under EG-79-C-03-1538.

5. APPENDIX A. Separatrices do Split [130]

Here I continue the discussion of section 2.4, below (2.26). Employing an (extremely) simple example it is shown that 2 separatrices, emanating from a hyperbolic point (cf. $\hat{y}$ in Fig. 6b, or $+\pi$ in Fig. 3), need not join each other (into one separatrix) via a finite loop, but *can intersect* (transversely). After we obtain one such 'homoclinic' intersection an infinity of other homoclinic points follows automatically, cf. section 2.4. For an integrable map on the other hand, we have a global (real-) analytic integral $I(y_t, y_{t+1})$, cf. section 2.1. The 2 separatrices (emanating from 1 point $\hat{y}$) are now level-lines of $I(y,y')$ at 1 level. Since most (real-)analytic functions $I(y,y')$ will not have level-lines that intersect themselves (transversely) in an infinity of points (which limit on $\hat{y}$) in general, we do not expect any homoclinic intersections for integrable systems in general [13] (exceptions are possible for some mechanical-"equilibrium" positions of a system of Hamiltonian differential equations [129]).

In the next paragraph I shall discuss a piecewise-linear mapping. Before you blame the (subsequent) homoclinic intersections of its separatrices solely on the singularities of that map, I want to briefly mention the piecewise-linear pendulum,

$$\ddot{x} = \begin{bmatrix} -x & , & |x| \le \frac{1}{2}\pi \\ +x-\pi & , & \frac{1}{2}\pi \le x \le 3\pi/2 \end{bmatrix} \quad \text{etc., mod. } 2\pi. \tag{A.1}$$

It has a simple (piecewise-quadratic) Hamiltonian $H(x,\dot{x})$ which is continuously differentiable everywhere. Its phase plane plot is analogous to that of the ordinary pendulum, in Fig. 3. Again, there are hyperbolic (mechanical equilibrium) points at $x = \pm\pi$, $\dot{x} = 0$. The separatrices emanating from them, cf. Fig. 3, i.e. level-lines of the simple Hamiltonian $H(x,\dot{x})$ above, do <u>not</u> have homoclinic intersections.

Consider the piecewise-linear mapping,

$$y_{t+1} + y_{t-1} = \begin{bmatrix} -\frac{1}{2}y_t + 4\frac{1}{2} & , & y_t \ge 1 \\ +4y_t & , & |y_t| \le 1 \\ -\frac{1}{2}y_t - 4\frac{1}{2} & , & y_t \le -1 \end{bmatrix} \equiv f(y_t). \qquad \begin{matrix} \text{(A.2a)} \\ \text{(A.2b)} \\ \text{(A.2c)} \end{matrix}$$

The function $\frac{1}{2}f(y_t)$ is plotted (dashed) in Fig. 29. The origin is hyperbolic with eigenvalues $\lambda_\pm = 2 \pm \sqrt{3}$ and eigenvectors $\vec{e}_\pm = (1/\lambda_\pm, 1)$, cf. eqs. (2.20-.26). We construct the unstable separatrix by mapping out a small piece of the unstable eigenvector, from the (hyperbolic) origin on. As the mapping (A.2b) is linear there, the first part of the separatrix is along the *straight* line: $(y/\lambda_+, y)$. Consider two particular points on this line, point A:$(1/\lambda_+, 1)$ and its image A': $(1, \lambda_+)$, cf. Fig. 29. Points between A and A' have coordinates $(y/\lambda_+, y)$, with $1 < y < \lambda_+$, and must be mapped by (A.2a) rather than (A.2b). Thus the image of AA' is the straight line: A'A" whose points have coordinates $(y, -y(2\frac{1}{2}-\sqrt{3})+4\frac{1}{2})$, with $1 < y < \lambda_+$. This line intersects the line "$y = x$", at some point D in Fig. 29, *but <u>not</u> perpendicularly!* Since the complete phase plane plot of our (reversible) map must be symmetric about the line "$y = x$" we can obtain the other, stable, separatrix (0 - B' - B"...) by mirroring the unstable one (0 - A' - A"...) about "$y = x$" (or re-do the calculation above, to obtain B'B" explicitly). *Hence, there is a homoclinic intersection of the two separatrices* at D, because A'A" does not intersect "$y = x$" perpendicularly. Additional applications of piecewise-linear conservative maps are found in refs. 56,39,202,201.

Approximating the piecewise-linear function in (A.2) - uniformly, for <u>every</u> point of some y interval, e.g. (-10,10) - by an *<u>infinitely differentiable</u>* (:C^∞) function $f(y)$ in (A.2) we can ensure that the latter's separatrices stay within some distance ε (small) from A'A" (and B'B"), in Fig. 29. Hence, all maps (A.2) with (C^∞-) $f(y)$, "sufficiently close" to our piecewise-linear function (A.2), have homoclinic intersections as well.

Considering (real-) *analytic* functions $f(y)$ in (A.2) we confine ourselves of course to ($f(y)$:) mapping whose separatrices have at least 1 point in common (but might still join). Any (real-) analytic function $f(y)$ can be approximated

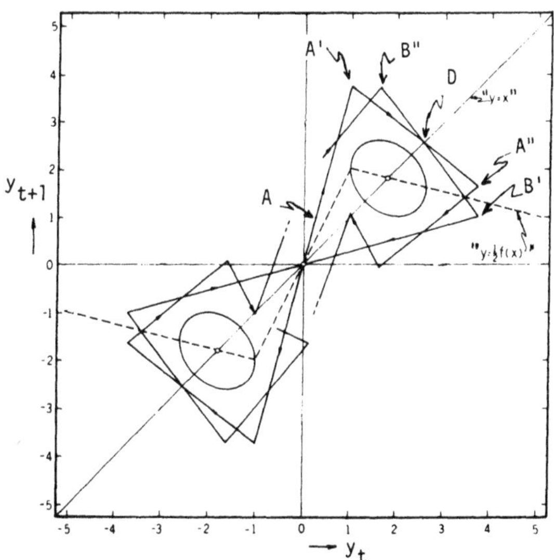

Figure 29
Intersecting-wild-Separatrices of the piecewise-linear mapping (A.2); from [56]. The unstable separatrix (piece), A'A", does not intersect "y = x" perpendicularly. Hence it has only one (homoclinic) point, D, in common with its mirror image, B'B".

arbitrarily well by piecewise-linear functions, with finitely many pieces (on some y-interval, e.g. (-10,10)). Some of these approximations can be "degenerate", in that their analogues of A'A" and B'B" may coincide (or may come within the small distance, 2ε, discussed above). Such degeneracies can easily be "perturbed away" by further approximation. Hence, we conclude that "sufficiently close" to "every" (real-) analytic mapping there is an infinity of C^∞-mappings with homoclinic intersections. In (physics-) practice this might suffice to indicate that "most" systems have homoclinic intersections. In (mathematics and) theory however this is much weaker than Siegel's results on analytic approximations, discussed in section 2.3 and note [89] [12,77,78].

If one wants to repeat the explicit calculation, above, for some analytic function f(y) in (A.2) one has to be extremely careful: Since I cannot obtain (exact) explicit expressions now, for the first two "folds" of the separatrices, I must approximate them. This is equivalent [56] to working with an another analytic f(y), for which the approximate separatrices, above, are exact. Even if this new f(y) does have intersecting separatrices it is very difficult to conclude anything about the original f(y): From section 2.3 we infer that 'most' "Taylor-developable" (:analytic) mappings do have intersecting separatrices and are *dense* among all analytic mappings. Hence 'most' *analytic*-approximations, no matter how close-, will yield intersecting separatrices, irrespective of whether the original map had them, or not.

6. REFERENCES

Non-Integrable Hamiltonian Mechanics
(Refs. 1-6,12,13(chapt. 1),14,17,19,20(Int.)-22(§1) contain reviews or introductory articles).

[1] *Topics in Nonlinear Dynamics*, ed. S. Jorna, Am.Inst.Phys.Conf.Proc., vol. 46, e.g. its Appendix (1978)
[2] M.V. Berry, Regular and Irregular Motion, ref. 1, p. 16-120 (1978)
[3] J. Moser, Nearly Integrable and Integrable Systems, ref. 1, p. 1-15 (1978)
[4] J. Moser, Stable and Unstable Motion in Dynamical Systems, ref. 200, p. 222-235 (1979)
[5] J. Moser, Is the Solar System Stable?, Math.Intelligencer 1, 65-71 (1978), in German: Neue Zürcher Zeitung (May 14, 1975)
[6] M.V Berry, Regularity and Chaos in Classical Mechanics, Illustrated by 3 Deformations of a Circular Billiard, Eur. J. Phys. 2, 91-102 (1981)

Classical Mechanics Texts

[7] H. Goldstein, *Classical Mechanics*, Addison-Wesley Publ.,Reading,Mass.,U.S.A. (1950)
[8] D. ter Haar, *Elements of Hamiltonian Mechanics*, North Holland Publ., Amsterdam (1961)
[9] H.C. Corben and P. Stehle, *Classical Mechanics*, (2nd enlarged edition) Krieger Publ., Huntington, N.Y. (1974)
[10] E.T. Whittaker, *Analytical Dynamics of Particles and Rigid Bodies*, Cambridge Univ. Press (1904-1964)
[11] L.A. Pars, *Analytical Dynamics*, Heinemann Publ., London (1965)

Texts, with chapters on Non-Integrable Systems

[12] J. Moser, Lectures on Hamiltonian Systems, Memoirs Am.Math.Soc. 81, 1-60 (1968); and its refs.
[13] J. Moser, *Stable and Random Motions in Dynamical Systems*, Princeton Univ. Press (1973); and its refs.
[14] V.I. Arnold, *Mathematical Methods of Classical Mechanics*, Springer Verlag, N.Y. and Heidelberg (1978); besides the original, in Russian (1974), there also exists a French translation, MIR, Moscow (1976); esp. its appendices and refs.
[15] V.I. Arnold and A. Avez, *Ergodic Problems of Classical Mechanics*, Benjamin Publ.,N.Y. and Amsterdam(1968); esp. its appendices and refs.
[16] R. Abraham and J.E. Marsden, *Foundations of Mechanics*, (2nd enlarged edition), Benjamin/Cummings Publ., Reading, Mass., U.S.A. (1978), [Chapter 8]; W. Thirring, *Classical Dynamical Systems*, Springer Verlag, Berlin (1978)
[17] A.J. Lichtenberg and M.A. Lieberman, *Regular and 'Stochastic' Motion*, Springer Verlag, Berlin (1983); J.H. Bartlett, *Classical and Modern Mechanics*, University of Alabama Press (1975)
[18] H. Poincaré, *Les Méthodes Nouvelles de la Mécanique Céleste*, Gauthier-Villars, Paris (1892); Dover Press (1957); (in English:) N.A.S.A. Translation TT F-450/452, U.S. Fed. Clearinghouse, Springfield, VA, U.S.A. (1967)
[19] V.I. Arnold, *Geometrical Methods in the Theory of Ordinary Differential Equations*, Springer Verlag, Berlin (1983); MIR, Moscow (French, 1980; Russian, 1978); also see [145,276/302].
[20] V.I. Arnold, Small Denominators and Problems of Stability of Motion in Classical and Celestial Mechanics, Russian Math. Surveys 18, 85-191 (1963)
[21] V.I. Arnold, The Classical Theory of Perturbations and the Problem of Stability of Planetary Systems, Sov.Math.Dokl. 3, 1008-1011 (1962)
[22] V.I. Arnold, Proof of a Theorem of A.N. Kolmogorov on the Invariance of Quasi-Periodic Motions under small Perturbations of the Hamiltonian, Russian Math.

Surv. 18, 9-36 (1963); Kolmogorov's original (1954) article and lecture are reprinted, in English, in ref. 100, p. 51-56 (1979), respectively ref. 16, Appendix (1978)

[23] N.N. Bogoljubov, Ju.A. Mitropolski and A.M. Samoilenko, *Methods of Accelerated Convergence in Nonlinear Mechanics*, Springer Verlag, Heidelberg (1976); also in Russian (1969)

Hamiltonian Systems, Examples

[24] G. Contopoulos, on the Existence of a Third Integral of the Motion, Astron. J. 68, 1-4 (1963); 70, 817-835 (1965); and, Integrable and 'Stochastic' Behavior in Dynamical Astronomy, ref. 100, p. 1-17 (1979)

[25] M. Hénon and C. Heiles, The Applicability of the Third Integral of the Motion, Some Numerical Experiments, Astron. J. 69, 73-79 (1964)

[26] F. Gustavson, On Constructing Formal Integrals of a Hamiltonian System near an Equilibrium Point, Astron. J. 71, 670-686 (1966)

[27] C. Froeschlé, A Numerical Study of the Stochasticity of Dynamical Systems with Two Degrees of Freedom, Astron. & Astrophys. 9, 15-23 (1970)

[28] When $N \geq 3$ we select (vanishing) initial conditions for all but 2 of the degrees of freedom, such that we end up with an $N = 2$ subsystem. If this is nonintegrable, so is the total system. If it is integrable we can separate the system into two *un*coupled ones with fewer degrees of freedom than N [17,11,18]. Then we repeat this procedure for each of these two [28]

[29] G. Casati, J. Ford, F. Vivaldi and W.M. Visscher, Energy Propagation in One-Dimensional Systems: Validity of Fourier's Law, preprint (1980)(Physics, Univ., 20133 Milano, Italy); and W.M. Visscher in: *Methods in Computational Physics*, Vol. 15, p. 371-408 (1976), Academic Press; A. Casher and J.L. Lebowitz, J.Math. Phys. 12, 1701 (1971); T.P. Valkering, J.Phys. 11A, 1885-1897 (1978); E.A. Jackson, Non Linearity and Irreversibility of Lattice Dynamics, Rocky Mountain J.Math. 8,126-196(1978), Review, in same volume as in [34] W.M. Visscher and J.E. Gubernatis, in *Dynamical Propts. Solids*, North Holland Co. (1980)

[30] P. Mazur and E. Braun, Physica 30,1973(1964); E. Braun, On the Statistical Mechanics Theory of Brownian Motion II, Physica 33,528-546(1967); P. Ullersma, Thesis Theor. Phys., U. of Utrecht, The Netherlands); U. Titulaer, Physica 70,470(1973); L. van Hemmen, Ref. 100, p. 232-240; also see refs. 125-127

[31] M.D. Kruskal, J.Math.Phys. 3,806(1962); See sectn. 2.3d and 2.5c of [17(1983)]; B. McNamara, Super-convergent Adiabatic Invariants with Resonant Denominators by Lie Transforms, preprint UCRL-79843, submitted to J.Math.Phys. (1978), and its refs.; H. Grad, Phys. Fluids 10,137-154(1967) ; Also see refs. 205 and 12.

Integrable N-Body Chains, Examples

[32] M. Adler, Some Finite Dimensional Integrable Systems and Their Scattering Behavior, Report MRC#1718, Math.Res.Center, Madison Wisconsin (Febr. 1977); J. Moser, Three Integrable Hamiltonian Systems Connected with Isospectral Deformations, Adv. in Math. 16, #2 (May 1975); M.A. Olshanetzky and A.N. Perelomov, Explicit Solution of the Calogero Model in the Classical Case and Geodesic Flows on Symmetric Spaces of Zero Curvature, Lett. Nuovo Cimento 16, 33 (1976); the quantized problem was solved earlier by F. Calogero, J.Math. Phys. 12, 419 (1971)

[33] E. Fermi, J. Pasta and S. Ulam, Los Alamos Report (1954), reprinted in Fermi's Collected Works (Univ. Chicago Press) (1965) and in: *Nonlinear Wave Motion*, ed. A.C. Newell, Lectures in Appl.Math. (A.M.S.) 15, p. 143 (1974); M.D. Kruskal and N.J. Zabusky, Progress on the Fermi-Pasta-Ulam Nonlinear String Problem, Plasma Physics Lab. Report MATT-Q-21, p. 301-308, (1963); N.J. Zabusky and M.D. Kruskal, Interactions of "Solitons" in a Collisionless Plasma and the Recurrence of Initial States, Phys.Rev.Lett. 15, 240-243 (1965); and M. Kruskal, The Birth of the Soliton, in *Research Notes in Mathematics*, Vol. 26, p. 1-8 (1978); ed. F. Calogero, Pitman Publ., London

[34] C.S. Gardner, J.M. Greene, M.D. Kruskal and R.M. Miura, Methods for Solving

the KdeV Equation, Phys.Rev.Lett. 19, 1095-1097; N.B. there exist 5 more articles by these authors, e.g. Comm. Pure Appl.Math. 27, 97 (1974); see G. Eilenberger, *Solitons*, Springer Verlag (1981); the Proceedings of the (1976-Tuscon) Conference on the Theory and Applications of Solitons, eds. H. Flaschka and D.W. McLaughlin, Rocky Mountain J.Math. 8, #1, #2 (1978); and: *Solitons*, eds. R. Bullough and P. Caudrey, Springer Verlag, Berlin (1980); The (integrable) Toda Chain has a comparable history of: a) numerical search for chaos (J. Ford et al.) and b) subsequent proofs of integrability and complete Soliton solutions [90] (Hénon; Flaschka, Kac and van Moerbeke, Proc.Natl.Acad. Sci. U.S.A.); The Toda Chain was recently quantized by Faddeev et al. [300]

Conservative Dynamical Mappings, Examples

[35] M. Hénon, Numerical Study of Quadratic Area-Preserving Mappings, Quart.Appl. Math. 27, 291-312 (1969); and its refs.; SEE REF. [56]
[36] J. Roels and M. Hénon, Recherche des Courbes Invariants d'une Transformation Ponctuelle Plane Conservant les Aires, Bulletin Astronomique (série 3) 2, 267-285 (1967); G. Servizi, et al., Phys.Lett., 95A, 11-14 (1983)
[37] F. Rannou, Numerical Study of Discrete Plane Area-Preserving Mappings, Astron. & Astrophys. 31, 289-301 (1974)
[38] A. Brahic, Numerical Study of a Simple Dynamical System, Astron. & Astrophys. 12, 98-110 (1971); See section 3.4 of ref. 17
[39] R.L. Devaney, A Piecewise Linear Model for the Zones of Instability, preprint (Math. Boston U.) (1983)

Periodic and Quasi-Periodic Orbits

[40] J.M. Greene, A Method for Determining a 'Stochastic' Transition, J. Math. Phys. 20, 1183-1201 (1979); R. MacKay thesis (Princeton, 1982); Springer Verlag (?1983)
[41] J.M. Greene, K.A.M. Surfaces Computed from the Hénon-Heiles Hamiltonian, ref. 200, p. 257-271 (1979)
[42] J.M. Greene, The Calculation of K.A.M. Surfaces, ref. 301 (1980)
[43] J.M. Greene, Two-Dimensional Measure-Preserving Mappings, J.Math.Phys. 9, 760-768 (1968)
[44] R.H.G. Helleman, Variational Solutions of Non-Integrable Systems, ref. 1, p. 264-285 (1978) [Addendum (below (2.37)): "When m_1, m_2 are not relative primes, we obtain the 'islands' about the (elliptic) primary periodic solutions"]
[45] R.H.G. Helleman and T. Bountis, Periodic Solutions of Arbitrary Period, Variational Methods, ref. 100, p. 353-375 (1979) (Misprint: 7 lines below Fig. 3, should read: "...is 10^6 larger...")
[46] C.R. Eminhizer, R.H.G. Helleman and E.W. Montroll, On a Convergent Nonlinear Perturbation Theory without Small Denominators or Secular Terms, J.Math.Phys. 17, 121-140 (1976) [esp. sections 4 and 5]; C. Eminhizer, Thesis, U. Rochester
[47] C.R. Eminhizer, P.A. Vuillermot, T.C. Bountis and R.H.G. Helleman, ref. 203, p. 25-59 (1979)
[48] R.H.G. Helleman, On the Iterative Solution of a 'Stochastic' Mapping, p. 343-370, in: *Statistical Mechanics and Statistical Methods in Theory and Application*, Plenum Publ.Co., N.Y. (1977); [better erase: 'ordered' in (2.10)]
[49] T.C. Bountis, Thesis, Physics, Univ. of Rochester, N.Y. (1978)(and University Microfilms Inc.); T.C. Bountis and R.H.G. Helleman, On the Stability of Periodic Orbits of Two-Dimensional Mappings, J.Math.Phys. 22, 1867-1877 (1981)
[50] G. Contopoulos, Astron.J. 75, 96-130 (1970); 76, 147-156 (1971)
[51] R.C. Churchill, G. Pecelli and D.L. Rod, ref. 100, p. 76-136 (1979)
[52] A. Weinstein, ref. 1, p. 260-263; Inventiones Math. 20, 47-57 (1973); and several preprints (Math., Berkeley, U.S.A.); J. Moser, Comm. Pure Appl.Math. 29, 727-747 (1976); P.H. Rabinowitz, preprint MRC #1783 (Math. Research Center, Madison, Wisconsin, U.S.A.) (1977)
[53] P. Deift and E. Trubowitz, Some Remarks on the Korteweg-de Vries and Hill's Equations, ref. 301 (1980)
[54] W. Magnus and S. Winkler, *Hill's Equation*, Interscience Publ., N.Y. (1966)
[55] H.P. McKean and E. Trubowitz, Comm. Pure Appl.Math. 24, 142-226 (1976); H.P. McKean and P. van Moerbeke, Inventiones Math. 30, 217-274 (1975)

[56] E.M. McMillan, A Problem in the Stability of Periodic Systems, p. 219-244, in: *Topics in Modern Physics - A tribute to Edward U.Condon*, eds. E. Britton and H. Odabasi, Colorado Assoc.Univ.Press, Boulder, CO (1971)
[57] D.F. Escande and F. Doveil, Renormalization Method for Computing a 'Stochastic' Threshold, [300], submitted to Phys.Rev.Lett., (Lab. P.M.I., Ecole Polytechnique, Palaiseau, France) (1980); and preprint P.M.I. 1053 (Oct. 1980)
[58] I.C. Percival, Variational Principles for Invariant Tori and Cantori, ref. 200, p. 302-310, (1979); A Variational Principle for Invariant Tori of Fixed Frequency, J.Phys. 12A, L57 - L60 (1979)
[59] E. Zehnder, Homoclinic Points near Elliptic Fixed Points, Comm. Pure Appl.Math. 26, 131-182 (1973)
[60] S. Channon and J.L. Lebowitz, Numerical Experiments in 'Stochasticity' and Heteroclinic Oscillation, ref. 301(Physics,Rutgers U.,New Brunswick,N.J.)(1980)
[61] G.H. Lunsford and J. Ford, J.Math.Phys. 13, 700-705, cf. Appendix A (1972); Phys.Rev. 1A, 59 (1970)
[62] J. Ford, ref. 1, p. 121-146 (1978); ref. 90,91; G.H. Walker and J. Ford, Phys. Rev. 188, 416 (1969)
[63] G. Benettin, C. Cercignani, L. Galgani and A. Giorgilli, Universal Properties in Conservative Dynamical Systems, (Physics, Univ. of Padova, Italy), Lett. Nuovo Cimento 28, 1-4 (1980); also see refs. 215-249
[64] J. Greene, private communication (Plasma Lab., Princeton University, N.J., U.S.A.) (1980); F. Vivaldi and J. Ford, private communication (Physics, Georgia Tech., Atlanta, GA, U.S.A.) (1980); also see ref. 244
[65] J.B.J. van Zeyts, Internal D-1 Report (Theor. Physics, Twente Univ. of Technology, Enschede, The Netherlands) (1980)
[66] G. Contopoulos and M. Zikides, Periodic Orbits and Ergodic Components of a Resonant Dynamical System, (Astron., Univ. of Athens, Greece), Astron. & Astrophys. (1980), in press
[67] G. Benettin, L. Galgani, A. Giorgilli and J.-M. Strelcyn, preprint, submitted to: Meccanica (cf. ref. 63) (1980); also see refs. 266, 267,198-199,244
[68] T.C. Bountis, Period Doubling, Bifurcations and Universality in Conservative Systems, subm. Physica D (1980) [80]

Quasi-Random References

[69] In practice these systems are damped, enabling us to observe a *final* outcome. If there were no damping we could take a flash-photograph after some fixed time interval. Its results would be just as random (but the game more expensive)
[70] C.L. Siegel and J.K. Moser, *Lectures on Celestial Mechanics*, Springer Verlag, Heidelberg (1971)
[71] V.M. Alexeev, Sur l'Allure Finale du Mouvement dans le Problème des Trois Corps, Actes Int.Congr.Math. 1970, Vol. 2, p. 893-907, Gauthier-Villars Publ., Paris (1971)
[72] V.M. Alexeev, Uspekhi Mat.Nauk. 24, 185-186 (1969)
[73] V.M. Alexeev, Quasi Random Dynamical Systems, Mat. USSR Sbornik 7, 1-43 (1969); 6, 505-560 (1968); 5, 73-128 (1968); Quasi Random Oscillations..., E.R.D.A.-TR-302, (Techn.Info.Serv., Oak Ridge Natl.Lab., Tennessee, U.S.A.) (1976)
[74] V.M. Alexeev, Math. Zametki 6, 489-498(1969); Sov.Math.Dokl. 8, 1421-1424(1967)
[75] K. Sitnikov, Dokl.Akad.Nauk. USSR 133, 303-306 (1960)
[76] S. Smale, Bull. A.M.S. 73, 747-817 (1967); and p. 63-80 in: *Differential and Combinatorial Topology*, Princeton Univ. Press (1965); and C. Conley, references see [13]
[77] C.L. Siegel, Math.Ann. 128, 144-170 (1954); R. Robinson, Amer.J.Math. 92, 562(1970); 897(1970)
[78] C.L. Siegel, Ann.Math. 42, 806-822 (1941)
[79] J. Ford, How Random is A Coin Toss?, Physics Today 40 (April), 40-47 (1983), especially his references 8 and 7

Arnold Diffusion
(Refs. 2-4,80,81,84 contain reviews or introductory articles)

[80] T.C. Bountis, The Role of Resonances, 'Stochasticity' and Arnold Diffusion in Models of the Beam-Beam Interaction, preprint (Math., Clarkson College,

Potsdam, N.Y., U.S.A.) (also reviews refs. 82,83, and parts of 84), in: Seminar on the Beam-Beam Interaction, ed. M. Month (B.N.L., Upton, N.Y.) (1980), S.L.A.C. Report of the May 22-23 Seminar (Stanford)

[81] M. Tabor, The Onset of Chaotic Motion in Dynamical Systems, preprint (La Jolla Institute, P.O.B. 1434, La Jolla, CA 92038, U.S.A.)(also reviews other "chaos estimates"), Adv. Chem. Phys. 46, 73 (1981)

[82] J.L. Tennyson, M.A. Lieberman and A.J. Lichtenberg, Diffusion in Near-Integrable Hamiltonian Systems with Three Degrees of Freedom, ref. 200, p. 272-301 (1979); C. Froeschlé, Astrophys. Space Sci. 14, 110 (1971); and with J.P. Scheidecker, 25, 373 (1973). Also see chapt. 6 of ref. 17 (1983)

[83] B.V. Chirikov, J. Ford and F. Vivaldi, Some Numerical Studies of Arnold Diffusion in a Simple Model, ref. 200, p. 323-340 (1979)

[84] B.V. Chirikov, A Universal Instability of Many-Dimensional Oscillator Systems, Physics Reports 52, 265 (1979); and some (as yet) unpublished preprints on Arnold Diffusion (Inst.Nucl.Phys., Novosibirsk, U.S.S.R.)

[85] Ph. Holmes and J. Marsden, J. Math. Phys. 23, 669 (1982); cf. Commun. Math. Phys. 82, 524 (1982)

[86] N.N. Nekhoroshev, Exponential Estimate of the Time of Stability for Nearly Integrable Hamiltonian Systems, (I)Russ.Math.Surv.32,1-63(1977);(II)Trudy Petrovsky Seminar 5,5-50(Russian)(1979);Funct.Anal.&Appl. 5,82(1971)

[87] V.I. Arnold, Instability of Dynamical Systems with several Degrees of Freedom, Sov.Math.Dokl. 5, 581-585 (1964)

[88] L. Brillouin, Poincaré and the Shortcomings of the Hamilton-Jacobi Method for Classical or Quantized Mechanics, reprinted in chapt. 9 of: *Scientific Uncertainty, and Information*, same author, Academic Press (1964); 1st version: Arch. Rational Mech. Anal. 5, 76-94 (1960)

[89] If we broaden the definition of neighborhood by prescribing (2.18) only for the lower order coefficients, and allowing *any* choice of higher order coefficients, the integrable Hamiltonians do become dense, while the nonintegrable ones remain dense. In that case the nonintegrable ones are more abundant than the integrable ones, in each neighborhood, analogous to the situation with the irrational and rational numbers on the real axis [12,77,78].

Transition to Statistical Mechanics
(Refs. 90-96 contain reviews or introductory articles)

[90] J. Ford, The Statistical Mechanics of Classical Analytic Dynamics, p. 215-255, in: *Fundamental Problems in Statistical Mechanics*, Vol. 3, ed. E.G.D. Cohen, North Holland Publ. Co., Amsterdam (1975)

[91] J. Ford, The Transition From Analytic Dynamics to Statistical Mechanics, Adv. Chem.Phys., 24, 155-185 (1973)

[92] J.L. Lebowitz and O. Penrose, Modern Ergodic Theory, Physics Today February, 26, 23-29 (1973)

[93] A.S. Wightman, Statistical Mechanics and Ergodic Theory - An Expository Lecture, p. 4-25 in: *Statistical Mechanics at the Turn of the Decade*, ed. E.G.D. Cohen, (Dekker) North Holland Publ. Co., Amsterdam (1971)

[94] R. Balescu, *Equilibrium and Nonequilibrium Statistical Mechanics*, Appendix A, Wiley - Interscience Publ., N.Y. (1975)

[95] Ya.G. Sinai, Ergodicity of Boltzmann's Equation, in *Statistical Mechanics, Foundations and Applications*, ed. T.A. Bak, Benjamin Publ. N.Y. (1967); Russian Math. Surveys 25, 137-189 (1970); and his article in ref. 301 (1980)

[96] Ya.G. Sinai, Development of Krylov's Ideas, p.239-281 in: *Works on the Foundations of Statistical Physics*, N.S. Krylov, Princeton U. Press (1979)

[97] Ya.B. Pesin, Characteristic Lyapunov Exponents and Smooth Ergodic Theory, Russian Math. Surveys 32, 55-114 (1977); M. Brin, in ref. 301 (1980)

[98] Ya.G. Sinai, *Introduction to Ergodic Theory*, Princeton University Press (1976)

Quantization of Non-Integrable Systems
(Refs. 99,100,105,111 and 118 contain reviews or introductory articles)

[99] M.V. Berry, Quantization of Mappings and Other Simple Classical Models, ref. 301 [reviews refs. 101-103 and others] (Physics,Univ. of Bristol,U.K.) (1980)

[100] *'Stochastic' Behavior in Classical and Quantum Hamiltonian Systems*, eds. G. Casati and J. Ford, Lecture Notes in Physics, Vol. 93, Springer Verlag (1979); Ph. Holmes, Proof of Non-Integrability for the Hénon-Heiles Hamiltonian..., Physica 5D, 335 (1982); Y. Chang, M. Tabor and J. Weiss, J.Math.Phys. 23, 531 (1982); T. Bountis, H. Segur and F. Vivaldi, Phys.Rev. 25A, 1257 (1982)
[101] M.V. Berry, N.L. Balazs, M. Tabor and A. Voros, Quantum Maps, Ann.Phys. 122, 26-63 (1979); first 2 authors: J. Phys. 12A, 625 (1979)
[102] J.H. Hannay and M.V. Berry, Quantization of Linear Maps on a Torus..., Physica 1D, 267-290 (1980)
[103] M.V. Berry, Quantizing a Classically Ergodic System: Sinai's Billiard and the KKR Method, Ann. Phys. (N.Y.) 131, 163 (1981)
[104] S.W. McDonald and A.N. Kaufman, Phys.Rev.Lett. 42, 1189-1191 (1979) (quantization of the 'stadium')
[105] R.A. Marcus, Molecular Behavior in the Quasi-Periodic and 'Stochastic' Regimes, ref. 301 (1980), espec. Figures 5 and 6; and his article in: *Horizons in Quantum Chemistry*, eds. K. Fukui and B. Pullman, Reidel Publ., Dordrecht, The Netherlands (1980); D.W. Noid and R.A. Marcus, J. Chem. Phys. 67, 599 (1977); 62, 2119-2124 (1975); and [on the Hénon-Heiles system (2.10):] with M.L. Koszykowski and M. Tabor, in press, J. Chem. Phys. 72, ... (1980); their articles in ref. 100; Also see: Proc. 1983-Como Conf., eds. Casati & Ford
[106] F.M. Izraelev and D.L. Shepelyanski, Quantum Resonance for the Rotator in a Nonlinear Periodic Field, Doklady 249, 1103-1107 (1979), in English: preprint I.Ya.Ph. 78-37 (Inst.Nucl.Phys., Novosibirsk, U.S.S.R)(1978); and the earlier article in ref. 100
[107] G.P. Berman and G.M. Zaslavsky, Condition of 'Stochasticity' in Quantum Nonlinear Systems, Physica 97A, 367-382 (1979); 91A, 450 (1978); Phys.Lett. 61A, 295 (1977); Zaslavsky, Sov. Phys. JETP, 46, 1094 (1977)

Semi-Classical Quantization

[108] M.C. Gutzwiller, p. 163-200, in: *Path Integrals and their Applications in Quantum Statistical and Solid State Physics*, Plenum Publ.Co., N.Y.(1978); his article in ref. 100; J.Math.Phys. 18, 806-823(1977); 14, 139-152(1973); 12, 343-358(1971); Phys.Rev.Lett. 45, 150-153(1980); N.B.: Physica 5D, 183-207 (1982)
[109] M. Tabor, The Role of Periodic Orbits in Semiclassical Quantization, ref. 100, p. 293-298(1979); with M. Berry, Proc. Roy. Soc. A356, 375-394 (1977)
[110] M.V. Berry, Philos. Trans. Roy. Soc. A287, 237-271 (1977); J. Phys. 10A, 2083-2091 (1977)
[111] I.C. Percival, Adv.Chem.Phys. 36, 1(1977); and ref. 100, p. 259-282 (1979); J.Phys. 6B, L229-L232 (1973); Adv.Chem.Phys. 36, 1-61 (1977); I.C. Percival an N. Pomphrey in: Molec.Phys., 35, 649(1978); Molec.Phys. 31, 97-114 (1976); J.Phys. 9B, 3131(1976)
[112] N. Pomphrey, J.Phys. 7B, 1909(1974) [Hénon-Heiles System (2.10)]
[113] N.C. Handy, S.M. Colwell and W.H. Miller, Faraday Disc.Chem.Soc. 62, 29(1977); and their paper in ref. 100

Related Q.M. References

[114] Quantized Integrable Chains are referred to in refs. 32,34
[115] A. Einstein, Zum Quantensatz von Sommerfeld und Epstein, Verh.Deut.Phys.Ges. 19, 82-92 (1917)
[116] V.P. Maslov, *Théorie des Perturbations et des Méthodes Asymptotiques*, Dunod Cie., Paris (1972)(in Russian: 1965); and with N.V. Fedoryuk, *The Quasi-Classical Approximation for Quantum Mechanic Equations*, Nauka, Moscow (1976); V.I. Arnold, ref. 14, Appendix 11 (1978); E.C. Titchmarsh, *Eigenfunction Expansions*, I,II, Oxford U. Press (1961-2nd. ed., 1958)
[117] J.B. Keller, Ann.Phys. 4, 180-188 (1958)
[118] D. Park, *Classical Dynamics and its Quantum Analogues*, Lecture Notes in Physics Vol. 110, Springer Verlag, Heidelberg (1979); S.I. Tomonaga, *Quantum Mechanics*, Vol. 1, North Holland Publ., Amsterdam (1962); B.L. van der Waerden, *Sources of Quantum Mechanics*, North Holland Publ., Amsterdam (1967): especially the translated articles with Heisenberg and/or Born and/or Jordan as authors
[119] F.T. Hioe, D. MacMillen and E.W. Montroll, Quantum Theory of Anharmonic

Oscillators: Energy Levels of a Simple and a Pair of Coupled Oscillators with Quartic Coupling, Physics Reports 43, 305-335 (1978), espec. Figure 2; and by F.T. Hioe: Phys.Rev. B16, 4112 (1977) D15, 488 (1977); S. Biswas, K. Datta, R. Saxena, P. Srivastava and V. Varma, J.Math.Phys. 14, 1190 (1973)
[120] K.S.J. Nordholm and S.A. Rice, J.Chem.Phys. 61,768(1974); 61,203(1974); and with J. Dancz, 67 1418 (1977)
[121] E.J. Heller, J.Chem.Phys. 72,1337-1347(1980); Chem.Phys.Lett. 60,338(1979); and with E.B. Stechel and M.J. Davis, J.Chem.Phys., in press (1980); and a preprint on the Hénon-Heiles model (2.10) (C.S.N.D., Los Alamos Natl. Lab.)
[122] C.M. Bender and T.T. Wu, Phys.Rev. 184,1231(1969); D7,1620(1973); with T. Banks: Phys.Rev. D8,3346(1973); D8,3366(1973)

Level Crossings

[123] F. Hund, Zur Deutung der Molekulspektren,(II:) Z.Phys. 42,93(1927);(I:) 40, 742(1927); Ueber Zuordnungsfragen, ins besondere über die Zuordnung von Multiplettermen zu Seriengrenzen, Z.Phys. 42,601(1929)
[124] J. von Neumann and E.P. Wigner, On the Behaviour of Eigenvalues in Adiabatic Processes, in: *Symmetry in the Solid State*, eds. R.S. Knox and A. Gold, p. 167, Benjamin Inc., New York (1964), in German: Physik.Z.30, 467-470(1929); E. Teller, J.Phys.Chem. 41,109-116(1937); V.I. Arnold, ref. 14, Appendix 10, especially sections B,C (1978); cf. §4 of ref. 103
[125] W.J. Caspers, Crossing of Energy Levels, Physica 40,125-138(1968); with H.P. van de Braak, P.W. Verbeek and J. Verstelle, Physica 53,210-224(1971)
[126] T.P. Valkering, Level Crossing and Constants of the Motion, Thesis, Physics, Twente U. Techn., Enschede, The Netherlands (1973), chapter 3; Physica 53, 117(1971); and with W.J. Caspers, Physica 63,113(1973); 78,516-526(1974)
[127] P. Mazur, Non-ergodicity of Phase Functions in Certain Systems, Physica 43, 533(1969); cf. ref. 30

Random References

[128] Z. Nitecki, *Differentiable Dynamics*, MIT Press (1971)
[129] R.L. Devaney, Transversal Homoclinic Orbits in an Integrable System, Amer. J. Math. 100, 631-642 (1978)
[130] The original Appendix A, entitled 'Separatrices Split', contained an error. I thank J. Duistermaat (Utrecht) and S. Ushiki (Kyoto) for pointing it out to me
[131] Changing variables to x + y and x - y we can separate the equations of motion.
[132] G.D. Birkhoff, *Dynamical Systems*, A.M.S. Colloqium Publs. (1927/1966)
[133] G.D. Birkhoff, Collected Works, Vols 1 and 2, A.M.S. Publ. (1960)
[134] D. ter Haar, *Elements of Statistical Mechanics*, Appendix I, The H Theorem and the Ergodic Theorem, esp. § 5, Holt, Rinehart and Winston Publ., N.Y. (1954/1961)
[135] E. Fermi, Beweis, das ein Mechanisches Normal system im Algemeinen Quasi-Ergodisch ist, Phys.Z. 24,261-264(1923)
[136] R.E. Peierls, Ann.Phys. 5,1055(1929), Leipzig; and in: *Quantum Theory of Solids*, Oxford Univ. Press (1956)
[137] J.C. Oxtoby and S.M. Ulam, Ann.Math. 42,874(1941)
[138] J.C. Oxtoby, *Measure and Category*, Springer Verlag, Heidelberg (1970)
[139] M. Casartelli, E. Diana, L. Galgani and A. Scotti, Phys.Rev. 13A,1921(1976); and with G. Casati, Theor.Math.Phys. 29,1022(1977)
[140] M. Casartelli, Phys.Rev. 19A,1741(1979); P. Bocchieri, A. Scotti, B. Bearzi and A. Loinger, Phys.Rev. 2A,2013(1970)
[141] Secular solutions, proportional to t, can arise when Q is a half integer [54,55,142-145]
[142] T. Fort, *Finite Differences and Difference Equations in the Real Domain*, Oxford Univ. Press (1948)
[143] H. Levy and F. Lessman, *Finite Difference Equations*, MacMillan Co., N.Y. (1961)
[144] R. de Vogelaere, p. 53 in Vol. 4 of *Contributions to the Theory of Nonlinear Oscillations*, ed. S. Lefschetz, Princeton Univ. Press (1958)
[145] A. Chenciner, Systémes Dynamiques Differentiables, in: Encyclopaedia Universalis, Paris (1983) (Math., U. Paris VII, 2 Place Jussieu)

[146] Equation (2.12) is of the type $d^2y/d\tau^2 = f(y) + p(\tau)g(y)$, with p a periodic function of the time τ. Hence its Hamiltonian is an explicitly periodic function. Most of the Hamiltonian results in this paper have versions which apply to periodic Hamiltonians [14,13]. We find many of the mappings (2.13-.16), derived from such equations, to be *nonintegrable* while we saw in section 2.1 that all 1-degree of freedom systems are *integrable*. This paradox is resolved by noting that p(t) itself will satisfy some other differential equation. Hence eq. (2.12) is the same as the "(1+1)-degree" of freedom *system* $d^2y/d\tau^2 = f(y) + pg(y)$ and $d^2p/d\tau^2 = h(p)$, with fixed $p(0)$, $\dot{p}(0)$. The unit time-step in difference equations [142-145] like (2.13-2.16) is due to this additional degree of freedom and reflects that we are describing a surface of section within a larger phase space

[147] Transformations Ponctuelles et Leurs Applications, (33 papers in English and French), Colloques Internat. du C.N.R.S., # 229(1973), Editions du C.N.R.S., Toulouse, France (1976)

[148] J. Moser, Nachr.Akad.Wiss.Göttingen, Math.Phys.IIa, 1-20(1962); H. Rüssmann (ibidem), 67-105(1970); and in ref. 301(1980);

[149] We employ y rather than Δy since $\hat{y} = 0$ here, cf. (2.23-24) or (3.11)

[150] When $Q = n/3$ or $n/4$ (,n/2 or n) the function $r\cos(t2\pi Q)$ contains at most 3 different numbers (t: integer). For our *second* difference equation (2.29) this is not enough to make one Fourier component functionally indepent of its higher harmonics as required to arrive at (2.34). In general, r values with $\sigma(r) = n/3$ or $n/4$ must be handled differently. This is not a mere technicality since qualitatively different behavior can arise in these cases [13,14,35,48]. The individual periodic orbits (no tori of course) with $\sigma = n/4, n/3, n/2$ or n can be obtained analytically [151,201,35,48]

[151] The series (2.32) cannot have a (finite) region of convergence, as discussed in section 2.1. If it did eq. (2.29) would be integrable. Hence (2.32) only approximates the solution of (2.29) over a time interval short enough for y_t to remain small. Note that in (2.35) we are not approximating the orbit for all time, but the mapping for small y. Yet, the series (2.32) can be made to converge locally at some $\vec{r}_0$ when $\sigma(r_0)$ is a *fixed rational* number. This yields the individual periodic orbits of a nonintegrable system [44,45-49, 52-55,201] which Poincaré conjectured to be dense among all bounded orbits [18,45,14]. presumably under some conditions on the equations of motion since some pathological counterexamples exist. At most fixed *irrational* values of $\sigma(r_0)$ the local convergence of such series can be guaranteed in principle by the K.A.M. theorem [12-15], as discussed in section 2.5. The approximations (2.35), (2.37) could also be obtained from the Birkhoff series [36,13,14]

[152] Note that the $2+\beta$ in (2.41) allows even more K.A.M. tori than expected [146] from (2.46)

[153] M. Herman, "A $C^{3-\varepsilon}$-Counterexample to the Twist Theorem", I.M.P.A. Lecture, Rio de Janeiro (1981) (Math.,Ecole Polytechnique,Palaiseau); v.v. see [156]; F. Takens, Indag. Math. 33, 379 (1971); M. Herman, Lectr. Notes, E.N.S.: Astérisque

[154] K.R. Meyer, Generic Bifurcation of Periodic Points, Trans.Am.Math.Soc. 149, 95-107(1970); esp. theorem 2.2, 2.1

[155] In section 3.2 we treat a more general mapping (3.7) which reproduces (2.47) for B=1. The α expression for that mapping is (3.24), derived in note [164]. Hence, in order to obtain the conservative expression (2.53) we merely set B=1 in (3.24). I thank R. MacKay [245] for correcting the algebra in my original note [155]

[156] The perturbations $\mathfrak{f}$, f, g, ∂ (H-$\bar{H}$) of sectns. 2.5, 2.8 are smooth enough for some proofs of the K.A.M. Theorem if their 3rd derivatives w.r. to A&A varbls. of the unperturbed system are "sufficiently"-continuous (Hölder-, cf. [13]); v.v. see [153]

[157] Transforming (2.47), with $u_t \equiv y_t + C - 1$ and $\mu \equiv 2 - C$ (2.49) yields $u_{t+1} + u_{t-1} = 2\mu u_t + 2u_t^2$. This is the same as (2.47), apart from $C \to \mu$. Thus there is a "mirror-Feigenbaum" with $C_{-k} = 2 - C_k$ and $\mu_k = 2 - C_k$ $(-\infty < k < \infty)$. The period-1 elliptic and hyperbolic orbits switch between $y = 0$ and $y = 1 - C$ when we go to the mirror tree, cf. [158]

[158] Transforming (3.7) with $u_t \equiv y_t + C - (1+B)/2$ and $\mu \equiv 1 + B - C$ yields $u_{t+1} + Bu_{t-1} = 2\mu u_t + 2u_t^2$. This is the same as (3.7), apart from $C \to \mu$. Thus there is a mirror-Feigenbaum with $C_{-k} = 1 + B - C_k$ and $\mu_k = 1 + B - C_k$ $(-\infty < k < \infty)$. The period-1

attractor and repellor switch between $y = 0$ and $y = (1+B)/2 - C$ when we go to the mirror tree, cf. [157]

[159] Consider $Ax_{t+1} + Bx_{t-1} = P + Qx_t + Rx_t^2$: No generality is lost by setting $A = 1$. The change of variables: $y_t \equiv R(x_t - a)/2$, with a determined from $Ra^2 + a(Q - 1 - B) + P = 0$, transforms our mapping into the standard form (3.7), with $2C = Q + 2aR$ real, *iff* $(Q - 1 - B)^2 > 4RP$, cf. also [158] and [245]

[160] The transformation $y_t \equiv a(e - x_t)/2$, with e solved from $-a^2e^2 + (b-1)ae + a = 0$, puts (3.8) into the form (3.7) with $C = -ae$ real, *iff* $(b-1)^2 > -4a$, cf. also [158]

[161] The transformation $y_t \equiv -ax_t/2$ puts (3.9) in the form (3.7) with $C = a/2$, cf. also [158]

[162] The transformation $x_t \equiv -2y_t/\mu + g$, with g solved from $-\mu^2g^2 - \mu g + \mu = 0$, puts (3.10) into the form (3.7); $C = -\mu g$ real, iff $4\mu > -1$ (see also [158])

[163] When $|B| > 1$ we divide (3.7) by B. The transformations y_t Bw_{-t}, $\zeta \equiv 1/B$ and $\Omega \equiv C/B$ yields $w_{-t+1} + \zeta w_{-t-1} = 2\Omega w_t^2$. Since $|\zeta| \le 1$ then, this is the "same" as (3.7), (apart from $t \to -t$) but yielding attracting trees as $t \to -\infty$

[164] With $r(\tau) \equiv \Delta y_{2\tau+1}/\Delta y_{2\tau-1}$, the square bracket term in (3.22) becomes $(r^2+B)\Delta y_{2\tau-1}^2$, and (3.16) yields $(r+B)\Delta y_{2\tau-1} = \beta\Delta y_{2\tau} + O(\Delta^2)$. Combining these we find the square bracket term of (3.22) to be $\Delta y_{2\tau}^2 \beta^2(r^2+B)/(r+B)^2$ + higher orders. Thus the *minimum* second order contribution from the square bracket term is $\Delta y_{2\tau}^2\ \beta^2/(1+B)$, (if $B < 0$, we start with the second bifurcation, that is, with the mapping applied twice since $B' > 0$, cf. (3.18). Thus we obtain the last term in the α-expression (3.24). In the conservative case, $B=1$, this yields (2.53). These expressions differ slightly from those in the original note [164], cf. note [155] and [245]

[165] C. Simó, On the Hénon-Pomeau Attractor, J.Stat.Phys. 21, 465 (1979)
J.H. Curry, On Some Systems motivated by the Lorenz Equations, Numerical Results, preprint (1980/81)(Math., U. Colorado, Boulder, U.S.A.)

[166] In order to arrive at (3.4) we take $B = 2(q_+ - q_-)/[\exp(2\pi q_+) - \exp(2\pi q_-)]$ in (3.2). This is not a limitation since we can still choose any nonlinear F(y) in (3.2-4) or transform variables after choosing F(y) [159]; See [245]

[167] P. Coullet, C. Tresser and A. Arneodo, Phys.Lett. 72A,268(1979);cf.[279]

[168] The standard example is the set of points obtained by removing all points between 1/3 and 2/3, of the interval (0,1) and by then removing the middle one-third parts of the remaining intervals, etc.. An infinite number of points, not dense on (0,1), remains

[169] R. Bowen, Equilibrium States and the Ergodic Theory of Anosov Diffeomorphisms, Lecture Notes Math.Vol. 470, Springer Verlag, Heidelberg (1975); C.B.M.S. 35(78)

(Refs. 170-199 added after completion of manuscript:)

Lorenz Equations and Extensions
cf. [253-257, 271-272,290,296,187-188] and cf. (3.35-37)

[170] K.A. Robbins, Periodic Solutions and Bifurcation Structure at high R in the Lorenz Model, SIAM J.Appl.Math. 36,457-472(1979); Proc.Cambr.Phil.Soc.82,309(77)

[171] I. Shimada and T. Nagashima, Prog.Theor.Phys. 58,1318-1319(1977); 59,1033 (1978); 61,1605-1616(1979); 1st. author, 62,61(1979)

[172] R. Graham and H.J. Scholz, Analytic Approximation of the Lorenz Attractor by Invariant Manifolds, Phys.Rev. 22A,1198-1204(1980)

[173] H. Yahata, Prog.Theor.Phys.Suppl. 64,176(1978); 61,791(1979)

[174] T. Shimizu and N. Moriaka, Phys.Lett. 66A,182(1978); 69A,148(1978), 66A,447(78)

[175] P. Manneville and Y. Pomeau, Physica 1D,219-226(1980)

[176] C. Boldrighini and V. Franceschini, Commun.Math.Phys. 64,159(1979)

[177] J.H. Curry, A Generalized Lorenz System, Commun.Math.Phys. 60,193-204(1978)

[178] J.H. Curry, SIAM J.Math.Anal. 10,71-77(1979); Phys.Rev.Lett. 43,1013-1016 (1979)

[179] K. Nakamura, Prog.Theor.Phys. 57,1874(1977); Prog.Theor.Phys.Suppl. 64,378 (1978); Proc. I.N.S. Nihon Univ. 14,9(1979)

[180] M. Lücke, J.Stat.Phys. 15,455(1976); E. Knobloch, 20,695(1979)
[181] G.M. Zaslavsky, The Simplest Case of a Strange Attractor, Phys.Lett. 69A,145-147(1978); with Kh.R.Ya. Rachko, Singularities of the Transition to a Turbulent Motion, Sov.Phys.JETP 49,1039-1044(1979)

Adding External Noise

[182] B. Shraiman, C. Wayne and P.C. Martin (1981), in this Vol.
[183] G. Mayer-Kress and H. Haken, The Influence of Noise on the Logistic Model, preprint (1980) and [300], cf. [194](Physics,U. of Stuttgart, W. Germany)
[184] J.P. Crutchfield and B.A. Huberman, Fluctuations and the Onset of Chaos, preprint (1980) (address: [251]); cf. [249]
[185] A. Zippelius and M. Lücke, The Effect of External Noise in the Lorenz Model of the Bénard Problem, in press, J.Stat.Phys. (1980)

Turbulence (cf. [290-302]),Attractors

[186] A.S. Monin, On the Nature of Turbulence, Sov.Phys.Usp. 21,429-442(1978)
[187] *Turbulence Seminar*, Lect. Notes Math., Vol. 615, Springer Verlag, Heidelberg (1977); e.g. the articles by Williams; Lanford; Marsden; and Bowen
[188] *Turbulence and the Navier Stokes Equation*, Lect. Notes Math., Vol. 565, Springer Verlag, Heidelberg (1976); e.g. the articles by Hénon and Pomeau; and by Ruelle; also see the Volume in [214]
[189] *Hydrodynamic Instabilities and the Transition to Turbulence*, eds. H.L. Swinney and J.P. Gollub, Springer Verlag, Heidelberg (1980); and their artls. in [301]
[190] G. Ahlers, Phys.Rev.Lett. 33,1185(1975); and: Low-Temperature Studies of the Rayleigh-Bénard Instabilities and Turbulence, in [301]; with R.P. Behringer, The R-B Instabilities and the Evolution of Turbulence, Progr.Theor.Phys.Suppl. 64,186-201; cf. G. Ahlers and R.W. Walden, preprint (1980)
[191] J.P. Gollub and H.L. Swinney, Onset of Turbulence... (in this Vol.), Phys. Rev.Lett. 35,927(1975); with S.V. Benson and P.R. Fenstermacher in [216]
[192] A. Libchaber and J. Maurer, J. Physique Lett. 39, L-369 (1978)
[193] M.I. Rabinovich, 'Stochastic' Self-Oscillations and Turbulence, Sov.Phys. Usp. 21,443-469(1978)
[194] H. Haken, Phys.Lett. 53A,77-78(1975); E. Bullard in [1]

---

[195] P.M. Kloeden, A.B. Deakin and A.Z. Tirkel, A Precise Definition of Chaos, Nature 264,295(1976)
[196] L. Markus and K. Meyer, Generic Hamiltonian Dynamical Systems are neither Integrable nor Ergodic, Mem.Am.Math.Soc., 144(1974)
[197] C.Froeschlé and R. Gonczi, Lyapunov Characteristic Numbers and Kolmogorov Entropy of a Four-Dimensional Mapping, Il Nuovo Cimento, Serie 11, 55B, 59-69[1980]
[198] G. Benettin, L. Galgani and A. Giorgilli, Further Results on Universal Properties in Conservative Systems, preprint (1980) (address: [63])
[199] B. Derrida and Y. Pomeau, Feigenbaum's Ratios of Two Dimensional Area Preserving Maps, to be subm. to Phys.Lett. A (1980)(address: [236])

Storage Rings and Accelerators
(Refs. 200, 204-208 contain reviews or introductory articles)

[200] *Nonlinear Dynamics and the Beam-Beam Interaction*, eds. M. Month and J.C. Herrera, Am.Inst.Phys.Conf.Proc., Vol. 57(1979)
[201] R.H.G. Helleman, Exact Results for Some Linear and Nonlinear Beam-Beam Effects, ref. 200, p. 236-256 (1979)
[202] L.J. Laslett, Some Illustrations of 'Stochasticity', ref. 1, p. 221-247 (1978)
[203] *A Review of Beam-Beam Phenomena*, ed. M. Month, Brookhaven National Laboratory Report B.N.L. 25703(1979)[Accelerator Dept., B.N.L., Upton,N.Y.,U.S.A.] and a new B.N.L. report, cf. ref. 80, ed. M. Month, (1980)
[204] Reference 9, Chapter 17 (1974)
[205] For Applications to Magnetic Confinement (Fusion) and Plasma Physics see: *Intrinsic 'Stochasticity' in Plasmas*, eds. G. Laval and D. Gresillon, Les

Editions de Physique Publ., Orsay, France (1980); ref. 255(1978); their refs.; references 207 and 212

[206] H. Bruck, *Circular Particle Accelerations*, Los Alamos Translation LA-TR-72-10 Ref. (1972) [Los Alamos,N.M.,U.S.A.]; in French: Presses Universitaires de France (1966)

[207] C. Horton, L. Reichl and V. Szebehely, eds., *Long-Time Prediction in Dynamics*, J. Wiley Publ., New York (1982)

[208] A.A. Kolomensky and A.N. Lebedev, *Theory of Cyclic Accelerators*, Wiley Publ. Co., N.Y. (1966)

[209] C.R. Eminhizer, A Preliminary Numerical Study of the Beam-Beam Interaction in Two Dimensions, in the S.L.A.C. report mentioned in ref. 80 (La Jolla Institute [81])(1980)

[210] F.M. Izraelev, S.I. Misnev and G.M. Tumaikin, Numerical Studies of 'Stochasticity' Limit in Colliding Beams (1-d. Model), preprint I.Ya.Ph.77-43 (address, cf. [106])(1977)

[211] F.M. Izraelev, Nearly Linear Mappings and their Applications, Physica 1D, 243 (1980)

[212] J.M. Wersinger, J.M. Finn and E. Ott, Phys.Rev.Lett. 44,453(1980)

Period-Doubling, Onset of Chaos

(Refs. 214-222) contain reviews or introductory articles; for Period-Doubling in conservative systems see section 2.6 and [63-68,244,198-199]

[213] N. Metropolis, M.L. Stein and P.R. Stein, (1973), in this Vol.

[214] E. Hopf, Bifurcation of a Periodic Solution from a Stationary Solution of a Differential System, p. 163-193 in: *The Hopf Bifurcation and Its Applications*, eds. J. Marsden and M. McCraken, Springer Verlag, Heidelberg (1976)Vol.19;(in German:) Ber.Math.Phys.Sachsische Acad.Wissensch.(Leipzig) 94,1-22(1942);[299]

[215] R.M. May, Simple Mathematical Models with Very Complicated Dynamics, Nature 261,459-467(1976); its references and his articles in ref. 301(1980) and 216 (1979), with their refs.; (1 st. artl.:) in this Vol.

[216] *Bifurcation Theory and Applications in Scientific Disciplines*, eds. O. Gurel and O.E. Rössler, Ann.N.Y. Acad.Sci., Vol. 316(1979)

[217] J. Guckenheimer, The Bifurcation of Quadratic Functions, ref. 216, p. 78-85 (1979); his article in ref. 301 (1980); Sensitive Dependence on Initial Conditions for One Dimensional Maps, (Math., Univ. of California at Santa Cruz, CA,USA); Commun.Math.Phys. 70,133(1979); his C.I.M.E. Lecture Notes on: Bifurcations of Dynamical Systems (1980) M. Jakobson, Topological and Metric Properties of One Dimensional Endomorphisms, Dokl.Acad.Nauk.S.S.S.R. 243, 866-869(1978); Commun. Math. Phys. 81, 39-88 (1981)

[218] M.J. Feigenbaum, The Onset Spectrum of Turbulence, Phys. Lett. 74A, 375(1979)

[219] P. Collet and J.P. Eckmann, Properties of Continuous Maps of the Interval to Itself, Proc.Lausanne Conf.Math.Phys., ed. P. Choquard, Lecture Notes Physics, Springer Verlag, Heidelberg (1980); On The Abundance of Ergodic Behaviour For Maps On The Interval, preprint (1979) (address: [237, 221]); article in [301]; Commun.Math.Phys. 73,115(1980)

[220] M.J. Feigenbaum, Metric Universality in Nonlinear Recurrence, ref. 100, p. 163-166(1979); and his article in [301]

[221] J.P. Eckmann, refs. 300,302 (Physics, Univ. of Geneva, Switzerland) (1980)

[222] O.E. Lanford III, Lectures: on Maps of the Interval [301,302] (Math., Univ. of California at Berkeley,CA,U.S.A.)(1980); his C.I.M.E. lectures (1976)

[223] M.J. Feigenbaum, The Universal Metric Properties of Nonlinear Transformations, J.Stat.Phys. 21,669-706(1979); 19,25(1978); and his lecture at the June-1976 Gordon Conference (N.H.); and his first artl. in this Vol.

[224] M.J. Feigenbaum, The Transition to Aperiodic Behavior in Turbulent Systems, Comm.Math.Phys. 77,65-86(1980); in this Vol.

[225] R.M. May and G.F. Oster, Period-Doubling and the Onset of Turbulence, An Analytic estimate of the Feigenbaum Ratio, Phys.Lett. 78A,1-3(1980)

[226] P. Collet and J.P. Eckmann, *Iterated Maps on the Interval as Dynamical Systems*,

Progress in Physics, Vol. 1, Birkhäuser Verlag, Basel and Boston (1980);
[227] P. Collet, J.P. Eckmann and H. Koch, Period Doubling Bifurcations for Families of Maps on R^n, J. Stat. Phys. (1980), in this Vol.
[228] Y. Oono, Prog.Theor.Phys. 59,1028(1978); cf. [240]
[229] R. Bowen, Commun.Math.Phys. 69,1(1979)
[230] D. Ruelle, Commun.Math.Phys. 55,47(1977); M. Misiurewicz, Absolutely Continuous Measures for Certain Maps of an Interval, I.H.E.S. preprint, (Bures sur Yvette, France) (1979); cf. [217] (1981)
[231] G. Julia, Mémoires sur l'Iteration des Fonctions Rationelles, J.Math.Pures, Appl. 4,47-245(1918)
[232] J. Crutchfield, D. Farmer, N. Packard, R. Shaw, G. Jones and R.J. Donnelly, Power Spectral Analysis ... (in this Vol.), Phys. Lett. 76A, 1-4 (1980)
[233] I. Gumowski and C. Mira, *Point Mappings, Order-Disorder Transition*, Springer Verlag, Heidelberg (1980); (in French:) Cepadues Publ., Toulouse (1980)
[234] V. Francescini and C. Tebaldi, Sequences of Infinite Bifurcations and Turbulence in a Five-Mode Truncation of the Navier-Stokes Equations, J.Stat. Phys. 21,707-726(1979); and: A Seven Mode Truncation of the Plane Incompressible Navier-Stokes Equations, preprint, (Math., Univ. of Modena, Italy) (1980); (1979:) in this Vol.
[235] V. Francescini, A Feigenbaum Sequence of Bifurcations in the Lorenz Model, J.Stat.Phys. 22,397-406(1980)
[236] B. Derrida, A. Gervois and Y. Pomeau, two preprints, (SPT/CEN de Saclay, Gif-sur-Yvette, France) (1977) and J.Phys. 12A,269(1979); first author, J.Physique C5,49(1978)
[237] P. Collet, J.P. Eckmann and O.E. Lanford III, Universal Properties of Maps on an Interval, Commun. Math. Phys. 76, 211-254 (1980); (1979); articles in ref. 301,302(1980);
[238] P. Coullet and C. Tresser, Critical Transition to Stochasticity for Some Dynamical Systems, to appear in J. de Physique, Lettres (1980); On the Existence of Hysteresis in a Transition to Chaos after a Single Bifurcation, with A. Arneodo, same ref. [Méc.Statistique,Univ. de Nice,France]; C.R. Acad. Sc.Paris 287A,577(1978); S. Grossmann and S. Thomae, (1977), in this Vol.
[239] E.N. Lorenz, The Problem of Deducing the Climate from the Governing Equations, Tellus 16, 1-11 (1964)
[240] T.Y. Li and J.A. Yorke, Period Three Implies Chaos, Am.Math.Monthly 82,985-992(1975); cf. [228]
[241] Ya.N. Belyaev, A.A. Monakhov, S.A. Scherbakov and I.M. Yavorshaya, J.E.T.P. Lett. 29,295(1979)
[242] J. Maurer and A. Libchaber, Rayleigh-Bénard Experiment in Liquid Helium: Frequency Locking and the Onset of Turbulence, J.Physique Lett. 40,L419-L423 (1979); Une Expérience de Rayleigh-Bénard de Géometrie Reduite..., preprint (1979), to appear J. Physique (?1980); cf. their artls. in this Vol.
[243] J.P. Gollub, S.V. Benson and J. Steinman, A Subharmonic Route to Turbulent Convection, ref. 301 (Physics, Haverford College, PA, U.S.A.)(1980); J.P. Gollub and S.V. Benson, Chaotic Response to Periodic Perturbation of a Convecting Fluid, Phys.Rev.Lett. 41,948-951(1978)
[244] J.B. McLaughlin, Period Doubling Bifurcations and Chaotic Motion for a Parametrically Forced Pendulum, J.Stat.Phys. 15,307-326(1975) (Chem.Eng., Clarkson Coll., Potsdam, N.Y.) (1980)
[245] R.H.G. Helleman and R.S. MacKay, One Mechanism for the Onsets of Large-Scale Chaos in Conservative and Dissipative Systems, pp. 95-126 in ref. 207 (1982); A. Zisook, Phys.Rev. A24,1640(1981), in this Vol.
[246] R.M. May and G.F. Oster, Bifurcations and Dynamic Complexity in Simple Ecological Models, Amer.Natur. 110,573-599(1976)
[247] D. Singer, Stable Orbits and Bifurcations of Maps on the Interval, S.I.A.M. J.Appl.Math. 35,260-269(1978); T. Li and J. Yorke, in: *Dynamical Systems*, ed. L. Cesari, Academic Press, N.Y. 2,203(1978)
[248] G. Iooss, *Bifurcations of Maps and Applications*, North Holland Publ. (1979); with W.F. Langford, Conjectures on the Routes to Turbulence via Bifurcations, ref. 301 (1980); G. Iooss and D.D. Joseph, *Elementary Stability and Bifurcation Theory*, to appear, North Holland (?) Publ. (1980)

[249] B.A. Huberman and J.P. Crutchfield, Chaotic States of Anharmonic Systems in Periodic Fields, Phys.Rev.Lett. 43,1743-1747(1979); B.A. Huberman and J. Rudnick, Phys.Rev.Lett. 45,154-157(1980)

Strange Attractors
(Refs. 250-256,268-273,301,302 contain reviews or introductory articles)

[250] D. Ruelle, Strange Attractors, Math. Intellegencer 2,126(1980) (in French: La Recherche 11,132-144(1980)), in this Vol.
[251] R. Shaw, Strange Attractors, Chaotic Behavior and Information Flow, Z. Naturforsch. 36a, 80 (1981); other articles of the Santa Cruz group, e.g.: D. Farmer, J. Crutchfield, H. Froehling, N. Packard and R. Shaw, Power Spectra and Mixing Properties of Strange Attractors, ref. 301 (1980); as well as the accompanying movies [301]
[252] D. Ruelle, Sensitive Dependence on Initial Conditions and Turbulent Behavior of Dynamical Systems, ref. 216, p. 408-416 (1979); Dynamical Systems with Turbulent Behavior, p. 341-360 in: *Mathematical Problems in Theoretical Physics*, eds. G. Dell 'Antonio, S. Doplicher and G. Jona-Lasinio, Lect. Notes Physics, Vol. 80, Springer Verlag, Heidelberg (1978); article in [301]
[253] J.L. Kaplan and J.A. Yorke, The Onset of Chaos in a Fluid Flow Model of Lorenz, ref. 216, p. 400-407(1979); Yorke, Commun.Math.Phys. 67,93(1979)
[254] O.E. Rössler, Continuous Chaos - Four Prototype Equations, ref. 216, p. 376-392(1979); An Equation for Continuous Chaos, Phys.Lett. 57A,397-398(1976); 71A,155(1979)
[255] Y.M. Trève, Chaotic Motions in Dissipative Systems: The Case of Strong Turbulence (= Chapter 2.), ref. 1, p. 183-202(1978)
[256] H.L. Swinney and J.P. Gollub, The Transition to Turbulence, Physics Today 31,41-49(1978)
[257] E.N. Lorenz, Deterministic Nonperiodic Flow, J.Atmos.Sci. 20,130-141(1963); cf. [170-188]; B. Saltzman, J. Atmos.Sci. 19, 329(1962)
[258] P.B. Wilson, An Interesting Mathematical Sequence ($x_{n+1}=c^{x_n}$) and its Relation to the Threshold for Turbulent Bunch Lengthening, Notes, PEP-232, SLAC, Stanford
[259] D. Ruelle and F. Takens, On the Nature of Turbulence, Commun.Math.Phys. 20, 167-192(1971); 23,343-344(1971); with S. Newhouse, Commun.Math.Phys. 64,35-40(1978); first 2 authors: Commun.Math.Phys. 64,35-40(1978)
[260] M. Hénon, A Two-Dimensional Mapping with a Strange Attractor, Commun.Math. Phys. 50,69-77(1976); in this Vol.
[261] M. Misiurewicz and B. Szewc, Existence of a Homoclinic Point for the Hénon Map, preprint (1980) (address: [267]); V. Franceschini and L. Russo, Stable and Unstable Manifolds of the Hénon Mapping, J.Stat.Phys. 25, 757 (1981) F.R. Marotto, Chaotic Behavior in the Hénon Mapping, Commun.Math.Phys. 68, 187-194(1979)
[262] C. Tresser, P. Coullet, and A. Arneodo, Topological Horseshoe and Numerically Observed Chaotic Behavior in the Hénon Mapping, J.Phys. 13A, L123-L127(1980); Transition to Turbulence for Doubly Periodic Flows, to appear in Phys.Lett. A (1980) [address, cf. ref. 238]
[263] J.H. Curry, On the Hénon Transformation, Commun.Math.Phys. 68,129-140(1979); See refs. 165,177,178; J.B. McLaughlin, Phys.Lett. 72A,271(1979)
[264] S.D. Feit, Characteristic Exponents and Strange Attractors, Commun.Math.Phys. 61,249(1978); cf. ref. 265
[265] G. Bennettin, L. Galgani and J.M. Strelcyn, Phys.Rev. 14A,2338(1976); and with A. Giorgilli, C.R. Acad.Sc. Paris 286A,431(1978); first 2 authors: in ref. 205 (1980); also compare refs. 97 and 197; V.I. Oseledec, Trans. Moscow Math.Soc. 19,197(1968)
[266] R. Lozi, Un Attracteur Etrange (?) du Type Attracteur de Hénon, Journ. de Physique 39(C5),9-10(1978)
[267] M. Misiurewicz, Strange Attractors for the Lozi Mapping, ref. 301 (Math., Warsaw University, Poland)(1980)
[268] J.-P. Eckmann, Rev. Mod. Phys. 53, 643 (1981), in this Vol.
[269] E. Ott, Rev. Mod. Phys. 53, 655 (1981)

[270] Ya.G. Sinai, The Stochastic Nature of Dynamic Systems, in: Proceedings of the Gorky Spring School on Nonlinear Waves (Landau Institute of Theor.Physics, 117334 Moscow, U.S.S.R.)(1977)

[271] L.A. Bunimovich and Ya.G. Sinai, The Stochastic Nature of the Attractor in the Lorenz Model, same ref. as above (1977)

[272] S.E. Newhouse, Asymptotic Behavior and Homoclinic Points in Nonlinear Systems, ref. 301 (Math., Univ. of North Carolina, Chapel Hill, N.C., U.S.A.); his I.H.E.S. lecture notes (1977); C.C. Pugh and M. Shub, Ergodic Attractors, to appear (Math., Univ. of California at Berkeley, CA, U.S.A.)(1980); and their paper in ref. 301 (1980); Newhouse, C.I.M.E. Lectures (1978)

[273] S. Smale, On the Problem of Reviving the Ergodic Hypothesis of Boltzmann and Birkhoff, ref. 301 (Math. Univ. of California, Berkeley ,CA, U.S.A.) (1980)

[274] D.V. Anosov, Trudy Instituta Steklova 90 (1967), (in English:) Am.Math.Soc. (1969); cf. ref. 14; and with Ya.G. Sinai, Some Smooth Ergodic Systems, Russ. Math.Surv. 22,103-172(1967)

[275] S.A. Orszag and J.B. McLaughlin, Evidence that Random Behavior is Generic for Nonlinear Differential Equations, Physica 1D,68-79(1980)

[276] A. Chenciner, Sections 2-6 of his article in ref. 302; M. Lieberman and A. Lichtenberg, (section 3.4 and) pp. 436-442 of ref. 17. B.V. Chirikov, lecturing on the "persistence" of chaotic behavior under some dissipative perturbations at the (1979) Clarkson Summer School (Inst.Nucl.Phys.,Novosibirsk, U.S.S.R.); and his articles with F.M. Izraelev in: Proc.Conf. on Programming & Math. Methods for Solving Physics, Inst.Nucl. Phys. Dubna, U.S.S.R. (1974); ref. 147, p. 409-428(1976); preprint 80-128(1980)

Duffing Equation and Nonlinear Oscillations

[277] P. Holmes, A Nonlinear Oscillator with a Strange Attractor, Phil.Trans.Roy. Soc. A292,420-447(1979); with J.E. Marsden, in [301]

[278] Y. Ueda, Explosion of Strange Attractors Exhibited by Duffing's Equation, in ref. 301 (Electr.Engin.,Kyoto University,Japan)(1980); his article in Proc. (1979) Engin. Foundation Conf.: *New Approaches to Nonlinear Problems in Dynamics*, ed. P. Holmes (Mechanics,Cornell Univ. Ithaca,N.Y.)(1980); Random Transitional Phenomena in the System Governed by Duffing's Equation, J.Stat.Phys. 20,181-196(1979); Trans.Inst.Elec.Eng. Japan 98A,167-173(1978); J. Ford, private communication (1980); J.B. McLaughlin, ref. 244

[279] A. Arneodo, P. Coullet and C. Tresser, A Forced Oscillator with Chaotic Behavior: An Illustration of a Theorem of Shil'nikov, submitted to J.Physics A (1980); and with W.F. Langford and J. Coste, A Mechanism for a Soft Mode Instability, Phys.Lett. 78A,11-14 (1980); Possible New Strange Attractors with Spiral Structure, preprint (July 1980); cf. [167] [address cf. ref. 238]; C. Hayashi, Selected Papers, Nippon Print Co. (1975)

[280] Other references to the Duffing Equation are in Refs. 249,46 and the following texts:

[281] J. Moser, Combination Tones for Duffing's Equation, Commun. Pure Appl.Math. 18,167-181(1965)

[282] J. Stoker, *Nonlinear Vibrations*, Interscience Publ. N.Y. (1950)

[283] A.H. Nayfeh and D.T. Mook, *Nonlinear Oscillations*, Interscience Publ., N.Y. (1979)

[284] R. Bellman, *Methods of Nonlinear Analysis*, Volumes 1 and 2, Academic Press, N.Y. (1970,1973)

[285] M. Roseau, *Vibrations Non-Lineaires et Theorie de la Stabilitē*, Springer Verlag, Heidelberg and N.Y. (1966)

[286] N. Minorski, *Nonlinear Oscillations*, van Nostrand Co., Princeton, N.J., U.S.A.(1962)

[287] N.V. Butenin, *Elements of the Theory of Nonlinear Oscillations*, Blaisdell Publ., N.Y. (1965)

[288] H.T. Davis, *Introduction to Nonlinear Differential and Integral Equations*, Dover Publ., N.Y. (1962); A. Arneodo, P. Coullet and C. Tresser, Occurrence of Strange Attractors in

Three-dimensional Volterra Equations, preprint (June 1980) [address cf. ref. 238]

[289] E.A. Coddington and N. Levinson, *Theory of Ordinary Differential Equations*, McGraw-Hill Co., N.Y. (1955)

Turbulent References

(References 291,293 contain reviews or introductory articles, also cf. [186-194])

[290] J.B. McLaughlin and P.C. Martin, Transition to Turbulence in a Statically Stressed Fluid System, Phys.Rev. 12A,186(1975)

[291] U. Frisch, Fully developed Turbulence and Intermittency, ref. 301 (Observatoire de Nice, France)(1980); E.D. Siggia and H. Aref, Scaling and Structures in Fully Turbulent Flows, ref. 301 (Atomic Physics, Cornell Univ., Ithaca, N.Y.)(1980)

[292] These references use some of the methods discussed in the section on Hamiltonian Mechanics: E.A. Novikov, Stochastization and Collapse of Vortex Systems, in ref. 301 (1980); H. Aref and N. Pomphrey, Integrable and Chaotic Motions of Four Vortices, Phys.Lett. 78A,297-300(1980); and a more detailed preprint (Lab. Solid State Physics, Cornell Univ., Ithaca, N.Y.)(1980)

[293] A.T. Winfree, Rotating Reactions, Sci.Am. 230, 82-95(1974); Springer V. (1980)

[294] N. Kopell, Waves, Shocks and Target Patterns in an Oscillating Chemical Reagent, in *Research Notes in Mathematics*, Vol. 14. p. 129-154, eds. W. Fitzgibbon and H. Walker, Pitman Publ., London (1977); article in ref. 301 (1980) and with L.N. Howard (to appear) in Studies in Appl.Math. (1980)

[295] C. Vidal, J.C. Roux, S. Bachelart and A. Rossi, Experimental Study of the Transition to Turbulence in the Belousov-Zhabotinsky Reaction, ref. 301 (1980) and (in French) C.R. Acad.Sci. Paris 289C,73-76(1979); K. Tomita and I. Tsuda, Chaos in the Belousov-Zhabotinsky Reaction in a Flow System, Phys.Lett., in press (1980)

---

[296] V.S. Afraymovich, V.V. Bykov and L.P. Shilnikov, Dok.Akad.Nauk. S.S.S.R. 234 (#2), 336 (1977)

[297] At a=4 (3.9) has the exact solution $x_t = \sin^2(2^t\phi_0)$; cf. also [215]

[298] L.D. Landau and E.M. Lifshitz, *Fluid Mechanics*, Pergamon Press, Oxford (1959); C.R. Acad.Sci. U.R.S.S. 44,311(1944)

[299] E. Hopf, A Mathematical Example Displaying Features of Turbulence, Comm. Pure Appl.Math. 1,303-322(1948); cf. [214]

[300] Private Communications during the 1980 and 1981 *Dynamics Days Twente* spring-conferences, Twente U. Technology, Programs reprinted in: *Nonlinear Science Abstracts*

[301] Proceedings of the 1979 '*International Conference on Nonlinear Dynamics*', ed. R.H.G. Helleman, Ann. N.Y. Acad.Sci., Vol. 357 (1980)

[302] Proceedings of the (July) 1981 "Les Houches" Summerschool on *Chaotic Behavior of Deterministic Systems*, eds. G. Iooss, R. Helleman and R. Stora, North Holland Publ., Amsterdam (1983)

Editor's note: We also recommend the excellent review by Berry (1978).

Part 8

Recent Developments

Russian Mathematical Surveys **39** 1–40 (1984)

Feigenbaum universality and the thermodynamic formalism

E.B. Vul, Ya.G. Sinai, and K.M. Khanin

CONTENTS

§1. Sequences of period-doubling bifurcations

1.1. This article is an exposition of a recent remarkable discovery in the theory of dynamical systems—the so-called Feigenbaum universality. This discovery lies on the boundary of mathematics and physics: the statement of the problem comes, in essence, from mathematics, while the method of approach to its solution is taken from physics and is based on the well known renormalization-group method in theoretical physics, first used in statistical mechanics and quantum field theory.

The problem with which Feigenbaum started bears a very general character and consists in studying how dynamical systems depending on a parameter pass from a stable type of motion, which it is natural to call laminar, to an unstable type that involves the appearance of strong statistical properties frequently associated with turbulence. The ensuing discovery goes even beyond the framework of this problem, since it provides a completely new method for investigating microscopic (that is, local) properties of dynamical systems.

The Feigenbaum universality refers directly to sequences of period-doubling bifurcations. In traditional bifurcation theory it is usual to consider the local behaviour of families of dynamical systems in a neighbourhood of a bifurcation value of the parameter. Here, however, we encounter a completely new problem: the local behaviour of a family of dynamical systems in a neighbourhood of a parameter value where infinitely many parameter bifurcation values accumulate. Historically, Metropolis, M.L. Stein, and P.R. Stein were among the first workers in this area [1]. They observed that the form of the trajectories becomes more complicated as the parameter increases for a broad class of one-parameter families of maps of a closed interval into itself, namely, a stable periodic trajectory becomes unstable as the parameter increases, and a stable periodic trajectory with twice the period is created, which attracts all points except for unstable cycles. We mention also the earlier paper [2]. Feigenbaum, working with a small pocket calculator, observed that the successive parameter values where such bifurcations take place for the family of maps $x \mapsto \mu x(1 - x)$ $(0 \leqslant \mu \leqslant 4)$ of $[0, 1]$ into itself converge to a limit at the rate of a geometric progression with the ratio $\delta = 4.6692$ ..., the famous Feigenbaum constant. He then made analogous calculations with the family $f(x; \mu) = \mu \sin(\pi x)$ and observed here a geometric progression with the same ratio. This led to the natural conjecture that δ does not depend at all on the form of the specific family of maps. Feigenbaum also proposed a theory explaining the universality of δ (see [3]-[6]).

It is useful qualitatively to form an intuitive picture of the phenomenon taking place when there is an infinite sequence of period-doubling bifurcations. We describe it separately for the cases of continuous and of discrete time .

Let us consider a smooth (C^∞-) system of ordinary differential equations in $\mathbf{R}^n$ depending on a single parameter μ:

$$\frac{dx_i}{dt} = f_i(x_1, \ldots, x_n; \mu). \tag{1.1}$$

We assume that for $\mu = \mu_0$ the system (1.1) has a stable periodic solution $x(t; \mu_0) = (x_1(t; \mu_0), \ldots, x_n(t; \mu_0))$ with period $T = T(\mu_0)$, that is, $x(t + T; \mu_0) = x(t; \mu_0)$. In this case stability means that the spectrum of the monodromy matrix lies strictly inside the unit disk. Suppose that a loss of stability of this periodic trajectory takes place as μ increases to some value μ_1: one of the eigenvalues of the monodromy matrix becomes equal to -1, while all the other eigenvalues stay less than 1 in modulus. In this case, if the higher derivatives satisfy certain simple inequalities, then there is a period-doubling bifurcation as the parameter goes through μ_1: the previously stable periodic trajectory becomes unstable, and at the same time a stable periodic trajectory with twice the period appears. As the parameter increases further, each of these periodic trajectories is somewhat displaced in the space, while keeping its own type of stability. We assume that the next period-doubling bifurcation happens when $\mu = \mu_2 > \mu_1$, with the result that two unstable periodic trajectories appear along with one stable periodic trajectory with period roughly twice that of the previous stable trajectory (see Fig. 1.1) and roughly four times that of the original periodic trajectory.

By now it is clear what happens aften n such bifurcations.

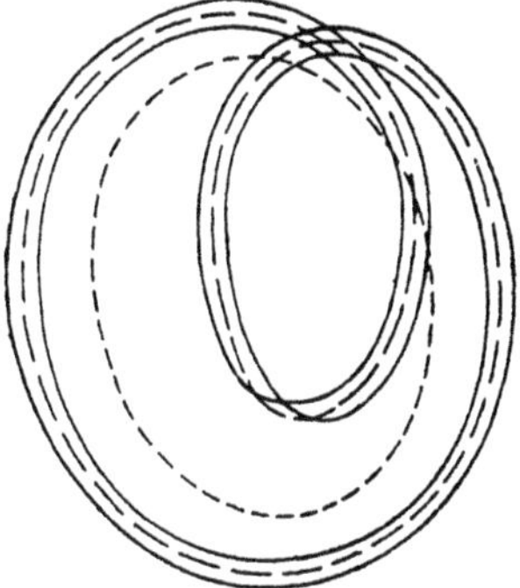

Fig. 1.1.

Namely, there is one stable periodic trajectory with the largest period, surrounded by n unstable periodic trajectories with periods decreasing roughly in a geometric progression with period 2^{-1}. We can now describe the object emerging as $n \to \infty$: we obtain an invariant set that is topologically a van Dantzig solenoid, and there are unstable periodic trajectories in any neighbourhood of it. In each such sufficiently small neighbourhood all the trajectories starting in an open dense subset are attracted to this invariant set. The invariant set described is an attractor in this sense. It is difficult to imagine that such an object can be found analytically.

1.2. We now describe the same phenomenon of an infinite sequence of period-doubling bifurcations for one-dimensional maps. The case of a one-parameter family of vector fields considered above can be reduced in the standard way to that of one-parameter families of maps with the help of the Poincaré succession map. Since period-doubling bifurcations bear a one-dimensional character (as will become clear), one-dimensional maps reflect the specific nature of the situation in a fairly complete manner. We consider a C^2-map f of the line $\mathbf{R}^1$ onto itself. This means that the value of $f(x)$ at a point x is the point into which x is mapped under the action of f. A point x_0 is called a stable fixed point of f if it has a neighbourhood U such that $\lim_{n\to\infty} f^{(n)}(y) = x_0$ for every $y \in U$. Here and later, $f^{(n)}$ denotes the n-th superposition of f. Sufficient conditions for a point x_0 to be stable are obviously the relations $f(x_0) = x_0$ and $|f'(x_0)| < 1$. Since the questions of interest to us are of a local nature, we assume that the domain of attraction U of x_0 is the whole line $\mathbf{R}^1$. Next, we assume that f depends on a real parameter μ, that is, we have a one-parameter family of maps. Then the fixed point x_0 also depends on μ, that is, $x_0 = x_0(\mu)$. Suppose that x_0 is stable for the initial value $\mu = \mu_0$, with $0 < f_x'(x_0;\ \mu) < 1$. We now start to increase μ. It can happen that $f_x'(x_0;\ \mu)$ then decreases, passes through the value 0 for $\mu = \bar{\mu}_0$, and then becomes negative and takes the value -1 for some $\mu = \mu_1$. The character of the transformation $f(x;\ \mu)$ in these cases is represented in Fig. 1.2 under the assumption that $f''_{xx}(x_0, \bar{\mu}_0) \neq 0$.

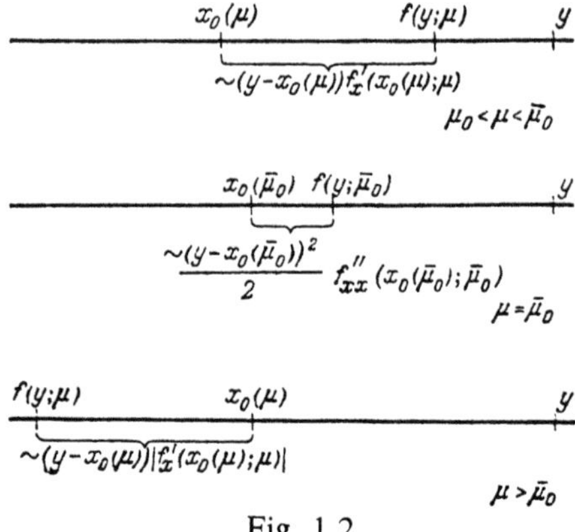

Fig. 1.2.

If the second-order derivatives satisfy fairly simply inequalities, then after a value $\mu = \mu_1$ is passed, the point $x_0(\mu)$ becomes unstable and there appear around it two points $x_{11}(\mu)$ and $x_{12}(\mu)$ forming a stable periodic trajectory of period 2. We depict this phenomenon graphically in two ways (Figs. 1.3 and 1.4).

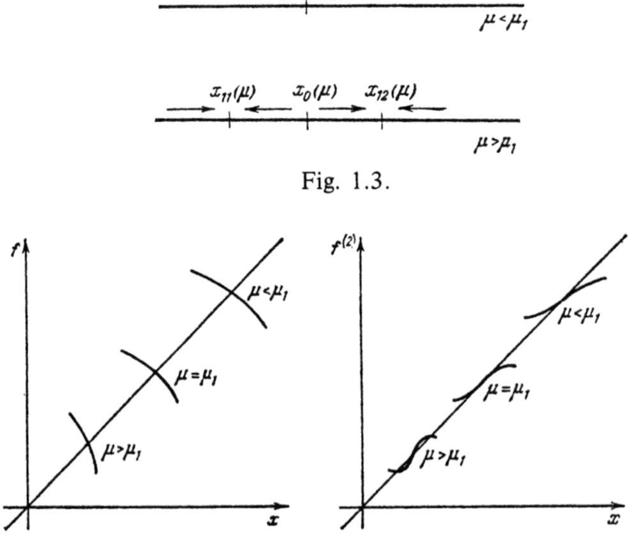

Fig. 1.3.

Fig. 1.4.

The differences $|x_{11}(\mu) - x_0(\mu_1)|$ and $|x_{12}(\mu) - x_0(\mu_1)|$ are of order $(\mu-\mu_1)^{1/2}$ as $\mu-\mu_1 \to 0$, while $|x_0(\mu) - x_0(\mu_1)| = O(\mu - \mu_1)$.

Thus, under one period-doubling bifurcation the previously stable fixed point x_0 becomes unstable, and a stable periodic trajectory of period 2 appears near it. As the parameter is further increased, the original fixed point x_0 continues to exist as an unstable fixed point, and all the remaining points are attracted to the stable periodic trajectory of period 2. This happens up to some value $\mu = \mu_2$, at which the periodic trajectory of period 2 loses stability in such a way that

$$\left.\frac{\partial f^{(2)}(x;\ \mu_2)}{\partial x}\right|_{x=x_{11}} = \left.\frac{\partial f^{(2)}(x;\ \mu_2)}{\partial x}\right|_{x=x_{12}} = -1.$$

We can then repeat the same arguments and the same pictures for $f^{(2)}$ and find that the periodic trajectory of period 2 becomes unstable and a periodic trajectory of period 4 appears near it. As μ is increased, an infinite sequence $\{\mu_n\}$ of parameter values emerges such that at $\mu = \mu_n$ there is a loss of stability of the periodic trajectory of period 2^{n-1}, and a periodic trajectory of period 2^n arises. By now we can imagine what happens when $\mu = \mu_\infty = \lim_{n\to\infty} \mu_n$: the map $f(x;\ \mu_\infty)$ has an invariant set F of Cantor type surrounded by infinitely many unstable periodic trajectories of periods 2^n. Moreover, all the points except those belonging to these unstable trajectories and their inverse images are attracted to F under the action of $f(x;\ \mu_\infty)$. Feigenbaum universality in its simplest form means that the sequence $\{\mu_n\}$ behaves in a universal manner, that is, $\mu_\infty - \mu_n \sim C\delta^{-n}$, where the constant C depends on the family f, while δ is universal and is the Feigenbaum constant mentioned above. In fact, we shall see that Feigenbaum universality is not exhausted by this: in a certain sense the structure of the attractor F, in particular, its Hausdorff dimension, and the behaviour of the iterates $f^{(n)}$ in a neighbourhood of $\mu = \mu_\infty$ do not depend on f.

We present some results of a numerical investigation of concrete examples of maps and vector fields; the machine computation points to the appearance of an infinite sequence of period-doubling bifurcations, as well as the universality.

The results of a numerical computation for the quadratic family $f(x;\ \mu) = 1-\mu x^2$ of transformations of $[-1, 1]$ into itself are given in Table 1.1 (see [8]).

In Table 1.2 we give bifurcation values of the parameter for the famous Lorenz model described by the following system of differential equations:

$$\begin{cases} \dot{x} = -\sigma x + \sigma y, \\ \dot{y} = -xz + rx - y, \\ \dot{z} = xy - bz. \end{cases}$$

Here $\sigma = 10$, $b = 8/3$, and r is regarded as a parameter (see [50]).

Table 1.1.

n	μ_n	$\Delta_n = \mu_n - \mu_{n-1}$	$\delta_n = \frac{\mu_{n-2} - \mu_{n-1}}{\mu_{n-1} - \mu_n}$
1	0.75		
2	1.25	0.5	
3	1.3680989394	0.1180989394	4.233738275
4	1.3940461566	0.0259472172	4.551506949
5	1.3996312389	0.0055850823	4.645807493
6	1.4008287424	0.0011975035	4.663938185
7	1.4010852713	0.0002565289	4.668103672
8	1.401140214699	0.000054943399	4.668966942
9	1.401151982029	0.000011767330	4.669147462
10	1.401154502237	0.000002520208	4.669190003

Table 1.2.

n	r_n	$\Delta_n = r_{n-1} - r_n$	$\delta_n = \frac{r_{n-2} - r_{n-1}}{r_{n-1} - r_n}$	n	r_n	$\Delta_n = r_{n-1} - r_n$	$\delta_n = \frac{r_{n-2} - r_{n-1}}{r_{n-1} - r_n}$
0	100.7952	—	—	3	99.54712	0.08139	4.319
1	99.9800	0.8152	—	4	99.52934	0.01778	4.578
2	99.62851	0.35149	2.319	5	99.525533	0.003807	4.670

The structure of this paper is as follows. In §2 we present a number of facts we need from the theory of one-dimensional maps. In §3 we give an explanation of Feigenbaum universality, mainly following [3]–[8]. The properties of the Feigenbaum attractor are studied in §4. A number of the results in this section are new. Small random perturbations of the Feigenbaum map are studied in §5 on the basis of results in §4. In §6 we present results of Eckmann, Collet, and Koch [9] referring to the multi-dimensional case, as well as some other generalizations.

§2. Facts from the general theory of continuous one-dimensional maps

2.1. A few years ago it seemed that the theory of one-dimensional continuous maps was a part of the general theory of dynamical systems of little interest. It was thought that, firstly, such maps have a fairly simple structure, and, secondly, the results then existing are characteristic of the one-dimensional case and do not have natural multi-dimensional generalizations. Both these convictions turned out to be erroneous: it became clear that the structure of one-dimensional continuous maps is highly non-trivial, and the facts discovered here are exceedingly beautiful and unexpected; in addition, a number of these facts admit natural extensions to the multi-dimensional case. It should be noted, however, that as far back as 20 years ago

Sharkovskii (see [10], [11]) observed the non-triviality and depth of the theory of one-dimensional maps, as well as the complexity of their structure. In this section we describe the results from the theory of one-dimensional continuous maps that we need for what follows. There are many publications on this theme; the text below is not an exhaustive survey of the whole range of questions, nor do the references pretend to be complete.

Singer [12] made the important discovery that the Schwarzian derivative is useful in the theory of one-dimensional maps. The Schwarzian derivative of a function $f \in C^3$ is defined to be $Sf = f'''/f' - (3/2)(f''/f')^2$. An essential property of Sf is that if $Sf_1 < 0$ and $Sf_2 < 0$, then $S(f_1 \circ f_2) < 0$. Deep properties of one-dimensional maps can be established by assuming that they are given by functions with negative Schwarzian derivatives. For example, Singer showed (see [12]) that for maps f of a closed interval into itself such that $Sf < 0$ each stable periodic trajectory attracts either the trajectory of one of the end-points of the interval or the trajectory of some critical point x_c, that is, a point at which $f'(x_c) = 0$. It is customary in the theory of one-dimensional maps to consider the so-called unimodal maps.

Definition 2.1. A continuous map of a closed interval into itself is said to be *unimodal* if the interior of the interval contains an extremum point x_c, and on both sides of it the map is strictly monotone (increasing and decreasing, respectively).

It follows from Singer's result above that unimodal maps with negative Schwarzian derivatives have at most one stable periodic trajectory. The last assertion is valid under certain additional assumptions, because, in general, another stable fixed point can exist. Guckenheimer [13] established that when there is a unique stable periodic trajectory, then the set of points not attracted to it has measure zero. There are diverse concrete examples showing that this set can have the structure of a Cantor set on which the action is unstable and thus stochastic.

A very important property of one-dimensional maps was discovered by Sharkovskii. We arrange the natural numbers in the following order:

$$1 \prec 2 \prec 2^2 \ldots \prec 2^n \prec \ldots \prec \ldots \prec 2^m \cdot 7 \prec 2^m \cdot 5 \prec 2^m \cdot 3 \prec \ldots$$
$$\ldots \prec 2 \cdot 7 \prec 2 \cdot 5 \prec 2 \cdot 3 \prec \ldots \prec \ldots 7 \prec 5 \prec 3.$$

This is called the Sharkovskii order. We see that it contains infinitely many transfinite numbers of the second kind corresponding to the subsequences $3 \cdot 2^m \succ 5 \cdot 2^m \succ 7 \cdot 2^m \succ \ldots$

Sharkovskii's Theorem (see [10]). *If a unimodal mappings of* $[a, b]$ *into itself has a periodic trajectory of period k, then it also has periodic trajectories of all periods preceding k in the Sharkovskii order.*

By now there are several proofs of Sharkovskii's theorem (see [4]), including some very elementary ones; one even awaits publication in the

Russian student journal "Quantum". We also direct attention to the paper [15] by Li and Yorke entitled "Period three implies chaos". In fact, here the proof is for a special case of Sharkovskii's theorem, namely, it is proved in [15] that if a unimodal map of a closed interval into itself has a periodic trajectory of period 3, then there are periodic trajectories of any period, which the authors associate with stochasticity.

The maps corresponding to the first transfinite number of the second kind in the Sharkovskii order plays a special role in the Feigenbaum theory. These maps have the property that there is a periodic trajectory of period 2^n for each $n \geqslant 0$. Such mappings have been the subject of special investigations. Here we should mention in the first place the papers of Barkovskii and Levin [16] and of Misiurewicz [17]. For example, the following class of one-dimensional maps is considered by Misiurewicz.

Definition 2.2. An even map of $[-1, 1]$ into itself belongs to the class G if:

1) $f \in C^1([-1, 1])$ and $f \in C^3((-1, 0) \cup (0, 1))$;
2) $f(-1) = -1$ and $f'(-1) > 1$;
3) $f'(x) \neq 0$ for $x \neq 0$, and $Sf < 0$ for $x \neq 0$;
4) f has a periodic trajectory of period 2^n for any $n \geqslant 0$;
5) f does not have other periodic trajectories.

For a map in the class G one can construct for all $n \geqslant 1$ a system of closed intervals $\Delta_k^{(n)}$ $(0 \leqslant k < 2^n)$ with the following properties.

I. The intervals $\Delta_k^{(n)}$ are pairwise disjoint for different k with $0 \leqslant k < 2^n$.

II. $$f(\Delta_k^{(n)}) = \Delta_{k+1}^{(n)}, \quad 0 \leqslant k < 2^n - 1, \quad f(\Delta_{2^n-1}^{(n)}) \subset \Delta_0^{(n)}.$$

III. Each interval $\Delta_k^{(n-1)}$ contains exactly two intervals of the n-th rank: $\Delta_k^{(n)}$ and $\Delta_{k+2^{n-1}}^{(n)}$.

We consider the set $F = \bigcap_{n \geqslant 1} \bigcup_{k=1}^{2^n-1} \Delta_k^{(n)}$. It is easy to verify that F is invariant under f and homeomorphic to the Cantor set. It turns out that for maps in the class G the set F is an attractor in the sense that $\operatorname{dist}(f^{(l)}(x), F) \to 0$ as $l \to \infty$ for almost all $x \in [-1, 1]$ (with the exception of the points that are carried under the action of $f^{(n)}$ into unstable periodic trajectories). We shall see later that the maps arising as a result of a sequence of period-doubling bifurcations belong, in essence, to the class G and also have attractors of Cantor type. Right now we observe that the existence of the system of closed intervals $\Delta_k^{(n)}$ enables us to describe the metric and spectral properties of the dynamical system $f|_F$ (see [17]).

Theorem 2.1. 1. *The transformation f has a unique invariant measure μ_0, which is concentrated on F. Moreover,* $\mu_0(\Delta_k^{(n)}) = 2^{-n}$.

2. *The transformation f of F with the measure μ_0 is ergodic and has a discrete spectrum consisting of the numbers of the form*

$$\exp\{2\pi i(2r+1)2^{-n}\}, \quad n \geqslant 1, \quad 0 \leqslant r \leqslant 2^{n-1} - 1.$$

Proof. 1. Since the intervals $\Delta_k^{(n)}$ are disjoint, $\mu_0(\Delta_k^{(n)})$ is the same for all k. Moreover, $\bigcup_{k=0}^{2^n-1} \Delta_k^{(n)} \supset F$. The first assertion follows from this.

2. For each $n \geqslant 1$ we construct an eigenfunction $e^{(n)}(x)$ with eigenvalue $\exp\{2\pi i 2^{-n}\}$. To do this, let $e^{(n)}(x) = \exp\{2\pi i k 2^{-n}\}$ for $x \in \Delta_k^{(n)}$ $(0 \leqslant k \leqslant 2^n - 1)$. Since $f^{(2n)}(\Delta_0^{(n)}) \subset \Delta_0^{(n)}$, this is well defined. It is not hard to show that the $e_r^{(n)} = (e^{(n)}(x))^{2r+1}$, $(0 \leqslant r \leqslant 2^{n-1} - 1)$ are eigenfunctions of U_f with eigenvalues $\exp\{2\pi i(2r+1)\cdot 2^{-n}\}$ and form a basis in $\mathscr{L}^2(F, \mu_0)$. This proves the theorem.

We take a function $\varphi \in C^1([-1, 1])$ with $\int \varphi \, d\mu_0 = 0$ and form its Fourier coefficients

$$c_n = (U_f^n \varphi, \varphi)_{\mu_0} = \int_0^1 e^{2\pi i \omega n} \, d\rho(\omega).$$

Here ρ is the spectral measure of φ (see [18]). It follows from Theorem 2.1 that

$$\rho = \sum_{n=1}^{\infty} \sum_{r=0}^{2^{n-1}-1} |\rho_r^{(n)}|^2 \delta\left(\omega - \frac{2r+1}{2^n}\right),$$

where the $\rho_r^{(n)}$ are constants. We now estimate the decrease of $\rho_r^{(n)}$.

***Theorem* 2.2.**

$$|\rho_r^{(n)}| \leqslant \frac{1}{2} \max_{x \in [-1, 1]} |\varphi'(x)| \cdot \max_{0 \leqslant k \leqslant 2^{n-1}-1} |\Delta_k^{(n-1)}|.$$

Proof. An explicit expression for $\rho_r^{(n)}$ is:

$$\rho_r^{(n)} = \int_F \varphi(x) \overline{e_r^{(n)}(x)} \, d\mu_0(x) =$$

$$= \sum_{k=0}^{2^{n-1}-1} \int_{\Delta_k^{(n-1)}} \varphi(x) \overline{e_r^{(n)}(x)} \, d\mu_0(x) =$$

$$= \sum_{k=0}^{2^{n-1}-1} \left[\int_{\Delta_k^{(n)}} \varphi(x) \overline{e_r^{(n)}(x)} \, d\mu_0(x) + \int_{\Delta_{k+2^{n-1}}^{(n)}} \varphi(x) \overline{e_r^{(n)}(x)} \, d\mu_0(x) \right] =$$

$$= \sum_{k=0}^{2^{n-1}-1} \{\varphi(x_k^{(n-1)}) [\overline{e_r^{(n)}}|_{\Delta_k^{(n)}} + \overline{e_r^{(n)}}|_{\Delta_{k+2^{n-1}}^{(n)}}] 2^{-n}\} +$$

$$+ \sum_{k=0}^{2^{n-1}-1} \left[\int_{\Delta_k^{(n)}} (\varphi(x) - \varphi(x_k^{(n-1)})) \overline{e_r^{(n)}(x)} \, d\mu_0 + \right.$$

$$\left. + \int_{\Delta_{k+2^{n-1}}^{(n)}} (\varphi(x) - \varphi(x_k^{(n-1)})) \overline{e_r^{(n)}(x)} \, d\mu_0 \right].$$

Here $x_k^{(n-1)}$ is the mid-point of $\Delta_k^{(n-1)}$ and $e_r^{(n)}|_{\Delta_k^{(n)}}$ is the value of $e_r^{(n)}(x)$ on $\Delta_k^{(n)}$. Now

$$e_r^{(n)}|_{\Delta_k^{(n)}} = (e^{(n)})^{2r+1}|_{\Delta_k^{(n)}} = e^{2\pi i(2r+1)k2^{-n}},$$

$$e_r^{(n)}|_{\Delta_{k+2^{n-1}}^{(n)}} = (e^{(n)})^{2r+1}|_{\Delta_{k+2^{n-1}}^{(n)}} = -e^{2\pi i(2r+1)k2^{-n}}.$$

Thus, the first sum is zero. For the second sum

$$|\varphi(x)-\varphi(x_k^{(n-1)})| \leqslant \frac{1}{2} \max_{x\in[-1,1]} |\varphi'(x)| \cdot \max_{0\leqslant k\leqslant 2^{n-1}-1} |\Delta_k^{(n-1)}|.$$

The theorem follows immediately from this.

This theorem describes the structure of the discrete spectrum of a smooth function: there is a δ-function amplitude at the point $1/2$, then two δ-function amplitudes at the points $1/2^2$ and $3/2^2$, then smaller amplitudes at the points $1/2^3$, $3/2^3$, $5/2^3$, and $7/2^3$, and so on. The total intensity of these amplitudes at the points of the form $(2r+1)2^{-n}$ satisfies the estimate

$$\sum_{r=1}^{2^{n-1}-1} |\rho_r^{(n)}|^2 \leqslant \text{const} \max_{0\leqslant k\leqslant 2^{n-1}-1} (|\Delta_k^{(n-1)}|^2)\cdot 2^n.$$

2.2. We now discuss in what sense a limit point μ_∞ of bifurcation parameter values μ_n can be regarded as the onset of stochastic behaviour. Let f be a map of the interval $[a, b]$ into itself. A measure ν is said to be invariant if $\nu(C) = \nu(f^{-1}C)$ for any $C \subset [a, b]$.

Definition 2.3. A map f has stochastic behaviour if it has an absolutely continuous invariant measure ν_0 with respect to which the σ-algebra $\bigwedge_n f^{-n}(\mathcal{F})$ consists of finitely many atoms. Here $\mathcal{F}$ is the σ-algebra of Borel subsets of $[a, b]$, and $f^{-n}(\mathcal{F})$ is the σ-algebra of sets of the form $f^{-n}C$, $C \in \mathcal{F}$.

Let us clarify this definition in a more intuitive way. To say that the number of atoms in the σ-algebra $\bigwedge_n f^{-n}(\mathcal{F})$ is r means that there are r subsets $C_1, \ldots, C_r$ such that $C_i \cap C_j = \varnothing$ for $i \neq j$, $f(C_i) = C_{i+1}$ for $i < r$, and $f(C_r) \subseteq C_1$. The map $f^{(r)}$ is non-ergodic and its ergodic components are the sets C_i $(1 \leqslant i \leqslant r)$. The map $f^{(r)}|_{C_i}$ has very nice mixing properties. It is an exact endomorphism (see [18]), which implies, in particular, that it has a Lebesgue spectrum of countable multiplicity, mixing of all degrees, and positive entropy. In many cases it is possible to estimate the rate of decrease of the correlations (see, for example, [19]-[21]). We remark that if a unimodal map f with stochastic behaviour has a negative Schwarzian derivative, then it cannot have stable periodic trajectories (see above).

The first example of a map with stochastic behaviour in which f has a critical point is due to von Neumann and Ulam [22], who proved that the mapping $f(x) = 4x(1-x)$ of $[0, 1]$ onto itself has stochastic behaviour. The density $p(x)$ of the corresponding invariant measure ν_0 has the form

$p(x) = \text{const}\,(x(1-x))^{1/2}$. We see that it has root singularities at the points 1 and 0, which form the trajectory of the critical point 1/2.

The next step was taken by Ruelle (see [23]), who showed that the map $f(x;\ \mu) = \mu x(1-x)$ of [0, 1] into itself also has stochastic behaviour for a special value $\mu = \mu_0$ of the parameter. This value μ_0 is chosen so that the critical point 1/2 is carried in two steps into an unstable periodic point of period 3. The density of the invariant measure also has root singularities along the trajectory of the critical point. An analogous result was obtained much earlier by Bunimovich (see [24]) for the map $x \to \{\lambda \sin 2\pi x\}$. Ognev (see [25]) and Misiurewicz (see [26]) obtained the following result fairly recently: suppose that a unimodal map f satisfies the condition $Sf < 0$ and, moreover, that the critical point x_c lands after finitely many iterations on an unstable periodic trajectory with arbitrary period. Then f has stochastic behaviour. (Misiurewicz actually proved a somewhat more general assertion.)

The most difficult and significant result in this whole range of questions is due to Jakobson. Let $f(x;\ \mu) = \mu x(1-x)$.

Jakobson's Theorem (see [27]). *The set $\mathcal{A}$ of values of the parameter μ such that f has stochastic behaviour is of positive Lebesgue measure.*

In fact, families of more general form are considered in [27]. The density of the invariant measure in Jakobson's theorem has a dense set of root singularities and is therefore extremely complicated.

We can now explain the sense in which μ_∞ can be regarded as the onset of stochastic behaviour. Any right half-neighbourhood U of μ_∞ contains values of μ such that the conditions of the Ognev-Misiurewicz theorem are satisfied, consequently, $f(x;\ \mu)$ has stochastic behaviour. Moreover, it follows from Jakobson's theorem that the Lebesgue measure of the intersection $U \cap \mathcal{A}$ is positive for any U, though it is not known whether μ_∞ is a point of density of $\mathcal{A}$. What we have said means that the stochastic behaviour for $\mu > \mu_\infty$ manifests itself as bursts between which there are structurally stable regimes with a stable periodic trajectory to which almost all points are attracted. In this connection we mention the famous conjecture that the parameter values corresponding to the stable periodic regimes are dense in the domain in which μ varies. Numerical experiments (see [28]) show that the part of the parameter values where there is stochastic behaviour grows as μ increases.

§3. The doubling transformation

In this section we explain the phenomenon of Feigenbaum universality.

3.1. We consider a one-parameter family of smooth maps $f(x;\ \mu)$ of [−1, 1] into itself. We confine ourselves to the case of families of unimodal maps. Moreover, we assume that $x_c(\mu) = 0$ is the unique critical point and is a maximum point for all μ. This condition is no restriction of generality, because any unimodal family $f(x;\ \mu)$ can be brought to this form by means of a suitable conjugation $\tilde{f}(\cdot;\ \mu) = S_\mu^{-1} \circ f(\cdot;\ \mu) \circ S_\mu$. The properties of existence or non-existence of periodic trajectories and the character of their stability (attracting, repelling) do not change under a conjugation $\tilde{f} = S^{-1} \circ f \circ S$, where S is a diffeomorphism of the interval onto itself.

Suppose that an infinite sequence of period-doubling bifurcations take place for the family $f(x;\ \mu)$ as μ increases; as before, let $\mu_1, \mu_2, \ldots$ denote the sequence of bifurcation values of the parameter. We recall that a stable periodic trajectory of period 2^n appears for $\mu = \mu_n$. Let us consider in a little more detail what happens in the interval $\mu_n < \mu < \mu_{n+1}$. The map $f^{(2^n)}(x;\ \mu)$ has a stable fixed point $x(\mu)$ for these values of the parameter. In fact there are 2^n such points. For μ close to μ_n the derivative $\frac{\partial}{\partial x} f^{(2^n)}(x;\ \mu)\,|_{x=x(\mu)}$ is close to $+1$, while for $\mu \uparrow \mu_{n+1}$

$$\frac{\partial}{\partial x} f^{(2n)}(x;\ \mu)\,|_{x=x(\mu)} \to -1$$

and at $\mu = \mu_{n+1}$ there is a loss of stability of the periodic trajectory. As μ varies, a value $\mu_n < \bar{\mu}_n < \mu_{n+1}$ appears such that $\frac{\partial}{\partial x} f^{(2^n)}(x;\ \bar{\mu}_n)\,|_{x=x(\bar{\mu}_n)} = 0$. In this case $f^{(2^n)}(x;\ \bar{\mu}_n)$ is said to have a superstable fixed point. By our assumptions $x(\bar{\mu}_n) = 0$. It is technically somewhat more convenient to follow the sequence of values $\bar{\mu}_n$.

We proceed to explain the main idea, due to Feigenbaum (see [3]), for proving universality. The graphs of the maps $f^{(2^{n-1})}(x;\ \mu)$ and $f^{(2^n)}(x;\ \mu)$ for $\bar{\mu}_n \leqslant \mu \leqslant \bar{\mu}_{n+1}$ are drawn in Fig. 3.1. The main argument is that for large n the maps $f^{(2^{n-1})}(x,\ \bar{\mu}_n)$ and $f^{(2^n)}(x;\ \bar{\mu}_{n+1})$ coincide asymptotically in a neighbourhood of 0 depending on n, to within a scale transformation and a reflection in the x-axis. Actually, the whole families of maps

$$f^{(2^{n-1})}(x;\ \mu),\ \bar{\mu}_n \leqslant \mu \leqslant \bar{\mu}_{n+1} \text{ and } f^{(2^n)}(x;\ \mu),\ \bar{\mu}_{n+1} \leqslant \mu \leqslant \bar{\mu}_{n+2}$$

coincide asymptotically after a shift and renormalization of μ, a reflection in the x-axis, and a scale transformation.

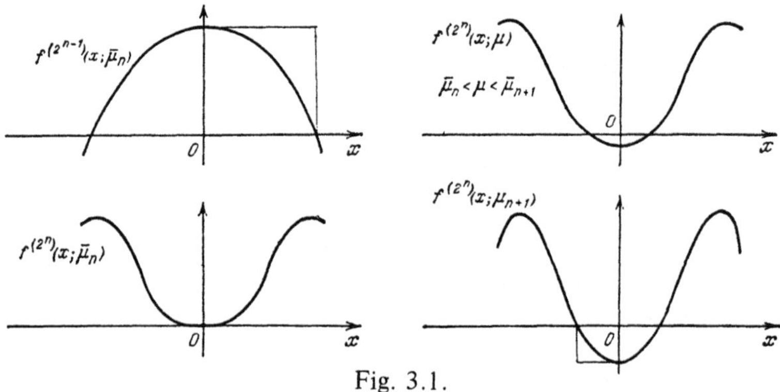

Fig. 3.1.

Thus, by successively doubling the family of maps and carrying out parameter renormalization and scale transformations, we obtain in the limit a family of maps that is invariant under the transformations. It is natural to think that the limiting family does not depend on the original one, but is entirely determined by the transformations listed above. A precise meaning is given to these arguments in the Feigenbaum theory.

If the assumption on the universal family is reasonable, then for maps of the class G we must obtain one and the same map independent of the choice of the original map after 2^n iterations and suitable renormalization of x as $n \to \infty$. The first step consists in determining this universal map, and we do that now.

3.2. We define the doubling transformation T acting in the space of maps of $[-1, 1]$ into itself. Let $f(x)$ be an even unimodal map of $[-1, 1]$ into itself, and $x = 0$ the maximum point of f. We write

$$\alpha = \alpha(f) = -\frac{f(0)}{f(f(0))}.$$

Under the action of f the interval $[-\alpha^{-1}, \alpha^{-1}]$ is mapped onto $[f(\alpha^{-1}), f(0)]$, and this, in turn, is mapped onto $[f(f(0)), f(f(\alpha^{-1}))]$.

We assume the following conditions:

$$\alpha > 0, \quad f(f(\alpha^{-1})) < \alpha^{-1}, \quad \alpha^{-1} < f(\alpha^{-1}), \quad f(0) > 0.$$

Then $[f(f(0)), f(f(\alpha^{-1}))] \subset [-\alpha^{-1}, \alpha^{-1}]$ and $[f(\alpha^{-1}), f(0)] \cap [-\alpha^{-1}, \alpha^{-1}] = \varnothing$. Thus, $h(x) = -\alpha f(f(\alpha^{-1}x))$ is again a unimodal map of $[-1, 1]$ into itself, and $h(0) = f(0)$. We now define the doubling transformation T by setting

$$(Tf)(x) = -\alpha f(f(\alpha^{-1}x)), \qquad \alpha = -\frac{f(0)}{f(f(0))}.$$

If T is defined for some map f, then also for maps close to f. In this section we study mainly the properties of T or of transformations similar to it. We now explain how T is connected with the universality phenomenon for a sequence of period-doubling bifurcations.

We consider the space of maps $f(x)$ (not necessarily even) of $[-1, 1]$ into itself such that $f(x) \in C^1([-1, 1])$, $x = 0$ is a maximum point, and $f(0) = \text{const}$, to be definite, const $= 1$. This space is invariant under T. It turns out that T has a fixed point $g(x)$ in it and the spectrum of the linearized transformation $DT(g)$ at the fixed point lies inside the unit disk except for a single eigenvalue greater than 1. This is the Feigenbaum constant $\delta = 4.6692\ldots$. Therefore, a one-dimensional unstable separatrix $\Gamma^u(g)$ passes through the fixed point g consisting of maps that recede from g under the action of T, and a stable separatrix $\Gamma^{(s)}(g)$ of codimension 1 consisting of maps attracted to g under the action of T. The unstable separatrix $\Gamma^u(g)$ corresponds to the eigenvalue δ.

We mentioned above that a period-doubling bifurcation takes place when the derivative at the fixed point of the mapping passes through -1. Let Σ_1 be the hypersurface of codimension 1 that consists of the maps in the function space having derivative -1 at the fixed point. If the family of maps intersects Σ_1 transversally, then a period-doubling bifurcation takes place. Let

$$\Sigma_2 = T^{-1}\Sigma_1, \ldots, \Sigma_k = T^{-1}\Sigma_{k-1}.$$

When the family of maps $f(x;\ \mu)$ intersects the surface Σ_k, a stable periodic trajectory of period 2^k is created from the stable periodic trajectory of period 2^{k-1}, that is, the parameter values corresponding to the intersections of the family of maps with the surfaces Σ_k are the bifurcation parameter values. For if the family $f(x;\ \mu)$ intersects Σ_k, then $T^{k-1}f$ intersects Σ_1, and this means that a period-doubling bifurcation takes place and a periodic trajectory of period 2^k appears.

The unstable separatrix $\Gamma^{(u)}(g)$ intersects Σ_1 transversally, hence, it intersects all the Σ_k ($k \geqslant 2$) transversally (see Fig. 3.2).

The surfaces Σ_k, as is easy to show, converge to $\Gamma^{(s)}(g)$, and asymptotically for large k the distance between Σ_{k+1} and $\Gamma^{(s)}(g)$ is δ times smaller than that between Σ_k and $\Gamma^{(s)}(g)$. Therefore, the bifurcation parameter values for any family of maps $f(x;\ \mu)$ in some neighbourhood of $\Gamma^{(u)}(g)$ satisfy the relation $\mu_\infty - \mu_k \sim \text{const}\ \delta^{-k}$, where const depends on the family of mappings and is determined by the derivative $\frac{\partial}{\partial\mu} f(x;\ \mu)|_{\mu=\mu_\infty}$. The map g is universal in the sense that $\lim_{k\to\infty} T^k f(x, \mu_\infty) = g$. Of course, the whole unstable separatrix $\Gamma^{(u)}(g)$ has an analogous universality property (see below).

In what follows we give a somewhat more formal explanation of Feigenbaum universality in terms of the behaviour of the unstable separatrix, but first we state the properties of T that lead to universality.

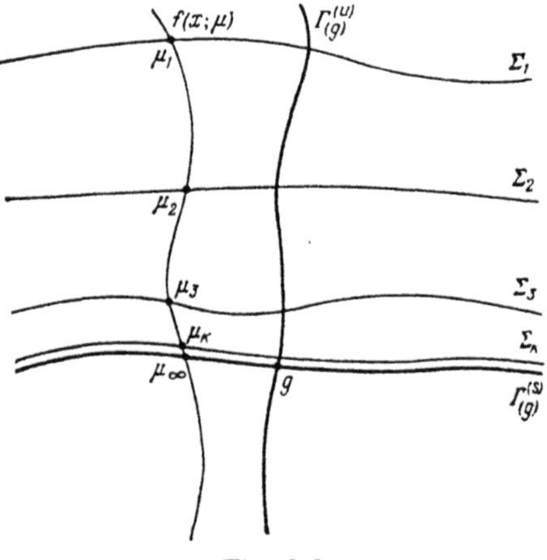

Fig. 3.2.

1. T has a fixed point g.
2. The linearized transformation $DT(g)$ has only one eigenvalue δ greater than 1 in modulus; $\delta = 4.6692\ldots$.
3. The unstable separatrix corresponding to δ intersects the surface Σ_1 transversally.

At the present time the properties 1, 2, and 3 of T can be regarded as proven.

Proofs were obtained by Lanford [29] and by Campanino and Epstein [30]. In a certain sense these proofs differ from the usual mathematical proofs. The fact is that they both are based on computer results, and there is no other way of verifying them than by repeating the computer calculations. More specifically, the course of the arguments is as follows. We assume that $\tilde{g}$ is an even polynomial of degree $2m$. Then $-\alpha\tilde{g}(\tilde{g}(\alpha^{-1}x))$ is a polynomial of degree $4m^2$. The problem is to find a polynomial $\tilde{g}$ close to $-\alpha\tilde{g}(\tilde{g}(\alpha^{-1}x))$. One approach is to consider a shortened polynomial $\tilde{g}_1(x)$, that is, a polynomial obtained by forcibly throwing out the higher powers in $-\alpha\tilde{g}(\tilde{g}(\alpha^{-1}x))$. Then one looks for a polynomial $\tilde{g}$ such that $\tilde{g} = \tilde{g}_1$. In another approach the values of $\tilde{g}(x)$ and $-\alpha\tilde{g}(\tilde{g}(\alpha^{-1}x))$ at previously chosen points are compared. It is this method that was realized by Lanford by means of a computer. In [29] he gave results of computations to an accuracy sometimes as great as 10^{-40}, that is, to 40 decimal places. After obtaining a polynomial $\tilde{g}(x)$ approximating the true fixed point of T with great accuracy we can linearize (in fact, rigorously) our problem in a neighbourhood of $\tilde{g}(x)$ and show, again rigorously, by Newton's method that there is a fixed point of the whole doubling transformation in a suitable neighbourhood. Even the singularities of $g(x)$ when it is extended analytically into the complex plane are studied in [31]. The first terms in the expansion of g are as follows:

$$g(x) \approx 1 - 1.52763x^2 + 0.104815x^4 - 0.0267057x^6 + \ldots,$$
$$\alpha = \alpha(g) = 2.50290 \ldots$$

The quantity α is also a universal constant, which characterizes the scale change associated with doubling of the maps. Below, α always denotes the constant $2.50290\ldots$.

Generally speaking, T has many other fixed points. For let us consider the space of maps of the form $f(|x|^\gamma)$, $\gamma > 1$, where f is an analytic function in some neighbourhood of the interval $[0, 1]$. This space is invariant under T and contains a fixed point. Apparently for all $\gamma > 1$ the transformation T has a fixed point of the indicated form, and the linearized transformation (in the space of functions of the form $f(|x|^\gamma)$) has only one eigenvalue greater than 1. Of course, this value $\delta(\gamma)$ characterizes the universal properties of the sequence of period-doubling bifurcations for families of maps having a singularity of the form $|x|^\gamma$ at the critical point. For $\gamma = 1 + \varepsilon$ (where ε is small) one can establish the properties 1–3 rigorously by the theory of perturbations with respect to ε, and this is what is done in [32]. In this case $\delta(1+\varepsilon) \to 2$ as $\varepsilon \to 0$. However, most important from the point of view of applications is the case $\varepsilon = 1$ or $\gamma = 2$ considered above, therefore, the theory of perturbations is difficult to carry through to the end.

3.3. The space of even maps is invariant under T. It turns out that the fixed point g and the eigenvector $g(\delta)$ corresponding to the eigenvalue δ are even functions. In this subsection we make a general study of some properties of the linearized transformation $DT(g)$ and show that the part of the spectrum of $DT(g)$ corresponding to the odd functions is trivial in the sense that it can be found explicitly.

Let us assume that T acts in the Banach space $M(\mathscr{D})$ of real analytic maps defined in some complex neighbourhood $\mathscr{D}$ of $[-1, 1]$.

The transformation T has a fixed point satisfying the conditions:

a) $g(x) = -\alpha g(g(\alpha^{-1}x))$, $\alpha = 2.50290...$;

b) $g(0) = 1$, $g'(x) > 0$ for $x < 0$, and $g'(x) < 0$ for $x > 0$;

c) $g(x) = g(-x)$.

We denote by $\widetilde{T}$ the transformation that is the same as T, but with fixed $\alpha(g) = 2.50290...$:

$$(\widetilde{T}f)(x) = -\alpha f(f(\alpha^{-1}x)), \qquad \alpha = 2{,}50290 \ldots$$

Of course, $\widetilde{g}(x)$ is also a fixed point of $\widetilde{T}$. We now find the explicit form of the linearized transformations $D\widetilde{T}(g)$ and $DT(g)$:

$$(D\widetilde{T}(g))\,h(x) = -\alpha g'(g(\alpha^{-1}x))\,h(-\alpha^{-1}x) - \alpha h(g(\alpha^{-1}x)),$$

$$(DT(g))\,h(x) = (D\widetilde{T}(g))\,h(x) + c(h)\,e_1,$$

$$c(h) = (\alpha^2 - 1)\,h(0) - \alpha h(1), \quad e_1 = g'(x)\cdot x - g(x).$$

As will be shown below, e_1 is an eigenvector with eigenvalue 1 both for $D\widetilde{T}(g)$ and for $DT(g)$ (it is easy to check that $c(e_1) = 0$). The difference $D\widetilde{T}(g) - DT(g)$ is the projection operator on the eigendirection e_1, therefore, the spectra of the two operators coincide.

We study the spectrum of $D\widetilde{T}(g)$ (see [9]).

It follows from Montel's theorem and the explicit form of $D\widetilde{T}(g)$ that it is compact and, consequently, has a discrete spectrum.

Let $S_n(\varepsilon)$ be the coordinate transformation

$$S_n(\varepsilon)\colon\ x \mapsto x + \varepsilon x^n, \qquad n \geqslant 0.$$

The inverse transformation $S_n^{-1}(\varepsilon)$ exists in a neighbourhood of $[-1, 1]$ for sufficiently small ε.

We define the set of vectors e_k ($k \geqslant 0$) as follows:

$$e_k = \frac{d}{d\varepsilon} S_k^{-1}(\varepsilon) \circ g \circ S_k(\varepsilon)\big|_{\varepsilon=0} = g'(x)\cdot x^k - (g(x))^k.$$

We claim that the e_k ($k \geqslant 0$) are eigenvectors of $D\widetilde{T}(g)$ with the eigenvalues $(-\alpha^{-1})^{k-1}$. We denote by A the linear transformation $Ax = -\alpha^{-1}x$.

Then

$$(D\widetilde{T}(g))e_k = \frac{d}{d\varepsilon}\widetilde{T}(S_k^{-1}(\varepsilon)\circ g\circ S_k(\varepsilon))\ |_{\varepsilon=0},$$

$$\begin{aligned}\widetilde{T}(S_k^{-1}(\varepsilon)\circ g\circ S_k(\varepsilon)) &= A^{-1}\circ(S_k^{-1}(\varepsilon)\circ g\circ S_k(\varepsilon))\circ(S_k^{-1}(\varepsilon)\circ g\circ S_k(\varepsilon))\circ A = \\ &= A^{-1}\circ S_k^{-1}(\varepsilon)\circ A\circ(A^{-1}\circ g\circ g\circ A)\circ A^{-1}\circ S_k(\varepsilon)\circ A = \\ &= A^{-1}\circ S_k^{-1}(\varepsilon)\circ A\circ g\circ A^{-1}\circ S_k(\varepsilon)\circ A.\end{aligned}$$

We have used the fact that $A^{-1}\circ g\circ g\circ A = g$. Observe that $A^{-1}\circ S_k(\varepsilon)\circ A = S_k(\varepsilon\cdot(-\alpha^{-1})^{k-1})$. Therefore,

$$(D\widetilde{T}(g))\,e_k = \frac{d}{d\varepsilon}S_k^{-1}(\varepsilon(-\alpha^{-1})^{k-1})\circ g\circ S_k(\varepsilon(-\alpha^{-1})^{k-1})|_{\varepsilon=0} = (-\alpha^{-1})^{k-1}e_k.$$

The eigenvectors e_k $(k \geqslant 0)$ are connected with coordinate transformations and therefore are inessential in the investigation of the universality properties. It is easy to show that the space spanned by the e_k $(k \geqslant 0)$ has infinite codimension (see [8], [9]). The vector $e_0 = g'(x) - 1$ has the eigenvalue $(-\alpha) = -2.5\ldots$. It is excluded by the condition $f'(0) = 0$. The vector e_1 with eigenvalue 1 is excluded by the condition $f(0) = \text{const}$. The remaining vectors e_k $(k \geqslant 2)$ have eigenvalues less than 1 in modulus.

We have already mentioned that the space $M^{(e)}(\mathscr{Z})$ of entire functions is invariant under $(D\widetilde{T})(g)$. It is this space that contains the whole essential part of the spectrum of the operator, in particular, the eigenvector $g(\delta)$ corresponding to the eigenvalue δ. For, we can show that any function $f \in M(\mathscr{Z})$ has a unique expansion

$$f = \sum_{k=0}^{\infty} f_{2k}\cdot e_{2k} + f^{(e)},$$

where $f^{(e)}$ is an even function. Let

$$f(x) = \sum_{k=0}^{\infty} c_k x^k.$$

Since

$$\sum_{k=0}^{\infty} f_{2k}e_{2k} = \Big(\sum_{k=0}^{\infty} f_{2k}x^{2k}\Big)\,g'(x) - \sum_{k=0}^{\infty} f_{2k}\,(g(x))^{2k},$$

we obtain

$$\Big(\sum_{k=0}^{\infty} f_{2k}x^{2k}\Big)\,g'(x) = \sum_{k=0}^{\infty} c_{2k+1}x^{2k+1},$$

$$f^{(e)} = \sum_{k=0}^{\infty} c_{2k}x^{2k} + \sum_{k=0}^{\infty} f_{2k}\,(g(x))^{2k}.$$

We make use of the fact that $g'(x)$ is an odd function. Thus, the coefficients f_{2k} $(k = 0, 1, \ldots)$ and the function $f^{(e)}$ are uniquely determined by the relations

$$\sum_{k=0}^{\infty} f_{2k}x^{2k} = \frac{x}{g'(x)}\sum_{k=0}^{\infty} c_{2k+1}\cdot x^{2k}$$

and

$$f^{(e)} = \sum_{k=0}^{\infty} c_{2k} x^{2k} + \frac{g(x)}{g'(h(x))} \sum_{k=0}^{\infty} c_{2k+1} (g(x))^{2k}.$$

3.4. As we saw above, to prove the Feigenbaum universality we have to find a fixed point of T, construct the unstable separatrix of the fixed point, and verify that the separatrix intersects the period-doubling surface.

Below we determine a way for constructing the unstable separatrix numerically (see [33]). To do this we consider a transformation T^* in a space of one-parameter families of one-dimensional maps. Let $f(x;\ \mu)$ be a family of even maps of an interval into itself depending on a parameter μ. We assume that for $\mu = 0$ the critical point $x = 0$ passes into 1, and 1 passes into 0, that is,

$$f(0;\ 0) = 1, \quad f(1;\ 0) = 0.$$

We set $f^{(2)}(x;\ \mu) = f(f(x;\ \mu);\ \mu)$ and find the smallest positive parameter value $\bar{\mu}(f)$ such that $x = 0$ is a periodic point with period 2 for the map $f^{(2)}(x;\ \mu)$, that is, $f^{(2)}(f^{(2)}(0;\ \bar{\mu}(f));\ \bar{\mu}(f)) = 0$. We can now define the transformation T^*:

$$(T^*f)(x;\ \mu) = -\alpha(f)\, f^{(2)}(\alpha^{-1}(f)\, x;\ \bar{\mu}(f)\,(1+\mu)),$$

where $\alpha(f) = -(f^{(2)}(0;\ \bar{\mu}(f)))^{-1}$ is chosen so that the normalization condition $(T^*f)(0;\ 0) = 1$ holds.

It is not hard to show that the unstable separatrix of the fixed point g of $\tilde{T}$ is a stable fixed point of T^* to within a scale transformation. We denote by $g(x;\ \mu)$ the fixed point of T^*, and by $\bar{\mu} = \bar{\mu}(g)$ and $\alpha = \alpha(g)$ the values of the constants $\bar{\mu}$ and α corresponding to $g(x;\ \mu)$. The parametrization on the unstable separatrix is chosen so that the parameter value $\mu = \bar{\mu}/(1-\bar{\mu})$ corresponds to the map $g(x)$. The new value of α is the same as the old one, and $\bar{\mu} = \delta^{-1}$.

The fixed point $g(x;\ \mu)$ can be constructed numerically. Moreover, $g(x;\ \mu)$ can be found as a polynomial in x and μ:

$$g(x;\ \mu) = \sum_{i=0}^{k} \sum_{j=0}^{m} g_{ij} x^{2i} \mu^{j}.$$

Instead of T^* we consider the transformation $T^*_{k,m}$ that differs from it in that the terms $x^{2p}\mu^q$ with $p > k$ or $q > m$ are discarded after the doubling. Iterating some initial family $f(x;\ \mu) = \sum_{i=0}^{k} \sum_{j=0}^{m} f_{ij} x^{2i} \mu^{j}$ with the help of $T^*_{k,m}$, we obtain in the limit an approximate fixed point $g^{(k,m)}(x;\ \mu)$. The coefficients $g^{(k,m)}_{ij}$ for the cases $k = 4$, $m = 5$ and $k = 5$, $m = 6$ are given in Table 3.1. We remark that for $k = 5$, $m = 6$ the values of the constants $(\bar{\mu};\ \alpha)$ in the finite-dimensional problem are (0.214115; 2.50306), while their precise values are (0.214169; 2.50290). A more thorough investigation

Table 3.1.

$i \backslash j$	0	1	2	3	4	5	6
0	1.000−0 1.000−0	1.334−0 1.282−0	3.040−1 2,811−1	8.675−3 6.771−3	−3,309−3 −3,456−3	−3,781−4 −5.274−4	— −3.345−5
1	−1,032−0 −1,046−0	−2.626−1 −2.569−1	9.840−3 9.685−3	9,577−3 8.878−3	1,515−3 1,371−3	1.096−4 1.061−4	— 2.526−6
2	4.441−2 4.577−2	−1.562−2 −1.582−2	−7,950−3 −7,684−3	−1,390−3 −1,209−3	−1.205−4 −4,435−5	−2,575−6 1,952−5	— 3.095−6
3	4.374−3 4.676−3	2,691−3 2.777−3	5,241−4 4.969−4	7,531−6 −2,561−5	−1,721−5 −3.567−5	−2.736−6 −8.755−6	— −8.744−7
4	−3.193−4 −3.515−4	−8.448−5 −8.995−5	2.037−5 2.576−5	1,655−5 2,205−5	4.072−6 6.912−6	4.072−7 1.242−6	— 1.095−7
5	— 4.539−6	— −8.033−6	— −6,907−6	— −2,594−6	— −6,235−7	— −1.024−7	— −8,975−9

(Explanation: the first number in each place of the table corresponds to the case $k = 4$, $m = 5$, and the second to $k = 5$, $m = 6$; the expression $-2.561-5$ means -2.561×10^{-5}.)

of the finite-dimensional transformations $T^*_{k,m}$ (including an estimate of the second derivatives) allows us to prove rigorously the existence of a fixed point for the infinite-dimensional transformation T^* and to obtain explicit criteria for the existence of an infinite sequence of period-doubling bifurcations for families of one-dimensional maps.

Using the transformation T^* and its stable fixed point $g(x;\ \mu)$ we can give a simple explanation of the universality phenomenon. For let $f(x;\ \mu)$ be a family of maps in the domain of attraction of $g(x;\ \mu)$. We consider the sequence of families of maps given by $f_0(x;\ \mu) = f(x;\ \mu)$, $f_1(x;\ \mu) = T^*f_0(x;\ \mu)$, $f_n(x;\ \mu) = T^*f_{n-1}(x;\ \mu) = -\alpha(n)f_{n-1}(f_{n-1}(\alpha(n)^{-1}x;\ \bar{\mu}(n)(1+\mu));\ \bar{\mu}(n)(1+\mu))$. As $n \to \infty$, $f^{(n)}(x;\ \mu) \to g(x;\ \mu)$, $\alpha(n) \to \alpha$ and $\bar{\mu}(n) \to \bar{\mu}$. Let $\bar{\mu}_n$ $(n = 1, 2, \ldots)$ be the sequence of values of μ corresponding to superstability of the periodic trajectory of period 2^n for the original family $f(x;\ \mu)$. An immediate consequence of the definition of T^* is that

$$\bar{\mu}_{n+1}-\bar{\mu}_n=\bar{\mu}(1)\,\bar{\mu}(2)\,\ldots\,\bar{\mu}(n).$$

Consequently, $\dfrac{\bar{\mu}_{n+1}-\bar{\mu}_n}{\bar{\mu}_n-\bar{\mu}_{n-1}} = \bar{\mu}(n) \to \bar{\mu} = \delta^{-1}$ as $n \to \infty$.

Another way of constructing the unstable separatrix was realized in [34] and [35].

§4. Properties of the Feigenbaum attractor

4.1. In the preceding section we have shown that the main role in the Feigenbaum theory is played by a fixed point of the transformation T, which satisfies the equation

$$g(x) = -\alpha g(g(\alpha^{-1}x)), \quad \alpha = g(1). \tag{4.1}$$

In what follows this is called the doubling equation. It was first obtained by Cvitanovic jointly with Feigenbaum ([3], [36]). A similar equation was presented in [37]. In accordance with (4.1), the graph of $g(g(x))$ on the interval $(-\alpha^{-1}, \alpha^{-1})$ passes into that of $g(x)$ after a reflection in the horizontal axis and a dilatation scale transformation with coefficient α. The approximate form of $g(x)$ is illustrated in Fig. 4.1.

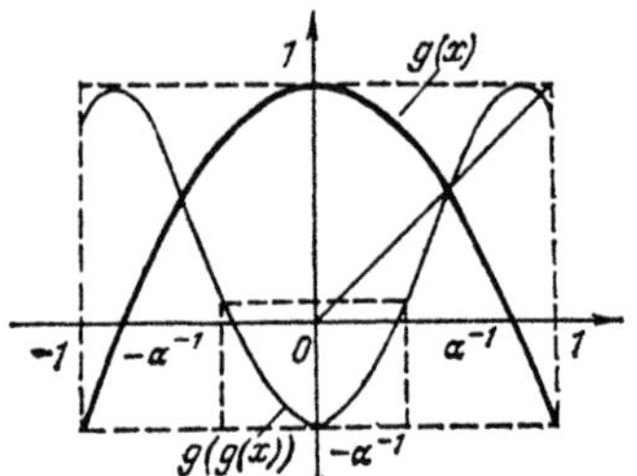

Fig. 4.1.

The same letter g denotes the map of $[-1, 1]$ into itself given by $g(x)$. We shall see presently that this map belongs, in essence, to the class G introduced in §2, and the corresponding closed intervals $\Delta_k^{(n)}$ $(0 \leqslant k < 2^n)$ can be constructed explicitly here and have some remarkable properties.

It is immediately clear that g has a fixed point $x_0^{(0)}$, which is unstable. Let $\Delta_0^{(1)}$ denote the interval $[-\alpha^{-1}, \alpha^{-1}]$. Then $g(\Delta_0^{(1)}) \equiv \Delta_1^{(1)} = [g(\alpha^{-1}), 1]$ does not intersect $\Delta_0^{(1)}$, and the unstable point $x_0^{(0)}$ lies between $\Delta_0^{(1)}$ and $\Delta_1^{(1)}$. Moreover, $g(\Delta_1^{(1)}) \subset \Delta_0^{(1)}$, as follows immediately from the doubling equation: if $x \in [-1, 1]$, then $\alpha^{-1}x \in [-\alpha^{-1}, \alpha^{-1}]$ and $g(g(\alpha^{-1}x)) = -\alpha^{-1}g(x) \in [-\alpha^{-1}, \alpha^{-1}]$, because $|g(x)| < 1$. Now we set $\Delta_0^{(n)} = [-\alpha^{-n}, \alpha^{-n}]$ and $\Delta_k^{(n)} = g^{(k)}(\Delta_0^{(n)})$ for $0 \leqslant k < 2^n$.

***Theorem* 4.1.** *The intervals $\Delta_k^{(n)}$ $(0 \leqslant k < 2^n)$ have the following properties:*

1°. $g^{(2^n)}(\Delta_0^{(n)}) \subset \Delta_0^{(n)}$.

2°. *Each $\Delta_k^{(n-1)}$ contains only the two intervals $\Delta_k^{(n)}$ and $\Delta_{k+2^{n-1}}^{(n)}$, $0 \leqslant k < 2^{n-1}$.*

3°. *The $\Delta_k^{(n)}$ are pairwise disjoint for different k, $0 \leqslant k < 2^n$.*

4°. $\Delta_{2k+1}^{(n)} \subset \Delta_0^{(1)}$ *and* $\Delta_{2k}^{(n)} \subset \Delta_0^{(1)}$.

5°. *Each $\Delta_{2k}^{(n+1)}$ is obtained from $\Delta_k^{(n)}$ by composing two transformations: the contraction with the coefficient α^{-1} and the reflection $x \to -x$.*

6°. *If $|\Delta|$ denotes the length of an interval, then*

$$|\Delta_0^{(n)}| = \max_{0 \leqslant k < 2^n} |\Delta_k^{(n)}|, \quad |\Delta_1^{(n)}| = \min_{0 \leqslant k < 2^n} |\Delta_k^{(n)}|.$$

Proof. 1° follows directly from the doubling equation, since $g^{(2^n)}(\alpha^{-n}x) = (-1)^n\alpha^{-n}g(x)$.

2° and 3° are proved simultaneously by induction on n. We assume that they have been established for n and verify them for $n+1$. The doubling equation and the fact that g is even give us

$$\underbrace{g\circ\ldots\circ g}_{2^n}(\alpha^{-(n+1)}x)=\underbrace{g\circ\ldots\circ g}_{2^n-2}(-\alpha^{-1}g(\alpha^{-n}x))=\underbrace{g\circ\ldots\circ g}_{2^n-2}(\alpha^{-1}g(\alpha^{-n}x))=\ldots$$
$$\ldots=-\alpha^{-1}\underbrace{g\circ\ldots\circ g}_{2^{n-1}}(\alpha^{-n}x)=\ldots=(-\alpha^{-1})^n g(\alpha^{-1}x).$$

From this we deduce immediately that $g^{(2^n)}(\Delta_0^{(n+1)}) = (-\alpha^{-1})^n\Delta_1^{(1)} \subset \Delta_0^{(n)}$ and $g^{(2^n)}(\Delta_0^{(n+1)}) \cap \Delta_0^{(n+1)} = \varnothing$. Thus, $\Delta_0^{(n+1)}$ and $\Delta_{2^n}^{(n+1)}$ satisfy 2°. Consequently, $\Delta_k^{(n+1)}$ and $\Delta_{k+2^n}^{(n+1)}$ are contained in $\Delta_k^{(n)}$. By the inductive hypothesis, the $\Delta_k^{(n)}$ are pairwise disjoint, and 2° and 3° follow easily from this.

4° follows from the fact that $g(\Delta_0^{(1)}) = \Delta_1^{(1)}$ and $g(\Delta_1^{(1)}) \subset \Delta_0^{(1)}$. To prove 5° we note that

$$\Delta_{2k}^{(n+1)}=\underbrace{g\circ\ldots\circ g}_{2k}(\Delta_0^{(n+1)})=\underbrace{g\circ\ldots\circ g}_{2k}(\alpha^{-1}\Delta_0^{(n)})=-\alpha^{-1}\underbrace{g\circ\ldots\circ g}_{k}(\Delta_0^{(n)})=-\alpha^{-1}\Delta_k^{(n)},$$

which is equivalent to 5°. Also, $\Delta_{2k+1}^{(n+1)} = g(\Delta_{2k}^{(n+1)})$, that is, it is obtained from $\Delta_{2k}^{(n+1)}$, which has the form described above, under the action of g.

It remains to prove 6°. Again we use induction on n. From 5° and the inductive hypothesis, $|\Delta_0^{(n+1)}| = \max_{0\leqslant k<2^n} |\Delta_{2k}^{(n+1)}|$. We assume that $|\Delta_{2k+1}^{(n+1)}| > |\Delta_0^{(n+1)}|$ for some k with $0 \leqslant k < 2^n$. Here $\Delta_{2k+1}^{(n+1)} \subset \Delta_1^{(1)}$ (by 4°), and that it can be seen directly from the form of g that $|g'(x)| > 1$ for $x \in \Delta_1^{(1)}$. Therefore, $|\Delta_{2k+2}^{(n+1)}| = |g(\Delta_{2k+1}^{(n+1)})| > |\Delta_{2k+1}^{(n+1)}| > |\Delta_0^{(n+1)}|$, which contradicts what was proved above.

We now establish that $\min_{0\leqslant k<2^{n+1}} |\Delta_k^{(n+1)}| = |\Delta_1^{(n+1)}|$ (again by induction on n). On the basis of 5°,

$$|\Delta_{2k}^{(n+1)}|=\alpha^{-1}|\Delta_k^{(n)}|\geqslant\alpha^{-1}|\Delta_1^{(n)}|>|\Delta_1^{(n+1)}|.$$

The last inequality can be verified directly. From it and 5°,

$$\min_{0\leqslant k<2^n}|\Delta_{2k}^{(n+1)}|=\alpha^{-1}\min_{0\leqslant k<2^n}|\Delta_k^{(n)}|=\alpha^{-1}|\Delta_1^{(n)}|=|\Delta_2^{(n+1)}|>|\Delta_1^{(n+1)}|.$$

We assume that $|\Delta_{2k+1}^{(n+1)}| < |\Delta_1^{(n+1)}|$ for some k with $0 \leqslant k < 2^n$. Since g is convex,

$$\min_{x\in\Delta_1^{(n+1)}}|g'(x)|>\max_{x\in\Delta_{2k+1}^{(n+1)}}|g'(x)|,$$

therefore, $|\Delta_{2k+2}^{(n+1)}| = |g(\Delta_{2k+1}^{(n+1)})| < |g(\Delta_1^{(n+1)})| = |\Delta_2^{(n+1)}|$, which contradicts what has been proved. The theorem is now proved.

Definition 4.1. The closed set $\bigcap_{n\geqslant 1}\bigcup_{k=0}^{2^n-1}\Delta_k^{(n)} = F$ is called the *Feigenbaum attractor*.

We see now that the system of intervals $\Delta_k^{(n)}$ has the same properties as the analogous system for maps in the class G. On the basis of Theorem 2.1 we draw the following conclusions from this.

1. The map $g|_F$ has a unique invariant measure μ_0 with respect to which g is ergodic, strictly ergodic, and has a discrete spectrum consisting of the numbers of the form

$$\exp\{2\pi i(2r+1)2^{-n}\},\quad n=1,2,\ldots,\quad 0\leqslant r\leqslant 2^{n-1}-1.$$

2. Suppose that $\varphi\in C^1([-1,1])$ and $\int\varphi\,d\mu_0=0$. We form the Fourier coefficients and their spectral representation

$$c_n=(U_g^n\varphi,\varphi)_{\mu_0}=\int_0^1 e^{2\pi i\omega n}\,d\rho(\omega).$$

Then

$$\rho=\sum_{n=1}^{\infty}\sum_{r=0}^{2^{n-1}-1}|\rho_r^{(n)}|^2\delta\left(\omega-\frac{2r+1}{2^n}\right).$$

From 6° and Theorem 2.2,

$$|\rho_r^{(n)}|\leqslant\max_{x\in[-1,1]}|\varphi'(x)|\alpha^{(1-n)}.$$

The proof of Theorem 2.2 actually gives us the stronger estimate

$$|\rho_r^{(n)}|\leqslant\frac{1}{2}\max|\varphi'(x)|\Big(\sum_{0\leqslant k<2^{n-1}}|\Delta_k^{(n-1)}|\Big)\cdot 2^{-(n-1)}.$$

It can be shown that

$$\Big(\sum_{0\leqslant k<2^n}|\Delta_k^{(n)}|\Big)\cdot 2^{-n}=O(\sigma^n),$$

where $\sigma\approx 0.29$ (see Theorem 4.2 below). The asymptotic behaviour of the amplitudes of the δ-functions is studied in more detail in [5], [38], and [68]. It is shown there that, for small r, the asymptotic behaviour is given by $\log|\rho_r^{(n)}|\sim n\log\bar{\sigma}$, where $\bar{\sigma}=0.15\ldots$, and $\log\Big(\sum_{r=0}^{2^{n-1}-1}|\rho_r^{(n)}|^2\Big)\sim n\log\tilde{\sigma}$, $\tilde{\sigma}=0.095\ldots$.

In what follows we need information about the structure of the unstable periodic points of g. We have already seen that there is a unique unstable fixed point $x_0^{(0)}$ between $\Delta_0^{(1)}$ and $\Delta_1^{(1)}$. It follows from the doubling equation that $g^{(2^n)}(\alpha^{-n}x)=(-\alpha^{-1})^n g(x)$, from which it is clear that each interval $\Delta_0^{(n)}$ contains an unstable periodic point $x_0^{(n)}$ of period 2^n separating $\Delta_0^{(n+1)}$ and $\Delta_{2^n}^{(n+1)}$. From this and 2° it follows that $x_k^{(n)}=g^{(k)}(x_0^{(n)})$ separates the intervals $\Delta_k^{(n+1)}$ and $\Delta_{k+2^n}^{(n+1)}$ inside $\Delta_k^{(n)}$. Consequently, if $\Delta_{k'}^{(n)}$ and $\Delta_{k''}^{(n)}$ are two adjacent intervals, then some periodic point $x_k^{(m)}$ lies between them, where $0\leqslant m<n$ and $0\leqslant k<2^m$. For if $\Delta_{k'}^{(n)}$ and $\Delta_{k''}^{(n)}$ lie in a single interval $\Delta_k^{(n-1)}$, then the assertion has already been proved. If not, then we pass to intervals $\Delta_{k_1'}^{(n-1)}$ and $\Delta_{k_1''}^{(n-1)}$ containing $\Delta_{k'}^{(n)}$ and $\Delta_{k''}^{(n)}$, respectively. Then either they lie inside a single $\Delta_{k_1'}^{(n-2)}$ and again everything is proved, or we must pass to $\Delta_{k_2'}^{(n-2)}$ and $\Delta_{k_2''}^{(n-2)}$ containing $\Delta_{k_1'}^{(n-1)}$ and $\Delta_{k_1''}^{(n-1)}$ and repeat the same argument.

We now show that g does not have other periodic points on $[-1, 1]$. From the form of g it follows at once that there cannot be periodic points inside $[-1, -\alpha^{-1}]$, because $g([-1, -\alpha^{-1}]) = [-\alpha^{-1}, g(\alpha^{-1})] \subset [-\alpha^{-1}, 1]$ and $g([-\alpha^{-1}, 1]) \subset [-\alpha^{-1}, 1]$. Next, the derivative $g'(x)$ is greater than 1 in absolute value on $[\alpha^{-1}, g(\alpha^{-1})]$, and $g([\alpha^{-1}, g(\alpha^{-1})]) \subset [-\alpha^{-1}, 1]$. In this case all points of $[\alpha^{-1}, g(\alpha^{-1})]$ except $x_0^{(0)}$ fall into $\Delta_0^{(1)}$ or $\Delta_1^{(1)}$ after finitely many steps. Consequently, there are no other periodic points inside $[\alpha^{-1}, g(\alpha^{-1})]$ except $x_0^{(0)}$. Repeating similar arguments for the interval $\Delta_0^{(n)}$ and the map $g^{(2^n)}$, we obtain the required result.

These arguments also indicate the sense in which the set F is an attractor. Suppose that $x \in [-1, 1]$ is such that $g^{(m)}(x) \neq x_k^{(n)}$ for any m, n, and k. Then for any $n \geqslant 0$ there is an r such that $g^{(r)}(x) \subset \Delta_k^{(n)}$ for some k with $0 \leqslant k < 2^n$, that is, dist $(g^{(r)}(x), F) \to 0$ as $r \to \infty$.

We show that by means of the system of intervals $\Delta_k^{(n)}$ we can construct a symbolic dynamics for the map g. (Of course, this can also be done for all maps of the class G in §2.) Suppose that $x \in F$ and $x \in \Delta_{k_n}^{(n)}$ $(n = 1, 2, \ldots)$. Then it follows from 2° in Theorem 4.1 that either $k_{n+1} = k_n$ or $k_{n+1} = k_n + 2^n$. We now write down the dyadic representation of k_n: $k_n = \varepsilon_0 + \varepsilon_1 \cdot 2 + \ldots + \varepsilon_{n-1} \cdot 2^{n-1}$, where the ε_i take the values 1 or 0. Then $k_{n+1} = k_n + \varepsilon_n \cdot 2^n$. Since $x = \bigcap_{n \geqslant 1} \Delta_{k_n}^{(n)}$, we obtain a one-to-one map $\varphi: x \leftrightarrow (\varepsilon_0, \varepsilon_1, \ldots, \varepsilon_n, \ldots)$ taking F into the space E of one-sided dyadic sequences $\varepsilon = (\varepsilon_0, \varepsilon_1, \ldots, \varepsilon_n, \ldots)$. This space can be regarded as the Abelian group of 2-adic integers. Since $g(\Delta_k^{(n)}) = \Delta_{k+1}^{(n)}$ for $0 \leqslant k < 2^n - 1$, φ takes g to the transformation that adds the element $\bar{\varepsilon} = (1, 0, \ldots, 0, \ldots)$. The image of μ_0 under φ is denoted by μ_0^*. It is easy to see that relative to μ_0^* the coordinates $\varepsilon_0, \varepsilon_1, \varepsilon_2, \ldots$ form a sequence of independent random variables taking the values 1 and 0 with probability 1/2.

4.2. We now define three parameters characterizing the metric properties of F.

***Theorem* 4.2.** I.[(1)] *There is a number $\gamma < 0$ such that for μ_0-almost every $x \in F$*

$$\lim_{n \to \infty} \frac{1}{n} \log |\Delta_s^{(n)}(x)| = \gamma,$$

where $\Delta_s^{(n)}(x)$ is the interval of the n-th rank that contains x.

II. *There is a number β_0 such that*

$$\lim_{n \to \infty} \frac{1}{n} \log \Big[\sum_{k=0}^{2^n - 1} |\Delta_k^{(n)}|^{\beta_0} \Big] = 0.$$

This number is equal to the Hausdorff dimension of F.

(1) As M. Misiurewicz has informed us, an assertion close to I has recently been obtained by himself and F. Ledrappier.

III. *There are positive numbers* λ *and* σ *such that*

$$\lim_{n\to\infty}\frac{1}{n}\log\Big[1+\sum_{k=1}^{2^n-1}\prod_{i=k}^{2^n-1}(g'(x_i))^2\Big]=2(\log\lambda-\log\alpha),$$

$$x_i=g^{(i)}(x_0),\quad x_0\in\Delta_0^{(n)};$$

$$\lim_{n\to\infty}\frac{1}{n}\log\Big[\frac{1}{2^n}\Big(\sum_{k=1}^{2^n-1}|\Delta_k^{(n)}|\Big)\Big]=\log\sigma.$$

The first constant shows how the lengths of the intervals $\Delta_k^{(n)}$ containing μ_0-typical points x decrease. As for II, we can assert that $\beta_0\geqslant-\log 2/\gamma$. However, equality does not hold in the last relation, and this shows that the Hausdorff dimension is attained at the expense of the set of non-typical points with respect to μ_0. The constant λ is essential for the analysis of small random perturbations of g (see §5). The constant σ is used in spectral estimates (see above).

Proof. 1. The doubling equation leads to an important asymptotic relation, which we shall call 1-universality. The length of the interval $\Delta_1^{(n)}$ adjoining 1 is of order α^{-2n}. In a neighbourhood of 1 we introduce a renormalized coordinate $X^{(n)}$ by setting $x=1-\alpha^{-2n}X^{(n)}$. As x varies in $\Delta_1^{(n)}$, so $X^{(n)}$ varies in $[0,d_n]$, where $d_n\to c$ as $n\to\infty$, and $c\approx1.52$ can be found from the relation $g(x)=1-cx^2+O(x^4)$ as $x\to0$. Next, $g^{(2^n)}(\Delta_1^{(n)})\subset\Delta_1^{(n)}$, and our aim is to describe this map in the renormalized variable $X^{(n)}$.
Given a value of $X^{(n)}$, we find an $X_0^{(n)}$ such that $X_0^{(n)}\alpha^{-n}\in\Delta_0^{(n)}$ and $g(X_0^{(n)}\alpha^{-n})=1-\alpha^{-2n}X^{(n)}$. It follows from the asymptotic expression for g that $X_0^{(n)}\sim\sqrt{X^{(n)}c^{-1}}$ as $n\to\infty$.
Further $g^{(2^n)}(1-\alpha^{-2n}X^{(n)})=g^{(2^n+1)}(\alpha^{-n}X_0^{(n)})=g(g^{(2^n)}(X_0^{(n)}\alpha^{-n}))$.
From the doubling equation it follows that $g^{(2^n)}(X_0^{(n)}\alpha^{-n})=(-\alpha^{-1})^n g(X_0^{(n)})$.
Therefore,

$$g(g^{(2^n)}(X_0^{(n)}\alpha^{-n}))=g((-\alpha^{-1})^n g(X_0^{(n)}))=g(\alpha^{-n}g(X_0^{(n)}))=$$
$$=1-c\alpha^{-2n}(g(X_0^{(n)}))^2+O(\alpha^{-4n}).$$

Thus, if

$$g^{(2^n)}(1-\alpha^{-2n}X^{(n)})=1-\alpha^{-2n}Y^{(n)},$$

then

$$Y^{(n)}=c(g(X_0^{(n)}))^2+O(\alpha^{-2n})=c(g(\sqrt{X^{(n)}c^{-1}}))^2+O(\alpha^{-2n}).$$

2. Let us prove the first assertion of the theorem. We fix a number k and consider the 2^k intervals $\Delta_{p\cdot2^{n-k}+1}^{(n)}$ $(0\leqslant p<2^k)$ in the neighbourhood of 1. In the renormalized variables $X^{(n-k)}$ the end-points of these intervals, hence, also their lengths converge to limits at an exponential rate as $n-k\to\infty$. Let $\Gamma_r^{(n)}=\Delta_1^{(n-r-1)}\setminus\Delta_1^{(n-r)}$. It is clear that $\Gamma_r^{(n)}$ contains 2^r intervals $\Delta_i^{(n)}$.

From a technical point of view it is convenient to consider the part of F inside $\Delta_1^{(1)}$. Let $x \in F \cap \Delta_1^{(1)}$ be fixed. We find intervals $\Delta_{p\cdot 2^{n-k}+1}^{(n)} \subset \Delta_1^{(n-k)}$ for which $\Delta_s^{(n)}(x) = g^{(t)}(\Delta_{p\cdot 2^{n-k}+1}^{(n)})$ $(0 \leqslant t < 2^{n-k})$. We introduce also $\Delta_{s'}^{(n-1)}(x) = g^{(t)}(\Delta_{p'\cdot 2^{n-k-1}}^{(n-1)}) \supset \Delta_s^{(n)}(x)$ and choose an arbitrary point $x \in \Delta_{p\cdot 2^{n-k}+1}^{(n)}$. The map $g^{(t)}$ is a homeomorphism of $\Delta_{p\cdot 2^{n-k}+1}^{(n)}$ onto $\Delta_s^{(n)}(x)$. Therefore,

$$|\Delta_s^{(n)}(x)| = \int\limits_{\Delta_{p\cdot 2^{n-k}+1}^{(n)}} \prod_{i=0}^{t-1} |g'(x_i)|\, dx_0 =$$

$$= \int\limits_{\Delta_{p\cdot 2^{n-k}+1}^{(n)}} \Big(\prod_{r=k}^{n-2}\Big(\prod_{i:\, x_i \in g^{-1}(\Gamma_r^{(n)})} |g'(x_i)|\Big)\Big)\Big(\prod_{i:\, x_i \in \Delta_1^{(1)}} |g'(x_i)|\Big) dx_0 =$$

$$= \prod_{i=0}^{t-1} |g'(\bar{x}_i)| \int\limits_{\Delta_{p\cdot 2^{n-k}+1}^{(n)}} \Big(\prod_{r=k}^{n-2}\Big(\prod_{i:\, x_i \in g^{-1}(\Gamma_r^{(n)})} (|g'(x_i)|/|g'(\bar{x}_i)|)\Big)\Big) \times$$

$$\times \prod_{i:\, x_i \in \Delta_1^{(1)}} (|g'(x_i)|/|g'(\bar{x}_i)|)\, dx_0.$$

Note that

$$\left|\frac{g'(x_i)}{g'(\bar{x}_i)}\right| = \left|1 + \frac{g'(x_i) - g'(\bar{x}_i)}{g'(\bar{x}_i)}\right|$$

and $|g'(x_i) - g'(\bar{x}_i)| \leqslant C_1 |x_i - \bar{x}_i| \leqslant 2C_1\alpha^{-n}$ (by 6° in Theorem 4.1), where $C_1 = \max\limits_{-1 \leqslant x \leqslant 1} |g''(x)|$. Now $|g'(\bar{x}_i)| \geqslant C_2\alpha^{-(n-r)}$ for $\bar{x}_i \in g^{-1}(\Gamma_r^{(n)})$. Moreover, $|g'(\bar{x}_i)| \geqslant C_3$ for $\bar{x}_i \in \Delta_1^{(1)}$. Therefore,

$$1 - 2C_1C_2^{-1}\alpha^{-r} \leqslant \left|\frac{g'(x_i)}{g'(\bar{x}_i)}\right| \leqslant 1 + 2C_1C_2^{-1}\alpha^{-r}$$

if $g(x_i) \in \Gamma_r^{(n)}$, and

$$1 - 2C_1C_3^{-1}\alpha^{-n} \leqslant \left|\frac{g'(x_i)}{g'(\bar{x}_i)}\right| \leqslant 1 + 2C_1C_3^{-1}\alpha^{-n}$$

if $x_i \in \Delta_1^{(1)}$. From this, using the fact that the number of i for which $g(x_i) \in \Gamma_r^{(n)}$ is at most 2^r, we see that

$$(1 - 2C_1C_3^{-1}\alpha^{-n})^{2^n} \prod_{r=k}^{n-2} (1 - 2C_1C_2^{-1}\alpha^{-r})^{2^r} \leqslant \frac{|\Delta_s^{(n)}(x)|}{\Big(\prod_{i=0}^{t-1} |g'(\bar{x}_i)|\Big) |\Delta_{p\cdot 2^{n-k}+1}^{(n)}|} \leqslant$$

$$\leqslant (1 + 2C_1C_3^{-1}\alpha^{-n})^{2^n} \prod_{r=k}^{n-2} (1 + 2C_1C_2^{-1}\alpha^{-r})^{2^r}.$$

From the fact that $2\alpha^{-1} < 1$ we find after a simple transformation that

$$\exp\{-\mathrm{const}\,(2\alpha^{-1})^k\} \leqslant \frac{|\Delta_s^{(n)}(x)|}{\left(\prod_{i=0}^{t-1}|g'(\bar{x}_i)|\right)|\Delta_{p\cdot 2^{n-k}+1}^{(n)}|} \leqslant \exp\{\mathrm{const}\,(2\alpha^{-1})^k\}.$$

Similar inequalities hold for $\Delta_{s'}^{(n-1)}(x)$, and the same points $\bar{x}_i$ can be chosen. Thus,

$$\exp\{-\mathrm{const}\,(2\alpha^{-1})^k\} \leqslant \frac{|\Delta_s^{(n)}(x)|}{|\Delta_{s'}^{(n-1)}(x)|} : \frac{|\Delta_{p\cdot 2^{n-k}+1}^{(n)}|}{|\Delta_{p'\cdot 2^{n-k-1}+1}^{(n-1)}|} \leqslant \exp\{\mathrm{const}\,(2\alpha^{-1})^k\}. \tag{4.2}$$

The next part of the arguments is fairly simple. Given a point $x \in F \cap \Delta_1^{(1)}$ and a sequence of intervals $\Delta_{s_n}^{(n)}(x)$ (we now indicate explicitly the dependence of s on n), we find a sequence of numbers $p_n(x) = p_n$ such that

$$\Delta_{s_n}^{(n)}(x) = g^{(t_n)}(\Delta_{p_n\cdot 2^{n-k}+1}^{(n)}),$$

$0 \leqslant t_n < 2^{n-k}$, $0 \leqslant p_n < 2^k$. The numbers $p_n(x)$ form a sequence of k-dependent random variables, that is, p_{n_1} and p_{n_2} are independent if $|n_1 - n_2| > k$. For suppose that x has the symbolic representation $(1, \varepsilon_1, \ldots, \varepsilon_{n-k}, \ldots, \varepsilon_{n-1}, \ldots)$. Then

$$s_n = 1 + \varepsilon_1\cdot 2 + \ldots + \varepsilon_{n-1}\cdot 2^{n-1}, \quad p_n = \varepsilon_{n-k} + \varepsilon_{n-k+1}\cdot 2 + \ldots + \varepsilon_{n-1}\cdot 2^{k-1}.$$

The assertion on the k-dependence follows from the independence of the random variables ε_i with respect to the invariant measure μ_0^*.

We rewrite $|\Delta_{s_n}^{(n)}(x)|$ in the form

$$|\Delta_{s_n}^{(n)}(x)| = |\Delta_{s_k}^{(k)}(x)| \prod_{m=k+1}^{n} \left(\frac{|\Delta_{s_m}^{(m)}(x)|}{|\Delta_{s_{m-1}}^{(m-1)}(x)|}\right).$$

We write

$$\frac{1}{n}\log|\Delta_{s_n}^{(n)}(x)| \equiv \gamma^{(n)}(x)$$

and

$$\frac{1}{n}\log\prod_{m=k+1}^{n}\left(\left|\frac{\Delta_{p_m\cdot 2^{m-k}+1}^{(m)}}{\Delta_{p_{m-1}\cdot 2^{m-k-1}+1}^{(m-1)}}\right|\right) \equiv \gamma_k^{(n)}(x).$$

Using (4.2) we have

$$|\gamma^{(n)}(x) - \gamma_k^{(n)}(x)| \leqslant \frac{\max_{0\leqslant i<2^k}|\log|\Delta_i^{(k)}||}{n} + \frac{\mathrm{const}\,(2\alpha^{-1})^k(n-k)}{n} \leqslant \mathrm{const}\,(2\alpha^{-1})^k. \tag{4.3}$$

We write $\gamma_k^{(n)}(x)$ in the form

$$\gamma_k^{(n)}(x) = \sum_{i=0}^{2^k-1}\frac{1}{n}\sum_{m:\,p_m=i}\log\left(\frac{|\Delta_{p_m\cdot 2^{m-k}+1}^{(m)}|}{|\Delta_{p_{m-1}\cdot 2^{m-k-1}+1}^{(m-1)}|}\right).$$

It follows from the 1-universality that the numbers $\log\left(\frac{|\Delta^{(m)}_{p_m\cdot 2^{m-k}+1}|}{|\Delta^{(m-1)}_{p_{m-1}\cdot 2^{m-k-1}+1}|}\right)$ converge to limits for fixed k at an exponential rate as $m\to\infty$, and these limits depend on the value $p_m=i$. We denote the corresponding limit values by $U_k(i)$. Moreover, by the strong law of large numbers, for μ_0-almost every $x\in F\cap\Delta^{(1)}_1$ the fraction of occurrances of the value $p_m=i$ $(0\leqslant i<2^k)$ tends to 2^{-k} for each k. Thus, for μ_0-almost every $x\in F\cap\Delta^{(1)}_1$ and all k

$$\gamma^{(n)}_k(x)\xrightarrow[n\to\infty]{}\frac{1}{2^k}\sum_{i=0}^{2^k-1}U_k(i)\equiv\gamma_k. \tag{4.4}$$

It follows from (4.3) and (4.4) that for μ_0-almost every $x\in F\cap\Delta^{(1)}_1$ the limit $\lim\limits_{n\to\infty}\gamma^{(n)}(x)$ exists and, furthermore, that $\lim\limits_{k\to\infty}\gamma_k$ exists and the two limits are equal. For

$$|\gamma^{(n)}(x)-\gamma^{(m)}(x)|\leqslant|\gamma^{(n)}(x)-\gamma^{(n)}_k(x)|+ \\ +|\gamma^{(n)}_k(x)-\gamma^{(m)}_k(x)|+|\gamma^{(m)}_k(x)-\gamma^{(m)}(x)|.$$

The first and third terms can be made as small as desired by choosing k suitably. For a fixed k the second term tends to zero as n, $m\to\infty$.

It can be proved similarly that $\lim\limits_{k\to\infty}\gamma_k$ exists and that the two limits are equal. We note that

$$|\gamma_k-\gamma_l|\leqslant\text{const}\,((2\alpha^{-1})^k+(2\alpha^{-1})^l),$$

thus, we obtain a constructive way of computing the constant γ. From the properties of the doubling equation it follows trivially that, since the limit exists for the μ_0-typical points in $F\cap\Delta^{(1)}_1$, it also exists for the μ_0-typical points in the whole attractor F. An approximate value of γ obtained by a numerical computation is -1.34.

3. We express the results obtained in the preceding part in terms of the symbolic representation given by the map φ. We consider the intervals $\Delta^{(n)}_{p\cdot 2^{n-k}+1}\subset\Delta^{(n-1)}_{p'\cdot 2^{n-k-1}+1}$ $(0\leqslant p<2^k)$. The map φ carries them into the respective words

$$(\underbrace{1,\ 0,\ \ldots,\ 0,}_{n-k}\ \varepsilon^{(k)},\ \varepsilon^{(k-1)},\ \ldots,\ \varepsilon^{(1)})\ \text{and}\ (\underbrace{1,\ 0,\ \ldots,\ 0,}_{n-k}\ \varepsilon^{(k)},\ \varepsilon^{(k-1)},\ \ldots,\ \varepsilon^{(2)})$$

Let $\log\left(\frac{|\Delta^{(n)}_{p\cdot 2^{n-k}+1}|}{|\Delta^{(n-1)}_{p'\cdot 2^{n-k-1}+1}|}\right)\equiv U^{(n)}_k(\varepsilon^{(1)},\ \varepsilon^{(2)},\ \ldots,\ \varepsilon^{(k)})$. For fixed $\varepsilon^{(1)},\ \ldots,\ \varepsilon^{(k)}$ the function $U^{(n)}_k(\varepsilon^{(1)},\ \varepsilon^{(2)},\ \ldots,\ \varepsilon^{(k)})$ has a limit as $n\to\infty$ in view of 1-universality. We denote the limit function by $U_k(\varepsilon^{(1)},\ \ldots,\ \varepsilon^{(k)})$. The same arguments as in the derivation of 1-universality enable us to prove the estimate

$$|U^{(n)}_k(\varepsilon^{(1)},\ \ldots,\ \varepsilon^{(k)})-U_k(\varepsilon^{(1)},\ \ldots,\ \varepsilon^{(k)})|\leqslant\text{const}\cdot\alpha^{-2(n-k)}\alpha^{2k}. \tag{4.5}$$

The functions $U_k(\varepsilon^{(1)},\ \ldots,\ \varepsilon^{(k)})$ converge as $k\to\infty$ to a limit, which we denote by $U(\varepsilon^{(1)},\ \varepsilon^{(2)},\ \ldots,\ \varepsilon^{(n)},\ \ldots)$. The existence of the limit follows easily from the results in the preceding part. There, in essence, we obtained the estimate

$$(4.6)\qquad |U(\varepsilon^{(1)},\ \ldots,\ \varepsilon^{(k)},\ \ldots) - U_k(\varepsilon^{(1)},\ \ldots,\ \varepsilon^{(k)})| \leqslant \text{const}\cdot(2\alpha^{-1})^k.$$

The main computation carried out in 2. shows that if $\Delta_s^{(n)} \subset \Delta_{s'}^{(n-1)} \subset \Delta_1^{(1)}$, $\varphi(\Delta_s^{(n)}) = (1,\ \varepsilon_2,\ \ldots,\ \varepsilon_n)$, and $\varphi(\Delta_{s'}^{(n+1)}) = (1,\ \varepsilon_2,\ \ldots,\ \varepsilon_{n-1})$, then

$$\exp\{-\text{const}\cdot(2\alpha^{-1})^k\} \leqslant$$

$$\leqslant \frac{|\Delta_s^{(n)}|}{|\Delta_{s'}^{(n-1)}|}\exp\{-U_k^{(n)}(\varepsilon_n,\ \ldots,\ \varepsilon_{n-k})\} \leqslant \exp\{\text{const}\cdot(2\alpha^{-1})^k\}.$$

Bearing (4.5) and (4.6) in mind, we obtain

$$\exp\{-\text{const}\cdot((2\alpha^{-1})^k + \alpha^{4k-2n})\} \leqslant$$

$$\leqslant \frac{|\Delta_s^{(n)}|}{|\Delta_{s'}^{(n-1)}|}\exp\{-U(\varepsilon_n,\ \ldots,\ \varepsilon_2,\ 1,\ 0,\ \ldots,\ 0,\ \ldots) \leqslant$$

$$\leqslant \exp\{\text{const}\,((2\alpha^{-1})^k + \alpha^{4k-2n})\},$$

where $1 \leqslant k \leqslant n$. Setting $k = [n/3]$ and multiplying by n we have

$$(4.7)\qquad \text{const} \leqslant \frac{|\Delta_s^{(n)}|}{\exp\left\{\sum_{s=1}^{n} U(\varepsilon_s,\ \varepsilon_{s-1},\ \ldots,\ \varepsilon_2,\ 1,\ 0,\ \ldots,\ 0,\ \ldots)\right\}} \leqslant \text{Const}.$$

Starting from (4.7) it is natural to use the language of statistical mechanics and call U the potential of the Feigenbaum attractor. The potential U is closely connected with the function $\sigma(t)$ studied in Feigenbaum's papers [5] and [6]. We show below how U can be used to determine the constants β_0, λ, and σ. The constant γ is obviously determined by the formula

$$\gamma = \int U(\varepsilon_1,\ \varepsilon_2,\ \ldots,\ \varepsilon_n,\ \ldots)\,d\mu^*(\varepsilon).$$

4. Let us now prove the second assertion in Theorem 4.2. The proof is based on some well known assertions from statistical mechanics. We can regard $U(\varepsilon^{(1)},\ \ldots,\ \varepsilon^{(k)},\ \ldots)$ as the interaction potential of the coordinate $\varepsilon^{(1)}$ with the remaining coordinates and thus apply standard theorems from the theory of Gibbsian limit distributions (see, for example, [39] and [40]). For any β we consider the new measure l_β given by $l_\beta(\Delta_k^{(n)}) = |\Delta_k^{(n)}|^\beta$ on the intervals $\Delta_k^{(n)}$.

By (4.7),

$$\text{const} \leqslant \frac{l_\beta(\Delta_k^{(n)})}{\exp\left\{\beta\sum_{s=1}^{n} U(\varepsilon_s,\ \ldots,\ \varepsilon_2,\ 1,\ 0,\ \ldots,\ 0,\ \ldots)\right\}} \leqslant \text{Const}.$$

The last relation justifies the notation β. From standard theorems in statistical mechanics (see [41] and [42]) we can conclude that the following limit exists:

$$f(\beta)=\lim_{n\to\infty}\frac{1}{n}\log\Big[\sum_{\varepsilon_2,\ldots,\varepsilon_n}\exp\Big\{\beta\sum_{s=1}^{n}U(\varepsilon_s,\varepsilon_{s-1},\ldots,\varepsilon_2,1,0,\ldots,0,\ldots)\Big\}\Big],$$

which is called the free energy. Since $U<0$, $f(\beta)$ is a smooth monotonically decreasing function of β. Moreover, $f(0)=\log 2$ and $\lim\limits_{\beta\to\infty} f(\beta)=-\infty$. Therefore, there is a unique $\beta_0>0$ such that $f(\beta_0)=0$. It can be shown that for this β_0 the sum

$$\sum_{\varepsilon_2,\ldots,\varepsilon_n}\exp\{\beta_0\sum_{s=1}^{n}U(\varepsilon_s,\varepsilon_{s-1},\ldots,\varepsilon_2,1,0,\ldots,0,\ldots)\}$$

is bounded uniformly in n by two positive constants, therefore,

$$\text{const}\leqslant\sum_{k=0}^{2^{n-1}-1} l_{\beta_0}(\Delta_{2k+1}^{(n)})\leqslant\text{Const}.$$

The total sum over all intervals Δ_k^n $(0\leqslant k<2^n)$ also lies between two constants, because

$$\text{const}\leqslant\frac{\sum\limits_{k=0}^{2^{n-1}-1} l_\beta(\Delta_{2k}^{(n)})}{\sum\limits_{k=0}^{2^{n-1}-1} l_\beta(\Delta_{2k+1}^{(n)})}\leqslant\text{Const}.$$

Hence it follows immediately that the Hausdorff dimension of F is at most β_0. A lower estimate is obtained by considering a subset of F consisting of typical points of F with respect to the Gibbsian limit distribution with potential $\beta_0 U$. For it is not hard to show that a subset of points of the attractor that are typical with respect to the Gibbsian limit distribution with potential βU has the Hausdorff dimension $(\beta-f(\beta)/f'(\beta))$. Consequently, for $\beta=\beta_0$ the Hausdorff dimension attains a maximum and coincides with β_0. Thus, the Gibbsian limit distribution with potential $\beta_0 U$ determines the part of F most massive in the sense of the Hausdorff dimension. A numerical computation gives the value $\beta_0\approx 0.54$. In this context we mention the article [43], where a more precise numerical estimate is obtained for the Hausdorff dimension of the Feigenbaum attractor.

5. Let us now prove the last two assertions of the theorem. We transform the sum defining λ:

$$S_n\equiv 1+\sum_{k=1}^{2^n-1}\prod_{i=k}^{2^n-1}(g'(x_i))^2=\prod_{i=1}^{2^n-1}(g'(x_i))^2\Big(1+\sum_{k=1}^{2^n-1}\prod_{i=1}^{k}(g'(x_i))^{-2}\Big)=$$

$$=|\Delta_1^{(n)}|^2\prod_{i=1}^{2^n-1}(g'(x_i))^2\Big[|\Delta_1^{(n)}|^{-2}+\sum_{k=1}^{2^n-1}|\Delta_1^{(n)}|^{-2}\prod_{i=1}^{k}(g'(x_i))^{-2}\Big].$$

In essence, it was proved in 2. that

$$\frac{\text{const}}{|\Delta_{k+1}^{(n)}|^2}\leqslant\frac{1}{|\Delta_1^{(n)}|^2}\prod_{i=1}^{k}(g'(x_i))^{-2}\leqslant\frac{\text{Const}}{|\Delta_{k+1}^{(n)}|^2}.$$

Therefore, const $\tilde{S}_n \leqslant S_n \leqslant$ Const $\tilde{S}_n$, where

$$\tilde{S}_n = |\Delta_{2^n}^{(n)}|^2 \cdot \sum_{k=1}^{2^n-1} |\Delta_k^{(n)}|^{-2}, \quad \Delta_{2^n}^{(n)} = g(\Delta_{2^n-1}^{(n)}).$$

As in the preceding part,

$$\lim_{n\to\infty} \frac{1}{n} \log\left[\sum_{k=1}^{2^n-1} |\Delta_k^{(n)}|^{-2}\right] = f(-2),$$

where $f(-2)$ is the free energy corresponding to the potential U for $\beta = -2$. Since $|\Delta_{2^n}^{(n)}| = O(\alpha^{-n})$, we obtain finally that

$$\lim_{n\to\infty} \frac{1}{n} \log S_n = \lim_{n\to\infty} \frac{1}{n} \log \tilde{S}_n = f(-2) - 2 \log \alpha, \quad \text{that is, } \lambda = e^{1/2 f(-2)}.$$

As we have already remarked, the constant λ, which is roughly 6.6, is related to the problem of small random perturbations of g. These questions will be treated in detail in the next section.

Finally, we consider the constant σ. The existence of the limit follows from (4.7). The same relation implies trivially that $\sigma = \frac{1}{2} e^{f(1)}$. A numerical computation yields the value $\sigma \approx 0.29$.

Theorem 4.2 is now proved.

We can also define a system of intervals $\Delta_k^{(n)}$ and a potential U for maps on the stable separatrix $\Gamma^{(s)}(g)$ corresponding to the fixed point g of the doubling transformation. It is easy to see that the potential coincides with that defined above for g. Thus, both U itself and the constants γ, β_0, λ, and σ found by means of it are universal.

§5. Small random perturbations of the map g

5.1. The problem of small random perturbations of the map g studied in detail in §4 is of great interest. The fact of the matter is that the observation of a sequence of period-doubling bifurcations in real systems inevitably happens in the presence of a certain amount of inherent noise, therefore, it is natural to investigate the influence of such noise on the properties of g. These questions have been studied by Crutchfield, Nauenberg, and Rudnick [44], Shraiman, Wayne, and Martin [45], and others ([46], [47]). Below we present a qualitative description of the relevant results.

Instead of g we now consider the Markov chain given by

$$x_{n+1} = g(x_n) + \xi_{n+1},$$

where the ξ_n are independent identically distributed random variables. We assume that the ξ_n are distributed in the interval $[-\varepsilon, \varepsilon]$, where $\varepsilon > 0$ is a parameter, and that the distribution is given by a density $p_\varepsilon(x)$, with

$$\int_{-\varepsilon}^{\varepsilon} x p_\varepsilon(x)\,dx = 0, \quad \int_{-\varepsilon}^{\varepsilon} x^2 p_\varepsilon(x)\,dx \sim C\varepsilon^2$$

as $\varepsilon \to 0$, where C is a constant. For example, it can be assumed that $p_\varepsilon(x) = 1/2\varepsilon$, corresponding to the uniform distribution on $[-\varepsilon, \varepsilon]$. This Markov chain has a stationary distribution with density $q(x;\ \varepsilon)$, which converges weakly to the measure μ_0 as $\varepsilon \to 0$ (see [48]). Our problem is to study the behaviour of $q(x;\ \varepsilon)$ for small but non-zero ε.

If the random component were not added, then for almost each $x_0 \in [-1, 1]$ the images $x_n = g^{(n)}(x_0)$ would tend to F, that is, $\operatorname{dist}(x_n, F) \to 0$ as $n \to \infty$. The random components spread the images of a point over the intervals $\Delta_k^{(n)}$. We now study this process in more detail.

Let $x_0 \in \Delta_0^{(n)}$. Then

$$x_{2^n} = \xi_{2^n} + g(\xi_{2^n-1} + g(\xi_{2^n-2} + \ldots + g(\xi_1 + g(x_0)\ \ldots)).$$

Since all the ξ_i are small, we can write a linear expansion with respect to ξ_i for the point x_{2^n}, that is,

$$x_{2^n} = \bar{x}_{2^n} + \xi_{2^n} + \sum_{i=1}^{2^n-1} \xi_i \prod_{s=i}^{2^n-1} g'(\bar{x}_s) + Q_{2^n}(\xi;\ x_0), \tag{5.1}$$

where $\bar{x}_s = g^{(s)}(x_0)$, and $Q_{2^n}(\xi;\ x_0)$ is the error. The sum

$$\xi_{2^n} + \sum_{i=1}^{2^n-1} \xi_i \prod_{s=i}^{2^n-1} g'(\bar{x}_s) \equiv \zeta_{2^n}(x_0)$$

is a random variable whose expectation is zero and whose variance is

$$\operatorname{var} \zeta_{2^n}(x_0) = \operatorname{var}\xi \left(1 + \sum_{i=1}^{2^n-1} \prod_{s=i}^{2^n-1} (g'(\bar{x}_s))^2\right). \tag{5.2}$$

The last sum was studied in §4, where it was shown that

$$\text{const}\ \alpha^{-2n}\lambda^{2n} \leqslant 1 + \sum_{i=1}^{2^n-1} \prod_{s=i}^{2^n-1} (g'(\bar{x}_s)^2) \leqslant \text{Const}\ \alpha^{-2n}\lambda^{2n}, \quad \lambda \approx 6.6.$$

Another approach to the study of the sum in (5.2) was presented in [44]. Let $x_0 = (-\alpha^{-1})^n z_0$, where $z_0 \in [-1, 1]$. We write

$$D_n(z_0) = 1 + \sum_{i=1}^{2^n-1} \prod_{s=i}^{2^n-1} (g'(\bar{x}_s))^2.$$

By using the doubling equation (4.1) it is then easy to obtain the following recursion relation:

$$D_{n+1}(z_0) = (g'(g(\alpha^{-1}z_0)))^2 D_n(\alpha^{-1}z_0) + D_n(g(\alpha^{-1}z_0)). \tag{5.3}$$

We denote by $\mathcal{L}$ the linear operator corresponding to (5.3), that is,

$$\mathcal{L}f(z) = (g'(g(\alpha^{-1}z)))^2 f(\alpha^{-1}z) + f(g(\alpha^{-1}z)).$$

Then $D_n(z_0) = \mathcal{L}^n\mathbf{1}$, where $\mathbf{1}$ is the function everywhere equal to 1. Since $\mathcal{L}$ is a positive operator, it has a non-degenerate positive maximal eigenvalue $\lambda(\mathcal{L})$ and a positive eigenfunction $l(z)$ corresponding to $\lambda(\mathcal{L})$. Therefore, as $n \to \infty$

$$D_n(z_0) \sim \lambda(\mathcal{L})^n \cdot l(z_0),$$

where $\lambda(\mathscr{L}) = \alpha^{-2}\lambda^2$. Of course, both approaches lead to the same result, but we obtain it directly in the framework of the thermodynamic formalism developed in §4.

We now find an $n_1 = n_1(\varepsilon)$ such that $\varepsilon \cdot \lambda^{n_1(\varepsilon)} \sim$ const as $\varepsilon \to 0$. This condition means that for an initial point $x_0 \in \Delta_0^{(n_1)}$ the point $\bar{x}_{2^{n_1}}$ is in $\Delta_0^{(n_1)}$, while the linear part of the random component spreads the image of the point over an interval whose length is of the same order as $|\Delta_0^{(n_1)}|$. In other words, in the course of 2^{n_1} steps there is a "loss of memory" about the initial condition on a scale of the order of the length of $\Delta_0^{(n_1)}$. The definition of the quantity $n_1(\varepsilon)$ is invariant with respect to the position of the initial point. For suppose that $x_0 \in \Delta_k^{(n_1)}$ $(0 \leqslant k < 2^{n_1})$. Then, as before, $x_{2^{n_1}} = \bar{x}_{2^{n_1}} + \zeta_{2^{n_1}}(x_0) + Q_{2^{n_1}}(\xi;\ x_0)$. Using the same arguments as in the proof of Theorem 4.2, III it is easy to show that

$$\operatorname{var}(\zeta_{2^{n_1}}(x_0)) = O(|\Delta_k^{(n_1)}|^2 \cdot \varepsilon^2 \cdot \lambda^{2n_1}) = O(|\Delta_k^{(n_1)}|^2),$$

that is, the variance of the random variable $\zeta_{2^{n_1}}(x_0)$ has the order of the square of the length of $\Delta_k^{(n_1)}$.

5.2. We now show that the stationary distribution of the original Markov chain is concentrated on $\bigcup_{k=0}^{2^{n_1}-1} \Delta_k^{(n_1)}$ in a certain sense. To do this, let us first estimate the error $Q_{2^n}(\xi;\ x_0)$.

Suppose, for simplicity, that $x_0 \in \Delta_0^{(n)}$, and $x_i = \bar{x}_i + L_i + Q_i$, where $\bar{x}_i = g^{(i)}(x_0)$, L_i is the part of the random perturbation linear in ξ, and Q_i is the error. We assume that $|L_i| \leqslant \tau\, |\Delta_i^{(n)}|$ for all $0 \leqslant i < 2^n$. We claim that if τ is sufficiently small, then $|Q_i| \leqslant C_1\tau^2\, |\Delta_i^{(n)}|$, where C_1 is an absolute constant that depends neither on the initial point nor on the realization of the random variables ξ. The recursion relations for L_i and Q_i are

$$\left\{ \begin{aligned} L_{i+1} &= L_i g'(\bar{x}_i) + \xi_{i+1}, \\ Q_{i+1} &= L_i (g'(\tilde{x}_i) - g'(\bar{x}_i)) + Q_i g'(\tilde{x}_i), \end{aligned} \right. \tag{5.4}$$

where $|\tilde{x}_i - \bar{x}_i| \leqslant |L_i + Q_i|$. It follows from (5.4) that

$$Q_j = \sum_{i=1}^{j-1} \left[L_i \cdot \frac{(g'(\tilde{x}_i) - g'(\bar{x}_i))}{g'(\tilde{x}_i)} \prod_{s=i}^{j-1} g'(\tilde{x}_s) \right].$$

We define points $x_i' \in \Delta_i^{(n)}$, $0 \leqslant i < 2^n$, such that $|g'(x_i')| \cdot |\Delta_i^{(n)}| = |\Delta_{i+1}^{(n)}|$ and assume that $|Q_i| \leqslant \chi\, |\Delta_i^{(n)}|$ $(0 \leqslant i < j)$.

Then $|g'(\tilde{x}_i) - g'(\bar{x}_i)| \leqslant M(\tau + \chi)\, |\Delta_i^{(n)}|$, where $M = \max_{x \in [-1,1]} g''(x)$, and we obtain

$$|Q_j| \leqslant M\tau(\tau+\chi) \sum_{i=1}^{j-1} \left[\frac{|\Delta_i^{(n)}|^2}{|g'(\tilde{x}_i)|} \left(\prod_{s=i}^{j-1} |g'(x_s')| \right) \left(\prod_{s=i}^{j-1} \frac{|g'(\tilde{x}_s)|}{|g'(x_s')|} \right) \right] =$$

$$= M\tau(\tau+\chi)\, |\Delta_j^{(n)}| \sum_{i=1}^{j-1} \left[\frac{|\Delta_i^{(n)}|}{|g'(\tilde{x}_i)|} \prod_{s=i}^{j-1} \frac{|g'(\tilde{x}_s)|}{|g'(x_s')|} \right].$$

We now estimate $\prod_{s=i}^{j-1} \frac{|g'(\tilde{x}_s)|}{|g'(x'_s)|}$:

$$\prod_{s=i}^{j-1} \frac{|g'(\tilde{x}_s)|}{|g'(x'_s)|} = \prod_{s=i}^{j-1} \left|1+\frac{g'(\tilde{x}_s)-g'(x'_s)}{g'(x'_s)}\right| \leqslant \prod_{s=i}^{j-1} \left|1+\frac{M\,|\Delta_s^{(n)}|\,(1+\tau+\chi)}{g'(x'_s)}\right| \leqslant$$

$$\leqslant \exp\left(\sum_{s=i}^{j-1} M(1+\tau+\chi)\frac{|\Delta_s^{(n)}|}{|g'(x'_s)|}\right) \leqslant \exp\left(MR_1(1+\tau+\chi)\right),$$

where $R_1 = \max_n \left(\sum_{s=0}^{2^n-1} \frac{|\Delta_s^{(n)}|}{|g'(x'_s)|} \right)$. Since $\sum_{i=1}^{j-1} \left(\frac{|\Delta_i^{(n)}|}{|g'(\tilde{x}_i)|} \right)$ $(2 \leqslant j \leqslant 2^n)$ does not exceed an absolute constant R_2, we obtain finally that

$$|Q_j| \leqslant MR_2\tau(\tau+\chi)\exp\left(MR_1(1+\tau+\chi)\right)|\Delta_j^{(n)}|.$$

Let $\chi = C_1\tau^2$, where $C_1 = 2MR_2 \exp(MR_1)$. Then for sufficiently small τ

$$MR_2\tau(\tau+\chi)\exp\left(MR_1(1+\tau+\chi)\right) \leqslant \chi$$

consequently,

$$|Q_j| \leqslant \chi\,|\Delta_j^{(n)}| = C_1\tau^2\,|\Delta_j^{(n)}|, \quad 0 \leqslant j \leqslant 2^n. \tag{5.5}$$

Thus, Q_j can be regarded as a small error as long as the quantities L_i $(1 \leqslant i \leqslant j)$ are small in comparison with $|\Delta_i^{(n)}|(1 \leqslant i \leqslant j)$. A similar assertion holds also when $x_0 \in \Delta_k^{(n)}$ $(0 \leqslant k < 2^n)$. Suppose now that $n = n_1 - m$. We write

$$\tilde{\Delta}_i^{(n_1-m)} = \{x:\ \mathrm{dist}(x, \Delta_i^{(n_1-m)}) \leqslant 4\alpha^{-2m}\,|\Delta_i^{(n_1-m)}|\}, \qquad 0 \leqslant i < 2^{n_1-m},$$

$$F(m) = \bigcup_{i=0}^{2^{n_1-m}-1} \tilde{\Delta}_i^{(n_1-m)}.$$

Theorem 5.1. *Let μ_ε be the stationary distribution of the Markov chain*

$$x_{i+1} = g(x_i) + \xi_{i+1},$$

and let $n_1 = [-\log\varepsilon/\log\lambda]$. Then for sufficiently large m

$$\mu_\varepsilon(F(m)) \geqslant 1 - \exp\left(-\mathrm{const}\cdot(\lambda\alpha^{-2})^m\right).$$

Remark. For the Feigenbaum attractor $\lambda\alpha^{-2} > 1$.

Proof of Theorem 5.1. We first estimate the probability that $\max_{0 \leqslant i \leqslant 2^{n_1-m}} \frac{|L_i|}{|\Delta_i^{(n_1-m)}|} > \alpha^{-2m}$. Using an inequality of Kolmogorov type,

$$\Pr\left(\max_{0 \leqslant i \leqslant 2^{n_1-m}} \frac{|L_i|}{|\Delta_i^{(n_1-m)}|} > \alpha^{-2m} \right) \leqslant \tag{5.6}$$

$$\leqslant 2\Pr\left(\frac{|L_{2^{n_1-m}}|}{|\Delta_{2^{(n_1-m)}}^{(n_1-m)}|} > \mathrm{const}\cdot\alpha^{-2m} \right).$$

Since the variance $\mathbf{var}\left(\frac{L_{2^{n_1-m}}}{|\Delta_{2^{n_1-m}}^{(n_1-m)}|} \right)$ is of order λ^{-2m} and $\alpha^{-2} > \lambda^{-1}$, we can use Bernstein's exponential estimates for the probability of large deviations (see [49]). It is easy to check that for sufficiently large m

$$\text{(5.7)} \qquad \Pr\left(\frac{|L_{2^{n_1-m}}|}{|\Delta^{(n_1-m)}_{2^{n_1-m}}|} > \text{const}\cdot\alpha^{-2m}\right) < \exp(-\text{const}\cdot(\lambda\alpha^{-2})^m).$$

We consider the transition probabilities p_N of the Markov chain for $N = B^m\cdot 2^{n_1-m}$ steps where B is a large constant. We show that for sufficiently large m

$$\text{(5.8)} \qquad \min_{x_0\in F(m+1)} p_N(x_0, F(m)) \geqslant 1-\exp(-\text{const}\,(\lambda\alpha^{-2})^m).$$

Let $x_0 \in \tilde{\Delta}^{(n_1-m-1)}_i \subset F(m+1)$. We write $y_s = x_{s\cdot 2^{n_1-m}}$ $(0 \leqslant s \leqslant B^m)$. It follows from (5.5)-(5.7) that with probability greater than $(1 - B^m \exp(-\text{const}\cdot(\lambda\alpha^{-2})^m))$

$$\text{(5.9)} \quad |y_s - g^{(2^{n_1-m})}(y_{s-1})| \leqslant \alpha^{-2m}\min(|\Delta^{(n_1-m)}_i|, |\Delta^{(n_1-m)}_{i+2^{n_1-m}-1}|), \qquad 1\leqslant s\leqslant B^m.$$

We assume that $y_{s'} \in \Delta^{(n_1-m)}_i(\Delta^{(n_1-m)}_{i+2^{n_1}-m-1})$ for some s'. Then it follows easily from (5.9) that $y_s \in \tilde{\Delta}^{(n_1-m)}_i(\tilde{\Delta}^{(n_1-m)}_{i+2^{n_1}-m-1})$ for all $s' \leqslant s < B^m$. Thus, it remains to show that with probability greater than $(1 - \exp(-\text{const}\cdot(\lambda\alpha^{-2})^m))$ there is an $s'(y_0) \leqslant B^m$ such that $y_{s'} \in \Delta^{(n_1-m)}_i \cup \Delta^{(n_1-m)}_{i+2^{n_1}-m-1}$. It suffices to consider the case $y_0 \in \Delta^{(n_1-m-1)}_i - \Delta^{(n_1-m)}_i - \Delta^{(n_1-m)}_{i+2^{n_1}-m-1} \equiv \bar{\Delta}^{(n_1-m)}_i$ (see Fig. 5.1).

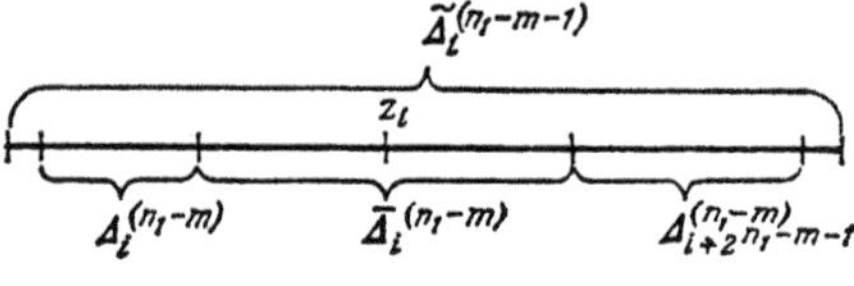

Fig. 5.1

There is an unstable fixed point z_i of the transformation $g^{(2^{n_1-m})}$ inside the interval $\bar{\Delta}^{(n_1-m)}_i$, therefore, for y_0 outside a neighbourhood of z_i that is exponentially small with respect to m, the exit time from $\bar{\Delta}^{(n_1-m)}_i$ is const.m, that is, $y_{\text{const}\cdot m} \in \Delta^{(n_1-m)}_i \cup \Delta^{(n_1-m)}_{i+2^{n_1}-m-1}$. On the other hand, the points y_s $(1 \leqslant s \leqslant B^m)$ leave any exponentially small neighbourhood of z_i because of fluctuations of the value of the random perturbation. More precisely, for sufficiently large B the probability of leaving an exponentially small (with respect to m) neighbourhood of z_i during the time B^m is greater than $(1 - \exp(-\text{const}\cdot(\lambda\alpha^{-2})^m))$, thus, (5.8) is proved. Using (5.8) we see that

$$\mu_\varepsilon(F(m)) = \int p_N(x, F(m))\,d\mu_\varepsilon(x) \geqslant (1-\exp(-\text{const}\cdot(\lambda\alpha^{-2})^m))\,\mu_\varepsilon(F(m+1)).$$

Since $\mu_\varepsilon(F(n_1)) = 1$, finally

$$\mu_\varepsilon(F(m)) \geqslant \prod_{i=m}^{\infty}(1-\exp(-\text{const}\cdot(\lambda\alpha^{-2})^i)) \geqslant 1-\exp(-\text{const}\cdot(\lambda\alpha^{-2})^m).$$

§6. Feigenbaum universality in the multi-dimensional case and some other generalizations

6.1. So far we have mainly studied families of one-dimensional maps. At the same time, results of numerical computations indicate that the constant $\delta = 4.6692$ characterizes the asymptotic properties of sequences of period-doubling bifurcations for various multi-dimensional dynamical systems not connected with one-dimensional maps. For example, sequences of period-doubling bifurcations subject to the universal asymptotic law $\mu_\infty - \mu_n \sim \sim \mathrm{const}\cdot\delta^{-n}$ were discovered in numerical investigations of the Lorenz model [50] (see §1), the five-mode approximation for the Navier-Stokes equations [51], the Hénon map [52], and for a number of other systems [53]. This fact was explained qualitatively in the article [9] of Eckmann, Collet, and Koch. Roughly speaking, their results consists in the following. Suppose that a family of multi-dimensional maps is structured so that in one direction it acts like the families of one-dimensional maps considered above, while in the other directions it realizes a strong contraction. As the parameter varies there is then an infinite sequence of period-doubling bifurcations subject to the same asymptotic law as in the one-dimensional case.

Let us pass to more precise formulations. We consider the space $\mathbf{C}^n$, presented as the direct sum $\mathbf{C} \oplus \mathbf{C}^{n-1}$. Vectors $z \in \mathbf{C}^n$ are written in the form $(z_0, \mathbf{z})$, where $z_0 \in \mathbf{C}^1$ and $\mathbf{z} \in \mathbf{C}^{n-1}$. Let D be an open subset of $\mathbf{C}^n$. We denote by $\mathscr{K}(D)$ the Banach space of bounded analytic maps from D to $\mathbf{C}^n$ with the norm

$$\|h\| = \sup\{\|h(z)\|,\ z \in D\}.$$

Let $D(\Delta)$, $\Delta > 0$, be the set $D(\Delta) = \{z \in \mathbf{C}^n : \|z - (y_0, 0)\| < \Delta$ for some $y_0 \in [-1, 1]\}$. We write the Feigenbaum map $g(x)$ in the form $g(x) = f(x^2)$ and put

$$\Phi(z) = (f(\zeta(z)),\ 0),$$

where $\zeta(z) = z_0^2 - \gamma\cdot\mathbf{z}$ and γ is an arbitrary vector in $\mathbf{C}^{n-1}$ with norm less than 2. The map $z \mapsto \Phi(z)$ is analytic in $D(\Delta)$ for sufficiently small Δ, that is, Φ belongs to the space $\mathscr{K}_\Delta \equiv \mathscr{K}(D(\Delta))$.

We denote by A the linear transformation

$$Az = (-\alpha^{-1}z_0,\ \alpha^{-2}\mathbf{z}),$$

where $\alpha = 2.50290...$, and define the doubling transformation by

$$T\colon G \mapsto A^{-1}\circ G\circ G\circ A,$$

where $G : \mathbf{C}^n \to \mathbf{C}^n$. It is easy to verify that T for sufficiently small Δ maps the ball in $\mathscr{K}_\Delta$ of radius $b\cdot\Delta$ with centre at Φ into $\mathscr{K}_\Delta$. Here $b > 0$ is an absolute constant. Moreover, it follows directly from the definition that Φ is a fixed point of T. To use the universality theory we must now study the spectrum of the differential of T at Φ. It is not hard to verify that $DT(\Phi)$

for sufficiently small $\Delta > 0$ is a compact operator on $\mathscr{K}_\Delta$. The main result in [9] is that the part of the spectrum of $DT(\Phi)$ outside the unit disk can be described explicitly. As in the one-dimensional case, there is a single eigendirection with the eigenvalue $\delta = 4.6692...$, while the remaining part of the spectrum outside the unit disk is inessential, because it is connected with changes of the variables in $\mathbf{C}^n$ (see §3). Thus, the multi-dimensional case reduces in a certain sense to one-dimensional maps.

6.2. In conclusion we dwell on some generalizations connected with the questions considered above. In [54] and [55] period-doubling bifurcations for families of area-preserving maps are studied, that is, for families of maps corresponding to Hamiltonian dynamical systems. In this case the bifurcation amounts to the following: as the parameter varies, an elliptic periodic trajectory becomes hyperbolic, but an elliptic periodic trajectory with twice the period appears in place of it. The bifurcation parameter values here are also described by a universal asymptotic law $\mu_\infty - \mu_n \sim \text{const}\cdot\tilde{\delta}^{-n}$, where $\tilde{\delta} \approx 8.72$. Conceptually, the situation here is completely analogous to that considered above. We have to study the corresponding doubling transformation, find a fixed point, and investigate the spectrum, which determines the constant $\tilde{\delta}$.

Another generalization that, in essence, uses the multi-dimensionality of the problem is the analysis of more complicated bifurcation sequences: period-tripling, period-quadrupling, and so on. In the multi-dimensional case such sequences of bifurcations are topologically stable for families of maps depending on two parameters ([56], [57], [58]). We consider in more detail the case of period-tripling bifurcations. Let $f(z;\ \mu)$ be a family of maps of $\mathbf{C}^1$ into $\mathbf{C}^1$ depending on a complex parameter μ. In the μ-plane there is the domain U_0 of the parameter values for which the map has a stable fixed point. The boundary of this domain consists of the parameter values for which the derivative at the fixed point lies on the unit circle. Touching U_0 are two domains $U_1^{(1)}$ and $U_2^{(1)}$ of smaller size corresponding to values of μ for which there are stable periodic trajectories of period 3. The points of contact are determined by the condition that for these parameter values the derivative at the fixed point is $-1/2 \pm (\sqrt{3}/2)i$. At this moment a bifurcation takes place in which a stable periodic trajectory of period 3 is created. Adjoining each of the domains $U_1^{(1)}$ and $U_2^{(1)}$ are two domains corresponding to stable trajectories of period 9, and so on. We consider the following sequence of bifurcation parameter values: for $\mu = \mu_n$ a stable periodic trajectory of period 3^n is created, and the derivative of the map $f^{(3^{n-1})}(z;\ \mu)$ at the fixed point is equal to a particular cube root of 1, say $-1/2+(\sqrt{3}/2)i$. Then $\mu_n \to \mu_\infty$ and $\mu_\infty - \mu_n \sim \text{const}\cdot(\delta^{(1)}(3))^n$, where $\delta^{(1)}(3) \approx 4.600+8.981i$ is a universal constant not depending on the family $f(z;\ \mu)$. The constant $\delta^{(2)}(3) = \overline{\delta^{(1)}(3)}$ corresponds to the sequence $\tilde{\mu}_n$ constructed for the other cube root of 1.

This paper does not cover the whole spectrum of problems on Feigenbaum universality. In particular, the interesting article [59], where properties of universality type are studied for other transfinite numbers in the Sharkovskii order, was left outside the boundaries of our survey. Moreover, there is a whole cycle of papers dealing with universality properties in the so-called intermittency phenomenon (see [60]-[63]), as well as a new development in the theory of smooth maps of the circle onto itself for which the inverse maps have singularities. The last problem seems to be connected with the collapse of invariant tori in the KAM theory (see [64]-[67]).

We thank A.I. Gol'berg, M. Misiurewicz, Ya.B. Pesin, M.I. Rabinovich, E.A. Sataev, M.J. Feigenbaum, P. Cvitanovic, A.N. Sharkovskii, and M.V. Jakobson for useful discussions of the questions treated in this article.

References

[1] M. Metropolis, M.L. Stein, and P.R. Stein, On finite limit sets for transformations of the unit interval, J. Combin. Theory Ser. A **15** (1973), 25-44. MR **47** # 5183.

[2] P.J. Myrberg, Iteration von Quadratwurzeloperationen, Ann. Acad. Sci. Fenn. Ser. AI **1958**, no. 259, 1-16. MR **21** # 7377.

[3] M.J. Feigenbaum, Quantitative universality for a class of non-linear transformations, J. Statist. Phys. **19** (1978), 25-52. MR **58** # 18601.

[4] ———, The universal metric properties of non-linear transformations, J. Statist. Phys. **21** (1979), 669-706. MR **82e**:58072.

[5] ———, The transition to aperiodic behaviour in turbulent systems, Comm. Math. Phys. **77** (1980), 65-86. MR **83e**:58053.

[6] ———, Universal behaviour in non-linear systems, Los Alamos Sci. **1** (1980), 4-27. MR **82h**:58031.
= Uspekhi Fiz. Nauk **141** (1983), 343-374.

[7] O.E. Lanford III, Smooth transformations of intervals, Lecture Notes in Math. **901** (1981), 36-54. MR **83k**:58066.

[8] P. Collet and J.-P. Eckmann, Iterated maps on the interval as dynamical systems, Birkhäuser, Basel-Boston-Stuttgart 1980. MR **82j**:58078.

[9] ———, ———, and H. Koch, Period-doubling bifurcations for families of maps on $\mathbf{R}^n$, J. Statist. Phys. **25** (1980), 1-14. MR **82i**:58052.

[10] A.N. Sharkovskii, Coexistence of cycles of a continuous map of the line into itself, Ukrain. Mat. Zh. **16** (1964), 61-71. MR **28** # 3121.

[11] ———, On cycles and the structure of a continuous mapping, Ukrain. Mat. Zh. **17** (1965), 104-111. MR **32** # 4213.

[12] D. Singer, Stable orbits and bifurcations of maps of the interval, SIAM J. Appl. Math. **35** (1978), 260-267. MR **58** # 13206.

[13] J. Guckenheimer, Sensitive dependence to initial conditions for one-dimensional maps, Comm. Math. Phys. **70** (1979), 133-160. MR **82c**:58037.

[14] L. Block, J. Guckenheimer, M. Misiurewicz, and L.S. Young, Periodic points and topological entropy of one-dimensional maps, Lecture Notes in Math. **819** (1980), 18-34. MR **82j**:58097.

[15] T.Y. Li and J.A. Yorke, Period three implies chaos, Amer. Math. Monthly **82** (1975), 985-992. MR **52** # 5898.

[16] Yu.S. Barkovskii and G.M. Levin, On a Cantor limit set, Uspekhi Mat. Nauk **35**:2 (1980), 201-202. MR **82c**:58032.
= Russian Math. Surveys **35**:2 (1980), 235-236.

[17] M. Misiurewicz, Structure of mappings of an interval with zero entropy, Inst. Hautes Études Sci. Publ. Math. **1981**, no. 53, 5-16. MR **83j**:58071.

[18] I.P. Kornfel'd, Ya.G. Sinai, and S.V. Fomin, *Ergodicheskaya teoriya,* Nauka, Moscow 1980. MR **83a**:28017.
Translation: Ergodic theory, Springer-Verlag, Berlin-Heidelberg-New York 1981. Zbl. **493** # 28007.

[19] L.A. Bunimovich and Ya.G. Sinai, The rate of decrease of correlations in one-dimensional ecological models, in: *Termodinamika i kinetika biologicheskikh protsessov* (The thermodynamics and kinetics of biological processes), Nauka, Moscow 1980.

[20] E. Hofbauer and G. Keller, Ergodic properties of invariant measures for piecewise monotonic transformations, Math. Z. **180** (1982), 119-140. MR **83h**:28028.

[21] M. Blank, An estimate of the rate of decrease of correlations in one-dimensional dynamical systems, Funktsional. Anal. i Prilozhen. **18**:1 (1984), 61-62.
= Functional Anal. Appl. **18**:1 (1984) (to appear).

[22] S.M. Ulam and J. von Neumann, On combinations of stochastic and deterministic processes, Bull. Amer. Math. Soc. **53** (1947), 1120.

[23] D. Ruelle, Applications conservant une mesure absolument continue par rapport à dx sur $[0, 1]$, Comm. Math. Phys. **55** (1977), 47-51. MR **57** # 7691.

[24] L.A. Bunimovich, On a certain transformation of the circle, Mat. Zametki **8** (1970), 205-206. MR **55** # 8309.
= Math. Notes **8** (1970), 587-592.

[25] A.I. Ognev, Metric properties of a certain class of maps of a closed interval, Mat. Zametki **30** (1981), 723-736. MR **83h**:58061.
= Math. Notes **30** (1981), 859-866.

[26] M. Misiurewicz, Absolutely continuous measures for certain maps of an interval, Inst. Hautes Études Sci. Publ. Math. **1981**, no. 53, 17-51. MR **83j**:58072.

[27] M.V. Jakobson, Absolutely continuous invariant measures for one-parameter families of one-dimensional maps, Comm. Math. Phys. **81** (1981), 39-88. MR **83j**:58070.

[28] R. Shaw, Strange attractors, chaotic behaviour, and information flow, Z. Naturforsch, A **36** (1981), 80-112. MR **82k**:58069.

[29] O.E. Lanford III, A computer-assisted proof of the Feigenbaum conjectures, Bull. Amer. Math. Soc. **6** (1982), 427-434. MR **83g**:58051.

[30] M. Campanino and H. Epstein, On the existence of Feigenbaum's fixed-point, Comm. Math. Phys. **79** (1981), 261-302. MR **82j**:58099.

[31] H. Epstein and J. Lascoux, Analyticity properties of the Feigenbaum function, Comm. Math. Phys. **81** (1981), 437-453. MR **83b**:58056.

[32] P. Collet, J.-P. Eckmann, and O.E. Lanford III, Universal properties of maps of an interval, Comm. Math. Phys. **76** (1980), 211-254. MR **83d**:58036.

[33] E.B. Vul and K.M. Khanin, On the unstable separatrix of Feigenbaum's fixed point, Uspekhi Mat. Nauk **37**:5 (1982), 173-174.
= Russian Math. Surveys **37**:5 (1982), 200-201.

[34] H. Daido, Theory of the period-doubling phenomenon of one-dimensional mappings based on the parameter dependence, Phys. Lett. A **83** (1981), 246-250. MR **82j**:58088.

[35] ———, Period-doubling bifurcations and associated universal properties including parameter dependence, Progr. Theoret. Phys. **67** (1982), 1698-1723. MR **84c**:58058.

[36] P. Cvitanovic, Universality in chaos (or, Feigenbaum for cyclists), Preprint NORDITA, Copenhagen, Denmark 1983.

[37] P. Coullet and C. Tresser, Itérations d'endomorphismes et groupe de renormalisation, J. Phys. Coll. **39**:C5 (1978), 25-28; supplément au **39**:8.
= C.R. Acad. Sci. Paris Ser. A-B **287** (1978), A577-A580. MR **80b**:58043.

[38] M.J. Feigenbaum, The onset spectrum of turbulence, Phys. Lett. A **74** (1979), 375-378. MR **82i**:76044.

[39] R.L. Dobrushin, Random Gibbsian fields for lattice systems with pairwise interaction, Funktsional. Anal. i Prilozhen. **2**:4 (1968), 31-43. MR **40** # 3862.
= Functional Anal. Appl. **2** (1968), 292-301.

[40] Ya.G. Sinai, *Teoriya fazovykh perekhodov. Strogie rezul'taty* (Theory of phase transitions. Rigorous results), Nauka, Moscow 1980. MR **82m**:82020.

[41] D. Ruelle, Statistical mechanics. Rigorous results, Benjamin, New York-Amsterdam 1969. MR **44** # 6279.
Translation: *Statisticheskaya mekhanika. Strogie rezul'taty,* Mir, Moscow 1971.

[42] ———, Thermodynamic formalism. The mathematical structures of classical equilibrium statistical mechanics, Addison-Wesley, Reading 1978. MR **80g**:82017.

[43] P. Grassberger, On the Hausdorff dimension of fractal attractors, J. Statist. Phys. **26** (1981), 173-179. MR **83i**:58063.

[44] J.P. Crutchfield, M. Nauenberg, and J. Rudnick, Scaling for external noise at the onset of chaos, Phys. Rev. Lett. **46** (1981), 933-935. MR **82f**:58059.

[45] B. Shraiman, E.C. Wayne, and P.C. Martin, Scaling theory for noisy period-doubling transitions to chaos, Phys. Rev. Lett. **46** (1981), 935-939. MR **82g**:70051.

[46] J.P. Crutchfield, J.D. Farmer, and B.A. Huberman, Fluctuations and simple chaotic dynamics, Phys. Rep. **92**:2 (1982), 45-82. PA **86** # 40497.

[47] T. Kai, Lyapunov number for a noisy 2^n cycle, J. Statist. Phys. **29** (1982), 329-343. PA **86** # 25812.

[48] R.Z. Khas'minskii, *Ustoichivost' sistem differentsial'nykh uravnenii pri sluchainykh vozmushcheniyakh ikh parametrov,* Nauka, Moscow 1969. MR **41** # 3925.
Translation: Stochastic stability of differential equations, Sijthoff and Noordhoff, Alphen aan den Rijn 1980. MR **82b**:60064.

[49] V.V. Petrov, *Summy nezavisimykh sluchainykh velichin,* Nauka, Moscow 1972. MR **48** # 1288.
Translation: Sums of independent random variables, Springer-Verlag, Berlin-New York 1975. MR **52** # 9335.

[50] V. Franceschini, A Feigenbaum sequence of bifurcations in the Lorenz model, J. Statist. Phys. **22** (1980), 397-406. MR **81g**:58023.

[51] ——— and C. Tebaldi, Sequences of infinite bifurcations and turbulence in a five-mode truncation of the Navier-Stokes equations, J. Statist. Phys. **21** (1979), 707-726. MR **81e**:35018.

[52] B. Derrida, A. Gervois, and Y. Pomeau, Universal metric properties of bifurcations of endomorphisms, J. Phys. A **12** (1979), 269-296. MR **80k**:58078.

[53] F.T. Arecchi and F. Lisi, Hopping mechanism generating $1/f$ noise in non-linear systems, Phys. Rev. Lett. **49** (1982), 94-98. PA **85** # 89627.

[54] P. Collet, J.-P. Eckmann, and H. Koch, On universality for area-preserving maps of the plane, Phys. D **3** (1981), 457-467. MR **83b**:58055.

[55] J.M. Greene, R.S. Mackay, F. Vivaldi, and M.J. Feigenbaum, Universal behaviour in families of area-preserving maps, Phys. D **3** (1981), 468-486. MR **82m**:58041.

[56] A.I. Gol'berg, Ya.G. Sinai, and K.M. Khanin, Universal properties for sequences of period-doubling bifurcations, Uspekhi Mat. Nauk **38**:1 (1983), 159-160.
= Russian Math. Surveys **38**:1 (1983), 187-188.

[57] P. Cvitanovic and J. Myrheim, Universality for period n-tuplings in complex mappings, Phys. Lett. A **94** (1983), 329-333. PA **86** # 49152.

[58] B. Mandelbrot, On the quadratic mapping $z \to z^2 - \mu$ for complex μ and z: the fractal structure of its $\mathcal{M}$-set, and scaling, Phys. D 7 (1983), 224-239. PA **86** # 111576.

[59] S. Kolyada and A.G. Sivak, Universal constants for one-parameter families of maps, in: *Ostsillyatsiya i ustoichivost' reshenii differentsial'no-funktsional'nykh uravnenii* (Oscillation and stability of solutions of differential-functional equations), Inst. Mat. Akad. Nauk Ukrain. SSR, Kiev 1982.

[60] Y. Pomeau and P. Manneville, Intermittent transition to turbulence in dissipative dynamical systems, Comm. Math. Phys. **74** (1980), 189-197. MR **81g**:58024.

[61] P. Manneville and Y. Pomeau, Different ways to turbulence in dissipative dynamical systems, Phys. D **1** (1980), 219-226. MR **81h**:58041.

[62] J.E. Hirsch, M. Nauenberg, and D.J. Scalapino, Intermittency in the presence of noise: a renormalization group formulation, Phys. Lett A **87** (1982), 391-393. MR **83f**:58060.

[63] B. Hu and J. Rudnick, Exact solutions to the Feigenbaum renormalization-group equations for intermittency, Phys. Rev. Lett. **48** (1982), 1645-1648. PA **85** # 84538.

[64] M.J. Feigenbaum, L.P. Kadanoff, and S.J. Shenker, Quasi-periodicity in dissipative systems: a renormalization group analysis, Phys. D **5** (1982), 370-386. MR **84f**:58101.

[65] S.J. Shenker, Scaling behaviour in a map of a circle onto itself: empirical results, Phys. D **5** (1982), 405-411. MR **84f**:58066.

[66] ——— and L.P. Kadanoff, Critical behaviour of a KAM surface: I. Empirical results, J. Statist. Phys. **27** (1982), 631-656. MR **83h**:58045.

[67] S. Ostlund, D. Rand, J. Sethna, and E. Siggia, Universal properties of the transition from quasi-periodicity to chaos in dissipative systems, Phys. D **8** (1983), 303-342. PA **86** # 111586.

[68] M. Nauenberg and J. Rudnick, Universality and the power spectrum at the onset of chaos, Phys. Rev. B **24** (1981), 493-495. PA **84** # 99025.

Translated by H. McFaden

Institute of Applied Mathematics
Academy of Sciences of the USSR

Institute of Theoretical Physics
Academy of Sciences of the USSR

Received by the Editors 23 November 1983

Physica Scripta **T9** 50–8 (1985)

Mode-Locking and the Transition to Chaos in Dissipative Systems

Per Bak

Physics Department, Brookhaven National Laboratory, Upton, New York 11973, USA

Tomas Bohr

Laboratory of Atomic and Solid State Physics, Cornell University, Ithaca, New York 14853, USA

and

Mogens Høgh Jensen

H. C. Ørsted Institute, Universitetsparken 5, Copenhagen, Denmark

Received July 25, 1984; accepted July 31, 1984

Abstract

Dissipative systems with two competing frequencies exhibit transitions to chaos. We have investigated the transition through a study of discrete maps of the circle onto itself, and by constructing and analyzing return maps of differential equations representing some physical systems. The transition is caused by interaction and overlap of mode-locked resonances and takes place at a critical line where the map looses invertibility. At this line the mode-locked intervals trace up a complete Devil's Staircase whose complementary set is a Cantor set with universal fractal dimension $D \sim 0.87$. Below criticality there is room for quasiperiodic orbits, whose measure is given by an exponent $\beta \sim 0.34$ which can be related to D through a scaling relation, just as for second order phase transitions. The Lebesgue measure serves as an order parameter for the transition to chaos.

The resistively shunted Josephson junction, and charge density waves (CDWs) in r.f. electric fields are usually described by the differential equation of the damped driven pendulum. The 2d return map for this equation collapses to 1d circle map at and below the transition to chaos. The theoretical results on universal behavior, derived here and elsewhere, can thus radily be checked experimentally by studying real physical systems. Recent experiments on Josephson junctions and CDWs indicating the predicted fractal scaling of mode-locking at criticality are reviewed.

1. Introduction

In the 17th century the Dutch physicist Christian Huyghens noted that two clocks hanging back-to-back on the wall tend to synchronize their motion [1]. This phenomenon is known as phase-locking and is generally present in dissipative systems with competing frequencies. The two frequencies may arise dynamically within the system (as for Huyghens coupled clocks, Fig. 1) or through the coupling of an oscillating or rotating motion to an external periodic force – as in a swing, Fig. 2. In many-dimensional systems the effective loss of degrees of freedom through dissipation may reduce the phase-locking phenomena to basically two coupled oscillators.

If some parameter is varied (as for instance the eigen frequency ω_1 of the swing) the system may pass through resonant regimes which are phase locked and regimes which are not. For weakly nonlinear coupling between the oscillators the phase locked intervals will have small measure. The motion is either (with small probability) periodic or, more likely, quasiperiodic, i.e. the ratio between the two frequencies ω_1/ω_2 is irrational. As the nonlinearity increases, the phase locked portions increase, and eventually chaotic motion may occur in addition to the periodic and quasiperiodic (incommensurate) motion. The mechanism, in these systems, leading eventually to chaotic behavior is interaction between the different resonances caused by non-linear couplings, and overlap between the resonant regions when the coupling exceeds a certain critical value. In some sense the mechanism is the analog for disspative systems of Chirikov's [2] instability of quasiperiodic orbits in Hamiltonian systems. In this article our recent work on the tansition to chaos caused by overlap of resonances [3–6] will be reviewed.

The layout of the article is as follows. In Section 2 we shall discuss briefly the Josephson junction in a microwave field, and sliding CDW's (as found for instance in niobium triselenide, $NbSe_3$) in applied d.c. plus a.c. electric fields, and the concept of a return map is introduced. It is shown that the return map exhibits "dimensional reduction" from $d = 2$ to $d = 1$, and acquires the form of a mapping of the circle onto itself. In Section 3 the mode-locking structure of the simple "sine" circle map is investigated, and critical exponents characterizing scaling of mode-locking at critically and the measure of quasiperiodic orbits below criticality and defined and estimated. The measure of

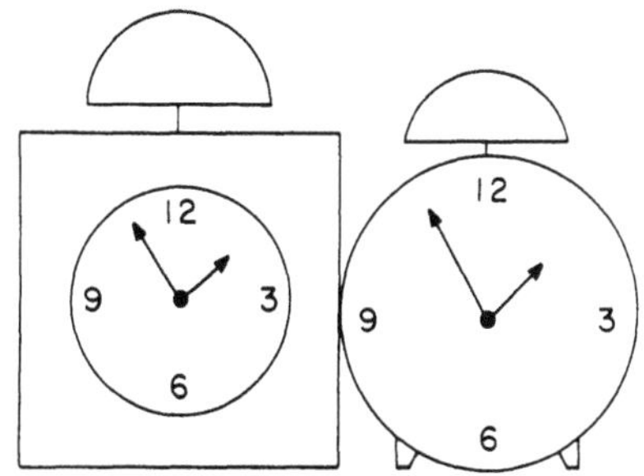

Fig. 1. The Huyghens clocks. Two weakly interacting clocks tend to synchronize each other.

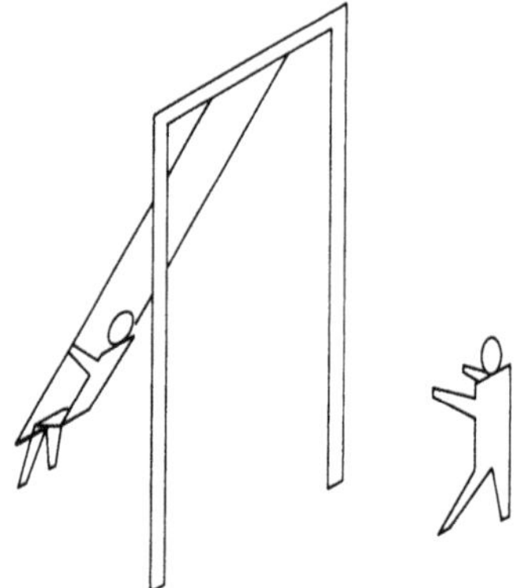

Fig. 2. The swing. A system with two coupled frequencies. If a torque A is applied to the pendulum the situation is very similar to the a.c.-driven Josephson junction.

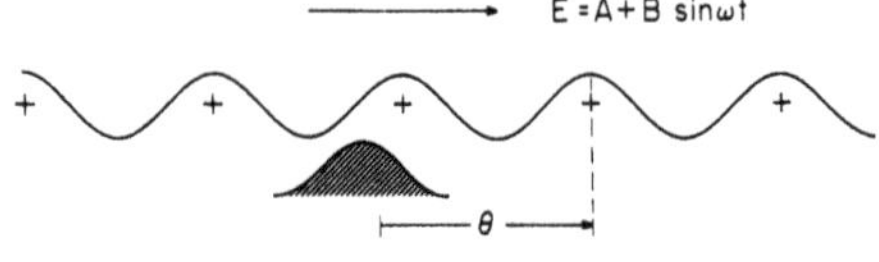

Fig. 4. Sliding charge density wave in a.c. plus d.c. electric field, and a pinning potential which could be a surface potential. The motion is that of a particle rolling down an oscillating washboard.

quasiperiodic orbits may serve as an order parameter for the transition to chaos. It is argued that the exponents are universal so the theory is predictive for real physical systems where the return map is a cricle map, although not necessarily the sine map.

In Section 4 we shall reconsider the return map for the damped driven pendulum in view of the theoretical results from the study of circle maps. It is shown that the return map remains one-dimensional up to and including the transition point. At criticality the curve "crinkles up" instead of just forming a third order inflection point (as the circle maps do). It will be shown that this is a common feature of 2d maps collapsing to 1d, and that criticality can be defined generally by the point where the return map looses its analytically. In Section 5, as a concrete example, the "Chirikov map with dissipation" will be studied. Numerical evidence is presented that the critical line is smooth and that there is scaling behavior of the mode-locked intervals as for the circle map.

Finally, in Section 6, specific experiments to investigate the transition to chaos are discussed and reviewed.

2. Phase locking in Josephson junctions and CDW systems

Figure 3 shows the equivalent electric diagram for the

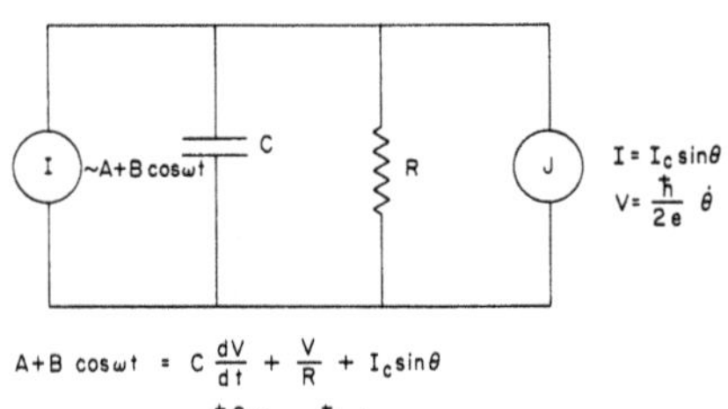

$$A+B\cos\omega t = C\frac{dV}{dt} + \frac{V}{R} + I_c\sin\theta$$
$$= \frac{\hbar C}{2e}\ddot{\theta} + \frac{\hbar}{2eR}\dot{\theta} + I_c\sin\theta$$

Fig. 3. The resistivity shunted Josephson junction, driven by a constant current A and a microwave current with amplitude B.

Josephson junction. Here, θ is the phase difference across the junction. The time dependence of the phase is governed by the differential equation.

$$\alpha\ddot{\theta} + \beta\dot{\theta} + \gamma\sin\theta = A + B\cos\omega t, \tag{1}$$

where $\alpha = \hbar/2eR$ and γ is the critical current I_c. Finally, A and B are the amplitudes of the d.c. and a.c. microwave component of the current through the junction. The model is usually referred to as the resistively shunted Josephson junction (RSJ) model, and a vast literature exists about it [7]. The differential equation is that of the forced, damped pendulum with mass α, damping coefficient β, and gravitational field γ. For certain values of the parameters the junction can be driven to a "noisy" state [8, 9] and indeed numerical simulations have indicated that the noise arises as chaotic solutions to the differential equation [10–12].

In the CDW systems, θ is the position of a sliding CDW relative to an "impurity" pinning potential (Fig. 4). The parameters α, β and γ are phenomenological constants representing the effective mass, damping, and pinning potential [13, 14]. A is a d.c. electric field which depins the CDW, and B is the amplitude of an oscillating r.f. electric field.

In those systems the phase-locking phenomenon shows up as a tendency of the average (angular) velocity $\langle\dot{\theta}\rangle$ to lock into rational multiples of the frequency of the external field,

$$\langle\dot{\theta}\rangle = \frac{N}{M}\omega \tag{2}$$

For small torque A on the pendulum (or d.c. current in the Josephson junction, or d.c. voltage in the CDW system) the pendulum stays near its downward position. When A exceeds a critical value the pendulum enters a running "rotating" mode with average velocity $\sim A/B$ (for γ and β not too large).

For the Josephson junction the voltage V is given by the Josephson relation

$$V = \frac{\hbar}{2e}\dot{\theta}, \tag{3}$$

so a locking of $\langle\dot{\theta}\rangle$ implies a locking of $\langle V\rangle$ and steps will be seen in the IV characteristics. For $M = 1$ these are the Shapiro steps [15], but in between them subharmonic steps (with $M > 1$) can often be seen. Figure 5 shows the striking experimental observation by Belykh et al. [9] of a multitude of such substeps in the IV characteristics of an Nb–Nb Josephson junction.

In the CDW systems the current carried by the sliding charge density wave is proportional to its velocity $\dot{\theta}$, so the average current is

$$I_{\mathrm{CDW}} \sim \langle\dot{\theta}\rangle. \tag{4}$$

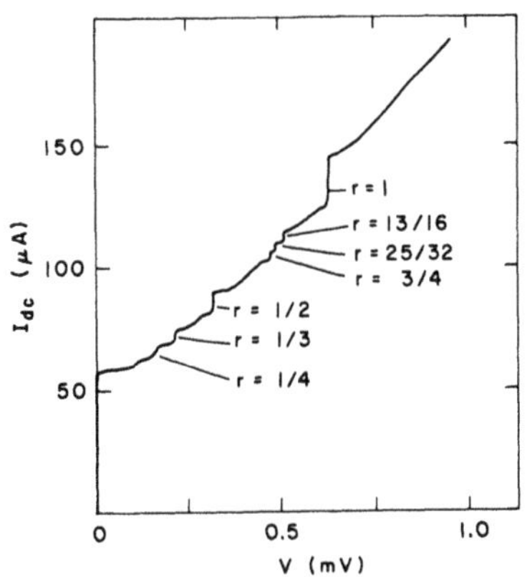

Fig. 5. IV characteristics of an Nb–Nb Josephson point junction in 295 GHz microwave field at $T = 4.2$ K. (Belykh et al., [9]).

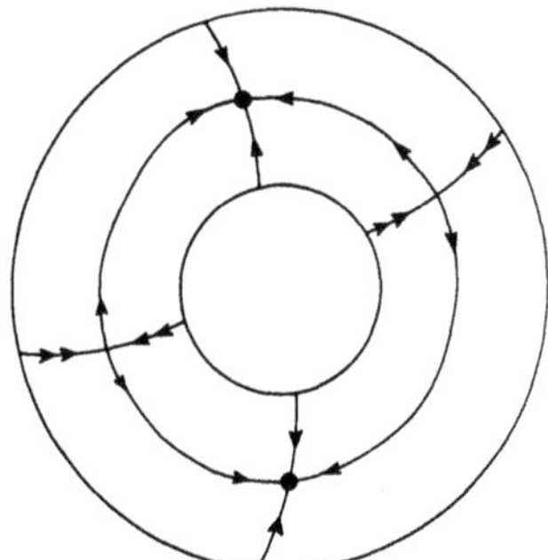

Fig. 6. Annular map (schematic) showing the invariant circle and the invariant manifolds. The simplest locked state, $P/Q = 1/2$ is shown with a stable and an unstable fixed point.

Hence, a locking of $\langle\dot{\theta}\rangle$ implies a locking of the current carried by the CDW. (The current carried by the normal electrons behaves in a smooth way, being proportional to the d.c. electric field.) The roles of currents and voltages are the reverse for Josephson junctions and CDW systems. Very recently, a multitude of subharmonic steps have been observed by Brown, Grüner, and Mozurkewich [16] in the sliding CDW system niobium-triselenlde, $NbSe_3$.

The most effective way of studying phase locking of differential equations such as (1) is through their return maps. Let us view the system with "stroboscopic light" at the discrete times $T_n = 2\pi n/\omega$, i.e. at the completion of the cycles of the external field. Since the differential equation is of second order the values of the phase θ_n and its derivative $\dot{\theta}_n$ at T_n contain all information about the system, and the values θ_{n+1} and $\dot{\theta}_{n+1}$ at T_{n+1} must be unique functions of θ_n and θ_n:

$$\begin{pmatrix}\theta_{n+1}\\ \\ \dot{\theta}_{n+1}\end{pmatrix} = \begin{pmatrix}G_1(\theta_n, \dot{\theta}_n)\\ \\ G_2(\theta_n, \dot{\theta}_n)\end{pmatrix} = R\begin{pmatrix}\theta_n\\ \\ \dot{\theta}_n\end{pmatrix} \tag{5}$$

Equation (5) defines the two dimensional return map for the differential equation. It is essential that the stroboscopic period be chosen as that of the external force. In general the functions G_1 and G_2 are not known analytically but must be determined numerically. Thus the map R can be thought of either as mapping the plane into itself or – identifying θ-values differing by multiples of 2π – as a mapping of cylinder (or an annulus) to itself as shown in Fig. 6.

The Jacobian of the mapping

$$J = \det\begin{pmatrix}\dfrac{\partial G_1}{\partial\theta_n} & \dfrac{\partial G_1}{\partial\dot{\theta}_n}\\ \\ \dfrac{\partial G_2}{\partial\theta_n} & \dfrac{\partial G_2}{\partial\dot{\theta}_n}\end{pmatrix} \tag{6}$$

is simply related to the parameters of the differential equation, namely

$$J = e^{-2\pi\beta/\omega\alpha} \equiv b \tag{7}$$

so that because of dissipation ($\beta > 0$) the map is area-contracting, $|J| < 1$. The initial condition will soon be forgotten due to the damping of the motion, and one might *hope* that the motion will asymptotically be confined to a unique invariant curve $\theta(t)$. This means that asymptotically $\dot{\theta}$ is just a given function of θ so the map R has a smooth invariant curve

$$\dot{\theta}_n = g(\theta_n) \tag{8}$$

on which the asymptotic behavior takes place. Going back to eq. (5) and inserting eq. (8) we find a unique relation

$$\theta_{n+1} = f(\theta_n) = G_1(\theta_n, g(\theta_n)), \tag{9}$$

where f maps the circle $0 < \theta_n \leqslant 2\pi$ to itself, i.e. $f(\theta_n)$ is a circle map.

Whether this dimensional reduction actually takes place depends crucially on the assumption (8) and for given values of the parameters we do not know whether or not it is satisfied, except in the limit $\alpha \ll 1$ where the connection with the circle map has been established analytically [17]. The best we can do is to generate the return map (9) by solving the differential equation (1) numerically. Figure 7 shows the reduced 1d map in a situation where the reduction does indeed happen. The curve seems to be "filled up" ergodically as the integration proceeds, so the asymptotic form is quasiperiodic and the map one-dimensional (see inset). Changing some parameters (for instance increasing A) we can generate plots that asymptotically display only a discrete set of points, namely the Q points $\theta_1^*, \theta_2^*, \ldots \theta_Q^*$ which are stable periodic points of the map R (or f in (9)). Increasing A still further leads again to quasi-periodic motion.

In general, $f(\theta)$ could be any periodic function of θ. However, it will be argued later that the specific form of f is not important for the critical behavior at the transition to chaos, so in the following chapter we shall consider a specific class of return maps, namely the sine circle maps.

3. Mode-locking and universality in circle maps

The "sine" circle maps are defined by the equation

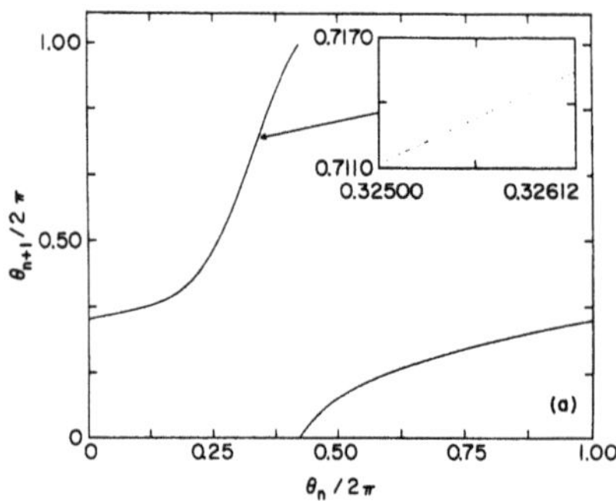

Fig. 7. Return map calculated numerically for $B = \alpha = \gamma = 1$, $\omega = 1.76$, $\beta = 1.576$, $A = 1.4$. The function $f(\theta_n)$ is monotonically increasing. The inset is a magnification, emphasizing the one-dimensionality of the map. Because of periodicity, only the square $0 < \theta_n, \theta_{n+1} \leqslant 1$ is shown.

$$\theta_{n+1} = f_\Omega(\theta_n) = \theta_n + \Omega - \frac{K}{2\pi} \sin 2\pi\theta_n \qquad (10)$$

where we have chosen the periodicity to be 1 instead of 2π. The function f is periodic since we identify the points θ and $\theta + \mathrm{n}$. The advantage of studying simple maps on this form is obvious. It is much easier to identify periodic, quasiperiodic and chaotic solutions by iterating the map than by a cumbersome numerical integration of the underlying differential equation. The map has a linear term θ_n and a bias term Ω representing the frequency of the system in the absence of nonlinear coupling K. To study the mode-locking in the circle map we consider iterations of the map, θ, $f(\theta)$, $f^2(\theta)\ldots$, or $\theta_1, \theta_2, \theta_3$. The frequency of the dynamical system is given by the winding number of the mapping,

$$W = \lim_{n\to\infty} \frac{f_\Omega^n - \theta}{n} \qquad (11)$$

Clearly, $W = \Omega$ in the absence of nonlinear coupling. Under iteration the variable θ_n may converge to a series which is either *periodic*, $\theta_{n+Q} = \theta_n + P$, with rational winding number P/Q, *quasiperiodic*, with irrational winding number $W = q$, or *chaotic* where the series behaves irregularly.

Although the question of the existence of smooth behaviour in circle maps has very much the flavor of the general problem of the existence of smooth invariant tori in dynamical systems (the KAM problem), much stronger theorems due to Arnold [18] and Herman [19] exist for the one-dimensional circle maps. As long as $f(\theta)$ is monotonic these theorems guarantee that no chaotic motion can occur. The transition to chaos, and the associated nontrivial scaling behavior that we shall discuss occurs precisely at the point in parameter space where $f(\theta)$ loses its invertibility. In that case the theorems above break down and not much is known in general. The first indication of interesting scaling behavior was given in a numerical investigation by Shenker [20] followed by renormalization group treatments by Feigenbaum, Kadanoff and Shenker [21] and Rand et al. [22]. These studies concentrate on quasiperiodic behaviour with specific well-behaved winding numbers.

In our work we focus on the mode-locking structure at and below the transition to chaos. The mapping (10) is sketched in Fig. 8(a) for $\Omega = 0.2$ and $K = 0.9$. Again, we have reduced it to the square $0 \leqslant \theta_n < 1$. We see two branches in the unit square. When $K < 1$ the map is strictly monotonic, just as for the return map calculated for pendulum equation, Fig. 7. At $K = 1$ the map develops a cubic inflection point at $\theta = 0$ so the map is still invertible but the inverse has a singularity.

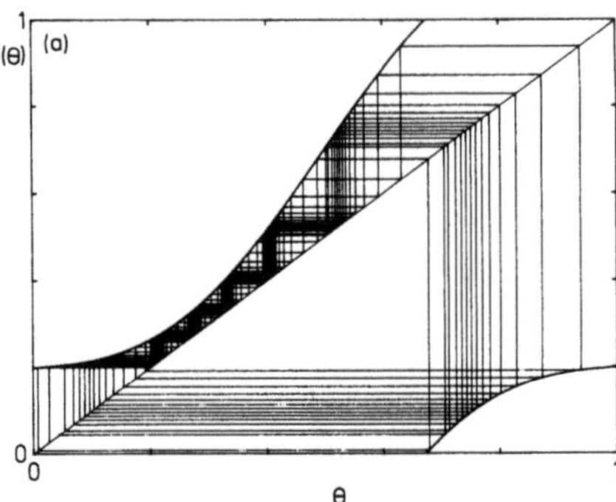

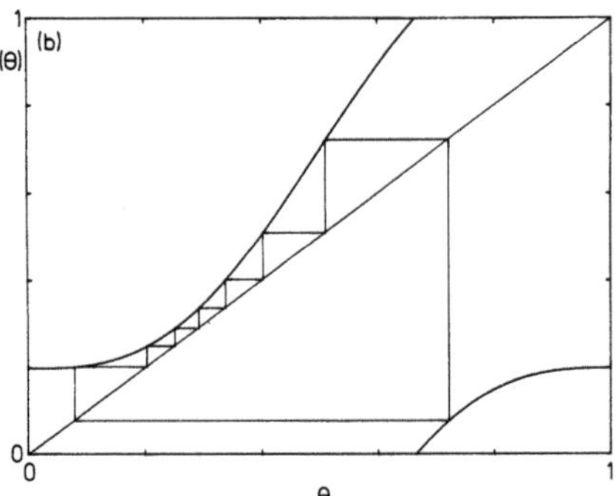

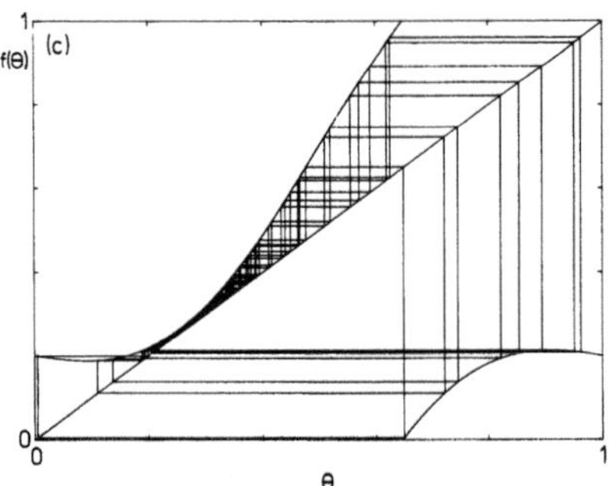

Fig. 8. Evolution of iterations of the circle map for $\Omega = 0.2$ and (a) $K = 0.9$ (quasiperiodic orbit), (b) $K = 1.0$ (limit cycle) and (c) $K = 1.1$ (chaotic orbit). For $K > 1$ the map develops local maxima and minima and chaotic behavior may occur.

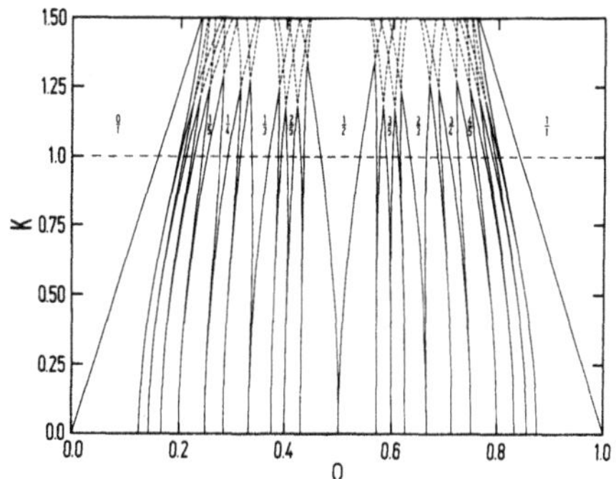

Fig. 9. Phase diagram for the "sine" circle map. Note the Arnold tongues where the winding number assumes locked rational values. The winding number assumes irrational values along one-dimensional curves ending at $K = 1$. The dotted lines indicate overlap of resonances.

For $K > 1$ (Fig. 8(c)) the map develops a local maximum and is no longer invertible. It will be seen later that a very similar scenario occurs for the return map of the differential equation (1) as the parameters are varied.

We shall here be concentrating on the situation for K equal to or slightly below 1. For $0 < K < 1$ it has been shown [19] that the winding number locks-in at every single rational number P/Q in a non-zero interval of Ω, $\Delta\Omega(P/Q)$. (See Fig. 9). For K close to zero all intervals are quite small so the probability that the winding number is rational is almost zero (as for the Huyghens clocks), and the probability of hitting an irrational winding number is almost 1. However, with increasing K the width of all intervals increase. The regimes in (Ω, K) space where W assumes rational values are called "Arnold tongues". Clearly, the widths of the resonances can not grow indefinitely; at some point they will interact and overlap.

One might speculate that at $K = 1$ the resonance will fill-up the critical line, and confine the quasiperiodic orbits to a Cantor set of zero measure. To test this conjecture we have caculated the widths $\Delta\Omega(P/Q)$ for all intervals with $Q \leqslant 95$. Figure 10 shows the widths of the "steps" vs P/Q. Note the self-similarity of the function under rescaling.

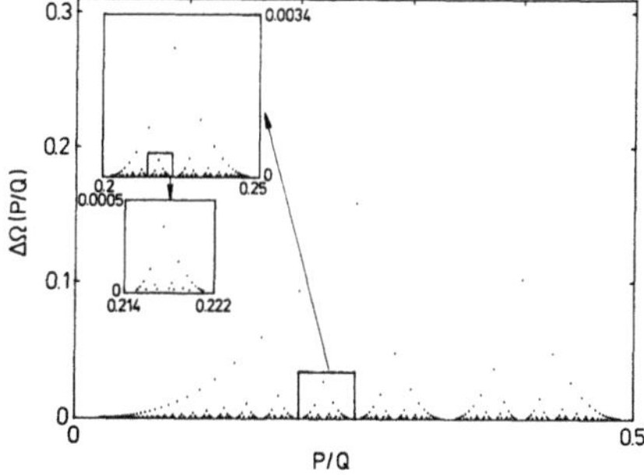

Fig. 10. Widths of intervals $\Delta\Omega(P/Q)$. The diagram is self-similar under magnification.

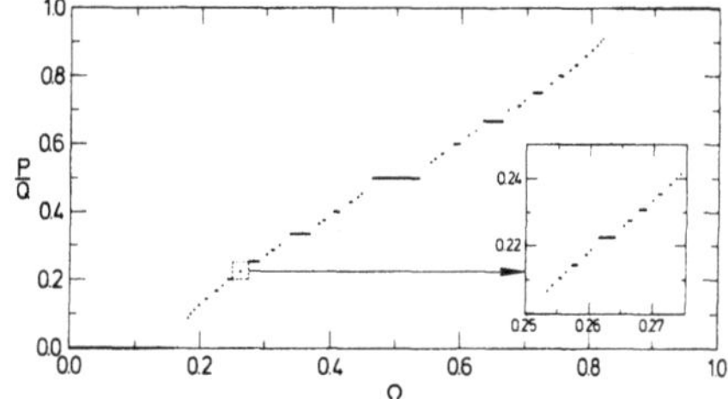

Fig. 11. Winding number W vs Ω for $K = 1$. Note the self-similar nature of the staircase.

Figure 11 shows the "staircase" function formed by plotting the winding number vs Ω. We conjecture that eventually $\Delta(P/Q) > 0$ for all P and all Q. By including more and more steps the Ω axis becomes more and more "filled up". To investigate whether or not the mode-locked intervals will eventually cover the entire Ω-axis we have calculated the total width of all steps which are larger than a given scale r. We are interested in the space between the steps, $1 - S(R)$ and have measured it on the scale r to find the number of holes, $N(r) = [1 - S(r)]/r$. On Fig. 12, log $N(r)$ has been plotted versus $1/r$ for 40 values of r in the interval [0.0009, 0.000017]. The points fall excellently on a straight line indicating a power law

$$N(r) \sim (1/r)^D \tag{12}$$

From the slope of the straight line we find $D = 0.8700 \pm 4 \times 10^{-4}$. The result (12) means that the space between the steps vanishes as

$$1 - S(r) \sim r^{1-D} \tag{13}$$

at $r \to 0$. Thus, there is no room for quasiperiodic motion at criticality, and the staicase Fig. 11 is called complete. The exponent D is the fractal dimension of the staircase, or rather the fractal dimension of the Cantor set which is the complementary set to the mode locked intervals on the Ω axis.

The fractal dimension can be determined by an alternative method simply by counting the number of steps $N_1(r)$ which are wider than a given scale r. This number is given by the equation

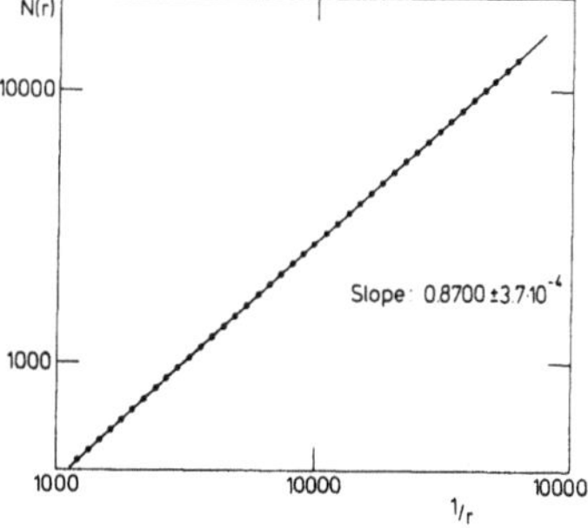

Fig. 12. Plot of log $N(r)$ vs log $(1/r)$ for the critical circle map. The slope of the straight line yields $D = 0.8700 \pm 3.7 \times 10^{-4}$.

$$\frac{\partial N_1}{\partial r} = \frac{1}{r}\frac{\partial S(r)}{\partial r} \sim \frac{1}{r}\frac{\partial r^{1-D}}{\partial r} \sim r^{-1-D} \tag{13a}$$

so

$$N_1(r) \sim r^{-D} \tag{13b}$$

The latter method of determining D seems easier since uncertainties in the determination of the stepwidth are not accumulated as when $S(r)$ is calculated. On the other hand, even if $N_1(r)$ obeys the simple power law eq. (13b) there is no guarantee that the staircase is complete. Integration of eq. (13b) lead to

$$S(r) \sim r^{1-D} + C, \tag{14}$$

where C is an integration constant. The two methods are equivalent only when the staircase is known to be complete, i.e. when $C = 0$.

When passing beyond the $K = 1$ line the steps continue to increase. Since they fill-up the whole Ω axis for $K = 1$ they must necessarily overlap for $K > 1$ (see Fig. 9). In an experimental situation the transition to chaos is most easily identified by locating the onset of hysteresis at the edges of the larger steps. The transition to chaos is basically caused by overlap of resonances and one can visualize the chaotic motion as an erratic jumping between the resonances. Most dissipative non-linear periodic systems perturbed by an external periodic field will probably exhibit a transition to chaos caused by overlap of resonances as described here.

It is important to know whether or not the critical behavior at the transition is universal, i.e.whether or not it depends on the specific function $f(\theta)$ in eq. (10). To check universality we have studied a class of mappings

$$f_{\Omega,a}(\theta) = \theta + \Omega - \frac{K}{2\pi}(\sin 2\pi\theta + a \sin^3 2\pi\theta). \tag{15}$$

Generally, the details are different from the staircase shown in Fig. 11. Some steps become wider, some become narrower. The scaling, however, remains the same, with $D \sim 0.87$. For $a = 1/6$ where the lowest order term in an expansion of $f(\theta)$ vs θ is of 5th order there is also a complete staircase, but the fractal dimension $D \approx 0.81$. Thus, the fractal dimension depends on the nature of the inflection point, but the generic critical exponent to be expected in an experiment is $D \sim 0.87$.

We would like to stress that although the dimension D was calculated by considering steps in a large Ω interval it is a well-defined and universal number at any infinitesimal interval. The self-similarity is a *local* property of the staircase.

The steps do not fill-up the entire Ω axis for $K < 1$ and the slope D in the log $N(r)$ vs log $(1/r)$ plot must necessarily converge towards $D = 1$ for small r. When K is slightly smaller than 1 it seems that the scaling follows $D \sim 0.87$ down to a certain scale (depending on $1 - K$) and then makes a smooth cross-over to the trivial scaling characterized by $D = 1$. We shall now define and estimate an exponent ν which characterizes this crossover, and another exponent β which governs the measure of the quasiperiodic orbits for $K < 1$. Since the measure, M, is zero at and above the transition, and nonzero below the transition, it may serve as an order-parameter for the transition to chaos. or rather for the transition away from chaos, with the regular phase playing the role of the ordered phase and the chaotic phase corresponding to the disordered phase at a thermodynamic second order transition. Hence, β is the order-parameter exponent just as the exponent characterizing the

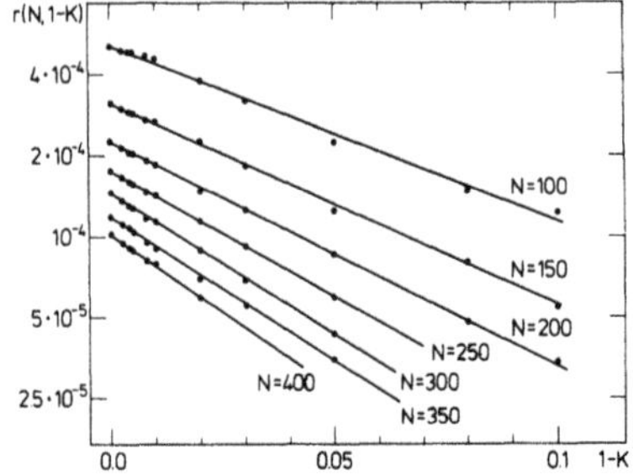

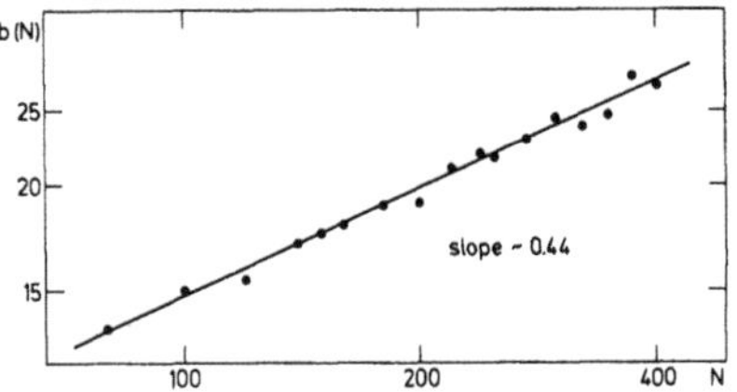

Fig. 13. (a) Scale r for which there are N intervals wider than r, plotted vs $(1 - K)$. (b) Plot of log $b(N_1)$ defined by eq. (16) vs log N_1.

magnetization in a magnetic system. The quantity $(1 - K)$ plays the role of the reduced temperature.

In order to quantify the cross-over we have calculated the scale $r(N_1, 1 - K)$ such that the number of resonances which are wider than r is precisely N_1. Figure 13(a) shows log $r(N_1, 1 - K)$ vs $(1 - K)$ for several values of N_1. The straight line indicates exponential behavior:

$$r(N_1, 1 - K) = r(N_1, 0) \exp(-b(N_1)(1 - K)). \tag{16}$$

Figure 13(b) shows log $b(N_1)$ vs log N_1. The linear behavior allows us to define an exponent ν:

$$b(N_1) \sim N_1^{1/D\nu}, \quad \frac{1}{D\nu} = 0.44 \pm 0.02 \tag{17}$$

Eqs. (16) and (17) yield a cut-off, N_0, in the number of resonances which give a contribution to the integrated measure of periodic orbits below criticality:

$$N_0 \sim (1 - K)^{-D\nu} \tag{18}$$

This cutoff in N_0 can be related to a cutoff scale r_0 through eq. (13b):

$$r_0 \approx (1 - K)^{\nu}, \quad \nu \sim 2.63. \tag{19}$$

For a given value of K the resonances which at $K = 1$ are narrower than the critical scale eq. (19) are effectively cut-off. The measure of quasiperiodic orbits is precisely the measure of the cut-off periodic orbits:

$$M(K) = \int_{r=0}^{(1-K)^{\nu}} \frac{\partial N_1}{\partial r} r \, dr$$
$$\sim (1 - K)^{\nu(1-D)} \equiv (1 - K)^{\beta}. \tag{20}$$

where $\beta \sim 0.34 \pm 0.02$.

Equation (20) defines a scaling relation,

$$D = 1 - \beta/\nu \qquad (21)$$

which is very similar to the relation

$$D = d - \beta/\nu$$

which has been derived for second order phase transitions. Here, d is the Euclidean dimension. It would be interesting to determine the measure of quasiperiodic orbits independently in order to check the scaling relation.

4. Transition to chaos in Josephson junctions, charge-density waves and damped driven pendula

Having conjectured the scaling behavior at the critical curve for the circle map, it is of interest to go back to the real physical systems considered in Section 2 (which are described by differential equations with 1d return circle maps) and investigate to what extent the results from the circle map can be taken over. First of all, it is important to establish the critical curve for the transition to chaos. For the circle map this line is just $K = 1$ but for higher dimensional systems no such simple line exists.

The return map depicted in Fig. 7 is monotonically increasing and thus corresponds to regular behavior only. To approach the regime with chaotic behavior we reduce the damping and vary A to keep the winding number roughly constant, $W \simeq 0.38$. Indeed it appears (Fig. 14) that the map acquires a zero slope at the transition to chaos so the transition seems closely related to the one occurring in the circle map. If the damping is reduced further (Fig. 15) the curve develops a local maximum allowing chaotic behavior.

However, as seen from the insets, the behavior of $f(\theta)$ is more complicated around the transition point: instead of just turning over to form isolated local maxima and minima, the curve "crinkles up" and seems to be filled-up in an uneven nonergodic way. This is not accidental but represents the typical critical behavior for return maps generated through the collapse of higher dimensional systems. Assuming a 2d return map of the form (5) it can be shown for a wide class of functions G that if $f'(\theta_n) \to 0^+$ then $f'(\theta_{n+1}) \approx -\infty$, as indicated numerically for the return map in Fig. 14. For a proof of this statement, see [6] and [23].

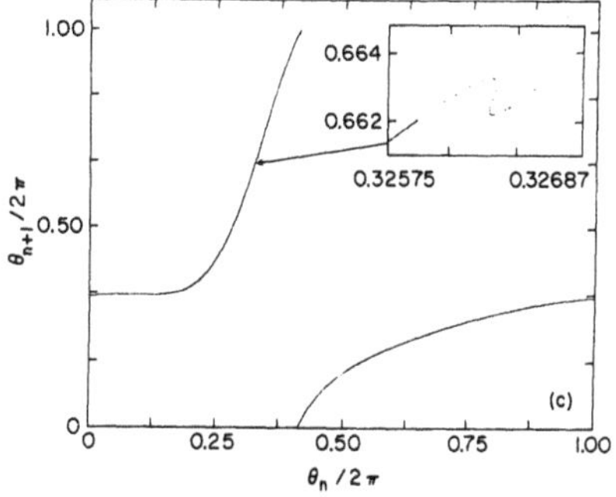

Fig. 14. Return map for eq. (7) calculated for $\beta = 1.253$, $A = 1.2$. The function develops a cubic inflection point indicating the transition to chaos. The inset shows an enlargement of the curve around $f^2(\theta_I)$ where θ_I is the inflection point ($f'(\theta_I) \to 0, f^{2\prime}(\theta_I) \to \infty$).

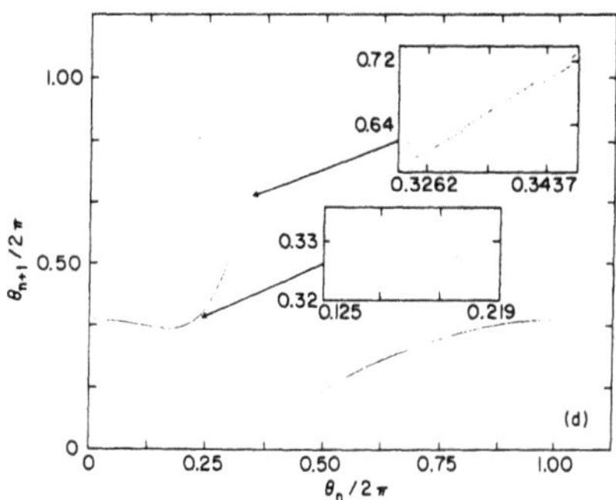

Fig. 15. Return map calculated for $\beta = 1.081$, $A = 1.094$. The map develops a local minimum and "wiggles" (insets) indicating chaotic behavior.

Consider now the graph of f as a function of some parameter and assume to start with that f is an increasing circle map. Varying the parameter decreases $f'(\theta)$ towards zero and therefore $f'(f(0))$ must approach minus infinity. But then $f'(\theta)$ must be zero somewhere in the interval $[0, f(0)]$ giving another infinite slope, etc. The whole curve crinkles up precisely as found in Figs. 14 and 15. Thus, the loss of monotonicity is tied to the loss of smoothness for the invariant circle: if there exists a value of θ for which $f'(\theta) = 0$ then the invariant circle has already broken up and the initial assumption of a smooth orbit $\dot{\theta} = g(\theta)$ is contradictory.

Then, when precisely does the map loose its smooth monotonic behavior? It can be shown that for limit-cycle fixed points this happens precisely when the eigenvalues of the product of the Jacobians eq. (6) at the Q limit cycle points become identical, $\lambda_1 = \lambda_2 = b^{Q/2}$, where b is defined by eq. (7). In the following section we shall see that this allows us to define a smooth transition curve by considering a specific two dimensional map.

5. Scaling behavior of the "Dissipative Standard Map"

Since we are looking for universal behavior we might as well choose a simple analytic map instead of doing the integration of the differential equation. For that particular map we shall try to locate the "critical line", i.e. the line in parameter space where the smoothness breaks down.

The fundamental question is whether the "critical line" in any sense defines a smooth curve. One might fear that since the definition of the criticality condition depends on the denominator Q of the winding number that the curve is fundamentally fractal. Our finding for the particular map which will be studied is that it is not fractal: The dissipation present in the map assures that the critical line defined through the very high order rational steps is smooth.

The map is defined by recursion relations in two variables θ and r:

$$\theta_{n+1} = \theta_n + \Omega - \frac{K}{2\pi} \sin 2\pi\theta_n + br_n \qquad (22)$$

$$r_{n+1} = br_n - \frac{K}{2\pi}\sin 2\pi\theta_n$$

where b is between 0 and 1. The equation has the required symmetry in θ and the variable r_n plays a role as $\dot{\theta}_n$ in eq. (6). The Jacobian matrix of the mapping is

$$D = \begin{pmatrix} 1 - K\cos 2\pi\theta & b \\ -K\cos 2\pi\theta & b \end{pmatrix} \tag{23}$$

with determinant b, showing indeed that eq. (22) is area contracting when $b < 1$. When $b \to 0$ we recover the "sine" circle map eq. (10) and when $b \to 1$ we obtain the so-called standard map. The conclusions derived in Section 4 are certainly valid for the effective reduced 1d map of eq. (23), if such a map exists.

The behavior of a limit cycle with period Q is characterized by the eigenvalues λ_1, λ_2 of the matrix

$$M = D(\theta_Q^*)D(\theta_{Q-1}^*)\ldots D(\theta_1^*) \tag{24}$$

and the condition for criticality is simply $\lambda_1 = \lambda_2 = b^{Q/2}$. This condition determines for large Q a curve in (K, Ω) space (equivalent to the (β, A) space for eq. (1)) as shown schematically in Fig. 16. Generally we wish to assign only one critical point to each cycle (and not a curve) and we chose this point to be the lowest point • on the hyperbola. Having these points for several P/Q cycles we define the critical curve as the curve determined by the accumulation points of these critical points. Figure 17 shows critical curves calculated for $b = 0.25$ and $b = 0.5$ We first of all notice that the critical points converge towards a smooth critical curve. A few points fall outside the curve but since they are associated with low Q-cycles they do not affect the limiting behavior.

Having identified a smooth critical curve the next problem is the scaling properties of P/Q steps. A particular limit cycle point or "step" is stable as long as the largest eigenvalue, say λ_1, is smaller than 1; the condition $\lambda_1 = 1$ is obtained when

$$\operatorname{Tr} M = 1 + b^Q \tag{25}$$

For high Q-cycles the quantity b^Q becomes very small so the condition for criticality is essentially $\lambda_1 = \lambda_2 = 0$, and the condition for stability, Tr $M = 1$, approach the corresponding conditions for the one-dimensional circle map. This gives a hint that the critical behavior is that of the circle map.

Figure 18 is a plot of the winding number verus Ω at criticality which again forms a staircase. Again, we measure the accumulated width of all steps larger than r in an Ω interval of length Ω_0. The "number of holes" plotted vs $1/r$ follows a power law

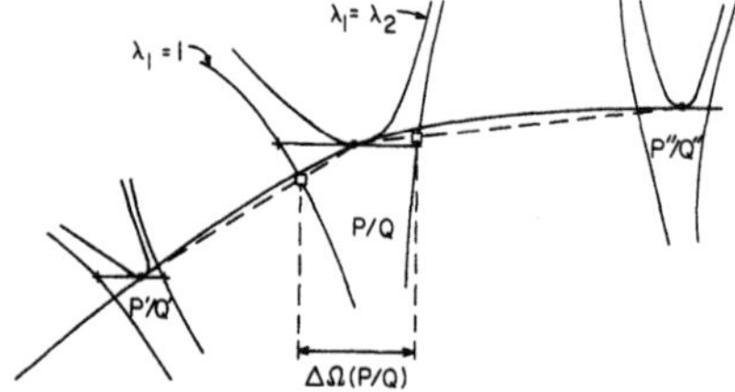

Fig. 16. Schematic diagram showing the hyerbolae where the invariant circle becomes critical at the stable periodic points. The critical curve is approximated by the minimum points •.

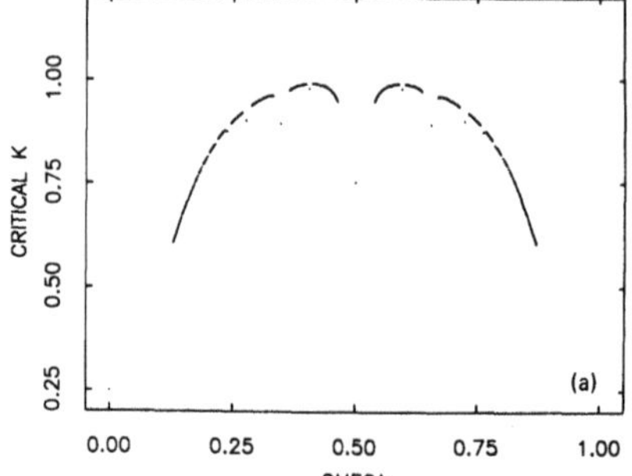

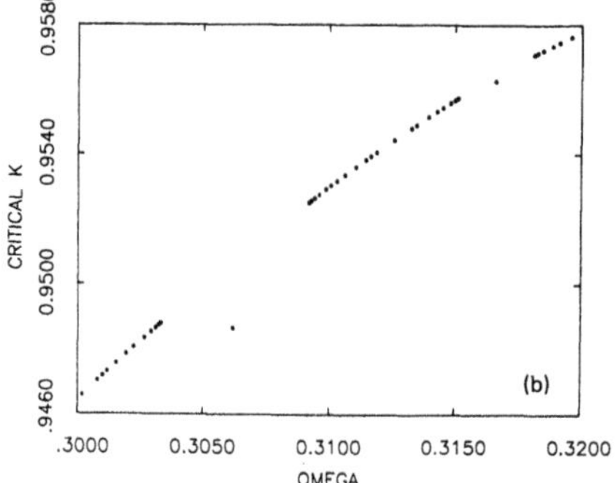

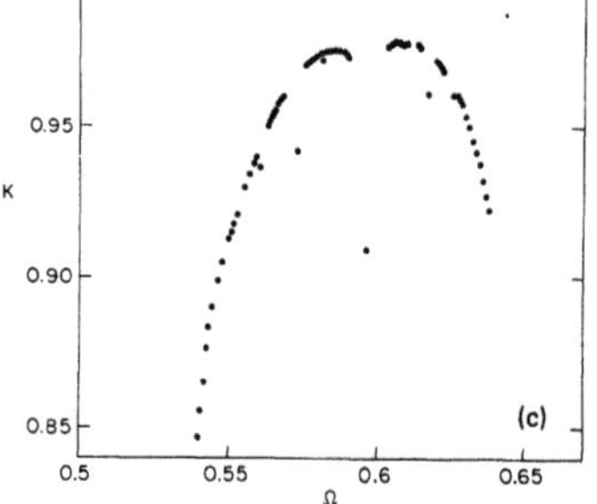

Fig. 17. Critical curves for 2d standard map. The curve (a) was calculated for $b = 0.25$; the curve (b) is a magnification of (a). The curve (c) was calculated for $b = 0.5$.

$$\frac{\Omega_0 - S(r)}{r} \sim \left(\frac{1}{r}\right)^D \tag{26}$$

with $D = 0.86 \pm 0.01$. The staircase is therefore complete and the dimension of the Cantor set is within the numerical accuracy identical to the universal exponent found for circle maps. We therefore believe that the critical behavior at the transition to chaos is that of the circle maps.

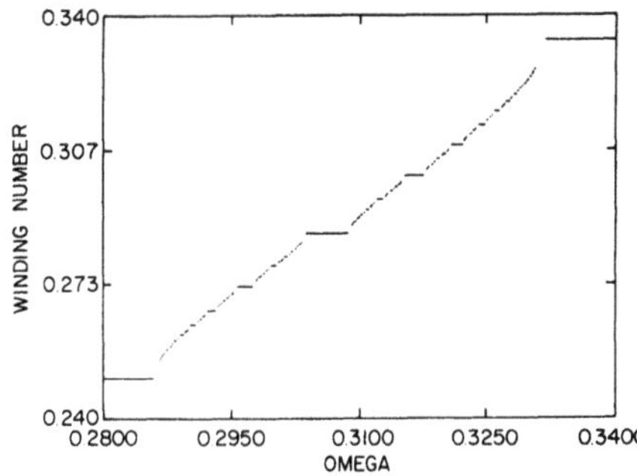

Fig. 18. The widths of phase locked steps at criticality for the map (22) with $b = 0.25$. Note the similarity with the circle map, Fig. 11.

6. Conclusion and discussion of experiments

We have investigated the transition to chaos by overlap of resonances by studying circle maps, differential equations representing real physical systems, and 2d discrete dissipative maps. Our conjecture is that the critical behavior of all these systems is the same, so theoretical results on circle maps, derived by us and others, can be taken over to explain and predict critical behavior at the transition to chaos for specific experimental systems. We urge that experiments on Josephson junctions, charge-density-wave systems and other systems with two competing periodicities be performed to check our predictions.

We are aware of three experiments which have been performed since we first announced our results: Kao et al. [24] and Alstrom et al. [25] have studied the differential equation (1) with a "Josephson junction simulator"; they find scaling behavior as predicted here, with $D \sim 0.91 \pm 0.04$ and $D \sim 0.87 \pm 0.02$, respectively. More importantly, Brown, Mozurkewich and Grüner [16] have measured subharmonic steps in the CDW system $NbSe_3$ in a.c. plus d.c. electric fields; they also find scaling behavior in agreement with our conjecture, with $D \sim 0.91 \pm 0.03$. We feel that the accuracy of the experiments could be improved; in particular more work should be done to locate the critical line, following, for instance, the ideas presented here. It would be of interest to measure the return map directly. The most precise measurements can probably be performed on Josephson junctions in microwave fields, but no experiment on the scaling behavior near the critical point has been reported so far.

At this symposium we have seen a number of other exciting experimental systems with competing frequencies for which the theory presented here can readily be applied. Jeffries [26] has found frequency locking and transitions to chaos in a resonantly driven p–n junction in silicon. Arecchi [27] has observed deterministic chaos and resonances for a homogeneously broadened laser with injected signal. Gollub has studied mode-locking and quasi periodic behavior in Rayleigh–Benard experiment [28]. The transition to chaos in this system is precisely of the type discussed here. Swinney [29] has found phase locking and a "devil's staircase" in the Belusov–Zhabotinsky chemical reaction. The staircase seems to have a different structure than the one found here, so it is not yet clear to what extent the theory applies in this case.

Acknowledgement

Work supported by the Division of Materials Sciences U.S. Department of Energy under contract DE-AC02-76CH00016, by National Science Foundation grant DMR-8314625 and by the Danish Natural Science Research Council.

References

1. Stated in footnote in Van der Pol, B., Phil. Mag. Ser. 7, Vol. 3, No. 13 (1927). We thank David Rand for making us aware of this.
2. Chirikov, B. V., Phys. Rev. **52**, 263 (1979).
3. Jensen, M. H., Bak, P. and Bohr, T., Phys. Rev. Lett. **50**, 1637 (1983).
4. Bak, P., Bohr, T., Jensen, M. H. and Christiansen, P. V., Solid State Commun. **51**, 231 (1984).
5. Jensen, M. H., Bak, P. and Bohr, T., Phys. Rev. A (October 1984).
6. Bohr, T., Bak, P. and Jensen, M. H., Phys. Rev. A (October 1984).
7. For reviews, see Lindelof, P. E., Rep. Prog. Phys. **44**, 949 (1981); Imry, Y., "Statics and Dynamics of Nonlinear Systems" (eds. Benedek, Bilz and Zeyher) p. 170. Springer, Blerin (1983).
8. Chiao, R. Y., Feldman, M. J., Peterson, D. W., Tucker, B. A. and Levinsen, M. T., "Future Trends in Superconducting Electronics", AIP Conf. Proc. **44**, 259 (1978).
9. Belykh, V. N., Pedersen, N. F. and Soerensen, O. H., Phys. Rev. **B16**, 4860 (1978).
10. Braiman, Y., Ben-Jacob, E. and Imry, Y., SQUID 80 (eds. H. D. Hahlbohm and H. Lubbig), Walter de Gruyter, Berlin (1980); Ben-Jacob, E., Braiman, Y., Shansky, R. and Imry, Y., Appl. Phys. Lett. **38**, 822 (1981); Ben-Jacob, E., Goldhirsch, I., Imry, Y. and Fishman, S., Phys. Rev. Lett. **49**, 1599 (1982); Kautz, R. L., J. Appl. Phys. **52**, 6241 (1981).
11. Huberman, B. A., Crutchfield, J. P. and Packard, N. H., Appl. Phys. Lett. **37**, 751 (1980); D'Humieres, D., Beasley, M. R., Huberman, B. A. and Libchaber, A., Phys. Rev. **A26**, 3483 (1982); Pedersen, N. F. and Davidson, A., Appl. Phys. Lett. **39**, 830 (1981); Cirillo, M. and Pedersen, N. F., Phys. Lett. **90A**, 150 (1982); Yeh, W. J. and Kao, Y. H., Phys. Rev. Lett. **49**, 1888 (1982); MacDonald, A. H. and Plischke, M., Phys. Rev. **B27**, 201 (1983); Yeh. W. J. and Kao, Y. H., Appl. Phys. Lett. **42**, 299 (1983). These papers deal with the case $A = 0$ which is less relevant here.
12. Levinsen, M. T., J. Appl. Phys. **53**, 4294 (1982).
13. Salam, F. M. A. and Sastry, S. S., preprint.
14. Grüner, G., Zawadowski, A. and Chaikin, P. M., Phys. Rev. Lett. **46**, 511 (1981); Bardeen, J. Ben-Jacob, E., Zettl, A. and Grüner, G., Phys. Rev. Lett. **49**, 493 (1982).
15. Shapiro, S., Phys. Rev. Lett. **11**, 80 (1963).
16. Brown, S. E., Mozurkewich, G. and Grüner, G., Phys. Rev. Lett. **52**, 2277 (1984).
17. Azbel, M. Ya. and Bak, P., Phys. Rev. B October 1984.
18. Arnold, V. I., American Mathematical Society Translations, Ser. 2, **46**, 213 (1965).
19. Herman, M. R., Geometry and Topology (ed. J. Palis) Lecture Notes in Mathematics **597**, 271. Springer, Berlin (1977).
20. Shenker, S. J., Physica **5D**, 405 (1982).
21. Feigenbaum, M. J., Kadanoff, L. P. and Shenker, S. J., Physica **5D**, 370 (1982).
22. Rand, D., Ostlund, S., Sethna, J. and Siggia, E., Phys. Rev. Lett. **49**, 132 (1982); Physica **6D**, 303 (1984).
23. Bohr, T., Phys. Lett. A (in press).
24. Yeh, W. J., He, D-R. and Kao, Y. H., Phys. Rev. Lett. **52**, 480 (1984).
25. Alstrom, P., Levinsen, M. T. and Jensen, M. H., Phys. Lett. A, **103**, 171 (1984).
26. Jeffries, C. D., these proceedings.
27. Arecchi, F. T., these proceedings.
28. Gollub, J. P., these proceedings.
29. Swinney, H., these proceedings.

Physical Review A **33** 1141–51 (1986)

Fractal measures and their singularities: The characterization of strange sets

Thomas C. Halsey, Mogens H. Jensen, Leo P. Kadanoff, Itamar Procaccia,* and Boris I. Shraiman†
The James Franck Institute, The Enrico Fermi Institute for Nuclear Studies, and Department of Chemistry, The University of Chicago, 5640 South Ellis Avenue, Chicago, Illinois 60637
(Received 26 August 1985)

We propose a description of normalized distributions (measures) lying upon possibly fractal sets; for example those arising in dynamical systems theory. We focus upon the scaling properties of such measures, by considering their singularities, which are characterized by two indices: α, which determines the strength of their singularities; and f, which describes how densely they are distributed. The spectrum of singularities is described by giving the possible range of α values and the function $f(\alpha)$. We apply this formalism to the 2^∞ cycle of period doubling, to the devil's staircase of mode locking, and to trajectories on 2-tori with golden-mean winding numbers. In all cases the new formalism allows an introduction of smooth functions to characterize the measures. We believe that this formalism is readily applicable to experiments and should result in new tests of global universality.

I. INTRODUCTION

Nonlinear physics presents us with a perplexing variety of complicated fractal objects and strange sets. Notable examples include strange attractors for chaotic dynamical systems,[1,2] configurations of Ising spins at critical points,[3] the region of high vorticity in fully developed turbulence,[4,5] percolating clusters and their backbones,[6] and diffusion-limited aggregates.[7,8] Naturally one wishes to characterize the objects and describe the events occurring on them. For example, in dynamical systems theory one is often interested in a strange attractor (the object) and how often a given region of the attractor is visited (the event). In diffusion-limited aggregation, one is interested in the probability of a random walker landing next to a given site on the aggregate.[8] In percolation, one may be interested in the distribution of voltages across the different elements in a random-resistor network.[6]

In general, one can describe such events by dividing the object into pieces labeled by an index i which runs from 1 up to N. The size of the ith piece is l_i and the event occurring upon it is described by a number M_i. For example, in critical phenomena, we can let M_i be the magnetization of the region labeled by i. Such a picture is natural in the droplet theory of the Ising model, where one argues that if the region i has a size of order l, the magnetization has a value of the order of

$$M_i \sim l^y , \tag{1.1}$$

where y (or y_σ) is one of the standard critical indices.[9] Since these droplets are imagined to fill the entire space, the density of such droplets is simply

$$\rho(l) \sim \frac{1}{l^d} , \tag{1.2}$$

where d is the Euclidean dimension of space. In fact, in critical phenomena we define a whole sequence of y_q's by saying that the typical values of $(M_i)^q$ vary with q and have the form[10]

$$(M_i)^q \sim l^{y_q}, \quad q=1,2,3,\ldots . \tag{1.3}$$

Typically, our attention focuses upon the values of y_q that are greater than zero and *we have only a few distinct values of these.*[11]

In this paper we are interested not in critical phenomena but instead in a broad class of strange objects. However, we specialize our treatment to the case in which M_i has the meaning of a probability that some event will occur upon the ith piece. For example, in experiments on chaotic systems one measures a time series $\{\mathbf{x}_i\}_{i=1}^N$. These points belong to a trajectory in some d-dimensional phase space. Typically, the trajectory does not fill the d-dimensional space even when $N\to\infty$, because the trajectory lies on a strange attractor of dimension D, $D<d$. One can ask now how many times, N_i, the time series visits the ith box. Defining $p_i=\lim_{N\to\infty}(N_i/N)$, we generate the measure on the attractor $d\mu(\mathbf{x})$, because

$$p_i=\int_{i\text{th box}} d\mu(\mathbf{x}) .$$

In many nonlinear problems, the possible scaling behavior is richer and more complex than is the case in critical phenomena. If a scaling exponent α is defined by saying that

$$p_i^q \sim l_i^{\alpha q} , \tag{1.4}$$

then α [roughly equivalent to y_q/q in Eq. (1.3)] can take on a range of values, corresponding to different regions of the measure. In particular, if the system is divided into pieces of size l, we suggest that the number of times α takes on a value between α' and $\alpha'+d\alpha'$ will be of the form

$$d\alpha' \rho(\alpha') l^{-f(\alpha')} , \tag{1.5}$$

where $f(\alpha')$ is a continuous function. The exponent $f(\alpha')$ reflects the differing dimensions of the sets upon which the singularities of strength α' may lie. This expression is roughly equivalent to Eq. (1.2), except that now, instead of the dimension d, we have a fractal dimension $f(\alpha)$

which varies with α. Thus, we model fractal measures by interwoven sets of singularities of strength α, each characterized by its own dimension $f(\alpha)$. The rest of our formalism attempts to unravel this complexity in a workable fashion.

The concept of a singularity strength α was stressed in the context of diffusion-limited aggregation in independent work of Turkevich and Scher[12] and of Halsey, Meakin, and Procaccia.[8] The latter group pointed out the significance of the density of singularities and expressed it in terms of f.

In order to determine the function $f(\alpha)$ for a given measure, we must relate it to observable properties of the measure. We relate $f(\alpha)$ to a set of dimensions which have been introduced by Hentschel and Procaccia, the set D_q defined by[13]

$$D_q=\lim_{l\to 0}\left[\frac{1}{q-1}\frac{\ln\chi(q)}{\ln l}\right], \tag{1.6}$$

where

$$\chi(q)=\sum_i p_i^q . \tag{1.7}$$

D_0 is just the fractal dimension of the support of the measure, while D_1 is the information dimension and D_2 is the correlation dimension.[14]

As q is varied in Eq. (1.7), different subsets, which are associated with different scaling indices, become dominant. Substituting Eqs. (1.4) and (1.5) into Eq. (1.7), we obtain

$$\chi(q)=\int d\alpha'\rho(\alpha')l^{-f(\alpha')}l^{q\alpha'} . \tag{1.8}$$

Since l is very small, the integral in Eq. (1.8) will be dominated by the value of α' which makes $q\alpha'-f(\alpha')$ smallest, provided that $\rho(\alpha')$ is nonzero. Thus, we replace α' by $\alpha(q)$, which is defined by the extremal condition

$$\frac{d}{d\alpha'}[q\alpha'-f(\alpha')]\bigg|_{\alpha'=\alpha(q)}=0 .$$

We also have

$$\frac{d^2}{d(\alpha')^2}[q\alpha'-f(\alpha')]\bigg|_{\alpha'=\alpha(q)}>0 ,$$

so that

$$f'(\alpha(q))=q , \tag{1.9a}$$

$$f''(\alpha(q))<0 . \tag{1.9b}$$

It then follows from Eq. (1.6) that[8]

$$D_q=\frac{1}{q-1}[q\alpha(q)-f(\alpha(q))] . \tag{1.10}$$

Thus, if we know $f(\alpha)$, and the spectrum of α values, we can find D_q. Alternatively, given D_q, we can find $\alpha(q)$ since

$$\alpha(q)=\frac{d}{dq}[(q-1)D_q] , \tag{1.11}$$

and, knowing $\alpha(q)$, $f(q)$ can be obtained from Eq. (1.10).

Equations (1.9)–(1.11) are the main formal results used in this paper. In the next section we develop the formalism outlined here in somewhat more detail and apply it to systems with strong self-similarity properties. In Sec. III we apply the formalism to some important examples of measures arising in dynamical systems. We examine the 2^∞ cycle of period doubling,[15] the devil's staircase of mode locking in circle maps,[16,17] and the elements of the critical cycle at the onset of chaos in circle maps with golden-mean winding number.[18–20] Although all of these cases have been examined previously, we are able to find a *smooth* function with which to characterize them. Furthermore, these characterizations are universal. Other attempts to study these measures have led to nowhere smooth scaling functions.[15,21] Since the characterizations are functions rather than numbers, they offer much more information than fractal dimensions. Unlike power spectra, these functions possess an immediate connection to the metric properties of the measures involved, and do not call for cumbersome interpretation. Therefore, we believe that experimental measurements of D_q, and thus of $f(\alpha)$, should replace more common tests of universality in the transition to chaos. We give many examples of the procedures employed, and we hope to encourage experiments to follow these lines.

II. EXACTLY SOLUBLE STRANGE SETS

A. Preliminaries

We begin by introducing a more general definition of the dimensions D_q. Consider a strange set S embedded in a finite portion of d-dimensional Euclidean space. Imagine partitioning the set into some number of disjoint pieces, $S_1,S_2,\ldots,S_N$, in which each piece has a measure p_i and lies within a ball of radius l_i, where each l_i is restricted by $l_i<l$. Then define a partition function

$$\Gamma(q,\tau,\{S_i\},l)=\sum_{i=1}^{N}\frac{p_i^q}{l_i^\tau} . \tag{2.1}$$

Eventually we shall argue that, for large N, this partition function is of the order unity only when

$$\tau=(q-1)D_q . \tag{2.2}$$

To make this argument, consider now two regions:

$$\text{region } A:\ q\ge 1,\ \tau\ge 0 , \tag{2.3a}$$

$$\text{region } B:\ q\le 1,\ \tau\le 0 . \tag{2.3b}$$

In region A, adjust the partition $\{S_i\}$ so as to maximize Γ. In region B, adjust it so that Γ is as small as possible. Then define

$$\Gamma(q,\tau,l)=\text{Sup}\ \Gamma(q,\tau,\{S_i\},l)\ \ (\text{region } A) , \tag{2.4a}$$

$$\Gamma(q,\tau,l)=\text{Inf}\ \Gamma(q,\tau,\{S_i\},l)\ \ (\text{region } B) . \tag{2.4b}$$

The supremum in region A will exist as long as there are constants $a>0$ and $\alpha_0>0$, so that for any possible subset of S, $\{S_i\}$, we have

$$p_i\le a(l_i)^{\alpha_0} . \tag{2.5}$$

Then $\Gamma(q,\tau,l)$ will exist and be less than infinity whenever

$$\alpha_0(q-1)>\tau . \tag{2.6}$$

Next define

$$\Gamma(q,\tau)=\lim_{l\to 0}[\Gamma(q,\tau,l)] . \tag{2.7}$$

Notice that $\Gamma(q,\tau)$ is a monotone nondecreasing function of τ and a monotone nonincreasing function of q. One can argue that there is a unique function $\tau(q)$ such that

$$\Gamma(q,\tau)=\begin{cases}\infty & \text{for } \tau>\tau(q),\\ 0 & \text{for } \tau<\tau(q) .\end{cases} \tag{2.8}$$

Equation (2.8) permits us to define D_q as

$$(q-1)D_q=\tau(q) . \tag{2.9}$$

Once D_q is known, Eqs. (1.10) and (1.11) will then give $\alpha(q)$ and $f(q)$. Notice that our definition of D_q is precisely the one which makes D_0 the Hausdorff dimension.

B. Connection to previously defined D_q

Hentschel and Procaccia[13] also defined a D_q, which we now denote as D_q^{HP}. To relate the two quantities, recall that the authors of Ref. 13 defined a partition in which all the diameters l_i had the same value l. We know that

$$\Gamma(q,\tau,l)\begin{cases}>l^{-\tau}\sum_{i=1}^{N}p_i^q & \text{(region } A)\\ <l^{-\tau}\sum_{i=1}^{N}p_i^q & \text{(region } B) .\end{cases} \tag{2.10}$$

If τ is chosen correctly, i.e., $\tau=\tau(q)$, the left-hand side of Eq. (2.10) will neither go to zero nor diverge very strongly as $l\to 0$. In particular, we guess that $\Gamma(l)$ is no worse than logarithmically dependent upon these quantities. Then

$$\lim_{l\to 0}[\ln\Gamma(q,\tau(q),l)/\ln l]\to 0 .$$

We have now

$$\frac{\tau}{q-1}\le\lim_{l\to 0}\left[\frac{\ln\left[\sum_{i=1}^{N}p_i^q\right]}{(\ln l)(q-1)}\right] . \tag{2.11}$$

The right-hand side of (2.11) is D_q^{HP}. We thus find

$$D_q\le D_q^{\mathrm{HP}} . \tag{2.12}$$

Since we believe that Eq. (2.10) will often be an order of magnitude equality when $\tau=\tau(q)$, we think that Eq. (2.12) will be an equality in most cases of interest.

At this point we turn to some simple examples to illustrate the quantities $\tau(q)$. These examples will enable us to gain intuition about the quantities $\alpha(q)$ and $f(\alpha)$.

C. Exactly soluble examples

1. Power-law singularity

One of the simplest possible applications of this formalism is to a probability measure with only one power-law singularity. Imagine a probability density $\rho(x)=\tilde{\alpha}x^{\tilde{\alpha}-1}$ on $x\in[0,1]$, where $0<\tilde{\alpha}<1$. Let us partition the interval into N segments $[x_i, x_i+\Delta x]$, with $\Delta x=N^{-1}$. The total probability measure on all of these intervals except for that adjoining zero is well approximated by $\rho(x_i)\Delta x$. The probability upon the segment adjoining zero possesses a probability $p_0=(\Delta x)^{\tilde{\alpha}}$. The partition function is therefore

$$\Gamma(q,\tau,\Delta x)\approx\frac{(\Delta x)^{\tilde{\alpha}q}}{(\Delta x)^{\tau}}+\sum_{i\neq 0}\frac{\tilde{\alpha}x_i^{\tilde{\alpha}-1}(\Delta x)^q}{(\Delta x)^{\tau}} . \tag{2.13}$$

There are $(\Delta x)^{-1}$ terms in the sum, so that

$$\Gamma(q,\tau,\Delta x)\sim(\Delta x)^{\tilde{\alpha}q-\tau}+(\Delta x)^{q-1-\tau} . \tag{2.14}$$

Thus, since we require that Γ neither go to zero nor infinity, we have that

$$\tau=\min\{q-1,\tilde{\alpha}q\} , \tag{2.15a}$$

or

$$D_q=\frac{1}{q-1}\min\{q-1,\tilde{\alpha}q\} . \tag{2.15b}$$

Thus for $q>q^*=1/(1-\tilde{\alpha})$, the dimensions correspond to a value of $\alpha=\tilde{\alpha}$ and of $f=0$, while for $q<q^*$ the dimensions correspond to $\alpha=1$ and $f=1$. Thus, in this example the f-α spectrum consists of two points, corresponding to the two types of behavior in the measure.

2. Cantor sets and generators

If a measure possesses an exact recursive structure, one can find its D_q. Suppose that the measure can be generated by the following process. Start with the original region which has measure 1 and size 1. Divide the region into pieces S_i, $i=1,2,\ldots,N$, with measure p_i and size l_i. Suppose that the maximum of l_i is given by l. Then at the first stage we can construct a partition function,

$$\Gamma(q,\tau,l)=\sum_i\frac{p_i^q}{l_i^{\tau}} . \tag{2.16}$$

Continue the Cantor construction. At the next stage each piece of the set is further divided into N pieces, each with a measure reduced by a factor p_j and size by a factor l_j. At this level the partition function will be

$$\Gamma(q,\tau,l^2)=[\Gamma(q,\tau,l)]^2 . \tag{2.17}$$

We see at once that, for this kind of measure, the first partition function $\Gamma(l)$ will generate all the others, and that $\tau(q)$ is defined by

$$\Gamma(q,\tau(q),l)=1 . \tag{2.18}$$

If a partition with finite N yields a Γ which obeys (2.17), that partition is called a generator.[22]

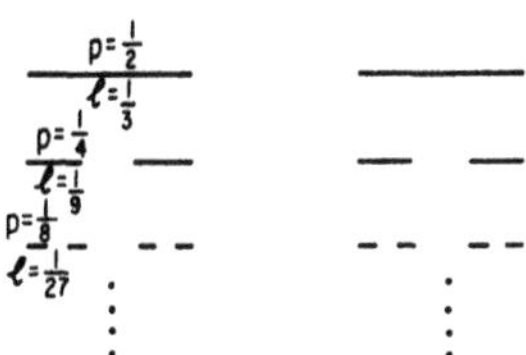

FIG. 1. The construction of the uniform Cantor set. At each stage of the construction the central third of each segment is removed from the set. Each segment has measure $p_0=(\frac{1}{2})^n$ and scale $l_0=(\frac{1}{3})^n$, where n is the number of generations.

3. *Uniform Cantor set*

A simple example is the classical Cantor set obtained by dividing the interval [0,1] as shown in Fig. 1. We initially replace the unit interval with two intervals, each of length $l=\frac{1}{3}$. Each of these intervals receives the same measure $p=\frac{1}{2}$. At the next stage of the construction of the measure this same process is repeated on each of these two intervals. Thus, for this measure we require

$$2\left[\frac{(\frac{1}{2})^q}{(\frac{1}{3})^\tau}\right]=1\,, \qquad (2.19)$$

which yields

$$\tau=(q-1)[\ln(2)/\ln(3)] \;\; [\text{or } D_q=\ln(2)/\ln(3)]\,. \qquad (2.20)$$

If l_0 is the length scale of the intervals at a particular level of the partitioning, and p_0 is the measure for such an interval, then

$$p_0=l_0^{\ln(2)/\ln(3)}\,. \qquad (2.21)$$

Calling the index of the singularity α, i.e., $p_0\sim l_0^\alpha$, we have here $\alpha=\ln(2)/\ln(3)$. If we further ask what is the density of these singularities, we find immediately that it is simply the density of the set,

$$\rho(l_0)=\frac{1}{l_0^{\ln(2)/\ln(3)}}\,, \qquad (2.22)$$

and Eq. (1.5) leads to $f=\ln(2)/\ln(3)$. Thus in this example, $\alpha=f$, and also

$$\tau(q)=q\alpha-f\,. \qquad (2.23)$$

Although Eq. (2.23) is trivial here, we shall see that its analog, Eq. (1.10), also holds in the most general cases.

4. *Two-scale Cantor set*

A somewhat less trivial example is obtained by constructing a Cantor set as in Fig. 2. Here we use two rescaling parameters l_1 and l_2 and two measures p_1 and p_2, and then continue to subdivide self-similarly. We assume that $l_2>l_1$. It is apparent that this example also has a generator, since the condition

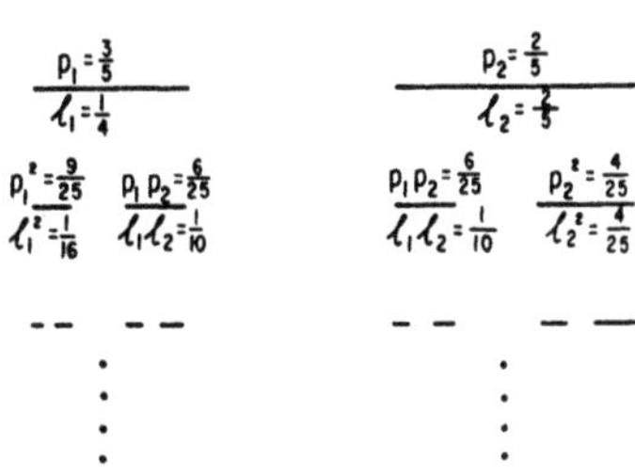

FIG. 2. A Cantor-set construction with two rescalings $l_1=0.25$ and $l_2=0.4$ and respective measure rescalings $p_1=0.6$ and $p_2=0.4$. The division of the set continues self-similarly.

$$\Gamma(q,\tau,l_2^n)=\left[\frac{p_1^q}{l_1^\tau}+\frac{p_2^q}{l_2^\tau}\right]^n=1 \qquad (2.24)$$

results in a τ that does not depend on n. The value of τ depends, however, on q. In Fig. 3 we show $D_q=\tau(q)/(q-1)$ as a function of q, as obtained numerically by solving Eq. (2.24). To further understand this curve, we can examine the quantity $\Gamma(l_2^n)$ for this case explicitly:

$$\Gamma(q,\tau,l_2^n)=\sum_m \binom{n}{m} p_1^{mq}p_2^{(n-m)q}(l_1^m l_2^{n-m})^{-\tau}=1\,. \qquad (2.25)$$

We expect that in the limit $n\to\infty$ the largest term in this sum should dominate. To find the largest term we compute

$$\frac{\partial \ln\Gamma(l_2^n)}{\partial m}=0\,. \qquad (2.26)$$

Using the Stirling approximation, we find that Eq. (2.26) is equivalent to

$$\tau=\frac{\ln(n/m-1)+q\ln(p_1/p_2)}{\ln(l_1/l_2)}\,. \qquad (2.27)$$

Since we expect that the maximal term dominates the sum, we have a second equation,

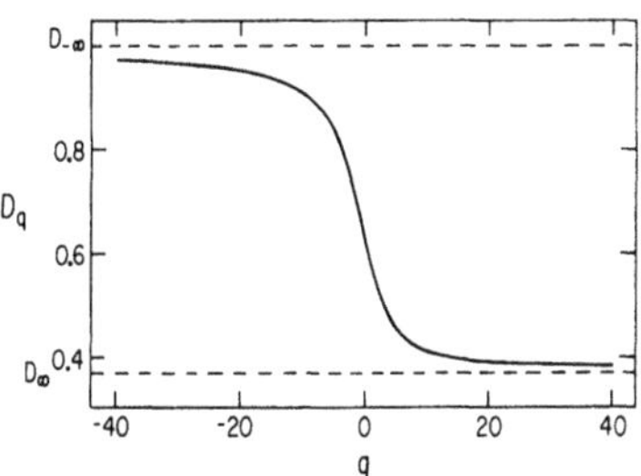

FIG. 3. D_q plotted vs q for the two-scale Cantor set of Fig. 2.

$$\binom{n}{m} p_1^{mq} p_2^{(n-m)} \left[\frac{1}{l_1^m l_2^{(n-m)}} \right]^{\tau} = 1 . \tag{2.28}$$

Inserting Eq. (2.27) into Eq. (2.28) leads to an equation for n/m. After some algebraic manipulation, one finds

$$\ln(n/m)\ln(l_1/l_2) - \ln(n/m-1)\ln l_1 = q(\ln p_1 \ln l_2 - \ln p_2 \ln l_1) . \tag{2.29}$$

We thus see that for any given q there will be a value of n/m which solves Eq. (2.29) and, in turn, determines τ from Eq. (2.27). This maximal term which determines τ actually comes from a set of $\binom{n}{m}$ segments, all of which have the same size $l_1^m l_2^{(n-m)}$. Their density exponent f is determined by

$$\binom{n}{m} (l_1^m l_2^{(n-m)})^f = 1 , \tag{2.30}$$

or

$$f = \frac{(n/m-1)\ln(n/m-1) - (n/m)\ln(n/m)}{\ln l_1 + (n/m-1)\ln l_2} . \tag{2.31}$$

The exponent determining the singularity in the measure, α, is determined by

$$p_1^m p_2^{(n-m)} = (l_1^m l_2^{(n-m)})^{\alpha} , \tag{2.32}$$

or

$$\alpha = \frac{\ln p_1 + (n/m-1)\ln p_2}{\ln l_1 + (n/m-1)\ln l_2} . \tag{2.33}$$

Thus, for any chosen q, the measure scales as $\alpha(q)$ on a set of segments which converge to a set of dimension $f(q)$. As q is varied, different regions of the set determine D_q. It can be shown that Eqs. (2.27), (2.29), (2.31), and (2.33) again lead to

$$\tau = (q-1)D_q = q\alpha(q) - f(q) . \tag{2.34}$$

We can also understand the spectrum of scaling indices α by considering the "kneading sequences" for the segments. In the first level of the construction there are two segments of sizes l_1 and l_2 and measures p_1 and p_2 which we can label L (left) and R (right). At the next level we have four segments, which we can reach by going left or right: LL, LR, RL, and RR. Thus the measure and the size of any segment are determined by its address, the kneading sequence of L's and R's. For example, the size of a segment is $l_1^m l_2^{(n-m)}$, where m and $n-m$ are, respectively, the numbers of L's and R's in the kneading sequence. Clearly, the sequence $LLL\ldots LLL\ldots$ is associated with $\alpha = \ln(p_1)/\ln(l_1) = D_\infty$, which lies on the edge of the spectrum, while the sequence $RRR\ldots RRR\ldots$ is associated with the singularity lying on the other edge of the spectrum. Other, less trivial kneading sequences lead to values of α between these two extremes. We note, however, that it is only the infinite "tail" of the sequence that determines the asymptotic scaling behavior. The number of sequences leading to the same singularity α may be simply found, and leads via Eq. (2.30) to exactly the same results for $f(\alpha)$ as the partition-function analysis above.

Finally, in Fig. 4 we display the curve $f(\alpha)$. The curve has been obtained for $l_1 = 0.25$, $l_2 = 0.4$ and $p_1 = 0.6$,

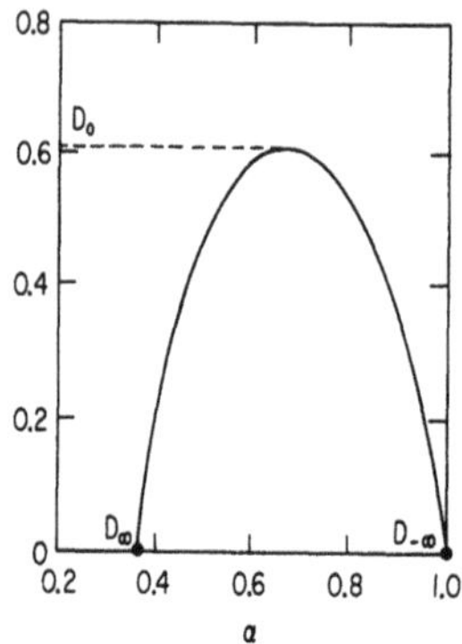

FIG. 4. The plot of f vs α for the set in Fig. 2. Note that $f=0$ corresponds to α values $D_{-\infty} = \ln(0.4)/\ln(0.4) = 1.0$ and $D_\infty = \ln(0.6)/\ln(0.25) = 0.3684$.

$p_2 = 0.4$. The leftmost point on the curve is $f=0$, $\alpha = \ln(0.6)/\ln(0.25)$. This is the value that in Eqs. (2.31) and (2.33) obtains for $n = m$. At any level of the construction there is exactly one such segment ($f=0$) and the singularity is

$$\ln p_1 / \ln l_1 = \mathrm{Inf}\{\ln p_1/\ln l_1, \ln p_2/\ln_2\} .$$

This value of α is also D_∞. The rightmost point on the graph again corresponds to $f=0$, but now

$$\alpha = \ln p_2/\ln l_2 = \mathrm{Sup}\{\ln p_1/\ln l_1, \ln p_2/\ln l_2\} .$$

This is also $D_{-\infty}$. Whereas D_∞ corresponds to the region in the set where the measure is most concentrated, $D_{-\infty}$ corresponds to that where the measure is most rarefied. For $q=0$ we simply obtain $f = D_0$, where D_0 is the Hausdorff dimension of the set. This is the maximum of the graph $f(\alpha)$.

Certain features of this curve are quite general, and follow from Eqs. (1.9)–(1.12). From Eq. (1.9) we find immediately that

$$\frac{\partial f}{\partial \alpha} = q , \tag{2.35a}$$

$$\frac{\partial^2 f}{\partial \alpha^2} < 0 . \tag{2.35b}$$

Thus, for any measure the curve $f(\alpha)$ will be convex, with a single maximum at $q=0$, and with infinite slope at $q = \pm\infty$. Also from Eq. (1.10) with $q=1$, we find that $\alpha(1) = f(1)$. The slope $\partial f/\partial\alpha$ there is unity. This general behavior of the curve $f(\alpha)$ will be seen in all cases where the measure possesses a continuous spectrum.

Although this example is rather simple, it contains many of the properties of the richer sets considered in Sec. III. In particular, we will not lose this intuitive view of the meaning of α and f.

5. Other types of spectra

We can obtain more insight into the meaning of the f-α spectrum for a measure by considering two examples of measures on continuous supports. Many of the most interesting measures encountered in applications lie on continuous supports, including the growth measure for diffusion-limited aggregates and strange attractors for systems of ordinary differential equations.

The first example is a simple generalization of the two-scale Cantor set defined by (2.24). A unit interval is subdivided into three segments, two of length l_2 and one of length l_1. The two former intervals each receive a proportion of the total measure given by p_2, and the latter interval receives a proportion given by p_1. We imagine that $l_1+2l_2=1$ and that $p_1+2p_2=1$. We also imagine, for the sake of the argument below, that $p_2/l_2>p_1/l_1$ and that $l_2>l_1$. Each of these three intervals is then subdivided in the same manner, and so forth. Although the measure on the line segment is rearranged at each step of the recursive process, the support for the measure remains at each step the original line segment. Thus we expect that D_0 for this measure will be 1. Furthermore, the densest intervals on the line segment contract not to one point (as was the case in the two-scale Cantor set), but to a set of points of finite dimension. Thus, we expect the lowest value of α, and hence the value of D_∞, to correspond to a nonzero value of f. Note that there is always only one segment at the lowest value of the density, so that we still expect $D_{-\infty}$ to correspond to a value of $f=0$. The condition (2.18) above on Γ requires that

$$\Gamma(q,\tau,l_2)=\frac{p_1^q}{l_1^\tau}+2\frac{p_2^q}{l_2^\tau}=1 . \tag{2.36}$$

The solution is simple and is displayed in Fig. 5. As predicted above, $f(q\rightarrow\infty)\neq 0$, so that the leftmost part of the f-α curve resembles a hook.

The second example is a set generated according to a different rule than the Cantor sets. The method is displayed in Fig. 6. At each stage, only the regions which have had their measure multiplied by a factor p_2 in the preceding stage are subdivided further, while the regions which have had their measure multiplied by a factor p_1 are not subdivided further. Thus the expression for the measure density of any region, at any stage of the iterative construction, will have, at most, one factor of p_1. The measure generated by this construction is much less self-similar than that considered in Sec. III C 4. For this measure the partition function is given for large n by

$$\Gamma(q,\tau,l_2^n)=(p_1^q/l_1^\tau)\Gamma^U(l_2^{n-1})+2(p_2^q/l_2^\tau)\Gamma(l_2^{n-1}) , \tag{2.37}$$

where Γ^U is the partition function for a uniform measure on a line segment. It is easy to show that

$$\tau(q)=\min\{q-1, q\tilde{\alpha}-\tilde{f}\} , \tag{2.38}$$

with $\tilde{\alpha}=\ln(p_2)/\ln(l_2)$, and $\tilde{f}=\ln(\frac{1}{2})/\ln(l_2)$. This example corresponds to a discrete, rather than a continuous, f-α curve, consisting of a point at $(\tilde{\alpha},\tilde{f})$ and a point at (1,1). This result should not surprise us, as this measure is properly described as a nonsingular background interrupted by singularities upon a Cantor set of dimension $\tilde{f}$.

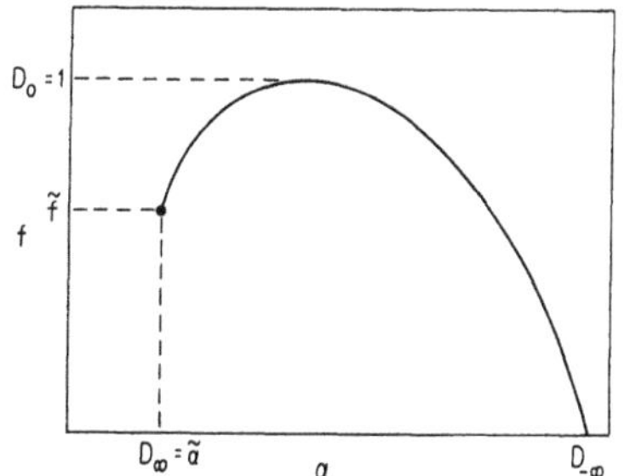

FIG. 5. The function $f(\alpha)$ for the measure defined by Eq. (2.36). Note that D_∞ corresponds to a nonzero value of f. Also, $D_0=1$.

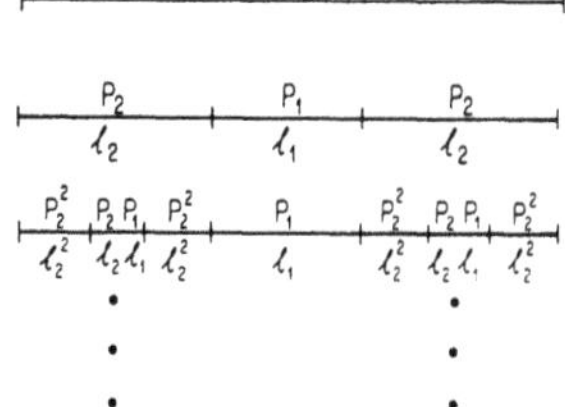

FIG. 6. The partitioning process for the measure yielding the partition function (2.37). Only those segments receiving a measure multiplied by p_2 at any stage of the construction are further subdivided. This measure is far less self-similar than those generated by the Cantor process.

III. EXAMPLES FROM DYNAMICAL SYSTEMS

In this section we examine the implications of the formalism of Sec. II for three examples: (i) the 2^∞ cycle at the accumulation point of period doubling, (ii) the set of irrational winding numbers at the onset of chaos via quasiperiodicity, and (iii) the critical cycle elements at the golden-mean winding number for the same problem. In all cases we calculate numerically the D_q, and use Eqs. (1.10) and (1.11) to extract $\alpha(q)$, $f(q)$, and a plot of $f(\alpha)$. In all three cases we can find theoretically $D_\infty, D_{-\infty}$, and thus $\alpha(q=\pm\infty)$.

A. The 2^∞ cycle of period doubling

Dynamical systems that period double on their way to chaos can be represented by one-parameter families of maps $M_\lambda(\mathbf{x})$, where M_λ: $R^F\rightarrow R^F$, and F is the number of degrees of freedom. At values of $\lambda=\lambda_n$ the system

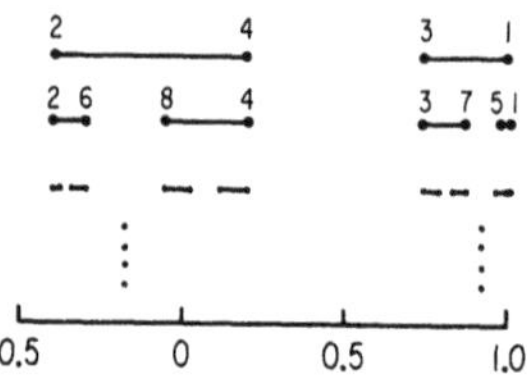

FIG. 7. The construction of the period-doubling attractor; the indices refer to the number of the iterate of $x=0$. The lines represent the scales l_i. Note the similarity with Fig. 2.

gains a stable 2^n-periodic orbit. This period-doubling cascade accumulates at λ_∞, where the system possesses a 2^∞ orbit. We generated numerically the set of elements of this orbit for the map $x'=\lambda(1-2x^2)$, with $\lambda_\infty \approx 0.837\,005\,134\ldots$.[15] The points making up the cycle are displayed in Fig. 7. The iterates of $x=0$ form a Cantor set, with half the iterates falling between $f(0)$ and $f^3(0)$ and the other half between $f^2(0)$ and $f^4(0)$. The most natural partition, $\{S_i\}$, for this case simply follows the natural construction of the Cantor set as shown in Fig. 7. At each level of the construction of this set, each l_i is the distance between a point and the iterate which is closest to it. The measures p_i of these intervals are all equal.

With 2^{11}-cycle elements we solved numerically $\Gamma=1$, thereby generating the D_q-versus-q curve shown in Fig. 8. From these results we calculated $\alpha(q)$ from Eq. (1.11) and $f(\alpha)$ from (1.10). The curve $f(\alpha)$ is displayed in Fig. 9.

To understand the shape of the curve in Fig. 9 we first consider the end points of the curve (for which $f=0$). As with the example solved in Sec. II C 4, we expect these two points to be determined by the most rarefied and the most concentrated intervals in the set. As has been shown by Feigenbaum,[15] these have scales $l_{-\infty}\sim\alpha_{PD}^{-n}$ and $l_{+\infty}\sim\alpha_{PD}^{-2n}$, respectively, where $\alpha_{PD}=2.502\,907\,875\ldots$ is the universal scaling factor.[15] Since the measures there

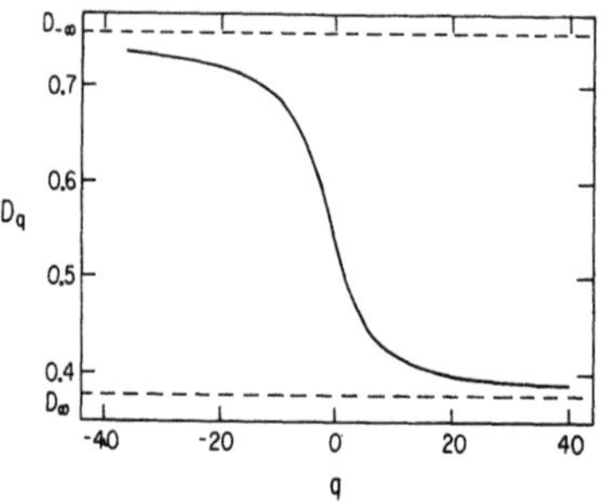

FIG. 8. D_q vs q calculated for the period-doubling attractor of Fig. 7.

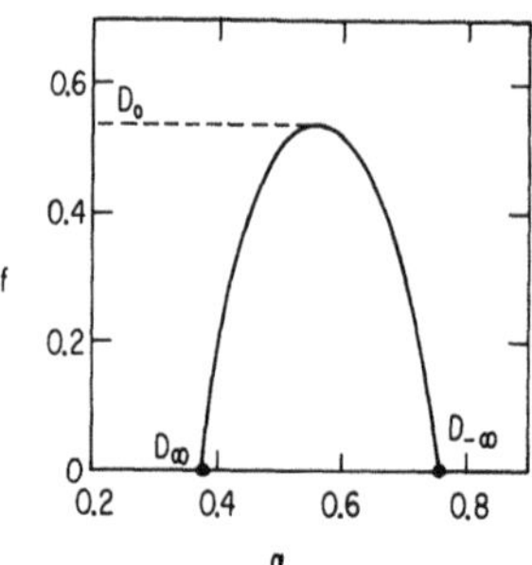

FIG. 9. The function $f(\alpha)$ for the period-doubling attractor of Fig. 7.

are simply $p_{-\infty}\sim 2^{-n}$, we expect these end points to be $\ln p_{-\infty}/\ln l_{-\infty}$ and $\ln p_\infty/\ln l_\infty$, respectively. These values are also $D_{-\infty}$ and D_∞, so that we find

$$D_{-\infty}=\frac{\ln 2}{\ln\alpha_{PD}}=0.755\,51\ldots, \tag{3.1a}$$

$$D_{\infty}=\frac{\ln 2}{\ln\alpha_{PD}^2}=0.377\,75\ldots. \tag{3.1b}$$

These values are in extremely good agreement with the numerically determined endpoints of the graph. The curve $f(\alpha)$ is perfectly smooth. The maximum is at $D_0=0.537\ldots$, in agreement with previous calculations of the Hausdorff dimension for this set. Since the slope of the curve $f(\alpha)$ is q, $\alpha(q)$ will be very close to $D_{\pm\infty}$ even for $|q|\sim 10$. However, Fig. 8 indicates that D_q is far from converged to $D_{\pm\infty}$ even for $q\sim\pm 40$. Thus, the transformations (1.10) and (1.11) lead more easily to good estimations of $D_{\pm\infty}$ than do direct calculations of the D_q's.

B. Mode-locking structure

Dynamical systems possessing a natural frequency ω_1 display very rich behavior when driven by an external frequency ω_2. When the "bare" winding number $\Omega=\omega_1/\omega_2$ is close to a rational number, the system tends to mode lock. The resulting "dressed" winding number, i.e., the ratio of the response frequency to the driving frequency, is constant and rational for a small range of the parameter Ω. At the onset of chaos the set of irrational dressed winding numbers is a set of measure zero, which is a strange set of the type discussed above. The structure of the mode locking is best understood in terms of the "devil's staircase" representing the dressed winding number as a function of the bare one. Such a staircase is shown in Fig. 10 as obtained for the map[16,17]

$$\theta_{n+1}=\theta_n+\Omega-\frac{K}{2\pi}\sin(2\pi\theta_n), \tag{3.2}$$

with $K=1$, which is the onset value above which chaotic orbits exist.

To calculate $D_{\pm\infty}$ analytically we make use of previous findings that the most extremal behaviors of this staircase are found at the golden-mean sequence of dressed winding numbers

$$F_n/F_{n+1} \to w^* = (\sqrt{5}-1)/2 \approx 0.6108\ldots ,$$

where F_n are the Fibonacci numbers, ($F_0=0$, $F_1=1$, and $F_n=F_{n-1}+F_{n-2}$ for $n \geq 2$) and at the harmonic sequence $1/Q \to 0$.[16,17] The most rarefied region of the staircase is located around the golden mean. Shenker found that the length scales l_i vary in that neighborhood as $l_{-\infty} \sim F_n^{-\delta} \sim (w^*)^{n\delta}$, where $\delta = 2.1644\ldots$ is a universal number.[18] The corresponding changes in dressed winding number are

$$p_{-\infty} \sim F_n/F_{n+1} - F_{n+1}/F_{n+2} \sim (w^*)^{2n} .$$

We thus conclude that

$$D_{-\infty} = \frac{\ln p_{-\infty}}{\ln l_{-\infty}} = \frac{2}{\delta} = 0.9240\ldots . \tag{3.3a}$$

For the $1/Q$ series it has been shown that changes in dressed winding number go as the square root of changes in bare winding number, i.e., that $p_i \sim l_i^{1/2}$.[16] This series determines the most concentrated portion of the staircase (Fig. 10), which means that $p_\infty \sim l_\infty^{1/2}$, leading to

$$D_\infty = \ln p_\infty / \ln l_\infty = \tfrac{1}{2} . \tag{3.3b}$$

To construct the curve $f(\alpha)$ we generated 1024 mode-locked intervals following the Farey construction, which also defines the partition $\{S_i\}$.[17] For each two neighboring intervals (see Fig. 10) we measured the change both in bare and in dressed winding numbers. The changes in bare winding numbers determined the scales l_i of the partition $\{S_i\}$, whereas the changes in dressed winding numbers were defined to be the measures p_i. Solving then the equation $\Gamma = 1$ we generated D_q as shown in Fig. 11 (for $q>0$ we accelerated the convergence as will be described shortly). Figure 12 shows $f(\alpha)$ for this case. Again the curve is smooth, in contrast to scaling functions found for

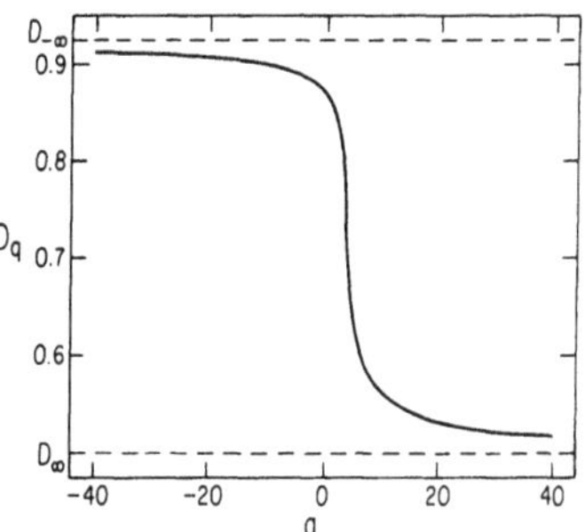

FIG. 11. D_q vs q for the staircase of Fig. 10.

the same problem by other authors.[15,21] Note that the maximum on Fig. 12 gives the fractal dimension D_0 of the mode-locking structure as $D_0 \approx 0.87\ldots$, in agreement with the predictions of Refs. 16 and 17. The rightmost branch of the curve $f(\alpha)$ in Fig. 12 (i.e., for $q<0$) converges vary rapidly within the Farey partition. This is, however, not the case for the leftmost branch (i.e., for $q>0$). To improve the convergence of this portion of the curve substantially, we made use of the following trick. In general, the partition function (2.1) will be of the form

$$\Gamma(l) = a e^{\gamma \ln l} , \tag{3.4}$$

where a and γ are constants. The convergence is often slowed down by the prefactor a and by the logarithmic dependence on l. However, by considering instead the ratio

$$\frac{\Gamma(l)}{\Gamma(2l)} = e^{-\gamma \ln 2} , \tag{3.5}$$

we find that a and l do not appear in the equation. We thus determine $\tau(q)$ by requiring that

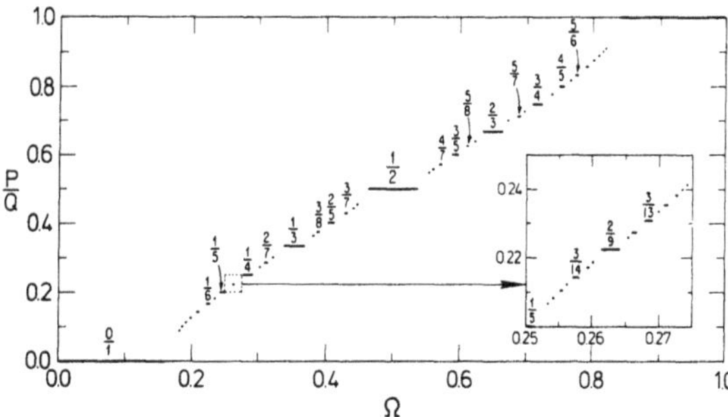

FIG. 10. The "devil's staircase" for the critical circle map of Eq. (3.2). The "dressed" winding number is plotted vs the "bare" winding number (Ref. 16).

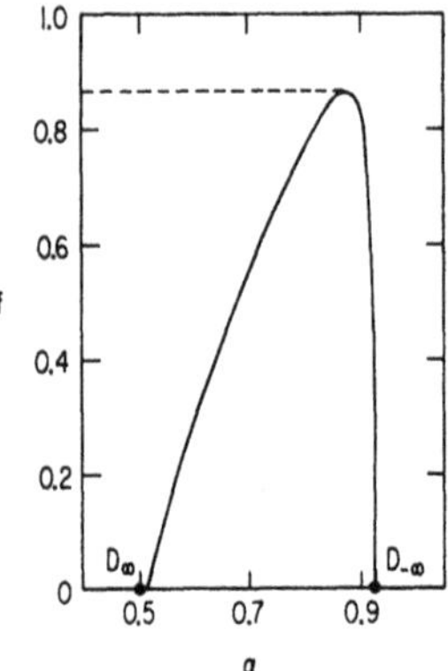

FIG. 12. A plot of f vs α for the mode-locking structure of the circle map. The left portion of the curve is found by accelerated convergence as described in Sec. III B.

$$\frac{\Gamma(l)}{\Gamma(2l)}(\tau,q)=1 .$$

In general, the denominator can be chosen to be of the form $\Gamma(bl)$, where b is a constant. The leftmost portion of the curve was generated with this method by calculating $\Gamma(l(1452))/\Gamma(l(886))=1$ [where $l(1452)$ and $l(886)$ are the maximal scales for partitions with 1452 and 886 intervals, respectively], and we observe that it passes through the point $(D_\infty,0)$. We found empirically that this method usually did not give reliable results for large values of $|q|$. Still, this method did successfully generate the entire curve in Fig. 12. We emphasize the ease of this measurement. The rightmost branch of the f-α curve of Fig. 12 converges very rapidly, even when only 8–16 mode-locked intervals are available.

C. Quasiperiodic trajectories for circle maps

Circle maps of the type (3.1) exhibit a transition to chaos via quasiperiodicity. A well-studied transition takes place at $K=1$ with dressed winding number equal to the golden mean, w^*. [18–20] We have at this point studied the structure of the trajectory $\theta_1,\theta_2,\ldots,\theta_i,\ldots$. To perform the numerical calculation we chose $\theta_1=f(0)$ and truncated the series θ_i at $i=2584=F_{17}$. The distances $l_i=\theta_{i+F_{16}}-\theta_i$ (calculated mod 1) define natural scales for the partition with measures $p_i=1/2584$ attributed to each scale. Figure 13 shows D_q versus q calculated for this set and Fig. 14 shows the corresponding function $f(\alpha)$. Again the curve is smooth. Shenker found for this problem that the distances around $\theta\sim 0$ scale down by a universal factor $\alpha_{\rm GM}=1.288\,5\ldots$ when the trajectory θ_i is truncated at two consecutive Fibonacci numbers, F_n,F_{n+1}.[18] This corresponds to the most rarefied region so that $l_{-\infty}\sim\alpha_{\rm GM}^{-n}$. The corresponding measure scales as $p_{-\infty}\sim 1/F_n\sim(\omega^*)^n$, leading to

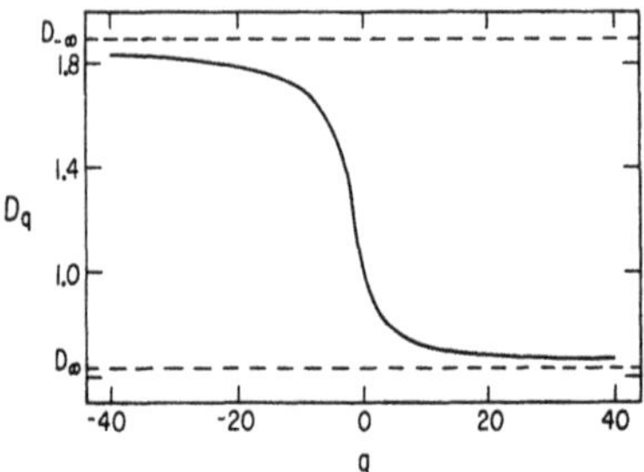

FIG. 13. D_q plotted vs q for the critical trajectory of a circle map with golden-mean winding number.

$$D_{-\infty}=\frac{\ln w^*}{\ln\alpha_{\rm GM}^{-1}}=1.898\,0\ldots . \tag{3.6a}$$

The map (3.2) for $K=1$ has at $\theta=0$ a zero slope with a cubic inflection and is otherwise monotonic. The neighborhood around $\theta=0$, which is the most rarefied region of the set, will therefore be mapped onto the most concentrated region of the set. As the neighborhood around $\theta=0$ scales as $\alpha_{\rm GM}$ when the Fibonacci index is varied, the most concentrated regime will scale as $\alpha_{\rm GM}^3$ due to the cubic inflection. This means that $l_\infty=\alpha_{\rm GM}^{-3n}$ and $p_\infty=(w^*)^n$, so that we obtain

$$D_\infty=\frac{\ln w^*}{\ln\alpha_{\rm GM}^{-3}}=0.632\,6\ldots . \tag{3.6b}$$

Figure 14 shows that the curve passes very close to the points $(D_\infty,0)$ and $(D_{-\infty},0)$. Again, however, we find

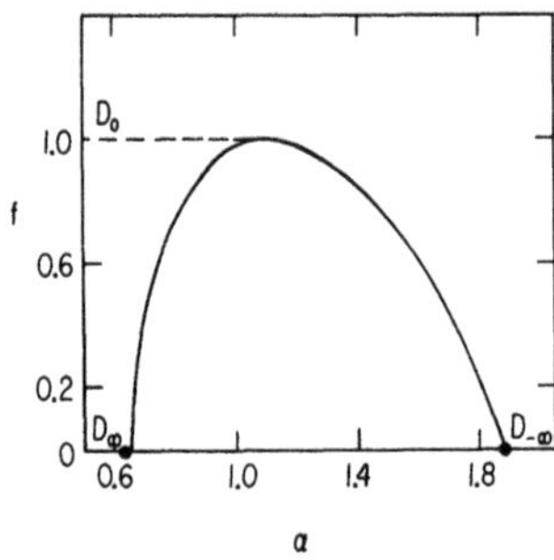

FIG. 14. A plot of f vs α for the golden-mean trajectory for the circle map.

that the dimensions D_q are far from $D_{\pm\infty}$ even for $q \sim \pm 40$.

To check for universality it is important to investigate $f(\alpha)$ for a higher-dimensional version of a circle map. We chose the dissipative standard map,

$$\theta_{n+1}=\theta_n+\Omega+br_n-\frac{K}{2\pi}\sin(2\pi\theta_n)\,,$$
$$r_{n+1}=br_n-\frac{K}{2\pi}\sin(2\pi\theta_n)\,, \tag{3.7}$$

and studied the critical cycle for $b=0.5$, again truncated at $i=2584=F_{17}$. We defined the scales by the Euclidean distances

$$l_i=[(\theta_{i+F_{16}}-\theta_i)^2+(r_{i+F_{16}}-r_i)^2]^{1/2}\,. \tag{3.8}$$

We found that the convergence for the two-dimensional (2D) case was slightly slower than for the one-dimensional (1D) case. This is, however, to be expected since it was found by Feigenbaum, Kadanoff, and Shenker that the convergence of the scaling number α_{GM} is slower for the 2D case than for the 1D case.[19] To improve the convergence we again made use of the ratio trick as embodied in Eqs. (3.4) and (3.5). For this case we calculated the partition function for two consecutive Fibonacci numbers, $F_{16}=1597$ and $F_{17}=2584$, and found τ from the requirement

$$\frac{\Gamma(l(F_{17}))}{\Gamma(l(F_{16}))}(\tau,q)=1$$

[$l(F_i)$ are the maximal scales for the partitions]. This improves the convergence significantly, and the f-α curve for this 2D case coincides almost completely with the curve found for the 1D case and displayed in Fig. 14.

IV. CONCLUSION

Most previous characterizations of strange sets arising in physics have followed the example of critical phenomena in relying upon a few universal numbers to characterize the physical systems generating these sets. Thus, strange attractors are characterized by their Hausdorff dimensions, or by the scaling exponents of particularly divergent regions of their measure. However, these numbers reflect only a small part of the universal scaling structure of these systems. Feigenbaum introduced scaling functions in order to describe the complex scaling properties of attractors at the onset of chaos.[15] These scaling functions contain all of the geometric information about the attractor, in contrast to the partial information furnished by local scaling exponents. These functions are, however, nowhere differentiable, and are thus very difficult to use. The full complexity of this scaling structure is more conveniently reflected by the continuous spectrum of exponents α and their densities $f(\alpha)$, of which previously investigated scaling exponents and Hausdorff dimensions comprise only a part.

Not only does this spectrum enrich our conceptual vocabulary, it should enrich our experimental vocabulary as well. The numerical studies of Sec. III were straightforward and did not require large investments of computer time in order to obtain extremely accurate results. Furthermore, this spectrum can be measured, and has been measured, in experiments upon physical realizations of dynamical systems.[23] The measurement of this spectrum should result in new tests of scaling theories of nonlinear systems.

ACKNOWLEDGMENTS

We would like to thank P. Jones and J. Rudnick for stimulating discussions. This work was supported by the U.S. Office of Naval Research, by the National Science Foundation through Grant No. DMR-83-16626, and through the Materials Research Laboratory of the University of Chicago. One of us (I.P.) wishes to thank Professor S. Berry for his warm hospitality and also acknowledge the support of the Minerva Foundation (Munich, West Germany).

*Permanent address: Department of Chemical Physics, The Weizmann Institute of Science, 76100 Rehovot, Israel.

†Present address: AT&T Bell Laboratories, 600 Mountain Avenue, Murray Hill, NJ 07974.

[1]R. Bowen, *Equilibrium States and the Ergodic Theory of Anisov Diffeomorphisms*, Vol. 470 of *Lectures Notes in Mathematics* (Springer, Berlin, 1975); M. Widom, D. Bensimon, L. P. Kadanoff, and Scott J. Shenker, J. Stat. Phys. **32**, 443 (1983).

[2]I. Procaccia, in Proceedings of the Nobel Symposium on Chaos and Related Problems [Phys. Scr. T 9, 40 (1985)].

[3]K. G. Wilson, Sci. Am. **241**, (2) 158 (1979).

[4]B. B. Mandelbrot, Ann. Isr. Phys. Soc. **225** (1977); H. Aref and E. D. Siggia, J. Fluid Mech. **109**, 435 (1981).

[5]I. Procaccia, J. Stat. Phys. **36**, 649 (1984).

[6]L. de Arcangelis, S. Redner, and A. Coniglio, Phys. Rev. B **31**, 4725 (1985).

[7]T. A. Witten, Jr. and L. M. Sander, Phys. Rev. Lett. **47**, 1400 (1981).

[8]T. C. Halsey, P. Meakin, and I. Procaccia (unpublished).

[9]B. Widom, J. Chem. Phys. **43**, 3892 (1965); M. E. Fisher, in Proceedings of the University of Kentucky Centennial Conference on Phase Transitions (March 1965) (unpublished); A. Z. Patashinskii and V. L. Prokovskii, Zh. Eksp. Teor. Fiz. **50**, 439 (1966) [Sov. Phys.—JETP **23**, 292 (1966)]; L. P. Kadanoff, W. Gotze, D. Hamblen, R. Hecht, E. A. S. Lewis, V. V. Palciauskas, M. Rayl, J. Swift, D. Aspnes, and J. Kane, Rev. Mod. Phys. **39**, 395 (1967).

[10]L. P. Kadanoff, Physics **2**, 263 (1966).

[11]In fact, for many two-dimensional critical phenomena we know [see, for instance, D. Friedan, Z. Qiu, and Stephen Shenker, Phys. Rev. Lett. **52**, 1575 (1984)] that there is only a finite set of dist t y's.

[12]L. Turkevich and H. Scher, Phys. Rev. Lett. **55**, 1024 (1985).

[13]H. G. E. Hentschel and I. Procaccia, Physica **8D**, 435 (1983).

[14]P. Grassberger and I. Procaccia, Physica **13D**, 34 (1984).

[15]M. J. Feigenbaum, J. Stat. Phys. **19**, 25 (1978); **21**, 669 (1979).

See also the two reprint compilations: P. Cvitanović, *Universality in Chaos* (Hilger, Bristol, 1984); Hao Bai-lin, *Chaos* (World Scientific, Singapore, 1984).

[16]M. H. Jensen, P. Bak, and T. Bohr, Phys. Rev. Lett. **50**, 1637 (1983); Phys. Rev. A **30**, 1960 (1984); **30**, 1970 (1984).

[17]P. Cvitanović, M. H. Jensen, L. P. Kadanoff, and I. Procaccia, Phys. Rev. Lett. **55**, 343 (1985).

[18]Scott J. Shenker, Physica **5D**, 405 (1982).

[19]M. J. Feigenbaum, L. P. Kadanoff, and Scott J. Shenker, Physica **5D**, 370 (1982).

[20]S. Ostlund, D. Rand, J. P. Sethna, and E. D. Siggia, Phys. Rev. Lett. **49**, 132 (1982); Physica **8D**, 303 (1983).

[21]P. Cvitanović, B. Shraiman, and B. Soderberg (unpublished).

[22]A. Cohen and I. Procaccia, Phys. Rev. A **31**, 1872 (1985), and references therein.

[23]M. H. Jensen, L. P. Kadanoff, A. Libchaber, I. Procaccia, and J. Stavans, Phys. Rev. Lett. **55**, 2798 (1985).

Journal of Statistical Physics **52** 527–69 (1988)

Presentation Functions, Fixed Points, and a Theory of Scaling Function Dynamics

Mitchell J. Feigenbaum[1]

Presentation functions provide the time-ordered points of the forward dynamics of a system as successive inverse images. They generally determine objects constructed on trees, regular or otherwise, and immediately determine a functional form of the transfer matrix of these systems. Presentation functions for regular binary trees determine the associated forward dynamics to be that of a period doubling fixed point. They are generally parametrized by the trajectory scaling function of the dynamics in a natural way. The requirement that the forward dynamics be smooth with a critical point determines a complete set of equations whose solution is the scaling function. These equations are compatible with a dynamics in the space of scalings which is conjectured, with numerical and intuitive support, to possess its solution as a unique, globally attracting fixed point. It is argued that such dynamics is to be sought as a program for the solution of chaotic dynamics. In the course of the exposition new information pertaining to universal mode locking is presented.

KEY WORDS: Scaling; thermodynamics; period doubling; mode locking; dynamical systems; chaos; renormalization group.

1. INTRODUCTION

The attempts to understand the full microscopic structure of chaotic dynamical motion succeeded through renormalization group-like treatments for a variety of transitional phenomena. The upshot of that work is a complicated delineation of the parameter space as well as the phase space marked by scaling phenomena. An important idea in that work was that the *dynamics* under the fixed-point map inherits from the fixed-point equation determining it a rich set of scaling symmetries. Indeed, the trajectory *scaling function* determined from the fixed point allows the full deter-

[1] Toyota Professor of the Rockefeller University, New York, N.Y.

mination of the orbit (and its Cantor set closure), or, Fourier transformed, the power spectrum given some low-resolution scale-fixing data. This information is embedding-free (an invariant—indeed the maximal invariant) and allows such computations whatever the embedding dimension of the phenomenon. The objects determined by these scaling functions are true "multifractals" in the present parlance, but of the richest variety with an infinity of scales.

As one pursues more chaotic, higher-dimensional strange attractors, it appears unlikely that a fully microscopic scaling function theory can be offered. At least provisionally, one has turned attention to various dimension-related notions of mere thermodynamic descrition. This raises questions as to just what these degenerate objects and their interrelations depend upon, how they are to be calculated, and their extensions and generalizations. A particular question of outstanding interest is the deduction of the numerically well-established dimension of the "gaps" in the set of all mode-locked intervals of quasiperiodic motion—the computation of average quantities from a highly complex microscopic distribution.

The crux of the above question is that, given a highly variegated set of microscopic scalings which determine average quantities as a subset of this full information, can one undo the labor of their extraction and find a better-behaved, simpler substrate from which the averages easily follow? The goal of this paper is to first present such machinery—that of so-called "presentation functions"—and then learn how fixed-point dynamics, scallings, and these new objects are all interrelated.

The plan of the paper is as follows. In the second section I extract the presentation function of period doubling dynamics from the fixed-point equation of the latter, and discern in it the schema of organizing arbitrary objects constructed on binary trees. In Section 3 I show how to immediately write down the thermodynamics given a presentation function by explicitly writing down a functional operator form of the "transfer matrix." I illuminate these notions by way of examples in Section 4 and present numerical evidence as to how mode locking thermodynamics might be derived. In Section 5 the notion of generalizations of the machinery to arbitrary trees, complete or "pruned," is developed. In Section 6 I invert the exposition and discover how the period doubling dynamics for a given tree can be obtained from its prior specified presentation function, thereby further exposing the "heart" of the mode locking problem and paving the way toward a new method of obtaining the fixed-point function. In Section 7 I relate presentation functions to scaling functions, the latter a special parameterization of the former. In Section 8 I explicitly write down all successive approximations to the computation of the scaling function. In Section 9 I present a dynamics in the space of scalings that (as a numerical observation in higher orders of approximation) possesses the solution of the equations of Section 8 as a globally stable fixed point and suggest that a new possibility for the rigorous proof of the isolated solution to the fixed-point equation is at hand. In a final section, I take stock of this offering and contemplate its meaning and possible future directions.

2. PRESENTATION FUNCTIONS

The dynamics of the period doubling fixed point[1] provides a scheme for the organization of objects definned by a regular binary tree. The fixed point obeys the equation

$$\alpha g^2 \alpha^{-1} = g; \qquad g(0) = 1 \tag{2.1}$$

where α denotes the linear transformation of multiplication by α. The function g is understood to possess (usually a quadratic) critical point at $x = 0$.

Denoting the critical value by x_0,

$$x_0 = g(0) = 1 \tag{2.2}$$

define the nth image under g of x_0 by x_n:

$$x_n = g^n(x_0) = g^{n+1}(0) \tag{2.3}$$

Observe that

$$x_{2n+1} = g^{2n+2}(0) = \alpha^{-1} g^{n+1} \alpha(0) = \alpha^{-1} x_n \tag{2.4}$$

and

$$x_{2n} = g^{2n+1}(0) = g^{-1}(x_{2n+1}) = g^{-1}\alpha^{-1} x_n = (\alpha g)^{-1}(x_n) \tag{2.5}$$

Since g is unimodal, there are *two* inverses of g. However, g alternately maps a central domain to a right domain, and g^2 maps within either of these domains. Thus, x_{2n} is always in the right domain that includes the critical value x_0. Thus, g^{-1} of (5) is the right inverse of g, that is, the inverse of the right half of g restricted to a domain strictly excluding the critical point, and so of bounded nonlinearity.

Let us denote the two results, (2.4) and (2.5), by

$$x_{2n+\varepsilon} = F_\varepsilon(x_n), \qquad \varepsilon = 0, 1 \tag{2.6}$$

so that

$$F_0(x) = (\alpha g)^{-1}(x) \tag{2.7}$$

$$F_1(x) = \alpha^{-1} x \tag{2.8}$$

Further, let us regard F_0 and F_1 as the two inverses of a "unimodal" map E

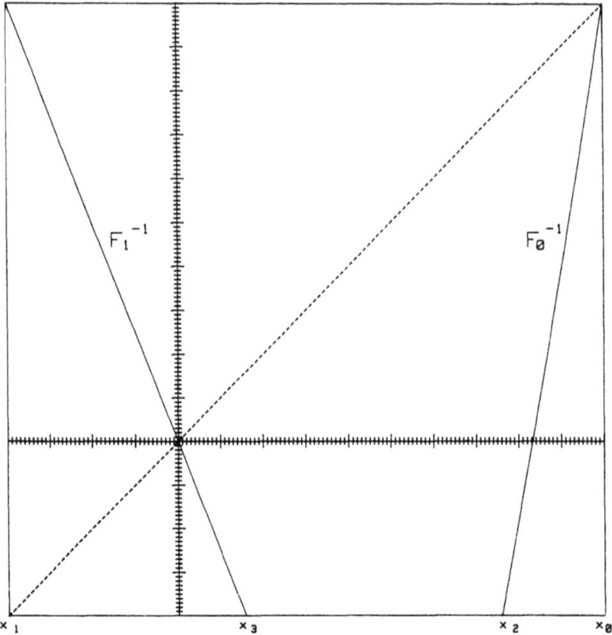

Fig. 1. The graph of the expanding map E for the quadratic period doubling fixed point. $F_1^{-1} = \alpha x$, $F_0^{-1} = \alpha g(x)$. The Cantor set is covered by the range $[x_1, x_3]$ of F_1 and $[x_2, x_0]$ of F_0.

(it can be verified that F_0 contracts, so that E is expanding). E is deicted in Fig. 1, and is defined on the two disjoint intervals $[x_1, x_3]$ and $[x_2, x_0]$. It is immediately verified from (2.1) and the content of (2.4) and (2.5) that

$$Eg^2 = gE \tag{2.9}$$

that is, E conjugates g^2 to g. Thus, the period doubling attractor, the closure of the orbit of x_0, is that Cantor set for which the center piece is mapped to the entire set by α and the right-hand piece is mapped to the entire set by αg.

Notice that E is defined on two intervals; E^2 is defined on four disjoint intervals, etc. The domain of E^n is a cover of the Cantor set by 2^n disjoint intervals. The intersection of the domains of E^n over all n is the Cantor set itself. E thus "presents" the Cantor set and is called its "presentation function."[2] Had we replaced αg by an approximating *linear* map, we see by the above that a two-scale Cantor set would have been constructed.

[2] The ideas of presentation functions are sprinkled in the literature. The name arose in discussions between D. Sullivan and myself. To my knowledge, its role here in carrying full dynamical information is new.

Let us consider E in the spirit of a dynamical system itself. Specifically, let us consider the set of all inverse images of x_0 under E. Denote the index r of an $x_r^{(n)}$ by its binary expansion

$$r = \varepsilon_n \cdots \varepsilon_1, \qquad \varepsilon_i = 0, 1 \tag{2.10}$$

so that $r = 0,..., 2^n - 1$ for any point in the set of nth inverse images. By (2.10) we immediately see that

$$x^{(n)}_{\varepsilon_n \cdots \varepsilon_1} = F_{\varepsilon_1} \circ F_{\varepsilon_2} \circ \cdots \circ F_{\varepsilon_n}(x_0) \tag{2.11}$$

Since

$$x_0 = x_{2x0} = F_0(x_0) \tag{2.12}$$

x_0 is a fixed point of F_0, so that, by (2.11), the first 2^{n-1} inverses of the nth set make up precisely the $(n-1)$th set of inverses. Thus,

$$x^{(n+r)}_{\underbrace{0 \cdots 0}_{r}\varepsilon_n \cdots \varepsilon_1} = x^{(n)}_{\varepsilon_n \cdots \varepsilon_1} \tag{2.13}$$

and the superscript (n) is superfluous: a point with the same evaluated index (leading 0 ε's) is the same at all levels possessing it.

But, x_n as defined in (3) is the nth image under g of x_0. *Thus, the inverse under E of a given index is precisely that iterate of the forward dynamics of g.* I will develop this observation later: namely, that to a given presentation function E, one can associate a period doubling fixed point, the dynamics of which agrees with the inverses under E. That is, to an *a priori* specified Cantor set, one can associate a dynamics that possesses the Cantor set as its attractor. Before doing so, however, let us investigate the thermodynamics of the Cantor set given its presentation function.

3. THE THERMODYNAMICS OF PRESENTATION FUNCTIONS

By thermodynamics I mean the deduction of variables and their relations to one another that in some well-defined sense are averages of exponential quantities defined microscopically on a set.[3] The exponential quantities here are the lengths of intervals covering a set.

Notice by (2.11) that an $x^{(n)}$ is obtained by n contractive mappings, so that $x^{(n)}\varepsilon_n \cdots \varepsilon_1$ is weakly dependent upon the higher indexed (leftmost) ε's.

[3] These ideas appear in the classical literature, and can be found in refs. 2 and 3. The notation here employed and the important focus on return times appears in ref. 4. The above work expositing Markov graphs for golden mean rotation should be understood as a serious elaboration upon ref. 5.

We are thus naturally led to approximations (not always actual covers in generalizations to follow) to the set by a set of 2^n intervals with each interval defined by the endpoints

$$x^{(n+1)}_{0\,\varepsilon_n\cdots\varepsilon_1} \quad \text{and} \quad x^{(n+1)}_{1\,\varepsilon_n\cdots\varepsilon_1} \tag{3.1}$$

Let us then denote an indexed nth-level interval length by

$$\Delta^{(n)}(\varepsilon_n\cdots\varepsilon_1)=x^{(n+1)}_{0\,\varepsilon_n\cdots\varepsilon_1}-x^{(n+1)}_{1\,\varepsilon_n\cdots\varepsilon_1} \tag{3.2}$$

By (2.11), we can estimate

$$\begin{aligned}\Delta^{(n)}(\varepsilon_n\cdots\varepsilon_1)&=F_{\varepsilon_1}\cdots F_{\varepsilon_n}F_0(x^{(0)}_0)-F_{\varepsilon_1}\cdots F_{\varepsilon_n}F_1(x^{(0)}_0)\\&\approx F'_{\varepsilon_1}(F_{\varepsilon_2}\cdots)\,F'_{\varepsilon_2}(F_{\varepsilon_3}\cdots)\cdots F'_{\varepsilon_r}(F_{\varepsilon_{r+1}}\cdots)\cdots\end{aligned} \tag{3.3}$$

The idea of (3.3) is that for n large, asymptotically each of the derivatives is taken at an argument insensitive to the highest indexed ε's so that each F_ε is asymptotically linear over the required range. This follows for F_ε that are differentiable, contractive, and of bounded nonlinearity. [A weakening of contractive is allowed to include $|F'(x)|=1$ at the boundary of definition.] The $\Delta^{(n)}$ will then be bounded by exponentials in n, that is, loosely,

$$|\Delta^{(n)}(\varepsilon_n\cdots\varepsilon_1)|\sim e^{-nh(\varepsilon_n\cdots\varepsilon_1)}=e^{-H(\varepsilon_n\cdots\varepsilon_1)} \tag{3.4}$$

where, by the above reasoning, it can easily be seen that h has increasingly weak dependence upon *low-index* ε's. (I shall return to this matter—generally the matter of scaling—later.)

Accordingly, in strict analogy to statistical mechanics, it is natural to consider the sum

$$\sum_{\{\varepsilon\}}|\Delta^{(n)}(\varepsilon_n\cdots\varepsilon_1)|^\beta\sim\sum_{\{\varepsilon\}}e^{-\beta H(\varepsilon_n\cdots\varepsilon_1)} \tag{3.5}$$

Since H is extensive in n, it is seen that the sum in (3.5) is the canonical partition sum for a statistical mechanical system of n "spins" ε_i one each at each lattice site i of a one-dimensional array. With "interactions" falling off with site separation, one expects the thermodynamic limit

$$\sum_{\{\varepsilon\}}|\Delta^{(n)}(\varepsilon_n\cdots\varepsilon_1)|^\beta\underset{n\to\infty}{\sim}2^{-nF(\beta)} \tag{3.6}$$

to go through (in the sense of logarithms). This "free energy" $F(\beta)$ expresses the thermodynamic relation between the variables F and β.

Notice that β_H such that

$$F(\beta_H)=0 \tag{3.7}$$

is "special" in that the sum has no exponential growth as the covering intervals are successively refined. Thus, β_H is to be identified with the Hausdorff dimension of the set. (If the "approximations" are as optimal as their definition suggests them to be, β_H *will* be the Hausdorff dimension.) Thus, β is related to "generalized" dimensions of the set.

Let us now proceed formally. Defining

$$\lambda(\beta) = 2^{-F(\beta)} \tag{3.8}$$

(3.6) asserts that

$$\sum |\Delta^{(n)}|^\beta \sim \lambda^n \tag{3.9}$$

and we seek an eigenvalue equation determining $\lambda(\beta)$. This is elementary to obtain from the F_ε; all we need do is use (3.3) in (3.9), writing the latter as an iterated sum:

$$\sum_{\{\varepsilon\}} |\Delta^{(n)}(\varepsilon_n \cdots \varepsilon_1)|^\beta \sim \cdots \sum_{\varepsilon_r} |F'_{\varepsilon_r}(F_{\varepsilon_{r+1}} \cdots)|^\beta \sum_{\varepsilon_{r-1}} |F'_{\varepsilon_{r-1}}(F_{\varepsilon_r} F_{\varepsilon_{r+1}} \cdots)|^\beta \cdots \tag{3.10}$$

Notice that the sum over ε_{r-1} and all lower-ε sums to its right depend only upon $F_{\varepsilon_r} F_{\varepsilon_{r+1}} \cdots$ defined by the outer sums. Accodingly, writing

$$\psi_{r-1}(F_{\varepsilon_r} F_{\varepsilon_{r+1}} \cdots) = \sum_{\varepsilon_{r-1}} |F'_{r-1}(F_{\varepsilon_r} F_{\varepsilon_{r+1}} \cdots)|^\beta \sum_{\varepsilon_{r-2}} |F'_{\varepsilon_{r-2}}(F_{\varepsilon_{r-1}}(F_{\varepsilon_r} \cdots))|^\beta \cdots \tag{3.11}$$

we then have

$$\psi_r(F_{\varepsilon_{r+1}} \cdots) = \sum_{\varepsilon_r} |F'_{\varepsilon_r}(F_{\varepsilon_{r+1}} \cdots)|^\beta \psi_{r-1}(F_{\varepsilon_r}(F_{\varepsilon_{r+1}} \cdots)) \tag{3.12}$$

Denoting any possible point $F_{\varepsilon_{r+1}} F_{\varepsilon_{r+2}} \cdots$ by x, we have that (3.12) reads

$$\psi_r(x) = \sum_\varepsilon |F'_\varepsilon(x)|^\beta \psi_{r-1}(F_\varepsilon(x)) \tag{3.13}$$

Since (3.13) is a linear transformation, $\psi_r(x)$ asymptotically in r behaves as

$$\psi_r(x) \sim \lambda^r \psi(x) \tag{3.14}$$

where λ is the largest eigenvalue obeying

$$\lambda\psi(x) = \sum_\varepsilon |F'_\varepsilon(x)|^\beta \psi(F_\varepsilon(x)) \tag{3.15}$$

Since the sum of (3.10) is $\psi_n \sim \lambda^n$, the λ of (3.15) is λ of (3.9) and (3.8), so that (3.15) is the desired eigenvalue equation for $\lambda(\beta)$.

[Although $\psi(x)$ by its definition was defined only for x in the "attractor" which can be proper (Cantor) subset of the interval, with F_ε differentiable enough, (3.15) determines its extension to all x.]

Thus, given the fixed point $g(x)$ for quadratic period doubling, (2.7) and (2.8) entered in (3.15) explicitly determines the equation for the thermodynamics of the period doubling attractor. It is useful, however, to write down and solve (3.15) for simpler presentation functions (i.e., when the F_ε are explicitly available). Accordingly, let us analyze both a trivial problem (a two-scale Cantor set) and highly nontrivial one (the parameter axis for subcritical quasiperiodic motion).[4]

4. EXAMPLES OF THERMODYNAMICS

1. Write $\alpha^{-1} = -\sigma_1$, so that

$$F_1(x) = -\sigma_1 x \tag{4.1}$$

and replace F_0 of Fig. 1 by

$$F_0(x) = 1 + \sigma_0(x-1) \tag{4.2}$$

with

$$0 < \sigma_0 < \sigma_1 < 1$$

Then (3.15) becomes

$$\lambda\phi(x) = \sigma_1^\beta \psi(-\sigma_1 x) + \sigma_0^\beta \psi(1 + \sigma_0(x-1)) \tag{4.3}$$

(4.3) clearly possesses the solution

$$\psi = \text{const}, \qquad \lambda = \sigma_1^\beta + \sigma_0^\beta \tag{4.4}$$

It is also easy to see that for each $n > 0$ there is a ψ_n, a polynomial of degree n in x, and $\lambda_n = \sigma_1^\beta(-\sigma_1)^n + \sigma_0^{\beta+n}$. However, the eigenvalue (4.4), λ_0, exceeds all these λ_n, and so (4.4) is the solution.

2. Consider the binary Farey tree shown in Fig. 2, constructed by placing

$$\frac{p}{q} \oplus \frac{p'}{q'} = \frac{p+p'}{q+q'}$$

between every pair of previously determined (*all* prior levels) fractions. It is easy to show that the nth layer of the tree consists precisely of all those fractions

$$[c_1, \ldots, c_k] = \cfrac{1}{c_1 + \cfrac{1}{c_2 + \cdots}}, \qquad c_k \geqslant 2 \tag{4.5}$$

[4] An initial exposition of these ideas can be found in ref. 6. More examples of this machinery and a discussion of the spectrum of (3.15) appear in ref. 7.

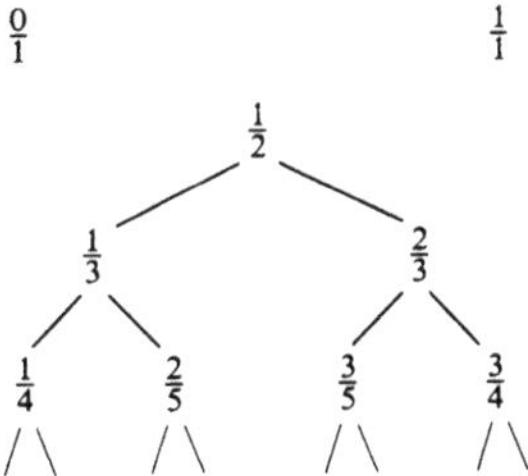

Fig. 2. Binary Farey tree.

for which

$$\sum_{1}^{k} c_i = n+2 \tag{4.6}$$

Define now a pair of transformations that determine the $(n+1)$th layer from the nth:

$$\begin{aligned} F_0&: \quad [c_1,..., c_k] \to [c_1+1, c_2,..., c_k] \\ F_1&: \quad [c_1,..., c_k] \to [1, c_1,..., c_k] \end{aligned} \tag{4.7}$$

It follows from (4.6) that (4.7) will inded accomplish its defined task. By (4.7),

$$F_1([c_1,..., c_k]) = \cfrac{1}{1+\cfrac{1}{c_1+\cdots}} = \frac{1}{1+[c_1,..., c_k]}$$

i.e.,

$$F_1(x) = \frac{1}{1+x} \tag{4.8}$$

and similarly,

$$F_0(x) = \frac{x}{1+x} \tag{4.9}$$

We now regard these F_ε as the inverses of E drawn in Fig. 3, with F_0 the left inverse and F_1 the right inverse.

Starting at $x_0^{(0)} = 1/2$, $x_0^{(1)} = F_0(1/2) = 1/3$, $x_1^{(1)} = F_1(1/2) = 2/3$, and, generally, the 2^n nth inverses of 1/2 are precisely the nth layer of the Farey tree. Notice that $F_0'(0) = 1$, so that each F_ε becomes marginally noncontractive at one endpoint of its domain. The intervals computed according to (3.2) are precisely the spacings between the two descendants at level n of a

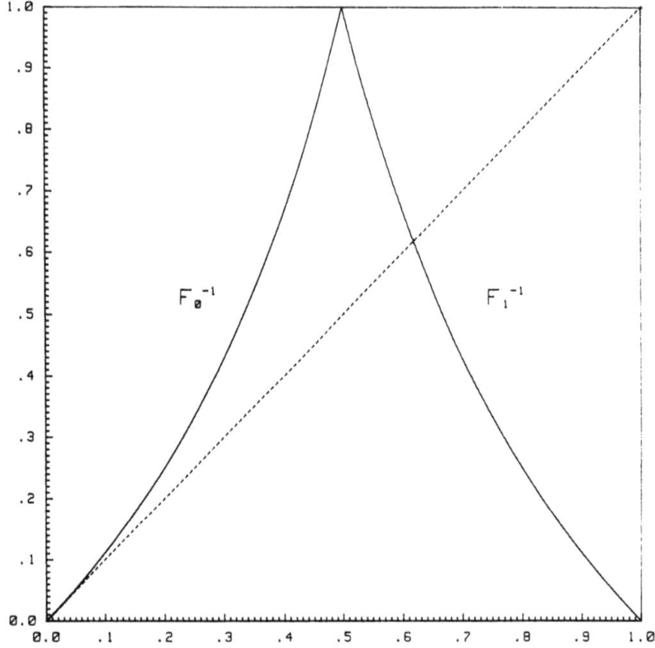

Fig. 3. The presentation function for the Farey model, Eqs. (4.8), (4.9).

level-$(n-1)$ parent (for example, $2/5-1/4$, the children of $1/3$ in Fig. 2). Since the tree determines all rationals, the "attractor" that E of Fig. 3 determines is the entire interval $[0, 1]$ partitioned at each level in a highly nontrivial fashion. The set of nth-level intervals now is a representative "subcovering" of the entire interval. As pointed out elsewhere, the study of these intervals is intimately related to the full devil's straircase of mode locking intervals for two coupled nonlinear ocillators, so that the results are of high physical interest.[5]

As mentioned at the end of Section 2, it is possible to construct a period doubling fixed point the *dynamics* of which is identical to the inverses of this E. I have elsewhere determined this so-called "Farey map," which has the property that it possesses a periodic orbit of length 2^n, one

[5] I first invented this "Farey model," itself a correct treatment of subcriticality, as a trial problem to understand the organization of the centroids of critical mode locking intervals. I learned of the critical problem and its suggestive scaling function from Cvitanovic. The essence of the idea that that organization is accomplished by a period doubling fixed point (and hence the organization of any complete binary tree) was developed in 1984 in my partially circulated and unpublished paper, "The renormalization of the Farey map." The present paper now renders that paper essentially defunct.

for each n, and of identical marginal stability -1, which are precisely the layers of the Farey tree. I shall explore this general matter in the next section.

Utilizing (4.8) and (3.15), we arrive at the equation

$$\lambda\psi(x) = (1+x)^{-2\beta}\left[\psi\left(\frac{x}{1+x}\right) + \psi\left(\frac{1}{1+x}\right)\right] \tag{4.10}$$

Despite the marginally contractive behavior of the F_ε, we know from two other independent (one harder, the other much harder) derivations that (4.10) is correct. There are some available solutions to (4.10).

(i) $\beta = 1 \rightarrow \lambda = 1$, $\psi = 1/x$. As anticipated, the set has Hausdorff dimension $\beta_H = 1$ and ψ is nontrivial.

(ii) $\beta = -n/2$, $n = 0, 1,...$; ψ now has polynomial solutions, with the leading ψ a polynominal of degree n. For example, $\lambda(+1/2) = 3$.

(iii) As $\beta \rightarrow -\infty$, $\lambda(\beta) \sim \rho^{2\beta}$, $\rho = (\sqrt{5}-1)/2$, intimately related to the fixed point $\rho = F_1(\rho)$ at which

$$F_1'(\rho) = -\rho^{-2}$$

(iv) For other $\beta < 1$, ψ has branch point at $x = 0$; it is shown in ref. 6 that

$$F \ln(-F) \sim 1 - \beta \qquad \text{as} \quad \beta \rightarrow 1 \quad \text{from below}$$

(v) $F(\beta) = 0$, $\beta > 1$.

By property (iv), this problem has an infinite-order phase transition at $\beta = 1$. (More information can be found in Appendix I of ref. 7. In particular, the statistical model is essentially that of a one-dimensional lattice gas with a one-particle saturating logarithmic interaction.)

In order to understand the statistical mechanics connection more fully and to better comprehend the period doubling dynamics equivalent to a given E, we shall have to turn to the notions of scaling functions. I first comment about generalizations.

5. OBJECTS ON COMPLETE AND REGULARLY PRUNED *n*-ARY TREES

Although we started with period doubling dynamics, it is clear that for any presentation function E we can construct a Cantor set and by (3.15) immediately write down the functional linear operator (the "transfer matrix" of the statistical mechanical analog) whose largest eigenvalue determines the thermodynamic relation of F to β. The simplest generalization away from usual period doubling is the replacement of $F_0=(\alpha g)^{-1}$ with some other function. More generally, we can write an E with n inverses, each then mapping the entire interval to an appropriate piece of the set. This is simply to replace a binary tree with a general n-ary tree. All of the formulas of the preceding sections remain unchanged, save for each ε, which now ranges from 0 to $n-1$.

However, another important class of objects is not yet covered. Consider, for example E of Fig. 4. The special property is that while any point on the F_1 arc has two inverses, a point on the F_0 arc has only an F_1 inverse. Thus, F_1F_1, F_1F_0 are allowed, but F_0F_0 is not. Thus, only $x_{\varepsilon_n\cdots\varepsilon_1}$ exist for which no two 0 epsilons can be consecutive. This situation is that of the incomplete tree of Fig. 5, which is easily seen to have F_n entries at level n, where $F_{n+1}=F_n+F_{n-1}$ are the Fibonacci numbers, and

$$F_n \sim \rho^{-n}; \qquad \rho^{-1}=\frac{\sqrt{5}+1}{2}$$

Observe that (3.9) implies that $\lambda(0)$ is the geometric growth rate of the number of pieces (or points) at level n, which for E of Fig. 4 should thus be

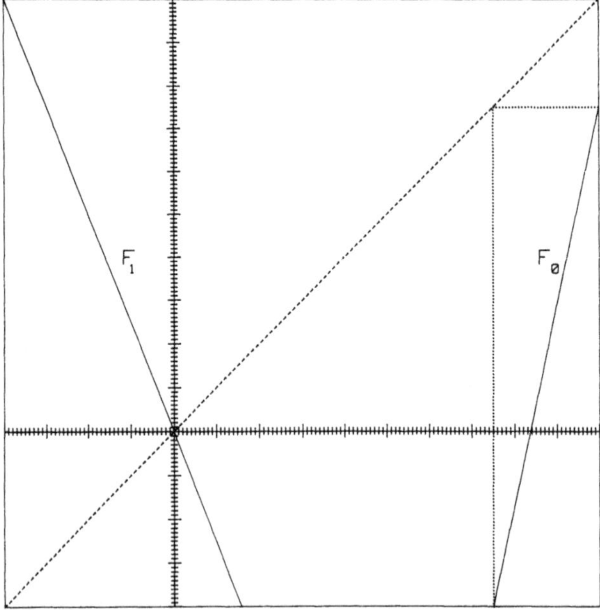

Fig. 4. A presentation function for a golden mean grammar incomplete binary tree: the range of F_0 is disjoint from its domain. (Section 5).

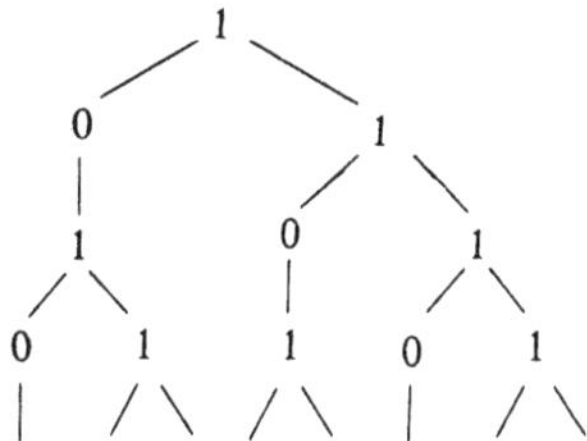

Fig. 5.

ρ^{-1}. For $\beta=0$, (3.15) always possesses $\psi=\text{const}$ as its leading eigenvalue with $\lambda=\sum_{\varepsilon} 1=2$, and so incompatible with the regularly "pruned" tree of Fig. 5 for E of Fig. 4.

It is easy to find the correct replacement for (3.15). Returning to (3.12), we immediately see that if $F_{\varepsilon_{r+1}}=F_0$, then ε_r cannot take on the value zero. Thus, the generic renaming of

$$F_{\varepsilon_{r+1}}F_{\varepsilon_{r+2}}\cdots=x$$

that led to (3.13) makes it impossible to determine the allowed range of ε summation in (3.13). To solve this problem, we simply backtrack one step and instead by x denote $F_{\varepsilon_{r+2}}F_{\varepsilon_{r+3}}\cdots$, so that (3.15) now reads

$$\lambda\phi(F_{\varepsilon_1}(x))=\sum_{\varepsilon_0}{}' \, |F'_{\varepsilon_0}(F_{\varepsilon_1}(x))|^{\beta}\,\psi(F_{\varepsilon_0}F_{\varepsilon_1}(x)) \tag{5.1}$$

where $\sum'_{\varepsilon_0}$ means that $F_{\varepsilon_0}F_{\varepsilon_1}$ must be an allowed composition.

Let us now denote the restriction of ψ to the range of F_{ε} by ψ_{ε}:

$$\psi_{\varepsilon}(x)=\psi(F_{\varepsilon}(x)) \tag{5.2}$$

so that (5.1) now reduces to

$$\lambda\psi_{\varepsilon_1}(x)=\sum_{\varepsilon_0}{}' \, |F'_{\varepsilon_0}(F_{\varepsilon_1}(x))|^{\beta}\,\psi_{\varepsilon_0}(F_{\varepsilon_1}(x)) \tag{5.3}$$

where again $\sum'_{\varepsilon_0}$ means that only those ε_0 such that $F_{\varepsilon_0}F_{\varepsilon_1}$ is allowed are summed. Equation (5.3) is the correct eigenvalue equation for a "grammar" of strings of ε's of length 2.

Specializing (5.3) to the tree of Fig. 5, we now have

$$\begin{aligned}\lambda\phi_0(x)&=|F'_1(F_0(x))|^{\beta}\,\psi_1(F_0(x))\\ \lambda\phi_1(x)&=|F'_0(F_1(x))|^{\beta}\,\psi_0(F_1(x))+|F'_1(F_1(x))|^{\beta}\,\psi_1(F_1(x))\end{aligned} \tag{5.4}$$

or,

$$\lambda^2\psi_1(x)=|F_0'(F_1(x))|^\beta|F_1'(F_0F_1(x))|^\beta\psi_1(F_0F_1(x))+\lambda|F_1'(F_1(x))|^\beta\psi_1(F_1(x))$$

which for $\beta=0$, $\psi_1=\text{const}$, reads

$$\lambda^2=1+\lambda\rightarrow\lambda_>=\frac{\sqrt{5}+1}{2}=\rho^{-1}$$

the correct growth rate.

So far we have treated $2-\varepsilon$ grammars. For restrictions among $n+1$ epsilons, we again turn to (5.12) and extend (5.1) to read

$$\lambda\phi(F_{\varepsilon_1}\cdots F_{\varepsilon_n}(x))=\sum_{\varepsilon_0}{}'\,|F_{\varepsilon_0}'(F_{\varepsilon_1}\cdot\cdot F_{\varepsilon_n}(x))|^\beta\,\psi(F_{\varepsilon_0}\cdots F_{\varepsilon_{n-1}}F_{\varepsilon_n}(x)) \tag{5.5}$$

where now (5.5) is written only for allowed $\varepsilon_1\cdots\varepsilon_n$ strings, and $\sum'$ means that only allowed strings are summed. Next, we define

$$\psi_{\varepsilon_1\cdots\varepsilon_n}(x)=\psi(F_{\varepsilon_1}\cdots F_{\varepsilon_n})(x) \tag{5.6}$$

$$\lambda\psi_{\varepsilon_1\cdots\varepsilon_n}(x)=\sum_{\varepsilon_0}{}'\,|F_{\varepsilon_0}'(F_{\varepsilon_1}\cdots F_{\varepsilon_n}(x))|^\beta\,\psi_{\varepsilon_0\cdots\varepsilon_{n-1}}(F_{\varepsilon_n}(x)) \tag{5.7}$$

The form of (5.7) suggests a Markov diagrammatic representation. One draws nodes for each allowed string of n ε's and unidirectionally links one node into another that agrees in its first $n-1$ ε's with last $n-1$ ε's of the former. Figure 6 represents one equation of the system (5.7) and is self-explanatory. To use Fig. 6, the sum of a $\psi(F_{\varepsilon_n}(x))$ timmes an F' link factor for all links into $\varepsilon_1\cdots\varepsilon_n$ produces λ times the $\psi(x)$ associated with the target node $\varepsilon_1\cdots\varepsilon_n$. Only "legal" nodes are drawn and "legal" links connect them. For example, the system (5.4) for the tree of Fig. 5 is determined by the graph of Fig. 7.

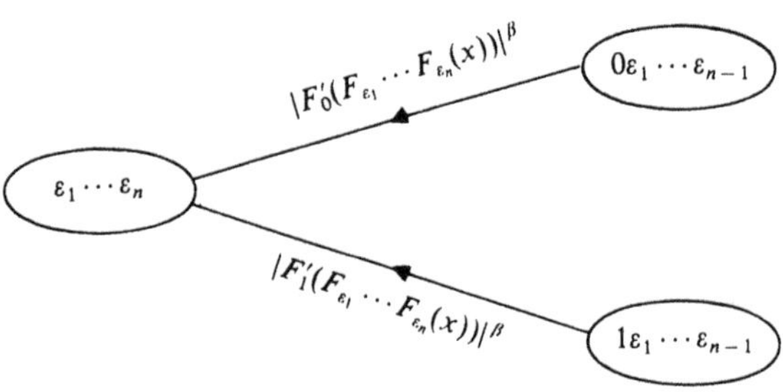

Fig. 6.

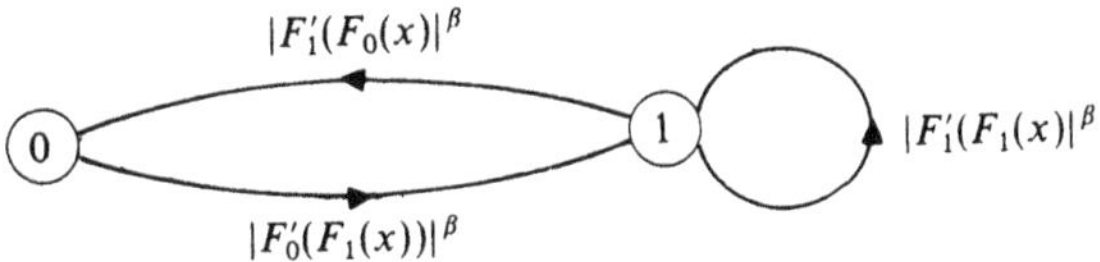

Fig. 7.

For $\beta = 0$ each link receives the factor $+1$; the largest λ is then the reciproval of the smallest positive zero of the determinant of the graph. By well-known methods, for Fig. 7 this determinant reads

$$0 = 1 - z - z^2 \rightarrow z_< = \frac{\sqrt{5} - 1}{2} \rightarrow \lambda_> = z_>^{-1} = \frac{\sqrt{5} + 1}{2}$$

Should the F's be linear, λ is again computed from the determinant of the graph, the links of which now are constants raised to the power β.

We thus see that the machinery of presentation functions easily extends to incomplete n-ary trees, and the thermodynamic information is readily available.

6. THE PERIOD DOUBLING DYNAMICS OF A PRESENTATION FUNCTION

I stated at the end of Section 2 that a dynamics can be associated with an *a priori* specified presentation function, in effect reversing the line of thought that led to presentation functions. By definition, this dynamics should generate the nth inverse images under E by the forward iteration of the dynamics. Since the dynamics will either possess orbits of length 2^n, one for each n, or a specified orbit of period 2^∞, the dynamics must be that of a period doubling fixed point. Let us now work out this connection. We shall see that a new method is afforded for determining these fixed points that perhaps can lead to a rigorous proof of their isolated existences.

In general there are 2^n nth inverses of E, labeled as $x_i^{(n)}$, $i = 0, \ldots, 2^n - 1$. By (2.6),

$$F_\varepsilon(x_i^{(n)}) = x_{2i+\varepsilon}^{(n+1)} \tag{6.1}$$

so that

$$F_1 F_0^{-1}(x_{2i}^{(n)}) = x_{2i+1}^{(n)} \tag{6.2}$$

Defining

$$g_0(x) = F_1 F_0^{-1} \tag{6.3}$$

g_0 provides the map that, applied to points of the right-hand piece of Fig. 1, images them to the central piece. Notice that g_0 images points at one level into others at the same level n.

Next observe that the range of F_1^r on the right-hand points is

$$F_1^r(x_{2i}^{(n)}) = x_{2^r(2i+1)-1}^{(x+r)} \tag{6.4}$$

Thus,

$$g_r(x) = F_0^r g_0 F_1^{-r}(x) \tag{6.5}$$

performs the mapping

$$g_r\colon \quad x_{2^r(2i+1)-1}^{(n)} \to x_{2^r(2i+1)}^{(n)} \tag{6.6}$$

so that the central (odd-indexed) points are mapped by the union of the restrictions of g, g_r, defined on intervals including just those x_i for which $i+1=2^r$ (odd). For this to make sense, the intersection of the interiors of the domains of any two distinct g_r must be empty. Moreover, the interval on which g_r is defined must include the relevant $x_i^{(n)}$ for *all* n. Since all x_{2i} lie within a fixed proper subinterval of the whole interval containing the support of E and since F_1 is a monotone contraction, this can be ascertained. In fact, provided E has a "gap" in its dommain of definition, as in Fig. 1, these intervals are totally disjoint; whereas, if there is no gap, as in Fig. 3, then intervals abut with empty interior intersections. Thus, the construction is as we have said. It should be noted that the domains of g_r as r diverges converge toward the fixed point of F_1, which, as we shall see, must then be the "critical point" of the map g.

By (6.3) and (6.6) we see that the range of g_0 includes the union of the domains of g_r; the range of each g_r is included within the domain of g_0. Notice by (6.3) that

$$g_{r-1} = F_0^{-1} g_r F_1 = F_1^{-1} g_0 g_r F_1 \tag{6.7}$$

that is, g^2 restricted to the domain of g_r for any $r \geqslant 1$ is smoothly conjugated by the F_1^{-1} part of E to g. Similarly,

$$g_{r-1} = F_0^{-1} g_r g_0 F_0 \tag{6.8}$$

so that g^2 restricted to the domain of g_0 is smoothly conjugate by the F_0^{-1} part of E to g. Thus, given any E, it serves the conjugating role of (2.9) for the g constructed by (6.3) and (6.5).

So long as F_1 is a smooth, monotone contraction with fixed point at x_c we can smoothly conjugate it to the linear transformation determined by its derivative at its fixed point. Calling this derivative α^{-1}, one has

$$F_1'(x_c) = \alpha^{-1} \tag{6.9}$$

and (6.7) in these new coordinates reads

$$g_{r-1} = \alpha g_0 g_r \alpha^{-1} \tag{6.10}$$

so that we have the usual period doubling fixed-point equation

$$g = \alpha g^2 \alpha^{-1}$$

which opened Section 2.

By way of example, if we turn to Example 2 of Section 4 for the Farey tree, $F_1(x)$ in (4.8) can be conjugated to homogeneous linear form through the fractional linear conjugacy

$$\tilde{F}_1 = hF_1h^{-1} \tag{6.11}$$

with

$$h(x) = \frac{1-\rho^{-1}x}{1+\rho x}, \qquad h^{-1}(x) = \frac{1-x}{\rho^{-1}+\rho x} \tag{6.12}$$

yielding

$$\tilde{F}_1(x) = -\rho^2 x, \qquad \alpha^{-1} = -\rho^2 \tag{6.13}$$

and

$$\tilde{F}_0(x) = \frac{1+2\rho x}{2\rho^{-1}-x} \tag{6.14}$$

This "canonical" form of E is depicted in Fig. 8.

By (6.3),

$$g_0(x) = -\rho^2 \tilde{F}_0^{-1}(x) = \frac{\rho^2 - 2\rho x}{2\rho + x} \tag{6.15}$$

with the important property that $g_0^2(x) = x$, so that the fixed point of g_0 has eigenvalue -1. Then by (6.5) all the other g_r can be explicitly computed. This g is the Farey map we have discussed elsewhere (see footnote 5), and is exhibited in Fig. 9.

The reason for considering the Farey tree lies in the numerical fact that critical mode locking of oscillators has the property that the complement of the mode-locked intervals has the universal dimension of 0.87. A theory for this numerical result is still lacking. Figure 8 pertains to the kindred but simpler subcritical problem, for which the dimension is 1. A careful analysis of the subcritical problem reveals that the thermodynamics

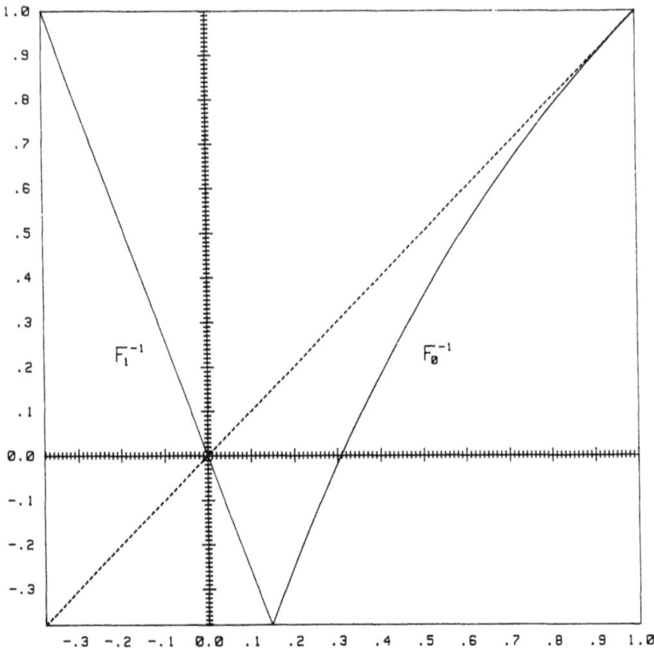

Fig. 8. The Farey presentation function of Fig. 3 after the conjugacy that brings F_1 to its canonical linear form, Eq. (6.13).

is saturated by the subtree at golden mean rotation. This part of the tree is universally determined by the unstable manifold of the renormalization-group fixed point of golden mean rotation. As we have just seen, for the subcritical case, h of (6.12) conjugates the entire tree to the golden mean subtree. However this works out for criticality, if the result is universal, it must also pertain to the unstable manifold of the critical golden mean fixed point. Thus there is an $\tilde{F}_1$ which is linear, but the derivative in criticality is the renormalized δ

$$\delta_r = -2.83361\ldots \tag{6.16}$$

instead of the subcritical $\delta = -\rho^2 = -2.618\ldots$. Thus, the arc $\tilde{F}_1$ for the critical problem is exactly known to be

$$F_1(x) = \delta_r^{-1} x$$

The corresponding F_0 is still not available. It suffices to say that it is a numerical fact that the critical case is smooth to within 10^{-8} (available precision) and is very well fit by

$$F_0(x) \approx x + k(1-x)^\nu, \qquad \nu \approx 1.37, \quad k \approx 0.408 \tag{6.17}$$

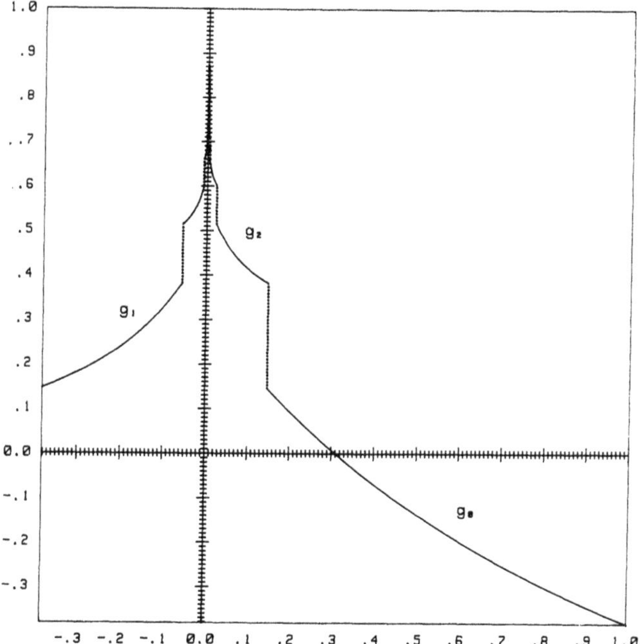

Fig. 9. The "Farey map" period doubling fixed point, the forward dynamics of which is the Farey tree.

and E is depicted in Fig. 10. The opening of the gap in E of course implies a dimension smaller than 1; the fit of (6.16) (theoretical) and (6.17) indeed reproduces the numerical 0.87 when entered in (3.15).

Let us end this section with the determination of g for the approximation (4.1) and (4.2) to the E of Fig. 1 for quadratic period doubling. We shall discover from this inquiry a general method for determing solutions to renormalization-group fixed-point equations.

We take

$$F_1 = \alpha^{-1} x \tag{6.18}$$

$$F_0 = 1 - \sigma_0(1 - x) \tag{6.19}$$

where (6.18) for the correct F_0 is exact. Notice by (6.5) that

$$g_r = 1 - \sigma_0^r[1 - g_0(\alpha^r x)] \tag{6.20}$$

where by (6.3)

$$g_0 = \alpha^{-1}[1 - \sigma_0^{-1}(1 - x)] \tag{6.21}$$

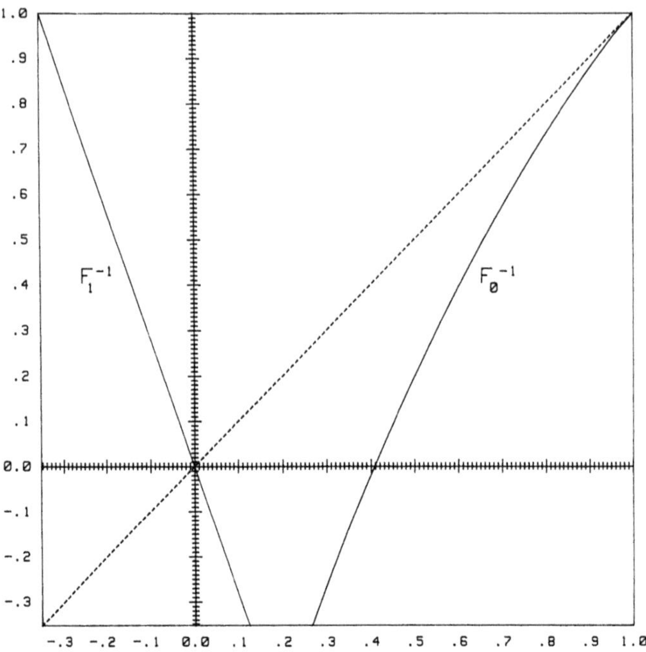

Fig. 10. An excellent *numerical* fit to F_0 of Eq. (6.17) for cubically critical mode locking. This presentation function, iterated, determines mode locking intervals and the well-known $\beta_H = 0.87$ to high accuracy. It is to be noted that since ν differs from 1.5, the intermittency argument Q^{-3} is misleadingly unimportant for β_H.

The requirement we impose on $\{g_r\}$ is that they should be the restrictions of some smooth function with a quadratic maximum at $x=0$ [which by (6.20) has the critical value of 1]. Notice by (6.20) that

$$\begin{aligned} g_r(\alpha^{-r}\hat{x}) &= 1 - [1 - g_0(\hat{x})]\sigma_0^r \\ &= 1 - \frac{1 - g_0(\hat{x})}{\hat{x}^2}\left(\frac{\sigma_0}{\alpha^{-2}}\right)^r (\alpha^{-r}\hat{x})^2 \end{aligned} \tag{6.22}$$

Thus, for any $\hat{x}$, if we fix

$$\sigma_0 = \alpha^{-2} \tag{6.23}$$

then (6.22) is compatible with

$$g = 1 - \mu x^2 \tag{6.24}$$

for

$$\mu = \frac{1 - g_0(\hat{x})}{\hat{x}^2} \tag{6.25}$$

Since by (6.21) g_0 entails the unknown constant α^{-1}, demanding that μ satisfies (6.25), at two distinct values of $\hat{x}$ will determine α^{-1}. However, choosing F_0 (as we have) to be the linear arc of (6.19), should the two choices of $\hat{x}$ be x_0 and x_2 of Fig. 1 [so that $x_0 = 1$, $x_2 = F_0(\alpha^{-1})$], then the linear arcs of g_r of (6.20) will be a linear polygonal period doubling fixed point with inscribing endpoints lying on the parabola (6.24) and so possessing an asymptotic quadratic critical point. Thus, set in (6.25)

$$\mu = \frac{1 - g_0(1)}{1} = 1 - \alpha^{-1} \tag{6.26a}$$

and

$$\mu = \frac{1 - g_0(x_2)}{x_2^2} = \frac{1 - g_0[F_0(\alpha^{-1})]}{x_2^2} = \frac{1 - \alpha^{-2}}{[1 - \alpha^{-2}(1 - \alpha^{-1})]^2} \tag{6.26b}$$

Eliminating μ in (6.26a), (6.26b) results in a quintic for α^{-1} with the root

$$\alpha = -2.48634\ldots$$

with an error of 0.66% from the actual result of $-2.502907875\ldots$. The polygonal fixed point is depicted in Fig. 11.

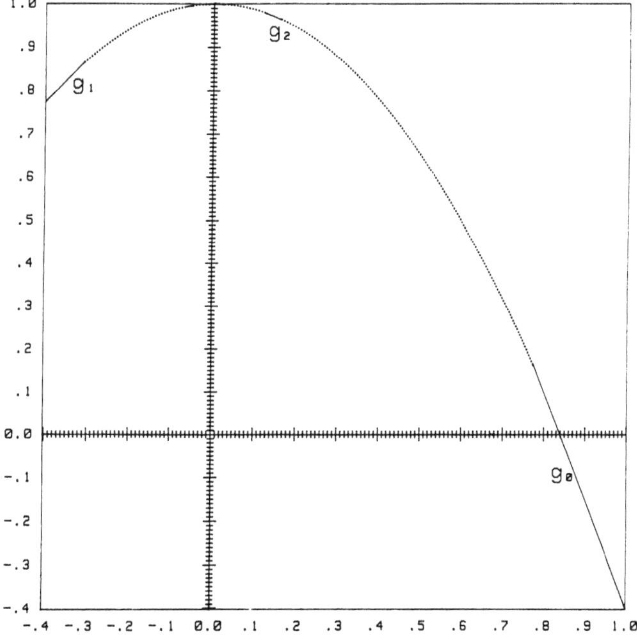

Fig. 11. The linear polygonal fixed point for "quadratic" period doubling of (6.22) that follows from F_0 of Fig. 1 replaced by that linear segment that renders g as quadratically smooth as possible.

Thus, in review, by requesting the $\{g_r\}$ (in some sense) to be the restrictions of a smooth, quadratically critical map, I have been led to an excellent linear polygonal solution to the fixed-point equation of the period doubling renormalization group. I now want to elevate this calulation to a systematically exact one. In order to do so, I must first explore the nature of approximations of F_0 by successively more linear segments, which I shall face in the next section through the idea of the scaling function. Having done that, in the final two sections I produce the general calculation for the fixed point and discover that to all orders (beyond the lowest, as a numerical observation) the solution is obtained as a globally stable fixed point of a natural dynamics in the space of all scalings. I shall mention my grander thoughts of the meaning of this dynamics in a final afterword.

To fortify the reader, Section 8 is technically arduous, with a rather simple final result. In erecting equations directly for the scalings, and moreover in discovering a natural and exceptionally well behaved flow on them, I believe new ground in being charted and, so beg the reader's indulgence.

7. THE RELATIONSHIP BETWEEN PRESENTATIONS AND SCALING FUNCTIONS

I presented a Markov graphical method of successive approximation to dynamical thrmodynamics in ref. 4. The links on those graphs were *constants* determined by the trajectory scaling function. In Section 5 (e.g., Fig. 7) I have drawn very low-order *exact* Markov graphs utilizing presentation functions. Should one proceed to construct successive higher-order duals of the F graphs, the links would have (F')'s of successively restricted arguments on them. With F_ε of bounded nonlinearity, these would approach constants, and so, obviously, would be identifiable with (constant) values of the scaling function σ. I now precisely work out that connection. In particular, it shall be possible to understand how an F_0 of a finite number of linear arcs is an expression of a σ of a finite number of constant values. Thus, from any smooth nonlinear F_ε one directly infers, in "modern" parlance, a multifractal of infinite scaling complexity. The ease with which one can deduce, for example, the thermodynamics of these objects should convice the reader that the machinery I have been erecting and discussing in these pages is very powerful.

The idea of scaling functions[6] is very simple. It is constituted of the observation that the quotient of small distances (asymptotically infinitesimals) is obviously invariant under smooth coordinate transfor-

[6] The scaling function is invented in ref. 8; see also refs. 9 and 10.

mations, and hence under the dynamics itself. That is, scaling properties are well ordered in *dynamical order* but not in space.

Using the ε notation of (2.6), define the scaling at a point on the orbit by

$$\sigma(\varepsilon_m \cdots \varepsilon_0) = \frac{x_{0\varepsilon_0 \cdots \varepsilon_0} - x_{1\varepsilon_m \cdots \varepsilon_0}}{x_{\varepsilon_m \varepsilon_{m-1} \cdots 0} - x_{\bar{\varepsilon}_m \varepsilon_{m-1} \cdots \varepsilon_0}} \tag{7.1}$$

where we are reexpressing the idea behind (3.3) that determines the smallest distance that can be identified with $m+1$ or m epsilons, and forming their quotient ($\bar{\varepsilon}$ denotes the complement of ε). Notice that the first of the numerator and denominator terms are evaluated at identical dynamical indexes, ensuring the invariance of σ under smooth coordinate transformations. (σ is an invariant, and generally a fuller one than the set of all periodic orbit eigenvalues,[7] the literature notwithstanding. Indeed, one cannot simply follow the Markov idea of the introduction to this section simply because thermodynamics and asymptotic exponential quantities are degenerate over the fuller information in σ.) Employing (2.6), one has that (7.1) relates σ to F:

$$\sigma(\varepsilon_m \cdots \varepsilon_0) = \frac{F_{\varepsilon_0} \cdots F_{\varepsilon_m} F_0(x_0) - F_{\varepsilon_0} \cdots F_{\varepsilon_m} F_1(x_0)}{F_{\varepsilon_0} \cdots F_{\varepsilon_m}(x_0) - F_{\varepsilon_0} \cdots F_{\bar{\varepsilon}_m}(x_0)} \tag{7.2}$$

The details to follow will implement the intuitive idea that after enough F's have been aplied to x_0, whether the first F is F_0, F_1, F_{ε_m}, or $F_{\bar{\varepsilon}_m}$, the resulting x's are all contracted enough toward one another as to lie within a single domain over which F is linear, so that the remaining lower-indexed F_ε's produce identical slope factors in numerator and denominator. That is, with F_0 a finite number of linear restrictions, there is only a finite number of distinct values of σ no matter how large m is: σ depends only upon a certain number of the highest indexed ε's. Let us see how this works out.

We begin by defining $F_0^{(n)}$ to be linear over each of the 2^n disjoint intervals of its domain. Accordingly, set

$$(F_{0,k}^{(n)})^{-1} \quad \text{linear on } (2k, 2k+2^{n+1}), \qquad k = 0,..., 2^n - 1 \tag{7.3}$$

$$F_{0,k}^{(n)}\colon \quad x_{l \cdot 2^n + k} \to x_{2(l \cdot 2^n + k)} \tag{7.4}$$

$$F_0^{(n)}(x_m) = F_{0,m \bmod 2^n}^{(n)}(x_m) \tag{7.5}$$

Notice in (7.3) that we use the dynamics to partition the domain of F^{-1} to be intervals whose endpoints are closest return neighbors. [The *order* of the endpoints might be reversed to $(2k+2^{n+1}, 2k)$ depending upon k.]

[7] Unpublished work of Dennis Sullivan, presented at the 1987 Noto Summer School, in which he establishes the scaling function as the maximal invariant for $C^{1+\alpha}$ geometries.

Condition (7.3) by (6.3) now determines restrictions of g_0:

$$g^{(n)}_{0,k} = F_1(F^{(n)}_{0,k})^{-1}: \qquad x_{2(k+l\cdot 2^n)} \to x_{2(k+l\cdot 2^n)+1} \tag{7.6}$$

and similarly by (6.5) for g_r:

$$g^{(n)}_{r,k} = F^r_0\, g^{(n)}_{0,k} F_1^{-r} \tag{7.7}$$

defined on the domain $F^r_1(\mathrm{dom}\, g_0)$ so that

$$g^{(n)}_{r,k}: \quad x_{2^r(2k+1+l\cdot 2^{n+1})-1} \to x_{2^r(2k+1+l\cdot 2^{n+1})} \tag{7.8}$$

The definition of (7.7) is completed with the understanding that each F_0 in it is to be that $F^{(n)}_{0,k}$ appropriate to the domain as given by (7.5). I want to explicitly record this. Since $\mathrm{dom}(g_r) = F^r_1(\mathrm{dom}\, g_0) = \alpha^{-r}(\mathrm{dom}\, g_0)$, we have

$$g^{(n)}_{r,k}(\alpha^{-r} x_{2k+\varepsilon 2^{n+1}}) = F^r_0\, g^{(n)}_{0,k}(x_{2k+\varepsilon 2^{n+1}}) = F^r_0(x_{2k+1+\varepsilon 2^{n+1}})$$

$$\overset{(r\geqslant n)}{=} [F^{(n)}_{0,0}]^{r-n} F^{(n)}_{0,2^{n-1}(2k+1)\bmod 2^n} \cdots F^{(n)}_{0,2(2k+1)\bmod 2^n} F^{(n)}_{0,2k+1}(x_{2k+1+\varepsilon 2^{n+1}}) \tag{7.9}$$

The most important aspect of (7.9) for the extension of the calculation of σ_1 at the end of Section 6 is that, apart from n other factors, g_r utilizes the one restriction $F_{0,0}$ that includes the fixed point of F_0 at $x = x_0 = 1$. I shall pick this up in the next section. For the moment, let us return to (7.2) now that the F's have been explicitly defined.

Note that the rightmost n of the F's of each denominator term in (7.2) produce

$$F^{(n)}_{\varepsilon_{m-n}} \cdots F^{(n)}_{\varepsilon_{m-1}} F^{(n)}_{\varepsilon}(x_0) = x_{\varepsilon 2^n + \varepsilon_{m-1}\cdots\varepsilon_{m-n}} = x_{\varepsilon 2^n + k} \tag{7.10}$$

for some $k < 2^n$. By (7.5) we see that (7.10) is in the domain of the *same* $F^{(n)}_{0,k}$ independent of ε. Identically, so, too, are the numerator terms

$$F^{(n)}_{\varepsilon_{m-n}} \cdots F^{(n)}_{\varepsilon_m} F^{(n)}_{\varepsilon}(x_0) \tag{7.11}$$

It follows further that the rest of the leftmost F's (of lower-indexed ε's) preserve the fact that each numerator and denominator x persists to lie in the domain of the same F restriction. Since each F is *linear*, we have

$$\begin{aligned} F^{(n)}_{\varepsilon_0} \cdots F^{(n)}_{\varepsilon_{m-1}} F^{(n)}_{\varepsilon_m}(x_0) - F^{(n)}_{\varepsilon_0} \cdots F^{(n)}_{\varepsilon_{m-1}} F^{(n)}_{\bar\varepsilon_m}(x_0) \\ = F^{(n)\prime}_{\varepsilon_0} \cdots F^{(n)\prime}_{\varepsilon_{m-n-1}}(x_{\varepsilon_m 2^n + k} - x_{\bar\varepsilon_m 2^n + k}) \end{aligned}$$

and

$$\begin{aligned} F^{(n)}_{\varepsilon_0} \cdots F^{(n)}_{\varepsilon_m} F^{(n)}_0(x_0) - F^{(n)}_{\varepsilon_0} \cdots F^{(n)}_{\varepsilon_m} F^{(n)}_1(x_0) \\ = F^{(n)\prime}_{\varepsilon_0} \cdots F^{(n)\prime}_{\varepsilon_{m-n-1}}(x_{\varepsilon_m 2^n + k} - x_{\varepsilon_m 2^n + k + 2^{n+1}}) \end{aligned}$$

Performing the quotient of (7.2), we find that the derivatives cancel, and we have

$$\sigma(\varepsilon_m \cdots \varepsilon_0) = \sigma(\varepsilon_m \cdots \varepsilon_{m-n}) = \frac{x_k - x_{k+2^{n+1}}}{x_k - x_{k+2^n \bmod 2^{n+1}}} = \sigma_n(k) \qquad (7.12)$$

where

$$k = 2^n \varepsilon_m + \cdots + 2^0 \varepsilon_{m-n} = 0, \ldots, 2^{n+1} - 1 \qquad (7.13)$$

Thus, with $F_0^{(n)}$ linearly defined on 2^n intervals, the set of all $\sigma(x_m)$ possesses 2^{n+1} distinct values constant over all but the highest indexed $n+1$ ε's. Also, 2^{n+1} independent parameters completely determine the 2^n linear functions $F_{0,k}^{(n)}$ (and *not* just the 2^n slopes $F_{0,k}^{(n)\prime}$), and, as we shall soon see, constitutes a most natural parameterization for dynamical purposes.

8. THE SCALING FUNCTION THEORY OF THE PERIOD DOUBLING FIXED POINT

Let us review what has been done so far. I started with the usual period doubling fixed point g and realized that its dynamics could also be determined by the backward dynamics (inverses) of an expanding map E, the presentation function. I next observed that the inverses of E, the functions F_ε, most naturally allow the determination of the thermodynamics of the dynamics of g. I next commented that forgetting g, the scheme of the F_ε is generally applicable to objects defined on trees. I then inverted the exposition, showing that the F_ε for any tree still determine the identical dynamics of some generalized period doubling fixed point that can be explicitly constructed from the F's. Moreover, it was shown at the end of Section 6 that requiring that fixed point to be smooth is a sufficient principle to determine the F_ε that give rise to it. In the last section it was realized that polygonal linear F's are equivalent to piecewise constant valued scaling functions, the latter a rich invariant under coordinate transformations. I am now prepared to again invert the exposition and discover how to frame the discussion of the underlying dynamics purely in the language of the scalings. I shall proceed to deduce the general equations that the scaling function satisfies, and then show in Section 9 that these equation can be solved by erecting a natural dynamics on the space of scalings which possesses as a globally stable fixed point the scaling function, hence the presentation function, and hence the period doubling fixed point that expresses the underlying dynamics.

Return to (7.9), and notice, as pointed out there, that the index r appears only as the power of the one restriction $F_{0,0}$. By (7.4), x_0 is its fixed point, so that with $x_0 = 1$ and $F'_{0,0} = \sigma_0$, we have

$$F_{0,0}^{(n)}(x) = 1 - \sigma_0(1 - x) \tag{8.1}$$

yielding

$$[F_{0,0}^{(0)}]^r(x) = 1 - \sigma_0^r(1 - x) \tag{8.2}$$

We can thus rewrite (7.9) explicitly in its r dependence for all $r \geqslant n$ as

$$g_{r,k}^{(n)}(\alpha^{-r} x_{2k+\varepsilon \cdot 2^{n+1}}) = 1 - \sigma_0^r U_{k,\varepsilon}^{(n)} \tag{8.3}$$

where

$$U_{k,\varepsilon}^{(n)} = \sigma_0^{-n}[1 - F_{0,2^n-1}^{(n)} \cdots F_{0,2(2k+1) \bmod 2^n}^{(n)} F_{0,2k+1}^{(n)}(x_{2k+1+\varepsilon 2^{n+1}})] \tag{8.4}$$

Now, just as we observed in(6.23), if we take

$$\sigma_0 = \alpha^2 \tag{8.5}$$

then (8.3) reads

$$g_{r,k}^{(n)}(\alpha^{-r} x_{2k-\varepsilon \cdot 2^{n+1}}) = 1 - (\alpha^{-r} x_{2k+\varepsilon \cdot 2^{n+1}})^2 \frac{U_{k,\varepsilon}^{(n)}}{x_{2k+\varepsilon \cdot 2^{n+1}}^2} \tag{8.6}$$

We now see, in analogy to (6.24)–(6.26), that by *requiring* that

$$U_{k,\varepsilon}^{(n)} = \mu_n x_{2k+\varepsilon \cdot 2^{n+1}}^2 \qquad \text{for some } \mu_n \tag{8.7}$$

we shall achieve

$$g_{r,k}^{(n)}(x) \text{ “=” } 1 - \mu_n x^2 \qquad (r \geqslant n, \text{ all } k) \tag{8.8}$$

where the quotes about the equals sign mean that (8.8) is satisfied just at the endpoints of the domains of definition of $g_{r,k}^{(n)}$. Notice that (8.8) is imposed only for $r \geqslant n$. Thus, the calculation of order n requires that all *but* the lowest 2^n restrictions of g lie on a parabola. As n increases, these lowest-order pieces determine the actual fixed point g, and not a parabola. However, μ_n converges to $1/2\, g''(0)$. Let us make sure that we understand what is to be determined.

The points

$$x_{2k+\varepsilon \cdot 2^{n+1}}, \qquad \varepsilon = 0, 1; \quad k = 0, \ldots, 2^n - 1 \tag{8.9}$$

are, by (7.3) the endpoints of the range of $F_{0,k}^{(n)}$. Using (8.4) in (8.7) should result in a sufficient number of equations to determine the x's of (8.9). However, as each $F_{0,k}^{(n)}$ is linear, and by (7.4) obeys

$$(F_{0,k}^{(n)})^{-1}: \quad x_{2k+\varepsilon\cdot 2^{n+1}} \to x_{k+\varepsilon\cdot 2^n} \tag{8.10}$$

while

$$x_{2k+1} = F_1(x_k) = \alpha^{-1}x_k$$

the $F_{0,k}^{(n)}$, *can* be parametrized by the x's of (8.9) *together* with the value of α^{-1}. Since, however, we have chosen the scale $x_0 = 1$, we still have between α^{-1} and the x_{2k} the required 2^{n+1} quantities to be determined by the system (8.7). More correctly, there is still the extra quantity μ_n of (8.7). However, the requirement that g possesses a *quadratic* critical point expressed by (8.5) provides the missing equation. We thus conjecture that the system of (8.7), (8.5), and $x_0 = 1$ possesses an isolated solution for the positive quantities x_{2k} of (8.9).

The reader should feel that this calculation is far from transparent. So it appears at this stage. However, recall that there are also precisely 2^{n+1} values $\sigma_n(k)$ of (7.12), one of which is $\sigma_n(0) = \sigma_0$ of (8.5). I promised that σ was a *natural* parametrization of the $F_{0,k}^{(n)}$. I shall now show why, and reduce the computation to one of systematic ease.

First, by (7.12),

$$\sigma_n(2^p(2k+1)) = \frac{x_{2^p(2k+1)} - x_{2^p(2k+1)+2^{n+1}}}{x_{2^p(2k+1)} - x_{2^p(2k+1)-2^n \bmod 2^{n+1}}} \qquad \text{for} \quad 2k+1 < 2^{n+1-p} \tag{8.11}$$

However, as also expressed by (7.12),

$$\sigma_n[2^p(2k+1)] = \sigma[2^{r+p}(2k+1)] \qquad \text{all} \quad r \geqslant 0 \tag{8.12}$$

which by (7.1) is again a quotient of differences of x's, but now for sufficiently large r ($\geqslant n-p$) the image (the index is even) of an endpoint of the domain of a $g_r^{(n)}$ obeying (8.8); that is,

$$x^{2^{r+p}(2k+1)} = g_r^{(n)}(x_{2^{r+p}(2k+1)-1}) = 1 - \mu_n x^2_{2^{r+p}(2k+1)-1} \tag{8.13}$$

By (8.12), (8.11) is now the quotient of the diferences of the x^2's of (8.13). However, by (6.4) and (6.18),

$$x^2_{2^{r+p}(2k+1)-1} = \alpha^{-2r} x^2_{2^p(2k+1)-1} \tag{8.14}$$

so that (8.11) is the quotient of the differences of the squares in (8.14). Finally, denoting $2^p(2k+1)-1=l$, we have

$$\sigma_n(l+1)=\frac{x_l^2-x_{l+2^{n+1}}^2}{x_l^2-x_{l+2^n \bmod 2^{n+1}}^2} \tag{8.15}$$

to be compared with the defining formula

$$\sigma_n(l)=\frac{x_l-x_{l+2^{n+1}}}{x_l-x_{l+2^n \bmod 2^{n+1}}}, \qquad l=0,\ldots,2^{n+1}-1 \tag{8.16}$$

There is one important proviso on (8.15): its range of applicability is

$$l+1=1,\ldots,2^n-1,\ 2^{n+1},\ldots,2^{n+1}-1 \quad \text{in} \quad (8.15) \tag{8.17}$$

This follows from the derivation of (8.15); should $p=n$ in (8.11), then k of (8.11) must equal zero. But then the second denominator index in that formula is $2^n+2^n \bmod 2^{n+1}=0$; however, x_0 is *never* the image of an x_m under $g_r^{(n)}$, whence the restriction in l of (8.17). Apart from the exceptions $l+1=0$ or 2^n (which I shall treat immediately), the content of (8.15) is that *the one-index advanced σ bears the identical relationship to the x^2's as does the σ of the unadvanced index to the same, but not squared, x's.* That is, the full content of the system (8.4), (8.7) is precisely this twice-defined encoding of the σ's.

To finish the deduction of the equations, let us turn to (8.16) for $l=2^n$:

$$\begin{aligned}\sigma_n(2^n)&=\frac{x_{2^n}-x_{3\cdot 2^n}}{x_{2^n}-x_0}\\&=\frac{g_{n,0}^{(n)}(x_{2^n-1})-g_{n,1}^{(n)}(x_{3\cdot 2^n-1})}{g_{n,0}^{(n)}(x_{2^n-1})-1}\\&=\frac{x_{2^n-1}^2-x_{3\cdot 2^n-1}^2}{x_{2^n-1}^2} && \text{[by (8.8)]}\\&=1-x_2^2 && \text{[by (6.4) and (6.18)]}\end{aligned}$$

That is,

$$1-x_2^2=\sigma_n(2^n) \tag{8.18}$$

Also, by (8.16) using (6.4) and (6.18),

$$\sigma_n(2^n-1)=\frac{x_{2^n-1}-x_{3\cdot 2^n-1}}{x_{2^n-1}-x_{2^{n+1}-1}}=\frac{1-x_2}{1-\alpha^{-1}} \tag{8.19}$$

However, with $l = 2^{n+1} - 1$ in (8.16), again using (6.4) and (6.18), we have

$$\sigma_n(2^{n+1} - 1) = \frac{x_{2^{n+1}-1} - x_{2^{n+2}-1}}{x_{2^{n+1}-1} - x_{2^n-1}} = -\alpha^{-1} \tag{8.20}$$

Thus, α^{-1} is also one of the σ values (the most obvious one) and, with (8.19), we have

$$1 - x_2 = \sigma_n(2^n - 1)[1 + \sigma_n(2^{n+1} - 1)] \tag{8.21}$$

Finally, noting (8.5) and (8.20), we can explicitly eliminate α^{-1} in

$$\sigma_n(0) = [\sigma_n(2^{n+1} - 1)]^2 \tag{8.22}$$

and write down the system of equations that determines σ:

$$\begin{aligned} 1 - [1 + \sigma_n(2^{n+1} - 1)]\, f_k^{(n)}[\sigma_n(i)] \\ = \{1 - f_k^{(n)}[\sigma_n(i+1)]\}^{1/2}, \qquad k = 1,..., 2^{n+1} - 1 \end{aligned} \tag{8.23}$$

where

$$f_k^{(n)}[\sigma_n(i+1)] = 1 - x_{2k}^2 \tag{8.24}$$

meaning that $f_k^{(n)}$ is that function of the set of one-index advanced σ's that is obtained by solving the set (8.15) for $1 - x_{2k}^2$ in terms of $1 - x_2^2$, which by (8.18) is simply $\sigma_n(2^n)$. The $f_k^{(n)}[\sigma_n(i)]$ on the left-hand side side of (8.23) is the same functional form as determined from (8.24) with each $\sigma_n(i+1)$ argument replaced by the argument $\sigma_n(i)$. The positive square root taken in (8.23) reflects the fact that $x_{2k} > 0$ in (8.24). Before going further, let us work out the two lowest orders, $n = 0$, 1, of this theory.

At $n = 0$, there are two values of σ, σ_0 and σ_1. With $k = 1$ in (8.24),

$$1 - x_2^2 = \sigma_1 = f_1(\sigma(i+1))$$

and (8.23) reads

$$1 - (1 + \sigma_1)\sigma_0 = (1 - \sigma_1)^{1/2}$$

which together with (8.22), $\sigma_0 = \sigma_1^2$, is easily seen to be the earlier result (6.26).

At order $n = 1$, we can see the machinery generally at work. By (8.22),

$$\sigma_0 = \sigma_3^2 \tag{8.25}$$

By (8.24),

$$k=1: \quad f_1(\sigma_{i+1})=1-x_2^2=\sigma_2 \qquad [\text{Eq. } (8.18)] \tag{8.26}$$

$$k=2: \quad f_2(\sigma_{i+1})=1-x_4^2=\frac{x_0^2-x_4^2}{x_0^2-x_2^2}(1-x_2^2)=\sigma_1\sigma_2 \tag{8.27}$$

$$k=3: \quad f_3(\sigma_{i+1})=1-x_6^2=(1-x_2^2)+\frac{x_2^2-x_6^2}{x_2^2-x_0^2}(x_2^2-x_0^2)$$

$$=\sigma_2(1-\sigma_3). \tag{8.28}$$

Entering these results in (8.23), we have

$$k=1: \quad 1-(1+\sigma_3)\sigma_1=(1-\sigma_2)^{1/2} \tag{8.29}$$

$$k=2: \quad 1-(1+\sigma_3)\sigma_0\sigma_1=(1-\sigma_1\sigma_2)^{1/2} \tag{8.30}$$

$$k=3: \quad 1-(1+\sigma_3)\sigma_1(1-\sigma_2)=[1-\sigma_2(1-\sigma_3)]^{1/2} \tag{8.31}$$

Together with (8.25), (8.29)–(8.31) determine a unique solution for $\sigma_0 \cdots \sigma_3 \in (0, 1)$ given numerically by

$$\sigma_0=0.1573393\ldots$$

$$\sigma_1=0.1759069\ldots$$

$$\sigma_2=0.4310046\ldots$$

$$\sigma_3=0.3966602\ldots$$

with $|\alpha|=\sigma_3^{-1}=2.521049\ldots$, an error similar to that of the $n=0$ solution of (6.26). As we proceed to higher n, the solutions geometrically converge toward the actual result.

There are several comments to make at this point. The first is that the extraction of $1-x_{2k}^2$ for each k from (8.15) is a straightforward task, so that any equation in any order of approximation can immediately be written down. However, as the solution to the full set of equations for a given order n becomes arduous and numerically problematic, I leave this exercise to the reader. Second, there is an important systematics that determines half of the order $n+1$ equations from all those of order n, so that the solution to order n, doubled up by pairs, exactly satisfies the first half of the order $n+1$ equations. By the bounded nonlinearity of F_0 follows the exponentially weak dependence of σ on lower-index ε's. This means that the remaining half to the order $n+1$ equations fail to be satisfied with exponentially small error for large n. Thus, there is good theoretical reason to believe that successive orders of approximation define a convergent theory. These results follow from the exact linearity of F_1:

$$\sigma_{n+1}(2k+1)=\frac{F_1(x_k)-F_1(x_{k+2^{n+1}})}{F_1(x_k)-F_1(x_{k+2_n \bmod 2^{n+1}})}$$

$$=\frac{x_k-x_{k+2^{n+1}}}{x_k-x_{k+2^n \bmod 2^{n+1}}}=\sigma_n(k)$$

It is then easy to see that by also replacing $\sigma_{n+1}(2k)$ by $\sigma_n(k)$, the first 2^{n+1} of Eqs. (8.23) at order $n+1$ becomes precisely the full set at order n.

However, as I have already alluded to, having written down the 2^{n+1} equations of order n in no way implies that we are close to a solution. Indeed, even using the trial approximation (the doubling up by pairs mentioned above)

$$\sigma_{n+1}(2k+1)=\sigma_{n+1}(2k)=\sigma_n(k)$$

while precisely satisfying half the order $n+1$ equations, and failing to satisfy the second half with exponentially decreasing errors, still fails to fall with the basin of attraction of Newton's method to the order $n+1$ solution. However, we can do much better.

9. DYNAMICS IN THE SPACE OF SCALINGS

I showed in the last section that the scalings σ can be written as a quotient of differences of either coordinates, or in critical image, the squares of coordinates. (I used this idea in the past to determine the discontinuities of σ, which, indeed, are exponentially graded accoding to closeness of approach to the critical point of g.) The formulas that fix σ are precisely the requirements that both evaluations agree.

Inspecting (8.15), one sees that

$$\sigma_n(l+1)=\frac{x_l^2-x_{l+2^{n+1}}^2}{x_l^2-x_{l+2^n \bmod 2^{n+1}}^2}=\sigma_n(l)\frac{x_l+x_{l+2^{2+1}}}{x_l+x_{l+2^n \bmod 2^{n+1}}} \tag{9.1}$$

or

$$\frac{\sigma_n(l+1)}{\sigma_n(l)}-1=\frac{x_{l+2^{n+1}}-x_{l+2^n \bmod 2^{n+1}}}{x_l+x_{l+2^n \bmod 2^{n+1}}}\sim\sigma_{\text{eff}}^n \tag{9.2}$$

so long as $|x_l|\sim 1$. Thus, provided that

$$l \bmod 2^r \neq 0 \qquad \text{for} \quad r=O(n) \tag{9.3}$$

$\sigma_n(l+1)$ is exponentially (in n) close to $\sigma_n(l)$. This suggests a simple strategy to (recursively) enfore the agreement of (8.15) and (8.16): Assume a given set of σ's and update them (by a dynamics in the 2^{n+1}-dimensional space of σ_n) by (i) inverting (8.15) to determine the x_{2k}^2 in terms of the $\sigma_n(i)$; (ii) taking square roots to produce the x_{2k}; and (iii) using the x_{2k} in (8.16) to determine new $\sigma_n'(i)$. That is, we have a transformation T in the space of scaling functions

$$\sigma_n' = T_n[\sigma_n] \tag{9.4}$$

which we now hope relaxes to the common solution of (8.15) and (8.16). For l's not "too" dyadically small [i.e., (9.3)], we anticipate from the discussion of (9.1) and (9.2) that

$$\sigma_n'(l-1) \approx \sigma_n(l) \tag{9.5}$$

Thus, the great bulk of the dynamics of T is the simple difference delay dynamics of (9.5) with more serious right-hand sides at the dyadically smallest values of l (e.g., 2^n). This implies in order n with 2^{n+1} σ's, after a major discontinuity, another 2^{n+1} steps are required for this significant modification to propagate "around" the set of σ's in order to be transformed again. Thus, if we denote by λ_n the eigenvalue of the T_n convergence to the stable fixed point σ_n^*, then we expect

$$\lambda_n^{2^{n+1}} \sim \Lambda \tag{9.6}$$

Indeed, after minor details are fixed [as before, $l+1 \neq 2^n$ in (8.15)], this program works perfectly. Thus, as I determine numerically, (9.6) is correct with

$$\Lambda \approx 0.70 e^{\pm 0.23 i} \tag{9.7}$$

However, not only does T possess σ^* as a stable fixed point, but so far I can determine numerically (although my search has not been exhaustive), T also possesses σ^* as a *global* attractor with basin $\{(x_1, \dots, x_{2^{n+1}}) \in (0,1)^{2^{n+1}}\}$. One can verify this in the lowest-order models where for all but 2^n sigmas, all other σ_N exactly satisfy the delay dynamics of (9.5). To fix ideas and to explain part of the last comment, let us now work out the $n=0, 1$ scaling dynamics using (9.5) taken with identity in I^N, where I is $(0, 1)$.

The first step is a simple piece of systematics. Return to (8.15) and evaluate

$$\sigma_n[2^{n-r}(2i+1)] = \frac{x^2_{2^{n-r}(2i+1)-1} - x^2_{2^{n-r}(2i+1)+2^{n+1}-1}}{x^2_{2^{n-r}(2i+1)-1} - x^2_{2^{n-r}(2i+1)+2^n-1 \bmod 2^{n+1}}}$$
$$n \geqslant r \geqslant 1, \qquad 2i < 2^{r+1} \tag{9.8}$$

Using (6.4) and (6.18) as in (8.14), we obtain

$$\sigma_n[2^{n-r}(2i+1)] = \frac{x_{2i}^2 - x_{2i+2^{r+1}}^2}{x_{2i}^2 - x_{2i+2^r \bmod 2^{r+1}}^2}, \qquad n \geqslant r \geqslant 1, \quad 2i < 2^{r+1} \tag{9.9}$$

and the related result

$$\sigma_n[2^{n-r}(2i+1)-1] = \frac{x_{2i} - x_{2i+2^{r+1}}}{x_{2i} - x_{2i+2^r \bmod 2^{r+1}}}, \qquad n \geqslant r \geqslant 0, \quad 2i < 2^{r+1} \tag{9.10}$$

The important content of these formulas is that the right-hand sides are *independent* of n: the equation that relates $\sigma_n(l-1)$ to $\sigma_n(l)$ is unchanged in all higher orders $n+p$ if l is replaced by $2^p l$. The notion of the set of σ_n interpreted as the nth-order step function approximant to the scaling *function* $\sigma(\tau)$ follows from this observation when τ is defined at level n as $l/2^{n+1}$:

$$\sigma_n(\tau) = \sigma_n(l) \qquad \text{for} \quad \tau \in (l/2^{n+1}, (l+1)/2^{n+1}) \tag{9.11}$$

The precise definition of T_n of (9.4) now follows. Throughout the rest of the discussion the subscript n on σ_n shall be implicitly understood. First, by (8.18)

$$x_2 = [1 - \sigma(2^n)]^{1/2} \tag{9.12}$$

Combining (8.19) and (8.20),

$$\sigma(2^n - 1) = \frac{1 - x_2}{1 + \sigma(2^{n+1} - 1)} \tag{9.13}$$

Now interpret (9.13) to determine the *new* (transformed under T) $\sigma(2^n - 1)$ by replacing x_2 in it by the square root of x_2^2 determined by the old $\sigma(2^n)$ as given in (9.12). Denoting transformed values by primes, we thus have

$$\text{(i)} \qquad \sigma'(2^n - 1) = \frac{1 - [1 - \sigma(2^n)]^{1/2}}{1 + \sigma(2^{n+1} - 1)} \tag{9.14}$$

This first equation required special treatment because of (8.17), now expressed by $r < 1$ in (9.9). As a (numerically, e.g.) systematic procedure, we save $(1 - x_2^2)$ $[=\sigma(2^n)]$ and $(1 - x_2)$ obtained by

$$(1 - x_2) = 1 - [1 - (1 - x_2^2)]^{1/2} = \frac{(1 - x_2^2)}{1 + [1 - (1 - x_2^2)]^{1/2}} \tag{9.15}$$

With these quantities known, we start with $r=1$ in (9.9), allowing $i=0, 1$, and rewrite it as

$$\text{(ii)} \qquad (1-x^2_{2i+2^{r+1}}) = (1-x^2_{2i}) + \sigma[2^{n-r}(2i+1)] \times [(1-x^2_{2i+2^r \bmod 2^{r+1}}) - (1-x^2_{2i})] \tag{9.16}$$

The $(1-x^2)$'s on the right-hand size are either $(1-x_0^2)=0$ or $(1-x_2^2)$ already saved. Thus, (9.16) determines the next two x_{2k}, $(1-x_4^2)$ and $(1-x_6^2)$, and the square root formula of (9.15) for general x_{2k} produces $(1-x_4)$ and $(1-x_6)$. Both of these quantities and the previous $(1-x_0)=0$ and $(1-x_2)$ now substituted in the right-hand side of (9.10) produces

$$\text{(iii)} \qquad \sigma'[2^{n-r}(2i+1)-1] = \frac{(1-x_{2i+2^{r+1}})-(1-x_{2i})}{(1-x_{2i+2^r \bmod 2^{r+1}})-(1-x_{2i})} \tag{9.17}$$

We now systematically increase $r=2,\ldots, n$, the last step at $r=n$ determining $\sigma'(2i)$, and, in particular, $\sigma'(0)$. The only σ' *not* produced by (9.10) is $\sigma'(2^{n+1}-1)$ (i.e., $-\alpha^{-1}$). But this, of course, is where we inject the required nature of the algebraic singularity of g (i.e., quadratic) at this critical point, (8.22):

$$\text{(iv)} \qquad \sigma'(2^{n+1}-1) = [\sigma'(0)]^{1/2} \tag{9.18}$$

This completes the precise implementation of T, and is an algorithm that simultaneously produces all the x_{2k} together with all the σ's. At the heart of this dynamics is an insistence upon taking square roots, an inherently stabilizing operation. Verified to be convergent in low order, the difference delay intuition suggests a rapid convergence with order n and persisting stability. Numerically this is fully borne out to the degree that choosing the initial σ's randomly within (0, 1) always leads to convergence to the unique fixed point σ^*. Let us finish this discussion with the $n=0$ and $n=1$ delay models.

Setting $r=1$ in (9.9), we have, using (9.12) for x_2,

$$i=0: \qquad \sigma(2^{n-1}) = \frac{1-x_4^2}{1-x_2^2} \to x_4 = [1-\sigma(2^{n-1})\,\sigma(2^n)]^{1/2} \tag{9.19}$$

and

$$i=1: \qquad \sigma(2\cdot 2^n) = \frac{x_2^2-x_6^2}{x_2^2-1} \to x_6 = \{1-\sigma(2^n)[1-\sigma(3\cdot 2^{n-1})]\}^{1/2} \tag{9.20}$$

Substituting in (9.10), we have the transformation formula

$$\sigma'(2^{n-1}-1) = \frac{1-x_4}{1-x_2} = \frac{1-[1-\sigma(2^n)\,\sigma(2^{n-1})]^{1/2}}{1-[1-\sigma(2^n)]^{1/2}} \tag{9.21}$$

and

$$\sigma'(3\cdot 2^{n-1}-1)=\frac{x_2-x_6}{x_2-1}=\frac{\{1-\sigma(2^n)[1-\sigma(3\cdot 2^{n-1})]\}^{1/2}-[1-\sigma(2^n)]^{1/2}}{1-[1-\sigma(2^n)]^{1/2}} \tag{9.22}$$

From these results, we can see how (9.2) works out. Rationalizing numerator and denominator in (9.21), we have

$$\sigma'(2^{n-1}-1)=\sigma(2^{n-1})\frac{1+[1-\sigma(2^n)]^{1/2}}{1+[1-\sigma(2^n)\,\sigma(2^{n-1})]^{1/2}} \tag{9.23}$$

and an analogous result for (9.22). Notice that the ratio of the *sums* of roots is of order 1. As we write down the formulas for σ's at dyadically larger indices [i.e., larger r in (9.9) and (9.10)], the analogous ratio of sums of roots will be exponentially closer to 1 as more and more products of σ's (smaller than 1) appear subtracted from 1 within the square roots. That is, (9.5) will become exponentially accurate for σ's of larger r. Thus, as the order n increases, the new equations, all for the largest r $(=n)$ become exponentially closer to (9.5), and the dynamics "saturates." The first approximation to $n\to\infty$ is obtained by using just (i) of (9.14) for $l=2^n$ and for all other $l=1,\ldots,2^{n+1}-1$,

$$\sigma'(l-1)=\sigma(l) \tag{9.24}$$

Thus, the $r=0$ model dynamics is

$$\sigma_{l-1,t+1}=\sigma_{l,t},\qquad l=1,\ldots,N-1,\,N+1,\ldots,2N-1 \tag{9.25a}$$

$$\sigma_{N-1,t+1}=\frac{1-(1-\sigma_{N,t})^{1/2}}{1+\sigma_{2N-1,t}} \tag{9.25b}$$

$$\sigma_{2N-1,t}=(\sigma_{0,t})^{1/2} \tag{9.25c}$$

where the time index t means $[\sigma_t]=T^t[\sigma_0]$. Let us compute the solution to (9.25). Write

$$X_t=\sigma_{2N-1,t},\qquad Y_t=\sigma_{N-1,t} \tag{9.26}$$

By (9.25a)

$$\sigma_{N,t-1+N}=\sigma_{2N-1,t}=X_t \tag{9.27}$$

so that (9.25b) becomes

$$\sigma_{N-1,t+N}=Y_{t+N}=\frac{1-(1-X_t)^{1/2}}{1+X_{t+N-1}} \tag{9.28}$$

By (9.25a) again,

$$\sigma_{0,t+2N-1} = \sigma_{N-1,t+N} = Y_{t+N} \tag{9.29}$$

which by (9.25c), produces

$$X_{t+2N-1} = (Y_{t+N})^{1/2}$$

or

$$X_{t+N-1} = (Y_t)^{1/2} \tag{9.30}$$

Substituting (9.30) in (9.28) produces the system

$$X_{t+N-1} = (Y_t)^{1/2} \tag{9.31a}$$

$$Y_{t+N} = \frac{1-(1-X_t)^{1/2}}{1+(Y_t)^{1/2}} \tag{9.31b}$$

Next, defining

$$x_k = X_{kN} \tag{9.32a}$$

$$y_k = Y_{kN} \tag{9.32b}$$

and ignoring the -1 in X_{t+N-1} (order $1/N$; we are interested in $N \to \infty$), we have

$$x_{k+1} = (y_k)^{1/2} \tag{9.33}$$

$$y_{k+1} = \frac{1-(1-x_k)^{1/2}}{1+(y_k)^{1/2}}$$

The fixed point of (9.33) is just $x = \sigma_1$, $y = \sigma_0$ of (6.26) and is the $\boldsymbol{n=0}$ fixed point. The eigenvalues at the fixed point are, however, different from the $n=0$ **dynamics** of (i)–(iv). The system (9.33) is the $\boldsymbol{r=0}$ $n \to \infty$ dynamics. The eigenvalues of (9.33) at its fixed point are

$$\lambda_1 = -0.832289\ldots, \qquad \lambda_2 = 0.688872\ldots \tag{9.34}$$

Since $x_k - x^* \sim \lambda_1^k$, by (9.32), $X_{kN} - x^* \sim \lambda_1^k$, and so

$$X_{2Nk} - x^* \sim (\lambda_1^2)^k \tag{9.35}$$

That is, each $2N = 2^{n+1}$ iterates of (9.25) constituting one entire passage through all σ's, converges with eigenvalue λ_1^2, so that Λ of (9.6) is

$$\Lambda_0 = \lambda_1^2 = 0.692705\ldots \tag{9.36}$$

Λ_0 turns out already to be an order 1 % result for the full $n \to \infty$ dynamics.

The $r=1$ $n\to\infty$ dynamics follows by adding (9.21) and (9.22) for the quarter-way σ's to (9.14) for the half-way σ and (9.25a) for all others. Denoting $N=2^{n-1}$ here, and $X_t=\sigma_{4N-1,t}$ with analogous quantities for the other σ's, and the N-step values by lowercase letters as in (9.32), we obtain

$$\begin{aligned} x' &= u^{1/2} \\ y' &= \frac{[1-y(1-x)]^{1/2}-(1-y)^{1/2}}{1-(1-y)^{1/2}} \\ z' &= \frac{1-(1-y)^{1/2}}{1+u^{1/2}} \\ u' &= \frac{1-(1-yz)^{1/2}}{1-(1-y)^{1/2}} \end{aligned} \tag{9.37}$$

The fixed point is $x=\sigma_3$, $y=\sigma_2$, $z=\sigma_1$, $u=\sigma_0$ of the $n=1$ dynamics, which is given below (8.31). The spectrum of (9.37) at its fixed point is

$$\begin{aligned} \lambda_c &= 0.90623457\,e^{\pm i(\pi/2-0.051493628)} \\ \lambda_1 &= 0.83050196 \\ \lambda_2 &= -0.81607177 \end{aligned} \tag{9.38}$$

Since $N=2^{n-1}$, λ^4 are now the full passage through the eigenvalues of σ, and

$$\Lambda_1=\lambda_c^4=0.6744698\,e^{\pm 0.2059745i} \tag{9.39}$$

The phase of Λ_1 means that convergence is a damped sinusoid of period ~ 30.50 (of full 2^{n+1} steps over the whole system). The modulus has changed $\sim 3\%$ from Λ_0. Indeed, the small departure of the phase of λ_c from $\pi/2$ is the very near agreement of λ_c^2 to the negative real λ_1 of (9.34). Also, $\lambda_1^2=0.68973\ldots$ in (9.38), to be compared to the subdominant $\lambda_2=0.688872$ of (9.34). That is, to surprising accuracy, the $r=0$ model faithfully follows the $r=1$ one. The actual $n\to\infty$ numerics are barely distinguishable (for n up to 8, or 512 σ's) from those of (9.37).

10. AFTERWORD AND CONCLUSIONS

In the last two sections we learned how the specification of the critical point singularity of the underlying dynamical map g determines the scaling function globally along the orbit. Since σ is invariant under smooth transformations, it is in particular almost a constant of the motion, since g is smooth away from its critical point, and hence σ everywhere is determined by the critical point singularity of g. The σ itself is a rich invariant encoding full knowledge of the temporal ordering along a strange set, and hence determining the refinement of a coarse-grained specification of the set by prologing the data x_t to successively larger ranges of t. Technically, as seen in the discussion surrounding (7.12), half of the information in σ is encoded in orbital eigenvalues; the other half of the information addresses the $O(1)$ coefficients of exponential quantities by determining the "finite"-size widths of the intervals obtained by the partitioning by the inverses under the presentation function. Let me be more precise.

By (3.3), the size of the intervals $\Delta^{(n)}(\varepsilon_n \cdots \varepsilon_1)$ is *estimated* by

$$\Delta^{(n)}(\varepsilon_n \cdots \varepsilon_1) \sim D(F_{\varepsilon_1} \cdots F_{\varepsilon_n}) \tag{10.1}$$

By choosing not x_0, but rather within $\Delta^{(n)}(0 \cdots 0)$ the periodic point

$$x_0^* = F_{\varepsilon_1} \cdots F_{\varepsilon_n}(x_0^*) \tag{10.2}$$

the derivative in (10.1) is effectively the eigenvalue of this orbit. This derivative in turn behaves as

$$\Delta^{(n)}(\varepsilon_n \cdots \varepsilon_1) \sim \sigma(\varepsilon_n \cdots \varepsilon_1)\, \sigma(\varepsilon_{n-1} \cdots \varepsilon_1) \cdots \sigma(\varepsilon_r \cdots \varepsilon_1) \cdots = k\sigma^n_{\text{eff}}(\varepsilon_n \cdots \varepsilon_1) \tag{10.3}$$

or,

$$\ln \sigma_{\text{eff}}(\varepsilon_n \cdots \varepsilon_1) \sim \frac{1}{n} \ln |D(F_{\varepsilon_1} \cdots F_{\varepsilon_n})| \tag{10.4}$$

It is precisely these asymptotic (in n) growth rate exponents that are determined by the slopes of $F_\varepsilon(x)$. Moreover, it is only this half of the information of σ that is "tested" by the thermodynamics. It follows that $f(\alpha)$ fails to encode the information of how the local linear segments of g are to be fitted together, and so is *far* from the full information invariantly available descriptive of the strange set (in addition to the complete loss of t-ordering information).

To put this differently, for $k = O(1)$ in (10.3), that equation asserts that the actual Δ's are of bounded variation from the Δ's *estimated* by $k = 1$. When we now reconsider the quotients leading to (7.12), we see that in the nth approximation to σ, asymptotically in m, the leading $m - n$ derivatives that determine σ_{eff} exactly cancel in numerator and denominator, so that σ is actually the quotient of the k's of (10.3) themselves. Thus, σ presents much finer information, inclusive of $O(1/n)$ terms, than the orbital eigenvalues and $f(\alpha)$, which determine information up to bounded variation only. This extra information allows orbit prolongation and attractor refinement not possible with the limited exponential information of (10.4).

Accordingly, a full and prescriptively useful end product of the "solution" of a dynamics $x_{t+1}=g(x_t)$ is the determination of its scaling function σ. But then, how can it be that σ is always determined from the equations that follow by eliminating the x_l from (8.15) and (8.16)? The answer to this query is the content of Section 6, specifically (6.10), which asserted that F_ε determines a period doubling fixed point. Inspection of the argument of Section 6 reveals that the "full" tree topology of inverses of F's as in Fig. 1, in contrast to an F of Fig. 4, is responsible for being led into period doubling. Indeed, an analysis like that of Section 6 applied to Fig. 4 would lead to a fixed-point dynamics of the golden mean renormalization group[11,12] and not that of period doubling. The analysis of Section 8 would then lead to σ's determined by the ratios of golden mean differences. I have presented the ideas of the golden mean scaling function in the context of circle maps elsewhere.[13] For the case of Fig. 4, the resulting σ is *not* that of a circle map, although of largely identical organization.

We thus see that the *topology* of intervals encoded in the topology of F (as in Fig. 4) determines the equations that fix σ. The golden mean topology *on a circle* is so much more important than that of Fig. 4 on an interval that I have not bothered here to present the scaling function theory for the problem whose thermodynamics is given by (5.4).

I have not done so because I have already worked out the theory for *circle map* topologies. That is, one can construct presentation functions on the circle. It turns out that there is a unique choice of the form of the golden mean fixed point that emerges whose scalings are identical to those of the original (i.e., not the fixed point) map for which E is expanding, and this choice is not that of Rand *et al.*[12] After F is known, the exact thermodynamic eigenvalue equation can be written down, which turns out to be subtly different from (5.4). There are a sufficient number of new ingredients in this circle map theory to make it inappropriate for this article. A sequel devoted to these new ideas is in preparation.

This brings us to a discussion of Sections 8 and 9, especially the latter. Section 8 is a culmination of a long effort to determine equations for the scaling function intrinsic to σ itself. (That is, not just determining σ as a detailed calculation available from the fixed-point dynamics g.) σ is the desired outcome of a calculation; g is the specification of the dynamics: it is the nature of the *orbits* of g that we want to know. Writing down g is writing down the *equations of motion*: since g is universal, at least we have written down a very generally applicable equation. But the real problem is to determine the *solution* to these equations, which are chaotic and derived from a highly nonlinear dynamics. As I have stressed, σ *is* the solution we seek in that it provides simple "genetic" building *knowledge* of the solution rather than a mindless enumeration of the ordered set $\{x_t\}$ of chaotically varying quantities. To generalize away from period doubling, we want to know how to generally write down equations for σ of a chaotic motion, and not just simulate the dynamics and "show" the solution in some complicated plot that merely reaffirms to the observer that the motion is chaotic—although at least not random. This is why I believe the ideas of the scaling function theory of Section 8 to be so important. But just so, I regard the dynamics in the space of scaling functions, Section 9, to be more important still.

As Section 9 stands, the σ dynamics is an "arbitrary" invention; as it might seem, merely a technical device to obtain the solution to the σ equations of Section 8. While true enough, I also believe otherwise.

As a mathematical point, the numerically observed good behavior of that dynamics (i.e., possessing a globally attracting fixed point), if proven, constitutes a proof of the uniqueness of the period doubling fixed point. (One would have to show that the solution for g just on the Cantor set is the restriction of an analytic function to prove existence by this line of thought. For the golden mean circle map fixed point, the "Cantor set" is now the entire circle, so a full solution is obtained now requiring proof of its analyticity.)

At least from this paper we know that nice scaling dynamics exists for a variety of highly nontrivial σ's. However, there is a conceptually deeper point: the equations of physics through the principles of inertia and causality are *local* (differential) equations of motion. Like the local (in t) dynamics $x_{t+1} = g(x_t)$, they fail to present in any transparent manner the inherently nonlocal "genetic" principle of a scaling function: the successive products of σ's as in (10.3) relate *distant* pieces of a solution to one another through common ancestors. What we must do is to implement that "change of variables" in the originally offered equations of motion to produce *dynamical equations* for σ itself. If we can do so, for σ's of just a few distinct values in some approximation, by easy calculation we can then compute the salient features of the solution and not the enumeration of the individually uninteresting x_t. Section 9 represents the first coming to grips with what such "intelligently" formulated dynamics should look like.

I want to conclude this paper with a conceptual analogy to (I hope) better illuminate the content of the last paragraph, and then exhibit an actual example of a true scalinglike dynamics drawn from the study of cellular automata.

Consider a cloudlike initial configuration of some fluid equation (a classical field theory). Imagine that the density of this configuration possesses rich scaling properties (e.g., a fixed spatial scale exponent over many decades.) Moreover, imagine that at successive moments of time it also possesses these scaling properties, although possibly variable in time. From this we should surmise that the instantaneous velocities should also possess similar scaling properties. Imagine that these scalings are easily specified, that is, we have discerned in this complex spatial object some prescriptive rules that if iterated would construct it. Now let us contemplate how we advance this structure in time. By the locality of the field equations we must *actually* spin out this iterative construction in order to provide the equations with the sort of initial data they require. Now we can advance the structure a step ahead in time. But what do we now have? Simply an immense list of local density and velocity values of high local irregularity. Of course, if we possess a good algorithm, we could now from this new pabulum of data again discern the scaling information—perhaps evolved—that we knew about anyway. This is obviously a foolish double regress. Since our informed understanding lay in the scaling description, we should obviously have transcribed our "true" local dynamics into one pertinent to these scalings, rather than mount a numerical program that strains the most powerful machines we possess. That is, the solution in the usual sense of our local field theories is apt to be a mindless enterprise when the solutions happen not to be simple. In this sense, our theories, while "true," are useful only to God, which seems not to be the hallmark of what humans adjudge to be truth.

As our last heuristic example, consider the time-dependent block probabilities of a one-dimension, nearest-neighbor, two-state-per-site cellular automaton. Denoting these probabilities in an nth approximation by $P_n(\varepsilon_m \cdots \varepsilon_1 \varepsilon_0)$, we will cast an analogy between the P_n and Δ_n. As with the Δ_n, the $P_n(\varepsilon_m \cdots \varepsilon_1 \varepsilon_0)$ are assembled from Markov transition probabilities $\sigma_n(\varepsilon_n \cdots \varepsilon_0)$ linking the 2^n nodes labeled by $\varepsilon_n \cdots \varepsilon_1$ of a strictly probabilistic Markov graph (the sum of the transition probabilities out of a node sums to one). Assigning nodal probabilities $\Pi_n(\varepsilon_n \cdots \varepsilon_1)$ determined by the stationarity of the process then produces the nth-order probabilities

$$P_n(\varepsilon_n \cdots \varepsilon_0) = \sigma_n(\sigma_n \cdots \varepsilon_0)\, \Pi_n(\varepsilon_n \cdots \varepsilon_1) \tag{10.5}$$

The point of this construction is that with the σ_n randomly assigned, these P_n are then a random *a priori* set of probabilities satisfying the Kolmogorov consistency conditions (i.e., that these P_n summed over right or left ε's consistently produces the lower-order P_{n-r} probabilities). The minimal-information *extension* of these P_n is now simply

$$\begin{aligned} &P_{n+r}(\varepsilon_{n+r},\dots, \varepsilon_{n+1}, \varepsilon_n,\dots, \varepsilon_0) \\ &\quad = \sigma_n(\varepsilon_{n+r},\dots, \varepsilon_r) \cdots \sigma_n(\varepsilon_{n+1},\dots, \varepsilon_1)\, P_n(\varepsilon_n,\dots, \varepsilon_0) \end{aligned} \tag{10.6}$$

so that the extensions P_{n+r} are constructed in precisely the same way as the $\Delta^{(n)}$ of (10.3). The important idea of Gutowitz *et al.*[14] (in addition to the Baysean extension method of the P_{n+r} equivalent to the Markov diagrammatic method just presented) is that the *dynamical* action of the automaton relates the P_n at time $t+1$ as an appropriate (depending upon the rule of the automaton) function of the $P_{n+2}(\varepsilon_{n+1}, \varepsilon_n, ..., \varepsilon_0, \varepsilon_{-1})$ for nearest neighbor rules (whence the additional ε_{n+1} and ε_{-1}). By (10.6) this becomes a dynamical rule for the evolution of the basic entities entities σ_n, Thus, in formal analogy, the kind of dynamics discussed in Section 9 is the actual temporal dynamics of these automata. That is, the idea of a dynamics *not* for the "obvious" variables $P(\cdots \varepsilon_m \cdots \varepsilon_0 \cdots)$, but rather for the "scaling" elaborative variables σ_n is the actual dynamics of these systems, so that the scheme of evolving "evolutionary" variables, with no requirement of approach to, say, a fixed point, is here realized.

Thus, in conclusion, I hope to have exposed some glimmers of a new program of dynamics for problems in which our accustomed partial differential fied equations lead us into a hopeless morass of boring numerical simulation. My examples to date are indeed too special, but perhaps suggestively illuminating.

ACKNOWLEDGMENTS

I have deeply profited from ongoing discussions with Dennis Sullivan over the last several years. This includes important points of understanding of presentation functions and his penetrations into the meaning of the period doubling scaling function, suggesting that its (σ's) smoothness should lead to a principle for its determination.

REFERENCES

1. M. J. Feigenbaum, Universality in Complex Discrete Dynamics, LA-6816-PR: Theoretical Division Annual Report, July 1975–September 1976, Los Alamos; *J. Stat. Phys.* **19**:25 (1978); **21**:669 (1979).
2. D. Ruelle, *Statistical Mechanics, Thermodynamic Formalism* (Addison-Wesley, 1978).
3. E. B. Vul, Ya. G. Sinai, and K. M. Khanin, *Uspekhi Mat. Nauk* **39**:3 (1984) [*Russ. Math. Surv.* **39**:1 (1984)].
4. M. J. Feigenbaum, *J. Stat. Phys.* **46**:919, 925 (1987).
5. T. C. Halsey, M. H. Jensen, L. P. Kadanoff, I. Procaccia, and P. I. Schraiman, *Phys. Rev. A* **33**:1141 (1986).
6. M. J. Feigenbaum, in *Proceedings of the 1987 Noto Summer School.*
7. M. J. Feigenbaum, I. Procaccia, and T. Tel, The scaling properties of multifractals as an eigenvalue problem, *Nonlinearity*, submitted.
8. M. J. Feigenbaum, *Phys. Lett.* **74A**:375 (1979).
9. M. J. Feigenbaum, *Commun. Math. Phys.* **77**:65 (1980).

10. M. J. Feigenbaum, in *Nonlinear Phenomena in Chemical Dynamics*, C. Vidal and A. Pecault, eds. (Springer-Verlag, 1981).
11. M. J. Feigenbaum, L. P. Kadanoff, and S. F. Shenker, *Physica D* **5**:370 (1982).
12. D. Rand, S. Ostlund, J. Sethna, and E. Siggia, *Physica D* **5** (1982).
13. M. J. Feigenbaum, in *Nonlinear Phenomena in Physics*, E. Claro, ed. (Springer-Verlag, 1984).
14. H. A. Gutowitz, J. D. Victor, and B. W. Knight, *Physica* **28D** (1987).

Nonlinearity **2** 305–10 (1989)

Fixed points of composition operators II

Henri Epstein†
Institut des Hautes Etudes Scientifiques, 35 route de Chartres, 91440 Bures-sur-Yvette, France

Received 6 May 1988
Accepted by D A Rand

Abstract. Analytic unicritical fixed points of composition operators of Feigenbaum type for interval and circle maps are shown to exist for every value of $r > 1$, where r is the order of the critical point.

AMS classification scheme numbers: 58F13, 30D05

1. Introduction

In this paper solutions are sought for the functional equations arising in the theories of period doubling and of golden circle maps due to Feigenbaum [6] (Cvitanović–Feigenbaum equation), Coullet and Tresser† [1], Feigenbaum *et al* [7] and Ostlund *et al* [8]. As in [3], we rewrite these equations in the common form

$$\phi(x) = -\frac{1}{\lambda}\phi\left(\frac{1}{\lambda^{\nu-1}}\phi(\lambda^\nu x)\right) \qquad \phi(0) = 1. \tag{1.1}$$

Here $\lambda \in (0, 1)$, $\nu \in [1, 2]$ are constants, and x varies in $[0, L)$ for some $L > \lambda^{1-\nu}$. The case $\nu = 1$ corresponds to *even* interval maps, and the case $\nu = 2$ to circle maps. We only consider solutions ϕ such that, on $[0, L)$, $\phi(x) = f(x^r)$, with f analytic, decreasing, and without critical points on $[0, L^r)$. It will be shown that such solutions exist for all $r > 1$ and $\nu \in [1, 2]$. This includes in particular the 'generic cases', $\nu = 1$ and $r = 2$ for the interval, $\nu = 2$ and $r = 3$ for the circle. (Different proofs already exist for these two cases.) While the remainder of this paper is self-contained, it is intended as a sequel to [3–5], to which we refer the reader for motivations, additional details, and a more complete bibliography (in particular [5] contains a list of pre-existing proofs). For a recent general review of the subject, see [9]. As in [3–5], instead of looking directly for a solution ϕ, we try to find $\psi(z) = U(z)/U(0)$ with $U = f^{-1}$. Section 2 contains those facts and definitions from the above references which are needed here, and a more precise statement of the problem. What is new in this paper is in §3.

† The functional equation used by these authors is actually slightly different, and no results concerning it are obtained here.

2. Preliminaries

Let $b \in (0, 1)$. We denote

$$\Omega(b) = \{z \in \boldsymbol{C}: \operatorname{Im} z \neq 0 \text{ or } z \in (-b^{-1}, b^{-2})\}. \tag{2.1}$$

Let $\mathcal{F}(b)$ denote the real Fréchet space of complex functions f holomorphic on $\Omega(b)$ and such that $f(z^*) = f(z)^*$, and $\boldsymbol{E}_0(b)$ be defined as

$$\boldsymbol{E}_0(b) = \{\psi_0 \in \mathcal{F}(b): \psi_0(\boldsymbol{C}_+) \subset \boldsymbol{C}_-,\ \psi_0(\boldsymbol{C}_-) \subset \boldsymbol{C}_+,\ \psi_0(0) = 1,\ \psi_0(1) = 0\} \tag{2.2}$$

with

$$\boldsymbol{C}_+ = -\boldsymbol{C}_- = \{z \in \boldsymbol{C}: \operatorname{Im} z > 0\}. \tag{2.3}$$

(The topology of $\mathcal{F}(b)$ is that of uniform absolute convergence on the compact subsets of $\Omega(b)$.)

If $\psi_0 \in \boldsymbol{E}_0(b)$, the function $-\log \psi_0$ is Herglotzian†, and

$$\operatorname{Im} z > 0 \Rightarrow 0 < -\operatorname{Im} \log \psi_0(z) < \pi. \tag{2.4}$$

Hence by Herglotz's theorem (see e.g. [2] or [10]), this function has the representation

$$-\log \psi_0(z) = \int \sigma(t)\, \mathrm{d}t \left(\frac{1}{t-z} - \frac{1}{t}\right) \qquad \forall z \in \boldsymbol{C}_+ \cup \boldsymbol{C}_- \cup (-b^{-1}, 1). \tag{2.5}$$

Here $\sigma \in L^\infty(\boldsymbol{R})$ vanishes on the interval $(-b^{-1}, 1)$, with $0 \leqslant \sigma \leqslant 1$ and $\sigma(t) = 1$ for all $t \in [1, b^{-2}]$. From this it follows that ψ_0 satisfies *a priori* bounds, e.g., on the reals:

$$\frac{1}{1+bz} \leqslant \frac{\psi_0(z)}{1-z} \leqslant \frac{1}{1-b^2 z} \qquad \forall z \in (0, b^{-2})$$

(reversed for $-b^{-1} < z < 0$). As a consequence of such inequalities it is easy to see that $\boldsymbol{E}_0(b)$ is compact in $\mathcal{F}(b)$ (this is done in detail in [5]). We shall also need the inequality

$$\frac{\psi_0''(z)}{\psi_0'(z)} \geqslant \frac{-2b}{1+bz} \qquad \forall z \in (-b^{-1}, b^{-2}) \tag{2.6}$$

which is easy to derive just from the positivity of the Schwarzian derivative of ψ_0 in $(-b^{-1}, b^{-2})$.

The problem we consider is the following: let ν be real and fixed in $[1, 2]$; find real constants $r > 1$, $\lambda \in (0, 1)$, $z_1 \in (0, 1)$, and a function $\psi \in \boldsymbol{E}_0(\lambda)$ such that, for all $z \in \Omega(\lambda)$,

$$\psi(z) = \lambda^{-r\nu} \psi(z_1 \psi(-\lambda z)^{1/r}). \tag{2.7}$$

We first study a preliminary exercise. Let ψ_0 be an element of $\boldsymbol{E}_0(\lambda)$ with $\lambda \in (0, 1)$. Determine $z_1 \in (0, 1)$ and $r > 1$ so that the function

$$V(\zeta) = \lambda^{-r\nu} \psi_0(z_1 \zeta^{1/r}) \tag{2.8}$$

† A complex function f is called a Herglotz, or Herglotzian function, if it is holomorphic in $\boldsymbol{C}_+ \cup \boldsymbol{C}_-$, $f(z^*) = f(z)^*$, and $f(\boldsymbol{C}_+) \subset \bar{\boldsymbol{C}}_+$; then $-f$ is an anti-Herglotz or anti-Herglotzian function. If such an f is non-constant and holomorphic at a real x, then $f'(x) > 0$ and $Sf(x) \geqslant 0$. ($Sf \equiv f'''/f' - \frac{3}{2}[f''/f']^2$ is the Schwarzian derivative of f.)

will satisfy

$$V(1)=1 \qquad V'(1)/V(1)=-1/\lambda. \tag{2.9}$$

This implies that z_1 should satisfy $q(z_1)=0$, where

$$q(z)=\frac{\psi_0'(z)}{\psi_0(z)}-\frac{A}{vz}\log\psi_0(z) \qquad A\equiv A(\lambda)=\frac{-1}{\lambda\log\lambda}\geqslant \mathrm{e}. \tag{2.10}$$

The function q has the representation

$$q(z)=\int\frac{\sigma(t)\,\mathrm{d}t}{(t-z)^2}\left(\frac{A(t-z)}{vt}-1\right) \tag{2.11}$$

which is valid for all $z\in \boldsymbol{C}_+\cup\boldsymbol{C}_-\cup(-\lambda^{-1},1)$. It is easy to show that q vanishes at a unique point z_1 in $(0,1)$, at which its derivative is negative. (The function zq vanishes at 0 and tends to $-\infty$ when $z\uparrow 1$. Its derivative can be written $z[\psi_0'/\psi_0]G(z)$, with $G'>0$ in $(0,1)$, and G has one zero in $(0,1)$. Hence in this interval zq has one maximum and just one root >0.) If $0\leqslant z\leqslant 1-v/A$, the integral in (2.11) is greater than zero; hence $z_1>1-v/A>\lambda^v$. Moreover the representations (2.5) and (2.11) respectively show that as ψ_0 varies over $\boldsymbol{E}_0(\lambda)$, for a fixed $z\in(0,1)$, $\psi_0(z)$ is a decreasing function, and $q(z)$ is an increasing function of σ. It follows that z_1 and $\lambda^{-rv}=1/\psi_0(z_1)$ are increasing functions of σ. Hence they are minimum when $\sigma=1$ on $[1,\lambda^{-2}]$ and $\sigma=0$ elsewhere, i.e. when ψ_0 coincides with ψ_2 given by

$$\psi_2(z,\lambda)=\frac{1-z}{1-\lambda^2 z}. \tag{2.12}$$

In particular (for fixed λ) $r\geqslant r_2(\lambda,v)$, where $r_2(\lambda,v)$, the value taken by r when $\psi_0=\psi_2$, is the unique zero in $(0,\infty)$ of the function

$$r\mapsto r\lambda^{rv-1}(1-\lambda^2)-(1-\lambda^{rv})(1-\lambda^{rv+2}) \tag{2.13}$$

and $r_2(\lambda,v)>1$ if and only if $\lambda>\lambda_0(v)$ (with $\lambda_0(1)=0$). For $\lambda>\lambda_0(v)$, the function $\lambda\mapsto r_2(\lambda,v)$ is strictly increasing and has a smooth inverse function denoted $r\mapsto b(r)$ defined on $(1,\infty)$ with values in $(0,1)$. (Details can be found in [5].)

3. Fixed-r method

Let $v\in[1,2]$ be fixed, as well as $r>1$. Denote $s=rv$. For $b\in(0,1)$, we will define an operator T_r by describing its action on an arbitrary element ψ_0 of $\boldsymbol{E}_0(b)$. (b is arbitrary for the moment, and later will be set equal to $b(r)$.) T_r will be constructed in three steps summarised as follows.

(i) Starting from ψ_0, define $V(\zeta)=\lambda^{-rv}\,\psi_0(z_1\zeta^{1/r})$ with $\lambda\in(0,1)$ and $z_1\in(0,1)$ chosen so that $V(1)=1$ and $V'(1)=-1/\lambda$.

(ii) Find an anti-Herglotz function ψ_1, holomorphic on $\boldsymbol{C}_+\cup\boldsymbol{C}_-\cup(-\lambda^{-1},1]$, such that $\psi_1(z)=V(\psi_1(-\lambda z))$, $\psi_1(0)=1$, $\psi_1(1)=0$ (i.e. ψ_1^{-1} is a lineariser of V).

(iii) Find an anti-Herglotz function $\psi\in\boldsymbol{E}_0(\lambda)$ such that $\psi(z)=\kappa^{-1}\psi(z_1\psi_1(-\lambda z)^{1/r})$, i.e. ψ is a lineariser of the function $z\mapsto\varphi_1(z)=z_1\psi_1(-\lambda z)^{1/r}$; this function has a fixed point at 1 with multiplier κ. Set $T_r\psi_0=\psi$.

It is in the third step that this method differs from the fixed-r method of [4, 5]. We now go through these steps in detail.

(i) The first step is to define a function V by

$$V(\zeta)=\lambda^{-r\nu}\psi_0(z_1\zeta^{1/r}) \tag{3.1}$$

where $\lambda\in(0,1)$ and $z_1\in(0,1)$ must be determined so that $V(1)=1$ and $V'(1)/V(1)=-1/\lambda$. This means

$$\psi_0(z_1)=\lambda^{r\nu} \qquad \frac{z_1\psi_0'(z_1)}{r\psi_0(z_1)}=-\frac{1}{\lambda} \tag{3.2}$$

and thus requires z_1 to satisfy

$$Y(z_1)=1 \qquad Y(z)\equiv-\frac{z\psi_0'(z)}{r\psi_0(z)^{1-1/s}} \qquad s=r\nu. \tag{3.3}$$

By the inequality (2.6), for $0<z<1$,

$$\begin{aligned}\frac{Y'(z)}{Y(z)}&=\frac{1}{z}+\frac{\psi_0''(z)}{\psi_0'(z)}-(1-1/s)\frac{\psi_0'(z)}{\psi_0(z)}\\ &\geqslant\frac{1}{z}\left(\frac{1-bz}{1+bz}\right)-(1-1/s)\frac{\psi_0'(z)}{\psi_0(z)}>0\end{aligned} \tag{3.4}$$

so that there is a unique $z_1\in(0,1)$ such that $Y(z_1)=1$. We define $\tau^\nu=\psi_0(z_1)$, $\lambda=\tau^{1/r}\in(0,1)$. The following inequality is essential:

$$z_1>1+\nu\lambda\log\lambda>\lambda^\nu. \tag{3.5}$$

To prove it we appeal to the exercise of §2: we note that $q(z_1)=0$, with q given by the same formula (2.10) as in §2, and with the representation (2.11) in $\boldsymbol{C}_+\cup\boldsymbol{C}_-\cup(-1/b,1)$. From this the inequality (3.5) again follows, by the same argument.

The function V is anti-Herglotzian, holomorphic in $\boldsymbol{C}_+\cup\boldsymbol{C}_-\cup(0,(z_1b^2)^{-r})$, and

$$V(1)=1 \qquad V'(1)=-\lambda^{-1} \qquad V(z_1^{-r})=0 \qquad V(0)=\tau^{-\nu}>z_1^{-r}>1. \tag{3.6}$$

Hence there is a $\zeta_0\in(0,1)$ such that $V(\zeta_0)=z_1^{-r}$. We define $W=V\circ V$, $\hat V(\zeta)=1-V(1-\zeta)$, and $\hat W=\hat V\circ\hat V$. Note that W is Herglotzian, holomorphic in a neighbourhood of 1, with $W(1)=1$ and $W'(1)=\lambda^{-2}$. It is also defined and continuous at z_1^{-r}, with $W(z_1^{-r})=V(0)=\tau^{-\nu}$. Moreover W is holomorphic at ζ_0, with $W(\zeta_0)=0$, i.e. $\hat W(1-\zeta_0)=1$.

(ii) As the second step let Ψ be defined as the solution of

$$\Psi(z)=\hat V(\Psi(-\lambda z)) \tag{3.7}$$

$$\qquad=\hat W(\Psi(\lambda^2 z)) \tag{3.8}$$

$$\Psi(0)=0 \qquad \Psi'(0)=1. \tag{3.9}$$

It is given by

$$\Psi(z)=\lim_{n\to\infty}\Psi_n(z) \qquad \Psi_n(z)=\hat V^n((-\lambda)^n z). \tag{3.10}$$

There exists a small open disk Δ around 0, depending only on b and r, such that the above sequence is uniformly bounded and converges absolutely and uniformly as ψ_0 varies in $\boldsymbol{E}_0(b)$. Since each Ψ_n is a Herglotz function, it follows from Vitali's theorem that the sequence also converges in $\boldsymbol{C}_+\cup\boldsymbol{C}_-\cup\Delta$, and that Ψ is a Herglotz function. On the real axis, there is a segment $[0, L) \subset \boldsymbol{R}_+$ on which Ψ is holomorphic, positive and strictly increasing, and a point $\hat{z} \in (0, L)$ such that $\Psi(\hat{z}) = 1 - \zeta_0$. Otherwise Ψ could be extended to the whole of $\boldsymbol{R}_+$ by (3.8), then to the whole of $\boldsymbol{R}_-$ by (3.7); hence it would have to coincide with z, a contradiction. But then $\Psi(\gamma) = 1$, where $\gamma = \hat{z}/\lambda^2$ and Ψ is holomorphic at γ. Now define

$$\psi_1(z) = 1 - \Psi(\gamma z). \tag{3.11}$$

Then ψ_1 is holomorphic in $\boldsymbol{C}_+ \cup \boldsymbol{C}_- \cup (-1/\lambda, 1]$, anti-Herglotzian, and satisfies

$$\psi_1(z) = V(\psi_1(-\lambda z)) \qquad \psi_1(0) = 1 \qquad \psi_1(1) = 0$$

$$\psi_1(-\lambda^{-1}) = V(0) = \tau^{-\nu} \qquad \psi_1(-\lambda) = z_1^{-r}. \tag{3.12}$$

The last equality follows from $V(\psi_1(-\lambda)) = \psi_1(1) = 0$. Note that for every compact K in the domain of analyticity of Ψ, there is an $n \geq 0$ such that $\Psi(z) = \hat{V}^{n+1}(\Psi((-\lambda)^{n+1}z))$ holds, with $(-\lambda)^n z \in \Delta$, for all $z \in K$. Therefore Ψ (and hence ψ_1) depend continuously on ψ_0.

(iii) The function φ_1 defined by

$$\varphi_1(z) = z_1 \psi_1(-\lambda z)^{1/r} \tag{3.13}$$

is holomorphic and Herglotzian in $\Omega(\lambda)$, continuous with infinite derivatives at $-\lambda^{-1}$ and λ^{-2}, and

$$\varphi_1(-\lambda^{-1}) = 0 \qquad \varphi_1(1) = 1 \qquad \varphi_1(\lambda^{-2}) = z_1 \lambda^{-\nu} < \lambda^{-2}. \tag{3.14}$$

It follows (by Schwarz's lemma) that every point of $\Omega(\lambda)$ is attracted to 1 under repeated application of φ_1, and that $\varphi_1'(1) = \kappa \in (0, 1)$. It is therefore possible to define a function H such that

$$H(z) = \kappa^{-1} H(\varphi_1(z)) \qquad H(1) = 0 \qquad H'(1) = 1 \tag{3.15}$$

by

$$H(z) = \lim_{n \to \infty} \kappa^{-n} [\varphi_1^n(z) - 1]. \tag{3.16}$$

H is a Herglotz function, holomorphic in $\Omega(\lambda)$. Since it is non-constant it is strictly increasing on $(-\lambda^{-1}, \lambda^{-2})$ and $H(0) < 0$. Thus we can define

$$\psi(z) = H(z)/H(0) \qquad T_r(\psi_0) = \psi \in \boldsymbol{E}_0(\lambda). \tag{3.17}$$

This completes the definition of T_r.

We now wish to show that b can be so chosen (for given ν and r) that T_r will map $\boldsymbol{E}_0(b)$ into itself. At this point we shall need the results of the preliminary exercise of §2. We will show that if $b = b(r)$, then $\lambda \leq b$ and hence $\boldsymbol{E}_0(\lambda) \subset \boldsymbol{E}_0(b)$. Indeed suppose that for some $\psi_0 \in \boldsymbol{E}_0(b)$, $\lambda > b$. Then $\psi_0 \in \boldsymbol{E}_0(\lambda)$ and since z_1, r, λ satisfy $q(z_1) = 0$, with q as in (2.10), it follows from the exercise of §2 that $r \geq r_2(\lambda, \nu)$, i.e. $\lambda \leq b(r)$, which is a contradiction.

The operator T_r is easily seen to be continuous on $\boldsymbol{E}_0(b(r))$ which it maps into itself, so that, by the Schauder–Tikhonov theorem, it has at least one fixed point $\psi_0 = \psi \in \boldsymbol{E}_0(b(r))$. These functions and the corresponding z_1, λ, ψ_1 and κ satisfy

$$\psi_1(z) = (1/\lambda^{r\nu})\psi(z_1\psi_1(-\lambda z)^{1/r})$$

$$\psi(z) = (1/\kappa)\psi(z_1\psi_1(-\lambda z)^{1/r})$$

which imply $\psi_1 = \kappa\lambda^{-r\nu}\psi$. But since $\psi_1(0) = \psi(0) = 1$ we have $\kappa = \lambda^{r\nu}$ and $\psi_1 = \psi$, so that ψ is a solution of our problem. Note that a solution to the functional equation (1.1) can be obtained by setting

$$\phi = u^{-1} \qquad u(z) = \frac{z_1}{\lambda^{\nu-1}}\,\psi(z)^{1/r}.$$

The result of this paper is thus expressed by the following theorem.

Theorem . For each $\nu \in [1, 2]$ and each $r > 1$ there is a $\lambda \in (0, 1)$, a $z_1 \in (0, 1)$, and a $\psi \in \boldsymbol{E}_0(\lambda)$ such that

$$\psi(z) = \lambda^{-r\nu}\psi(z_1\psi(-\lambda z)^{1/r})$$

holds for all $z \in \Omega(\lambda)$.

As shown in [5], any such ψ is univalent in $\Omega(\lambda)$.

Acknowledgment

I wish to thank J-P Eckmann and D Sullivan for many useful discussions.

References

[1] Coullet P and Tresser C 1978 Itération d'endomorphismes et groupe de renormalisation *J. Physique Colloq.* **539** C5–25; 1978 *C. R. Acad. Sci., Paris* A **287** 577–80

[2] Donoghue W F Jr 1974 *Monotone Matrix Functions and Analytic Continuation* (Berlin: Springer)

[3] Eckmann J-P and Epstein H 1986 On the existence of fixed points of the composition operator for circle maps *Commun. Math. Phys.* **107** 213–31.

[4] Epstein H 1986 New proofs of the existence of the Feigenbaum functions *Commun. Math. Phys.* **106** 395–426.

[5] Epstein H 1988 Fixed points of composition operators *Non-linear Evolution and Chaotic Phenomena* ed G Gallavotti and P Zweifel (New York: Plenum)

[6] Feigenbaum M J 1978 Quantitative universality for a class of non-linear transformations *J. Stat. Phys.* **19** 25–52; 1979 Universal metric properties of non-linear transformations *J. Stat. Phys.* **21** 669–706.

[7] Feigenbaum M J, Kadanoff L P and Shenker S J 1982 Quasi-periodicity in dissipative systems: a renormalization group analysis *Physica* **5D** 370–86.

[8] Ostlund S, Rand D, Sethna J and Siggia E 1983 Universal properties of the transition from quasi-periodicity to chaos in dissipative systems *Physica* **8D** 303–42.

[9] Rand D 1987 Universality and renormalization in dynamical systems *New Directions in Dynamical Systems* ed T Bedford and J W Swift (Cambridge: Cambridge University Press).

[10] Valiron G 1954 *Fonctions Analytiques* (Paris: Presses Universitaires de France).

BOUNDED STRUCTURE OF INFINITELY RENORMALIZABLE MAPPINGS

by

Dennis Sullivan

The period doubling universality observed by Feigenbaum is one example of an apparently large theory of geometric rigidity of critical orbits of dynamical systems. Even in the folding mappings of the interval there are uncountably many different topological forms illustrating the following phenomenon :

The orbit of critical point of f up to closest return times picks out collections of intervals permuted by f. (see § 3). The intersection of these interval collections is a labeled Cantor set *whose geometry at fine scales is conjecturally determined only by the sequence of permutations and the critical exponent of the mapping* f.

Here we describe how to get geometric bounds on the Poincaré return maps to these intervals in the C^2, power law singularity case. There are two ideas. Koebe distortion §1 and combinatorics § 2, 3, 4. These bounds show that limits of these Poincare return maps or renormalizations always exist §5 and have a certain analytic form : their inverse branches have extensions to the upper half plane which are limits of composition of roots $\sqrt[k]{z}$ and linear mappings § 6, 7.

The bounds which only depend on the critical exponent asymptotically, show the rigidity conjecture implies strong convergence results for iterated renormalization § 7.

Acknowledgements : The rigidity conjecture was suggested to me by the work of M. Herman and by conversations with M. Feigenbaum and C. Tressor. The work on the bounds was inspired by a special case first obtained by Guckenheimer. The Koebe technique was inspired by work of de Melo and Van Strien.

There is a second part of this work using holomorphic methods and quasiconformal mappings which depends on numerous collaborations with Curt Mc Mullen. It owes much to one of H. Epstein's papers, and proves the rigidity conjecture in the real analytic case.

§1 Koebe principle

If four consecutive equal intervals are mapped topologically so that the middle pair have widely disparate lengths then the image of three of the intervals (either the first three or the last three) looks like

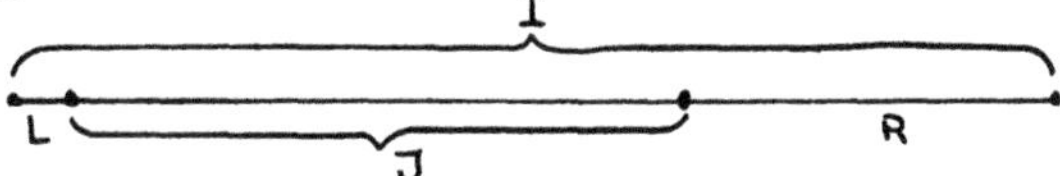

Namely, J is much longer than its distance to the boundary of I.

This means the projective length $[J \subset I]$ of J in I is large. $[J \subset I] = \log\left(1 + \frac{|J|\,|I|}{|L|\,|R|}\right)$ (see appendix 1). Now maps of positive schwarzian derivative *reduce projective* length. Observation of de Melo and van Strien. This happens for backwards branches near the critical value, for power law singularities $x \to |x|^r + c$, $r > 1$.

Away from the critical point the distortion of projective length in a composition $(J_1 \subset I_1) \xrightarrow{f} (J_2 \subset I_2) \xrightarrow{f} \dots \xrightarrow{f} (J_n \subset I_n)$ is controlled (for uniformly C^2 maps) by

$$\exp\left[\varepsilon\left(\max_\alpha I_\alpha\right)\sum_\alpha I_\alpha\right].$$

$\varepsilon(x)$ is the modulus of continuity of the gradient of $\log\left(\frac{fx - fy}{x - y}\right)$. See Appendix 1.

Thus we have the Koebe principle for C^2 maps with say power law singularities : *in any backwards composition with disjoint images of four consecutive intervals of equal length the middle two stay roughly the same size* (cf. de Melo and Van Strien [1]).

The outer pair of intervals impedes the distortion of the inner two intervals, and we refer to these buffer zones near the boundary as the Koebe space. We find this Koebe space in the next three sections.

§2 Backward branches and directed critical values.

Figure 1B, depicts graphically the backward branches of a high iterate of an interval mapping with one fold (figure 1A).

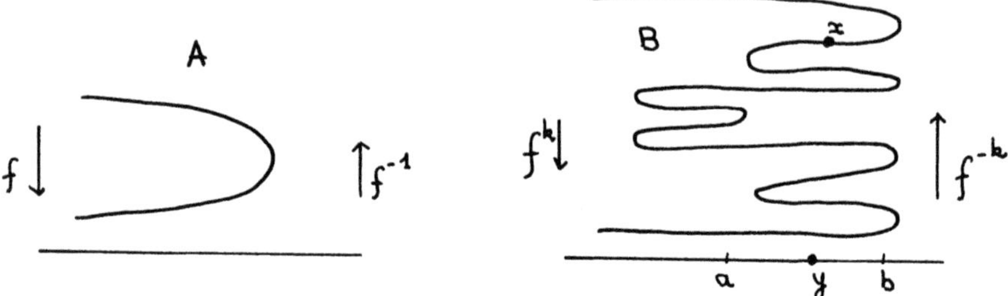

It shows that a backwards branch of f^{-k} (eg such that $f^k x = y$) is defined on an interval (a, b) limited by forward images of turning points of f^k. These points are just the forward images of the single critical point of f, which we denote $\{1, 2, 3, \ldots\}$.

Key observation : The limiting points a and b *point inward toward* the point y, where we say a point of the forward critical orbit $\{1, 2, 3, \ldots\}$ points in the direction to the side where the folded strands land. For example for f itself the points $\{1, 2\}$ are directed like :

(fig 2)A

because f and f^2 look like :

(fig 2)B

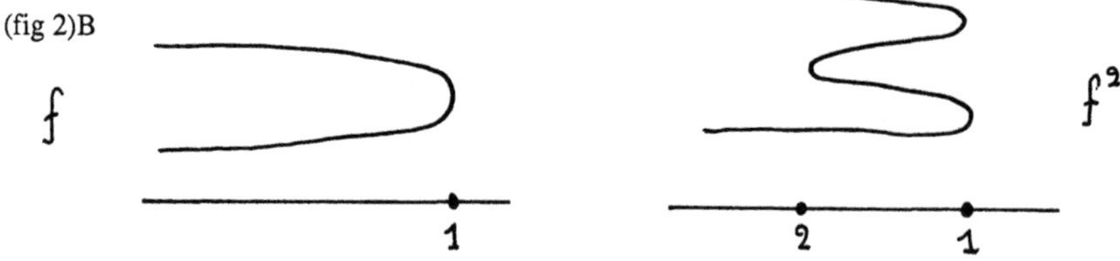

§ 3 Critical value directions for renormalizable mappings

The figure 3 shows the directions of $\{1, 2, \ldots, 2k\}$ (defined in §2) for a mapping whose k^{th} iterate is renormalizable

(Fig 3) → ← → ← ... → ← ... → ← → ←

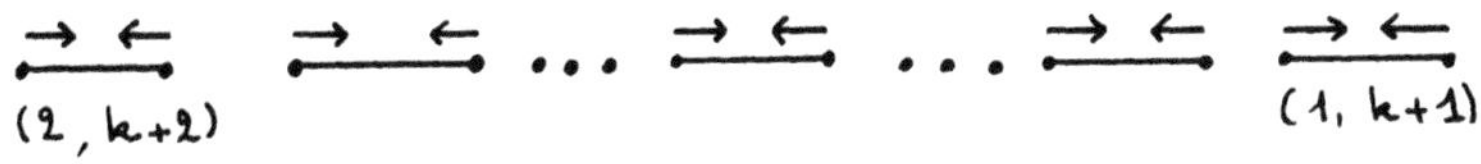

(2, k+2) (1, k+1)

Here we assume the intervals

$$I_V = (k + 1, 1), f(k + 1, 1), \ldots, f^{k-1}(k + 1, 1) = (2k, k) = I_C$$

are disjoint and the last one contains the turning point of f. Then f^k mapping any $(j, k + j)$ to itself is called a renormalization of f. Anyone of these is a simple one fold mapping with directions as in Figure 2. This explains figure 3.

Three intervals lemma : *All backwards branches on* $I_C = (2k, k)$ *up to time around* k (precisely $j \le k - 2$) *are defined on* I_C *together with its left and right neighboring intervals.* (assume $k > 2$, and see remark below).

Proof : by figure 1B the limiting points of the maximal interval of definition of a backward branch of f^s point toward I_C and lie in $\{1, 2, \ldots, 2j\}$. for $s \le j - 2$.

What we want follows from figure 3. The limiting points are not $(2k, k)$ because these points are too far forward in the orbit. *The limiting points are also not the next pair point out because these point the wrong way.* They might be the next pair out or further. Anyway there are three intervals in the domain of definition.

Remark : There is a similar statement for the interval $(r, k + r)$. *Any backwards branch of length* $\le r - 2$ *is defined on* $(r, k + r)$ *and its immediate neighboring intervals.*

§ 4 Koebe space around the critical value interval coming from the smallest interval

Consider a smallest interval $(s, k + s) = I_s$ among the collection $(j, k + j)$ $j = 1, 2, \ldots, k$. Suppose first $s > 2$. Then I_s has left and right neighbors as least as large as itself and any branch of f^{-j} is defined on the span of these for $j = 1, 2, \ldots, s - 2$ by § 3.

Thus by §1 these branches cannot distort I_s and its position in the span too much. It follows that $(2, k + 2)$ has one sided Koebe space in the sense that its size is not much larger than the distance to the far endpoint of its immediate neighbor. Applying f^{-1} once we obtain one sided space in the same sense for the interval $(k + 1, 1)$ (Here we lose a factor depending on the the non linearity of f^{-1} near $(k + 2, 2)$. In the argument later this will become negligible as these intervals become small.)

We apply f^{-1} once again and look at the figure 4

(fig4)

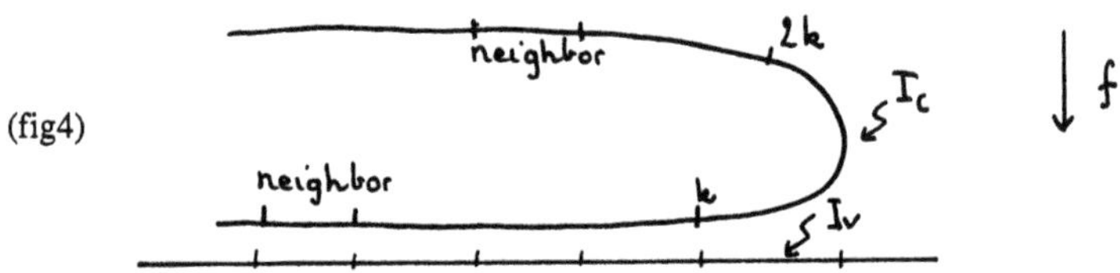

to get *2-sided space around* I_c in the sense that I_c is not much bigger than the distance to the far endpoints of its two neighboring intervals.

We note the Koebe space we get depends on the nature of the turning point of f. For example if in some C^2 coordinate chart φ we have a power law $x \to |x|^r$ $r > 1$, then the estimate depends on r and the non linearity of φ. The effect of the coordinate chart will become negligible as the size of the intervals become small.

<u>Remark</u> : If the smallest interval above is one of the end intervals $(1, k + 1)$ or $(2, k + 2)$ we pick up the argument above at the appropriate moment for one already would have then the one sided space needed.

§ 5 Koebe space around all the intervals

We have found one sided space around $(1, k + 1)$ and $(2, k + 2)$ and 2-sided space around $I_c = (k, 2k)$ in § 4. We pull back this space around I_c (using Koebe § 1) to obtain 2 sided space around all the interior intervals $(j, k + j)$, $j \geq 3$. Applying Koebe again we find controlled non linearity for all backwards branches on $(j, k + j)$ up to $j - 2$. This means that all the renormalizations of f are constructed from diffeormorphisms of bounded non linearity (large compositions of f) composed with f on a neigborhood of its critical point, that is

$$\text{renormalized map} = \binom{\text{bounded}}{\text{diffeo}} \binom{\text{power}}{\text{law}} \binom{\text{bounded}}{\text{diffeo}}.$$

The bound (in whatever C^k f lives in, see § 6) depends on f in various ways :

a) the nature of the turning point

b) the constants of the Koebe argument §1

c) the non linearity of f near 2.

As the depth of renormalization increases one knows the lengths of the intervals tends to zero (See Appendix 3). Thus a) becomes $x \to |x|^r$ and only r matters. Also c) becomes negligible. For $C^{1,1}$ (f' Lipschitz) folding maps b) becomes negligible exponentially fast in the depth of renormalization (Appendix 3).

§6 Renormalization Limits

In the composition $f_1 \circ f_2 \circ \ldots \circ f_n$ defining a high renormalization some factors occur on intervals near the critical point and are really non linear while others occur on intervals far away and produce little effect.

We can move the almost linear factors to one end of the composition at the cost of conjugating them by the non linear factors. When the total non linearity is bounded these conjugations are bounded operators in each C^k for all k . The total non linearity is bounded by Appendix 3.

Thus one sees that any limit of renormalizations (in any C^k, $k \geq 1$) is a limit of compositions where only the non linear terms corresponding to intervals near the critical point need be taken into account. For example, choose a C^k coordinate system so f is a power law $x \to |x|^r$ there. Take for the non linear factors only those intervals appearing in this neighborhood. Then in C^k, any limit of renormalization is a limit of compositions of powers $x \to |x|^r$ restricted to various intervals near the critical point.

Corollary : i) Any limit of renormalizations (in C^k) is a composition $h(|x|^r)$ where the inverse of h is a limit of finite compositions of diffeomorphisms with extensions to the upper half plane of the form $z \to az^{1/r}+b$, a,b real $a > 0$.

ii) The structure of this limit only depends on the fine structure of the forward critical orbit $\{1,2,3,...\}$ arbitrarily near the critical point of f.

Remark : The above is valid for arbitrary infinitely renormalizable mappings, even those where the combinatorics of the renormalizations is unbounded as the depth tends to infinity. (See Appendix 2, 3).

Corollary : Any renormalization limit has real analytic inverse branches which are each a composition of a schlicht mapping of the upper half plane with $z \to z^{1/r}$. In particular they belong to the Herglotz class considered by H. Epstein .

§7 Rigidity Conjecture implies Renormalization Convergence.

For a k-renormalizable map, the closure of the orbit of the critical point is contained in the collection of k-intervals. For an infinitely renormalizable mapping the intersection of all these interval collections is a Cantor set of measure zero. This follows from § 5 (Appendix 3). This Cantor set is the closure of the critical orbit and its fine geometry near the critical point determines all limits of renormalizations of f. § 6.

The rigidity conjecture concerns the homeomorphism between these Cantor sets for two mappings of the same combinatorics. We say the Cantor set is rigid if the non linearity of this homeomorphism restricted to smaller and smaller interval neighborhoods of the critical point tends to zero. (The non linearity of a homeomorphism on an interval is the maximum absolute value of the log of the ratio of the lengths of the images of two consecutive subintervals of equal length). By what we said above *if the Cantor set is rigid, the same sequence of iterated renormalization of two mappings brings them together in every* C^k *topology* $k \geq 2$ *, that makes sense*. § 6. This statement works even when the combinatorics of the successive renormalizations is unbounded.

Appendix 1 : The projective length of $J \subset I = [a,b] \subset [c,d]$ is $\int_a^b \left(\frac{dx}{x-c} + \frac{dx}{d-x}\right)$. The infinitesimal distortion of this length by a map f is $\left(\left(\frac{f'x}{fx-fc}\right) + \left(\frac{f'x}{fd-fx}\right)\right) / \left(\left(\frac{1}{x-c}\right) + \left(\frac{1}{d-x}\right)\right) =$ $f'x\left(\frac{fd-fc}{d-c}\right)\left(\frac{x-c}{fx-fc}\right)\left(\frac{d-x}{fd-fx}\right)$. Taking the logarithm and writing $\log\left(\frac{fx-fy}{x-y}\right) = \Delta(x,y)$, which becomes $\log f'x$ on the diagonal, we obtain a difference of differences of Δ over the vertical (or horizontal) sides of the square $[(c,d),(x,c),(d,x),(x,x)]$. Dividing by the appropriate side length of the square one obtains a difference of difference quotients of Δ . This quantity is controlled by the modulus of continuity of the gradient of $\Delta(x,y)$, which exists if the original map is C^2 .

Thus the distortion by an f in C^2 of the projective length $[J \subset I]$ is estimated by $1 + o(|I|)$ where the "small o" comes from the modulus of continuity of the gradient of $\log\left(\frac{fx-fy}{x-y}\right)$. (cf [1]).

Appendix 2 : (boundedness of total non linearity).

As we pull back $I_c = (k,2k)$ (k-1) times its projective length in the interval with definite buffer zones on each side only *changes* a bounded amount by Koebe and § 5. On the other hand each passage near the critical value creates a compression, whereas the C^2 calculation (appendix 1) shows the total change at other moments is negligible. If we calculate the compression part and invoke the first statement above, we prove the

Proposition. Write $[j,k+j] = [a_j,b_j]$ with an origin at the critical value. Then $\sum_{j=2}^{k} |\log \frac{b_j}{a_j}|$ is bounded independent of k . This means the *total non linearity* of the diffeomorphic part of renormalizations is bounded universally.

Appendix 3 : The figure 4 show a gap whose size is comparable to the size of I_C. For all other intervals the mirror image about the critical point of f of an interval is contained in a gap or it lies ouside the interval [2,1]. As we go deep into renormalization the ratio of the length of (2,c) to the length of (1,c) cannot go to zero, where c is the critical point, because the shape of the renormalized f is bounded and the combinatorics stays away from that of a period two critical point. These remarks provide sufficient reason for the assertion :

The total length of the intervals tends to zero exponentially fast in the depth of renormalization. In particular the Cantor set has measure zero and the contribution to Koebe distortion away from the critical point tends exponentially fast to zero in the depth if f' *is Lipschitz.*

[1] W de Melo and S J Van Strien. Schwarzian derivative and beyond. Bull of Amer Math Soc 18, 1988, pp 159-162.

[2] H. Epstein and J. Lascoux. Analyticaly Properties of the Feigenbaum Function. Comm. in Math. Physics 81, 413-453 (1981).

Part 9

References

References (articles included in this reprint selection are marked with a star)

G. Ahlers and R. Behringer (1982) Evolution of turbulence from the Rayleigh-Bénard instability, Phys. Rev. Lett. 40, 712.

V.M. Alekseev and M.V. Yakobson (1981) Symbolic dynamics and hyperbolic dynamic systems, Phys. Rep. 75, 287.

V.I. Arnold (1978) Mathematical methods of classical mechanics (Springer, New York).

F.T. Arecchi and F. Lisi (1982) Hopping mechanism generating 1/f noise in nonlinear systems, Phys. Rev. Lett. 49, 94.

* F.T. Arecchi, R. Meucci, G. Puccioni and J. Tredicce (1982) Experimental evidence for subharmonic bifurcations, multistability and turbulence in a Q-switched gas laser, Phys. Rev. Lett. 49, 1217.

A. Arneodo, P. Coullet, C. Tresser, A. Libchaber, J. Maurer and D. d'Humières (1983) On the observation of an uncompleted cascade in a Rayleigh-Bénard experiment, Physica 6D, 385.

D. Baiye and V. Franceschini (1981) Symmetry breaking on a model of five-mode truncated Navier-Stokes equations, J. Stat. Phys. (USA) 26, 471.

P. Bak (1981) Chaotic behaviour and incommensurate phases in the anisotropic Ising model with competing interactions, Phys. Rev. Lett. 46, 791.

P. Bak and M. Høgh Jensen (1982) Bifurcations and chaos in the ϕ^4 theory on a lattice, J. Phys. A15, 1893.

J.D. Barrow (1981) Chaos in the Einstein equations, Phys. Rev. Lett. 46, 963.

J.R. Beddington, C.A. Free and J.H. Lawton (1975) Dynamics complexity in predator-prey models framed in difference equations, Nature 255, 58.

J. Bellisard, D. Bessis and P. Moussa (1982) Chaotic states of almost periodic Schrödinger operators, Phys. Rev. Lett. 49, 701.

E. Ben-Jacob, I. Goldhirsch, Y. Imry and S. Fishman (1982) Intermittent chaos in Josephson junction, Phys. Rev. Lett. 49, 1599.

G. Benettin, L. Galgani and A. Giorgilli (1980) Further results on universal properties in conservative dynamical systems, Lett. Nuovo Cimento 29, 163.

G. Benettin, C. Cercignani, L. Galgani and A. Giorgilli (1980) Universal properties in conservative dynamical systems, Lett. Nuovo Cim. 28, 1.

G. Benettin, L. Galgani and J.-M. Strelcyn (1976) Kolmogorov entropy and numerical experiments, Phys. Rev. A14, 2338.

* P. Bergé, M. Dubois, P. Manneville and Y. Pomeau (1980) Intermittency in Rayleigh-Bénard convection, J. Phys. Lett. 41, L341.

M.V. Berry (1978) Regular and irregular motion, in: Topics in nonlinear dynamics, ed. S. Jorna (AIP Conf. Proc. 46, New York).

T.C. Bountis (1981) Period doubling bifurcations and universality in conservative systems, Physica 3D, 577.

T.C. Bountis and R.H.G. Helleman (1981) On the stability of periodic orbits of two-dimensional mappings, J. Math. Phys. 22, 1867.

H. Brolin (1965) Invariant sets under iteration of rational functions, Ark. Mat. 6, 103.

M. Campanino and H. Epstein (1981) On the existence of Feigenbaum's fixed point, Commun. Math. Phys. 79, 261.

G. Casati and I. Guarneri (1983) Chaos and special features of quantum systems under external perturbations, Phys. Rev. Lett. 50, 640.

J. Cascais, R. Dilão and A. Noronha da Costa (1983) Chaos and reverse bifurcations in a RCL circuit, Phys. Lett. 93A, 213.

Y.F. Chang, M. Tabor and J. Weiss (1982) Analytic structure of the Hénon-Heiles Hamiltonian in integrable and nonintegrable regimes, J. Math. Phys. 23, 531.

B.V. Chirikov (1979) A universal instability of many-dimensional oscillator systems, Phys. Rep. 52, 263.

B.V. Chirikov and D.L. Shepelyanskii (1981) Stochastic oscillations of classical Yang-Mills fields, JETP Lett. 34, 163.

P. Collet, J.P. Crutchfield and J.-P. Eckmann (1983) Computing the topological entropy of maps, Commun. Math. Phys. 88, 257.

P. Collet and J.-P. Eckmann (1980a) Iterated maps on the interval as dynamical systems (Birkhäuser, Boston).

P. Collet and J.-P. Eckmann (1980b) On the abundance of aperiodic behaviour for maps on the interval, Commun. Math. Phys. 73, 115.

* P. Collet, J.-P. Eckmann and H. Koch (1980) Period doubling bifurcations for families of maps on R^n, J. Stat. Phys. 25, 1.

P. Collet, J.-P. Eckmann and H. Koch (1981) On universality for area-preserving maps of the plane, Physica 3D, 457.

P. Collet, J.-P. Eckmann and O.E. Lanford III (1980) Universal properties of maps on an interval, Commun. Math. Phys. 76, 211.

P. Collet, J.-P. Eckmann and L. Thomas (1981) A note on the power spectrum of the iterates of Feigenbaum's function, Commun. Math. Phys. 81, 261.

G. Contopoulos (1981) Do successive bifurcations in Hamiltonian systems have the same universal ratio?, Lett. Nuovo Cim. 30, 498.

M. Cosnard (1981) Étude du comportement d'itérations d'un opérateur de renormalisation, C.R. Acad. Sci. (Paris) 293, 619.

P. Coullet and J. Tresser (1978a) Itérations d'endomorphismes et groupe de renormalisation, C.R. Acad. Sci. (Paris) 287, 577.

P. Coullet and C. Tresser (1978b) Itérations d'endomorphismes et groupe de renormalisation, J. de Phys. C5, 25.

P. Coullet and C. Tresser (1980) Critical transition to stochasticity for some dynamical systems, J. de Phys. Lett. 41, L255.

J.P. Crutchfield, J.D. Farmer and B.A. Huberman (1982) Fluctuations and simple chaotic dynamics, Phys. Rep. 92, 45.

* J.P. Crutchfield, J.D. Farmer, N.H. Packard, R. Shaw, G. Jones and R. Donnely (1980) Power spectral analysis of a dynamical system, Phys. Lett. 76A, 1.

* J.P. Crutchfield and B.A. Huberman (1980) Fluctuations and the onset of chaos, Phys. Lett. 77A, 407.

* J.P. Crutchfield, M. Nauenberg and J. Rudnick (1981) Scaling for external noise at the onset of chaos, Phys. Rev. Lett. 46, 933.

V. Croquette and C. Poitou (1981) Cascade of period doubling bifurcations and large stochasticity in the motions of a compass, J. de Phys. Lett. 42, L537.

* P. Cvitanović, Universality in chaos, Acta Phys. Pol. (to appear).

P. Cvitanović and M. Høgh Jensen (1982) Universalitet i overgang til chaos, Fysisk Tidsskrift 80, 82.

P. Cvitanović and J. Myrheim (1983) Universality for period n-tuplings in complex mappings, Phys. Lett. 94A, 329.

H. Daido (1981a) Theory of the period-doublings of 1-D mappings based on the parameter dependence, Phys. Lett. 83A, 246.

H. Daido (1981b) Universal relation of a band-splitting sequence to a preceding period-doubling one, Phys. Lett. 86A, 259.

B. Derrida (1980) Critical properties of one-dimensional mappings, in: Bifurcation phenomena in mathematical physics and related topics, eds. C. Bardos and D. Bessis (Reidel, Dordrecht) p. 137.

B. Derrida, A. Gervois and Y. Pomeau (1978) Iteration of endomorphisms on the real axis and representation of numbers, Ann. Inst. Henri Poincaré 29, 305.

B. Derrida, A. Gervois and Y. Pomeau (1979) Universal metric properties of bifurcations of endomorphisms, J. Phys. A12, 269.

B. Derrida and Y. Pomeau (1980) Feigenbaum's ratios of two-dimensional area preserving maps, Phys. Lett. 80A, 217.

D. D'Humières, M.R. Beasley, B.A. Huberman and A. Libchaber (1982) Chaotic states and routes to chaos in the forced pendulum, Stanford University preprint G.L.-3429 (May).

A. Douady (1982) Systèmes dynamiques holomorphes, Sem. Bourbaki 599 (November, 1982).

A. Douady and J.H. Hubbard (1982) Itération des polynômes quadratiques complexes, C.R. Acad. Sci. (Paris) 294, 123.

* J.-P. Eckmann (1981) Roads to turbulence in dissipative dynamical systems, Rev. Mod. Phys. 53, 643.

J.-P. Eckmann, H. Koch and P. Wittwer (1982) Existence of a fixed point of the doubling transformation for area-preserving maps of the plane, Phys. Rev. A26, 720.

J.-P. Eckmann, L. Thomas and P. Wittwer (1981) Intermittency in the presence of noise, J. Phys. A14, 3153.

H. Epstein and J. Lascoux (1981) Analyticity properties of the Feigenbaum function, I.H.E.S. preprint P/81/27 (May, 1981).

M. Faraday (1831) Phil. Trans. R. Soc. London 299, sect. 103.

* J.D. Farmer (1981) Spectral broadening of period-doubling bifurcation sequences, Phys. Rev. Lett. 47, 179.

J.D. Farmer (1982) Dimensions, fractal measures and chaotic dynamics, Los Alamos preprint LA-UR-82-1719.

J.D. Farmer, J. Hart and P. Weidman (1982) A phase space analysis of a baroclinic flow, Phys. Lett. 91A, 22.

J.D. Farmer, E. Ott and J.A. Yorke, The dimension of chaotic attractors, Physica D (to appear).

P. Fatou (1906) Comptes Rendus Notes (Paris) 143, 546.

P. Fatou (1919) Bull. Soc. Math. France 47, 161; ibid. 48, 33 and 208 (1920).

M.J. Feigenbaum (1978) Quantitative universality for a class of nonlinear transformations, J. Stat. Phys. 19, 25.

* M.J. Feigenbaum (1979a) The universal metric properties of nonlinear transformations, J. Stat. Phys. 21, 669.

M.J. Feigenbaum (1979b) The onset of turbulence, Phys. Lett. 74A, 375.

* M.J. Feigenbaum (1980a) Universal behaviour in nonlinear systems, Los Alamos Science 1, 4.

* M.J. Feigenbaum (1980b) The transition to aperiodic behaviour in turbulent systems, Commun. Math. Phys. 77, 65.

M.J. Feigenbaum (1982) Periodicity and the onset of chaos, in: Mathematical problems in theoretical physics, eds. R. Schrader et al., Lecture notes in physics 153 (Springer-Verlag, Berlin).

M.J. Feigenbaum and B. Hasslacher (1982) Irrational decimations and path integrals for external noise, Phys. Rev. Lett. 49, 605.

M.J. Feigenbaum, L.P. Kadanoff and S.J. Shenker (1982) Quasi-periodicity in dissipative systems: A renormalization group analysis, Physica 5D, 370.

P.R. Fenstermacher, H.L. Swinney and J.P. Gollub (1979) Dynamical instabilities and the transition to chaotic Taylor vortex flow, J. Fluid Mech. 94, 103.

V. Franceschini (1979) A Feigenbaum sequence of bifurcations in the Lorenz model, J. Stat. Phys. 22, 397.

* V. Franceschini and C. Tebaldi (1979) Sequences of infinite bifurcations and turbulence in a five-mode truncation of the Navier-Stokes equations, J. Stat. Phys. 21, 707.

H. Froehling, J.P. Crutchfield, J.D. Farmer, N.H. Packard and R. Shaw (1981) On determining the dimension of chaotic flows, Physics 3D, 605.

J. Frøyland (1982) Multifurcations in SU(2) Yang-Mills theory, University of Oslo preprint (May, 1982).

T. Geisel and J. Nierwetberg (1981) Onset of diffusion and universal scaling in chaotic systems, Phys. Rev. Lett. 48, 7.

H.M. Gibbs, F.A. Hopf, D.L. Kaplan and R.L. Shoemaker (1981) Observation of chaos in optical bistability, Phys. Rev. Lett. 46, 474.

* M. Giglio, S. Musazzi and V. Perini (1981) Transition to chaotic behaviour via a reproducible sequence of period-doubling bifurcations, Phys. Rev. Lett. 47, 243.

L. Glass and R. Perez (1982) Fine structure of phase locking, Phys. Rev. Lett. 48, 1772.

A.I. Golberg, Ya.G. Sinai and K.M. Khanin (1983) Universal properties of sequences of period triplings, Usp. Mat. Nauk 38, 159.

J.P. Gollub and S.V. Benson (1980) Many routes to turbulent convection, J. Fluid Mech. 100, 449.

J.P. Gollub, S.V. Benson and J. Steinman (1981) Ann. N.Y. Acad. Sci. 357, 22.

* J.P. Gollub and H.L. Swinney (1975) Onset of turbulence in a rotating fluid, Phys. Rev. Lett. 35, 927.

D.L. Gonzales and O. Piro (1983) Chaos in a nonlinear driven oscillator with exact solution, Phys. Rev. Lett. 50, 870.

P. Grassberger (1981) On the Hausdorff dimension of fractal attractors, J. Stat. Phys. 26, 173.

P. Grassberger and M. Scheunert (1981) Some more universal scaling laws for critical mappings, J. Stat. Phys. 26, 697.

P. Grassberger and H. Procaccia (1983) Characterization of strange attractors, Phys. Rev. Lett. 50, 346.

C. Grebogi, E. Ott and J.A. Yorke (1982) Chaotic attractors in crisis, Phys. Rev. Lett. 48, 1507.

C. Grebogi, E. Ott and J.A. Yorke (1983) Fractal basin boundaries, long-lived chaotic transients and unstable-unstable pair bifurcations, Phys. Rev. Lett. 50, 935.

J.M. Greene (1979) A method for determining a stochastic transition, J. Math. Phys. 20, 1183.

J.M. Greene, R.S. MacKay, F. Vivaldi and M.J. Feigenbaum (1981) Universal behaviour in families of area-preserving maps, Physica 3D, 468.

* S. Grossmann and S. Thomae (1977) Invariant distributions and stationary correlation functions of one-dimensional discrete process, Z. Naturforsch. 32A, 1353.

J. Guckenheimer (1977) On the bifurcation of maps of the interval, Inventions Math. 39, 165.

J. Guckenheimer (1979) Sensitive dependence to initial conditions for one-dimensional maps, Commun. Math. Phys. 70, 133.

J. Guckenheimer (1980) Dynamics of the van der Pol equation, IEEE Trans. Circuits and Systems CS-27, 983.

J. Guckenheimer (1980) Bifurcations of dynamical systems, in: Dynamical systems (Birkhäuser, Boston).

J. Guckenheimer, G. Oster and A. Ipaktchi (1977) Dynamics of density dependent population models, J. Math. Biol. 4, 101.

* M.R. Guevara, L. Glass and A. Shrier (1981) Phase locking, period-doubling bifurcations and irregular dynamics in periodically stimulated cardiac cells, Science 214, 1350.

I. Gumowski and C. Mira (1980a) Dynamique chaotique. Transformations ponctuelles. Transition ordre-désordre (Editions Cépadues, Toulouse).

I. Gumowski and C. Mira (1980b) Recurrences and discrete dynamic systems, Lecture notes in math. 809 (Springer, Berlin).

H. Haken (1978) Synergetics, an introduction (Springer, Berlin).

H. Haken, ed. (1981) Chaos and order in nature, Proc. of the Int. Symposium on synergetics at Schloss Elmau, Bavaria, (Springer, New York).

H. Haken (1983) At least one Lyapunov exponent vanishes if the trajectory of an attractor does not contain a fixed point, Phys. Lett. 94A, 71.

Bai-lin Hao and Shyu-yu Zhang (1982a) Subharmonic stroboscopy as a method to study period-doubling bifuractions, Phys. Lett. 87A, 267.

Bai-lin Hao and Shyu-yu Zhang (1982b) Hierarchy of chaotic bands, J. Stat. Phys. 28, 769.

* R.H.G. Helleman (1980a) Self-generated chaotic behaviour in nonlinear mechanics, in: Fundamental problems in statistical mechanics, vol.5, ed. E.G.D. Cohen (North-Holland, Amsterdam) p.165.

R.H.G. Helleman, ed. (1980b) Nonlinear dynamics, Ann. N.Y. Acad. Sci. 357, 1.

R.H.G. Helleman (1981) One mechanism for the onsets of large-scale chaos in conservative and dissipative systems, in: Nonequilibrium problems in statistical mechanics, vol.2, eds. W. Horton, L. Reichland and V. Szebehely (John Wiley, New York).

M. Hénon (1969) Quart. Appl. Math. 27, 291.

* M. Hénon (1976) A two-dimensional mapping with a strange attractor, Commun. Math. Phys. 50, 69.

J.E. Hirsch, B.A. Huberman and D.J. Scalapino (1982) Theory of intermittency, Phys. Rev. A25, 519.

* J.E. Hirsch, M. Nauenberg and D.J. Scalapino (1982) Intermittency in the presence of noise: a renormalization group formulation, Phys. Lett. 87A, 391.

D.R. Hofstadter (1981) Strange attractors: mathematical patterns delicately poised between order and chaos, Sci. Am. 245, 5.

M. Høgh Jensen, P. Bak and T. Bohr (1983) Complete devil's staircase, fractal dimension, and universality of mode-locking structure in the circle map, Phys. Rev. Lett. 50, 1637.

P.J. Holmes and J.E. Marsden (1982) Melnikov's method and Arnold diffusion for perturbations of integrable Hamiltonian systems, J. Math. Phys. 23, 669.

F.C. Hoppensteadt (1978) in: Mathematics Today, ed. L.A. Steen (Springer, New York).

F.A. Hopf, D.L. Kaplan, H.M. Gibbs and R.L. Shoemaker (1982) Bifurcations to chaos in optical bistability, Phys. Rev. A25, 2172.

B. Hu (1981) Dissipative bifurcation ratio in the area-non-preserving Hénon map, J. Phys. A14, L423.

B. Hu (1982) Introduction to the real-space renormalization-group methods in critical and chaotic phenomena, Phys. Rep. 91, 233.

B. Hu and J.M. Mao (1982a) Third-order renormalization-group calculation of the Feigenbaum number, Phys. Rev. A25, 1196.

B. Hu and J.M. Mao (1982b) Period doubling: universality and critical-point order, Phys. Rev. A25, 3259.

B. Hu and J. Rudnick (1982) Exact solutions to the Feigenbaum renormalization-group equations for intermittency, Phys. Rev. Lett. 48, 1645.

B.A. Huberman, J.P. Crutchfield and N. Packard (1980) Noise phenomena in Josephson junctions, Appl. Phys. Lett. 37, 750.

B.A. Huberman and J.P. Crutchfield (1979) Chaotic states of anharmonic systems in periodic fields, Phys. Rev. Lett. 43, 1743.

* B.A. Huberman and J. Rudnick (1980) Scaling behaviour of chaotic flows, Phys. Rev. Lett. 45, 154.

* B.A. Huberman and A.B. Zisook (1981) Power spectra of strange attractors, Phys. Rev. Lett. 46, 626.

* J.L. Hudson and J.C. Mankin (1981) Chaos in the Belousov-Zhabotinsky reaction, J. Chem. Phys. 74, 6171.

A. Ichimura and T. Shimizu (1982) Asymptotic solution of a chaotic motion, Phys. Lett. 91A, 52.

K. Ikeda, H. Daido and O. Akimoto (1980) Optical turbulence: chaotic behaviour of transmitted light from a ring cavity, Phys. Rev. Lett. 45, 709.

T. Janssen and J.A. Tjon (1983a) Bifurcations of lattice structures, J. Phys. A16, 673.

T. Janssen and J.A. Tjon (1983b) Universality in multi-dimensional maps, J. Phys. A16, 697.

D.J. Jefferies (1982) Period multiplication and chaotic behaviour in a driven nonlinear piezo-electric resonator, Phys. Lett. 90A, 316.

C. Jeffries and J. Pérez (1982) Observation of a Pomeau-Manneville intermittent route to chaos in a nonlinear oscillator, Phys. Rev. A26, 2117.

S. Jorna, ed. (1978) Topics in nonlinear dynamics, AIP Conf. Proc. no. 46 (Am. Inst. of Physics, New York).

G. Julia (1918) Mémoire sur l'itération des fonctions rationelles, J. Math. Pures et Appl. 4, 47.

T. Kai (1982) Lyapunov number for a noisy 2^n cycle , J. Stat. Phys. 29, 329.

T. Kai (1981) Universality of power spectra of a dynamical system with period-doubling sequence, Phys. Lett. 86A, 263.

R.L. Kautz (1981) Chaotic states of rf-biased Josephson junctions, J. Appl. Phys. 52, 6241.

R.L. Kautz (1981) J. Appl. Phys. 52, 3528.

R. Keolian, L.A. Turkevich, S.J. Putterman, I. Rudnick and J.A. Rudnick (1981) Subharmonic sequences in the Faraday experiment: departures from period doubling, Phys. Rev. Lett. 47, 1133.

H.J. Korsch and M.V. Berry (1981) Evolution of Wigner's phase-space density under a nonintegrable quantum map, Physica 3D, 627.

Y. Kuramoto and S. Koga (1982) Anomalous period-doubling bifurcations leading to chemical turbulence, Phys. Lett. 92A, 1.

O.E. Lanford (1980) Remarks on the accumulation of period-doubling bifurcations, in: Mathematical problems in theoretical physics, Lecture notes in Physics, vol.116 (Springer, Berlin) p.340.

O.E. Lanford (1981) Smooth transformations of intervals, in: Séminaire Bourbaki 563, Lecture notes in Math., vol.901 (Springer, Berlin) p.36.

* O.E. Lanford (1982) A computer-assisted proof of the Feigenbaum conjectures, Bull. Am. Math. Soc. 6, 427.

A. Lasota and P. Rusek (1970) Problems of the stability of the motion in the process of rotary drilling with cogged bits, Arch. Gornictua 15, 205.

W. Lauterborn and E. Cramer (1981) Subharmonic route to chaos observed in acoustics, Phys. Rev. Lett. 47, 1445.

M.T. Levinsen (1982) Even and odd subharmonic frequencies and chaos in Josephson junctions: impact on parametric amplifiers?, J. Appl. Phys. 53, 4294.

T.Y. Li and J.A. Yorke (1975) Period three implies chaos, Am. Math. Monthly 82, 985.

Tien-Yien Li, M. Misiurewicz, G. Pianigiani and J.A. Yorke (1982) Odd chaos, Phys. Lett. 87A, 271.

A. Libchaber (1982) Convection and turbulence in liquid helium I, Physica 109B and 110B, 1583.

* A. Libchaber, C. Laroche and S. Fauve (1982) Period doubling in mercury, a quantitative measurement, J. de Phys. Lett. 43, L211.

A. Libchaber and J. Maurer (1980) Une experience de Rayleigh-Bénard de géométrie réduite; multiplication, accrochage, et démultiplication de fréquencies, J. de Phys. 41, Colloq. C3, 51.

* A. Libchaber and J. Maurer (1981) A Rayleigh-Bénard experiment: helium in a small box, in: Nonlinear phenomena at phase transitions and instabilities, ed. T. Riste (Plenum, New York) p.259.

P.S. Linsay (1981) Period doubling and chaotic behaviour in a driven anharmonic oscillator, Phys. Rev. Lett. 47, 1349.

* E.N. Lorenz (1963) Deterministic nonperiodic flow, J. Atmos. Sci. 20, 130.

E.N. Lorenz (1964) The problem of deducing the climate from the governing equations, Tellus 16, 1.

E.N. Lorenz (1979) On the prevalence of aperiodicity for simple systems, in: Lecture notes in math. 755 (Springer, Berlin) p.53.

* E.N. Lorenz (1980) Noisy periodicity and reverse bifurcation, Ann. N.Y. Acad. Sci. 357, 282.

V.S. L'vov, A.A. Predtechenskii and A.I. Chernykh (1981) Bifurcation and chaos in a system of Taylor vortices: a natural and numerical experiment, Sov. Phys. JETP 53, 562.

G.A. McCreadie and G. Rowlands (1982) An analytic approximation to the Lyapunov number for 1-D maps, Phys. Lett. 91A, 146.

* R.S. Mackay (1983) Period-doubling as a universal route to stochasticity, in: Long-time prediction in dynamics, eds. C.W. Horton, L.E. Reichl and A.G. Szebehely (Wiley, New York) p.127.

J. McLaughlin (1981) Period-doubling bifurcations and chaotic motion for a parametrically forced pendulum, J. Stat. Phys. 24, 375.

B.B. Mandelbrot (1980) Fractal aspects of the iteration of $z \to \lambda z(1-z)$ for complex λ and z, Ann. N.Y. Acad. Sci. 357, 249.

B.B. Mandelbrot (1982) The fractal geometry of nature (Freeman, San Francisco).

P. Manneville and Y. Pomeau (1979) Intermittency and the Lorenz model, Phys. Lett. 75A, 1.

P. Manneville and Y. Pomeau (1980) Different ways to turbulence in dissipative dynamical systems, Physica 1D, 219.

N.S. Manton and M. Nauenberg (1983) Universal scaling behaviour for iterated maps in the complex plane, Commun. Math. Phys. 89, 555.

S.G. Matinyan, G.K. Savvidy and N.G. Ter-Arutyunyan-Savvidy (1981) Stochastic classical mechanics of Yang-Mills and its elimination by the Higgs mechanism, Pisma v. Zh. Eksp. Teor. Fiz. 34, 613.

S.G. Matinyan, G.K. Savvidy and N.G. Ter-Arutyunyan-Savvidy (1981) Classical Yang-Mills mechanics, nonlinear color oscillations, Zh. Eksp. Teor. Fiz. 80, 830.

* R.M. May (1976) Simple mathematical models with very complicated dynamics, Nature 261, 459.

R.M. May and G.F. Oster (1976) Bifurcations and dynamical complexity in simple ecological models, Am. Nat. 110, 573.

R.M. May and G.F. Oster (1980) Period-doubling and the onset of turbulence: an analytic estimate of the Feigenbaum ratio, Phys. Lett. 78A,1.

G. Mayer-Kress and H. Haken (1981) The influence of noise on the logistic model, J. Stat. Phys. 26, 149.

* M. Metropolis, M.L. Stein and P.R. Stein (1973) On finite limit sets for transformations of the unit interval, J. Combinatorial Theory (A) 15, 25.

C. Mira (1980) Complex dynamics in two-dimensional endomorphisms, Nonlinear Anal. 4, 1167.

R.F. Miracky, J. Clarke and R.H. Koch (1983) Chaotic noise observed in a resistivity shunted self-resonant Josephson tunnel junction, Phys. Rev. Lett. 50, 856.

M. Misiurewicz (1981) Absolutely continuous measures for certain maps of an interval, Publ. Math. I.H.E.S. 53, 17.

M. Misiurewicz and W. Szlenk (1980) Entropy of piecewise monotone mappings, Studie Math. 67, 45.

J. Moser (1968) Lectures on Hamiltonian systems, Memoirs Am. Math. Soc. 81, 1.

P.J. Myrberg (1958-1964) Iteration der réellen polynome zweiten grades, Ann. Acad. Sci. Fennicae, A.I. 253; ibid. 259 (1958), 268 (1959), 292 (1960), 336/3 (1964), 348 (1964).

P.J. Myrberg (1962) Sur l'itération des polynomes réels quadratiques, J. de Math. Pures et Appl., Sér. 9.41, 339.

K. Nakamura, S. Ohta and K. Kawasaki (1982) Chaotic states of ferromagnets in strong parallel pumping fields, J. Phys. C15, L143.

* M. Nauenberg and J. Rudnick (1981) Universality and the power spectrum at the onset of chaos, Phys. Rev. B24, 493.

E.A. Neppiras (1968) J. Acoust. Soc. Am. 46, 587.

S. Ostlund, D. Rand, J. Sethna and E. Siggia (1983) Universal properties of the transition from quasi-periodicity to chaos in dissipative systems, Physica D (to be published).

E. Ott (1981) Strange attractors and chaotic motions of dynamical systems, Rev. Mod. Phys. 53, 655.

A.M. Ozorio de Almeida and J.H. Hannay (1982) Geometry of two-dimensional ion in phase space: projections, sections and the Wigner function, Ann. Phys. (N.Y.) 138, 115.

N.H. Packard, J.P. Crutchfield, J.D. Farmer and R.S. Shaw (1980) Geometry from a time series, Phys. Rev. Lett. 45, 712.

P.O. Pedersen (1935) J. Acoust. Soc. Am. 6, 227.

I.C. Percival (1973) Regular and irregular spectra, J. Phys. B6, L229.

R. Perez and L. Glass (1982) Bistability, period-doubling bifurcations and chaos in a periodically forced oscillator, Phys. Lett. 90A, 441.

J. Pérez and C. Jeffries (1982) Direct observation of a tangent bifurcation in a nonlinear oscillator, Phys. Lett. 92A, 82.

H. Poincaré (1892) Les methodes nouvelle de la mécanique céleste (Gauthier-Villars).

* Y. Pomeau and P. Manneville (1980) Intermittent transition to turbulence in dissipative dynamical systems, Commun. Math. Phys. 74, 189.

* Y. Pomeau, J.C. Roux, A. Rossi, S. Bachelart and C. Vidal (1981) Intermittent behaviour in the Belousov-Zhabotinsky reaction, Phys. Lett. 42, L271.

I. Prigogine (1980) From being to becoming (Freeman, San Francisco).

R.A. Pullen and A.R. Edmonds (1981) Comparison of classical and quantal spectra for totally bound potential, J. Phys. A14, L477.

M.I. Rabinovich (1978) Stochastic self-oscillations and turbulence, Sov. Phys. Usp. 21, 443.

D. Rand, S. Ostlund, J. Sethna and E.D. Siggia (1982) Universal transition from quasi-periodicity to chaos in dissipative systems, Phys. Rev. Lett. 49, 132.

G. Riela (1982) Universal spectral property in higher dimensional dynamical systems, Phys. Lett. 92A, 157.

R.K. Ritala and M.M. Salomaa (1983) Odd and even subharmonics and chaos in rf SQUIDs, J. Phys. C16, L477.

A.L. Robinson (1982) Physicists try to find order in chaos, Science 218, 554.

O.E. Rössler (1976) Z. Naturforsch. 31a, 259, 1168, 1664.

O.E. Rössler (1979) Continuous chaos - four prototype equations, Ann. N.Y. Acad. Sci. 316, 376.

* J.-C. Roux, A. Rossi, S. Bachelart and C. Vidal (1980) Representation of a strange attractor from an experimental study of chemical turbulence, Phys. Lett. 77A, 391.

D. Ruelle (1977) Applications conservant une mesure absolument continue par rapport à dχ sur [0,1], Commun. Math. Phys. 55, 47.

* D. Ruelle (1980) Strange attractors, Math. Intelligencer 2 126.

D. Ruelle and F. Takens (1971) On the nature of turbulence, Commun. Math. Phys. 20, 167.

A.N. Šarkovskii (1964) Coexistence of cycles of a continuous map of a line into itself, Ukr. Mat. Z. 16, 61.

G.K. Savvidy (1983) Yang-Mills classical mechanics as a Kolmogorov K-system, Yerevan preprint (February, 1983).

M. Schell, S. Fraser and R. Kapral (1982) Diffusive dynamics in systems with translational symmetry: a one-dimensional map model, Phys. Rev. A26, 504.

R.A. Schmitz, K.R. Graziani and J.L. Hudson (1977) J. Chem. Phys. 67, 3040.

H.J. Scholz, T. Yamada, H. Brand and R. Graham (1981) Intermittency and chaos in a laser system with modulated inversion, Phys. Lett. 82A, 321.

I. Schreiber and M. Marek (1982) Strange attractor in coupled reaction-diffusion cells, Physica 5D, 258.

R.S. Shaw (1981) Strange attractors, chaotic behavior and information flow, Z. Naturforsch. 36a, 80.

* S.J. Shenker (1982) Scaling behaviour in a map of a circle onto itself: empirical results, Physica 5D, 405.

S.J. Shenker and L.P. Kadanoff (1982) Critical behaviour of a KAM surface (I). Empirical results, J. Stat. Phys. 27, 631.

* B. Shraiman, C.E. Wayne and P.C. Martin (1981) Scaling theory for noisy period-doubling transitions to chaos, Phys. Rev. Lett. 46, 935.

C.L. Siegel (1942) Iteration of analytic functions, Ann. Math. 43, 607.

* R.H. Simoyi, A. Wolf and H.L. Swinney (1982) One-dimensional dynamics in a multicomponent chemical reaction, Phys. Rev. Lett. 49, 245.

D. Singer (1978) Stable orbits and bifurcation of maps of the interval, SIAM J. Appl. Math. 35, 260.

C.W. Smith, M.J. Tejwani and D.A. Farris (1982) Bifurcation universality for first-sound subharmonic generation in superfluid helium-4, Phys. Rev. Lett. 48, 492.

P.R. Stein and S. Ulam (1973) in: Lectures, Nato Advanced Studies Institute, Istanbul (Reidel, Dordrecht).

D. Sullivan (1982) Conformal dynamical systems, IHES, Bures-sur-Yvette preprint M/82/50 (September, 1982).

H.L. Swinney and J.P. Gollub, eds. (1981) Hydrodynamic instabilities and the transition to turbulence (Springer, New York).

L. Tedeschini-Lalli (1982) Truncated Navier-Stokes equations: continuous transition from a five-mode to a seven-mode model, J. Stat. Phys. 27, 365.

* J.S. Testa, J. Pérez and C. Jeffries (1982) Evidence for universal chaotic behaviour of a driven nonlinear oscillator, Phys. Rev. Lett. 48, 714.

S. Thomae and S. Grossmann (1981) A scaling property in critical spectra of discrete systems, Phys. Lett. 83A, 181.

J.M.T. Thompson and R. Ghaffari (1982) Chaos after period-doubling bifurcations in the resonance of an impact oscillator, Phys. Lett. 91A, 5.

J.M.T. Thompson and R.J. Thompson (1980) Numerical experiments with a strange attractor, Bull. Inst. Math. Appl. 16, 150.

K. Tomita (1982) Chaotic response of nonlinear oscillators, Phys. Rep. 86, 113.

K. Tomita and T. Kai (1978) Stroboscopic phase portrait and strange attractors, Phys. Lett. 66A, 91.

C. Tresser and P. Coullet (1978) C.R. Acad. Sci. (Paris) 287, 577.

J.S. Turner, J.-C. Roux, W.D. McCormick and H.L. Swinney (1981) Alternating periodic and chaotic regimes in a chemical reaction - experiment and theory, Phys. Lett. 85A, 9.

Y. Ueda (1979) Randomly transitional phenomena in the system governed by Duffing's equation, J. Stat. Phys. 20, 181.

S.M. Ulam (1964) Computers, Sci. Am. 211 (3), 203.

S.M. Ulam and J. von Neumann (1947) On combinations of stochastic and deterministic processes, Bull. Am. Math. Soc. 53, 1120.

S. Utida (1957) Population fluctuation, an experimental and theoretical approach, Cold Spring Harbor Symposia on Quantitative Biology 22, 139.

M.G. Velarde and C. Normand (1980) Convection, Sci. Am. 243, July, p.78.

R. Villela Mendés (1981) Critical-point dependence of universality in maps of the interval, Phys. Lett. 84A, 1.

E.B. Vul and K.M. Khanin (1982) On the unstable manifold of Feigenbaum fixed point, Usp. Mat. Nauk. 37, 173.

J. Weiland and H. Wilhelmsson (1983) Transition to chaos for ballooning modes stabilized by finite Larmor radius effects, Physica Scripta (to be published).

C.O. Weiss, A. Godone and A. Olafsson (1983) Routes to chaotic emission in a cw He-Ne laser, Phys. Rev. A (to be published).

J.M. Wersinger, J.M. Finn and E. Ott (1980) Bifurcations and strange behaviour in instability saturation by nonlinear mode coupling, Phys. Rev. Lett. 44, 453.

* A. Wolf and J. Swift (1981) Universal power spectra for the reverse bifurcation sequence, Phys. Lett. 83A, 184.

T. Yamada and R. Graham (1980) Chaos in a laser system under a modulated external field, Phys. Rev. Lett. 45, 1322.

W.J. Yeh and Y.H. Kao (1982a) Universal scaling and chaotic behaviour of a Josephson junction analog, Phys. Rev. Lett. 49, 1888.

W.J. Yeh and Y.J. Kao (1982b) Intermittent transition to turbulence in a Josephson oscillator, preprint.

A. Zardecki (1982) Noisy Ikeda attractor, Phys. Lett. 90A, 274.

G.M. Zaslavsky (1981) Stochasticity in quantum systems, Phys. Rep. 80, 157.

S.L. Ziglin (1981) Dokl. Akad. Nauk. SSSR 257, N1.

* A.B. Zisook (1981) Universal effects of dissipation in two-dimensional mappings, Phys. Rev. A24, 1640.

A.B. Zisook (1982) Intermittency in area-preserving mappings, Phys. Rev. A25, 2289.

A.B. Zisook and S.J. Shenker (1982) Renormalization group for intermittency in area-preserving mappings, Phys. Rev. A25, 2824.

Author Index

For Product Safety Concerns and Information please contact our EU representative GPSR@taylorandfrancis.com
Taylor & Francis Verlag GmbH, Kaufingerstraße 24, 80331 München, Germany

www.ingramcontent.com/pod-product-compliance
Ingram Content Group UK Ltd.
Pitfield, Milton Keynes, MK11 3LW, UK
UKHW052031210425
457613UK00032BA/1052

* 9 7 8 0 8 5 2 7 4 2 6 0 0 *